Lecture Notes in Computer Science　　10532

Commenced Publication in 1973
Founding and Former Series Editors:
Gerhard Goos, Juris Hartmanis, and Jan van Leeuwen

More information about this series at http://www.springer.com/series/7410

Roberto Avanzi · Howard Heys (Eds.)

Selected Areas in Cryptography – SAC 2016

23rd International Conference
St. John's, NL, Canada, August 10–12, 2016
Revised Selected Papers

Springer

Editors
Roberto Avanzi
ARM, Systems Architecture Group
Grasbrunn
Germany

Howard Heys
Memorial University of Newfoundland
St. John's, NL
Canada

ISSN 0302-9743   ISSN 1611-3349   (electronic)
Lecture Notes in Computer Science
ISBN 978-3-319-69452-8   ISBN 978-3-319-69453-5   (eBook)
https://doi.org/10.1007/978-3-319-69453-5

Library of Congress Control Number: 2017956782

LNCS Sublibrary: SL4 – Security and Cryptology

Printed on acid-free paper

This Springer imprint is published by Springer Nature
The registered company is Springer International Publishing AG
The registered company address is: Gewerbestrasse 11, 6330 Cham, Switzerland

Preface

The Conference on Selected Areas in Cryptography (SAC) series started as a workshop in 1994, when it first was held at Queen's University in Kingston. SAC has been held annually since 1994 in several Canadian provinces and it is the only international conference series on cryptography that is held annually in Canada. Since 2006, the conference has been organized in co-operation with the International Association for Cryptologic Research (IACR).

During the past 23 years, SAC has established itself as an internationally reputed venue for researchers in cryptography to present and discuss new work on selected areas of current interest in a relaxed and friendly atmosphere. This past year, SAC was held at Memorial University of Newfoundland (MUN), in St. John's, Newfoundland, Canada. This was the second time that SAC was hosted in St. John's, and the fourth time in an Atlantic Canadian province.

To keep the SAC conference series focused, each year presents papers in four *selected areas*, of which three are fixed, and the fourth one is specially chosen every year. The areas for SAC 2016 were:

1. Design and analysis of symmetric key primitives and cryptosystems including block and stream ciphers, hash functions, MAC algorithms, and authenticated encryption schemes
2. Efficient implementations of symmetric and public key algorithms
3. Mathematical and algorithmic aspects of applied cryptology
4. Side channel, fault and related attacks on symmetric and asymmetric cryptographic primitives and their countermeasures

A total of 100 submissions were reviewed for SAC 2016, with the Program Committee and the chairs selecting 28 papers for presentation. In addition to these 28 papers, two speakers were invited to give presentations at the conference. Douglas Stebila presented the Stafford Tavares Lecture on "Post-Quantum Key Exchange for the Internet" and Francesco Regazzoni talked on "Physical Attacks and Beyond."

The Program Committee for SAC 2016 comprised a total of 46 members. The review process was thorough – each submission received the attention of at least three reviewers, with almost all accepted papers being reviewed by four. Submissions involving a Program Committee member required at least five reviews. A total of 441 reviews were uploaded, of which 154 were written by 115 external subreviewers. The reviews were then followed by in-depth discussions on the papers, which contributed in a decisive way to the quality of the final selection. Despite the huge amount of work, and the occasional difference in opinion, the atmosphere in the Program Committee was always friendly and cooperative. For us, it was an honor to work with these Program Committee members, and we thank them sincerely for their engagement.

For the second time, the SAC Summer School (S3) took place just before the conference. In line with the selected topics of the latter, the overall theme of the

Summer School was "Secure and Efficient Implementation of Cryptographic Algorithms" and comprised the following talks and speakers:

- "Hardware Implementation of Public Key Cryptography"
  Tim Güneysu, University of Bremen and DFKI, Germany
- "Software Implementation of Public Key Cryptography"
  Patrick Longa, Microsoft Research, USA
- "Secure Hardware Implementation of Symmetric Key Ciphers (Including Side Channel Resistance)"
  Francesco Regazzoni, ALaRI–USI, Lugano, Switzerland
- "Implementation and Analysis of Cryptographic Protocols"
  Douglas Stebila, McMaster University, Canada

We thank the speakers for accepting our invitation to present and for providing a coherent yet varied program for the Summer School.

Further, we would like to thank Saeed Samet (eHealth Research Unit, Faculty of Medicine, MUN), Jonathan Anderson and Jiming Xu (both with the Department of Electrical and Computer Engineering, MUN), and Mary Garnier (MUN Conference Services) for assistance with the local organization.

Lastly, we are very appreciative of the sponsorship provided to SAC 2016 from Microsoft Research, the Atlantic Association for Research in Mathematical Sciences (AARMS), MUN Faculty of Engineering, MUN Department of Electrical and Computer Engineering, and Memorial University of Newfoundland.

August 2017   Roberto Avanzi
                                                     Howard Heys

Organization

Program Chairs

Roberto Avanzi   ARM, Germany (Qualcomm, Germany at the time
.of the event)

Howard Heys   Memorial University of Newfoundland, Canada

Program Committee

Carlisle Adams	University of Ottawa, Canada
Elena Andreeva	COSIC, KU Leuven, Belgium
Kazumaro Aoki	NTT Secure Platform Laboratories, Japan
Jean-Philippe Aumasson	Kudelski Security, Switzerland
Roberto Avanzi (Co-chair)	ARM, Germany (Qualcomm, Germany at the time of the event)
Manuel Barbosa	HASLab–INESC TEC and DCC–FCUP, Portugal
Paulo S.L.M. Barreto	University of Washington Tacoma, USA and University of São Paulo, Brazil
Olivier Benoit	Qualcomm, USA
Céline Blondeau	Aalto University, Finland
Andrey Bogdanov	Technical University of Denmark, Denmark
Billy Bob Brumley	Tampere University of Technology, Finland
Joan Daemen	STMicroelectronics, Belgium and Radboud University, Nijmegen, The Netherlands
Itai Dinur	Ben-Gurion University, Israel
Pierrick Gaudry	Loria, France
Guang Gong	University of Waterloo, Canada
Johann Großschädl	University of Luxembourg, Luxembourg
Tim Güneysu	University of Bremen and DFKI, Germany
Anwar Hasan	University of Waterloo, Canada
Philip Hawkes	Qualcomm, Australia
Howard Heys (Co-chair)	Memorial University of Newfoundland, Canada
Laurent Imbert	LIRMM, CNRS, Université Montpellier 2, France
Kimmo Järvinen	Aalto University, Finland
Michael John Jacobson, Jr.	University of Calgary, Canada
Marc Joye	NXP Semiconductors, USA and École Normale Supérieure, Paris, France
Elif Bilge Kavun	Infineon Technologies, Germany
Liam Keliher	Mount Allison University, Canada
Nathan Keller	Bar-Ilan University, Israel, Einstein Institute of Mathematics, Hebrew University of Jerusalem, and Weizmann Institute, Revohot, Israel

Gregor Leander	HGI, Ruhr University Bochum, Germany
Gaëtan Leurent	Inria, France
Petr Lisonek	Simon Fraser University, Canada
Victor Lomné	ANSSI, France
Nicky Mouha	Katholieke Universiteit Leuven, Belgium and Inria, France
María Naya-Plasencia	Inria, France
Christian Rechberger	IAIK, Graz University of Technology, Austria
Francesco Regazzoni	ALaRI–USI, Lugano, Switzerland
Palash Sarkar	Indian Statistical Institute, Kolkata, India
Jörn-Marc Schmidt	secunet, Germany
Kyoji Shibutani	Sony Corporation, Japan
Francesco Sica	Nazarbayev University, Kazakhstan
Meltem Sönmez Turan	NIST, USA
François-Xavier Standaert	Université catholique de Louvain, Belgium
Michael Tunstall	Cryptography Research, USA
Vanessa Vitse	Institut Fourier, Université Grenoble Alpes, France
Damian Vizár	EPFL, Switzerland
Bo-Yin Yang	Academia Sinica, Taiwan
Amr Youssef	Concordia University, Canada

Local Arrangements Committee

Howard Heys
Saeed Samet
Jonathan Anderson
Jiming Xu
Mary Garnier

Additional Reviewers

Ahmed Abdelkhalek	Marc Blanc-Patin
Simon Abelard	Christina Boura
Toru Akishita	Cagdas Calik
Erdem Alkim	Rosario Cammarota
José Bacelar Almeida	Wouter Castryck
Florian Bache	Yao Chen
Subhadeep Banik	Ming-Shing Chen
Morgan Barbier	Ronald Cramer
Lejla Batina	Parthasarathi Das
Christof Beierle	Thomas De Cnudde
Shivam Bhasin	Jean-Paul Degabriele
Jean-François Biasse	Jérémie Detrey
Karim Bigou	Daniel Dinu
Peter Birkner	Benedikt Driessen

Orr Dunkelman
Sébastien Duval
Nevine Ebeid
Maria Eichelseder
Kirsten Eisentraeger
Jean-Pierre Flori
Benoît Gérard
Youssef Gahi
Vincent Grosso
Aurore Guillevic
Michael Hamburg
Harunaga Hiwatari
James Howe
Michael Hutter
Ilia Iliashenko
Takanori Isobe
Eliane Jaulmes
Anthony Journault
Antoine Joux
Sabyasachi Karati
Eunkyung Kim
Thorsten Kranz
Praveen Kumar Vadnala
Po-Chun Kuo
Tancrède Lepoint
Wen-Ding Li
Sebastian Lindner
Kalikinkar Mandal
Pedro Maat Costa Massolino
Filippo Melzani
Florian Mendel
Alfred Menezes
Lingchuan Meng
Santos Merino Del Pozo
Oliver Mischke
Dustin Moody
Carlos Moreno
Christophe Negre
Tobias Oder
Thomas Pöppelmann
Paolo Palmieri
Louiza Papachristodoulou
Kostas Papagiannopoulos
Sylvain Pelissier

Hervé Pelletier
Bo-Yuan Peng
Cesar Pereida Garcia
Leo Perrin
Jérôme Plût
Romain Poussier
Emmanuel Prouff
Ciara Rafferty
Somindu Ramanna
Damien Robert
Sujoy Sinha Roy
Markku-Juhani Olavi Saarinen
Pascal Sasdrich
Renate Scheidler
Tobias Schneider
Peter Schwabe
Yannick Seurin
Shashank Singh
Daniel Slamanig
Pierre-Jean Spaenlehauer
Valentin Suder
Ruggero Susella
Atsushi Takayasu
Yin Tan
Adrian Thillard
David Thomson
Arnaud Tisserand
Mohamed Tolba
Sohaib ul Hassan
Felipe Valencia
Kerem Varici
Serge Vaudenay
Philip Vejre
Vesselin Velichkov
Bastien Vialla
Karine Villegas
Qingju Wang
Friedrich Wiemer
Alexander Wild
Jong-Shian Wu
Shang-Yi Yang
Takanori Yasuda
Jens Zumbrägel

Sponsoring Institutions

Microsoft Research
Atlantic Association for Research in Mathematical Sciences (AARMS)
Memorial University of Newfoundland

Contents

Lattice-Based Cryptography

Efficient Classical Public Key Cryptography

Cryptanalysis of Asymmetric Primitives

Invited Lectures

Physical Attacks and Beyond

Francesco Regazzoni[(✉)]

ALaRI - USI, Lugano, Switzerland
regazzoni@alari.ch

Abstract. Physical attacks have been subject of extensive research since more than twenty years. Nevertheless, several problems still have to be solved. This paper, after recalling the most popular physical attacks, introduces three (of the many) possible research directions in the area: the methodological study of the interaction between countermeasures against one type of attack and the resistance against another attack, the development of automated techniques for applying and verifying the correct application of countermeasures, and the study of physical attacks in the novel and changed scenario of cyber-physical systems.

1 Introduction

Physical attacks exploit weaknesses of an implementation to reveal the secret information. These attacks are possible since very often an adversary has physical access to the target device and can easily record its activity. Among the physical attacks, side channel attacks have been demonstrated to be extremely powerful, since they allow to e.g., extract the secret key from a cryptographic circuit with minimal efforts. In a nutshell, side-channel attacks collect information leaked from the target device while data is being processed, and exploit the dependence between this leakage and the processed data. Information can leak through several "channels", including power consumption [15], the time needed to complete an operation [14], and the chip's electromagnetic emissions [1].

Researchers dedicated significant efforts to defeat these attacks. However, developing general, reliable and effective countermeasures against physical attacks remains an extremely challenging task. Countermeasures are often considered only in the later stages of the full design flow, and applied manually by designers with strong security expertise. Very little is known about the interaction between different physical attacks and about the role which a countermeasure against one physical attack would play on the robustness of the device against another attack. The problem will be further complicated in the near future when cyber-physical systems will pervade several areas of our daily lives, including numerous safety-critical or privacy-relevant ones. These devices will have to provide strong security, but they should also often provide safety, real time computation capabilities, and achieve an extremely little energy footprint.

In view of this increasingly relevant problem, it is crucial to have the complete awareness of the security threats which cyber-physical system will have to face and to address the design challenges associated with the deployment of systems

© Springer International Publishing AG 2017
R. Avanzi and H. Heys (Eds.): SAC 2016, LNCS 10532, pp. 3–13, 2017.
https://doi.org/10.1007/978-3-319-69453-5_1

secure against physical attacks. This paper summarizes the main physical attacks and discusses three possible future research direction in the area.

2   Overview of Physical Attacks

In cryptography, a physical attack is an attack where the adversary, instead of focussing on the mathematical structure of a cryptographic routine, tries to extract secret information by exploiting the weaknesses of its implementation. Physical attacks are usually divided in two groups: active attacks and passive attacks. During an active attack, the adversary has to actively manipulate the device, by modifying its inputs or its operating environment, to force it to behave abnormally. This abnormal behavior is then exploited to perform the attack. During a passive attack, the adversary observes the normal behavior of a device and analyzes some side effects to gain information of the secret key.

Side channel attacks are very powerful passive attacks. Informally, a side channel is an information (often unintentionally leaked) which, indirectly, allows to infer knowledge about a different, and often more interesting, event. Side channels have been used and are used in several fields. For instance, the amount of pizza delivery over night in offices was used to infer if some important activity was under planning, the use of electric power was used to determine if a person was actually residing in the declared house. A field where side channels are deeply used is biology. Biological tests and medical exams often do not check directly the presence of a particular virus or of a specific disease, but they examine a side consequence which indicates with good approximation if the searched disease is present (for instance, positron emission tomography checks the concentration of light emitted by radioactive sugar, to infer, in a less invasive way, the possible presence of cancer).

Side channel have been also used in security for several years. A well known example of of the use of side channel for security application is the opening of safes. The mechanical locks of the safes which were used in past centuries were producing a slightly different noise when the pin of the combination was aligned to the correct digit. Exploiting this difference in noise, bank robbers were able to infer the secret combination and open the safe without trying all possible combinations of the lock. Nowadays, most of our security systems are controlled by electronic components. Instead of mechanical noise, attackers exploit other side information, such as power consumption or time needed for computation, but the principle is the same: use some side information to extract the secret data.

To take advantage of physical vulnerabilities however, the adversary needs to have physical access to the device. Such access was not always available. At the beginning of the digital era, when the computation was carried out in insulated mainframes, it was almost impossible to access the devices. As a result, physical attacks were not possible. Years later, with the diffusion of personal computers, it became easier to have physical access to the devices. However, personal computers are usually located in a rather safe environment (inside an office or an apartment), which is still reasonably protected from an adversary. As a result, the main security threats for personal computers were mainly coming from viruses and unprotected network connections.

The situation dramatically changed with the creation of the internet of things (IoT) and the pervasive diffusion of the embedded systems which are populating it. These devices are often operating in a hostile environment, and very often they are easily accessible by adversaries. In this scenario, where the devices are available to the adversary, resistance against physical attacks has become of primary importance.

Physical attacks as we know today appeared in the open literature in the late Nineties, but the recent declassification of the project Transient Electromagnetic Pulse Emanations Standard (TEMPEST) [12] demonstrates that there was awareness of the problem at least since fifty years before.

Timing analysis, the first physical attack that was published, was presented in 1996 by Kocher et al. [14]. Timing analysis attacks exploit the different time required by a device to process different data and to carry out different computations. Such time difference is due to several factors, such as the time needed to fetch the data (cache or memory hit or miss), the program behavior (branch taken or not), or the speed of difference components (a multiplier is generally slower than a shifter). Although these timing characteristics are often extremely small, the work of Kocher et al. demonstrates that they are sufficient to infer the entire secret key.

Soon after the first timing analysis attack was presented, Boneh et al. [7] proposed fault attacks. Fault attacks are attacks in which an adversary voluntarily induces a fault into a circuit and exploits the erroneous behavior to gain information about the secret key. The first step of a fault attack is the introduction of an error, possible transient, in the device. There are several ways to induce a fault. The adversary usually trades the cost of the equipment for mounting the fault with the precision of the injection (and thus the power of the attack). Common methods to inject faults are: variation in supply voltage, variation of the external clock, variation of operating temperature, exposure to X-rays, or precise laser illumination.

Power analysis was presented in 1999 [15]. The instantaneous power consumption of a cryptographic device strongly depends on the processed data and on the performed operation. Power analysis attacks essentially exploit this fact. These attacks are very powerful and can be performed with pretty inexpensive equipment. Several variations of power analysis attacks have been proposed in the past, the two most common are simple power analysis and differential power analysis. In simple power analysis attacks, an adversary attempts to derive the secret directly interpreting a set of power traces collected during the computation of the cryptographic routine. To be effective, simple power analysis requires often a detailed knowledge about the implementation of the cryptographic algorithm under attack. Differential power analysis attacks allows to retrieve the secret key with the only knowledge of the algorithm used in the device under attack. DPA attacks are based on a divide and conquer approach: the general

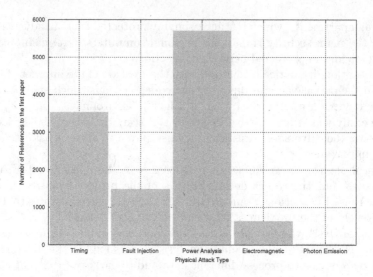

Fig. 1. Number of references to the first paper presenting each of the most common physical attacks (data collected the 8th of August 2016). This plot gives the intuition of the amount of research activity per attack.

idea is that the attacker, instead of attacking the whole key at once, targets a small portion of it, makes a hypothesis on possible values of the key and verifies these hypothesis using the power traces. The full key is recovered iterating this process.

After these pioneering works, several other physical attacks and channels have been discover and presented, including attacks exploiting the electromagnetic emission of a device [1] and attacks exploiting the photons emitted by electronic components during the computations [23]. The scientific community devoted significant efforts to the study of the problem. As usual, research activities focused on attacks and countermeasures against attacks. On the one hand, researchers tried to develop countermeasures to defeat physical attacks (or, at least, to complicate as much as possible the task of the attacker). Hardware implementations and software routines capable of computing cryptographic operations in constant time [13], hiding the power consumption using power analysis resistant logic styles [8, 26–28] or masking it using randomization [17, 19], or efficient error detection and correction codes [6] are possible examples of countermeasures developed over the years. On the other hand, researchers tried to improve the effectiveness of the attacks to better understand their potential and limits. Template attacks [9] and fault sensitivity [16] are two possible example of this improvement. Furthermore, especially for power analysis, researchers also developed metrics for fairly evaluate the robustness against attacks [25].

The amount of carried out research dealing with physical attacks is visible from Fig. 1, which depicts the number of references, as reported on Google scholar [24] the 8th of August 2016, to the first papers discussing each of the most common physical attacks. Although not being an exact and precise measure, this figure gives an intuition of the large impact which physical attacks had (and still continue to have). The physical attack more deeply studied is power analysis. This is probably caused by the power of the attack and by the relatively inexpensive equipment needed to mount it. Electromagnetic attacks and photon emission attacks did not get the same exposure as the other physical attacks simply because they appeared only recently. Surprisingly, fault attacks were not investigated with the same effort as timing and power analysis attacks. This fact is unexpected because, at least in their low cost version, fault attacks are extremely simple to be carried out.

Despite such a vast effort however, the problem of physical attack is still on scientific agenda, since some issues are still open. We need a better understanding of some physical attacks (as visible from Fig. 1, only power analysis and timing attacks have been explored in depth), we need to develop effective countermeasures against some other attacks, in particular photon emission, and several other problems have to be addressed to ensure the robustness of cyber-physical and embedded systems. The next section will focus on three of these problems, presenting their main challenges and highlighting possible research directions.

3   Challenge 1: Interaction Between Physical Attacks

Physical attacks, so far, have been mainly analyzed in isolation. This fact is even more evident when it comes to the design of countermeasures. Researchers often concentrate on one type of attack, developing a countermeasure against it and evaluating how the protected design behaves compared to the original one. However, the goal of the attacker is just to get access to the secret information stored in the device (and not get access to the secret key using a specific attack). Thus, application of a countermeasure against one attack without considering the global effect on security of the countermeasure is extremely dangerous. In fact, countermeasures against one attack might harm the robustness of the system against another type of attack.

An example of this risk reported in the past is the negative effect which countermeasures against fault attacks have on the resistance of a circuit against power analysis [21]. Several error-detecting and correcting codes have been used to harden the non-linear transformation of the AES algorithm and have been analyzed. Each error-detecting and correcting code is characterized by its coverage and its error recovery capability. As a result, some codes where more suitable than others to protect against fault attacks. After this exploration, the resistance against power analysis attacks of each of the considered error-detecting and correcting codes was analyzed in detail, using the information theory metric [25]. The results, reported in Fig. 2, demonstrate that the circuit characterized by the highest resistance against power analysis attacks is the one without any error-detecting and correcting codes (basically the one which could be easily attacked by fault attacks).

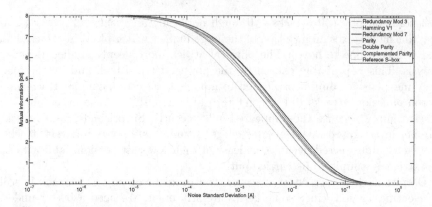

Fig. 2. Resistance against power analysis attacks analyzed using information theory (from [21]). Several error detection and correction circuits have been analyzed (Parity, Complemented Parity, Double Parity, Residue Modulo 3, Residue Modulo 7, Hamming Code, and a reference version without any additional circuit). The most resistant against power analysis attacks is the left most curve, the blue one, which is the reference circuit where no support for resistance against fault attacks was added. (Color figure online)

This example shows that, even though the intention of the designer was to increase the resistance of the circuit by making it more robust against fault attacks, the achieved result was to help the attacker, since the added circuit significantly simplified the procedure for extracting the secret key using a different type of attack. Currently we have a pretty good knowledge of some physical attacks, but we still know very little about the possible interaction between them and we know even less about the about interactions between the different countermeasures which we apply. Exploring these problems in much more depth is of crucial importance for designing much more resistant and much more secure embedded and cyber-physical systems.

4   Challenge 2: Automatic Application of Countermeasures

Despite the pervasive diffusion of electronic systems also in extremely private and critical aspects of our live, security is often considered only at the end of the whole design process, after other goals (such as performance and cost) are achieved. This is not a good approach for designing secure systems in general, but is even less effective for tackling the problem of physical attacks, since these attacks are strictly depending on the underlining architecture and on the specific implementation. Thus, a much more effective way of achieving robustness from physical attacks is to consider security since the beginning of the whole design flow, and to use security related metrics as forefront design variables as now are area or memory occupation, performance, and power consumption.

Furthermore, implementations of countermeasures against physical attacks require engineers and designers with strong security expertise and good knowledge of state of the art in the field. Currently designers have to rely only on their experience and on good practices for finshing the implementation. Once the design is completed, it is evaluated by laboratories which test the device against a number of known attacks. If problems are encountered, the design has to be corrected and re-evaluated.

A parallel can be made between today's techniques for achieving physical attacks resistance and the design process of electronic circuits as it was decades ago. At the beginning, design of electronic circuits was carried out by teams of expert designers, who were sometimes manually drawing the layout of the fabrication masks. Then electronic design automation arrived to support designers in their tasks. The boosted productivity (together with the progresses of technology) allowed us to achieve the level of integration and to handle the level of complexity which made possible the existence of extremely powerful personal computers, smart devices and all other electronic components which are currently populating our lives.

In the same way, security would significantly benefit from the development of design tools allowing designers to automatically apply countermeasures against physical attacks, to evaluate their effects, to early estimate the impact of these countermeasures on other design parameters and to verify their correct application. An automation tool would take an unprotected design and apply a set of existing countermeasures, as would have been done by a designer. Such automation tools would not replace the work of researchers studying and designing novel and more effective countermeasures (as electronic design automation did not replace designers manually implementing extremely optimized blocks), but would provide an essential support for implementing systems which are robust against physical attacks by constructions and, ultimate, overall more secure.

Despite the importance of these topics, design automation for security did not receive significant attention. Previous works have addressed the topic mainly from the hardware point of view, proposing design flows for power analysis resistant logic styles [20,28,29]. More recently, the topic of automatic application of physical attacks countermeasures tackled also software aspects, proposing tools for power analysis aware instruction set extensions [22], and compilers for the automatic application of software countermeasure as hiding and masking [3,4,18]. The focus was still mainly on power analysis attacks. Verification tools for asserting the correct applications of countermeasures have also subject of research [5,10,11]. Verification tools are extremely important since they would allow to immediately identify not only errors introduced by designers during the implementation, but also several other security pitfalls, such as intrinsic weaknesses of the applied countermeasure and the involuntary removal of protections caused by various optimizations carried out in the tool chain.

These works represent however only the begin of a research direction, the one of automation tools for security, still in infancy, which would, once more developed, enable the design of more physical resistant, and thus overall more secure, embedded and cyber-physical systems.

5   Challenge 3: Physically Secure Cyber-Physical Systems

Embedded systems are becoming more and more intelligent and connected. Together with network connectivity, these devices began to integrate sensors since several years. Now, these devices integrates also some support for autonomous decision and actuators for putting these decisions in place. Systems composed by an analysis and decision-making part (cyber) and by a sensing and actuating parts (physical) take the names of cybper-physical systems (CPSs). The block representation of such systems is depicted in Fig. 3.

The presence of actuators dramatically increase the consequences of misuse of such systems, since a malicious attack can cause much more damage than the ones cause by a leak of private data. Cyber-physical systems are often used in critical applications, e.g., to automatically monitor patients or to control our smart grid. The security of these applications should be guaranteed, since a breach in such systems might have also catastrophic consequences and cause also the loss of human lives. For these reasons, it is of utmost importance that the cyber part of CPSs is resistant against physical attacks.

However, this is not sufficient. Cyber-physical systems are composed of two parts, a cyber part, very similar to the computational part of embedded systems, and a physical part. We know what are the threats to the cyber-part, and we know what can be the defense mechanisms for it. However, the goal of the attacker is to take control of the system, not necessarily take control of the system attacking the cyber part. An attack directed to the physical part, could

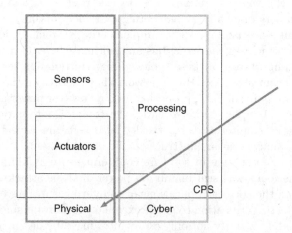

Fig. 3. Simplified schema of a Cyber-Physical system. It composes a cyber part, which analyzes the data and take the decisions, and a physical part, which usually consist in two parts: one devoted to sensing and one devoted to actuating the decision taken by the cyber part. Currently, security researchers are mainly focusing on securing the cyber part of the CPS. Almost no attention is devoted yet to the security of the physical part of the system. (Color figure online)

be much simpler while allowing the adversary to reach his goal. In the past, security was only concentrating on cyber attacks carried out against electronic components.

As discussed in Sect. 2, the discovery of physical attacks against the electronic components was devastating for embedded systems. Now, with the addition of a physical part to systems, the game changes again. The physical portion of CPSs will be exposed, exactly as the cyber part, to cyber and physical attacks. However, we are not prepared to address this new situation, since little or nothing is known about attacks and countermeasures against the physical portion of a system. Few works addressed the problems so far (physical attacks to the physical portion of CPSs were analyzed, for instance, in the context of additive manufacturing [2]). Future security research should definitely address, as indicated by the red arrow in Fig. 3, security threats and possible countermeasure devoted to the physical portion of systems, since the adversary will attack through the weakest point, and the physical part is much likely to be the weakest point of CPSs.

6   Conclusions

Approximately 20 years have passed since physical attacks were published in the open literature. Since then, researchers have deeply studied the subject, aiming on the one side at discovery of new and much more powerful ways for carrying out the attacks, and on the other attempting to increase the robustness of the implementations. Nevertheless, several problems are still open. This paper presented three (of the many) possible directions for future research in the area, namely the study of the interaction of different physical attacks (and the effects which a countermeasure against one attack might have on the robustness against another physical attack), the study of techniques for automatically applying countermeasures against physical attacks (and to verify the proper applications of them), and, finally, the study of the robustness of cyber-physical systems, where the presence of a physical part could completely change the rules of the game.

Acknowledgements. The author thanks Ilia Polian for his constructive comments.

References

1. Agrawal, D., Archambeault, B., Rao, J.R., Rohatgi, P.: The EM side-channel(s). Cryptographic Hardware and Embedded Systems - CHES 2002. LNCS, pp. 29–45. Springer, Heidelberg (2003). doi:10.1007/3-540-36400-5_4
2. Abdullah, M., Faruque, A., Chhetri, S.R., Canedo, A., Wan, J.: Acoustic side-channel attacks on additive manufacturing systems. In 2016 ACM/IEEE 7th International Conference on Cyber-Physical Systems (ICCPS), pp. 1–10 (2016). http://ieeexplore.ieee.org/xpl/articleDetails.jsp?arnumber=7479068
3. Bayrak, A.G., Regazzoni, F., Brisk, P., Ienne, P.: A first step towards automatic application of power analysis countermeasures. In: Proceedings of the 48th Design Automation Conference, San Diego, California, pp. 230–235, June 2011

4. Bayrak, A.G., Regazzoni, F., Novo, D., Brisk, P., Standaert, F.-X., Ienne, P.: Automatic application of power analysis countermeasures. IEEE Trans. Comput. **64**(2), 329–341 (2015)
5. Bayrak, A.G., Regazzoni, F., Novo, D., Ienne, P.: Sleuth: automated verification of software power analysis countermeasures. In: Bertoni, G., Coron, J.-S. (eds.) CHES 2013. LNCS, vol. 8086, pp. 293–310. Springer, Heidelberg (2013). doi:10.1007/978-3-642-40349-1_17
6. Bertoni, G., Breveglieri, L., Koren, I., Maistri, P., Piuri, V.: Error analysis and detection procedures for a hardware implementation of the advanced encryption standard. IEEE Trans. Comput. **52**(4), 492–505 (2003)
7. Boneh, D., DeMillo, R.A., Lipton, R.J.: On the importance of eliminating errors in cryptographic computations. J. Cryptol. **14**(2), 101–119 (2001)
8. Cevrero, A., Regazzoni, F., Schwander, M., Badel, S., Ienne, P., Leblebici, Y.: Power-gated MOS current mode logic (PG-MCML): a power aware DPA-resistant standard cell library. In Proceedings of the 48th Design Automation Conference, San Diego, California, pp. 1014–1019, June 2011
9. Chari, S., Rao, J.R., Rohatgi, P.: Template attacks. In: Kaliski, B.S., Koç, K., Paar, C. (eds.) CHES 2002. LNCS, vol. 2523, pp. 13–28. Springer, Heidelberg (2003). doi:10.1007/3-540-36400-5_3
10. Eldib, H., Wang, C.: Synthesis of masking countermeasures against side channel attacks. In: Biere, A., Bloem, R. (eds.) CAV 2014. LNCS, vol. 8559, pp. 114–130. Springer, Cham (2014). doi:10.1007/978-3-319-08867-9_8
11. Eldib, H., Wang, C., Schaumont, P.: Formal verification of software countermeasures against side-channel attacks. ACM Trans. Softw. Eng. Methodol. (TOSEM) **24**(2), 11 (2014)
12. McNamara, J.: The Complete, Unofficial TEMPEST Information Page (1996). http://www.eskimo.com/joelm/tempest.html
13. Käsper, E., Schwabe, P.: Faster and timing-attack resistant AES-GCM. In: Clavier, C., Gaj, K. (eds.) CHES 2009. LNCS, vol. 5747, pp. 1–17. Springer, Heidelberg (2009). doi:10.1007/978-3-642-04138-9_1
14. Kocher, P.C.: Timing attacks on implementations of Diffie-Hellman, RSA, DSS, and other systems. In: Koblitz, N. (ed.) CRYPTO 1996. LNCS, vol. 1109, pp. 104–113. Springer, Heidelberg (1996). doi:10.1007/3-540-68697-5_9
15. Kocher, P., Jaffe, J., Jun, B.: Differential power analysis. In: Wiener, M. (ed.) CRYPTO 1999. LNCS, vol. 1666, pp. 388–397. Springer, Heidelberg (1999). doi:10.1007/3-540-48405-1_25
16. Li, Y., Sakiyama, K., Gomisawa, S., Fukunaga, T., Takahashi, J., Ohta, K.: Fault sensitivity analysis. In: Mangard, S., Standaert, F.-X. (eds.) CHES 2010. LNCS, vol. 6225, pp. 320–334. Springer, Heidelberg (2010). doi:10.1007/978-3-642-15031-9_22
17. Messerges, T.S., Dabbish, E.A., Sloan, R.H.: Examining smart-card security under the threat of power analysis attacks. IEEE Trans. Comput. **51**(5), 541–552 (2002)
18. Moss, A., Oswald, E., Page, D., Tunstall, M.: Compiler assisted masking. In: Prouff, E., Schaumont, P. (eds.) CHES 2012. LNCS, vol. 7428, pp. 58–75. Springer, Heidelberg (2012). doi:10.1007/978-3-642-33027-8_4
19. Nikova, S., Rechberger, C., Rijmen, V.: Threshold implementations against side-channel attacks and glitches. In: Ning, P., Qing, S., Li, N. (eds.) ICICS 2006. LNCS, vol. 4307, pp. 529–545. Springer, Heidelberg (2006). doi:10.1007/11935308_38
20. Popp, T., Mangard, S.: Masked dual-rail pre-charge logic: DPA-resistance without routing constraints. In: Rao, J.R., Sunar, B. (eds.) CHES 2005. LNCS, vol. 3659, pp. 172–186. Springer, Heidelberg (2005). doi:10.1007/11545262_13

21. Regazzoni, F., Breveglieri, L., Ienne, P., Koren, I.: Interaction between fault attack countermeasures and the resistance against power analysis attacks. In: Joye, M., Tunstall, M. (eds.) Fault Analysis in Cryptography, pp. 257–272. Springer, Heidelberg (2012). doi:10.1007/978-3-642-29656-7

22. Regazzoni, F., Cevrero, A., Standaert, F.-X., Badel, S., Kluter, T., Brisk, P., Leblebici, Y., Ienne, P.: A design flow and evaluation framework for DPA-resistant instruction set extensions. In: Clavier, C., Gaj, K. (eds.) CHES 2009. LNCS, vol. 5747, pp. 205–219. Springer, Heidelberg (2009). doi:10.1007/978-3-642-04138-9_15

23. Schlösser, A., Nedospasov, D., Krämer, J., Orlic, S., Seifert, J.-P.: Simple photonic emission analysis of AES. In: Prouff, E., Schaumont, P. (eds.) CHES 2012. LNCS, vol. 7428, pp. 41–57. Springer, Heidelberg (2012). doi:10.1007/978-3-642-33027-8_3

24. Google Scholar. http://scholar.google.com/

25. Standaert, F.-X., Malkin, T.G., Yung, M.: A unified framework for the analysis of side-channel key recovery attacks. In: Joux, A. (ed.) EUROCRYPT 2009. LNCS, vol. 5479, pp. 443–461. Springer, Heidelberg (2009). doi:10.1007/978-3-642-01001-9_26

26. Tiri, K., Akmal, M., Verbauwhede, I.: A dynamic and differential CMOS logic with signal independent power consumption to withstand differential power analysis on Smart Cards. In: Proceedings of the 28th European Solid-State Circuits Conference, Florence, pp. 403–406, September 2002

27. Tiri, K., Verbauwhede, I.: Securing encryption algorithms against dpa at the logic level: next generation smart card technology. In: Walter, C.D., Koç, Ç.K., Paar, C. (eds.) CHES 2003. LNCS, vol. 2779, pp. 125–136. Springer, Heidelberg (2003). doi:10.1007/978-3-540-45238-6_11

28. Tiri, K., Verbauwhede, I.: A logic level design methodology for a secure DPA resistant ASIC or FPGA implementation. In: Proceedings of the Design, Automation and Test in Europe Conference and Exhibition, Paris, pp. 246–251, February 2004

29. Tiri, K., Verbauwhede, I.: A digital design flow for secure integrated circuits. IEEE Trans. Comput. Aided Des. Integr. Circuits Syst. 25(7), 1197–1208 (2006)

Post-quantum Key Exchange for the Internet and the Open Quantum Safe Project

Douglas Stebila[1]([⊠]) and Michele Mosca[2,3,4]

[1] Department of Computing and Software, McMaster University,
Hamilton, ON, Canada
stebilad@mcmaster.ca
[2] Institute for Quantum Computing and Department of Combinatorics
and Optimization, University of Waterloo, Waterloo, ON, Canada
mmosca@uwaterloo.ca
[3] Perimeter Institute for Theoretical Physics, Waterloo, ON, Canada
[4] Canadian Institute for Advanced Research, Toronto, ON, Canada

Abstract. Designing public key cryptosystems that resist attacks by
quantum computers is an important area of current cryptographic
research and standardization. To retain confidentiality of today's communications against future quantum computers, applications and protocols must begin exploring the use of quantum-resistant key exchange and
encryption. In this paper, we explore post-quantum cryptography in general and key exchange specifically. We review two protocols for quantum-
resistant key exchange based on lattice problems: BCNS15, based on the
ring learning with errors problem, and Frodo, based on the learning with
errors problem. We discuss their security and performance characteristics, both on their own and in the context of the Transport Layer Security
(TLS) protocol. We introduce the Open Quantum Safe project, an open-
source software project for prototyping quantum-resistant cryptography,
which includes liboqs, a C library of quantum-resistant algorithms, and
our integrations of liboqs into popular open-source applications and protocols, including the widely used OpenSSL library.

1 Introduction

All Internet security protocols that use cryptography, such as the Transport
Layer Security (TLS, a.k.a. the Secure Sockets Layer (SSL)) protocol [18] have
the same basic structure: *public key cryptography* is used to authenticate the
communicating parties to each other and to establish a shared secret key, which

Based on the Stafford Tavares Invited Lecture by D. Stebila.

D. Stebila—Supported in part by Australian Research Council (ARC) Discovery
Project grant DP130104304, Natural Sciences and Engineering Research Council of
Canada (NSERC) Discovery grant RGPIN-2016-05146, and an NSERC Discovery
Accelerator Supplement grant.

M. Mosca—Supported by NSERC, CFI, and ORF. IQC and the Perimeter Institute
are supported in part by the Government of Canada and the Province of Ontario.

© Springer International Publishing AG 2017
R. Avanzi and H. Heys (Eds.): SAC 2016, LNCS 10532, pp. 14–37, 2017.
https://doi.org/10.1007/978-3-319-69453-5_2

is then used in *symmetric cryptography* to provide confidentiality and integrity to their communication. The security of most public key cryptosystems depends on the difficulty of solving some mathematical problem, such as factoring large numbers or computing discrete logarithms in finite field or elliptic curve groups. The best known solutions to these problems run in exponential (or sub-exponential) time, making it infeasible for attackers to break the schemes.

Quantum mechanics allows for devices that operate on quantum bits, known as *qubits*, which are two-state quantum systems that can be in any quantum superposition of 0 and 1. Such devices are called quantum computers, and could solve certain types of problems much faster than "classical" (non-quantum) computers. Shor's algorithm [48] could efficiently (i.e., in polynomial time) factor large numbers and compute discrete logarithms, breaking all widely deployed public key cryptosystems. Most symmetric key schemes, such as the Advanced Encryption Standard (AES) cipher, would not be broken by quantum algorithms, although would generally need bigger keys. While large-scale quantum computers do not yet exist, building quantum computers is an active area of research. And Schoelkopf [17] identify seven stages in the development of quantum computers: so far, physicists can perform operations on single and multiple physical qubits, perform non-destructive measurements for error correction, and are making progress on constructing logical memories with longer lifetime than physical qubits; to achieve large-scale quantum computation, we will require the ability to perform operations on single and multiple logical qubits with fault-tolerant computation. Regarding the million-dollar question of when a large-scale quantum computer will be built, in 2015 Mosca [38] stated "I estimate a 1/7 chance of breaking RSA-2048 by 2026 and a 1/2 chance by 2031."

Any attacker who records present-day communications would be able to decrypt it once a quantum computer is built; and there is evidence that governments are storing vast quantities of encrypted Internet traffic. This motivates the urgent use of cryptography that is designed to be safe against quantum attackers—called "post-quantum" or "quantum-safe" or "quantum-resistant" cryptography. In August 2015, the United States National Security Agency (NSA) issued a memo regarding its Suite B cryptographic algorithms for government use, advising that it plans to "transition to quantum resistant algorithms in the not too distant future" [39]. In August 2016, the United States National Institute of Standards and Technology (NIST) launched its post-quantum crypto project[1], a multi-year process with the goal of evaluating and standardizing one or more quantum-resistant public key cryptosystems.

Post-quantum Cryptography. There are several classes of mathematical problems that are conjectured to resist attacks by quantum computers and have been used to construct public key cryptosystems, several of which date from the early days of public key cryptography. These include:

- *Code-based cryptography.* The McEliece public key encryption scheme [36] was one of the first public key schemes, and is based on error-correcting codes, in

---

[1] http://www.nist.gov/pqcrypto.

particular, the hardness of decoding a general linear code. Niederreiter [40] subsequently proposed a digital signature scheme based on error correcting codes.

- *Hash-based cryptography.* Merkle [37] first proposed the use of hash functions for digitally signing documents; Lamport [30] and Winternitz then showed how to convert Merkle's one-time signature scheme into a many-time signature scheme. These schemes are based entirely on standard hash function properties, and thus are believed to be among the most quantum-resistant. Modern variants include SPHINCS [7] and XMSS [13].
- *Multivariate cryptography.* These cryptosystems are based on the difficulty of solving non-linear, usually quadratic, polynomials, over a field [35,41].
- *Lattice-based cryptography.* Ajtai [1] proposed the first cryptographic schemes directly based on lattices. Regev [46] then introduced the related *learning with errors (LWE) problem*, the security of which is based on lattice problems, and which now forms the basis of a variety of public key encryption and signature schemes [31]. The *ring learning with errors (ring-LWE)* problem [33] uses additional structure which allows for smaller key sizes. Another scheme whose security relates to lattices is the NTRU scheme [26], which also allows for fairly small key sizes.
- *Supersingular elliptic curve isogenies.* One of the newest candidates for quantum-resistant public key cryptography is based on the difficulty of finding isogenies between supersingular elliptic curves [20].

In addition, quantum information can be used directly to create cryptosystems; this is called *quantum cryptography*. For example, quantum key distribution allows two parties to establish a shared secret key using quantum communication and an authenticated classical channel. While this can provide very strong security, it is not yet a candidate for widespread usage since it requires physical infrastructure capable of transmitting quantum states reliably over long distances, so in the rest of this paper we focus solely on quantum-resistant cryptography using classical (non-quantum) computers.

Existing quantum-resistant schemes generally have several limitations. Compared with traditional RSA, finite field, and elliptic curve discrete logarithm schemes, all quantum-resistant schemes have either larger public keys, larger ciphertexts/signatures, or slower runtime. Many quantum-resistant schemes are also based on mathematical problems that are, from a cryptographic perspective, quite new, and thus have received comparably less cryptanalysis. There remain many open questions in post-quantum cryptography, making it an exciting and active research area: the design of better public key encryption and signature schemes with smaller keys and ciphertexts/signatures; improved cryptanalysis leading to better parameter estimates; development of fast, secure implementations suitable for high-performance servers and small embedded devices; and integration into existing network infrastructure and applications.

(It is worth noting that research into post-quantum cryptography is valuable even if large-scale quantum computers are never built: it is possible that the factoring, RSA, or discrete logarithms problems will be solved by some (non-quantum) mathematical breakthrough. Having a diverse family of cryptography assumptions on which we can base public key cryptography protects against such a scenario. Furthermore, the cryptographic agility that will help prepare for a transition to yet-to-be-determined quantum-resistant cryptographic algorithms will also enable the ability to respond quickly to other unexpected weaknesses in cryptographic algorithms.)

This Paper. In this paper, we discuss two research projects in the area of lattice-based key exchange: the "BCNS15" protocol [10] based on the ring-LWE problem, and the "Frodo" protocol [9] based on the LWE problem. We will explain the basic mathematics of these protocols, and our results on the performance of these protocols and their integration into the TLS protocol. We will introduce the Open Quantum Safe project, an open-source software project designed for evaluating post-quantum cryptography candidates and prototyping their use in applications and protocols such as TLS.

This line of work focuses initially on key exchange, with digital signatures to follow closely. As noted above, any attacker who records present-day communications protected using non-quantum-resistant cryptography would be able to decrypt it once a quantum computer is built. This implies that information that needs to remain confidential for many years needs to be protected with quantum-resistant cryptography even before quantum computers exist. In communication protocols like TLS, digital signatures are used to authenticate the parties and key exchange is used to establish a shared secret, which can then be used in symmetric cryptography. This means that, for security against a future quantum adversary, authentication in today's secure channel establishment protocols can still rely on traditional primitives (such as RSA or elliptic curve signatures), but we should incorporate post-quantum key exchange to provide quantum-resistant long-term confidentiality. This has the benefit of allowing us to introduce new post-quantum ciphersuites in TLS while relying on the existing RSA-based public key infrastructure for certificate authorities. However, applications which require long-term integrity, such as signing contracts and archiving documents, will need to begin considering quantum-resistant signature schemes.

Notation. Let χ be a distribution; $a \xleftarrow{\$} \chi$ denotes sampling a randomly according to χ. The uniform distribution is denoted by $\mathcal{U}$. Vectors are denoted in lower-case bold, like $\mathbf{a}$; matrices are denoted in upper-case bold, like $\mathbf{A}$. The inner product of two vectors $\mathbf{a}$ and $\mathbf{b}$ is $\langle \mathbf{a}, \mathbf{b} \rangle$. Sampling each component of the length-n vector $\mathbf{a}$ independently at random from χ is denoted by $\mathbf{a} \xleftarrow{\$} \chi^n$. If $\mathcal{A}$ is a probabilistic algorithm, then $y \xleftarrow{\$} \mathcal{A}(x)$ denotes running $\mathcal{A}$ on input x with fresh randomness and storing the output in variable y, and $y \xleftarrow{\$} \mathcal{A}^O(x)$ denotes running $\mathcal{A}$ with oracle access to procedure O.

2 Lattice-Based Cryptography and the LWE Problems

In a seminal 1996 work, Ajtai [1] first proposed a cryptographic construction (in that case, a hash function) that relied on the hardness of a computational problem on lattices (the Short Integer Solution (SIS) problem). A subsequent work by Ajtai and Dwork [2] presented a public key encryption scheme based on another lattice problem. Concurrently, Hoffstein, Pipher, and Silverman [26] created the NTRU public key encryption scheme with can be viewed as involving algebraically structured lattices. A variety of research on the use of lattices in constructing cryptosystems continued during that era, and a detailed chronology is outside the scope of this paper; see one of the many surveys of lattice-based cryptography, such as Peikert's [44].

2.1   The Learning with Errors Problem

In 2005, Regev [46] introduced the *learning with errors (LWE) problem*, showed that LWE is related to the hardness of a lattice problem (the Gap Shortest Vector Problem (GapSVP)), and gave a public key encryption scheme based on LWE. Being a more abstract algebraic problem, LWE can be easier to work with in terms of building cryptosystems, and a large amount of research into the hardness of LWE and its use in cryptography has followed; again, see a survey such as [44] for a detailed chronology.

The search learning with errors problem is like a noisy version of solving a system of linear equations: given a matrix $\mathbf{A}$ and a vector $\mathbf{b} = \mathbf{A}\mathbf{s} + \mathbf{e}$, find $\mathbf{s}$.

Definition 1 (Search LWE problem). *Let n, m, and q be positive integers. Let χ_s and χ_e be distributions over $\mathbb{Z}$. Let $\mathbf{s} \xleftarrow{\$} \chi_s^n$. Let $\mathbf{a}_i \xleftarrow{\$} \mathcal{U}(\mathbb{Z}_q^n), e_i \xleftarrow{\$} \chi_e$, and set $b_i \leftarrow \langle \mathbf{a}_i, \mathbf{s} \rangle + e_i \mod q$, for $i = 1, \dots, m$. The search LWE problem for $(n, m, q, \chi_s, \chi_e)$ is to find $\mathbf{s}$ given $(\mathbf{a}_i, b_i)_{i=1}^m$. In particular, for algorithm $\mathcal{A}$, define the advantage*

$$\mathrm{Adv}^{\mathrm{lwe}}_{n,m,q,\chi_s,\chi_e}(\mathcal{A}) = \Pr\left[\mathbf{s} \xleftarrow{\$} \chi_s^n; \mathbf{a}_i \xleftarrow{\$} \mathcal{U}(\mathbb{Z}_q^n); e_i \xleftarrow{\$} \chi_e; \right.$$
$$\left. b_i \leftarrow \langle \mathbf{a}_i, \mathbf{s}_i \rangle + e \mod q : \mathcal{A}((\mathbf{a}_i, b_i)_{i=1}^m) = \mathbf{s})\right].$$

For appropriate distributions χ_s and χ_e, not only is it conjectured to be hard to find the secret vector $\mathbf{s}$, it is even conjectured that LWE samples $(\mathbf{a}, \langle \mathbf{a}, \mathbf{s} \rangle + e)$ look independent and random: this is the decision LWE problem.

Definition 2 (Decision LWE problem). *Let n and q be positive integers. Let χ_s and χ_e be distributions over $\mathbb{Z}$. Let $\mathbf{s} \xleftarrow{\$} \chi_s^n$. Define the following two oracles:*

- *$O_{\chi_e,\mathbf{s}}$: $\mathbf{a} \xleftarrow{\$} \mathcal{U}(\mathbb{Z}_q^n), e \xleftarrow{\$} \chi_e$; return $(\mathbf{a}, \langle \mathbf{a}, \mathbf{s} \rangle + e \mod q)$.*
- *U: $\mathbf{a} \xleftarrow{\$} \mathcal{U}(\mathbb{Z}_q^n), u \xleftarrow{\$} \mathcal{U}(\mathbb{Z}_q)$; return $(\mathbf{a}, u)$.*

The decision LWE problem for (n, q, χ_s, χ_e) is to distinguish $O_{\chi,\mathbf{s}}$ from U. In particular, for algorithm $\mathcal{A}$, define the advantage

$$\mathrm{Adv}^{\mathrm{dlwe}}_{n,q,\chi_s,\chi_e}(\mathcal{A}) = \left| \Pr(\mathbf{s} \xleftarrow{\$} \mathbb{Z}_q^n : \mathcal{A}^{O_{\chi_e,\mathbf{s}}}() = 1) - \Pr(\mathcal{A}^U() = 1) \right|.$$

Choice of Distributions. The error distribution χ_e is usually a discrete Gaussian distribution of width αq for "error rate" $\alpha < 1$.

LWE was originally phrased involving a uniform distribution on the secret $\mathbf{s}$ ($\chi_s^n = \mathcal{U}(\mathbb{Z}_q^n)$). Applebaum et al. [5] showed that the *short secrets* (or "normal form") variant, in which $\chi_s = \chi_e$, has a tight reduction to the original uniform secrets variant. In what follows, we use the short secrets variant throughout, and abbreviate to a single error distribution χ (using shorthand notation $\mathrm{Adv}^{\text{lwe}}_{n,m,q,\chi}$ and $\mathrm{Adv}^{\text{dlwe}}_{n,q,\chi}$).

Difficulty. Difficulty of both search and decision LWE problems depends on the size of n, m, and q, as well as the distributions χ_s and χ_e. Regev [46] showed that, for appropriate parameters, search and decision LWE are worst-case hard assuming the (average case) hardness of a lattice problem. In particular, he showed first that search-LWE is at least as hard as solving the worst-case lattice problems GapSVP_γ and SIVP_γ (for a parameter γ depending on n and α) using a quantum reduction; then that decision-LWE is at least as hard as the search version using a classical reduction. A sequence of later results have improved various aspects (making the first reduction classical, not quantum; handling more moduli); see Peikert's survey [44, Sect. 4.2.2] for a list.

Extracting Secret Bits. The decision LWE problem effectively yields an element $\langle \mathbf{a}, \mathbf{s} \rangle + e \in \mathbb{Z}_q$ that is indistinguishable from random. Parties using LWE to establish a shared secret for public key encryption (like in Regev's scheme) or key agreement (as we will see in the next section) will only approximately agree on the same value modulo q, so they will have to apply some reconciliation function and extracting a small number of bits (maybe even just 1 bit) from a single element of $\mathbb{Z}_q$. In order to establish a multi-bit shared secret with LWE, the parties will hence need to send many samples, which we can then think of in matrix form: a matrix $\mathbf{A} \xleftarrow{\$} \mathbb{Z}_q^{m \times n}$, and an error $\mathbf{e} \xleftarrow{\$} \chi^n$, to obtain $\mathbf{b} \leftarrow \mathbf{As} + \mathbf{e} \in \mathbb{Z}_q^m$. This increases communication sizes m-fold, and requires approximately $O(mn \log q)$ bits of communication to obtain an m-bit secret. To reduce communication sizes, one could try to introduce some structure to the matrix $\mathbf{A}$, for example making each row the cyclic shift of the previous row. However, rather than working in matrix form, we can shift our representation to a polynomial ring, leading us to the ring-LWE problem.

2.2   The Ring Learning with Errors Problem

In 2010, Lyubashevsky et al. [34] introduced the ring-LWE problem. Let $R = \mathbb{Z}[X]/\langle X^n + 1 \rangle$, where n is a power of 2. Let q be an integer, and define $R_q = R/qR$, i.e., $R_q = \mathbb{Z}_q[X]/\langle X^n + 1 \rangle$. In other words, R_q consists of polynomials of degree at most $n - 1$, with coefficients in $\mathbb{Z}_q$, and the wrapping rule that $X^n \equiv -1 \mod q$. The search and decision ring-LWE problems are analogues of the corresponding LWE problems, except with ring elements rather than vectors.

Definition 3 (Search ring-LWE problem). *Let n and q be positive integers. Let χ_s and χ_e be distributions over R_q. Let $s \xleftarrow{\$} \chi_s$. Let $a \xleftarrow{\$} \mathcal{U}(R_q), e \xleftarrow{\$} \chi_e$, and set $b \leftarrow as + e$. The search ring-LWE problem for (n, q, χ_s, χ_e) is to find s given (a, b). In particular, for algorithm $\mathcal{A}$ define the advantage*

$$\mathrm{Adv}_{n,q,\chi_s,\chi_e}^{\mathrm{rlwe}}(\mathcal{A}) = \Pr\left[s \xleftarrow{\$} \chi_s; a \xleftarrow{\$} \mathcal{U}(R_q); e \xleftarrow{\$} \chi_e; b \leftarrow as + e : \mathcal{A}(a, b) = s\right].$$

Again, for appropriate distributions χ_s and χ_e, not only is it conjectured to be hard to find the secret s, it is even conjectured that ring-LWE samples $(a, as + e)$ look independent and random: this is the decision ring-LWE problem.

Definition 4 (Decision ring-LWE problem). *Let n and q be positive integers. Let χ_s and χ_e be distributions over R_q. Let $s \xleftarrow{\$} \chi_s$. Define the following two oracles:*

- *$O_{\chi_e,s}$: $a \xleftarrow{\$} \mathcal{U}(R_q), e \xleftarrow{\$} \chi_e$; return $(a, as + e)$.*
- *U: $a, u \xleftarrow{\$} \mathcal{U}(R_q)$; return (a, u).*

The decision ring-LWE problem for (n, q, χ_s, χ_e) is to distinguish $O_{\chi_e,s}$ from U. In particular, for algorithm $\mathcal{A}$, define the advantage

$$\mathrm{Adv}_{n,q,\chi_s,\chi_e}^{\mathrm{drlwe}}(\mathcal{A}) = \left|\Pr(s \xleftarrow{\$} R_q : \mathcal{A}^{O_{\chi_e,s}}() = 1) - \Pr(\mathcal{A}^U() = 1)\right|.$$

Choice of Distributions. The error distribution χ_e is usually a discretized Gaussian distribution in the canonical embedding of R; for an appropriate choice of parameters, we can sample ring elements from χ_e by sampling each coefficient of the polynomial independently from a related distribution.

As with LWE, ring-LWE can be formulated using either a uniform secret ($\chi_s = \mathcal{U}(R_q)$) or with short secrets ($\chi_s = \chi_e$), which has a tight reduction to the original secrets variant. In what follows, we use the short secrets variant throughout, and abbreviate to a single error distribution χ (using shorthand notation $\mathrm{Adv}_{n,q,\chi}^{\mathrm{rlwe}}$ and $\mathrm{Adv}_{n,q,\chi}^{\mathrm{drlwe}}$).

Difficulty. Difficulty of both search and decision ring-LWE depends on the parameters n and q and the distributions χ_s and χ_e. Lyubashevsky et al. [34] showed that search ring-LWE as hard as quantumly solving approximate shortest vector problem on an *ideal* lattice in R; and then the classical search-to-decision reduction applies.

Because of the additional structure present in ring-LWE, the choice of n and q requires greater care than the unstructured LWE problem [45]. There is also the risk that the ring-LWE problem may be easier than the LWE problem. Currently, the best known algorithms for solving hard problems in ideal lattices [14,29] are the same as those for regular lattices (ignoring small polynomial speedups); and in some sieving algorithms, the ideal case enables one to save a small constant factor of time or space [11,47]. Very recently Cramer et al. [16] gave a quantum polynomial time algorithm algorithm for ideal-SVP with certain parameters, but this is not currently applicable to ring-LWE. In summary, some view LWE as a more "conservative" security choice than ring-LWE, though there is no appreciable security difference at present.

Extracting Secret Bits. The decision ring-LWE problem effectively yields a ring element that is indistinguishable from random. Being an element of $R_q = \mathbb{Z}_q[X]/\langle X^n + 1 \rangle$, we have n coefficients each of which is an element of $\mathbb{Z}_q$. As with LWE, cryptographic constructions using this will need to reconcile approximately equal shared secrets, and thus can extract only a small number of bits (maybe even just 1 bit) from each coefficient. But since there are n (independent-looking) coefficients, one can extract n random-looking bits from a single ring element. Thus, one needs approximately $O(n \log q)$ bits of communication to obtain an n-bit secret, a substantial reduction compared to LWE. Thus, in practice, one must decide between the decreased communication of ring-LWE versus the potentially more conservative security of LWE.

3 Key Exchange Protocols from LWE and Ring-LWE

Regev [46] was the first to give a public key encryption scheme from the learning with errors problem, and Lyubashevsky et al. [33] were the first to give a public key encryption scheme from ring-LWE. Like ElGamal public key encryption, both these schemes implicitly contain a key encapsulation mechanism and then one-time-mask the KEM shared secret with (an encoded form of) the message. Peikert [42] describes a corresponding approximate LWE key agreement protocol. In 2010, Lindner and Peikert [31] gave an improved LWE-based public key encryption scheme, and a ring-LWE analogue, and described how to view it as an approximate key agreement protocol. This was followed by detailed LWE- and ring-LWE-based key agreement protocols by Ding et al. [19] (including a single-bit reconciliation mechanism to obtain exact key agreement); a sketch of an LWE-based key agreement scheme by Blazy et al. [8, Figs. 1, 2]; and detailed ring-LWE-based key encapsulation mechanisms by Fujioka et al. [22, Sect. 5.2] and Peikert [43] (with an alternative single-bit reconciliation mechanism). In addition to basic unauthenticated key exchange, there have been works on using LWE to create *password-authenticated* key exchange [28] and using ring-LWE to create *authenticated* key exchange [49] (though the security proof of the latter is questioned [24]).

In this section, we will examine two unauthenticated key agreement protocols in which this paper's first author was involved. Frodo [9], an LWE-based key exchange protocol, is an instantiation of the Lindner–Peikert LWE approximate key agreement scheme using a generalization of Peikert's reconciliation mechanism in which multiple bits are extracted from a single element of $\mathbb{Z}_q$. BCNS15 [10], a ring-LWE-based key exchange protocol, is an instantiation of the key exchange scheme corresponding to the KEM in the Lyubashevsky–Piekert–Regev public key encryption scheme from ring-LWE using Peikert's reconciliation mechanism.

3.1 Common Tools: Reconciliation

In both Frodo and BCNS15, the parties will establish an approximately equal shared secret, then exchange some "hints" that allow them to perform a reconciliation operation on the approximately equal shared secret to extract some secret bits that are, with high probability, the same for both parties. The reconciliation technique of Ding et al. [19] sends a single bit "hint" for each key bit and relies on the low-order bits of the shared secret; Peikert's technique [43] also sends a single bit hint but relies on the high-order bits of the shared secret. The explanation below generalizes Peikert's approach [43] to extract multiple bits.

Let $B \in \mathbb{N}$ be the number of bits we aim to extract from one element of $\mathbb{Z}_q$. Assume $B < (\log_2 q) - 1$. Let $\overline{B} = \lceil \log_2 q \rceil - B$. Let $v \in \mathbb{Z}_q$, represented canonically as an integer in $[0, q)$. Define the *rounding* function

$$\lfloor \cdot \rceil_{2^B} : \mathbb{Z}_q \to \mathbb{Z}_{2^B} : v \mapsto \left\lfloor 2^{-\overline{B}} v \right\rceil \bmod 2^B,$$

where $\lfloor \cdot \rceil : \mathbb{R} \to \mathbb{Z}$ rounds real number x to the closest integer. When q is a multiple of 2^B, $\lfloor \cdot \rceil_{2^B}$ outputs the B most significant bits of $(v + 2^{\overline{B}-1}) \bmod q$, thereby partitioning $\mathbb{Z}_q$ into 2^B intervals of integers with the same B most significant bits (up to a cyclic shift of the values that centres these intervals around 0).

Define the *cross-rounding* function

$$\langle \cdot \rangle_{2^B} : \mathbb{Z}_q \to \mathbb{Z}_2 : v \mapsto \left\lfloor 2^{-\overline{B}+1} v \right\rfloor \bmod 2,$$

where $\lfloor \cdot \rfloor : \mathbb{R} \to \mathbb{Z}$ takes the floor of the real number x. When q is a multiple of 2^{B+1}, $\langle \cdot \rangle_{2^B}$ partitions $\mathbb{Z}_q$ into two intervals based according to their $(B + 1)$th most significant bit.

On input of $w \in \mathbb{Z}_q$ and $c \in \{0, 1\}$, the *reconciliation* function $\mathrm{rec}_{2^B}(w, c)$ outputs $\lfloor v \rceil_{2^B}$, where v is the *closest* element to w such that $\langle v \rangle_{2^B} = c$.

If Alice and Bob have approximately equal values $v, w \in \mathbb{Z}_q$, they can use the following process to derive B bits that are, with high probability, equal. Suppose q is a multiple of 2^B. Bob computes $c \leftarrow \langle v \rangle_{2^B}$ and sends c to Alice. Bob computes $k' \leftarrow \lfloor v \rceil_{2^B}$. Alice computes $k \leftarrow \mathrm{rec}_{2^B}(w, c)$.

Security of this technique follows from the following fact: if $v \in \mathbb{Z}_q$ is uniformly random, then $\lfloor v \rceil_{2^B}$ is uniformly random given $\langle v \rangle_{2^B}$.

Correctness follows if v and w are sufficiently close. Namely, if $|v - w \pmod{q}| < 2^{\overline{B}-2}$, then $\mathrm{rec}_{2^B}(w, \langle v \rangle_{2^B}) = \lfloor v \rceil_{2^B}$. Parameters must be chosen so that v and w are sufficiently close.

For our parameters in the ring setting, we will want to extract 1 bit from each element of $\mathbb{Z}_q$, but q will not be a multiple of 2. Peikert suggested the following technique: Bob computes $\overline{v} \xleftarrow{\$} \mathrm{dbl}(v)$, where $\mathrm{dbl} : \mathbb{Z}_q \to \mathbb{Z}_{2q} : x \mapsto 2x - e$, where e is sampled from $\{-1, 0, 1\}$ with probabilities $p_{-1} = p_1 = \frac{1}{4}$ and $p_0 = \frac{1}{2}$. Bob computes $c \leftarrow \langle \overline{v}/2 \rangle_2$ and sends c to Alice. Bob computes $k' \leftarrow \lfloor \overline{v}/2 \rceil_2$. Alice computes $k \leftarrow \mathrm{rec}_2(2w, c)$.

For ring-LWE, these functions are extended from $\mathbb{Z}_q$ to the ring $R_q = \mathbb{Z}_q[X]/\langle X^n + 1 \rangle$ coefficient-wise. For matrix forms of LWE, these functions can be extended to vectors component-wise.

3.2   Ring-LWE-Based Key Exchange: BCNS15

Protocol. The BCNS15 protocol [10], based on the ring-LWE problem, is shown in Fig. 1. Alice and Bob exchange ring-LWE samples $b = as + e$ and $b' = as' + e'$. They can then compute an approximately equal shared secret:

$$sb' = sas' + se' \approx sas' + s'e = bs' \in R_q = \mathbb{Z}_q[X]\langle X^n + 1 \rangle.$$

From each coefficient of the approximately equal shared secret, they extract a single secret bit.

Public parameters		
Decision ring-LWE parameters n, q, χ		
$a \xleftarrow{\$} \mathcal{U}(R_q)$		

Alice		Bob
$s, e \xleftarrow{\$} \chi$		
$b \leftarrow as + e \in R_q$	$\xrightarrow{\quad b \quad}$	$s', e' \xleftarrow{\$} \chi$
		$b' \leftarrow as' + e' \in R_q$
		$e'' \xleftarrow{\$} \chi$
		$v \leftarrow bs' + e'' \in R_q$
		$\bar{v} \xleftarrow{\$} \mathrm{dbl}(v) \in R_{2q}$
	$\xleftarrow{\quad b', c \quad}$	$c \leftarrow \langle \bar{v}/2 \rangle_2 \in \{0,1\}^n$
$k_A \leftarrow \mathrm{rec}_2(2b's, c) \in \{0,1\}^n$		$k_B \leftarrow \lfloor \bar{v}/2 \rceil_2 \in \{0,1\}^n$

Fig. 1. BCNS15: unauthenticated Diffie–Hellman-like key exchange from ring-LWE

Security. Assuming the decision ring-LWE problem is hard for the parameters chosen, the BCNS15 key exchange protocol is a secure unauthenticated key exchange protocol. The argument follows [31,43] by using two applications of the decision ring-LWE assumption: first, on Alice's computations involving s (so b becomes independent from s), and second on Bob's computations involving s' (so b' and v become independent from s'). This makes the approximately equal shared secret v uniformly random from the adversary's perspective, and as noted above the hint c reveals no information about extracted key k'.

Parameters. The BCNS15 protocol is instantiated with $n = 1024$ and $q = 2^{32} - 1$. The error distribution χ is a discrete Gaussian distribution; because n is a power of 2, this can be achieved by sampling each coefficient from a discrete Gaussian $D_{\mathbb{Z},\sigma}$ with has $D_{\mathbb{Z},\sigma}(x) = \frac{1}{S}e^{-x^2/(2\sigma^2)}$ where $S = 1 + 2\sum_{k=1}^{\infty} e^{-k^2/(2\sigma^2)}$. With these parameters, the probability that reconciliation yields $k \neq k'$ is much less than 2^{-128}. Total communication required for two parties to establish a shared secret is 8,320 bytes.

Based on hardness estimates by Albrecht et al. [3], breaking the system with these parameters would require $2^{163.8}$ operations on a classical computer with at least $2^{94.4}$ memory usage. Assuming a square-root speedup for quantum computers via Grover's algorithm (though it is not known how to achieve a full square-root speedup), this suggests at least $2^{81.9}$ quantum security. Based on the same difficulty estimates for the subsequent NewHope protocol [4], Alkim et al. list BCNS15 as having 86-bit classical security and 78-bit quantum security.

Subsequent Works. Alkim et al. [4] subsequently published the so-called "NewHope" protocol, making several improvements to the BCNS15 protocol. NewHope uses different parameters and a different error distribution (which was easier to sample), resulting in substantially improved performance and smaller communication (3,872 bytes). NewHope also uses a pseudorandomly generated a, rather than a fixed public parameter. In July 2016, Google announced that they were deploying a two-year experiment in the alpha version of their Chrome web browser (called "Canary") that uses the NewHope key exchange protocol in a hybrid ciphersuite with elliptic curve Diffie–Hellman [12]. Further improvements to NewHope have been given by several papers [25,32].

3.3 LWE-Based Key Exchange: Frodo

Protocol. The Frodo key exchange protocol [9], based on the LWE problem, is shown in Fig. 2. It uses a matrix form of the LWE problem: Alice uses m secrets $\mathbf{s}_1, \ldots, \mathbf{s}_m$, represented as a matrix $\mathbf{S}$; similarly for Bob. Alice and Bob exchange matrix LWE samples $\mathbf{B} = \mathbf{AS} + \mathbf{E}$ and $\mathbf{B}' = \mathbf{S}'\mathbf{A}' + \mathbf{E}'$. They can then compute an approximately equal shared secret:

$$\mathbf{B}'\mathbf{S} = \mathbf{S}'\mathbf{AS} + \mathbf{S}'\mathbf{E} \approx \mathbf{S}'\mathbf{AS} + \mathbf{SE}' = \mathbf{S}'\mathbf{B} \in \mathbb{Z}_q^{m \times m}.$$

From each entry of the approximately equal shared secret, they extract B secret bits. Frodo follows NewHope's idea of using a pseudorandomly generated $\mathbf{A}$.

Security. Assuming the decision LWE problem is hard for the parameters chosen, and PRF is a pseudorandom function, the Frodo key exchange protocol is a secure unauthenticated key exchange protocol. A hybrid argument goes from the standard decision-LWE problem to a matrix form of it, then the same argument as for BCNS15 above yields the indistinguishability of the session key.

Parameters. The Frodo paper contains several parameter sets, including a "recommended" parameter set, which uses $n = 752, q = 2^{15}, m = 8$, and $B = 4$. The error distribution χ is a concrete distribution specified in the paper, which is close in Renyi divergence to a rounded continuous Gaussian distribution (but requires fewer bits to sample). With these parameters, the probability that reconciliation yields $\mathbf{k} \neq \mathbf{k}'$ is $2^{-38.9}$. Total communication required for two parties to establish a shared secret is 8,320 bytes. The claimed security level is 140 bits of security against a classical adversary, and 130 bits against a quantum adversary. The paper also includes a higher-security "paranoid" parameter set, which conjectures a certain lower bound on lattice sieving for any adversary.

Public parameters

Decision LWE parameters n, q, χ; integer m

Alice	Bob

$seed \xleftarrow{\$} \{0,1\}^\lambda$
$\mathbf{A} \leftarrow \mathrm{PRF}(seed) \in \mathbb{Z}_q^{n \times n}$
$\mathbf{S}, \mathbf{E} \xleftarrow{\$} \chi(\mathbb{Z}_q^{n \times m})$
$\mathbf{B} \leftarrow \mathbf{AS} + \mathbf{E} \in \mathbb{Z}_q^{n \times m}$ $\xrightarrow{\;\;b, seed\;\;}$ $\mathbf{A} \leftarrow \mathrm{PRF}(seed) \in \mathbb{Z}_q^{n \times n}$
$\mathbf{S}', \mathbf{E}' \xleftarrow{\$} \chi(\mathbb{Z}_q^{m \times n})$
$\mathbf{B}' \leftarrow \mathbf{S}'\mathbf{A} + \mathbf{E}' \in \mathbb{Z}_q^{m \times n}$
$\mathbf{E}'' \xleftarrow{\$} \chi(\mathbb{Z}_q^{m \times m})$
$\mathbf{V} \leftarrow \mathbf{S}'\mathbf{B} + \mathbf{E}'' \in \mathbb{Z}_q^{m \times m}$

$\xleftarrow{\;\;\mathbf{B}', \mathbf{C}\;\;}$ $\mathbf{C} \leftarrow \langle \mathbf{V} \rangle_{2^B} \in \mathbb{Z}_{2^B}^{m \times m}$
$\mathbf{k} \leftarrow \mathrm{rec}_{2^B}(\mathbf{B}'\mathbf{S}, \mathbf{C}) \in \mathbb{Z}_{2^B}^m$ $\mathbf{k}' \leftarrow \lfloor \mathbf{V} \rceil_{2^B} \in \mathbb{Z}_{2^B}^m$

Fig. 2. Frodo: unauthenticated Diffie–Hellman-like key exchange from LWE

3.4 Performance of Post-quantum Key Exchange

Table 1 (copied from [9]) shows the performance characteristics of several post-quantum key exchange protocols:

- BCNS ring-LWE key exchange, C implementation [10];
- NewHope ring-LWE key exchange, C implementation [4];
- NTRU public key encryption key transport using parameter set EES743EP1, C implementation;[2] and
- SIDH (supersingular isogeny Diffie–Hellman) key exchange, C implementation [15].

The table also includes non-quantum-secure algorithms at the 128-bit classical security level for comparison: OpenSSL's implementation of ECDH (on the nistp256 curve) and RSA with a 3072-bit modulus. Results were measured on a single hardware hyper-thread on a 2.6 GHz Intel Xeon E5 (Sandy Bridge); see [9] for details. Although some implementations included optimizations using the AVX2 instruction set, the computer used for measurements did not support AVX2.

In the table, Alice0 denotes Alice's procedure for constructing her outgoing message, and Alice1 is her procedure for processing Bob's incoming message and deriving the shared secret.

The NewHope protocol has the best computational performance of the post-quantum key exchange algorithms tested, even outperforming traditional RSA and ECDH. However, all structured lattice schemes (ring-LWE and NTRU) have larger communication than RSA and ECDH, around 2–8 KiB round-trip.

---

[2] https://github.com/NTRUOpenSourceProject/ntru-crypto.

Table 1. Performance of standalone cryptographic operations, showing mean runtime in milliseconds of standalone cryptographic operations, communication sizes (public key/messages) in bytes, and claimed security level in bits. Table from [9].

Scheme	Alice0 (ms)	Bob (ms)	Alice1 (ms)	Communication (bytes)		Claimed security	
				A→B	B→A	Classical	Quantum
RSA 3072-bit	—	0.09	4.49	387/0*	384	128	—
ECDH nistp256	0.37	0.70	0.33	32	32	128	—
BCNS	1.01	1.59	0.17	4,096	4,224	86	78
NewHope	0.11	0.16	0.03	1,824	2,048	229	206
NTRU EES743EP1	2.00	0.28	0.15	1,027	1,022	256	128
Frodo recomm.	1.13	1.34	0.13	11,377	11,296	144	130
Frodo paranoid	1.25	1.64	0.15	13,057	12,976	177	161
SIDH	135	464	301	564	564	192	128

*In TLS, the RSA public key is already included in the server's certificate message, so RSA key transport imposes no additional communication from server to client.

Unstructured lattice schemes (LWE) also achieve good performance, on the order of 1 ms, but require even more communication, around 22 KiB round-trip. Supersingular isogeny Diffie–Hellman has much smaller keys (1 KiB round-trip, not much larger than RSA 3072), but orders of magnitude slower performance. (Note, however, that the AVX2 optimized implementation of SIDH was an order of magnitude faster than its C implementation). No code-based post-quantum protocol was included in the tests above. In particular, the implementation of Bernstein et al.'s "McBits" high-speed code-based cryptosystem [6] was not publicly available at the time of writing, but their paper reports speeds of 0.005ms (on a 3.4 GHz CPU) for decryption at the 128-bit quantum security level, but at the cost of 216 KiB public keys.

These trade-offs leave no clear post-quantum winner: the smallest key sizes come from SIDH but it has slow performance (though performance usually improves!); ring-LWE gives a decent tradeoff with fast performance and not-too-big keys; LWE's performance remains good, and avoids the use of a structured lattice, but requires larger communication. Though these larger public keys may be too big for embedded devices, it should be remembered that the average webpage is over 1 MB: if we had to switch the Internet to post-quantum cryptography today, the communication costs from post-quantum key exchange would not be much more than an extra emoticon on a webpage.

3.5   From Unauthenticated to Authenticated Key Exchange

Both the BCNS15 and Frodo protocols are for unauthenticated key exchange: they assume the adversary is passive. Of course in practice one must achieve security against an active network adversary. Peikert [43] noted the challenges that are faced in securing LWE and ring-LWE based protocols against an active adversary, and Fluhrer [21] described an explicit attack on ring-LWE protocols that reuse ephemeral key shares against an active adversary. Peikert suggested the use of a transform such as the Fujisaki–Okamoto transform [23] which converts a passively secure (IND-CPA) key encapsulation mechanism (KEM) into an actively secure (IND-CCA) KEM. For integration with TLS, there is also the possibility of using signatures in a signed-DH-like protocol to first authenticate the keyshares; see [10].

4   Integrating Post-quantum Key Exchange into TLS

All the post-quantum key exchange candidates explored in the previous section incur some penalty (either slower computation, or bigger communication, or both) compared to existing RSA or elliptic curve public key cryptography. It is therefore important to understand the impact of these penalties in a practical setting. Both the BCNS15 and Frodo papers integrate the corresponding key exchange scheme into the Transport Layer Security (TLS) protocol, the dominant protocol used securing Internet communications. In particular, they create new TLS version 1.2 ciphersuites which use traditional RSA or ECDSA certificates for signature, but use post-quantum key exchange to derive a shared secret, and then continue to use standard TLS authenticated encryption constructions (e.g., AES in GCM mode). (Due to the message flow in the TLS 1.2 handshake, the TLS server plays the role of "Alice" in the key exchange, and the TLS client plays the role of "Bob".) This is achieved by modifying OpenSSL, a common open-source library for SSL/TLS, which is used by applications such as the Apache httpd web server for securing web server communication.

Hybrid Ciphersuites. The experiments involving post-quantum ciphersuites in TLS also included hybrid ciphersuites, where the TLS handshake uses two key exchange algorithms: one post-quantum algorithm, and one traditional algorithm (in this case, ECDH). While the use of two key exchange algorithms does impact performance, it allows early adopters to retain the (current) security of traditional algorithms like ECDHE while obtaining (potential) security against quantum computers: since many post-quantum algorithms have had comparatively less cryptanalysis, there is an increased chance that parameter sizes for post-quantum algorithms will evolve more rapidly over the next few years in the face of new classical or quantum cryptanalytic advances. Interestingly, Google, in its recent NewHope experiment in Chrome, decided to use solely hybrid ciphersuites [12].

Security. As noted above, BCNS15 and Frodo were shown to be secure *unauthenticated* key exchange protocols, i.e., assuming a passive adversary. For security against an active adversary, we showed in the BCNS paper [10] how to achieve the standard security notion for TLS ("authenticated and confidential channel establishment" (ACCE) [27]) if the server signs both the client and server key share. Note that this would require reordering some of the messages in TLS. An alternative, as noted above, is to use a KEM transform to obtain an actively-secure key exchange protocol.

Table 2. Performance of Apache httpd web server, measured in connections per second, connection time in milliseconds, and handshake size in bytes. Table from [9].

Ciphersuite		Connections/second			Connection time (ms)		Handshake size (bytes)
Key exchange	Signature	1B	10 KiB	100 KiB	w/o load	w/load	
ECDHE nistp256	ECDSA	1187	1088	961	14.2	22.2	1, 264
	RSA	814	790	710	16.1	24.7	1,845
BCNS15	ECDSA	922	893	819	18.8	35.8	9,455
	RSA	722	716	638	20.5	36.9	9,964
NewHope	ECDSA	1616	1351	985	12.1	18.6	5,005
	RSA	983	949	771	13.1	20.0	5,514
NTRU EES743EP1	ECDSA	725	708	612	20.0	27.2	3,181
	RSA	553	548	512	19.9	29.6	3,691
Frodo Recomm.	ECDSA	923	878	843	18.3	31.5	23,725
	RSA	703	698	635	20.7	32.7	24,228
Hybrid ciphersuites							
BCNS15+ECDHE	ECDSA	736	728	664	23.1	37.7	9,595
	RSA	567	559	503	24.6	40.2	10,177
NewHope+ECDHE	ECDSA	1095	1017	776	16.5	25.2	5,151
	RSA	776	765	686	18.1	28.0	5,731
NTRU+ECDHE	ECDSA	590	578	539	22.5	34.3	3,328
	RSA	468	456	424	24.2	36.8	3,908
Frodo Rec.+ECDHE	ECDSA	735	701	667	22.9	36.4	23,859
	RSA	552	544	516	24.5	39.9	24,439

All TLS ciphersuites used AES256-GCM authenticated encryption with SHA384 in the MAC and KDF. Note that different key exchange methods are at different security levels; see Table 1 for details.

4.1 Performance of Post-quantum Key Exchange in TLS

Table 2 (copied from [9]) shows the performance of a TLS-protected Apache web server using various key exchange mechanisms and signature schemes. It measures:

- *throughput* (connections/second): number of connections per second at the server before server latency spikes, measured with requests of different payload sizes (1 B, 10 KiB, 100 KiB);
- *handshake latency* (milliseconds): time from when client sends first TCP packet till client receives first application data packet, measured on an unloaded server and a loaded server (with sufficiently many connections to achieve 70% CPU load).

Performance was measured on a 4-CPU server with the same CPU as in Sect. 3.4. See [9] for the detailed methodology.

Unsurprisingly, the additional overhead of other cryptographic and network operations in a TLS connection mutes the performance differences between key exchange protocols. For example, while the standalone performance of NewHope is 9× better than that of Frodo recommended, throughput of a NewHope-based ciphersuite is only 1.75× better than Frodo recommended when the server

returns 1 byte of application data, and the gap narrows further to just 1.12×
when the server returns 100 KiB of application data. Similarly, the latency of a
Frodo-based ciphersuite is only 1.5× slower than a NewHope-based ciphersuite.
When hybrid ciphersuites are used, the performance difference between slow and
fast post-quantum ciphersuites narrows even further.

5 Interlude: Programming is Hard

In the BCNS15 work on ring-LWE-based key exchange, we did a performance
evaluation at two levels: the standalone cryptographic operations of the ring-
LWE key exchange protocol, and its performance when run in the TLS protocol.
The first is a fairly common practice in cryptographic research: implement your
algorithms in C, then use some cycle counting or microsecond-accurate timing
code to determine the runtime of your algorithms.

Evaluating performance in the TLS protocol is less common due in part to
the difficulty of doing so: either one has to implement a network protocol from
scratch (which is painful and usually not the main purpose of the research), or
integrate the cryptographic algorithms into an existing cryptographic library,
such as OpenSSL. These libraries are often quite complex. When we wanted to
add our BCNS15 ring-LWE key exchange protocol to OpenSSL for testing purposes, we had to first "wrap" our core ring-LWE library inside of OpenSSL's
data structures inside the `crypto` directory, then modify OpenSSL's `ssl` directory to make use of those new data structures. Table 3 shows the number of files
and lines of code that were added or changed. While the core ring-LWE library
consisted of only 6 (standalone) C files totalling just under 900 lines of code,
integrating it into OpenSSL required touching 23 files and changing or adding
another 2143 lines of code.

Table 3. Source code changes to add BCNS15 ring-LWE key exchange to OpenSSL

Component	New files	Existing files	Lines of code*
Core ring-LWE library	6	0	896
Ring-LWE "wrapper" for OpenSSL	6	5	1229
SSL integration	0	12	914

*Lines of code excludes Makefiles and automatically generated files, but includes
comments and whitespace, and counts both lines added and deleted. Calculated
from https://github.com/dstebila/openssl-rlwekex/commit/f80719bf.

For the Frodo work on LWE-based key exchange, we again wanted to evaluate
the performance of our algorithms both in a standalone setting and in the context
of TLS, but we also wanted to compare with other post-quantum key exchange
candidates. Writing 2100 lines of wrapper/integration code for each algorithm we
wanted to add was an unappealing prospect. For the Frodo project, we developed

an intermediate API that allowed us to more easily integrate different post-quantum key exchange algorithms into OpenSSL for performance comparison. This not-publicly-released intermediate API was the predecessor of and partial motivation for some of the features added to the Open Quantum Safe framework.

6   Open Quantum Safe: A Software Framework for Post-quantum Cryptography

The goal of our Open Quantum Safe (OQS) project (https://openquantumsafe.org) is to support the development and prototyping of quantum-resistant cryptography. OQS consists of two main lines of work: liboqs, an open source C library for quantum-resistant cryptographic algorithms; and prototype integrations into protocols and applications, including the widely used OpenSSL library.

As an example of where the OQS framework can assist with the grand challenge of moving quantum-resistant cryptography towards reliable widespread deployment, consider a small- or medium-sized enterprise that understands the need to integrate quantum-resistant cryptography into its products. Perhaps their products protect information that requires long-term confidentiality. Perhaps their products will be deployed in the field for many years with no easy opportunity for changing the cryptographic algorithms later. Or perhaps they or their customers are worried about the small but non-negligible chance that today's algorithms will be broken, by quantum computers or otherwise, much earlier than expected.

Whatever their reason for wishing to integrate quantum-safe cryptography into their products sooner rather than later, this would not be an easy path for them to take. In-house implementation of quantum-safe primitives requires advanced specialized expertise in order to understand the research literature, choose a suitable scheme, digest the new mathematics, choose suitable parameters, and develop robust software or hardware implementations. This is an enormous, expensive, and risky endeavour to undertake on one's own, especially for a small- or medium-sized enterprise.

Commercially available alternatives, especially back in 2014 when this project started taking shape, were few, and also potentially problematic from a variety of perspectives: cost, patents, transparency, maintenance, degree of external scrutiny, etc.

Companies who would like to offer a quantum-safe option today do not have an easy or robust path for doing so.

OQS gives such organizations the option of prototyping an available quantum-resistant algorithm in their applications. Since these are still largely experimental algorithms that have not yet received the intense scrutiny of the global cryptographic community, our recommendation is to use one of the available post-quantum algorithms in a "hybrid" fashion with a standard algorithm that has received intense scrutiny with respect to classical cryptanalysis and robust implementation.

Since we fully expect that ongoing developments and improvements in the design, cryptanalysis, and implementation of quantum-safe algorithms, OQS is designed so improvements and changes in the post-quantum algorithm can be adopted without major changes to application software.

Organizations who do not wish or need to use open source in their products can still benefit from:

− reference implementations that will guide them in their own implementations
− benchmark information that will guide their choice of algorithm
− the ability to test alternatives in their products before deciding which algorithms to choose.

OQS was thus designed with the goal of both facilitating the prototyping and testing of quantum-resistant algorithms in a range of applications, and of driving forward the implementation, testing, and benchmarking of quantum-resistant primitives themselves.

The high-level architecture of the OQS software project is shown in Fig. 3.

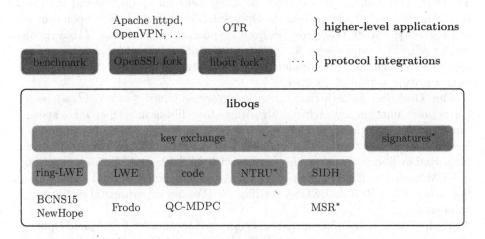

Fig. 3. Architecture of the Open Quantum Safe project. (* denotes future plans.)

6.1   liboqs

liboqs (https://github.com/open-quantum-safe/liboqs) provides a common interface for key exchange and digital signature schemes, as well as implementations of a variety of post-quantum schemes. Some implementations are based on existing open source implementations, either adapting the implementation or putting a thin "wrapper" around the implementation. Other implementations have been written from scratch directly for the library. As of writing, liboqs includes key exchange based on:

- ring-LWE using the BCNS15 protocol (adaptation of existing implementation) [10];
- ring-LWE using the NewHope protocol (wrapper around existing implementation) [4];
- LWE using the Frodo protocol (adaptation of existing implementation) [9];
- error correcting codes – quasi-cyclic medium-density parity-check codes using the Niederreiter cryptosystem (new implementation).

liboqs also includes common routines available to all liboqs modules, including a common random number generator and various symmetric primitives such as AES and SHA-3.

liboqs includes a benchmarking program that enables runtime comparisons of all supported implementations. The library and benchmarking program build and have been tested on Mac OS X 10.11.6, macOS 10.12, and Ubuntu 16.04.1 (using clang or gcc), and Windows 10 (using Visual Studio).

6.2   Application/Protocol Integrations

The OQS project also includes prototype integrations into protocols and applications. Our first integration is into the OpenSSL library,[3] which is an open source cryptographic library that provides both cryptographic functions (libcrypto) and an SSL/TLS implementation (libssl). OpenSSL is used by many network applications, including the popular Apache httpd web server and the OpenVPN virtual private networking software.

Our OpenSSL fork (https://github.com/open-quantum-safe/openssl) integrates post-quantum key exchange algorithms from liboqs into OpenSSL's speed command, and provides TLS 1.2 ciphersuites using post-quantum key exchange based on primitives from liboqs. For each post-quantum key exchange primitive supported by liboqs, there are ciphersuites with AES-128 or AES-256 encryption in GCM mode (with either SHA-256 or SHA-384, respectively), and authentication using either RSA or ECDSA certificates. (We use experimental ciphersuite numbers.)

Each of these four ciphersuites is also mirrored by another four *hybrid* ciphersuites which use both elliptic curve Diffie–Hellman (ECDHE) key exchange *and* the post-quantum key exchange primitive.

Our OpenSSL integration also includes *generic* ciphersuites. liboqs includes interfaces for each key exchange algorithm so it can be selected by the caller at runtime, but it also includes a generic interface that can be configured *at compile time*. Our OpenSSL integration does include ciphersuites for each individual key exchange algorithm in liboqs, but it also includes a set of ciphersuites that call the generic interface, which will then use whatever key exchange algorithm was specified at compile time. This means that a developer can add a new algorithm to liboqs and immediately prototype its use in SSL/TLS without changing a single line of code in OpenSSL, simply by using the generic OQS ciphersuites in OpenSSL and compiling liboqs to use the desired algorithm.

---

[3] https://www.openssl.org.

6.3   Case Study: Adding NewHope to liboqs and OpenSSL

As mentioned earlier, one of the goals of the Open Quantum Safe project is to make it easier to prototype post-quantum cryptography. It should be easy to add a new algorithm to liboqs, and then easy to use that algorithm in an application or protocol that already supports liboqs.

Recently, we added the NewHope ring-LWE-based key exchange to liboqs and our OpenSSL fork. It is interesting to compare the amount of work required to add NewHope to liboqs and our OpenSSL fork with the figures in Table 3 on adding BCNS15 directly to OpenSSL.

In liboqs, the wrapper around NewHope is 2 new files, totalling 163 lines of code, and requires 5 lines of code to be changed in 2 other files (plus changes in the Makefile).

As noted above, liboqs includes a "generic" key exchange method which can be hard-coded at compile time to any one of its implementations, and our OpenSSL fork already includes a "generic OQS" key exchange ciphersuite that calls liboqs' generic key exchange method. Thus, once NewHope has been added to liboqs, it is possible to test NewHope in OpenSSL with zero changes to the OpenSSL fork via the generic key exchange method and recompiling. However, to explicitly add named NewHope ciphersuites to OpenSSL, we are able to reuse existing data structures, resulting in a diff that touches 10 files and totals 222 lines of code. Moreover, the additions can very easily follow the pattern from previous diffs,[4] making adding a new OQS-based ciphersuite a 15-min job.

7   Conclusion and Outlook

The next few years will be an exciting time in the area of post-quantum cryptography. With the forthcoming NIST post-quantum project, and with continuing advances in quantum computing research, there will be increasing interest from government, industry, and standards bodies in understanding and using quantum-resistant cryptography. Lattice-based cryptography, in the form of the learning with errors and the ring-LWE problems, is particularly promising for quantum-resistant public key encryption and key exchange, offering high computation efficiency with reasonable key sizes. More cryptanalytic research will be essential to increase confidence in any standardized primitive. Since each post-quantum candidate to date has trade-offs between computational efficiency and communication sizes compared to existing primitives, it is also important to understand the how applications and network protocols behave when using different post-quantum algorithms. The Open Quantum Safe project can help rapidly compare post-quantum algorithms and prototype their use in existing protocols and applications, and experiments like Google's use of NewHope in its Chrome Canary browser will give valuable information about how post-quantum cryptosystems behave in real-world deployments.

---

[4] https://github.com/open-quantum-safe/openssl/commit/cb91c708 and https://github.com/open-quantum-safe/openssl/commit/3a04b822.

For cryptographers interested in designing new public key encryption, digital signature schemes, and key exchange protocols—for cryptanalysts looking to study new mathematical problems—for cryptographic engineers building new systems—and for standards bodies preparing for the future—exciting times lie ahead!

Acknowledgements. Research on LWE and ring-LWE based key exchange discussed in this paper includes joint work with Joppe W. Bos (NXP Semiconductors), Craig Costello (Microsoft Research), Léo Ducas (CWI), Ilya Mironov (Google), Michael Naehrig (Microsoft Research), Valeria Nikolaenko (Stanford University), and Ananth Raghunathan (Google) and was published as [9,10].

The Open Quantum Safe project grew out of discussions with a number of colleagues over the past few years, especially during early 2014, around the challenges of taking post-quantum cryptography closer to widespread practical application. These colleagues include: Scott Vanstone and Sherry Shannon Vanstone (Trustpoint); Matthew Campagna (Amazon Web Services); Alfred Menezes, Ian Goldberg, and Guang Gong (University of Waterloo); William Whyte and Zhenfei Zhang (Security Innovation); as well as the research colleagues in the paragraph above; we are grateful for all the valuable discussions. The Open Quantum Safe project has been supported in part by the Tutte Institute for Mathematics and Computing. Contributors to OQS are listed at https://github.com/open-quantum-safe/liboqs/graphs/contributors and as of writing include Jennifer Fernick, David Jao, Tancrède Lepoint, Shravan Mishra, Christian Paquin, Alex Parent, John Schanck, and Sebastian Verschoor.

D.S. thanks the chairs of SAC 2016 for the invitation to give the Stafford Tavares Invited Lecture at Canada's annual cryptography conference.

References

1. Ajtai, M.: Generating hard instances of lattice problems (extended abstract). In: 28th ACM STOC, pp. 99–108. ACM Press, May 1996
2. Ajtai, M., Dwork, C.: A public-key cryptosystem with worst-case/average-case equivalence. In: 29th ACM STOC, pp. 284–293. ACM Press, May 1997
3. Albrecht, M.R., Player, R., Scott, S.: On the concrete hardness of learning with errors. Cryptology ePrint Archive, Report 2015/046 (2015). http://eprint.iacr.org/2015/046
4. Alkim, E., Ducas, L., Pöppelmann, T., Schwabe, P.: Post-quantum key exchange - a new hope. In: USENIX Security 2016. USENIX Association, August 2016. https://eprint.iacr.org/2015/1092
5. Applebaum, B., Cash, D., Peikert, C., Sahai, A.: Fast cryptographic primitives and circular-secure encryption based on hard learning problems. In: Halevi, S. (ed.) CRYPTO 2009. LNCS, vol. 5677, pp. 595–618. Springer, Heidelberg (2009). doi:10.1007/978-3-642-03356-8_35
6. Bernstein, D.J., Chou, T., Schwabe, P.: McBits: fast constant-time code-based cryptography. In: Bertoni, G., Coron, J.-S. (eds.) CHES 2013. LNCS, vol. 8086, pp. 250–272. Springer, Heidelberg (2013). doi:10.1007/978-3-642-40349-1_15
7. Bernstein, D.J., et al.: SPHINCS: practical stateless hash-based signatures. In: Oswald, E., Fischlin, M. (eds.) EUROCRYPT 2015. LNCS, vol. 9056, pp. 368–397. Springer, Heidelberg (2015). doi:10.1007/978-3-662-46800-5_15

8. Blazy, O., Chevalier, C., Ducas, L., Pan, J.: Exact smooth projective hash function based on LWE. Cryptology ePrint Archive, Report 2013/821 (2013). http://eprint.iacr.org/2013/821

9. Bos, J.W., Costello, C., Ducas, L., Mironov, I., Naehrig, M., Nikolaenko, V., Raghunathan, A., Stebila, D.: Frodo: take off the ring! practical, quantum-secure key exchange from LWE. In: ACM CCS 2016. ACM Press, October 2016. https://eprint.iacr.org/2016/659

10. Bos, J.W., Costello, C., Naehrig, M., Stebila, D.: Post-quantum key exchange for the TLS protocol from the ring learning with errors problem. In: 2015 IEEE Symposium on Security and Privacy, pp. 553–570. IEEE Computer Society Press, May 2015

11. Bos, J.W., Naehrig, M., van de Pol, J.: Sieving for shortest vectors in ideal lattices: a practical perspective. Cryptology ePrint Archive, Report 2014/880 (2014). http://eprint.iacr.org/2014/880

12. Braithwaite, M.: Google security blog: experimenting with post-quantum cryptography, July 2016. https://security.googleblog.com/2016/07/experimenting-with-post-quantum.html

13. Buchmann, J., Dahmen, E., Hülsing, A.: XMSS - a practical forward secure signature scheme based on minimal security assumptions. In: Yang, B.-Y. (ed.) PQCrypto 2011. LNCS, vol. 7071, pp. 117–129. Springer, Heidelberg (2011). doi:10.1007/978-3-642-25405-5_8

14. Chen, Y., Nguyen, P.Q.: BKZ 2.0: better lattice security estimates. In: Lee, D.H., Wang, X. (eds.) ASIACRYPT 2011. LNCS, vol. 7073, pp. 1–20. Springer, Heidelberg (2011). doi:10.1007/978-3-642-25385-0_1

15. Costello, C., Longa, P., Naehrig, M.: Efficient algorithms for supersingular Isogeny Diffie-Hellman. In: Robshaw, M., Katz, J. (eds.) CRYPTO 2016. LNCS, vol. 9814, pp. 572–601. Springer, Heidelberg (2016). doi:10.1007/978-3-662-53018-4_21

16. Cramer, R., Ducas, L., Wesolowski, B.: Short Stickelberger class relations and application to Ideal-SVP. Cryptology ePrint Archive, Report 2016/885 (2016). http://eprint.iacr.org/2016/885

17. Devoret, M.H., Schoelkopf, R.J.: Superconducting circuits for quantum information: an outlook. Science 339(6124), 1169–1174 (2013)

18. Dierks, T., Rescorla, E.: The Transport Layer Security (TLS) Protocol Version 1.2. RFC 5246 (Proposed Standard), August 2008. http://www.ietf.org/rfc/rfc5246.txt

19. Ding, J., Xie, X., Lin, X.: A simple provably secure key exchange scheme based on the learning with errors problem. Cryptology ePrint Archive, Report 2012/688 (2012). http://eprint.iacr.org/2012/688

20. Feo, L.D., Jao, D., Plût, J.: Towards quantum-resistant cryptosystems from supersingular elliptic curve isogenies. J. Math. Cryptol. 8(3), 209–247 (2014)

21. Fluhrer, S.: Cryptanalysis of ring-LWE based key exchange with key share reuse. Cryptology ePrint Archive, Report 2016/085 (2016). http://eprint.iacr.org/2016/085

22. Fujioka, A., Suzuki, K., Xagawa, K., Yoneyama, K.: Practical and post-quantum authenticated key exchange from one-way secure key encapsulation mechanism. In: Chen, K., Xie, Q., Qiu, W., Li, N., Tzeng, W.G. (eds.) ASIACCS 13, pp. 83–94. ACM Press, May 2013

23. Fujisaki, E., Okamoto, T.: How to enhance the security of public-key encryption at minimum cost. In: Imai, H., Zheng, Y. (eds.) PKC 1999. LNCS, vol. 1560, pp. 53–68. Springer, Heidelberg (1999). doi:10.1007/3-540-49162-7_5

24. Gong, B., Zhao, Y.: Small field attack, and revisiting RLWE-based authenticated key exchange from Eurocrypt'15. Cryptology ePrint Archive, Report 2016/913 (2016). http://eprint.iacr.org/2016/913

25. Gueron, S., Schlieker, F.: Speeding up R-LWE post-quantum key exchange. Cryptology ePrint Archive, Report 2016/467 (2016). http://eprint.iacr.org/2016/467

26. Hoffstein, J., Pipher, J., Silverman, J.H.: NTRU: a ring-based public key cryptosystem. In: Buhler, J.P. (ed.) ANTS 1998. LNCS, vol. 1423, pp. 267–288. Springer, Heidelberg (1998). doi:10.1007/BFb0054868

27. Jager, T., Kohlar, F., Schäge, S., Schwenk, J.: On the security of TLS-DHE in the standard model. In: Safavi-Naini, R., Canetti, R. (eds.) CRYPTO 2012. LNCS, vol. 7417, pp. 273–293. Springer, Heidelberg (2012). doi:10.1007/978-3-642-32009-5_17

28. Katz, J., Vaikuntanathan, V.: Smooth projective hashing and password-based authenticated key exchange from lattices. In: Matsui, M. (ed.) ASIACRYPT 2009. LNCS, vol. 5912, pp. 636–652. Springer, Heidelberg (2009). doi:10.1007/978-3-642-10366-7_37

29. Laarhoven, T.: Sieving for shortest vectors in lattices using angular locality-sensitive hashing. In: Gennaro, R., Robshaw, M. (eds.) CRYPTO 2015. LNCS, vol. 9215, pp. 3–22. Springer, Heidelberg (2015). doi:10.1007/978-3-662-47989-6_1

30. Lamport, L.: Constructing digital signatures from a one-way function. Technical report SRI-CSL-98, SRI International Computer Science Laboratory, October 1979

31. Lindner, R., Peikert, C.: Better key sizes (and attacks) for LWE-based encryption. In: Kiayias, A. (ed.) CT-RSA 2011. LNCS, vol. 6558, pp. 319–339. Springer, Heidelberg (2011). doi:10.1007/978-3-642-19074-2_21

32. Longa, P., Naehrig, M.: Speeding up the number theoretic transform for faster ideal lattice-based cryptography. Cryptology ePrint Archive, Report 2016/504 (2016). http://eprint.iacr.org/2016/504

33. Lyubashevsky, V., Peikert, C., Regev, O.: On ideal lattices and learning with errors over rings. In: Gilbert, H. (ed.) EUROCRYPT 2010. LNCS, vol. 6110, pp. 1–23. Springer, Heidelberg (2010). doi:10.1007/978-3-642-13190-5_1

34. Lyubashevsky, V., Peikert, C., Regev, O.: A toolkit for ring-LWE cryptography. In: Johansson, T., Nguyen, P.Q. (eds.) EUROCRYPT 2013. LNCS, vol. 7881, pp. 35–54. Springer, Heidelberg (2013). doi:10.1007/978-3-642-38348-9_3

35. Matsumoto, T., Imai, H.: Public quadratic polynomial-tuples for efficient signature-verification and message-encryption. In: Barstow, D., et al. (eds.) EUROCRYPT 1988. LNCS, vol. 330, pp. 419–453. Springer, Heidelberg (1988). doi:10.1007/3-540-45961-8_39

36. McEliece, R.: A public-key cryptosystem based on algebraic coding theory. DSN Progress Report 42-44, January and February 1978. http://ipnpr.jpl.nasa.gov/progress_report2/42-44/44N.PDF

37. Merkle, R.C.: A certified digital signature. In: Brassard, G. (ed.) CRYPTO 1989. LNCS, vol. 435, pp. 218–238. Springer, New York (1990). doi:10.1007/0-387-34805-0_21

38. Mosca, M.: Cybersecurity in an era with quantum computers: will we be ready? Cryptology ePrint Archive, Report 2015/1075 (2015). http://eprint.iacr.org/2015/1075

39. National Security Agency: NSA Suite B cryptography: Cryptography today, August 2015. https://www.nsa.gov/ia/programs/suiteb_cryptography/

40. Niederreiter, H.: Knapsack-type cryptosystems and algebraic coding theory. Prob. Control Inf. Theory. Problemy Upravlenija i Teorii Informacii **15**, 159–166 (1986)

41. Ong, H.; Schnorr, C.P.: Signatures through approximate representation by quadratic forms. In: Chaum, D. (ed.) CRYPTO 1983, pp. 117–131. Plenum Press, New York (1983)
42. Peikert, C.: Some recent progress in lattice-based cryptography. In: Reingold, O. (ed.) TCC 2009. LNCS, vol. 5444, pp. 72–72. Springer, Heidelberg (2009). doi:10.1007/978-3-642-00457-5_5
43. Peikert, C.: Lattice cryptography for the internet. In: Mosca, M. (ed.) PQCrypto 2014. LNCS, vol. 8772, pp. 197–219. Springer, Cham (2014). doi:10.1007/978-3-319-11659-4_12
44. Peikert, C.: A decade of lattice cryptography. Found. Trends Theor. Comput. Sci. **10**(4), 283–424 (2016)
45. Peikert, C.: How (not) to instantiate ring-LWE. Cryptology ePrint Archive, Report 2016/351 (2016). http://eprint.iacr.org/2016/351
46. Regev, O.: On lattices, learning with errors, random linear codes, and cryptography. In: Gabow, H.N., Fagin, R. (eds.) 37th ACM STOC, pp. 84–93. ACM Press, May 2005
47. Schneider, M.: Sieving for shortest vectors in ideal lattices. In: Youssef, A., Nitaj, A., Hassanien, A.E. (eds.) AFRICACRYPT 2013. LNCS, vol. 7918, pp. 375–391. Springer, Heidelberg (2013). doi:10.1007/978-3-642-38553-7_22
48. Shor, P.W.: Algorithms for quantum computation: discrete logarithms and factoring. In: 35th FOCS, pp. 124–134. IEEE Computer Society Press, November 1994
49. Zhang, J., Zhang, Z., Ding, J., Snook, M., Dagdelen, Ö.: Authenticated key exchange from ideal lattices. In: Oswald, E., Fischlin, M. (eds.) EUROCRYPT 2015. LNCS, vol. 9057, pp. 719–751. Springer, Heidelberg (2015). doi:10.1007/978-3-662-46803-6_24

Side Channels and Fault Attacks

Detecting Side Channel Vulnerabilities in Improved Rotating S-Box Masking Scheme—Presenting Four Non-profiled Attacks

Zeyi Liu[1,2,3], Neng Gao[1,2], Chenyang Tu[1,2(✉)], Yuan Ma[1,2], and Zongbin Liu[1,2]

[1] Data Assurance and Communication Security Research Center, Beijing, China
{liuzeyi,gaoneng,tuchenyang,mayuan,liuzongbin}@iie.ac.cn
[2] State Key Laboratory of Information Security,
Institute of Information Engineering, CAS, Beijing, China
[3] University of Chinese Academy of Sciences, Beijing, China

Abstract. Improved Rotating S-box Masking (RSM2.0 for short) is a well-known countermeasure designed and implemented by DPA Contest V4.2 committee to provide security protection for AES-128. By combining both 1st-order masking and shuffling techniques, improved RSM claims to offer at least non-profiled resistance for its software implementation and up to now no systematic research has been published to challenge such security claim yet. To study the practical security of RSM2.0 against non-profiled attacks, we first propose an analytical methodology to guide the detection of the exploitable vulnerabilities in RSM2.0. On the basis of the methodology, several potential flaws hidden in both the algorithm design and detailed implementation of RSM2.0 are discovered and we make use of them to design six attacking schemes in total, all of which belong to non-profiled attacks. Four representative attacks are eventually implemented and submitted to DPA Contest V4.2 for official evaluation and the results show that all the submitted attacks are both practical and feasible. Among them, the best attack scheme requires only 257 power traces to crack the complete 128-bit master key with 80% success rate. To further improve the security level of RSM2.0, we also discuss some possible strategies to eliminate or mitigate the threats proposed by us.

Keywords: Side-channel analysis · 1st-order masking schemes · Shuffling · Non-profiled attack · Second order CPA · DPA Contest V4.2

1 Introduction

Adding side channel resistances is indispensable for modern cryptographic devices to thwart the potential attack first proposed by Kocher et al. in [1].

C. Tu—The work is supported by a grant from the National High Technology Research and Development Program of China (863 Program, No. 2013AA01A214) and the National Basic Research Program of China (973 Program, No. 2013CB338001). Besides, the work is also supported by a grant from the National Natural Science Foundation of China (No. 61402470).

R. Avanzi and H. Heys (Eds.): SAC 2016, LNCS 10532, pp. 41–57, 2017.
https://doi.org/10.1007/978-3-319-69453-5_3

The basic idea of the attack is to collect the observable leakages derived from the operations of sensitive intermediate values and make use of them with the help of statistical methods to deduce the hidden secret in the devices, generally the cryptographic key. Masking and shuffling are two classic countermeasures most extensively studied to enhance the security level of cryptographic devices. By bringing in random numbers, masking schemes [2–4] divide each sensitive intermediate value into several individual parts while keeping each part random. This scheme cuts off the relationship between the hidden secret and the direct leakage from sensitive intermediate value, thus efficiently resisting the common statistical analysis methods in side channel areas such as [1,5,6]. On the other hand, shuffling schemes [7,8] provide the side channel protection from another perspective, namely time dimension. With the help of randomized index table, shuffling schemes either disorder the execution path of cryptographic algorithm or randomly insert the dummy operations, thus randomizing the leakage position of each sensitive intermediate value and putting up obstacles to most of the analysis methods that mainly rely on the constant leakage instant.

To counteract possible attacks against one single defense strategy and achieve a higher security level, combining both masking and shuffling has been a tendency [3,9–11] in the design of side channel countermeasures. Among them, improved RSM [11] is a most recent and well-known countermeasure proposed by DPA Contest V4.2 committee to provide security protection for AES-128. Improved from the original RSM [12], RSM2.0 updates the original masking strategy with newly introduced offset array and performs shuffled operations with the help of the shuffle array, thus aiming to counteract the existing non-profiled attacks proposed in V4.1. An attack is considered to be non-profiled when adversaries don't have the chance to build the precise leakage model of the device they target in a previous training phase, such as DPA [1], CPA [5] etc. Thus, compared with classic profiled method, such as template attack [13] or stochastic model [14], non-profiled attacks usually require more power trace for secret extraction but show a higher security risk.

The combined countermeasure in RSM2.0 shows its resistance to non-profiled attacks in the following way. On the one hand, one byte of offset index is superseded by an offset array of sixteen bytes which determine the mask usage for all state[1] bytes independently. By this means, the second order attack proposed by Zhou et al. [15] which combines S-box output with input mask m_i and the masked plaintext with mask m_{i+1} doesn't work anymore. It's also impossible to exploit the significant power difference when operating on mask 0xFF and 0x00 for offset recovery [16] or to perform two kinds of constructive collision attacks proposed also in [16] since the relationship of mask usage within the first and last round has been cut off. On the other hand, the employment of shuffle array completely disorders the predictable sequence of mask S-box execution both in the first and last round, causes changeable execution window of the concerned S-boxes and results in the obstacle to perform constant instant related attacks, such as 1st-order attack [17] or second order attack which relies on the leakage

---

[1] State denotes the basic data unit of 4 by 4 matrix as defined in standard AES-128.

preprocessing of two instantaneous moments. Furthermore, V4.2 committee also rewrites the implementation in assembly code and precharges the specific state registers before overwritten by new numbers. Thus the constructive first-order attack proposed in [18] by exploiting hamming distance leakage between the input and output of masked S-box doesn't work anymore.

According to the official evaluation, we are the first to launch the non-profiled attacks against RSM2.0, and this is also the first paper to systematically analyze the potential non-profiled vulnerabilities hidden in RSM2.0. Although some other attack schemes have also been submitted to official website [19], almost all of them belong to profiled schemes where attackers have to perform a training phase with large quantities of power traces in order to characterize the real leakage model of the targeted device. Such kind of attacks are capable to recover secret key within several power traces but require a stronger assumption for the abilities of the attackers. In this paper, we are only dedicated in the attacks of non-profiled type.

The contribution of this paper mainly lies in the following aspects. We make use of an analytical methodology to guide the search of exploitable flaws both in the algorithm design and implementation of RSM2.0. Then, on the basis of the discovered non-profiled flaws, we come up with several attack schemes, more precisely second order schemes and its variants. Four of the attacks are eventually implemented as examples to validate the usability of the flaws. Official results show that all of our uploaded attacks are both feasible and practical. And the best scheme require only 257 traces to recover the AES-128 master key with 80% global success rate(GSR), thus breaking the security claim of RSM2.0 for the first time. Furthermore, in order to eliminate or mitigate the threats proposed by us, some possible countermeasures are also discussed in this paper.

The rest of the paper is organized as follows. In Sect. 2 we review the detailed algorithm design of RSM2.0, especially the countermeasures newly added to prevent the enhanced implementation from some known attacks. Then, in Sect. 3, we first explain the analytical methodology used to restrict the range of vulnerability detection and then point out several potential flaws hidden in either the design or implementation of RSM2.0. Besides, the reasons of flaw generation are also clearly explained in this section. Afterwards, in Sect. 4, we show our practical attack processes together with the official evaluation results of our four exemplary attacks. Furthermore, the discussion of some possible countermeasures is presented in Sect. 5. Finally, we conclude our work in the last section of the article.

2   Improved RSM Scheme

In this section, the algorithm details of improved RSM are described explicitly. The description focuses on the mask usage and tracking in the algorithm flow and also on some important features newly brought in, including the shuffling countermeasure and the offset array. What's worth mentioning is that, for the simplicity of description, we omit "modulo 16" after all the addition operation used hereafter unless special explanation is made.

2.1 Algorithm Description

Mask array, offset array and shuffle array, denotes as M[], O() and Sf[] respectively, are three core components to build up the whole RSM2.0 scheme.

Mask array is designed to be a fixed and publicly known array of 16 bytes. The latest values are chosen meticulously with the goal to not only minimize the mutual information leakage [20] but also take side-channel indistinguishability [21] into consideration. We denote each individual value in the array as M[i], $i \in [0, 15]$, and the whole mask array can be specified as $\{0x03, 0x0c, 0x35, 0x3a, 0x50, 0x5f, 0x66, 0x69, 0x96, 0x99, 0xa0, 0xaf, 0xc5, 0xca, 0xf3, 0xfc\}$.

Offset array contains 16 four-bit random numbers which range from 0 to 15 and are refreshed in each encryption. It cooperates with Mask array to randomly and independently select mask values which provide the initial protection for the input state in each encryption round. That is, according to each offset byte, noted as O(i), $i \in [0, 15]$, mask $M[O(i)+r]$ is picked out and later Xored with the sensitive input variable at the start of each encryption round, where $r \in [0, 9]$ represents the round index. We denote such input masks in each round as $Mask_r$, and it can be represented in the form of a mask state:

$$Mask_r = \begin{pmatrix} M[O(0)+r] & M[O(4)+r] & M[O(8)+r] & M[O(12)+r] \\ M[O(1)+r] & M[O(5)+r] & M[O(9)+r] & M[O(13)+r] \\ M[O(2)+r] & M[O(6)+r] & M[O(10)+r] & M[O(14)+r] \\ M[O(3)+r] & M[O(7)+r] & M[O(11)+r] & M[O(15)+r] \end{pmatrix}$$

Shuffle array is a new feature introduced in RSM2.0. It is refreshed trace by trace and kept secret to analysts as offset array does. What's different is that Sf[] is a random permutation of [0, 15] and is deployed to disorder the non-linear transformation of 16 S-boxes together with the subsequent linear layer operation, namely ShiftRows, both in the first and last round. In fact, two separate shuffle arrays are deployed which are defined as Sf0[] and Sf10[] by us to distinguish the position of their usage.

After the description of the three fundamental arrays, the round functions in RSM2.0 are explained below. Apart from the unchanged AddRoundKey(AR) function, the other round functions can be divided into two categories, namely the non-linear and linear layer functions.

- **Non-linear layer function:**
  The only one function that belongs to the non-linear layer is MaskedSub-Bytes(MS). Unlike standard SubBytes function in AES, MaskedSubBytes consists of sixteen different and reconstructed S-boxes corresponding to both the input mask and the output mask. Each masked S-box can be defined as $MaskedSubByte_{mm'}[x] = SubByte[x \oplus m] \oplus m'$, where m and m' represent the input and output mask byte respectively. Specifically in RSM2.0, m and m' are designed to be two successive masks in $M[]$. That is, each new S-box can be denoted as $MaskedSubByte_{M[i]M[i+1]}[]$ ($MaskedSubByte_i[]$ for short), $i \in [0, 15]$, and can be previously computed due to the already

known $M[]$. Thus the input mask state $Mask_r$ becomes traceable (switch to $Mask_{r+1}$) when going through the non-linear layer and such special and circular way of mask usage for S-box reconstruction is called the Rotating S-boxes Masking (RSM).

- **Linear layer functions:**
Linear layer is composed of three functions in total, namely ShiftRows(SR) and MixColumns(MC), and also the additionally introduced MaskCompensation(MCP). On the one hand, the first three functions keep unchanged as in the standard AES. The only difference is that all of their inputs and outputs are protected with masks to randomize all the sensitive intermediate values in the practical encryption and more importantly, these masks are naturally traceable due to the linear property of all these functions.

On the other hand, the MaskCompensation function is newly introduced to eliminate the derived output masks after the MixColumns function and simultaneously re-mask the intermediate variable with the input masks of the next round. To achieve this goal, the compensation mask, denoted as $MaskCompensation_r$ (MCP_r for short), $r \in [0, 8]$, should be first generated and can be expressed as:

$$MCP_r = MC(SR(Mask_{r+1})) \oplus Mask_{r+1}$$

Then the MaskCompensation happening in the first nine rounds can be described in the following derivation process:

$$\begin{aligned}
&MC(SR(MS(K_r \oplus X_r \oplus Mask_r))) \oplus MCP_r \\
&= MC(SR(SubBytes(K_r \oplus X_r) \oplus Mask_{r+1})) \oplus MCP_r \\
&= MC(SR(SubBytes(K_r \oplus X_r))) \oplus MC(SR(Mask_{r+1})) \\
&\quad \oplus MC(SR(Mask_{r+1})) \oplus Mask_{r+1} \\
&= MC(SR(SubBytes(K_r \oplus X_r))) \oplus Mask_{r+1}
\end{aligned}$$

The only change in MaskCompensation happens in the last round where MixColumns is omitted and no next round input masks are needed due to the requirement of the unmasked and correct ciphertext output. Thus $MaskCompensation_9$ satisfies:

$$MaskCompensation_9 = ShiftRows(Mask_{10})$$

With the explanation of functions in both the linear layer and the non-linear layer, all the major process of RSM2.0 has been clearly demonstrated. And more detailed and complete algorithm of RSM2.0 is presented in Appendix 1.

2.2   Acquisition Platform and Measurements

The measurement of all the official power traces is completed on a 8-bit AVR microcontroller Atmega163 embedded in a smartcard. It contains 16 Kb of in-system programmable flash, 512 bytes of EEPROM, and 32 general purpose working registers. The acquisition of traces is performed through a LeCroy WaveRunner 6100 A oscilloscope by the use of an EM probe. The sampling rate FS equals to 500 MS/s and the acquisition bandwidth is 200 MHz.

3   Detecting Non-profiled Vulnerabilities in RSM2.0

In this section, we explain the analytical methodology we comply with to lead the search of the non-profiled vulnerabilities and show the discovered exploitable leakages and the reason of their generation. Our discoveries validate that the improved countermeasure are far from perfect to counteract the type of non-profiled attacks. What's worse, some of the newly added defense mechanisms even directly result in the attacks presented in this section.

3.1   Analytical Methodology for Vulnerability Detection

Although 1st-order masking schemes can not resist 2nd-order attack theoretically, RSM2.0 puts targeted obstacles to such kind of attack. On the one hand, by protecting each state byte with an independent and randomly indexed mask, none of the two masked intermediates in the algorithm share the same mask part. On the other hand, the common attacking points, namely non-linear layer transformation, and also the subsequent ShiftRows operation are both implemented with shuffled order thus making it difficult to collect the power leakage from those parts.

In order to follow the traditional second order idea and achieve a better performance, we are not only expected to discover the new attacking point for mask elimination but also expected to bypass the shuffle countermeasure in order to avoid the costly *integrated-and-combined* strategy [9,22] for second order attack.

To accomplish the above target, we first perform the vulnerability detection process in both the algorithm design and the code implementation of RSM2.0. Our detection mainly comply with the following guidelines aiming at obtaining optimal attacking performance:

1. Restricting the range of detection in the intermediate values that contain only 8-bit subkey. Although the direct side channel attack against larger subkey block is possible [23], the expensive resource overheads, such as GPU acceleration and huge memory usage make it inefficient and unsuitable in the official evaluation platform. Thus, to acquire better attacking performance we focus on 8-bit subkey recovery at a time.
2. The predictable intermediates utilized by the attackers should be the results of the non-linear layer transformation. The characteristics of the non-linear transformation lead to the fact that each single bit guessing error of the target subkey would influence as much bits of the prediction result as possible. Thus, making it easier to distinguish the correct key from others with less power traces.
3. The security of the newly introduced MaskCompensation process should be taken into consideration additionally.

Based on the guidelines above, the vulnerability detection can be simply limited in the following sensitive regions (Fig. 1).

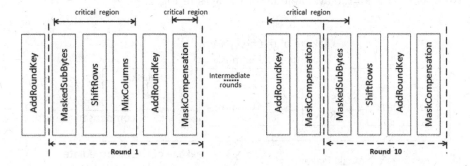

Fig. 1. Critical regions for vulnerability detection

3.2 Flaws in the Algorithm Design

Online Derivation of Compensation Mask. The first vulnerability we discover in the critical regions lies in the entire process of MCP_r derivation which must be implemented online in RSM2.0. The inevitable online feature is caused by the replacement of the offset index by the offset array $O()$. In the original RSM, the input mask state of round r, which we denote as $Mask_{idx,r}$, is uniquely determined by a 4-bit index denoted by idx. More accurately, such input mask state satisfies the following formula:

$$\begin{pmatrix} M[idx+0+r] & M[idx+4+r] & M[idx+8+r] & M[idx+12+r] \\ M[idx+1+r] & M[idx+5+r] & M[idx+9+r] & M[idx+13+r] \\ M[idx+2+r] & M[idx+6+r] & M[idx+10+r] & M[idx+14+r] \\ M[idx+3+r] & M[idx+7+r] & M[idx+11+r] & M[idx+15+r] \end{pmatrix}$$

Due to the fact that $M[]$ is a fixed mask array, there are only 16 possible values for $Mask_{idx,r}$ state. Therefore, the compensation mask state which is completely dependent on $Mask_{idx,r}$, is actually derived offline and stored previously for later use. However, the input mask state $Mask_r$ in RSM2.0 utilizes offset array $O()$ to index the selected mask for each input mask byte as stated in Sect. 2. Each element $O(i)$ in the offset array is independent and identically distributed, thus resulting in 16^{16} possible values for the $Mask_r$ state. In order to store all these compensation values derived from $Mask_r$, $16 * 16^{16}$ bytes of storage space is required which is unreachable in the embedded devices with constrained resources. Thus, online compensation mask derivation becomes indispensable and all of its related power consumption would be recorded in the power traces.

The other significant cause that leads to this online process further exploitable is the omission of shuffling protection during this stage. This vulnerability is serious since it means all the steps during the derivation of compensation mask, including the loading of $Mask_r$, the derivation for the first part of compensation mask and the compensation mask generation, are processed at the constant time instants and leak the power consumptions corresponding to the original input $Mask_r$ or its intermediate state during the transformation. We briefly show all the available mask leakages in Fig. 2.

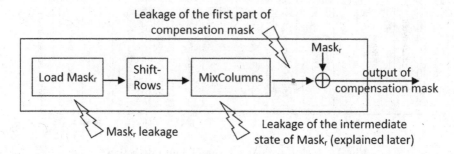

Fig. 2. Exploitable mask leakages in the derivation of compensation mask (Color figure online).

AddRoundKey Function Flaw. In addition to compensation mask derivation, AddRoundKey operation in RSM2.0 also lacks shuffling protection, thus causing potential second order threats. The exploitable loophole caused by this sequential process appears at the end of the ninth round, where the AddRound-Key function is performed right after the MixColumns of the current round. That is to say, each output byte of AddRoundKey is protected by the same mask as in MixColumns output, as shown in Fig. 3, where the X_i represents the input bytes of the ninth round and K_i, K_i' are the subkey bytes of the eighth and ninth round respectively. Special note is that such output masks (in red) are also generated as the first part of the compensation mask in the online mask derivation phase (the third leakage point in Fig. 2).

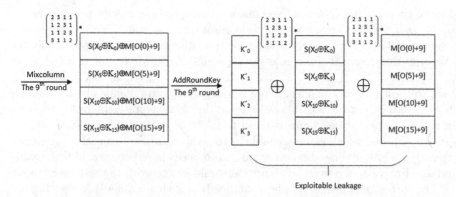

Fig. 3. Exploitable leakages in AddRoundKey, taking the first column as an example.

3.3   Flaws in the Implementation Level

Location of MaskCompensation. Flaws in the implementation level appear because of the inserted position of the online MaskCompensation. More precisely, in the source code of RSM2.0 implementation, the MaskCompensation of the current round starts operating after AddRoundKey has finished, just before the MaskedSubBytes of the next round, as shown in Fig. 4. This seemingly negligible implementation order does matter since all the masked variables after the ninth round AddRoundKey can be derived reversely from the known ciphertext by only guessing 8-bit subkey of the last round key, which is a proper subkey guessing space for side channel attacks. Besides, the MaskCompensation here would switch the protection masks of its output state to $Mask_9$ which has also leaked its power consumption at the start of the ninth round MaskCompensation (as shown in Fig. 2).

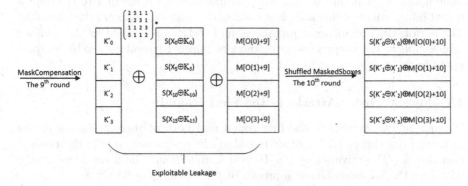

Fig. 4. Exploitable leakages at the output of MaskCompensation, taking the first column as an example.

Flaws in Linear Layer Function. More critical security flaws appear at linear layer. Since almost all of the proposed attacks against original RSM select the non-linear function, i.e. masked S-box, as their attacking point, designers of RSM2.0 pay too much attention to the protection of S-box execution while on the other hand ignore the potential security risk in the linear layer transformation.

The flaw we find in the linear layer appears in the process of MixColumns operation in the first encryption round. The essential reason that gives rise to this flaw actually lies in two aspects. The first one is that both the MixColumns function used for encryption and the MixColumns included in the compensation mask derivation stage share the same assembler code. The only difference is the input variables, namely $ShiftRows(X \oplus K \oplus Mask_1)$ and $ShiftRows(Mask_1)$ respectively. Also due to the linear characteristics of the MixColumns operation, such shared implementation implies that whenever a masked intermediate value is used or generated in the MixColumns for encryption, the corresponding mask itself, which is used for the protection of that intermediate value, will also appear in the exactly same position of the MixColumns for compensation mask derivation. The second reason that makes the flaws further feasible is that the entire MixColumns code is implemented in a fixed sequence, which means that attackers are able to find out leakages in constant time instant, thus meeting the prerequisite of classic second order attacks.

4   Practical Attacks and Official Evaluation Results

With the clear explanation of all the exploitable vulnerabilities in the section above, six non-profiled attacks, more precisely second order attacks and the variants, can be launched. We selectively perform four of them as examples to validate the usability of the discovered flaws and show the effectiveness of our attack schemes by citing the official evaluation data.

The basic idea of classic second order attacks, as present in [22,24,25], is to preprocess the leakages from two parts of the power traces and both parts respectively correspond to the intermediate variables protected with the exact same mask. After preprocessing, the combined leakage is relevant to the unprotected intermediate value, and thus making the later correlation attack feasible. The performance of different preprocessing functions is studied in [26] and our attacks follow the preprocessing method of improved product combining proposed in this paper.

4.1   Second Order Attacks in the First Round[2]

We present two attacks in the first round and both of them make use of the leakages from the execution of shared MixColumns source code in the encryption and MCP_r derivation stage. To better understand the attacks, we briefly introduce the implementation approach of MixColumns in RSM2.0.

Suppose $(V_{0,j}, V_{1,j}, V_{2,j}, V_{3,j})^T$ is a column of 4 input bytes, which serves as a basic unit for MixColumns transformation. Then the output of the transformation, where $i \in [0,3]$, $j \in [0,3]$, can be formalized and recombined as:

$$V_{i,j} = (2 * V_{i,j}) \oplus (3 * V_{(i+1)\%4,j}) \oplus V_{(i+2)\%4,j} \oplus V_{(i+3)\%4,j}$$
$$= (2 * V_{i,j} \oplus 2 * V_{(i+1)\%4,j}) \oplus V_{(i+1)\%4,j} \oplus V_{(i+2)\%4,j} \oplus V_{(i+3)\%4,j}$$
$$= V_{i,j} \oplus (V_{1,j} \oplus V_{2,j} \oplus V_{3,j} \oplus V_{4,j}) \oplus 2 * (V_{i,j} \oplus V_{(i+1)\%4,j})$$

The implementation follows the process of the calculation in the last line. That is, $V_{1,j} \oplus V_{2,j} \oplus V_{3,j} \oplus V_{4,j}$ is firstly generated and shared for all the bytes in column j. Then, to derive a specific byte $V_{i,j}$, the generation of $(V_{i,j} \oplus V_{(i+1)\%4,j})$ is subsequently completed. This value is later used as the input of the lookup table Xtime which is previously calculated to store 2*X (under $GF(2^8)$) in the position of X, where X ranges from 0 to 255. Thus, after going through Xtime table, $2 * (V_{i,j} \oplus V_{(i+1)\%4,j})$ is acquired. Finally, by Xoring together $(V_{1,j} \oplus V_{2,j} \oplus V_{3,j} \oplus V_{4,j})$, $2 * (V_{i,j} \oplus V_{(i+1)\%4,j})$ and the original value $V_{i,j}$ stored in the register, the newly updated $V_{i,j}$ comes into being.

---

[2] For the need of expression, all the "mod" operation would be explicitly added in this subsection.

Attacking Xtime Input. During the generation of the Xtime input $(V_{i,j} \oplus V_{(i+1)\%4,j})$, $V_{i,j}$ is first loaded into the register and simultaneously leaks instantaneous power consumption. When it happens in the encryption process, $V_{i,j}$ actually equals to one of the output variables of MaskedSubBytes function. Thus all the MaskedSubBytes outputs in the form of $S(X_i \oplus K_i) \oplus M[(O(i)+1)\%16]$, $i \in [0, 15]$, leaks information. On the other hand, when it goes to MaskCompensation function, $V_{i,j}$ in fact represents one of the mask bytes in $Mask_1$ state. Thus, $M[(O(i) + 1)\%16]$ leaks its information as well. By preprocessing both of the leakages, the combined leakage would be relevant to the following intermediate variable:

$$(S(X_i \oplus K_i) \oplus M[(O(i) + 1)\%16]) \oplus M[(O(i) + 1)\%16] = S(X_i \oplus K_i), i \in [0, 15]$$

which is an unprotected and predictable value, appropriate for the traditional CPA attack.

Official evaluation results show that, the second order attack proposed here are both feasible and practical. Based on the published performance parameter, merely 258 traces are required (Fig. 5(a)) to recover all of the 128-bit master key with 80% success rate (the so-called 80% global success rate, GSR [27]) and only 210 traces would suffice to reduce the maximum partial guessing entropy(PGE [27]) under 10, which means that the remaining key guessing space is less than 10^{16}.

Optimized Chained Attack. Unlike the attack above, here we utilize the leakage exposed when generating each complete input byte of the Xtime, namely $(V_{i,j} \oplus V_{(i+1)\%4,j})$. After the same combination of the leakages in MixColumns of different stages, the united leakage would also be related to a predictable intermediate value which is involved with 16 bits of the master key. Taking the first column as an example, four predictable values are $S(X_0 \oplus K_0) \oplus S(X_5 \oplus K_5)$, $S(X_5 \oplus K_5) \oplus S(X_{10} \oplus K_{10})$, $S(X_{10} \oplus K_{10}) \oplus S(X_{15} \oplus K_{15})$ and $S(X_{15} \oplus K_{15}) \oplus S(X_0 \oplus K_0)$ respectively, where X_i and K_i are the ith plaintext and master key respectively. Direct attacks by guessing the first and the third predictable value or guessing the second and the fourth could recover all the subkeys in this column but the key guessing space is $2 * 2^{16}$ in total.

In order to further reduce the computation overhead, we optimized the attack process in a chained way. That is, we first attack K_0 by using the leakages and method mentioned first in this subsection, then with the most probable guessing K_0 revealed, predictable value $S(X_0 \oplus K_0) \oplus S(X_5 \oplus K_5)$ is utilized to extract K_5 only. The same approach then goes on for the second predictable value, where this time, K_5 is fixed to the most probable value obtained in the last step and K_{10} becomes the only key which needs to be guessed in this step. Finally, K_{15} is also revealed in the same way. By this means, the key guessing space for extracting four subkeys in a column is now reduced to $4 * 2^8$ and the subkeys in other three columns can be inferred similarly as in the first column.

Figure 5(b) depicts the official evaluation result of such chained second-order attack. It shows that such method does work, and 565 traces are needed to uniquely determine the master key with 80% success rate(80% GSR).

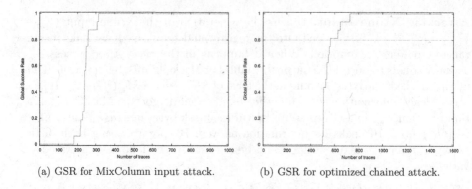

(a) GSR for MixColumn input attack.   (b) GSR for optimized chained attack.

Fig. 5. Official evaluation results of Global Success Rate, GSR.

Note also that the similar second order and chained attack can also be launched at the position of Xtime output, namely $2 * (V_{i,j} \oplus V_{(i+1)\%4,j})$, and the generation of shared $(V_{1,j} \oplus V_{2,j} \oplus V_{3,j} \oplus V_{4,j})$ respectively. The only difference lies in the form of the predictable intermediate value, which is selected and utilized by the attackers to infer the subkey. To avoid repetition, we don't describe them here anymore.

4.2   Second Order Attacks in the Ninth Round

Attacking MaskCompensation Output. The attack here utilizes the implementation level flaw explained in Sect. 3.3. Due to the inserted position of MaskCompensation, each individual output byte here contains only 8-bit subkey when derived from the ciphertext and is actually protected by the corresponding mask in $Mask_9$. Besides, $Mask_9$ has also been loaded byte by byte when generating the ninth round compensation masks as depicted in Fig. 2. Thus by combining the leakages from both positions, the preprocessed leakage has relevance to the following values which are predictable from the perspective of side channel opponents:

$$(Sbox^{-1}ShiftRows^{-1}[C \oplus K_L] \oplus Mask_9) \oplus Mask_9 = Sbox^{-1}ShiftRows^{-1}[C \oplus K_L]$$

where C refers to the output ciphertext state and K_L denotes the last round key. After 16 bytes of K_L is revealed completely, a reversed key expansion process of AES should be performed in order to fetch our final target, namely the 128-bit master key.

Figure 6(a) depicts the GSR tendency of this attack evaluated by DPA Contest committee. It shows that the attack by exploiting the implementation level flaw is of high efficiency that only 257 traces can meet the requirement for 80% GSR. Besides, to reduce the Max PGE under 10, only 205 traces are required.

Attacking AddRoundKey Output. The only difference between the output of the ninth round AddRoundKey and MaskCompensation lies in the masks for protection. For AddRoundKey function, these masks are obtained by transforming the input mask state of the ninth round, namely $Mask_8$, through Masked-SubBytes, ShiftRows and MixColumns in sequence. Thus, the AddRoundKey output in the ninth round can be derived from the ciphertext C:

$$Sbox^{-1}ShiftRows^{-1}[C \oplus K_L] \oplus MixColumns[ShiftRows[Mask_9]]$$

On the other hand, during the construction of the compensation mask in the current round, the first compensation part $- MixColumns[ShiftRows[Mask_9]]$ should be generated first as shown in Fig. 2.

Likewise, by performing second order attacks at both parts, the last round key K_L involved in the predictable variable $Sbox^{-1}ShiftRows^{-1}[C \oplus K_L]$ can be recovered and the master key would also be deduced instantly with the help of the key expansion process but in a reversed way. The official evaluation of GSR in Fig. 6(b) shows that such attack is also feasible in RSM2.0, and the number of traces needed to acquire 80% GSR is 698.

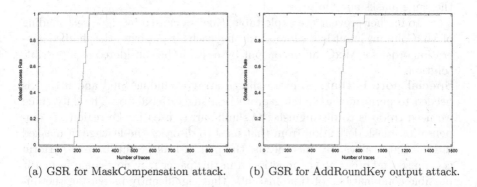

(a) GSR for MaskCompensation attack.   (b) GSR for AddRoundKey output attack.

Fig. 6. Official evaluation results of Global Success Rate, GSR.

5   Discussion of Possible Countermeasures

In order to mitigate or even get rid of the non-profiled threats presented in this article, we propose the following coping strategies. All of the strategies follow the basic principle that both leakages exploited in a single second order attack should be protected individually by either eliminating the source of the leakage or by randomizing the position of its appearance.

1. **Adding resistance in the ninth round:** exchanging the order of the AddRoundKey and MaskCompensation function in the ninth round can be the first step to enhance the security level of RSM2.0. In fact, compared to the original attack aiming at the MaskCompensation output leakage, now attackers have to deduce each byte of hypothetical intermediate state, namely $Sbox^{-1}ShiftRows^{-1}[C \oplus K_L] \oplus K_P$, by searching 16-bit subkey space, where K_L and K_P represent the last and penultimate round key respectively.
   After exchanging, the output states of both functions are protected with $Mask_9$. In order to further prevent the second order threat, the sequential

operations of loading $Mask_9$, Xoring compensation masks and AddRound-Key execution should all be respectively shuffled. Fortunately, such protections could be easily added since every operation mentioned above deals with 16 state bytes independently.

2. **Adding resistance in the first round:** the shared MixColumns code in the compensation mask derivation and the original encryption part leads to our attacks in the first round. To plug such leakages, one possible method is to implement MixColumns in different ways. For example, alter the MixColumns implementation in the encryption part by following the process of the original transformation formula, namely $V_{i,j} = 2 * V_{i,j} \oplus 3 * V_{(i+1)\%4,j} \oplus V_{(i+2)\%4,j} \oplus V_{(i+3)\%4,j}$. Thus all of the chained attacks mentioned can be surely counteracted since the derived intermediates in different implementations don't share the same masks any more.

   To further prevent the exploitable leakages caused by the direct loading of MixColumns input bytes for calculating each output byte, the shuffled generation order for MixColumns output bytes could be considered as a possible solution.

3. **Special note is that:** no extra shuffle array(excluding Sf0[] and Sf10[]) is needed to perform the shuffling protection suggested above. The only thing we need to do is to distinguish the shuffle array used in the region of compensation mask derivation from that used to disorder the leakage of masked intermediate values. By this means, the pair of leakages utilized to perform certain second order attack mentioned in this paper may now come from 16^2 possible combinations of time instants, thus significantly increasing the difficulty for target leakage location and even causing huge overhead (number of power traces) when using advanced *integrated-and-combined* technique to accumulate the shuffled leakages as mentioned in [9,22].

6   Conclusion

This is the first paper to systematically analyze the non-profiled vulnerabilities in RSM2.0. To achieve the goal, we first propose the analytical methodology to guide the vulnerability detection and then scrutinize both the algorithm design and implementation details of the RSM2.0 countermeasure. Based on all the vulnerabilities newly found, several attacking schemes are proposed and four of them are finally implemented as examples and submitted for official evaluation.

The evaluation reports show that all of our proposed attacks are both feasible and practical. Thus, this study validates in the first time that the currently used countermeasures and implementation are still insufficient to provide RSM2.0 with non-profiled resistance.

To further fix the vulnerabilities described in this paper, we come up with several corresponding suggestions which either try to eliminate the second order leakages or to extend the protection region of shuffle countermeasure. These improvements are not unique but they can be considered as a general direction to set up obstacles for potential attackers and to further improve the security level of RSM2.0, especially against non-profiled attacks.

A   Algorithm of Improved Rotating S-Boxes Masking

Algorithm 1. Improved rotating S-boxes masking scheme

Input:
  16-bytes plaintext $X = [X0, X1, \ldots, X15]$;
  One offset array: $O(i), i \in [0, 15]$,
  (uniformly random, unknown);
  Two shuffle arrays: $Sf0[i], Sf10[i], i \in [0, 15]$,
  (uniformly random permutations, unknown);
Output:
  16-bytes ciphertext $X = [X0, X1, \ldots, X15]$;

1: On-the-fly key expansion $RoundKey_r, r \in [0, 10]$,
2: $RoundKey_0 \leftarrow RoundKey_0 \oplus Mask_0$
3: $X = X \oplus RoundKey_0$
  /*All rounds but the last one*/
4: **for** each $i \in [0, 8]$ **do**
5:     **if** $r = 0$ **then**
6:         **for** $i \in Sf0[0, 15]$ **do**
7:             $X_i = MaskedSubBytes_{O(i)+r}(X_i)$
8:             $X_i = SR(X_i)$
9:         **end for**
10:    **else**
11:        **for** $i \in [0, 15]$ **do**
12:            $X_i = MaskedSubBytes_{O(i)+r}(X_i)$
13:            $X_i = SR(X_i)$
14:        **end for**
15:    **end if**
16:    $X = MC(X)$
  /* AddRouondKey of the next round */
17:    $X = X \oplus RoundKey_{r+1}$
18:    $MCP_r = MC(SR(Mask_{r+1})) \oplus Mask_{r+1}$
19:    $X = X \oplus MCP_r$;
20: **end for**
  /* Last round */
21: **for** $i \in Sf10[0, 15]$ **do**
22:    $X_i = MaskedSubBytes_{O(i)+9}(X_i)$
23:    $X_i = SR(X_i)$
24: **end for**
25: $X = X \oplus RoundKey_{10}$ /* Ciphertext unmasking */
26: $MCP_9 = SR(Mask_{10})$
27: $X = X \oplus MCP_9$;

References

1. Kocher, P., Jaffe, J., Jun, B.: Differential power analysis. In: Wiener, M. (ed.) CRYPTO 1999. LNCS, vol. 1666, pp. 388–397. Springer, Heidelberg (1999). doi:10.1007/3-540-48405-1_25

2. Coron, J.-S., Goubin, L.: On Boolean and arithmetic masking against differential power analysis. In: Koç, Ç.K., Paar, C. (eds.) CHES 2000. LNCS, vol. 1965, pp. 231–237. Springer, Heidelberg (2000). doi:10.1007/3-540-44499-8_18

3. Herbst, C., Oswald, E., Mangard, S.: An AES smart card implementation resistant to power analysis attacks. In: Zhou, J., Yung, M., Bao, F. (eds.) ACNS 2006. LNCS, vol. 3989, pp. 239–252. Springer, Heidelberg (2006). doi:10.1007/11767480_16

4. Rivain, M., Prouff, E.: Provably secure higher-order masking of AES. In: Mangard, S., Standaert, F.-X. (eds.) CHES 2010. LNCS,, vol. 6225, pp. 413–427. Springer, Heidelberg (2010). doi:10.1007/978-3-642-15031-9_28

5. Brier, E., Clavier, C., Olivier, F.: Correlation power analysis with a leakage model. In: Joye, M., Quisquater, J.-J. (eds.) CHES 2004. LNCS, vol. 3156, pp. 16–29. Springer, Heidelberg (2004). doi:10.1007/978-3-540-28632-5_2

6. Gierlichs, B., Batina, L., Tuyls, P., Preneel, B.: Mutual information analysis. In: Oswald, E., Rohatgi, P. (eds.) CHES 2008. LNCS, vol. 5154, pp. 426–442. Springer, Heidelberg (2008). doi:10.1007/978-3-540-85053-3_27

7. Mangard, S., Oswald, E., Popp, T.: Power Analysis Attacks: Revealing the Secrets of Smart Cards. Springer Science & Business Media, Heidelberg (2008)

8. Veyrat-Charvillon, N., Medwed, M., Kerckhof, S., Standaert, F.-X.: Shuffling against side-channel attacks: a comprehensive study with cautionary note. In: Wang, X., Sako, K. (eds.) ASIACRYPT 2012. LNCS, vol. 7658, pp. 740–757. Springer, Heidelberg (2012). doi:10.1007/978-3-642-34961-4_44

9. Tillich, S., Herbst, C., Mangard, S.: Protecting AES software implementations on 32-bit processors against power analysis. In: Katz, J., Yung, M. (eds.) ACNS 2007. LNCS, vol. 4521, pp. 141–157. Springer, Heidelberg (2007). doi:10.1007/978-3-540-72738-5_10

10. Rivain, M., Prouff, E., Doget, J.: Higher-order masking and shuffling for software implementations of block ciphers. In: Clavier, C., Gaj, K. (eds.) CHES 2009. LNCS, vol. 5747, pp. 171–188. Springer, Heidelberg (2009). doi:10.1007/978-3-642-04138-9_13

11. Bhasin, S., Bruneau, N., Danger, J.-L., Guilley, S., Najm, Z.: Analysis and improvements of the DPA contest v4 implementation. In: Chakraborty, R.S., Matyas, V., Schaumont, P. (eds.) SPACE 2014. LNCS, vol. 8804, pp. 201–218. Springer, Cham (2014). doi:10.1007/978-3-319-12060-7_14

12. Nassar, M., Souissi, Y., Guilley, S., Danger, J.-L.: RSM: a small and fast countermeasure for AES, secure against 1st and 2nd-order zero-offset SCAs. In: Design, Automation & Test in Europe Conference & Exhibition (DATE 2012), pp. 1173–1178. IEEE (2012)

13. Chari, S., Rao, J.R., Rohatgi, P.: Template attacks. In: Kaliski, B.S., Koç, K., Paar, C. (eds.) CHES 2002. LNCS, vol. 2523, pp. 13–28. Springer, Heidelberg (2003). doi:10.1007/3-540-36400-5_3

14. Schindler, W., Lemke, K., Paar, C.: A stochastic model for differential side channel cryptanalysis. In: Rao, J.R., Sunar, B. (eds.) CHES 2005. LNCS, vol. 3659, pp. 30–46. Springer, Heidelberg (2005). doi:10.1007/11545262_3

15. Attacks in the v4.1 phase. http://www.dpacontest.org/v4/rsm_hall_of_fame.php

16. Kutzner, S., Poschmann, A.: On the security of RSM - presenting 5 first- and second-order attacks. In: Prouff, E. (ed.) COSADE 2014. LNCS, vol. 8622, pp. 299–312. Springer, Cham (2014). doi:10.1007/978-3-319-10175-0_20

17. Ye, X., Eisenbarth, T.: On the vulnerability of low entropy masking schemes. In: Francillon, A., Rohatgi, P. (eds.) CARDIS 2013. LNCS, vol. 8419, pp. 44–60. Springer, Cham (2014). doi:10.1007/978-3-319-08302-5_4

18. Moradi, A., Guilley, S., Heuser, A.: Detecting hidden leakages. In: Boureanu, I., Owesarski, P., Vaudenay, S. (eds.) ACNS 2014. LNCS, vol. 8479, pp. 324–342. Springer, Cham (2014). doi:10.1007/978-3-319-07536-5_20

19. Other attacks submitted in the officail website. http://www.dpacontest.org/v4/42_hall_of_fame.php

20. Nassar, M., Guilley, S., Danger, J.-L.: Formal analysis of the entropy/security trade-off in first-order masking countermeasures against side-channel attacks. In: Bernstein, D.J., Chatterjee, S. (eds.) INDOCRYPT 2011. LNCS, vol. 7107, pp. 22–39. Springer, Heidelberg (2011). doi:10.1007/978-3-642-25578-6_4

21. Carlet, C., Guilley, S.: Side-channel indistinguishability. In: Proceedings of the 2nd International Workshop on Hardware and Architectural Support for Security and Privacy, p. 9. ACM (2013)

22. Tillich, S., Herbst, C.: Attacking state-of-the-art software countermeasures—a case study for AES. In: Oswald, E., Rohatgi, P. (eds.) CHES 2008. LNCS, vol. 5154, pp. 228–243. Springer, Heidelberg (2008). doi:10.1007/978-3-540-85053-3_15

23. Moradi, A., Kasper, M., Paar, C.: Black-box side-channel attacks highlight the importance of countermeasures. In: Dunkelman, O. (ed.) CT-RSA 2012. LNCS, vol. 7178, pp. 1–18. Springer, Heidelberg (2012). doi:10.1007/978-3-642-27954-6_1

24. Oswald, E., Mangard, S., Herbst, C., Tillich, S.: Practical second-order DPA attacks for masked smart card implementations of block ciphers. In: Pointcheval, D. (ed.) CT-RSA 2006. LNCS, vol. 3860, pp. 192–207. Springer, Heidelberg (2006). doi:10.1007/11605805_13

25. Standaert, F.-X., Veyrat-Charvillon, N., Oswald, E., Gierlichs, B., Medwed, M., Kasper, M., Mangard, S.: The world is not enough: another look on second-order DPA. In: Abe, M. (ed.) ASIACRYPT 2010. LNCS, vol. 6477, pp. 112–129. Springer, Heidelberg (2010). doi:10.1007/978-3-642-17373-8_7

26. Prouff, E., Rivain, M., Bevan, R.: Statistical analysis of second order differential power analysis. IEEE Trans. Comput. 58(6), 799–811 (2009)

27. Standaert, F.-X., Malkin, T.G., Yung, M.: A unified framework for the analysis of side-channel key recovery attacks. In: Joux, A. (ed.) EUROCRYPT 2009. LNCS, vol. 5479, pp. 443–461. Springer, Heidelberg (2009). doi:10.1007/978-3-642-01001-9_26

Bridging the Gap: Advanced Tools for Side-Channel Leakage Estimation Beyond Gaussian Templates and Histograms

Tobias Schneider[1]([✉]), Amir Moradi[1], François-Xavier Standaert[2], and Tim Güneysu[3]

[1] Horst Görtz Institute for IT Security, Ruhr-Universität Bochum, Bochum, Germany
{Tobias.Schneider-a7a,amir.moradi}@rub.de
[2] ICTEAM/ELEN/Crypto Group, Université catholique de Louvain, Louvain-la-Neuve, Belgium
fstandae@uclouvain.be
[3] University of Bremen and DFKI, Bremen, Germany
tim.gueneysu@uni-bremen.de

Abstract. The accuracy and the fast convergence of a leakage model are both essential components for the efficiency of side-channel analysis. Thus for efficient leakage estimation an evaluator is requested to pick a Probability Density Function (PDF) that constitutes the optimal trade-off between both aspects. In the case of parametric estimation, Gaussian templates are a common choice due to their fast convergence, given that the actual leakages follow a Gaussian distribution (as in the case of an unprotected device). In contrast, histograms and kernel-based estimations are examples for non-parametric estimation that are capable to capture any distribution (even that of a protected device) at a slower convergence rate.

With this work we aim to enlarge the statistical toolbox of a side-channel evaluator by introducing new PDF estimation tools that fill the gap between both extremes. Our tools are designed for parametric estimation and can efficiently characterize leakages up to the fourth statistical moment. We show that such an approach is superior to non-parametric estimators in contexts where key-dependent information in located in one of those moments of the leakage distribution. Furthermore, we successfully demonstrate how to apply our tools for the (worst-case) information-theoretic evaluation on masked implementations with up to four shares, in a profiled attack scenario. We like to remark that this flexibility capturing information from different moments of the leakage PDF can provide very valuable feedback for hardware designers to their task to evaluate the individual and combined criticality of leakages in their (protected) implementations.

1 Introduction

Physical attacks are known to pose a major threat to the cryptographic components and security services in many embedded devices. An attacker obtaining

© Springer International Publishing AG 2017
R. Avanzi and H. Heys (Eds.): SAC 2016, LNCS 10532, pp. 58–78, 2017.
https://doi.org/10.1007/978-3-319-69453-5_4

side-channel leakages such as the power consumption or electromagnetic emissions from a cryptographic implementation can extract the secret cryptographic key by applying suitable statistical tools on the collected data. A number of reports have demonstrated that such attacks are not just a theoretical concern but that also real-world devices can be compromised [18,28,38,51]. As a consequence, the seminal Differential Power Analysis (DPA) paper by Kocher *et al.* [21] has been followed by a vast literature on solutions for a wide range of contexts to mitigate these attacks. For example, the inclusion of random delays [10], or shuffling [49] are a frequently used heuristic to improve the physical protection of software implementations. In contrast to this, re-keying strategies, formalized under the name of leakage-resilient cryptography, provide theoretical tools that enable reducing the security of multiple iterations to a single one (cf. [17] for an early result and [47] for a recent one). In this context, one of the most investigated and best understood protection against side-channel attacks is masking [7,13,41] that bridges theory and practice. Its underlying principle is to encode any sensitive variable in an implementation into d shares, and to perform the computations on these shares only. Given that the leakage of all the shares is independent and that the measurements are sufficiently noisy, it ensures that the smallest key-dependent (mixed) moment in the leakage distribution is d. Therefore, any adversary trying to extract information from a masked implementation should (ideally) estimate this (mixed) higher-order moment, a task of which the complexity increases exponentially in d.

A drawback with all these solutions is the significant performance overhead. As a result, the development of methodologies enabling a fair assessment of their security level has evolved in parallel with the development of countermeasures so that designers can discuss security and performance implications for their implementations on a sound basis [46]. Since side-channel analysis is essentially based on the comparison of key-dependent leakage models with actual measurements, these methodological developments have led to a central division between *profiled* and *non-profiled* evaluation tools and attacks [50]. In the first case, the adversary/evaluator is allowed to build an accurate (yet not perfect [16]) model for his target device that generally corresponds to an estimation of the leakage Probability Density Function (PDF)[1]. As depicted in the upper left part of Fig. 1, Gaussian Template Attacks (TA) are the most common tool for this purpose [8]. In this (here: exhaustive) approach, one builds a Gaussian model for the leakage of every target intermediate value in the implementation. The main limitation of Gaussian templates is that they are bound to the analysis of the first two moments in a leakage distribution (i.e., unprotected implementations and masking with $d = 2$). According to the state-of-the-art, the canonical way to analyze higher-order masked implementations would be to switch to non-parametric PDF estimation, e.g., based on histograms and kernels. But this comes at the cost of two important drawbacks. First, these tools imply a more complex (hence measurement intensive) estimation problem. Second, they estimate all the statistical

---

[1] Profiled attacks can also be referred to when the adversary possesses a device with a biased randomness source (as masks).

		PROFILED EVALUATIONS & ATTACKS			NON-PROFILED ATTACKS			
		tool	moments	estim. cost				
EXHAUSTIVE	PDF-BASED	Gaussian-TA	1,2	*				
		Histogram-TA	all	***				
		Kernel-TA	all	***				
		EMG-TA	1,2,3	**				
		Pearson-TA	1,2,3,4	**				
		SGL-TA	1,2,3,4	**				
	PER MOMENT	MCP-DPA	any d (one by one)	exp(d)				
					tool	moments	estim. cost	
					CPA (and equiv)	1	*	PER MOMENT
					HO-CPA CEPACA MC-DPA	any d (one by one)	exp(d)	
					Gaussian-MIA	1,2	*	A PRIORI / PDF-BASED
					Histogram-MIA	all	***	
					Kernel-MIA	all	***	
					EMG-MIA	1,2,3	**	
					Pearson-MIA	1,2,3,4	**	
					SGL-MIA	1,2,3,4	**	
		tool	moments	estim. cost	tool	moments	estim. cost	
SIMPLIFYING		linear regression	any d (one by one)	exp(d)	on-the-fly regr. stepwise regr.	any d (one by one)	exp(d)	PDF-BASED & PER MOMENT

Fig. 1. Summary of side-channel evaluation tools and attacks.

moments at once, meaning that one loses the detailed intuition that could be obtained from the separate examination of all moments. Alternatively, one could use the Moments-Correlating Profiled DPA (MCP-DPA) introduced in [31] that suffers from the complementary drawback. Namely, since MCP-DPA is essentially a "per moment" approach, the intuitions extracted now only correspond to moment taken separately, and it is unclear how one could extend these attacks towards the joint exploitation of multiple moments at the same time.

A comprehensive understanding of how the information leakage of a masked cryptographic implementation is spread among different statistical moments is essential to interpret the results of its security evaluation. That is, in general a $(d-1)$th-order secure implementation is defined as an implementation for which the smallest key-dependent moment in the leakage distribution is d, and this is ideally expected to occur for d shares. But in practice, it frequently happens that glitches (i.e., non-independent leakages) contradict this expectation, leading to informative moments of smaller orders than d, both in hardware and software case studies [9, 26]. Significant research efforts have been dedicated to the design of glitch-free implementations, e.g., based on multiparty computation [42] or threshold implementations [30, 32]. However, in the latter case the number of shares is larger than the claimed order. This, however, highlights the demand for the ability to determine the exact moment that actually leaks [3]. Simple leakage detection tests (e.g., t-test [44]) can be used for this, however they provide only limited information and merely show the existence of leakage (for a more detailed discussion of the limitations of t-test based leakage detection see [15]). Eventually, the recent results in [14] showed that by quantifying the informativeness of each statistical moment in a side-channel attack, one can extrapolate the security level of an implementation in function of the noise in its measurements (i.e., a parameter that is typically easier to adapt for HW engineers).

Contribution. Based on this state-of-the art, our contribution is threefold.

First, we extend the evaluation toolbox for profiled side-channel analysis with three new PDF estimation tools, based on Exponentially Modified Gaussian (EMG) distributions, Pearson distribution system and Shifted Generalized Lognormal (SGL) distributions. As illustrated in the upper left part of Fig. 1, they allow characterizing statistical moments up to the fourth one, which captures all most relevant masked implementations published so far.

Second, we show that these tools enable the computation of the information leakage in each statistical moment of a leakage distribution (up to the fourth one). We further illustrate that based on such computations, we can design efficient attacks that are able to exploit the information in all the leaking moments jointly, and that the efficiency of these attacks is proportional to the sum of the information provided by each moment.

Eventually, we observe that our tools also have applications in the context of non-profiled side-channel analysis, where the adversary assumes some a-priori model for his target implementation (e.g., typically Hamming weights, Hamming distances). In this context as well, one can divide existing solutions between "per moment" and "PDF-based" distinguishers (see the middle right part of Fig. 1). Usual representatives of the first category include Correlation Power Analysis (CPA) [6] or its equivalents [25] for first-order moments, and higher-order DPA [37], Correlation-Enhanced Power Analysis Collision Attacks (CEPACA) [27] or Moments Correlating Collision-DPA (MCC-DPA) [31] for higher-order moments. The most common representative of the second category is Mutual Information Analysis (MIA) [19], which usually relies on (non-parametric) histograms or kernels [2], although any PDF estimation tool is in principle eligible[2]. We show that MIA based on the previously mentioned PDF estimation tools (EMG, Pearson, SGL) leads to interesting efficiency tradeoffs for implementations leaking in moments up to four.

The combination of these tools and methods are valuable inputs for the evaluation of the masking countermeasure, since they allow a more accurate understanding of its implementation weaknesses due to glitches (or any other physical default). Furthermore, they are not limited to analysis techniques and also lead to new attacks exploiting a (practically relevant) combination of moments. Eventually, we remark that our results raise relevant questions regarding the so-called simplifying distinguishers in the bottom of Fig. 1. In this context, the adversary/evaluator does not build a model for every target intermediate value

---

[2] Such as cumulants which are used in [22] to estimate the mutual information.

but for a combination of them (or of their bits). All the published simplifying distinguishers (e.g., linear regression in the profiled case [43], its on-the-fly extension [12] and stepwise regression [50] in the non-profiled case) mix a "per moment" approach [11] with simple (typically Gaussian) PDF estimations. Hence, finding whether one could combine a simplifying distinguisher (that provides useful intuitions regarding the parts of the computations that leak more) with more complex PDF estimation tools as in this paper (that provide similarly useful intuitions regarding which moments are leaking) remains an interesting open problem.

2 Background

Generally, *density estimation* – as a well-studied field in statistics – refers to two major categories, namely non-parametric and parametric methods. Histograms and kernels are amongst the well-known non-parametric ones, which do not make any assumptions about the form of the distribution and use only the sampled data to estimate the distribution. By contrast, Gaussian density estimation, which is the most popular parametric PDF estimator, assumes a symmetric form for the distribution, and characterizes it based on its (sample) mean and standard deviation only. As mentioned in the introduction, our focus in this paper is side-channel evaluation, which is commonly based on PDF estimation for building the leakage models. In this section, we shortly recall some frequently-applied PDF estimation techniques in the field of side-channel analysis. We only consider a univariate scenario, which is motivated by our experimental case study in Sect. 5, that is based on a threshold implementation in which all the shares are manipulated in parallel.

Notations. The parametric PDF estimators make use of statistical moments that we specify as follows. Let X be a (univariate) random variable. The dth-order raw statistical moments are defined as $\mathsf{E}(X^d)$, with $\mu = \mathsf{E}(X)$ the mean of the distribution and $\mathsf{E}(.)$ the expectation operator. The dth-order central moments are defined as $\mathsf{E}\left((X - \mu)^d\right)$, with $\sigma^2 = \mathsf{E}\left((X - \mu)^2\right)$ the variance of the distribution. The dth-order standardized moments are defined as $\mathsf{E}\left(\left(\frac{X-\mu}{\sigma}\right)^d\right)$, with $\gamma_1 = \mathsf{E}\left(\left(\frac{X-\mu}{\sigma}\right)^3\right)$ the skewness (a measure of the *asymmetry* of the distribution, also known as the first shape parameter), and $\beta_2 = \mathsf{E}\left(\left(\frac{X-\mu}{\sigma}\right)^4\right)$ the kurtosis (a measure of the *peakedness* of the distribution, also known as the second shape parameter). Unless otherwise stated, for simplicity we denote first raw, second central, third (and fourth) standardized moments by first, second, third (and fourth) moments respectively.

Gaussian Density Estimation. In this case, it is assumed that the leakages follow a Gaussian (normal) distribution, and the PDF is given by:

$$F(x) = \frac{1}{\sigma\sqrt{2\pi}} e^{-\frac{(x-\mu)^2}{2\sigma^2}},$$

with μ and σ the estimated mean and standard deviation of the samples. Since a Gaussian distribution considers only the first two moments, it generally leads to a more efficient estimation compared to the non-parametric histograms or kernels (as long as the actual distribution is close enough to a Gaussian one). In other words, if the higher (>2nd) statistical moments of the underlying distribution of the samples are negligible, Gaussian density estimation is going to be extremely efficient. Gaussian Templates and regression-based models are part of the widely-used tools exploiting such an assumption [16].

Gaussian Mixtures. We mention that yet another approach to PDF estimation for masked implementations would be to consider mixture distributions. As demonstrated in [48], this solution is especially efficient when the profiling phase assumes the knowledge of the shares. By contrast, it becomes heuristic – since based on the Expectation Maximization (EM) algorithm – if they are not [23], which will be our running scenario in this work. In particular, we will consider contexts where the different modes of the mixture distributions are well inter-leaved (i.e. when the noise is large enough for masking to enforce good security guarantees), which makes the EM algorithm hard(er) to apply and stands in contrast with contexts where the modes can be trivially identified by the adversary (for example see [29]). That is, our goal is to investigate simple(r) tools that apply to masking when it delivers its promises and are guaranteed to converge without any need to guess about the number of shares in the target device.

3 New Proposals

We now describe three alternative parametric distributions that can cover moments up to the fourth one. We discuss their advantages as well as the challenges one may face to set the parameters to use them.

3.1 Exponentially Modified Gaussian

Since the Gaussian distribution is symmetric, its skewness is always zero. The exponentially Modified Gaussian (EMG) is another parametric distribution which additionally includes this first shape parameter. The PDF of such a distribution, that covers the first three moments, is defined by [20]:

$$F(x) = \frac{\lambda_3}{2} e^{\frac{\lambda_3}{2}(2\lambda_1 + \lambda_3 \lambda_2^2 - 2x)} erfc\left(\frac{\lambda_1 + \lambda_3 \lambda_2^2 - x}{\sqrt{2}\lambda_2}\right), \tag{1}$$

where $\lambda_1, \lambda_2, \lambda_3$ are the parameters of the distribution and $erfc(.)$ refers to the complementary error function defined as:

$$erfc(x) = \frac{2}{\sqrt{\pi}} \int_x^\infty e^{-t^2} dt.$$

By means of the sample mean μ, standard deviation σ and skewness γ_1 of the data, these three parameters can be estimated as:

$$\lambda_1 = \mu - \sigma \left(\frac{\gamma_1}{2}\right)^{1/3}, \quad \lambda_2^2 = \sigma^2 \left(1 - \left(\frac{\gamma_1}{2}\right)^{2/3}\right), \quad \lambda_3 = \frac{1}{\sigma \left(\frac{\gamma_1}{2}\right)^{1/3}}.$$

It should be noted that EMG does not cover symmetric distributions, i.e., $\gamma_1 = 0$. However, it usually causes no issue in practice (and in particular for side-channel attacks) as the estimated skewness is never exactly zero. Nevertheless, if the underlying skewness is zero, the estimated skewness might be very small. These cases can lead to numerical problems, which can be solved by using libraries for higher precision computations or switching to a distribution which covers zero skewness (Gaussian, Pearson). Besides, note that for a negative skewness $\gamma_1 < 0$, the distribution is parametrized with the absolute value $|\gamma_1|$, and then mirrored around the mean.

3.2 Pearson Distribution System

The Pearson distribution system is a collection of probability distributions that can be parametrized using the first four moments. In total twelve different distributions (cf. [33–35]) are defined in such a way that depending on the estimated moments one type is preferred, and the corresponding PDF estimation technique is applied. In our experiments we noticed that types I, IV and VI (which are presented in detail below) are the only necessary ones. For further descriptions of the other types, the interested reader is referred to the original articles [33–35].

Cautionary Note. Distribution systems like Pearson's are in general very flexible as they allow characterizing a broad range of combinations of moments. However, they require the estimation of several PDFs, and may face stability problems at the transitions between the different types of distributions (which may occur, e.g., by increasing the number of side-channel samples). Hence, in these cases, it is preferable to rely on a single distribution.

3.3 Shifted Generalized Lognormal

In [24], Low introduced the Shifted Generalized Lognormal distribution (SGL). It can be parametrized with the first four moments and covers a large interval of possible combinations of skewness and kurtosis. Both of these properties are desirable in side-channel evaluations, and therefore this distribution can be an interesting alternative to the Pearson's distribution system. The realm covered by the SGL is vast and we found it to be sufficient for all our practical experiments.

Concretely, the PDF of the SGL is given by:

$$F(x) = \frac{1}{2\lambda_3^{1/\lambda_3} \lambda_4 \Gamma(1 + 1/\lambda_3)(x - \lambda_1)} e^{-\frac{1}{\lambda_3 \lambda_4^{\lambda_3}} \left| ln\left(\frac{x - \lambda_1}{\lambda_2}\right)\right|^{\lambda_3}}, \tag{2}$$

for $\lambda_1 < x < \infty$, where λ_1, λ_2, λ_3, and λ_4 are the distribution parameters and $\Gamma(.)$ denotes the gamma function. These parameters can be estimated using the first four moments. For conciseness, we only give a brief overview of the resulting estimation problem in the full version of the paper [45], and refer the interested readers to [24].

3.4   Computational Complexity

The presented parametric methods have all different PDFs with different computation complexities. For SGL, the computation of the parameters from the first four moments takes considerably longer than for all other discussed distributions. To present some intuitions on the run time of the different PDFs, we performed experiments using 100 randomly generated sets of moments and run each PDF[3] 100 times for each of these sets. Then we computed the average over all 1000 executions of each PDF. The Gaussian distribution is used as a reference value and has an average of 0.0034 s on an Intel i5-4200M CPU. The averages increase with the number of moments considered in the distribution: 0.0082 s (EMG), 0.029 s (Pearson), 1.70 s (SGL).

4   Simulated Experiments

In order to better understand the interest of the tools proposed in Sect. 3 in the context of side-channel analysis, we present a couple of simulated experiments. In the following we use mathematically-generated leakages derived from:

$$l = \mathrm{HW}(s \oplus c_1 \oplus c_2) + \mathrm{HW}(c_1) + \mathrm{HW}(c_2), \qquad (3)$$

where $\mathrm{HW}(.)$ denotes the Hamming weight function, s a sensitive (secret) 4-bit variable, and c_1 and c_2 uniformly distributed random masks in $\{0,1\}^4$. Note that this example is related to any nibble-oriented cipher, e.g., PRESENT [4], and the basic evaluation procedure presented in this paper does not change for larger bit sizes. The only adjustment is the number of possible different classifications, i.e., 2^n instead of 2^4 for n-bit variables. In this simulation it is supposed that the target is a hardware design where the shares are processed at the same time. This scenario essentially emulates a second-order Boolean masking scheme, where we only focus on the encoding of a single variable s in a noise-free situation. In this context, the first and second moments of the leakage distribution are expected to be independent of s. For each $s \in \{0,1\}^4$, we estimate the PDF using both non-parametric (kernels) and parametric (Gaussian, EMG, Pearson, SGL) tools. The first four moments for each s, plotted in Fig. 2(a), reveal that there is indeed no dependency between s and the first two moments (i.e., they remain constant for all s). Hence, the only way that s can be distinguished is by observing the third moment. Since kernel-based density estimation considers all possible moments, it can be used to distinguish s as shown in Fig. 2(b).

---

[3] We implemented three distributions in MATLAB and used the publicly available pearspdf [5].

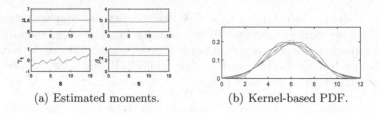

(a) Estimated moments.   (b) Kernel-based PDF.

Fig. 2. The estimated moments for each possible $s \in \{0,1\}^4$ (a) and kernel-estimated PDFs (b) for mathematically-generated leakages corresponding to a 2nd-order masking.

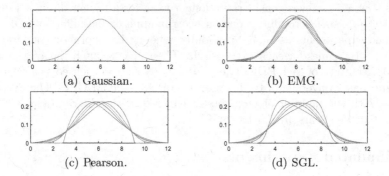

(a) Gaussian.   (b) EMG.

(c) Pearson.   (d) SGL.

Fig. 3. The estimated PDFs for mathematically-generated leakages corresponding to a 2nd-order masking, obtained with various parametric tools from Sects. 2 and 3.

By contrast, the third moment is not used to parametrize the Gaussian distribution and thus each s results in the same distribution in this case (as per Fig. 3(a)). This example shows why Gaussian density estimation cannot be used to analyze the leakages that reside in an order higher than two. Eventually, our newly proposed estimators consider moments up to the fourth one, and therefore they can be used to quantify the information leakage of our simulated masking experiment (this can be seen in the remaining part of Fig. 3).

5 Practical Case Studies

To examine the application and efficiency of the above-mentioned solutions, we consider a threshold implementation of the PRESENT cipher [4] on an FPGA platform. More precisely, the target design is the *Profile 2* presented in [36] that follows a serialized architecture, i.e., using one instance of the S-box for the whole SLayer. Such a masked hardware implementation has been selected for the practical investigations due to its second- and third-order univariate leakages which allow us to examine our proposed tools. If we would have no leakage at order three and higher, examining the difference between our tools and Gaussian would not be possible.

In the target implementation, the data state is represented by $d = 3$ Boolean shares, and the SLayer is based on the 2-stage masked S-box described in [32]. In other words, each S-box on a 4-bit data is implemented in a pipeline fashion and needs two clock cycles to be computed. For more details on the design architecture we refer the interested reader to [31,36].

The leakage traces are collected from a Xilinx Virtex-II Pro FPGA embedded on SASEBO [1]. The sampling rate was set to 1 GS/s and the target FPGA clock was driven at a frequency of 3 MHz.

We collected 100,000,000 traces to be used in our experiments. During the measurements, the PRNG that provides random data (masks) for the sharing of the plaintext was kept active. We also examined and confirmed the uniform distribution of the masks.

A former analysis of MCP-DPA by Moradi and Standaert in [31] on the same implementation revealed that the first pipeline stage of the target S-box exhibits the most informative leakages. It confirms that no first-order leakage can be exploited from this implementation, whereas the second and third moments are indeed informative. It also suggests that second-order leakages are more informative than third ones. By contrast, and as exhaustively discussed in the introduction, two important questions remain open. First, can we quantify the informativeness of the different moments on a (somewhat) more formal basis? Second, and given that more than a single moment provides information, can we design an attack that jointly exploits these moments? (which is in contrast with MCP-DPA that only exploits moments one by one).

Both questions can be answered in the affirmative by the following discussion. In order to make our results comparable with [31], we focus on the same parts of the leakage traces. Namely, we analyze the most informative clock cycle in the S-box execution. Based on this case study, we show that the newly introduced PDF estimation tools are powerful ingredients for the information theoretic analysis of a threshold implementation. First, they are able to extract an amount of information from the traces comparable to a kernel density estimation. Second, they are useful to estimate the informativeness of each moment, and to perform attacks based on the best combination of moments carrying significant information. Eventually, they can naturally and efficiently be embedded in PDF-based non-profiled attacks such as MIA.

5.1   Profiled Evaluations and Attacks

First, we examine the information leakage of the target device using an information theoretic approach. The idea to use Mutual Information (MI) as an evaluation metric was introduced in [46]. It was later refined in [39] to incorporate the fact that the leakage distribution is only estimated, which can potentially bias the estimation of the MI. The so-called Perceived Information (PI) is used to reflect this bias and can be computed as:

$$\hat{PI}(S; L) = H[S] - \sum_{s \in \mathcal{S}} Pr[s] \sum_{l \in \mathcal{L}} Pr_{\mathsf{chip}}[l|s] \cdot log_2 \hat{Pr}_{\mathsf{model}}[s|l], \qquad (4)$$

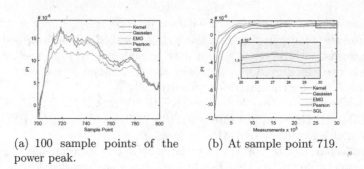

(a) 100 sample points of the power peak.

(b) At sample point 719.

Fig. 4. Kernel-, Gaussian-, EMG-, Pearson- and SGL-based PI estimation with all covered moments (a) using 100,000,000 meas., (b) over the number of meas.

where Pr_{chip} denotes the chip's true distribution (which is unknown but can be sampled) and $\hat{Pr}_{model}$ refers to the adversary's estimated model (for which we have an analytical formula). Computing the PI essentially requires an estimated $\hat{Pr}_{model}$, which is exactly what our PDF estimation tools provide. In our experiments, we followed the procedure presented in [16] for computing this metric. In particular, we used 10-fold cross-validation and report the mean of the resulting PI estimates. We start by looking at the information extracted using all the moments enabled by each PDF estimation tool. We then analyze (subsets of) these moments separately.

Combined Moments. In order to compare our proposed solutions (EMG, Pearson, SGL) with the established ones (kernels, Gaussian), we first compute the PI using all the covered moments. We estimate $\hat{Pr}_{model}$ using the different estimators and compare the results. As previously mentioned, this experiment only covers 100 sample points corresponding to the power peak of the targeted clock cycle. The 100,000,000 traces are divided into 10 sets. For each of the 10 runs we use one of these 10 sets (each with 10,000,000 measurements) as samples of the chip's true distributions, and the remaining 9 sets (90,000,000 measurements) to estimate the model distribution ($\hat{Pr}_{model}$). Figure 4(a) contains the results.

At the first glance, it can be observed that the achieved PI using the Gaussian distribution to estimate $\hat{Pr}_{model}$ is the lowest. This implies that not all available information is contained in the first two moments (that are the only ones captured by a Gaussian distribution). More interestingly, kernel-based density estimation is non-parametric and therefore is expected to provide the highest PI if its bandwidth is well adapted and enough samples are available. Yet, we observe that this is not exactly the case in our experiments. As depicted in Fig. 4(a) (where we focus on the most informative sample 719), this is most likely due to an estimation issue (i.e., a lack of samples). As expected, the non-parametric kernel density estimation is the slowest to converge in this case. This suggests an interesting feature of our new parametric tools. Namely, whereas Gaussian estimation is very fast but limited to the exploitation of two moments (hence leads to less efficient attacks, as will be discussed next), EMG-, Pearson- and SGL-based estimations combine a faster convergence than kernels with a similar informativeness.

Summarizing, we can conclude that PDFs covering the right combination of moments lead to the best tradeoff between a fast convergence towards a well estimated model, and a well-informative model once properly estimated (i.e., a model for which the PI should be close to the MI [16]). By contrast, the previous results do not allow to deduce about the relative informativeness of each moment (which could possibly be used to further speed up the model estimation and attacks), which motivates the following analysis.

Separate Moments. An interesting property of the parametric estimators is the ability to consider only selected moments instead of trying to characterize any possible moment (as in non-parametric estimations). Using the Gaussian distribution as an example, we can compute the information contained exclusively in the first two moments, as this distribution only considers the mean and variance. Similarly, it is also possible to compute the PI for the first three moments (with EMG distributions) and the first four moments (with Pearson's distribution system and SGL distributions). In the following, we present an approach that enables us to compute the PI both for each moment taken separately and for any combination of those.

For this purpose, and taking the case where we focus on a single moment, we simply have to set all but one of the moments to a fixed value. For example, suppose that we want to consider the information contained in the first moments of a Gaussian distribution only. We achieve this by considering a Gaussian model where the means are estimated as in the previous section, but the variances are set to a fixed value, which essentially removes any secret-dependent information they could carry from the templates through the second moments. Since changing the variances affects the shape of the distributions, the fixed value can be chosen as the average of the variances (over the 16 templates) to minimize the distance between the original distributions and the ones with a fixed variance[4]. A similar technique actually works for any of our parametric estimators, and for any (combination of) moments.

As an illustration, let us first recall the influence of the first four statistical moments on the shape of the resulting distribution. The third moment, called skewness, measures the asymmetry of the distribution. Therefore, distributions with positive skewness tend to left while distributions with a negative skewness tend to the right. The fourth moment, called kurtosis, measures the "peakedness" (sharpness) of the distribution. As a consequence, the higher the kurtosis, the sharper is the distribution. Note that we consider only the first four moments in our analysis, hence we omitted definitions for moments of any further orders.

---

[4] Instead, one can also consider the variance of whole trace set. Here we need only a fixed value which is not too different from the variance of each template. Such an approach is not valid in case of Gaussian mixtures as stated in Sect. 2.

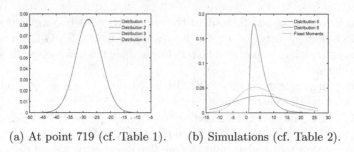

(a) At point 719 (cf. Table 1).   (b) Simulations (cf. Table 2).

Fig. 5. The PDFs of the six distributions from Tables 1 and 2.

Table 1. The first four statistical moments of four distributions at sample point 719.

	Dist. 1	Dist. 2	Dist. 3	Dist. 4	Average
Mean	−27.9734310	−27.9811494	−27.9827913	−27.9782609	−27.9789082
Variance	22.3624316	21.9979663	22.2165081	22.2660171	22.2107308
Skewness	0.0075083	0.0053184	0.0131009	−0.0000767	0.0064627
Kurtosis	3.0177549	3.0202503	3.0219293	3.0183596	3.0195735

When we set specific higher-order moments (as in our approach) to specific values, we actually fix the width of the distributions (i.e., variance), or their right-left tendency (i.e., skewness) or their sharpness (i.e., kurtosis). Hence, information sitting in the corresponding moments does not contribute in the information-theoretic-based evaluation, e.g., mutual information. We like to emphasize that the estimated higher-order moments in real side-channel measurements (categorized, for example, based on the processed data) are very slightly different. Consider for example the PDFs of four exemplary distributions shown in Fig. 5(a), taken from the most leaking point of the measurements of our case study (see Fig. 4(a)). The first four moments of each distribution are given in Table 1. All moments of the four distributions are very similar to each other, e.g., the skewness of all these four distributions is only slightly different. Hence, fixing the skewness of all of them to a specific value (e.g., the average of all skewnesses given by 0.0064627) does not significantly change the shape of the distributions.

Here we consider four different cases:

1. All moments except the first are fixed to their average (evaluation through means).
2. All moments except the second are fixed to their average (evaluation through variances).
3. All moments except the third are fixed to their average (evaluation through skewnesses)
4. All moments except the fourth are fixed to their average (evaluation through kurtoses).

Table 2. The first four statistical moments of two simulated distributions.

	Dist. 5	Dist. 6	Average
Mean	4.9997939	7.400773	6.2002834
Variance	10.0032941	149.017440	79.5103671
Skewness	1.7063003	0.377136	1.0417184
Kurtosis	7.8417563	3.648649	5.7452030

For each case, the shape of the resulting distributions is very close to the original shape in Fig. 5(a). The resulting PDFs of the modified distributions for each case is provided in the full version of the paper [45].

It should be noted that in case of simulated data with significantly different moments for each distribution the resulting shapes of each distribution would be also dramatically different to each other. Therefore in this case, setting the corresponding moments to a fixed (average) value does not make the distributions to roughly follow the same shape. If such a huge difference between the moments of the (categorized) distributions exists in practice by any (rare) chance, the corresponding implementation is significantly vulnerable to certain attacks. Obviously, this makes the necessity of performing per-moment evaluations questionable. As an example, we show in Fig. 5(b) two simulated distributions formed by the moments from Table 2. It is obvious that the shape of the distribution with fixed moments is considerably different than that of the original two distributions. In this case, a per-moment approach would not be easily possible with an information-theoretic evaluation tool.

We analyze this moment-based investigation based on the same case study as for the previous information theoretic analysis. Hence, we repeat the previous experiments (of Fig. 4(a)) with the same parametric estimators (Gaussian, EMG, Pearson, SGL), but this time we consider each possible moment separately. The results are depicted in Fig. 6 where the PI curves are categorized based on the employed estimator. Each part of the figure contains the PI curves obtained for each moment separately. For example, in Fig. 6(a) the curve labeled *Gaussian (1st)* shows the PI achieved for the first moments (and the curve *Gaussian (2nd)* depicts the same for the second moments, etc.). Further, we included the PI curve of the combined moments (taken from Fig. 4(a)) and the sum of the PI curves of the separate moments (e.g., *Gaussian Sum* as the sum of the PI curves of *Gaussian (1st)* and *Gaussian (2nd)*).

As expected, the first moment does not contain any exploitable information as the implementation is first-order secure. It is also noticeable that the chosen estimator does not affect the PI for the first moment. The second moment leads to the highest PI, and therefore is the most informative moment. As similarly indicated by MCP-DPA, the third moment is informative but not as much as the second one. Furthermore, using two estimators (Pearson, SGL) that also cover the fourth moment, we are not able to detect any significant information leakage in the fourth moment. Therefore, a combination of the second and third moments should suffice to capture most of the available information in the underlying measurements.

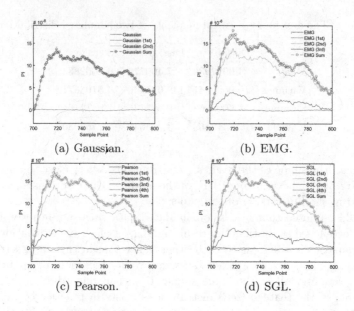

(a) Gaussian. (b) EMG.

(c) Pearson. (d) SGL.

Fig. 6. PI estimates for the separate moments.

Most interestingly, we observe that the sum of the PI values obtained for the separate moments is actually close to the PI estimated with the combined moments. Although informal, this observation is particularly interesting in view of the recent results by Duc *et al.* in [14] where the PI values are connected with the success rate of a (worst-case) template attack using the same model. Indeed, since the sum of the PI values obtained per moment is essentially the same as the PI value obtained with the non-parametric kernel method, it implies that *in our case study*, the separation between moments did not lead to any significant information loss. This suggests that a (simple and intuitive) moment-based side-channel evaluation could be well-founded, at least in certain contexts that would be worth formalizing. And very concretely, it also means that an attack exploiting out two informative (i.e., second and third) moments will be close to optimal in our case.

Profiled Attacks. The results in [14] prove that (under sufficiently noisy leakages) the success rate of a profiled template attack is inversely proportional to the PI value estimated with the same model. In view of the previous discussions, it means that our proposed estimation tools (EMG, Pearson, SGL) should lead to more effective profiled attacks than their counterparts with Gaussian estimation (because of modeling errors) and kernels (because of assumption errors). Furthermore, the attacks exploiting the second moment should lead to a higher success rate than attacks exploiting the other three moments. Eventually, the best attack should exploit the combination of second and third moments. For completeness, we ran experiments to confirm these expectations. We built univariate templates

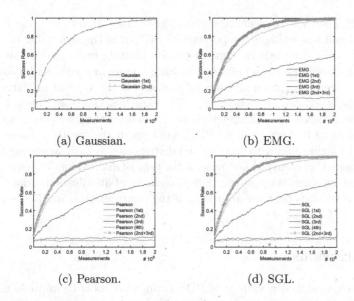

(a) Gaussian.   (b) EMG.

(c) Pearson.   (d) SGL.

Fig. 7. Success rate of several univariate template attacks exploiting separate and combined moments, for the most informative sample point 719 in our traces.

(for the most informative sample point 719) from 90,000,000 measurements and, for each given number of measurements, repeated an attack 1000 times for different measurements (excluding those used for profiling) to compute a subkey recovery success rate. The results of this last experiment are depicted in Fig. 7 and are well in line with theoretical predictions. In this respect, the most interesting curves are the ones corresponding to the combination of second and third moments, since they correspond to the best tradeoff between model complexity and attack efficiency, and could not have been reached with existing side-channel evaluation tools.

Non-profiled Attacks. A detailed discussion with experimental results is provided in the full version of the paper [45].

5.2   Selection of Tools

We have discussed multiple parametric tools, each with its own advantages and disadvantages. Compared to the traditional non-parametric tools, they offer a higher flexibility and convergence. Therefore, they should be preferred if the number of samples is too small or a special case (e.g., only two moments) should be evaluated. The PDF of EMG can be computed very efficiently compared SGL and Pearson. However, it considers only the first three moments instead of four. The Pearson distribution system includes the kurtosis and its PDF is still relatively efficient compared to SGL. Nevertheless, it is made up of multiple different distributions which can be problematic in certain cases as pointed out in Sect. 3.2. Therefore, in scenarios where the computation time of the PDF can be ignored and the leakages are covered by SGL, it is the preferable tool.

However, the computation time is often a limiting factor and it can be significantly reduced in certain cases by choosing a more limited distribution which is still sufficient to capture all relevant leakage. If the type of implementation and leakage is known, this choice is easily possible. Gaussian (resp. EMG) is the preferred choice for leakage which is exclusive in the first two (resp. three) moments due to its very efficient PDF. Leakage in the fourth moment can be also efficiently captured with the Pearson distribution system, assuming that the aforementioned problems do not arise. If the type of masked implementation, i.e., the order of masking, is unknown, then this choice of distribution cannot be that easily made. SGL is the best approach, if the distribution is inside the plane of existence of SGL.

6 Conclusions

This paper introduced a variety of PDF estimation tools to improve the evaluation of leaking devices, both in the profiled and non-profiled settings. Their main interest is their flexibility: our proposals can indeed capture information lying in different moments of the leakage PDF. As a result, we can easily analyze masked implementations and extract useful feedback to hardware designers, i.e. in terms of how much information is lying in every moment and how to combine it. This brings a concrete and more founded counterpart the recent evaluations of implementations with non-independent leakages in [14], where this quantity of information "per moment" is required. More generally, our findings provide efficient tradeoffs between the cost of profiling and the efficiency of the resulting attacks, since they allow adversaries and evaluators to build models that are tailored to their implementations. These results naturally open various interesting research challenges for future work. As mentioned in introduction, combining an analysis of moments as in this work with simplifying approaches to leakage modeling (e.g. based on linear regression) would be even more convenient to evaluators. Besides, investigating the "summing rule" of Sect. 5.1 more formally is certainly worth further efforts as well. Eventually, our current tools are limited to univariate leakages. While this was sufficient to analyze our hardware case study, it naturally suggests the extension to multivariate case studies as yet another important question. This is especially interesting given that even hardware designs with univariate d-order security may include a multivariate vulnerability for which less than d points are combined [40]. A starting point for this purpose would be to exploit some popular "combining" functions from the side-channel literature (which would allow us to exploit our univariate tools directly).

Acknowledgments. This work is partly supported by the DFG Research Training Group GRK 1817 Ubicrypt and the ERC project 280141. François-Xavier Standaert is a research associate of the Belgian Fund for Scientific Research (FNRS-F.R.S.).

References

1. Side-Channel Attack Standard Evaluation Board (SASEBO). http://satoh.cs.uec. ac.jp/SAKURA/index.html
2. Batina, L., Gierlichs, B., Prouff, E., Rivain, M., Standaert, F.-X., Veyrat-Charvillon, N.: Mutual information analysis: a comprehensive study. J. Cryptol. **24**(2), 269–291 (2011)
3. Bilgin, B., Gierlichs, B., Nikova, S., Nikov, V., Rijmen, V.: A more efficient AES threshold implementation. In: Pointcheval, D., Vergnaud, D. (eds.) AFRICACRYPT 2014. LNCS, vol. 8469, pp. 267–284. Springer, Cham (2014). doi:10.1007/978-3-319-06734-6_17
4. Bogdanov, A., Knudsen, L.R., Leander, G., Paar, C., Poschmann, A., Robshaw, M.J.B., Seurin, Y., Vikkelsoe, C.: PRESENT: an ultra-lightweight block cipher. In: Paillier, P., Verbauwhede, I. (eds.) CHES 2007. LNCS, vol. 4727, pp. 450–466. Springer, Heidelberg (2007). doi:10.1007/978-3-540-74735-2_31
5. Pierce Brady. pearspdf. http://www.mathworks.com/matlabcentral/fileexchange/ 26516-pearspdf
6. Brier, E., Clavier, C., Olivier, F.: Correlation power analysis with a leakage model. In: Joye, M., Quisquater, J.-J. (eds.) CHES 2004. LNCS, vol. 3156, pp. 16–29. Springer, Heidelberg (2004). doi:10.1007/978-3-540-28632-5_2
7. Chari, S., Jutla, C.S., Rao, J.R., Rohatgi, P.: Towards sound approaches to counteract power-analysis attacks. In: Wiener, M. (ed.) CRYPTO 1999. LNCS, vol. 1666, pp. 398–412. Springer, Heidelberg (1999). doi:10.1007/3-540-48405-1_26
8. Chari, S., Rao, J.R., Rohatgi, P.: Template attacks. In: Kaliski, B.S., Koç, K., Paar, C. (eds.) CHES 2002. LNCS, vol. 2523, pp. 13–28. Springer, Heidelberg (2003). doi:10.1007/3-540-36400-5_3
9. Coron, J.-S., Giraud, C., Prouff, E., Renner, S., Rivain, M., Vadnala, P.K.: Conversion of security proofs from one leakage model to another: a new issue. In: Schindler, W., Huss, S.A. (eds.) COSADE 2012. LNCS, vol. 7275, pp. 69–81. Springer, Heidelberg (2012). doi:10.1007/978-3-642-29912-4_6
10. Coron, J.-S., Kizhvatov, I.: An efficient method for random delay generation in embedded software. In: Clavier, C., Gaj, K. (eds.) CHES 2009. LNCS, vol. 5747, pp. 156–170. Springer, Heidelberg (2009). doi:10.1007/978-3-642-04138-9_12
11. Dabosville, G., Doget, J., Prouff, E.: A new second-order side channel attack based on linear regression. IEEE Trans. Comput. **62**(8), 1629–1640 (2013)
12. Doget, J., Prouff, E., Rivain, M., Standaert, F.-X.: Univariate side channel attacks and leakage modeling. J. Cryptogr. Eng. **1**(2), 123–144 (2011)
13. Duc, A., Dziembowski, S., Faust, S.: Unifying leakage models: from probing attacks to noisy leakage. In: Nguyen, P.Q., Oswald, E. (eds.) EUROCRYPT 2014. LNCS, vol. 8441, pp. 423–440. Springer, Heidelberg (2014). doi:10.1007/ 978-3-642-55220-5_24
14. Duc, A., Faust, S., Standaert, F.-X.: Making masking security proofs concrete. In: Oswald, E., Fischlin, M. (eds.) EUROCRYPT 2015. LNCS, vol. 9056, pp. 401–429. Springer, Heidelberg (2015). doi:10.1007/978-3-662-46800-5_16
15. Durvaux, F., Standaert, F.-X.: From improved leakage detection to the detection of points of interests in leakage traces. In: Fischlin, M., Coron, J.-S. (eds.) EUROCRYPT 2016. LNCS, vol. 9665, pp. 240–262. Springer, Heidelberg (2016). doi:10. 1007/978-3-662-49890-3_10
16. Durvaux, F., Standaert, F.-X., Veyrat-Charvillon, N.: How to certify the leakage of a chip? In: Nguyen, P.Q., Oswald, E. (eds.) EUROCRYPT 2014. LNCS, vol. 8441, pp. 459–476. Springer, Heidelberg (2014). doi:10.1007/978-3-642-55220-5_26

17. Dziembowski, S., Pietrzak, K.: Leakage-resilient cryptography. In: Foundations of Computer Science, pp. 293–302. IEEE Computer Society (2008)

18. Eisenbarth, T., Kasper, T., Moradi, A., Paar, C., Salmasizadeh, M., Shalmani, M.T.M.: On the power of power analysis in the real world: a complete break of the KEELOQ code hopping scheme. In: Wagner, D. (ed.) CRYPTO 2008. LNCS, vol. 5157, pp. 203–220. Springer, Heidelberg (2008). doi:10.1007/978-3-540-85174-5_12

19. Gierlichs, B., Batina, L., Tuyls, P., Preneel, B.: Mutual information analysis. In: Oswald, E., Rohatgi, P. (eds.) CHES 2008. LNCS, vol. 5154, pp. 426–442. Springer, Heidelberg (2008). doi:10.1007/978-3-540-85053-3_27

20. Grushka, E.: Characterization of exponentially modified Gaussian peaks in chromatography. Anal. Chem. **44**(11), 1733–1738 (1972). PMID: 22324584

21. Kocher, P., Jaffe, J., Jun, B.: Differential power analysis. In: Wiener, M. (ed.) CRYPTO 1999. LNCS, vol. 1666, pp. 388–397. Springer, Heidelberg (1999). doi:10.1007/3-540-48405-1_25

22. Le, T.-H., Berthier, M.: Mutual information analysis under the view of higher-order statistics. In: Echizen, I., Kunihiro, N., Sasaki, R. (eds.) IWSEC 2010. LNCS, vol. 6434, pp. 285–300. Springer, Heidelberg (2010). doi:10.1007/978-3-642-16825-3_19

23. Lemke-Rust, K., Paar, C.: Gaussian mixture models for higher-order side channel analysis. In: Paillier, P., Verbauwhede, I. (eds.) CHES 2007. LNCS, vol. 4727, pp. 14–27. Springer, Heidelberg (2007). doi:10.1007/978-3-540-74735-2_2

24. Low, Y.M.: A new distribution for fitting four moments and its applications to reliability analysis. Struct. Saf. **42**, 12–25 (2013)

25. Mangard, S., Oswald, E., Standaert, F.-X.: One for all - all for one: unifying standard differential power analysis attacks. IET Inf. Secur. **5**(2), 100–110 (2011)

26. Mangard, S., Popp, T., Gammel, B.M.: Side-channel leakage of masked CMOS gates. In: Menezes, A. (ed.) CT-RSA 2005. LNCS, vol. 3376, pp. 351–365. Springer, Heidelberg (2005). doi:10.1007/978-3-540-30574-3_24

27. Moradi, A.: Statistical tools flavor side-channel collision attacks. In: Pointcheval, D., Johansson, T. (eds.) EUROCRYPT 2012. LNCS, vol. 7237, pp. 428–445. Springer, Heidelberg (2012). doi:10.1007/978-3-642-29011-4_26

28. Moradi, A., Barenghi, A., Kasper, T., Paar, C.: On the vulnerability of FPGA bitstream encryption against power analysis attacks: extracting keys from xilinx Virtex-II FPGAs. In: Computer and Communications Security, CCS 2011, pp. 111–124. ACM (2011)

29. Moradi, A., Kirschbaum, M., Eisenbarth, T., Paar, C.: Masked dual-rail precharge logic encounters state-of-the-art power analysis methods. IEEE Trans. VLSI Syst. **20**(9), 1578–1589 (2012)

30. Moradi, A., Poschmann, A., Ling, S., Paar, C., Wang, H.: Pushing the limits: a very compact and a threshold implementation of AES. In: Paterson, K.G. (ed.) EUROCRYPT 2011. LNCS, vol. 6632, pp. 69–88. Springer, Heidelberg (2011). doi:10.1007/978-3-642-20465-4_6

31. Moradi, A., Standaert, F.-X.: Moments-correlating DPA. IACR Cryptology ePrint Archive 2014:409 (2014)

32. Nikova, S., Rijmen, V., Schläffer, M.: Secure hardware implementation of nonlinear functions in the presence of glitches. J. Cryptol. **24**(2), 292–321 (2011)

33. Pearson, K.: Contributions to the mathematical theory of evolution. II. Skew variation in homogeneous material. R. Soc. Lond. Philos. Trans. Ser. A **186**, 343–414 (1895)

34. Pearson, K.: Mathematical contributions to the theory of evolution. X. Supplement to a memoir on skew variation. R. Soc. Lond. Philos. Trans. Ser. A **197**, 443–459 (1901)

35. Pearson, K.: Mathematical contributions to the theory of evolution. XIX. Second supplement to a memoir on skew variation. R. Soc. Lond. Philos. Trans. Ser. A **216**, 429–457 (1916)

36. Poschmann, A., Moradi, A., Khoo, K., Lim, C.-W., Wang, H., Ling, S.: Side-channel resistant crypto for less than 2, 300 GE. J. Cryptol. **24**(2), 322–345 (2011)

37. Prouff, E., Rivain, M., Bevan, R.: Statistical analysis of second order differential power analysis. IEEE Trans. Comput. **58**(6), 799–811 (2009)

38. Rao, J.R., Rohatgi, P., Scherzer, H., Tinguely, S.: Partitioning attacks: or how to rapidly clone some GSM cards. In: IEEE Symposium on Security and Privacy 2002, pp. 31–41. IEEE Computer Society (2002)

39. Renauld, M., Standaert, F.-X., Veyrat-Charvillon, N., Kamel, D., Flandre, D.: A formal study of power variability issues and side-channel attacks for nanoscale devices. In: Paterson, K.G. (ed.) EUROCRYPT 2011. LNCS, vol. 6632, pp. 109–128. Springer, Heidelberg (2011). doi:10.1007/978-3-642-20465-4_8

40. Reparaz, O., Bilgin, B., Nikova, S., Gierlichs, B., Verbauwhede, I.: Consolidating masking schemes. In: Gennaro, R., Robshaw, M. (eds.) CRYPTO 2015. LNCS, vol. 9215, pp. 764–783. Springer, Heidelberg (2015). doi:10.1007/978-3-662-47989-6_37

41. Rivain, M., Prouff, E.: Provably secure higher-order masking of AES. In: Mangard, S., Standaert, F.-X. (eds.) CHES 2010. LNCS, vol. 6225, pp. 413–427. Springer, Heidelberg (2010). doi:10.1007/978-3-642-15031-9_28

42. Roche, T., Prouff, E.: Higher-order glitch free implementation of the AES using secure multi-party computation protocols - extended version. J. Cryptogr. Eng. **2**(2), 111–127 (2012)

43. Schindler, W., Lemke, K., Paar, C.: A stochastic model for differential side channel cryptanalysis. In: Rao, J.R., Sunar, B. (eds.) CHES 2005. LNCS, vol. 3659, pp. 30–46. Springer, Heidelberg (2005). doi:10.1007/11545262_3

44. Schneider, T., Moradi, A.: Leakage assessment methodology. In: Güneysu, T., Handschuh, H. (eds.) CHES 2015. LNCS, vol. 9293, pp. 495–513. Springer, Heidelberg (2015). doi:10.1007/978-3-662-48324-4_25

45. Schneider, T., Moradi, A., Standaert, F.-X., Güneysu, T.: Bridging the gap: advanced tools for side-channel leakage estimation beyond Gaussian templates and histograms. Cryptology ePrint Archive, Report 2016/719 (2016). http://eprint.iacr.org/2016/719

46. Standaert, F.-X., Malkin, T.G., Yung, M.: A unified framework for the analysis of side-channel key recovery attacks. In: Joux, A. (ed.) EUROCRYPT 2009. LNCS, vol. 5479, pp. 443–461. Springer, Heidelberg (2009). doi:10.1007/978-3-642-01001-9_26

47. Standaert, F.-X., Pereira, O., Yu, Y.: Leakage-resilient symmetric cryptography under empirically verifiable assumptions. In: Canetti, R., Garay, J.A. (eds.) CRYPTO 2013. LNCS, vol. 8042, pp. 335–352. Springer, Heidelberg (2013). doi:10.1007/978-3-642-40041-4_19

48. Standaert, F.-X., Veyrat-Charvillon, N., Oswald, E., Gierlichs, B., Medwed, M., Kasper, M., Mangard, S.: The world is not enough: another look on second-order DPA. In: Abe, M. (ed.) ASIACRYPT 2010. LNCS, vol. 6477, pp. 112–129. Springer, Heidelberg (2010). doi:10.1007/978-3-642-17373-8_7

49. Veyrat-Charvillon, N., Medwed, M., Kerckhof, S., Standaert, F.-X.: Shuffling against side-channel attacks: a comprehensive study with cautionary note. In: Wang, X., Sako, K. (eds.) ASIACRYPT 2012. LNCS, vol. 7658, pp. 740–757. Springer, Heidelberg (2012). doi:10.1007/978-3-642-34961-4_44

50. Whitnall, C., Oswald, E., Standaert, F.-X.: The myth of generic DPA...and the magic of learning. In: Benaloh, J. (ed.) CT-RSA 2014. LNCS, vol. 8366, pp. 183–205. Springer, Cham (2014). doi:10.1007/978-3-319-04852-9_10

51. Zhou, Y., Yu, Y., Standaert, F.-X., Quisquater, J.-J.: On the need of physical security for small embedded devices: a case study with COMP128-1 implementations in SIM cards. In: Sadeghi, A.-R. (ed.) FC 2013. LNCS, vol. 7859, pp. 230–238. Springer, Heidelberg (2013). doi:10.1007/978-3-642-39884-1_20

Uniform First-Order Threshold Implementations

Tim Beyne[✉] and Begül Bilgin

ESAT/COSIC, KU Leuven and iMinds, Leuven, Belgium
tim.beyne@student.kuleuven.be, begul.bilgin@esat.kuleuven.be

Abstract. Most masking schemes used as a countermeasure against side-channel analysis attacks require an extensive amount of fresh random bits on the fly. This is burdensome especially for lightweight cryptosystems. Threshold implementations (TIs) that are secure against first-order attacks have the advantage that fresh randomness is not required if the sharing of the underlying function is uniform. However, finding uniform realizations of nonlinear functions that also satisfy other TI properties can be a challenging task. In this paper, we discuss several methods that advance the search for uniformly shared functions for TIs. We focus especially on three-share implementations of quadratic functions due to their low area footprint. Our methods have low computational complexity even for 8-bit Boolean functions.

Keywords: Boolean functions · Correction terms · Masking · Randomness · The threshold implementations · Uniformity

1 Introduction

Side channel attacks (SCA), which are shown to be a great threat to today's cryptosystems [1,14,15], derive sensitive information (e.g. secret key) by correlating various characteristics of the device such as timing, power consumption and electromagnetic emanation leakages with intermediate values of the cryptographic algorithm during execution [15,16]. In this paper, we consider adversaries that can only use first-order SCA, i.e. can use only first-order statistical moments of the side-channel information or equivalently can use information from a single wire [13]. The threshold implementation (TI) method is a countermeasure that is proven to be secure with minimal adversarial and implementation assumptions [4,19,21] and is used for symmetric-key algorithms. Being a masking scheme, its essence lies in splitting the sensitive data into s uniformly distributed shares and adopting the (round) functions to operate on these shares in a way that the correct output is calculated. Unlike other masking schemes, first-order TI additionally requires each output share of a function to be independent of at least one of its input shares. This enables security on demanding non-ideal (such as glitchy) circuits and is called the non-completeness property. The uniformly shared input combined with non-completeness randomizes the calculation, and hence breaks the linear relation between the side-channel information and the sensitive data for each function.

© Springer International Publishing AG 2017
R. Avanzi and H. Heys (Eds.): SAC 2016, LNCS 10532, pp. 79–98, 2017.
https://doi.org/10.1007/978-3-319-69453-5_5

In the most generic case, the non-completeness property implies the bound $s \geq td + 1$ on the number of shares where t is the algebraic degree of the function [4,19] and d is the attack order. Hence, in our setting s increases only with the degree of the underlying function to be calculated. Fortunately, any high degree permutation can be represented by a combination of quadratic functions, by means of sequential combination alone [2] or parallel and sequential combinations together [10,17]. Since more shares typically imply an increase in required resources such as area, it is desired to keep s as low as possible. Therefore, we mainly target implementations with three shares while keeping the discussions generic.

Related Work. When the round-based nature of symmetric-key algorithms and the uniformly shared input requirement are considered, it is useful to construct the sharing of nonlinear functions in each round such that their output, which is the input of the following round, is also uniform. A sharing of a function (realization) satisfying this property is called a uniform sharing (realization).

So far, the strategy for finding uniform realizations has been to exhaustively check uniformity for all possible non-complete realizations. For some Sboxes [20], this strategy yields positive results rather quickly. However, even for small, low-degree Sboxes with few shares the search space of possible realizations is very large [7]. Therefore, proving the (non-)existence of a non-complete uniform sharing for a particular nonlinear function is a difficult task [2,3,18].

Alternative to finding a uniform realization, fresh randomness can be added to the output shares of a nonuniform realization. This operation, which makes the sharing uniform, is commonly referred to as *remasking*. The increased cost of high throughput fresh random number generation is undesired and sometimes even unaffordable for a lightweight system.

Contribution. Even though there may not exist a known uniform realization of a given vectorial Boolean function, it is beneficial to find a subset of outputs for which the realization can be made jointly uniform since this reduces the randomness cost significantly. Starting from this partially uniform realization idea, which is described in Sect. 2.5, we focus on finding uniform realizations for Boolean functions, then combine them appropriately. Finding uniform realizations has two main challenges. First, no efficient method to check the uniformity of a realization has been presented so far. Second, if the realization under test is not uniform, another realization needs to be checked and a systematic way to reduce the search space has not been presented yet.

In this paper, we tackle both of these challenges. In Sect. 3, we introduce an efficient method to check uniformity. In Sects. 4 and 5, we respectively discuss adding linear and quadratic terms to output shares in order to make the realization uniform and provide examples. We prove that any realization which uses a bent function as an output share can not be made uniform by adding only linear terms. We also re-prove that there exists no uniform realization of a nonlinear function with two inputs and one output with three shares. This result was previously shown by exhaustive search [19].

2   Threshold Implementations

2.1   Notation

We denote the vector space of dimension n over the Galois field of order 2 by $\mathbb{F}^n$. We use lower case characters for elements of $\mathbb{F}^n$ and vectorial Boolean functions from $\mathbb{F}^n$ to $\mathbb{F}^m$. Superscripts refer to each bit and each coordinate function, i.e. $x = (x^1, \ldots, x^n)$ where $x^i \in \mathbb{F}$ and $f = (f^1, \ldots, f^m)$ where f^i is a Boolean function. We omit the superscript when $n = 1$ for elements and $m = 1$ for functions. The ring of $n \times m$ matrices over $\mathbb{F}$ is written as $\mathbb{F}^{n \times m}$. The dot-product and the field addition of x, y are denoted by $x \cdot y$ and $x + y$ respectively. $\overline{x}$ represents the bitwise complement of x. $|\mathcal{S}|$ denotes the cardinality of the set $\mathcal{S}$.

The notation used for TIs is similar to [2,4,5,19,21]. A correct s share vector $x^i = (x^i_1, \ldots, x^i_s)$ of x^i has the property that $x^i = \sum_{j=1}^{s} x^i_j$. In particular, $\mathrm{Sh}(x^i)$ is the set of correct sharings for the variable x^i. This notation can be readily extended to elements of $\mathbb{F}^n$ and (vectorial) Boolean functions. The sharing $f = f^1_1, f^1_2, \ldots, f^m_{s_{\mathrm{out}}}$ defined from $\mathbb{F}^{ns_{\mathrm{in}}}$ to $\mathbb{F}^{ms_{\mathrm{out}}}$ with s_{in} input and s_{out} output shares is called a *realization*. The realization is correct if $f^i = \sum_{j=1}^{s_{\mathrm{out}}} f^i_j$ for all i. Each share f^i_j of a coordinate function f^i is called a component function. Constructing a uniform and non-complete realization for a linear function is trivial [19]. Therefore, we focus only on nonlinear functions.

2.2   Non-completeness

Non-completeness is the key property that makes TI secure even on glitchy circuits. Without loss of generality, a first-order non-complete realization has the property that its ith output share is independent of its ith input share [19]. This independence implies that leakage of a single share is independent of the unmasked input, proving the security [19]. As described in Sect. 1, $s_{\mathrm{in}}, s_{\mathrm{out}} \geq t+1$ due to this property [19]. A non-complete three-share realization $y = (y_1, y_2, y_3)$ of an AND gate ($y = f(x) = x^1 x^2$) is provided in Eq. (1) as an example.

$$y_1 = f_1(x^1_2, x^1_3, x^2_2, x^2_3) = x^1_2 x^2_2 + x^1_2 x^2_3 + x^1_3 x^2_2$$
$$y_2 = f_2(x^1_1, x^1_3, x^2_1, x^2_3) = x^1_3 x^2_3 + x^1_1 x^2_3 + x^1_3 x^2_1 \qquad (1)$$
$$y_3 = f_3(x^1_1, x^1_2, x^2_1, x^2_2) = x^1_1 x^2_1 + x^1_1 x^2_2 + x^1_2 x^2_1$$

2.3   Uniformity

A sharing of a variable is uniform if, for each unshared value $x \in \mathbb{F}^n$, every $x \in \mathrm{Sh}(x)$ occurs with equal probability. A realization f is called uniform if, for uniformly generated input, its output is also uniformly generated. Namely, f is uniform if and only if

$$\mathcal{N_U} = |\{x \in \mathrm{Sh}(x) | f(x) = y\}| = \frac{2^{n(s_{\mathrm{in}}-1)}}{2^{m(s_{\mathrm{out}}-1)}}$$

for each input $x \in \mathbb{F}^n$ and $y = f(x) \in \mathbb{F}^m$ [5].

Definition 1 (Uniformity table $\mathcal{U}$). *Let f be a shared realization from $\mathbb{F}^{ns_{in}}$ to $\mathbb{F}^{ms_{out}}$ and $\mathcal{U}_{x,y}$ be the cardinality of $\{x \in Sh(x)|f(x) = y\}$ for the unshared input x and shared output y. The $2^n \times 2^{ms_{out}}$ table which has $\mathcal{U}_{x,y}$ as its $(x, y)^{th}$ element is called the uniformity table $\mathcal{U}$ of f.*

Here, we assume that the rows and columns of $\mathcal{U}$ are ordered lexicographically by each unshared then shared output. If f is uniform, the elements of $\mathcal{U}$ are equal to either 0 or $\mathcal{N}_\mathcal{U}$. We provide the uniformity table of Eq. (1) in Table 1 for completeness [5].

Table 1. The uniformity table of Eq. (1).

(x^1, x^2)	(y_1, y_2, y_3)							
	000	011	101	110	001	010	100	111
(0,0)	7	3	3	3	0	0	0	0
(0,1)	7	3	3	3	0	0	0	0
(1,0)	7	3	3	3	0	0	0	0
(1,1)	0	0	0	0	5	5	5	1

Note that Table 1 shows that the aforementioned realization is not uniform since the table contains elements different from 0 and $\mathcal{N}_\mathcal{U} = 4$. Since we want to limit the randomness requirement to minimize resources, we focus on methods to find partially or completely uniform sharings directly. We also keep the number of shares as small as possible for performance considerations.

2.4 Correction Terms

When a realization is not uniform, it is nevertheless possible that a different construction of the component functions yields a uniform realization. One possible way of generating alternative realizations is adding *correction terms* (CTs) to an even number of component functions without breaking the non-completeness property [19]. We assume CTs are generated using only the input shares of the realization.

Consider a realization of a quadratic Boolean function with n variables with three output and input shares. The number of linear and quadratic CTs is $3(n + \binom{n}{2})$. Therefore, there exist $2^{3(n+\binom{n}{2})}$ possible non-complete three share realizations for this function. If we consider such realizations for 3- and 4-bit (quadratic) Sboxes, we get 2^{18} and 2^{30} possibilities for each coordinate function and $(2^{18})^3$ and $(2^{30})^4$ possible realizations for the Sbox [6].

2.5 Partial Uniformity

Definition 2 (Partial Uniformity). *Consider the function f with m coordinate functions f^i. A realization that is uniform in at least one l-combination of its coordinate functions, i.e. without loss of generality with uniform $f_1^1, f_2^1, \ldots f_{s_{out}}^l$, is called a partially uniform realization of f.*

The case where $l = m$ implies that f has a uniform realization. The motivation to find a partially uniform function is that, if l output variables are jointly uniform, they do not need to be remasked to make the joint distribution of output shares uniform [3]. Hence, the required randomness for remasking can be reduced from $m \cdot (s_{out} - 1)$ bits to $(m - l) \cdot (s_{out} - 1)$ bits. Note that by using this method alone, the authors of [3] gained 60% efficiency on fresh randomness.

A straightforward way to find partially uniform realizations, which we apply, starts by finding uniform realizations for each coordinate function of f. These realizations are then combined iteratively until it is not possible to combine any more component functions uniformly[1]. Therefore, we mainly focus on finding uniform realizations of a Boolean function efficiently in the rest of the paper and use their combinations for partial uniformity only on examples.

There are two main obstacles in this approach:

1. It is relatively expensive to check whether a given realization is uniform. So far, the only proposed way to check uniformity is generating the uniformity table $\mathcal{U}$ of the realization completely and checking if its nonzero elements are equal to $\mathcal{N}_{\mathcal{U}}$. This requires $2^{ns_{in}}$ evaluations of the realization (for all possible s_{in} shares of each of the n input variables) in the worst case.
2. Going through all possible realizations, i.e. trying all possible CTs, can be extremely expensive due to the large amount of CTs as discussed at the end of Sect. 2.3. Even if we focus only on the linear CTs, there exist $2^{ns_{in}}$ different realizations implying $O(2^{ns_{in}} 2^{ns_{in}}) = O(2^{2ns_{in}})$ complexity to check uniformity for all of them.

Therefore, both decreasing the search space of possible realizations and reducing the complexity of checking uniformity for each realization would have significant impact on the overall complexity of finding uniform realizations for Sboxes.

3   Fast Uniformity Check for Boolean Functions

This section aims to reduce the complexity O_U of checking whether a given realization of a Boolean function is uniform. In order to do that, we first analyze the dependencies between the elements $\mathcal{U}_{x,y}$ of the uniformity table. We observe that if the realization has three output shares, it is sufficient to calculate only one row of $\mathcal{U}$ due to the dependency between $\mathcal{U}_{x,y}$, reducing O_U. For this reason, we consider the case $s_{out} = 3$ in the second half of this section. Note that using three output shares limits the degree of the Boolean function to two. However, any high-degree function can be decomposed into quadratic Boolean functions and using a small number of shares typically reduces the implementation cost [5, 7, 17].

---

[1] One possible algorithm to find a (partial) uniform realization is provided in Appendix A for completeness. Note that this algorithm returns a uniform realization if it exists.

3.1   Observations on the Rows of $\mathcal{U}$

Consider a non-complete realization f with s_{in} input and s_{out} output shares, represented by x and y respectively. For each unshared input x, let $N_{x,i}$ denote the number of inputs of f for which the component y_i has a fixed value $b \in \{0, 1\}$. That is

$$N_{x,i} = \begin{cases} \sum_y y_i \cdot \mathcal{U}_{x,y} & \text{if } b = 1 \\ \sum_y \overline{y_i} \cdot \mathcal{U}_{x,y} & \text{if } b = 0. \end{cases} \tag{2}$$

Lemma 1. $N_{x,i}$ is independent of x.

Proof. Due to non-completeness of f, y_i is (without loss of generality) independent of $(x_i^1, x_i^2, \ldots, x_i^n)$. Hence, $N_{x,i}$ is also independent of $(x_i^1, x_i^2, \ldots, x_i^n)$. Since the input sharing x is uniform, $N_{x,i}$ is independent of x. □

Hence, we will write N_i rather than $N_{x,i}$. The lemma implies that the entries of any row of $\mathcal{U}$ corresponding to some unshared input x are related by the same set of equations for a constant binary value b:

- For $1 \leq i \leq s_{\text{out}}$, N_i satisfies Eq. (2)
- The sum of the values must be equal to $2^{n(s_{\text{in}}-1)}$

$$\sum_y \mathcal{U}_{x,y} = 2^{n(s_{\text{in}}-1)}. \tag{3}$$

If $y \in \text{Sh}(\overline{f(x)})$ with $\overline{f(x)}$ the complement of $f(x)$, then $\mathcal{U}_{x,y} = 0$. Hence, we will say that the system of Eqs. (2) and (3) has only $2^{s_{\text{out}}-1}$ unknowns.

Lemma 2. *Given Eq. (3), the equations given in Eq. (2) for $b = 1$ and $b = 0$ are linearly dependent.*

Proof. Form the coefficient matrix of the system of Eqs. (2) and (3) such that each row of the matrix represents an equation for which the columns are the coefficients of $\mathcal{U}_{x,y}$. Clearly, for each i, the rows p and r of this matrix corresponding to N_i when $b = 0$ and $b = 1$ are binary complements. Let j be the row of ones corresponding to Eq. (3). Then $p = j - r$ describes the linear dependence among these rows for each i. □

Lemma 2 implies that there are at most $s_{\text{out}} + 1$ linearly independent equations describing the unknowns $\mathcal{U}_{x,y}$. Since there are $2^{s_{\text{out}}-1}$ unknowns, the values N_i completely determine each row of $\mathcal{U}$ only if $2^{s_{\text{out}}-1} \leq s_{\text{out}} + 1$. This inequality holds only for $s_{\text{out}} \leq 3$. Since s_{out} must be greater than the degree of the function and we focus on nonlinear operations, $s_{\text{out}} = 3$. Note that fixing the number of output shares to three has no implication on the number of input shares, nor on the amount of input variables. In what follows, we investigate the case $s_{\text{out}} = 3$ further. For this case, Appendix B lists the four linearly independent equations for each unshared input x that describe the relation between elements in a single row of $\mathcal{U}$.

3.2   Observations on $\mathcal{U}$ when $s_{\text{out}} = 3$

Theorem 1. *Let f be a realization of a Boolean function with $s_{out} = 3$. Then any row of its uniformity table $\mathcal{U}$ uniquely determines all elements of $\mathcal{U}$.*

Proof. Recall that the rows of $\mathcal{U}$ correspond to the unshared inputs. For any two inputs x, x', the systems of Eqs. (2) and (3) will be identical provided that we choose the same constant value for b. If the elements of some row are known, one can easily deduce the values N_i and hence the system of equations. The proof is completed by the fact that for $s_{\text{out}} = 3$ the system of equations completely determines the elements of any row. $\square$

Note that the theorem does not imply that all rows are equal, since the unknowns in the system of equations for x and x' are different if $f(x) \neq f(x')$. Namely, they are $\mathcal{U}_{x,y}$ with $y \in \text{Sh}(f(x))$ and $\mathcal{U}_{x',y}$ with $y \in \text{Sh}(f(x'))$ respectively. Hence, the rows can in general take only two different values.

Corollary 1. *If the realization f of a Boolean function with $s_{out} = 3$ has a uniform distribution for one unshared input value (one row of $\mathcal{U}$), then it has a uniform distribution for all unshared input values (all rows).*

Proof. If the distribution of output shares is uniform for input x, then all nonzero elements in that row are equal to $\mathcal{N}_\mathcal{U}$. Hence, by Theorem 1, all values of $\mathcal{U}$ are fixed. Since the uniformity table of a uniform realization is a possible solution for $\mathcal{U}$, and the solution must be unique, it follows that f is a uniform realization. $\square$

Using Corollary 1, the computational complexity of the uniformity check (O_U) when $s_{\text{out}} = 3$ is reduced to $O(2^{n(s_{\text{in}}-1)})$, i.e. computing a single row of $\mathcal{U}$. To be able to compare with the results of the following section, we note that at most $O(2^{n(s_{\text{in}}-1)}2^{ns_{\text{in}}}) = O(2^{2ns_{\text{in}}-n})$ evaluations are required if checking all linear CTs is desired. We conclude this section with the following theorem from which we will benefit in the remainder of the paper.

Theorem 2. *A realization with one output variable and three output shares is uniform if and only if each of its component functions is a balanced Boolean function.*

Proof. According to Theorem 1, it is enough to solve the system of Eqs. (2) and (3) for a single row of $\mathcal{U}$ to determine all the elements $\mathcal{U}_{x,y}$. By Lemma 2, the equations for either $b = 0$ or $b = 1$ suffice. Hence, we consider the system of equations for $b = 0$ which is provided in Eq. (10) in Appendix B and simplified as the following extended coefficient matrix with columns ordered lexicographically by each shared, then unshared output:

$$\begin{pmatrix} 1 & 1 & 0 & 0 & 1 & 1 & 0 & 0 & | & N_1 \\ 1 & 0 & 1 & 0 & 1 & 0 & 1 & 0 & | & N_2 \\ 1 & 0 & 0 & 1 & 0 & 1 & 1 & 0 & | & N_3 \\ 1 & 1 & 1 & 1 & 1 & 1 & 1 & 1 & | & 2^{(s_{\text{in}}-1)n} \end{pmatrix}$$

Depending on whether the output $(y = \sum_i y_i)$ is 0 or 1, the elements $\mathcal{U}_{x,y}$ corresponding to either the first or the second four coefficients of the matrix (equivalently either the first or the second line of each equation in Eq. (10)) are non-zero. Here, we only provide the solution for the system $y = 0$ given in Eq. (4). The solution for $y = 1$ is similar.

$$\mathcal{U}_{x,(0,0,0)} = -2^{(s_{in}-1)n-1} + \frac{1}{2}(N_1 + N_2 + N_3), \quad \mathcal{U}_{x,(0,1,1)} = 2^{(s_{in}-1)n-1} + \frac{1}{2}(N_1 - N_2 - N_3),$$

$$\mathcal{U}_{x,(1,0,1)} = 2^{(s_{in}-1)n-1} + \frac{1}{2}(-N_1 + N_2 - N_3), \quad \mathcal{U}_{x,(1,1,0)} = 2^{(s_{in}-1)n-1} + \frac{1}{2}(-N_1 - N_2 + N_3)$$

$$\tag{4}$$

$$\mathcal{U}_{x,(0,0,1)} = \mathcal{U}_{x,(0,1,0)} = \mathcal{U}_{x,(1,0,0)} = \mathcal{U}_{x,(1,1,1)} = 0$$

($\Rightarrow$): For a uniform realization, all non-zero expressions in Eq. (4) must be equal to each other and have the value $\mathcal{N}_{\mathcal{U}} = 2^{n(s_{in}-1)-m(s_{out}-1)} = 2^{n(s_{in}-1)-2}$. This uniquely determines N_1, N_2 and N_3 for a uniform realization. In particular, we have

$$N_1 = N_2 = N_3 = 2^{n(s_{in}-1)-1}, \tag{5}$$

implying that each component function is balanced.

($\Leftarrow$): If each output bit is uniform satisfying Eq. (5), then each $\mathcal{U}_{x,y}$ is either 0 or $\mathcal{N}_{\mathcal{U}}$. This implies the uniformity of the realization.   □

Note that one side of the proof stating that if the realization is uniform, each of the component functions must be balanced has already been proven in [7] and is independent of the number of shares or the degree of the function.

4   Using Linear Correction Terms Efficiently to Satisfy Uniformity

In this section, we show how to avoid trying all the linear correction terms one by one in order to find uniform realizations of Boolean functions. We benefit from the Walsh-Hadamard transformation (WHT) to directly see which linear correction terms can lead to uniform realizations and eliminate a significant portion of the search space. Even though we describe our method for $s_{out} = 3$ to benefit from Theorem 2, the idea can be used for efficient uniformity checks of component functions with $s_{out} > 3$.

Definition 3. *The Walsh-Hadamard transformation of f is denoted by W_f. For $\omega \in \mathbb{F}^n$, it is given by*

$$W_f(\omega) = \sum_{x \in \mathbb{F}^n} (-1)^{f(x)+\omega \cdot x},$$

i.e. the discrete Fourier transform of $(-1)^{f(x)}$.

Here, the addition in the exponent is in $\mathbb{F}$, and the summation is in the integers. This transformation can be efficiently calculated with $O(n2^n)$ computational complexity using fast WHT.

Definition 4. *A Boolean function f is called bent if and only if for all vectors* $\omega \in \mathbb{F}^n$, $W_f(\omega) = 2^{n/2}$.

Bent functions only exist for even n [22]. In our study, we will treat bent and non-bent component functions separately for reasons that will be clarified later in this section. Moreover, we will make use of the following well-known result.

Fact. f is balanced if and only if $W_f(0) = 0$. Moreover, $f(x) + a \cdot x$ is balanced if and only if $W_f(a) = 0$.

4.1 Realizations Without Bent Component Functions

Adding linear correction terms to a nonuniform realization $f = (f_1, f_2, f_3)$ is described by the following equations, where a and b are binary correction vectors.

$$f_1' = f_1 + a_{\hat{1}} \cdot x \quad f_2' = f_2 + b_{\hat{2}} \cdot x \quad f_3' = f_3 + (a_{\hat{1}} + b_{\hat{2}})_{\hat{3}} \cdot x.$$

The notation $a_{\hat{i}}$ indicates that the bits corresponding to every i^{th} share are zero. Due to the restrictions implied by this notation, the new sharing is non-complete.

By Theorem 2, (f_1', f_2', f_3') is uniform if and only if f_i's are balanced. Therefore, $W_{f_1}(a_{\hat{1}})$, $W_{f_2}(b_{\hat{2}})$ and $W_{f_3}((a_{\hat{1}} + b_{\hat{2}})_{\hat{3}})$ must be zero which can easily be checked by using fast WHT. We use Algorithm 1.

Algorithm 1. Find linear correction terms for the realization $f = (f_1, f_2, f_3)$.

1: Compute W_{f_1}, W_{f_2} and W_{f_3} using WHT.
2: **for** $a_{\hat{1}} \in \mathbb{F}^{n s_{\text{in}}}$ **do**
3:     **if** $W_{f_1}(a_{\hat{1}}) \neq 0$ **then**
4:         **continue**
5:     **end if**
6:     **for** $b_{\hat{2}} \in \mathbb{F}^{n s_{\text{in}}}$, $(a_{\hat{1}} + b_{\hat{2}})_{\hat{3}}$ **do**
7:         **if** $W_{f_2}(b_{\hat{2}}) = 0$ and $W_{f_3}((a_{\hat{1}} + b_{\hat{2}})_3) = 0$ **then**
8:             **yield** $(a_{\hat{1}}, b_{\hat{2}})$
9:         **end if**
10:    **end for**
11: **end for**

The computational complexity of the three Walsh-Hadamard transformations in this algorithm is $O(n(s_{\text{in}} - 1) \cdot 2^{n(s_{\text{in}}-1)})$. The outer for-loop iterates over $2^{n(s_{\text{in}}-1)}$ values, the inner over $2^{n(s_{\text{in}}-2)}$ values. Hence, the for loop has complexity $O(2^{n(2s_{\text{in}}-3)})$. It follows that the total worst-case complexity is

$$O\left(n(s_{\text{in}} - 1) \cdot 2^{n(s_{\text{in}}-1)} + 2^{n(2s_{\text{in}}-3)}\right) = O\left(2^{n(2s_{\text{in}}-3)}\right).$$

To find a single solution the best-case complexity is $O\left(n(s_{\text{in}} - 1) \cdot 2^{n(s_{\text{in}}-1)}\right)$.

The table below summarizes the complexities of each uniformity-check method presented so far when only linear correction terms are considered. The efficiency of using the WHT is clear as the input size of the Boolean function increases. We emphasize that to find uniform realizations of vectorial Boolean functions all the aforementioned methods should be repeated for each coordinate function. Hence the complexity gain observed for a single Boolean function in the following table gains in significance for Sboxes.

Method	Worst-case complexity	$s_{in} = 3$	$s_{in} = 4$
Naive	$O(2^{2ns_{in}})$	$O(2^{6n})$	$O(2^{8n})$
Fast uniformity check	$O(2^{2ns_{in}-n})$	$O(2^{5n})$	$O(2^{7n})$
Using WHT	$O(2^{2ns_{in}-3n})$	$O(2^{3n})$	$O(2^{5n})$

Application to $\mathcal{Q}_{300}^4$. It has been shown in [7,10,17] that 4-bit permutations can be decomposed into quadratic functions in order to enable three-share realization of cryptographic algorithms. There exist six 4-bit quadratic permutation classes [7] up to affine equivalence that can be used for decomposition. For all quadratic permutation classes except one (namely $\mathcal{Q}_{300}^4$ as denoted by [7]) a uniform realization with three-shares has been found. However, for class $\mathcal{Q}_{300}^4$ the (non-)existence result was inconclusive so far since the search space is too big for practical verification. By using Algorithm 1 together with Algorithm 3 on the representative of $\mathcal{Q}_{300}^4$, we found that two out of four coordinate functions have jointly uniform realizations as described in Appendix C. This implies a 50% reduction when a permutation from $\mathcal{Q}_{300}^4$ is instantiated, which shows the relevance of this section. We note that no further improvements are possible for this permutation using only linear correction terms.

4.2   Realizations with Bent Component Functions

Theorem 3. *If any component function of a realization—seen as a function on $\mathbb{F}^{n(s_{in}-1)}$—is bent, then this realization is not uniform and it can not be made uniform by using only linear correction terms.*

Proof. Take one of the component functions f_i of the realization of f, viewed as a function on $\mathbb{F}^{n(s_{in}-1)}$. Further assume f_i is bent. Since the Walsh spectrum of f_i does not contain any zeros, it is clear that neither f_i is a balanced function nor any linear correction term makes f_i balanced. Hence, the realization cannot be made uniform with only linear CTs. Note that for $s_{out} \geq 4$, balancedness is still a necessary condition. Thus, the theorem also holds for more than three output shares.   □

We discuss two ad-hoc solutions to remedy this problem for any nonlinear function in Appendix D. More generally, it is easier to find linear correction terms if the number of zeros of the Walsh-Hadamard transform of each of the components is high. Section 5 provides further insight into this matter.

5 Finding Uniform Realizations of Quadratic Functions

So far we only focused on using linear CTs to find uniform realizations. In this section we benefit from quadratic forms to find quadratic CTs to enable uniform sharing on quadratic Boolean functions even if they have bent component functions.

5.1 Quadratic Forms

Any function $f : \mathbb{F}^n \to \mathbb{F}$ composed of only quadratic terms can be described by its *quadratic form* as $f(x) = x\,S\,x^T$ with S an upper triangular coefficient matrix.

Similarly, its *bilinear form* $B_f(x,y) = f(x+y) + f(x) + f(y)$ can be described by the equation $B_f(x,y) = y\,(S + S^T)\,x^T$. This bilinear map defines a subspace of $\mathbb{F}^n$, given by $\mathrm{rad}(f) = \{x \in \mathbb{F}^n \mid \forall y \in \mathbb{F}^n : B_f(x,y) = 0\}$, i.e. the radical or kernel of f. It follows from the rank-nullity theorem that $\dim(\mathrm{rad}(f)) = n - \mathrm{rank}\,(S + S^T)$.

Proposition 1. [22] *f is bent if and only if* $\dim(\mathrm{rad}(f)) = 0$.

Let L be an $n \times n$ invertible matrix. Then $(S + S^T)$ and $L^T(S + S^T)L$ are called *cogredient*. The cogredience relation divides the set of $n \times n$ matrices into mutually disjoint classes of cogredient matrices with the same rank.

It is well known that any symmetric matrix over $\mathbb{F}$ has the following normal form [23]:

$$
N = \begin{pmatrix}
\begin{smallmatrix} 0 & 1 \\ 1 & 0 \end{smallmatrix} & & & \\
& \begin{smallmatrix} 0 & 1 \\ 1 & 0 \end{smallmatrix} & & \\
& & \ddots & \\
& & & 0
\end{pmatrix}. \tag{6}
$$

That is, there exists an invertible matrix T such that $S + S^T = T N T^T$. For more information on quadratic forms over fields of characteristic two, see [11,23].

5.2 Quadratic Forms in TI Context

Let $f(x)$ where $x \in \mathbb{F}^n$ be a quadratic Boolean function to be shared with the realization $\boldsymbol{f} = (f_1, \ldots, f_{s_{\mathrm{out}}})$. In addition, let M_i be the matrices associated with the bilinear form of f_i, that is, $B_{f_i}(\boldsymbol{x}, \boldsymbol{y}) = \boldsymbol{x} M_i \boldsymbol{y}^T$ where

$$
M_i = \begin{pmatrix}
0 & X_i^{12} & X_i^{13} & \cdots & X_i^{1\,n} \\
X_i^{12} & 0 & X_i^{23} & \cdots & X_i^{2\,n} \\
\vdots & \vdots & \vdots & \ddots & \vdots \\
X_i^{1\,n} & X_i^{2\,n} & X_i^{3\,n} & \cdots & 0
\end{pmatrix}. \tag{7}
$$

Each of the X_i^{kj} are $s_{\mathrm{in}} \times s_{\mathrm{in}}$ matrices, with zeros in the ith row and column to satisfy non-completeness. These matrices are zero when $x^k x^j$ does not appear in the unshared function[2].

Similarly, let M be a block-matrix constructed from the matrix $S + S^T$ of the bilinear form B_f. Each block is of size $s_{\mathrm{in}} \times s_{\mathrm{in}}$ and its values equal the value

---

[2] We assume that there are no *superfluous terms* that appear in an even number of M_i and hence can be canceled out from the realization.

of the corresponding element of $S + S^T$. Hence, the correctness requirement can be stated as $\sum_{i=1}^{s_{out}} M_i = M$.

Corollary 2. *If $rank(M_i) = n(s_{in} - 1)$ for any i, then using linear CTs on the realization f does not make it uniform.*

Proof. Proof follows from Theorem 3 and Proposition 1.   □

The proof of the following theorem clarifies the quadratic form notation for TI.

Theorem 4. *No nonlinear Boolean function with two inputs and one output can be uniformly shared using three shares.*

Proof. Consider the following direct sharing for the product $x^1 x^2$ (AND gate):

$$f_1 = \underline{x_2^1 x_3^2 + x_3^1 x_2^2 + x_2^1 x_2^2}, \quad f_2 = \underline{x_1^1 x_3^2 + x_3^1 x_1^2 + x_3^1 x_3^2}, \quad f_3 = \underline{x_1^1 x_2^2 + x_2^1 x_1^2 + x_1^1 x_1^2}$$

The underlined terms cannot be moved due to the non-completeness property of TIs, hence their corresponding coefficients are fixed to be 1 in M_i whereas the other terms are flexible. It will be shown that any realization of $x^1 x^2$ contains at least one bent Boolean function. Equivalently, by Corollary 2, there exists at least one M_i that is of full rank $n(s_{in} - 1)$.

We have the following matrices M_i associated with $f_i : \mathbb{F}^{n(s_{in}-1)} \to \mathbb{F}$ ($i \in \{1, 2, 3\}$):

$$\begin{pmatrix} 0 & 0 & A_i & 1 \\ 0 & 0 & 1 & B_i \\ A_i & 1 & 0 & 0 \\ 1 & B_i & 0 & 0 \end{pmatrix}$$

Note that the zero rows and columns of M_i—corresponding to the unused share due to non-completeness—have been removed in the above notation, since they have no influence, leaving $n(s_{in} - 1) \times n(s_{in} - 1)$ matrices.

Due to the orthogonality of the columns and rows, the above matrix is always of rank four, unless $A_i = B_i = 1$. It follows from Proposition 1 that every condition other than $A_i = B_i = 1$ implies that f_i is a bent function. However, the only remaining configuration is not possible since it corresponds to the sharing

$$f_1' = x_2^1 x_3^2 + x_3^1 x_2^2 + x_2^1 x_2^2 + x_3^1 x_3^2$$
$$f_2' = x_1^1 x_3^2 + x_3^1 x_1^2 + x_3^1 x_3^2 + x_1^1 x_1^2$$
$$f_3' = x_1^1 x_2^2 + x_2^1 x_1^2 + x_1^1 x_1^2 + x_2^1 x_2^2,$$

which is not correct. Note also that any correction must have degree less than three to preserve non-completeness. By Theorem 3, there exist no linear correction terms that makes the realization uniform. Hence, an AND gate has no uniform sharing with three input and output shares. Further, since linear terms have no influence on B_{f_i}, it follows that no nonlinear Boolean function with two inputs can be uniformly shared using three shares.   □

The correctness of the above theorem has been shown in [19] by enumeration of all possible correction terms. Our proof indicates that the matrix representation of quadratic forms is a useful tool to study the uniformity of quadratic realizations.

5.3 Using Quadratic Forms to Find Uniform Realizations

Proposition 2. *A uniform realization of a quadratic function can be found only if there exist s_{out} matrices M_i formed as in Eq. (7), satisfying the following properties:*

1. *$\sum_{i=1}^{s_{out}} M_i = M$.*
2. *$\forall i \in \{1, \ldots, s_{out}\} : \text{rank}(M_i) < n(s_{in} - 1)$.*
3. *Linear correction terms can be found (e.g. by using Algorithm 1)[3]*

Moreover, by Theorem 2, the above requirements are also sufficient if $s_{out} = 3$.

Note that the proposition applies not only to quadratic forms but also quadratic functions in general, since linear terms do not influence the block matrix M of the bilinear form. In what follows, we discuss how the second requirement of Proposition 2 can be met, which is non-trivial. Moreover, we mainly focus on bent functions.

Recall that the matrix $S + S^T$ of the bilinear form of a bent Boolean function f is cogredient to its normal form N, given by Eq. (6). Note that if $S + S^T$ is cogredient to N, then there also is a transformation T such that $M = T N' T^T$. The matrix T is obtained by replacing ones in the original cogredience transformation matrix with identity matrices, and zeros with zero-blocks of the appropriate size. N' is the following block matrix:

$$N' = \begin{pmatrix} \begin{smallmatrix} 0 & J \\ J & 0 \end{smallmatrix} & & & \\ & \begin{smallmatrix} 0 & J \\ J & 0 \end{smallmatrix} & & \\ & & \ddots & \\ & & & 0 \end{pmatrix}. \tag{8}$$

The matrix J is an $s_{in} \times s_{in}$ matrix of ones. It now suffices to select matrices N'_i such that $N' = \sum_{i=1}^{s_{out}} N'_i$ with $\text{rank}(N'_i)$ as low as possible for each i. In particular, since the transformation T preserves the rank and it does not act on individual shares, if one can choose N'_i such that $\text{rank}(N'_i) < n(s_{out} - 1)$, then for $M_i = T N'_i T^T$, the first and second requirements from Proposition 2 are satisfied. One possible way of constructing the matrices N'_i, is by decomposing each of the block matrices J occurring in the normal form of Eq. (8). The decomposition of each J must be chosen such that it induces a linear dependence relation among the rows of at least one of the matrices N'_i, and hence reduces the rank of one of the matrices M_i. Eq. (9) provides three such decompositions of J for $s_{in} = s_{out} = 3$:

$$\begin{aligned} J &= \begin{pmatrix} 0&0&0 \\ 0&1&1 \\ 0&1&1 \end{pmatrix} + \begin{pmatrix} 1&0&1 \\ 0&0&0 \\ 1&0&0 \end{pmatrix} + \begin{pmatrix} 0&1&0 \\ 1&0&0 \\ 0&0&0 \end{pmatrix} \\ &= \begin{pmatrix} 0&0&0 \\ 0&0&1 \\ 0&1&0 \end{pmatrix} + \begin{pmatrix} 1&0&1 \\ 0&0&0 \\ 1&0&1 \end{pmatrix} + \begin{pmatrix} 0&1&0 \\ 1&1&0 \\ 0&0&0 \end{pmatrix} \\ &= \begin{pmatrix} 0&0&0 \\ 0&0&1 \\ 0&1&0 \end{pmatrix} + \begin{pmatrix} 0&0&1 \\ 0&0&0 \\ 1&0&1 \end{pmatrix} + \begin{pmatrix} 1&1&0 \\ 1&1&0 \\ 0&0&0 \end{pmatrix}. \end{aligned} \tag{9}$$

---

[3] The possibility of finding linear CTs increases as the rank of the matrix decreases since a low-rank matrix typically has more zeros in the Walsh spectrum.

Notice that the ith decomposition (from top to bottom) has identical rows in the ith term of the decomposition. The choice of the ith decomposition reduces the rank of M_i by two[4], since J is the only nonzero block in a row in N' and N' is symmetric. Hence, to ensure that each $\operatorname{rank}(M_i) < n(s_{\text{in}} - 1)$, each decomposition from Eq. (9) must be used at least once. This implies that this method can be used effectively only when $n \geq 6$. The method to generate M_i's using the decomposition of J for $s_{\text{out}} = 3$ is formalized in Algorithm 2. Note that lines 2–9 are merely intended to construct the block matrix T from the corresponding matrix L. The computational complexity of the algorithm is as low as $O(n^3)$ since finding the normal form can be done using a particular type of simultaneous row and column reduction, see for example Wan [23] for a description. Since the search space for $n \leq 5$ is feasible, we opted for a generic search algorithm to generate the matrices N_i' for these cases. Specifically, we focused on $n = 4$ since there exist no odd-sized bent functions and Theorem 4 completes the work for $n = 2$. The following theorem formalizes our findings.

Algorithm 2. Low-rank decomposition of the matrix M for $s_{\text{out}} = s_{\text{in}} = 3$.

Input: $S + S^T \in \mathbb{F}^{n \times n}$ ▷ The matrix representation of B_f.
Output: M_1, M_2, M_3 ▷ Matrices of the bilinear forms of the output shares.

1: Find T such that $S + S^T = TNT^T$ with N the normal form as in Eqn. (6).
2: Let $L \in \mathbb{F}^{ns_{\text{in}} \times ns_{\text{in}}}$.
3: **for** $1 \leq i, j \leq n$ **do**
4:      **if** $T[i, j] = 1$ **then**
5:          $L[i : i + s_{\text{in}} - 1, j : j + s_{\text{in}} - 1] \leftarrow I_{s_{\text{in}}}$
6:      **else**
7:          $L[i : i + s_{\text{in}} - 1, j : j + s_{\text{in}} - 1] \leftarrow 0$
8:      **end if**
9: **end for**
10: Let $M_1, M_2, M_3 \in \mathbb{F}^{ns_{\text{in}} \times ns_{\text{in}}}$.
11: **for** $1 \leq i, j \leq n$ **do**
12:      **if** $N[i, j] = 1$ **then**
13:          ▷ Choose any decomposition from Eqn. (9).
14:          ▷ Use each decomposition at least once (only possible if $n \geq 6$).
15:          Let $J = J_1 + J_2 + J_3$.
16:          $M_l[i : i + s_{\text{in}} - 1, j : j + s_{\text{in}} - 1] \leftarrow J_l$ for $l = 1, 2, 3$.
17:      **end if**
18: **end for**
19: **return** $L(M_1 + M_1^T)L^T, L(M_2 + M_2^T)L^T, L(M_3 + M_3^T)L^T$

Theorem 5. *Let f be any quadratic Boolean function on $\mathbb{F}^n, n \geq 4$. Then there is a sharing $\boldsymbol{f}$ with $s_{in} = s_{out} = 3$ shares, such that none of the output shares of $\boldsymbol{f}$ are bent functions.*

---

[4] We consider the matrix derived from f_i of size $n(s_{\text{in}} - 1) \times n(s_{\text{in}} - 1)$ without the zero rows and columns.

Furthermore, we conjecture that if the first two requirements of Property 2 hold, then a quadratic Boolean function f can always be made uniform using three shares with linear correction terms. We conclude this section with an example.

Application to an $\mathbb{F}_4$ Multiplier. The AES S-box can be decomposed into several multiplications in $\mathbb{F}_4$, additions and rotations [8]. No three share uniform realization of $\mathbb{F}_4$ has been found so far, which can be explained with the fact that both coordinate functions of this multiplication which are given in Eq. (12) in Appendix E are bent. Since $n = 4$, we used a generic algorithm to find the matrices M_i leading to a realization with non-bent coordinate functions which is provided in Eq. (13) in Appendix E. We then performed a search on linear correction terms as described in Algorithm 1 to make the realization uniform. We found several uniform realizations for both coordinate functions. Details of this investigation leading to an implementation with 50% lower randomness requirements are given in Appendix E.

6   Conclusion

In this paper, we provided methods to find uniform realizations of nonlinear (vectorial) Boolean functions efficiently. We limit ourselves to first-order TIs because the uniformity property is insufficient to provide theoretical security against higher-order attacks. We started by discussing how the uniformity check of especially three output share realizations of Boolean functions can be performed efficiently. We then described how the Walsh-Hadamard transformation can be used to find all linear correction terms that lead to uniform realizations without the need for an exhaustive search. This method can be applied to any n-bit Boolean function with worst-case complexity $O(2^{2ns_{in}-3n})$ where s_{in} is the number of input shares of the threshold implementation. We proved that if the shared realization has a bent component function, this share can not be made uniform by using only linear correction terms. On the other hand, we showed that we can use the theory of quadratic forms to find uniform realizations for many quadratic functions. We demonstrated the applicability of the theory by providing partially uniform three-share realizations for a representative of the problematic quadratic 4-bit permutation class $\mathcal{Q}_{300}^4$ and a $\mathbb{F}_4$ multiplier that requires 50% less randomness compared to their naive implementations.

Acknowledgement. The authors are especially grateful to Vincent Rijmen for his contributions to this work. Additionally, we would like to thank the anonymous reviewers for providing constructive and valuable comments and Faruk Gologlu for fruitful discussions. This work was supported in part by NIST with the research grant 60NANB15D346, in part by the Research Council KU Leuven (OT/13/071 and C16/15/058) and in part by the Flemish Government through FWO Thresholds G0842.13. Begül Bilgin is a Postdoctoral Fellow of the Fund for Scientific Research - Flanders (FWO).

A   Algorithm to Find Partial Uniform Realizations

Algorithm 3. Find (partially) uniform realizations.

Input: $f = (f^1, \ldots, f^m)$ s.t. f^i is the realization of the coordinate function f^i of f; The initial set S^0 of all possible correction functions c^i;

Output: The set Σ_l of all sets $S^{t_1, \ldots, t_l}$ with elements $(c^{t_1}, \ldots, c^{t_l})$ s.t. $f' = f^{t_1, \ldots, t_i} + c^{t_1, \ldots, t_l} = (f^{t_1} + c^{t_1}, \ldots, f^{t_l} + c^{t_l})$ is a uniform realization.

1: **function** IsUNIFORM(f)
2:    **return true** if f is uniform, **false** otherwise
3: **end function**

4: **function** GENERATECORRECTIONFUNCTIONS(f^i, S^0)
5:    $S^i \leftarrow \emptyset$
6:    **for** $c^i \in S^0$ **do**
7:       **if** IsUNIFORM($f^i + c^i$) **then**
8:          $S^i \leftarrow S^i \cup \{c^i\}$
9:       **end if**
10:    **end for**
11:    **return** S^i
12: **end function**

13: **function** COMBINECORRECTIONFUNCTIONS(f, Σ_{l-1})
14:    $\Sigma_l \leftarrow \emptyset$
15:    ▷ Denote the set of l-combinations from $\{1, \ldots, m\}$ by $\mathcal{S}$.
16:    **for** $\{t_1, \ldots, t_l\} \in \mathcal{S}$ **do**
17:       $S^{t_1, \ldots, t_l} \leftarrow \emptyset$
18:       **for** $c^{t_2, \ldots, t_l} \in S^{t_2, \ldots, t_l}, c^{t_1} \in \{c^{t_1} | c^{t_1, t_3, \ldots, t_l} \in S^{t_1, t_3, \ldots, t_l}\}$ **do**
19:          $c^{t_1, \ldots, t_l} \leftarrow c^{t_2, \ldots, t_{l-1}} + c^{t_1}$
20:          **if** $\forall 3 \leq i \leq l : c^{t_1, \ldots, t_{i-1}, t_{i+1}, \ldots, t_l} \in S^{t_1, \ldots, t_{i-1}, t_{i+1}, \ldots, t_l}$
               and IsUNIFORM($f^{t_1, \ldots, t_l} + c^{t_1, \ldots, t_l}$) **then**
21:             $S^{t_1, \ldots, t_l} \leftarrow S^{t_1, \ldots, t_l} \cup \{c^{t_1, \ldots, t_l}\}$
22:          **end if**
23:       **end for**
24:       $\Sigma_l \leftarrow \Sigma_l \cup \{S^{t_1, \ldots, t_l}\}$
25:    **end for**
26:    **return** Σ_l
27: **end function**

28: **function** FINDPARTIALLYUNIFORMREALIZATION(S^0, g)
29:    **for** $1 \leq i \leq m$ **do**
30:       $S^i = $ GENERATECORRECTIONFUNCTIONS(f^i, S^0)
31:    **end for**
32:    $\Sigma_1 \leftarrow \{S^1, \ldots, S^m\}$
33:    $l \leftarrow 2$
34:    **while** $l \leq m$ **and** $\exists S^{t_1, \ldots, t_{l-1}} \in \Sigma_{l-1} : S^{t_1, \ldots, t_{l-1}} \neq \emptyset$ **do**
35:       $\Sigma_l \leftarrow$ COMBINECORRECTIONFUNCTIONS(f, Σ_{l-1})
36:       $l \leftarrow l + 1$
37:    **end while**
38:    **return** Σ_{l-1}
39: **end function**

B   Fast Uniformity Check for $s_{out} = 3$

For each unshared input x, the four linearly independent equations describing each row of the uniformity table $\mathcal{U}$ of the Boolean function f with $s_{out} = 3$ are as follows:

$$
\begin{aligned}
&1 \cdot \mathcal{U}_{x,(0,0,0)} + 1 \cdot \mathcal{U}_{x,(0,1,1)} + 0 \cdot \mathcal{U}_{x,(1,0,1)} + 0 \cdot \mathcal{U}_{x,(1,1,0)} + \\
&1 \cdot \mathcal{U}_{x,(0,0,1)} + 1 \cdot \mathcal{U}_{x,(0,1,0)} + 0 \cdot \mathcal{U}_{x,(1,0,0)} + 0 \cdot \mathcal{U}_{x,(1,1,1)} = N_1 \\
&1 \cdot \mathcal{U}_{x,(0,0,0)} + 0 \cdot \mathcal{U}_{x,(0,1,1)} + 1 \cdot \mathcal{U}_{x,(1,0,1)} + 0 \cdot \mathcal{U}_{x,(1,1,0)} + \\
&1 \cdot \mathcal{U}_{x,(0,0,1)} + 0 \cdot \mathcal{U}_{x,(0,1,0)} + 1 \cdot \mathcal{U}_{x,(1,0,0)} + 0 \cdot \mathcal{U}_{x,(1,1,1)} = N_2 \\
&1 \cdot \mathcal{U}_{x,(0,0,0)} + 0 \cdot \mathcal{U}_{x,(0,1,1)} + 0 \cdot \mathcal{U}_{x,(1,0,1)} + 1 \cdot \mathcal{U}_{x,(1,1,0)} + \\
&0 \cdot \mathcal{U}_{x,(0,0,1)} + 1 \cdot \mathcal{U}_{x,(0,1,0)} + 1 \cdot \mathcal{U}_{x,(1,0,0)} + 0 \cdot \mathcal{U}_{x,(1,1,1)} = N_3 \\
&1 \cdot \mathcal{U}_{x,(0,0,0)} + 1 \cdot \mathcal{U}_{x,(0,1,1)} + 1 \cdot \mathcal{U}_{x,(1,0,1)} + 1 \cdot \mathcal{U}_{x,(1,1,0)} + \\
&1 \cdot \mathcal{U}_{x,(0,0,1)} + 1 \cdot \mathcal{U}_{x,(0,1,0)} + 1 \cdot \mathcal{U}_{x,(1,0,0)} + 1 \cdot \mathcal{U}_{x,(1,1,1)} = 2^{n(s_{in}-1)}
\end{aligned} \tag{10}
$$

Depending on whether the output $(y = \sum_i y_i)$ is 0 or 1, either the first or the second line of each equation in Eq. (10) will have non-zero terms $\mathcal{U}_{x,y}$.

C   Finding Uniform Realizations Using Fast WHT

Partial uniform realization for $\mathcal{Q}_{300}^4$. Here, we describe definitive results regarding the use of linear correction terms on the representative permutation of $\mathcal{Q}_{300}^4$ with truth table $[0, 1, 2, 3, 4, 5, 8, 9, 6, 7, 12, 13, 14, 15, 10, 11]$. Namely, it is possible to find multiple uniform realizations for each coordinate function of the permutation using the contribution from this section. However, this does not imply that the realization for the permutation is also uniform. Our algorithm revealed that we can make two out of four coordinate functions jointly uniform. We provide the algebraic description of one such realization where the unshared permutation is described as $(y^1, y^2, y^3, y^4) = f(x^1, x^2, x^3, x^4)$ in Eq. (11).

$$
\begin{aligned}
y_1^1 &= x_2^2 x_2^3 + x_2^2 x_3^3 + x_2^2 x_2^4 + x_2^2 x_3^4 + x_3^2 x_2^3 + x_3^2 x_3^3 + x_3^2 x_2^4 + x_3^2 x_3^4 + x_2^3 x_2^4 + x_2^3 x_3^4 + x_3^3 x_2^4 + x_3^3 x_3^4 + x_1^2 \\
y_2^1 &= x_1^2 x_1^3 + x_1^2 x_3^3 + x_1^2 x_1^4 + x_1^2 x_3^4 + x_3^2 x_1^3 + x_3^2 x_1^4 + x_1^3 x_1^4 + x_1^3 x_3^4 + x_3^3 x_1^4 \\
y_3^1 &= x_1^2 x_2^3 + x_1^2 x_2^4 + x_2^2 x_1^3 + x_2^2 x_1^4 + x_1^3 x_2^4 + x_2^3 x_1^4 + x_1^2 \\
y_1^2 &= x_2^2 x_2^3 + x_2^2 x_3^3 + x_3^2 x_2^3 + x_3^2 x_3^3 + x_2^2 x_2^4 + x_2^2 x_3^4 + x_2^3 + x_3^2 x_2^4 + x_3^2 x_3^4 + x_3^3 + x_2^4 + x_3^4 \\
y_2^2 &= x_1^2 x_1^3 + x_1^2 x_3^3 + x_3^2 x_1^3 + x_1^2 x_1^4 + x_1^2 x_3^4 + x_1^3 + x_3^2 x_1^4 + x_1^4 \\
y_3^2 &= x_1^2 x_2^3 + x_2^2 x_1^3 + x_1^2 x_2^4 + x_2^2 x_1^4 \\
y_1^3 &= x_2^2 x_2^3 + x_2^2 x_3^3 + x_2^2 + x_3^2 x_2^3 + x_3^2 x_3^3 + x_3^2 + x_2^4 + x_3^4 \\
y_2^3 &= x_1^2 x_1^3 + x_1^2 x_3^3 + x_1^2 + x_3^2 x_1^3 + x_1^4 \\
y_3^3 &= x_1^2 x_2^3 + x_2^2 x_1^3 \\
y_1^4 &= x_3^1, \quad y_2^4 = x_1^1, \quad y_3^4 = x_2^1.
\end{aligned} \tag{11}
$$

This particular realization makes the joint realization of the pair (y_1, y_4) uniform. The component functions corresponding to the coordinate functions (y_2, y_3)

should be remasked for uniformity of the permutation's realization. Hence, the required randomness is reduced by 50% compared to remasking every bit. No further improvements are impossible using only linear correction terms.

D Constructions to Avoid Bent Component Functions

Two generic constructions for avoiding bent functions are listed below:

1. Add a term of degree higher than $n(s_{\text{in}} - 1)/2$ which is the maximum degree of a bent function [9]. If $n(s_{\text{in}} - 1)/2 < n(s_{\text{in}} - 2)$, we must add an additional share due to non-completeness. Hence, this is mainly useful for $s_{\text{in}} \geq 4$.
2. It can be shown that the derivative $D_\omega f(\boldsymbol{x}) = f(\boldsymbol{x}) + f(\boldsymbol{x} + \boldsymbol{\omega})$ is a balanced Boolean function if f is bent [9]. Hence, adding $f_i(\boldsymbol{x} + \boldsymbol{\omega})$ to both share i and a new share makes share i balanced if f is bent. The new share can be avoided if some component f is independent of two input shares.

E Using Quadratic Correction Terms For Uniformity

Partial uniform realization for $\mathbb{F}_4$ *multiplier.* It has been shown in [8] that the AES S-box can be decomposed into several multiplications in $\mathbb{F}_4$, additions and rotations. This decomposition has been used for TIs of AES in [12,18]. Since, no uniform realization of $\mathbb{F}_4$ has been found so far, these TIs relied heavily on adding fresh randomness. This can be explained with the fact that both coordinate functions of this multiplication which are given in Eq. (12) are bent.

$$f^1(x) = x^1 x^4 + x^2 x^3 + x^2 x^4 \qquad f^2(x) = x^1 x^3 + x^1 x^4 + x^2 x^3. \qquad (12)$$

Since $n = 4$, we used a generic algorithm to find the matrices M_i leading to a realization with non-bent coordinate functions which is provided in Eq. (13). Note that this realization is not uniform. Hence, we performed a search on linear correction terms as described in Algorithm 1. This gave several uniform realizations for both coordinate functions such as Eq. (15) corresponding to y_1.

$$y_1^1 = x_2^1 x_2^2 + x_2^1 x_2^3 + x_2^1 x_3^4 + x_3^1 x_2^4 + x_3^1 x_3^4 + x_2^2 x_3^3 + x_2^2 x_3^4 + x_3^2 x_2^3 + x_3^2 x_2^4$$

$$y_2^1 = x_1^1 x_3^4 + x_3^1 x_1^4 + x_1^2 x_1^3 + x_1^2 x_1^3 + x_1^2 x_3^4 + x_3^2 x_1^3 + x_3^2 x_3^3 + x_3^2 x_1^4 + x_3^2 x_3^4$$

$$y_3^1 = x_1^1 x_1^4 + x_1^1 x_2^4 + x_2^1 x_2^2 + x_2^1 x_2^3 + x_2^1 x_1^4 + x_2^1 x_2^4 + x_1^2 x_1^3 + x_1^2 x_1^4 + x_1^2 x_2^4 \qquad (13)$$
$$+ x_2^2 x_1^3 + x_2^2 x_2^3 + x_2^2 x_1^4 + x_2^2 x_2^4$$

$$y_1^2 = x_2^1 x_3^3 + x_2^1 x_3^4 + x_3^1 x_2^3 + x_3^1 x_3^3 + x_3^1 x_2^4 + x_2^2 x_3^3 + x_2^2 x_2^4 + x_3^2 x_3^3$$

$$y_2^2 = x_1^1 x_3^3 + x_1^1 x_1^4 + x_1^1 x_3^4 + x_3^1 x_1^3 + x_3^1 x_1^4 + x_3^1 x_3^3 + x_1^2 x_1^3 + x_3^2 x_1^3 + x_3^2 x_3^3$$

$$y_3^2 = x_1^1 x_1^3 + x_1^1 x_2^3 + x_1^1 x_2^4 + x_2^1 x_2^3 + x_2^1 x_2^3 + x_2^1 x_1^4 + x_2^1 x_2^4 + x_1^2 x_1^3 + x_1^2 x_2^3 \qquad (14)$$
$$+ x_2^2 x_1^3 + x_2^2 x_2^3 + x_2^2 x_2^4$$

Since no combination of possible uniform realizations for coordinate functions yielded a uniform result, we conclude that the sharing of either one of the coordinate functions should still be remasked. This requires two bits of randomness.

$$f_1^1 = y_1^1 + x_2^1 \qquad f_2^1 = y_2^1 + x_1^2 + x_1^3 \qquad f_3^1 = y_3^1 + x_1^2 + x_1^3 \qquad (15)$$

References

1. Balasch, J., Gierlichs, B., Verdult, R., Batina, L., Verbauwhede, I.: Power analysis of atmel cryptomemory – recovering keys from secure EEPROMs. In: Dunkelman, O. (ed.) CT-RSA 2012. LNCS, vol. 7178, pp. 19–34. Springer, Heidelberg (2012). doi:10.1007/978-3-642-27954-6_2
2. Bilgin, B.: Threshold Implementations as Countermeasure Against Higher-Order Differential Power Analysis. PhD thesis, KU Leuven and University of Twente (2015)
3. Bilgin, B., Daemen, J., Nikov, V., Nikova, S., Rijmen, V., Van Assche, G.: Efficient and first-order dpa resistant implementations of KECCAK. In: Francillon, A., Rohatgi, P. (eds.) CARDIS 2013. LNCS, vol. 8419, pp. 187–199. Springer, Cham (2014). doi:10.1007/978-3-319-08302-5_13
4. Bilgin, B., Gierlichs, B., Nikova, S., Nikov, V., Rijmen, V.: Higher-order threshold implementations. In: Sarkar, P., Iwata, T. (eds.) ASIACRYPT 2014. LNCS, vol. 8874, pp. 326–343. Springer, Heidelberg (2014). doi:10.1007/978-3-662-45608-8_18
5. Bilgin, B., Gierlichs, B., Nikova, S., Nikov, V., Rijmen, V.: Trade-offs for threshold implementations illustrated on AES. IEEE Trans. Comput.-Aided Des. Integr. Circ. Syst. **34**, 1–13 (2015)
6. Bilgin, B., Nikova, S., Nikov, V., Rijmen, V., Stütz, G.: Threshold implementations of all 3×3 and 4×4 S-boxes. In: Prouff, E., Schaumont, P. (eds.) CHES 2012. LNCS, vol. 7428, pp. 76–91. Springer, Heidelberg (2012). doi:10.1007/978-3-642-33027-8_5
7. Bilgin, B., Nikova, S., Nikov, V., Rijmen, V., Tokareva, N., Vitkup, V.: Threshold Implementations of Small S-boxes. Crypt. Commun. **7**(1), 3–33 (2015)
8. Canright, D.: A very compact S-box for AES. In: Rao, J.R., Sunar, B. (eds.) CHES 2005. LNCS, vol. 3659, pp. 441–455. Springer, Heidelberg (2005). doi:10.1007/11545262_32
9. Carlet, C.: Boolean Functions for Cryptography and Error Correcting Codes (2006)
10. Carlet, C., Prouff, E., Rivain, M., Roche, T.: Algebraic decomposition for probing security. In: Gennaro, R., Robshaw, M. (eds.) CRYPTO 2015. LNCS, vol. 9215, pp. 742–763. Springer, Heidelberg (2015). doi:10.1007/978-3-662-47989-6_36
11. Carlet, C., Yucas, J.L.: Piecewise constructions of bent and almost optimal Boolean functions. Des. Codes Crypt. **37**, 449–464 (2005)
12. De Cnudde, T., Bilgin, B., Reparaz, O., Nikov, V., Nikova, S.: Higher-order threshold implementation of the AES S-box. In: Homma, N., Medwed, M. (eds.) CARDIS 2015. LNCS, vol. 9514, pp. 259–272. Springer, Cham (2016). doi:10.1007/978-3-319-31271-2_16
13. Duc, A., Dziembowski, S., Faust, S.: Unifying leakage models: from probing attacks to noisy leakage. In: Nguyen, P.Q., Oswald, E. (eds.) EUROCRYPT 2014. LNCS, vol. 8441, pp. 423–440. Springer, Heidelberg (2014). doi:10.1007/978-3-642-55220-5_24
14. Eisenbarth, T., Kasper, T., Moradi, A., Paar, C., Salmasizadeh, M., Shalmani, M.T.M.: Physical cryptanalysis of KeeLoq code hopping applications. Cryptology ePrint Archive, Report 2008/058 (2008). http://eprint.iacr.org/
15. Kocher, P., Jaffe, J., Jun, B.: Differential power analysis. In: Wiener, M. (ed.) CRYPTO 1999. LNCS, vol. 1666, pp. 388–397. Springer, Heidelberg (1999). doi:10.1007/3-540-48405-1_25
16. Kocher, P.C.: Timing attacks on implementations of Diffie-Hellman, RSA, DSS, and other systems. In: Koblitz, N. (ed.) CRYPTO 1996. LNCS, vol. 1109, pp. 104–113. Springer, Heidelberg (1996). doi:10.1007/3-540-68697-5_9

17. Kutzner, S., Nguyen, P.H., Poschmann, A.: Enabling 3-share threshold implementations for any 4-bit S-box. Cryptology ePrint Archive, Report 2012/510 (2012). http://eprint.iacr.org/
18. Moradi, A., Poschmann, A., Ling, S., Paar, C., Wang, H.: Pushing the limits: a very compact and a threshold implementation of AES. In: Paterson, K.G. (ed.) EUROCRYPT 2011. LNCS, vol. 6632, pp. 69–88. Springer, Heidelberg (2011). doi:10.1007/978-3-642-20465-4_6
19. Nikova, S., Rijmen, V., Schläffer, M.: Secure hardware implementations of nonlinear functions in the presence of glitches. J. Cryptology 24, 292–321 (2010)
20. Poschmann, A., Moradi, A., Khoo, K., Lim, C.-W., Wang, H., Ling, S.: Side-channel resistant crypto for less than 2,300 GE. J. Cryptology 24(2), 322–345 (2011)
21. Reparaz, O., Bilgin, B., Nikova, S., Gierlichs, B., Verbauwhede, I.: Consolidating masking schemes. In: Gennaro, R., Robshaw, M. (eds.) CRYPTO 2015. LNCS, vol. 9215, pp. 764–783. Springer, Heidelberg (2015). doi:10.1007/978-3-662-47989-6_37
22. Rothaus, O.: On bent functions. J. Comb. Theory Ser. A 20(3), 300–305 (1976)
23. Wan, Z.-X.: Lectures on Finite Fields and Galois Rings. World Scientific, Singapore (2003)

Attacking Embedded ECC Implementations Through cmov Side Channels

Erick Nascimento[1]([⊠]), Łukasz Chmielewski[2], David Oswald[3], and Peter Schwabe[4]

[1] Institute of Computing, University of Campinas, Campinas, Brazil
enascimento.pub@gmail.com
[2] Riscure BV, Delft, The Netherlands
Chmielewski@riscure.com
[3] School of Computer Science, University of Birmingham, Birmingham, UK
d.f.oswald@cs.bham.ac.uk
[4] Digital Security Group, Radboud University, Nijmegen, The Netherlands
peter@cryptojedi.org

Abstract. Side-channel attacks against implementations of elliptic-curve cryptography have been extensively studied in the literature and a large tool-set of countermeasures is available to thwart different attacks in different contexts. The current state of the art in attacks and countermeasures is nicely summarized in multiple survey papers, the most recent one by Danger et al. [21]. However, any combination of those countermeasures is ineffective against attacks that require only *a single trace* and directly target a conditional move (cmov) – an operation that is at the very foundation of all scalar-multiplication algorithms. This operation can either be implemented through arithmetic operations on registers or through various different approaches that all boil down to loading from or storing to a secret address. In this paper we demonstrate that such an attack is indeed possible for ECC software running on AVR ATmega microcontrollers, using a protected version of the popular μNaCl library as an example. For the targeted implementations, we are able to recover 99.6% of the key bits for the arithmetic approach and 95.3% of the key bits for the approach based on secret addresses, with confidence levels 76.1% and 78.8%, respectively. All publicly available ECC software for the AVR that we are aware of uses one of the two approaches and is thus in principle vulnerable to our attack.

Keywords: ECC · Montgomery ladder · Power analysis · AVR · Conditional move

E. Nascimento—The author was supported by the Brazilian National Council for Scientific and Technological Development (CNPq), under the Science Without Borders program, process 206508/2014-0. This work was done while the author was visiting Radboud University.

P. Schwabe—This work was supported by the Netherlands Organisation for Scientific Research (NWO) through Veni 2013 project 13114. Permanent ID of this document: bb3c834d7cc8ffbe7e7520f1c21bd408. Date: July 18, 2016.

R. Avanzi and H. Heys (Eds.): SAC 2016, LNCS 10532, pp. 99–119, 2017.
https://doi.org/10.1007/978-3-319-69453-5_6

1 Introduction

For many years, efficient software implementations of cryptographic algorithms for constrained embedded processors were mainly restricted to symmetric ciphers. However, in recent years, various libraries for elliptic curve cryptography (ECC) have been published that offer acceptable runtime and code size also on microcontrollers with very limited computational resources, e.g., the 8-bit AVR ATmega series of processors. Notable examples for these ECC implementations are summarized in Table 1.

Table 1. Overview of ECC implementations for embedded AVR processors.

Name	Description	SCA countermeasures
micro-ecc [43]	8/32/64-bit C impl. for NIST curves	Not documented; apparently randomized projective coordinates
nano-ecc [33]	Derivate of micro-ecc	Same as micro-ecc
μNaCl [23,32,49]	Curve25519 for 8/16/32-bit processors	Constant-time
AVR-Crypto-Lib [53]	ECDSA with NIST P-192	None
FLECC_IN_C [59]	8/16/32/64-bit C impl. for various curves	Constant time, randomized projective coordinates
RELIC [2]	Various curves and fields supported	Constant-time
WM-ECC [58]	Impl. for sensor networks	None
TinyECC [42]	Impl. for sensor networks	None
MIRACL [13]	Lib. supporting multiple curves	None
WolfSSL [60]	Support for AVR unclear	None
Wiselib [1]	Lib. for distributed systems	None
CRS ECC [56]	Commercial, closed source	None

Due to the fact that an adversary often has physical access to an embedded device performing ECC operations, implementation attacks and in particular side-channel analysis (SCA) are severe threats in this scenario. Consequently, several libraries comprise countermeasures against SCA, for example, by performing computations in constant-time, or by using randomized projective coordinates. The protected implementations are further detailed in Table 1.

Many common SCA countermeasures assume that the adversary needs access to multiple traces (with identical scalar) to recover the secret key, which inherently protects protocols with ephemeral scalars. In this paper, we challenge this assumption and target fundamental building blocks of any ECC implementation, namely *conditional moves* and loads/stores from/to secret memory addresses.

We show that template attacks allow to recover most of the secret scalar with a single trace of elliptic-curve scalar multiplication (ECSM) in both cases, which in turn renders all currently published ECC implementations for the AVR (and likely other, similar architectures) insecure.

Note that although this paper focuses on implementations of ECC, our attacks also apply to exponentiation algorithms as used in, e.g., RSA, classical Diffie-Hellman, DSA, or ElGamal. We actually expect the attacks to work even better there, because group elements are larger and thus require more loads (or conditional moves). We leave this investigation for future work.

Related work. Carefully combining countermeasures like uniformity of modular operations, (re-)randomization of the projective representation of points, scalar blinding, point blinding, and random field (or curve) isomorphisms prevent classical side-channel attacks like timing [38], SPA [20], DPA [39], CPA [11] or collision attacks [25,31]. These attacks require a fixed scalar for multiple measured power or electromagnetic traces. The main protection relies on the full randomization of intermediate data, including input point, scalar and group, during the execution of an ECSM [4,19,24]. In this work we consider implementations based on the Montgomery ladder algorithm, protected by scalar randomization (SR) and projective-coordinate randomization[1].

To overcome the aforementioned countermeasures two kinds of attacks have emerged: template and horizontal attacks. Although in general template attacks [14] can be used to attack multiple traces that share the same scalar, we need to attack ECSM traces independently, because of the SR. Template attacks combine statistical modeling and power-analysis, and consist of two phases. In the first phase, called *profiling*, the attacker builds templates by executing a sequence of instructions using a fixed scalar (with SR turned off). The second phase is called *matching*, in which the attacker matches the templates to attacked single traces (with SR turned on). The assumption is that the attacker possesses a *profiling* device, in order to build templates, that behaves the same as the target device, and runs the same implementation.

Template attacks on ECC trace back to an attack on ECDSA demonstrated by Medwed and Oswald [44]. However, this attack requires an offline DPA on the ECSM during profiling, in order to select the points of interest. Moreover, since the attack exploits data-dependent leakage it requires profiling with multiple templates (i.e., 33) while for our attacks two templates are enough. Furthermore, the attack only needs to recover a few bits of the multiple ephemeral scalars and can then employ ECDSA-specific lattice techniques to recover the long-term secret key [10]. This is not possible in the context of our work, since we do not target ECDSA: an attacker has only a single trace to recover sufficiently many bits of the randomized scalar using SCA to be able to compute the remaining bits.

---

[1] The implementations actually attacked apply only projective coordinates randomization, however, our attack also works on an implementation with SR enabled, because we do not make any assumption about the secret scalar, i.e., it may be different from one execution to another.

Another template attack on ECC is presented in [30]. This attack follows a similar approach to our attack, but instead of exploiting address-dependent leakage, it exploits register location based leakage using a high-resolution inductive EM probe. As a result the attack is considerably expensive to execute. A template attack on a wNAF ECC algorithm is presented in [61]. However, this attack is applied to an implementation that is not protected with either, scalar randomization or base-point randomization. Another approach to attack ECC are the so called online template attacks [5,22]. These attacks work if SR is enabled, but not when point randomization is enabled.

The template attack from [16] targets load instructions. However, multiple traces are required in the attack phase. Therefore, this attack does not work against implementations protected by SR. The template attack from [28] aims to extract a random multiplicative mask (base-blinding) out of a single measurement exploiting data leakage; then it is possible to unmask all intermediate values and run DPA.

Horizontal attacks on RSA [6,8,9,15,17,18,29,54,55,57] and ECC [7,27] are emerging forms of side-channel attacks on exponentiation-based or scalar-multiplication-based algorithms. Their methodology allows recovering the exponent bits through the analysis of individual traces. Therefore, these attacks are efficient against SR even when combined with point and group randomization. The attacks employ different common distinguishers: SPA, horizontal correlation analysis [18], Euclidean distance [57], horizontal collision-correlation [6–8,17], horizontal cross-correlation [27], or clustering [29,55].

An interesting horizontal address-based DPA attack on Montgomery multiplications is presented in [15]. The approach is similar to ours, but this attack exploits Hamming weight leakage of addresses. Furthermore, the analysis in [15] lacks the results for a full modular exponentiation (only a few iterations are attacked) and success rates.

The main issue of horizontal attacks is that extracting leakage from a single unlabeled trace is usually heavily limited by noise. Therefore, we have decided to attack our state-of-the art implementations, that contains scalar and point randomizations, using a more powerful attack paradigm, from the point of view of the attacker setting, namely, template attacks.

Contributions. The main contributions of this paper are threefold:

1. First, by the example of a protected version of μNaCl, we show that the single-trace leakage of conditional moves within the Montgomery ladder can be exploited to recover the scalar.
2. Second, we show that a similar attack applies to loads and stores from/to secret-dependent addresses. In doing so, we show that even implementations on embedded devices *without* cache cannot tolerate secret-dependent memory accesses.
3. Finally, we generalize the method from [26] to tolerate a certain number of incorrectly recovered scalar bits without relying on normal or side-channel-enhanced exhaustive search. Furthermore, we present experimental results for our algorithm.

Organization of the paper. The remainder of this paper is structured as follows: in Sect. 2, we review the use of conditional moves in scalar multiplication algorithms, together with possible countermeasures against side-channel analysis. Then, in Sect. 3, we describe the measurement setup and target implementation used for our attacks presented subsequently: while Sect. 4 deals with template attacks on the (arithmetic) conditional swap within the Montgomery ladder, Sect. 5 applies similar methods to recover the scalar by exploiting the leakage of secret load addresses. Section 6 discusses how to tolerate a certain number of incorrectly recovered scalar bits more efficiently than by simple exhaustive search. Finally, we conclude in Sect. 7 with directions for future work, in particular regarding countermeasures.

2   Scalar Multiplication and Conditional Moves

The most basic scalar-multiplication algorithm is the double-and-add algorithm, which scans through the bits of the scalar and performs a double operation for each zero bit and a double-and-add operation for each one bit. This algorithm is well known to be vulnerable to all kind of side-channel attacks, including power analysis and timing attacks.

The first step to side-channel protection is to always perform the same sequence of finite-field operations, independent of the scalar. The most common approaches to achieve such a structure are either to use (fixed-window) double-and-add-always scalar multiplication or ladder-based approaches (typically the Montgomery ladder [45] or, for general Weierstrass curves, the Brier-Joye ladder [12]). Another layer of side-channel protection then adds randomization of the scalar (through one of various blinding methods), and the internal representation of points (for example through projective randomization, field isomorphisms, or curve isomorphisms). By re-randomizing before or after each ECSM loop iteration, most horizontal collision or cross-correlation attacks are thwarted.

Interestingly, even with all those countermeasures in place, scalar-multiplication algorithms contain operations that *choose one out of two (or more) curve points* depending on bit(s) of the scalar. An attacker who learns all of these choices from side-channel information from just one trace, learns all of the scalar bits used in this scalar multiplication and thus obtains the secret key. On microcontrollers with restricted register space, there are essentially two different ways to implement this *conditional move* (cmov): either by loading from (or storing to) addresses that depend on the secret scalar, or by using arithmetic operations to perform a conditional register-to-register move. The latter approach is very common on large processors with cache, where the former approach leaks through cache-timing information. Essentially, the idea is to replace a computation of the form $R \leftarrow P[s]$, where s is a secret scalar bit, by a computation of the form $R \leftarrow sP[1] + (1-s)P[0]$. Note that this approach does not require actual multiplications; it is much easier to expand s to a bit mask of all ones or all zeros and use bit-logical instructions.

Most implementations of ECSM contain considerably more than just one secretly-indexed load, store, or conditional move. Sometimes this is a choice made by the implementors to improve performance (by avoiding otherwise unnecessary loads and stores); sometimes it is an inherent property of the ECSM algorithm. For example, the Montgomery ladder needs a conditional swap (cswap) of two points instead of a conditional move, which requires significantly more operations that involve the secret scalar bit than a simple cmov (for details, see Sect. 4).

The side-channel attacks described in the remainder of this paper attack both implementations that make use of secretly indexed memory accesses (in Sect. 5) and implementations that use the arithmetic cmov operation (or more specifically, the cswap operation) in Sect. 4. The idea of attacking loads from secret positions through side-channel information is not new: it is not only used in various cache-timing attacks (that do not apply to simple architectures such as the AVR), but it is also the underlying principle of address-bit-DPA [34]. What is novel is the fact that we need only a single trace. This renders countermeasures such as scalar blinding and address randomization [35,36] ineffective.

3   Attack Setup

In this section, we describe the targeted implementations, the utilized micro-controller, our measurement setup. The trace pre-processing, frequency filtering and alignment, are described in the full paper [48].

3.1   Target Implementations

We target two protected ECSM implementations based on [49]. Both employ the Montgomery ladder, with the pseudocode given in Algorithm 1. The main difference between the two variants is the realization of the cmov (i.e., the function CSWAP_COORDS): The first implementation, described in more detail in Sect. 4.1, consists of applying an arithmetic conditional swap of the respective coordinates values of the working points $P_1 = (X_1 : Z_1)$ and $P_2 = (X_2 : Z_2)$. The second, described in Sect. 5.1, replaces the arithmetic conditional swap by a conditional swap of pointers to the coordinate values. Both implementations utilize projective-coordinate re-randomization as the main side-channel counter-measure. A randomly generated $\lambda \in \mathbb{F}_p$ is multiplied with the coordinates of $P_1 = (X_1 : Z_1)$ and $P_2 = (X_2 : Z_2)$ at the beginning of every ECSM iteration. We make publicly available the source code for both implementations [47].

3.2   Target Device and Measurement Setup

We carried out our experiments with an ATmega328P 8-bit microcontroller placed on the target board of the ChipWhisperer [51] side-channel evaluation platform. While the ChipWhisperer also provides the possibility to capture ana-log signals (e.g., power consumption or electro-magnetic emanation), we used a separate oscilloscope (Picoscope 5203) due to the limited bandwidth, memory, and sample rate of the ChipWhisperer.

Algorithm 1. Montgomery ladder with arithmetic cswap and randomized projective coordinates.

// ... initialization omitted ..
$bprev \leftarrow 0$
for $i = 254 \dots 0$ **do**
    RE_RANDOMIZE_COORDS($work$)
    $b \leftarrow$ bit i of scalar
    $s \leftarrow b \oplus bprev$
    $bprev \leftarrow b$
    CSWAP_COORDS($work$, s)
    LADDERSTEP($work$)
end for

The targeted ATmega328P has a 32 KB of Flash, 2 KB of SRAM, and 1 KB of EEPROM. The register file contains 32 registers (R0–R31), among which 6 serve as pointers for indirect 16-bit addressing and have the following aliases: X (R27:R26), Y (R29:R28) and Z (R31:R30). Arithmetic instructions take 1 cycle, with the exception of multiplication instructions, which take 2 cycles. Loads and stores from/to SRAM take 2 cycles. Loads from Flash take 3 cycles. More technical details about the target device are given in the full paper [48].

4   Attacking Arithmetic Cswaps

In this section, we describe a template attack on conditional swaps (cswaps) in the Montgomery ladder step. In our case, the cswap is implemented using Boolean and arithmetic operations in constant time.

4.1   Target Implementation

In the Montgomery ladder (Algorithm 1), the function CSWAP_COORDS implements the cswap (based on input bit s) by first creating a mask m, which is either 0x00 or 0xFF for $s = 0$ and $s = 1$, respectively, by setting $m = -s$ (assuming m, s are 8-bit values). Then, a (conditional) XOR swap is executed as follows:

Listing 1.1. Conditional XOR swap.

```
1  ld  xx, X    ; X register points to first value
2  ld  yy, Z    ; Z register points to second value
3  mov tt, xx
4  eor tt, yy
5  and tt, m    ; tt = (xx XOR yy) AND m
6  eor xx, tt   ; xx = xx XOR tt
7  eor yy, tt   ; yy = yy XOR tt
8  st  X+, xx   ; Store first value
9  st  Z+, yy   ; Store second value
```

In other words, if $m = 0x00$ ($s = 0$), $tt = 0$ and the XORs $xx = xx \oplus tt$ and $yy = yy \oplus tt$ leave the values unchanged. Otherwise, if $m = 0xFF$ ($s = 1$), we have a standard XOR swap, i.e., $xx = xx \oplus xx \oplus yy = yy$ (equivalent for yy).

4.2 Template Generation and Matching

We generated templates for the and instruction (line 5 of Listing 1.1), grouping the traces in the profiling set into two sets V_0 and V_1. Traces in V_0 represent those where $m = 0$ (i.e., an AND with 0x00), while V_1 are traces where $m = 0xFF$. Note that the traces were cut to only contain the clock cycle for the targeted and instruction, i.e., each trace is $64 \cdot 67 = 4288$ samples long (cf. Appendix 2 of the full paper [48]). For V_i, $i = 0, 1$, we subsequently computed templates consisting of the pointwise mean vector $\boldsymbol{\mu}^{(i)}$ and the covariance matrix $\boldsymbol{\Sigma}^{(i)}$ [14]. Note that the two possible leakages 0x00 (all bits zero) and 0xFF (all bits one) can be expected to be maximally (or at least to a large degree) different, which should facilitate template attacks in this particular case.

We matched the templates to the traces in the test set with the standard approach, i.e., computing the respective probabilities using the multivariate normal distribution pdf and identifying the template with the highest probability to recover the respective bit of the scalar. The respective success rates wrt the size of the profiling set are given in Sect. 4.3.

Classification. For each template we computed the Euclidean distance between the sample vector and the template mean vector. The template (T_0 or T_1) that results in the smallest distance is considered the best match for the sample vector. In this attack, the index of the closest template (0 or 1) corresponds to the swap bit.

Confidence score and confidence level. For the first classification method we derived a simple confidence score on the recovered bit value based on the distances (d_0 and d_1) to each template. It varies linearly for a particular $d_0 + d_1$ value, ranging from 0 (no confidence) and 1 (full confidence):

$$\text{conf\_score} = 2 \cdot \left| 0.5 - \frac{min(d_0, d_1)}{d_0 + d_1} \right| \tag{1}$$

We furthermore define the *confidence level* of a given trace (in the test set) as follows: Let us call a recovered bit *suspicious* if its confidence level is less than the greatest confidence score of any falsely identified bit (whereas this threshold is determined experimentally in the profiling phase). Then, the confidence level is the percentage of bits that are not suspicious, i.e., that can be unambiguously recovered. Note that the average confidence level (over all number of traces in the test set) is always less than or equal to the average success rate, since an incorrectly recovered bit is always suspicious.

4.3 Attack Results

Figure 1 shows the average and best case success rates (computed over all 255 scalar bits), together with the respective confidence levels over the number of traces used for template generation and matching. Note that each full trace comprises 255 ECSM iterations, which were all used for generating the templates – in other words, each full trace contributes 255 "effective" traces to the profiling set.

The traces used for template generation and matching were taken from different trace sets (coming from different capture sessions). The same number of traces was used for profiling and testing, i.e., a given value on the horizontal axis of Fig. 1 is the same for profiling and testing.

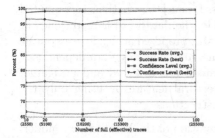

Fig. 1. Success rates for the template attack on cswap for different number of full traces.

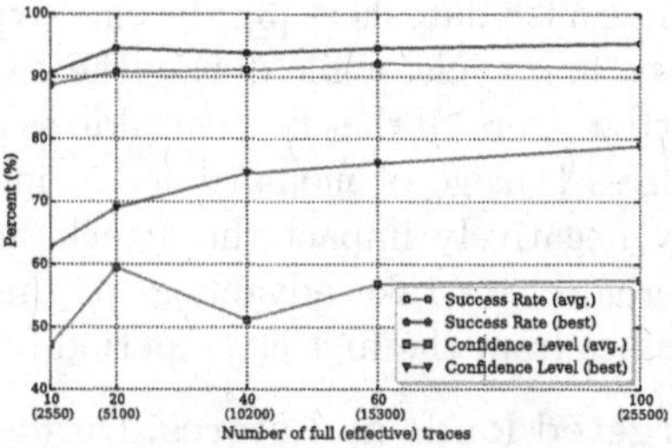

Fig. 2. Results for the template attack on loads/stores for different number of full traces.

As evident in Fig. 1, already for 10 full traces (i.e., about 2,550 effective traces), the average success rate reaches 96.71%, i.e., we can recover most of the bits of the scalar. Furthermore, the best success rate reaches 99.6% with the confidence level 76.1%. By increasing the number of traces, both success rate and confidence level change only minimally; due to the strong leakage of the targeted device, most information can be already extracted with a low trace count.

5   Attacking Secret-Dependent Memory Accesses

In general, ECC (and in particular NaCl-derived) implementations avoid loads from secret-dependent addresses altogether due to the possibility of cache-timing attacks. However, for embedded implementations without caches, secret load addresses are sometimes deemed acceptable. In this section, we show that template attacks can be employed to exploit this leakage.

5.1   Target Implementation

The targeted implementation replaces the cswap of the $(X_1 : Z_1)$ and $(X_2 : Z_2)$ coordinates values used in the targeted implementation in Algorithm 1 by working with pointers to those coordinates, and conditionally swapping these pointers. Besides being slightly faster, this implementation also potentially exhibits less leakage, because it uses the secret-dependent mask m in an AND operation only twice for each pointer cswap[2], rather than 32 times as in the ECSM implementation based on arithmetic cswap (cf. Sect. 4.1).

However, in implementations of finite-field operations both input and output operands are pointers. The values of these pointers are addresses to the memory holding the actual field element value, and those addresses directly depend on whether the swap occurred or not, which in turn depends on the value of the secret mask bit.

AVR memory access instructions internals. Memory access instructions (loads and stores) on an AVR take 2 clock cycles to execute. According to the ATmega328 datasheet [3], the effective address for such instructions is computed in the first cycle, while during the second cycle, the data word is read (load) or written (store) if the effective address is valid. Our proposed attack focuses on the address leakage of memory access instructions, and thus any data-dependency may negatively impact the attack success rate if not detected and mitigated. Therefore, we take advantage of this architectural feature by using only the samples from the first clock period of such instructions.

Targeted loads and stores. During each iteration of the Montgomery ladder, the actual field arithmetic occurs in the so-called LADDERSTEP function (cf. Algorithm 1). We target the loads and stores addresses in the first three field operations in LADDERSTEP, i.e., addition, subtraction, and addition. Each of these operations has two $\mathbb{F}_p$ inputs (a and b) and one output r.

Finite-field addition and subtraction are implemented with reduction modulo $2^{256} - 38$. The reduction step also execute loads and stores, of which the samples are also used for template creation and matching. Listing 1.2 shows a small segment of the execution trace containing the loads of the first operands bytes and the store of the first byte of the result (before reduction):

Listing 1.2. Segment of the execution trace for a field addition.

```
1 0x171a:  fp_add+0x5    LD R20, X+       ; first byte of a
2 0x171a:  fp_add+0x5    CPU-waitstate
3 0x171c:  fp_add+0x6    LD R21, Y+       ; first byte of b
4 0x171c:  fp_add+0x6    CPU-waitstate
5 0x171e:  fp_add+0x7    ADD R20, R21
6 0x1720:  fp_add+0x8    ST Z+, R20       ; first byte of r
7 0x1720:  fp_add+0x8    CPU-waitstate
```

Our oscilloscope's memory is divided into 255 segments, each of which is 65 kSample in length. A memory segment holds the samples captured from a single ECSM iteration. Due to the 65 kSample limit for each ECSM iteration, we were able to capture the samples from all the loads and stores from the first field addition and the first field subtraction, but only half of the loads and stores from the arithmetic part of the second field addition. Note that the memory limitation

---

[2] For the AVR architecture, pointers are 16 bit wide and one AND with the secret-dependent bit is required to cswap a byte. Thus a pointer cswap requires two ANDs.

is due to the relatively low-cost oscilloscope we used—high-end equipment would further facilitate the presented attack.

Table 2 shows the number of executed instructions of each type that are used in the attack. We used a total of 372 instructions, which are concatenated into a single sample vector. After trace preprocessing, 67 power samples are available per clock cycle, and as only the first clock period of a memory access instruction is used, the sample vector per ECSM iteration has $n_v = 24,924$ samples.

Table 2. Number of executed instructions of each type that are used in the attack.

Type	1st fp_add	fp_sub	2nd fp_add	Total
LD R20, X+	32	32	16	80
LD R21, Y+	32	32	16	80
LD R20, Z+0	33	33	0	66
ST Z+, R20	65	65	16	146

5.2 Template Generation

Each load or store instruction accesses at most two possible addresses. If it always accesses the same address, then it does not provide useful leakage relevant for the attack. Considering only those loads and stores that may access two addresses, during any execution of the LADDERSTEP, only two distinct sequences of addresses can be accessed: A_{noswap}, containing the addresses accessed before the first pointers swap has taken place[3], i.e., an even state (noswap state); and A_{swap} containing the addresses accessed in an odd state (swap state).

First, we grouped the sample vectors into two sets. The first set, V_0, consists of the load/store sample vectors for addresses in the set A_{noswap}, while the second set, V_1, contains those originating from addresses in set A_{swap}. Then, we computed various statistics for each sample index of V_i, $i = 0, 1$: mean $\mu^{(i)}$, standard deviation $\sigma^{(i)}$, median $md^{(i)}$, as well as lower $l^{(i)}$ and upper $u^{(i)}$ percentiles (the actual percentiles used are discussed in Sect. 5.3). The collection of these statistics for V_0 and V_1, called T_0 and T_1, are the two possible templates.

5.3 Point-of-Interest Selection

The POI selection consists of using the lower and upper percentile vectors $l^{(i)}$ and $u^{(i)}$ (i=0,1) to compute the intersection of the pair of intervals $[l_j^{(0)}, u_j^{(0)}]$ and $[l_j^{(1)}, u_j^{(1)}]$ for each sample index $j = 1, \ldots, n_v$. The sample indices where the intersection is empty are the considered POIs.

---

[3] These addresses are the same as those accessed after the 2nd but before the 3rd swap, or after the 4th but before the 5th swap, and so on.

Intuitively, the sample indices with an empty intersection are those that are good distinguishers for the two templates, because in these points the samples tend to be clustered around the median (and also typically around the mean) of one template, rather than being scattered.

Different values for the lower and upper percentiles may give a different number of POIs, and that directly affects the success rate and confidence level of the attack. Thus, we tested the attack for different pairs of values for these parameters, ranging from wider and more selective percentiles $(12.5, 87.5)$[4] to narrow, less selective $(40, 60)$. We emphasize that the POI selection is completely based on the samples of the traces used for the generation—it does not depend on the samples of the trace being attacked (i.e., the sample vector to classify). In fact, the POIs are represented as a Boolean vector used during template matching to select the samples from the target trace vector to be classified.

POI selection refinements. To improve the confidence level of the attack, we tested two POI selection refinements, as explained above. First, we noticed that when using more selective percentile parameters, the current selection method returned sample indices that were clustered in a few instructions, while most of the remaining instructions were not covered by any sample, although they should in theory contribute some leakage. To make the POIs more evenly distributed and exploit leakage from all useful instructions, we forced a minimum of one sample index per instruction to be included in the POI vector. If there was no sample index for a given instruction in the current POI vector, one was randomly selected. Second, also due to the clustering of the POIs in a few instructions, we limit the number of samples per instruction to one. In the case that sample indices had to be removed, we selected those randomly as well.

5.4   Template Matching

At first, without using any POI selection, we tried to use the standard multivariate Gaussian model, taking advantage of both the mean vector and covariance matrix computed from V_0 and V_1 (also known as *complete templates*) similar to the approach of Sect. 4. However, in contrast to Sect. 4, the sample vectors to be classified and the mean template vectors are relatively long $(24, 924$ samples) and relatively similar to each other (i.e., their Euclidean distance is very small), numerical instability issues due to almost singular matrices arose during the computation of the probability density function. For those reasons, we decided to use *reduced templates* instead, which uses only the mean vectors.

After applying POI selection, the matched sample vectors are much smaller, and thus full templates could then in principle be applied, as the covariance matrices would not lead to numerical instability. However, due to the high success rates achieved using the reduced templates, we decided to not use full templates to avoid increasing storage and computational requirements.

We also evaluated the effect on the attack success rate and confidence level of compressing the sample vector using normal and absolute sum for different

---

[4] I.e., the lower is the 12.5-percentile and the upper is the 87.5-percentile.

window lengths. In addition, we applied a straightforward outlier detection to remove samples that have likely been subject to larger distortions: In the matching phase, we discarded all samples that have a distance of more than a multiple of standard deviations to the mean trace at the respective point in time. Using reduced templates, template matching boils down to computing the (squared) Euclidean distance between the sample vector to match and the template mean vectors. The lower that distance is, the stronger is the match. In this case, other distinguishers can be used in a straightforward way, and thus we also tested the attack using the Pearson correlation coefficient.

Classification methods and confidence score. As a first classification method to test, we selected the template closer to the sample vector (cf. Sect. 4.2). We also tested majority voting classification, where each sample is individually classified, also based on its distance to the corresponding element in the templates mean vectors, and the majority vote wins. In both cases, as each template directly corresponds to a scalar bit value, the classification output is the recovered bit value. The confidence score was computed in the same way as in Sect. 4.2.

5.5   Attack Results

Figure 2 depicts average and best case success rates for the template attack on secret-dependent memory accesses for the best and average cases. Again, as in Sect. 4.3, the trace sets used for template generation and matching were recorded in different capture sessions, and the same number of traces was used for each set. Again, only a limited number of profiling traces was sufficient to reach success rates exceeding 90%; the best success rate reaches 95.3% (there are only 12 errors) with the confidence level 78.8% (the 12 errors are included in the 54 suspicious bits). To investigate the effect of various pre-processing steps and attack parameters, using 10 traces we investigated the average success rate and confidence level depending on various attack parameters. In particular, we investigated various signal frequency filtering options, POI selection methods, classification and compression methods, outlier filtering, and distinguishers; the result of the investigation are described in the full paper [48]. The best parameters that we discovered, were used to perform the main attack described in this section.

6   Error Detection and Correction

Due to noise, data leakage (note that we are aiming at exploiting the address leakage only), and other aspects that interfere with the side-channel analysis (misalignment, clock jitter, etc.), the derivation of the final scalar for a single trace likely contains errors. If the amount of wrong bits is sufficiently small, then a brute-force attack may still be feasible. However, first the attacker needs a metric to indicate the location of the possible wrong bits in the recovered scalar. The notion of suspicious bits (cf. Sect. 4.2) can be used as a reference for the scalar bits selection with respect to a brute-force attack.

Let us consider the trace with smallest amount of suspicious bits from the experiment from Sect. 5; for this trace there are 54 suspicious bits that comprise all falsely identified bits. Unfortunately, to recover a full randomized scalar, even in this case, the attacker needs 2^{54} operations, which is generally impractical. Note, that we consider only the worst-case complexity and not the average case.

To improve upon the brute-force search complexity, there are two options. The first approach is to try to exploit the distribution of suspicious bits for incorrectly (red) and correctly (blue) recovered bits (Fig. 3). While there is a clear trend for incorrect bits to have lower confidence score, the intersection between correct and incorrect bits is large. Still, it may possible to exploit the trend with an informed brute force attack [40], prioritizing bits with the lowest confidence score. Unfortunately this attack works well if the bits containing errors are adjacent to each other and that is not the case in our setting.

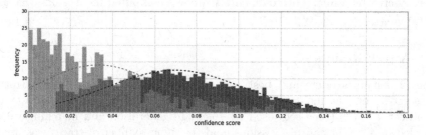

Fig. 3. Distribution of confidence scores over all traces for suspicious bits. Red: incorrectly recovered bits, blue: correctly recovered but suspicious bits. (Color figure online)

Alternatively (or combined with the informed brute-force search), we apply the second algorithm from [26], which is originally designed for *square-and-multiply chains*, to the Montgomery ladder. We describe how the algorithm works using the aforementioned example trace, which contains $s = 54$ suspicious bits, as an example. Let us represent the indices of these bits as a list sorted in descending order: $i_s, \ldots i_1$, where each $i_j \in \{0, \ldots 254\}$ and $s \geq j \geq 1$; note that there are 255 bits in total. Let x denote the bit index $i_{\lfloor \frac{s}{2}+1 \rfloor}$ (namely, i_{28} for the example trace). Let a be the number represented by the bit string corresponding to the left part of the scalar from x (including i_x) and let b be the number corresponding to the bit string of the (least significant) right part. Furthermore, we know that $R = [k]P$, where R is the resulting point, k the scalar to be recovered, and P the input point. Then, clearly $R = [k]P = [a \cdot 2^{i_x} + b]P = [a]([2^{i_x}]P) + [b]P$. If we denote $[2^{i_x}]P$ by H, then the above equation reduces to

$$R - [b]P = [a]H \tag{2}$$

We can use Eq. 2 to check correctness of our guess. Now, following [26], we use a time-memory trade-off technique to speed up an exhaustive search: Consider all different possible guesses for a. For each guess, we compute $[a]H$ and store all pairs $(a, [a]H)$. We then sort all pairs based on the value of $[a]H$ and store them in an ordered table.

Next, we make a guess for b and compute $z = R - [b]P$. If our guess for b is correct, then z is present in the second column of some row in the table we built—the first column is the corresponding a. Finding such a pair can be done using binary search, as the table is sorted as per the second column. If z is present, we are done since we have determined the scalar. Otherwise, we make a new, different guess for b and continue. Since there are approximately $2^{\frac{s}{2}}$ guesses for a and b, the time complexity is $O(2^{\frac{s}{2}})$ operations. As there are $2^{\frac{s}{2}}$ guesses for a, the table has that many entries and the space complexity is $O(2^{\frac{s}{2}})$ points. This way, we limit the time complexity to $O(2^{\frac{s}{2}})$ (cf. [26] for a detailed complexity analysis), which is 2^{27} for the example trace.

We do not know which trace contains the smallest number of suspicious bits since we do not know the maximum confidence score of a falsely identified bit. However, to use the above algorithm we assume that we know the number of suspicious bits to be bruteforced to recover the correct scalar. This can be determined by using templates to attack some traces, for which we know the randomized key. Furthermore, note that if the attack fails, we can extend the execution to the second most likely suspicious bit and reuse the previously obtained data. Based on our experiments, we determined that the number 54 of suspicious bits should cover all falsely identified bits for at least one trace. Our complete attack works as follows: we run the above algorithm sequentially for each of the n traces. We stop the attack as soon as the time-memory trade-off technique succeeds for one trace.

Since we are running the attack n times, the complexity of the complete attack is multiplied by n. It totals to $O(n \cdot 2^{\frac{s}{2}})$ operations and $O(n \cdot 2^{\frac{s}{2}})$ points in memory. For the attack from the previous section, this corresponds to $100 \cdot 2^{27} = 2^{32}$ operations. Therefore, we conclude that the scalar can be recovered successfully and efficiently even in the presence of multiple errors and uncertain bits (for experimental results see Sect. 6.1). Furthermore, we believe that the above technique may be of independent interest since it can be applied to a commonly used ECSM algorithm, i.e., Montgomery ladder, even if errors are randomly spread across the scalar recovered by the SCA attack.

6.1 Algorithm Implementation and Experimental Results

The first challenge we faced is how to compute the point subtraction in Eq. 2. Curve25519 is a curve in the Montgomery form, and as such, there is an efficient formula for differential point addition using XZ coordinates, but no efficient formula to compute a standard point addition, as far as we know. For that reason, we decided to do the point addition in affine coordinates, which costs a field inversion and a few multiplications. However, to use them we need to know the y-coordinates $y(R)$ and $y([b]P)$. The attack assumes that $x(R)$ (the ECSM output) is known, but $y(R)$ is not, and thus has to be computed. To do so, we use the curve formula directly to compute the two possible values for $y(R)$, at the cost of a field square root, an expensive operation, but it has to be done only once for each value of R. In the case of $y([b]P)$, an efficient algorithm by Okeya and Sakurai [52] costs one field inversion.

To generate the table of precomputed points $A = [a]H$ and to compute $B = [b]P$ in Eq. (2), the naive approach is to compute a full ECSM for each value of a and b. A more efficient method is to apply Gray coding to the suspicious bits in scalars a and b. One property of such a code is that consecutive code words differ in just a single bit, which means that, in our context, we can generate $[k']P$ from $[k]P$ using a single point addition (if the bit changed from 0 to 1) or point subtraction (if the change is from 1 to 0), where k and k' are scalars whose unknown bits are represented as Gray code words, and the code word in k' is the successor of the respective code word in k. To compute the sequence of points $[k_i]P$ $(i = 0, 1 \ldots)$, we first construct the scalar k_0, by setting the unknown bits to zero and the (assumed correct) recovered bits from the output of the SCA attack to their respective values. Then, we apply the full ECSM algorithm to compute $[k_0]P$, and from there we use the aforementioned method to generate the sequence of points $[k_1]P, [k_2]P \ldots$, which costs essentially a point addition per each computed point.

We implemented the key recovery algorithm with the aforementioned arithmetic-level optimizations as a single-threaded program. We tested our implementation in a smaller scale, to recover 40 suspicious bits of a scalar on a PC with 8 GB of RAM total, but only 5 GB available for the program, a i7-3740QM CPU, running at 2.7 GHz. It took 1h23 to recover the correct scalar, where about 1.5 ms is spent to add a single entry to the table and about 3 ms to test a possible value of b. By using these time values as a reference, we estimate that the time for the recovery of a scalar with 60 suspicious bits using the current implementation is around 18 days. The source code of the key recovery implementation is publicly available [46].

7   Conclusions and Possible Countermeasures

In this paper we show that the single-trace data leakage of conditional moves can be exploited to recover the scalar using a template attack. We also show that a similar attack applies to address leakage due to loads and stores from/to secret-dependent addresses. Furthermore, we generalize the method from [26] to tolerate a certain number of incorrectly recovered scalar bits without relying on normal exhaustive search.

Now we discuss possible countermeasures against our attack. We consider evaluating or improving our attack to work against these countermeasures as future work. First of all, note that any countermeasure based on modifying the base point before or during the scalar multiplication does not protect against our attacks, since they aim at exploiting address-dependent and the cswap leakage. Similarly, scalar blinding or splitting does not affect the attack, since we require only one trace and could hence recover the blinded or split scalar. The knowledge of the randomized scalar (or the split scalars) is sufficient to either recover the original scalar or to compute the correct scalar multiplication result. A potential countermeasure against our attack is presented in [50], performing online data randomization during the exponentiation to prevent horizontal collision-correlation attacks. The main idea is to the split scalar to two parts and to randomly interleave two scalar multiplications. However, we believe that our attack might still be mounted if four templates are used to recognize which bit is processed and during which ECSM.

The idea behind Itoh et al. [34] memory-address countermeasure is to store sensitive variables at different memory addresses, but with the same Hamming weight. We believe that although this would cause our attack to be less effective, the addresses leakage may still be identified by template matching. Randomization of memory addresses of the coordinates used in the Montgomery ladder before the ECSM might lead to our attack being less effective, since the templates are prepared assuming fixed addresses. The above countermeasure can be improved by randomizing not only the addresses but also the memory accesses [35–37].

The countermeasure of [30] protects against localized EM template attacks on the ECC Montgomery ladder. The main idea is to randomly swap the ladder registers at the end of a ladder iteration; the addressing of the registers within the loop is inverted according to whether the registers have been swapped. The countermeasure is uniform in its operation sequence, and hence, our template attacks would be infeasible in principle. In addition, several randomization techniques protecting the Montgomery ladder are presented in [41]. Similarly to the countermeasure of [30], these techniques generate operation sequences independent from the scalar. Thus we assume that our attack would be less effective or ineffective against them. We therefore regard as future work evaluating and improving our attacks with respect to the three latter countermeasures.

References

1. Amaxilatis, D.: A generic algorithms library for heterogeneous, distributed, embedded systems. https://github.com/ibr-alg/wiselib
2. Aranha, D.F., Gouvêa, C.P.L.: RELIC is an Efficient LIbrary for Cryptography. https://github.com/relic-toolkit/relic
3. Atmel. Atmega328P datasheet (2016). http://www.atmel.com/devices/atmega32 8p.aspx
4. Bajard, J.-C., Imbert, L., Liardet, P.-Y., Teglia, Y.: Leak resistant arithmetic. In: Joye, M., Quisquater, J.-J. (eds.) CHES 2004. LNCS, vol. 3156, pp. 62–75. Springer, Heidelberg (2004). doi:10.1007/978-3-540-28632-5_5
5. Batina, L., Chmielewski, Ł., Papachristodoulou, L., Schwabe, P., Tunstall, M.: Online template attacks. In: Meier, W., Mukhopadhyay, D. (eds.) INDOCRYPT 2014. LNCS, vol. 8885, pp. 21–36. Springer, Cham (2014). doi:10.1007/978-3-319-13039-2_2
6. Bauer, A., Jaulmes, É.: Correlation analysis against protected SFM implementations of RSA. In: Paul, G., Vaudenay, S. (eds.) INDOCRYPT 2013. LNCS, vol. 8250, pp. 98–115. Springer, Cham (2013). doi:10.1007/978-3-319-03515-4_7

7. Bauer, A., Jaulmes, É., Prouff, E., Reinhard, J., Wild, J.: Horizontal collision correlation attack on elliptic curves - extended version -. Cryptogr. Commun. **7**, 91–119 (2015)
8. Bauer, A., Jaulmes, E., Prouff, E., Wild, J.: Horizontal and vertical side-channel attacks against secure RSA implementations. In: Dawson, E. (ed.) CT-RSA 2013. LNCS, vol. 7779, pp. 1–17. Springer, Heidelberg (2013). doi:10.1007/978-3-642-36095-4_1
9. Bauer, S.: Attacking exponent blinding in RSA without CRT. In: Schindler, W., Huss, S.A. (eds.) COSADE 2012. LNCS, vol. 7275, pp. 82–88. Springer, Heidelberg (2012). doi:10.1007/978-3-642-29912-4_7
10. Benger, N., van de Pol, J., Smart, N.P., Yarom, Y.: "Ooh aah... just a little bit": a small amount of side channel can go a long way. In: Batina, L., Robshaw, M. (eds.) CHES 2014. LNCS, vol. 8731, pp. 75–92. Springer, Heidelberg (2014). doi:10.1007/978-3-662-44709-3_5
11. Brier, E., Clavier, C., Olivier, F.: Correlation power analysis with a leakage model. In: Joye, M., Quisquater, J.-J. (eds.) CHES 2004. LNCS, vol. 3156, pp. 16–29. Springer, Heidelberg (2004). doi:10.1007/978-3-540-28632-5_2
12. Brier, É., Joye, M.: Weierstraß elliptic curves and side-channel attacks. In: Naccache, D., Paillier, P. (eds.) PKC 2002. LNCS, vol. 2274, pp. 335–345. Springer, Heidelberg (2002). doi:10.1007/3-540-45664-3_24
13. CertiVox. MIRACL Cryptographic SDK. https://github.com/CertiVox/MIRACL
14. Chari, S., Rao, J.R., Rohatgi, P.: Template attacks. In: Kaliski, B.S., Koç, K., Paar, C. (eds.) CHES 2002. LNCS, vol. 2523, pp. 13–28, Springer, Heidelberg (2003). doi:10.1007/3-540-36400-5_3
15. Chen, C.-N.: Memory address side-channel analysis on exponentiation. In: Lee, J., Kim, J. (eds.) ICISC 2014. LNCS, vol. 8949, pp. 421–432. Springer, Cham (2015). doi:10.1007/978-3-319-15943-0_25
16. Choudary, O., Kuhn, M.G.: Efficient template attacks. In: Francillon, A., Rohatgi, P. (eds.) CARDIS 2013. LNCS, vol. 8419, pp. 253–270. Springer, Cham (2014). doi:10.1007/978-3-319-08302-5_17
17. Clavier, C., Feix, B., Gagnerot, G., Giraud, C., Roussellet, M., Verneuil, V.: ROSETTA for single trace analysis. In: Galbraith, S., Nandi, M. (eds.) INDOCRYPT 2012. LNCS, vol. 7668, pp. 140–155. Springer, Heidelberg (2012). doi:10.1007/978-3-642-34931-7_9
18. Clavier, C., Feix, B., Gagnerot, G., Roussellet, M., Verneuil, V.: Horizontal correlation analysis on exponentiation. In: Soriano, M., Qing, S., López, J. (eds.) ICICS 2010. LNCS, vol. 6476, pp. 46–61. Springer, Heidelberg (2010). doi:10.1007/978-3-642-17650-0_5
19. Coron, J.-S.: Resistance against differential power analysis for elliptic curve cryptosystems. In: Koç, Ç.K., Paar, C. (eds.) CHES 1999. LNCS, vol. 1717, pp. 292–302. Springer, Heidelberg (1999). doi:10.1007/3-540-48059-5_25
20. Courrège, J.-C., Feix, B., Roussellet, M.: Simple power analysis on exponentiation revisited. In: Gollmann, D., Lanet, J.-L., Iguchi-Cartigny, J. (eds.) CARDIS 2010. LNCS, vol. 6035, pp. 65–79. Springer, Heidelberg (2010). doi:10.1007/978-3-642-12510-2_6
21. Danger, J.-L., Guilley, S., Hoogvorst, P., Murdica, C., Naccache, D.: A synthesis of side-channel attacks on elliptic curve cryptography in smart-cards. J. Cryptogr. Eng. **3**(4), 1–25 (2013)
22. Dugardin, M., Papachristodoulou, L., Najm, Z., Batina, L., Danger, J., Guilley, S., Courrège, J., Therond, C.: Dismantling real-world ECC with horizontal and vertical template attacks. Cryptology ePrint Archive, Report 2015/1001 (2015)

23. Düll, M., Haase, B., Hinterwälder, G., Hutter, M., Paar, C., Sánchez, A.H., Schwabe, P.: High-speed curve25519 on 8-bit, 16-bit and 32-bit microcontrollers. Des. Codes Crypt. **77**(2), 493–514 (2015)
24. Dupaquis, V., Venelli, A.: Redundant modular reduction algorithms. In: Prouff, E. (ed.) CARDIS 2011. LNCS, vol. 7079, pp. 102–114. Springer, Heidelberg (2011). doi:10.1007/978-3-642-27257-8_7
25. Fouque, P.-A., Valette, F.: The doubling attack – *why upwards is better than downwards*. In: Walter, C.D., Koç, Ç.K., Paar, C. (eds.) CHES 2003. LNCS, vol. 2779, pp. 269–280. Springer, Heidelberg (2003). doi:10.1007/978-3-540-45238-6_22
26. Gopalakrishnan, K., Thériault, N., Yao, C.Z.: Solving discrete logarithms from partial knowledge of the key. In: Srinathan, K., Rangan, C.P., Yung, M. (eds.) INDOCRYPT 2007. LNCS, vol. 4859, pp. 224–237. Springer, Heidelberg (2007). doi:10.1007/978-3-540-77026-8_17
27. Hanley, N., Kim, H.S., Tunstall, M.: Exploiting collisions in addition chain-based exponentiation algorithms using a single trace. In: Nyberg, K. (ed.) CT-RSA 2015. LNCS, vol. 9048, pp. 431–448. Springer, Cham (2015). doi:10.1007/978-3-319-16715-2_23
28. Herbst, C., Medwed, M.: Using templates to attack masked montgomery ladder implementations of modular exponentiation. In: Chung, K.-I., Sohn, K., Yung, M. (eds.) WISA 2008. LNCS, vol. 5379, pp. 1–13. Springer, Heidelberg (2009). doi:10.1007/978-3-642-00306-6_1
29. Heyszl, J., Ibing, A., Mangard, S., Santis, F., Sigl, G.: Clustering algorithms for non-profiled single-execution attacks on exponentiations. In: Francillon, A., Rohatgi, P. (eds.) CARDIS 2013. LNCS, vol. 8419, pp. 79–93. Springer, Cham (2014). doi:10.1007/978-3-319-08302-5_6
30. Heyszl, J., Mangard, S., Heinz, B., Stumpf, F., Sigl, G.: Localized electromagnetic analysis of cryptographic implementations. In: Dunkelman, O. (ed.) CT-RSA 2012. LNCS, vol. 7178, pp. 231–244. Springer, Heidelberg (2012). doi:10.1007/978-3-642-27954-6_15
31. Homma, N., Miyamoto, A., Aoki, T., Satoh, A., Shamir, A.: Comparative power analysis of modular exponentiation algorithms. IEEE Trans. Comput. **59**(6), 795–807 (2010)
32. Hutter, M., Schwabe, P.: NaCl on 8-bit AVR microcontrollers. In: Youssef, A., Nitaj, A., Hassanien, A.E. (eds.) AFRICACRYPT 2013. LNCS, vol. 7918, pp. 156–172. Springer, Heidelberg (2013). doi:10.1007/978-3-642-38553-7_9
33. iSec Partners. nano-ecc - a very small ECC implementation for 8-bit microcontrollers (2016). https://github.com/iSECPartners/nano-ecc
34. Itoh, K., Izu, T., Takenaka, M.: Address-bit differential power analysis of cryptographic schemes OK-ECDH and OK-ECDSA. In: Kaliski, B.S., Koç, K., Paar, C. (eds.) CHES 2002. LNCS, vol. 2523, pp. 129–143. Springer, Heidelberg (2003). doi:10.1007/3-540-36400-5_11
35. Itoh, K., Izu, T., Takenaka, M.: A practical countermeasure against address-bit differential power analysis. In: Walter, C.D., Koç, Ç.K., Paar, C. (eds.) CHES 2003. LNCS, vol. 2779, pp. 382–396. Springer, Heidelberg (2003). doi:10.1007/978-3-540-45238-6_30
36. Izumi, M., Ikegami, J., Sakiyama, K., Ohta, K.: Improved countermeasure against address-bit DPA for ECC scalar multiplication. In: 2010 Design, Automation & Test in Europe Conference and Exhibition (DATE 2010), pp. 981–984. IEEE (2010)
37. Izumi, M., Sakiyama, K., Ohta, K.: A new approach for implementing the MPL method toward higher SPA resistance. In: International Conference on Availability, Reliability and Security, ARES 2009, pp. 181–186. IEEE (2009)

38. Kocher, P.C.: Timing attacks on implementations of Diffie-Hellman, RSA, DSS, and other systems. In: Koblitz, N. (ed.) CRYPTO 1996. LNCS, vol. 1109, pp. 104–113. Springer, Heidelberg (1996). doi:10.1007/3-540-68697-5_9

39. Kocher, P., Jaffe, J., Jun, B.: Differential power analysis. In: Wiener, M. (ed.) CRYPTO 1999. LNCS, vol. 1666, pp. 388–397. Springer, Heidelberg (1999). doi:10.1007/3-540-48405-1_25

40. Lange, T., Vredendaal, C., Wakker, M.: Kangaroos in side-channel attacks. In: Joye, M., Moradi, A. (eds.) CARDIS 2014. LNCS, vol. 8968, pp. 104–121. Springer, Cham (2015). doi:10.1007/978-3-319-16763-3_7

41. Le, D.-P., Tan, C.H., Tunstall, M.: Randomizing the montgomery powering ladder. In: Akram, R.N., Jajodia, S. (eds.) WISTP 2015. LNCS, vol. 9311, pp. 169–184. Springer, Cham (2015). doi:10.1007/978-3-319-24018-3_11

42. Liu, A., Ning, P.: TinyECC: A Configurable Library for Elliptic Curve Cryptography in Wireless Sensor Networks (Version 1.0). http://discovery.csc.ncsu.edu/software/TinyECC/ver1.0/index.html

43. Mackay, K.: micro-ecc – ECDH and ECDSA for 8-bit, 32-bit, and 64-bit processors (2016). https://github.com/kmackay/micro-ecc

44. Medwed, M., Oswald, E.: Template attacks on ECDSA. In: Chung, K.-I., Sohn, K., Yung, M. (eds.) WISA 2008. LNCS, vol. 5379, pp. 14–27. Springer, Heidelberg (2009). doi:10.1007/978-3-642-00306-6_2

45. Montgomery, P.L.: Speeding the Pollard and elliptic curve methods of factorization. Math. Comput. **48**(177), 243–264 (1987)

46. Nascimento, E.: SAC 2016 - Implementation of algorithm for ECDLP with errors based on a time-memory tradeoff (2016). https://github.com/enascimento/SCA-ECC-keyrecovery

47. Nascimento, E.: SAC 2016 - Targeted Curve25519 implementations for AVR (2016). https://github.com/enascimento/sac2016-avr-target-impls

48. Nascimento, E., Chmielewski, L., Oswald, D., Schwabe, P.: Attacking embedded ECC implementations through cmov side channels (2016). https://eprint.iacr.org/2016/923

49. Nascimento, E., López, J., Dahab, R.: Efficient and secure elliptic curve cryptography for 8-bit AVR microcontrollers. In: Chakraborty, R.S., Schwabe, P., Solworth, J. (eds.) SPACE 2015. LNCS, vol. 9354, pp. 289–309. Springer, Cham (2015). doi:10.1007/978-3-319-24126-5_17

50. Negre, C., Perin, G.: Trade-off approaches for leak resistant modular arithmetic in RNS. In: Foo, E., Stebila, D. (eds.) ACISP 2015. LNCS, vol. 9144, pp. 107–124. Springer, Cham (2015). doi:10.1007/978-3-319-19962-7_7

51. O'Flynn, C., Chen, Z.D.: ChipWhisperer: an open-source platform for hardware embedded security research. In: Prouff, E. (ed.) COSADE 2014. LNCS, vol. 8622, pp. 243–260. Springer, Cham (2014). doi:10.1007/978-3-319-10175-0_17

52. Okeya, K., Sakurai, K.: Efficient elliptic curve cryptosystems from a scalar multiplication algorithm with recovery of the y-coordinate on a montgomery-form elliptic curve. In: Koç, Ç.K., Naccache, D., Paar, C. (eds.) CHES 2001. LNCS, vol. 2162, pp. 126–141. Springer, Heidelberg (2001). doi:10.1007/3-540-44709-1_12

53. Otte, D.: Avr-crypto-lib (2016). https://git.cryptolib.org/avr-crypto-lib.git

54. Perin, G., Chmielewski, Ł.: A semi-parametric approach for side-channel attacks on protected RSA implementations. In: Homma, N., Medwed, M. (eds.) CARDIS 2015. LNCS, vol. 9514, pp. 34–53. Springer, Cham (2016). doi:10.1007/978-3-319-31271-2_3

55. Perin, G., Imbert, L., Torres, L., Maurine, P.: Attacking randomized exponentiations using unsupervised learning. In: Prouff, E. (ed.) COSADE 2014. LNCS, vol. 8622, pp. 144–160. Springer, Cham (2014). doi:10.1007/978-3-319-10175-0_11

56. Sigma. ECDSA and ECDH cryptographic algorithms for 8-bit AVR microcontrollers. http://www.cmmsigma.eu/products/crypto/crs_avr010x.en.html

57. Walter, C.D.: Sliding windows succumbs to big mac attack. In: Koç, Ç.K., Naccache, D., Paar, C. (eds.) CHES 2001. LNCS, vol. 2162, pp. 286–299. Springer, Heidelberg (2001). doi:10.1007/3-540-44709-1_24

58. Wang, H.: WM-ECC is an Elliptic Curve Cryptography (ECC) primitive suite developed exclusively for wireless sensor motes. http://cis.csuohio.edu/~hwang/WMECC.html

59. Wenger, E., Unterluggauer, T., Werner, M.: 8/16/32 shades of elliptic curve cryptography on embedded processors. In: Paul, G., Vaudenay, S. (eds.) INDOCRYPT 2013. LNCS, vol. 8250, pp. 244–261. Springer, Cham (2013). doi:10.1007/978-3-319-03515-4_16

60. wolfSSL. Embedded Web Server for AVR. https://www.wolfssl.com/wolfSSL/Blog/Entries/2010/11/16_Embedded_Web_Server_for_AVR.html

61. Zhang, Z., Wu, L., Mu, Z., Zhang, X.: A novel template attack on wNAF algorithm of ECC. In: 2014 Tenth International Conference on Computational Intelligence and Security (CIS), pp. 671–675. IEEE (2014)

Lattice Attacks Against Elliptic-Curve Signatures with Blinded Scalar Multiplication

Dahmun Goudarzi[1,2]([✉]), Matthieu Rivain[1], and Damien Vergnaud[2]

[1] CryptoExperts, Paris, France
{dahmun.goudarzi,matthieu.rivain}@cryptoexperts.com
[2] ENS, CNRS, Inria and PSL Research University, Paris, France
damien.vergnaud@ens.fr

Abstract. Elliptic curve cryptography is today the prevailing approach to get efficient public-key cryptosystems and digital signatures. Most of elliptic curve signature schemes use a *nonce* in the computation of each signature and the knowledge of this nonce is sufficient to fully recover the secret key of the scheme. Even a few bits of the nonce over several signatures allow a complete break of the scheme by lattice-based attacks. Several works have investigated how to efficiently apply such attacks when partial information on the nonce can be recovered through side-channel attacks. However, these attacks usually target unprotected implementation and/or make ideal assumptions on the recovered information, and it is not clear how they would perform in a scenario where common countermeasures are included and where only noisy information leaks via side channels. In this paper, we close this gap by applying such attack techniques against elliptic-curve signature implementations based on a blinded scalar multiplication. Specifically, we extend the famous Howgrave-Graham and Smart lattice attack when the nonces are blinded by the addition of a random multiple of the elliptic-curve group order or by a random Euclidean splitting. We then assume that noisy information on the blinded nonce can be obtained through a template attack targeting the underlying scalar multiplication and we show how to characterize the obtained likelihood scores under a realistic leakage assumption. To deal with this scenario, we introduce a filtering method which given a set of signatures and associated likelihood scores maximizes the success probability of the lattice attack. Our approach is backed up with attack simulation results for several signal-to-noise ratio of the exploited leakage.

1 Introduction

In 1985, Koblitz [Kob87] and Miller [Mil86] independently proposed to use the algebraic structure of elliptic curves in public-key cryptography. Elliptic curve cryptography requires smaller keys and it achieves faster computation and memory, energy and bandwidth savings. It is therefore well suited for embedded devices. There exists several digital signature schemes based on the *discrete logarithm problem* in the group of points of an elliptic curve (e.g. [ElG84, Sch91, Nat00]). These schemes use a *nonce k* for each signed message

© Springer International Publishing AG 2017
R. Avanzi and H. Heys (Eds.): SAC 2016, LNCS 10532, pp. 120–139, 2017.
https://doi.org/10.1007/978-3-319-69453-5_7

and compute $[k]P$ for some public point P on the elliptic curve. It is well known that the knowledge of partial information on the nonces used for the generation of several signatures may lead to a total break of the scheme [HS01,Ble00].

Side-channel attacks are a major threat against implementations of cryptographic algorithms [Koc96,KJJ99]. These attacks consists in analyzing the physical leakage of a cryptographic hardware device, such as its power consumption or its electromagnetic emanations. Elliptic curves implementations have been subject to various side-channel attacks. In order to prevent the leakage of partial information on the nonce k from the run of the algorithm that computes scalar multiplication $[k]P$, many countermeasures have been proposed. To thwart simple side-channel analysis, it is customary to ensure a constant (or secret-independent) operation flow (see for instance [Cor99,JY03,IMT02]). To prevent more complex attacks it is necessary to use a probabilistic algorithm to encode the sensitive values such that the cryptographic operations only occur on randomized data. In [Cor99], Coron proposed notably to randomize the scalar k and the projective coordinates of the point P. These countermeasures are nowadays widely used and it is not very realistic to assume that a specific set of bits from the nonces could be recovered in clear by a side-channel attacker.

Related works. Two famous attacks have been designed against elliptic-curve signature schemes that exploit partial information on the nonces of several signatures: the Bleichenbacher's attack [Ble00] and the Howgrave-Graham and Smart's attack [HS01]. Nguyen and Shparlinski [NS02,NS03] proposed a proven variant of Howgrave-Graham and Smart's attack when a single block of consecutive bits is unknown to the adversary. Very few results on the security of elliptic-curve signatures with noisy partial information on the nonces are known. In [LPS04], Leadbitter, Page and Smart considered adversaries that can determine some relation amongst the bits of the secret nonces rather than their specific values (but this relation is known with certainty). This work was recently extended by Faugère et al. in [FGR13]. In [BV15], Bauer and Vergnaud designed an attack where the adversary learns partial information on the nonces but not with perfect certainty (but their attack does not apply to ECDSA or Schnorr signatures).

In [CRR03], Chari et al. introduced the so-called *template attacks* which aim at exploiting all the available side-channel information when the adversary can only obtain a limited number of leakage traces (which is the case in our discrete logarithm setting since a nonce is used only once). Template attacks require that the adversary is able to perform a profiling of the side-channel leakage (*e.g.* based on a copy of the target device under her control). Template attacks against ECDSA were proposed in [MO09,HMHW09,HM09,MHMP13] but none of them considered a blinded implementation of the scalar multiplication.

Our contributions. We consider *practical* attack scenario, where the target implementation is protected with usual countermeasures and where the adversary recovers some noisy information from a signature computation. We consider an elliptic curve signature based on a regular scalar multiplication algorithm which is protected using classic randomization techniques, such as the masking of projective coordinates and the scalar blinding [Cor99,CJ03]. Our contributions are three-fold.

Firstly, we adapt the lattice-based attack proposed by Howgrave-Graham and Smart [HS01] to the setting where the adversary gets partial information on *blinded nonces* of the form $k + r \cdot q$ where r is a small random integer (typically of 32 bits) and q is the elliptic curve group order [Cor99]. We show that the attack works essentially in the same way than the original one but the number of known bits per nonce must be increased by the bit-length of the random r and the number of unknown blocks of consecutive bits.

Afterwards, we consider a scenario where some noisy information is leaked on the bits of the blinded nonces. Under a realistic leakage assumption, the widely admitted *multivariate Gaussian assumption*, we show how to model the information recovered by a template attacker. Specifically, we characterize the distribution of the obtained likelihood scores with respect to a *multivariate signal-to-noise ratio* parameter. We then introduce a filtering method which, given a set of signatures and associated likelihood scores, select the blinded nonce bits to construct the lattice a way to maximize the success probability of the attack. The method relies on a criteria derived from the analysis of the Howgrave-Graham and Smart's attack and techniques from *dynamic programming*.

Finally, we consider a second implementation setting in which the scalar multiplication is protected by the random Euclidean splitting method [CJ03]. In this setting, the nonces k are split as $k = \lceil k/r \rceil \cdot r + (k \bmod r)$ for a (small) random r and the adversary gets partial information on r, $\lceil k/r \rceil$ and $(k \bmod r)$. We show how this partial information can be use to directly get likelihood scores on the nonce bits and we adapt our filtering method to this scenario. For both blinding schemes, we provide some experimental results for our attack based on several values for the multivariate SNR parameter. Our experiments are simulation-based but one could equally use practically-obtained score vectors from a template attack against an actual implementation. The obtained results would be the same for similar multivariate signal-to-noise ratios.

2   Implementation and Leakage Model

2.1   ECDSA Signature Scheme

The ECDSA signature scheme [Nat00] relies on an elliptic curve E defined over some finite field $\mathbb{K}$ where $E(\mathbb{K})$, the group of $\mathbb{K}$-rational points of E, has (almost) prime order q. The public parameters of the scheme include a description of $E(\mathbb{K})$, a base point $\boldsymbol{P} \in E(\mathbb{K})$ that is a generator of the group (or of the large prime-order subgroup), and a cryptographic hash function $H : \{0,1\}^* \mapsto [\![0, 2^{\ell-1}]\!]$, where $\ell := \lceil \log_2 q \rceil$. The secret key x is randomly sampled over $[\![0, q-1]\!] = [0, q) \cap \mathbb{Z}$ and the corresponding public key is set as the point $\boldsymbol{Q} = [x]\boldsymbol{P}$, that is the scalar multiplication of $\boldsymbol{P}$ by x.

A signature $\sigma = (t, s)$ of a message $m \in \{0,1\}^*$ is then computed from the secret key x as $t = \text{xcoord}([k]\boldsymbol{P})$ and $s = k^{-1}(h + t \cdot x) \bmod q$, where k is a random nonce sampled over $[\![1, q-1]\!]$ and $h = H(m)$. One can then verify the signature from the public key $\boldsymbol{Q}$ by computing $u = s^{-1}h \bmod q$ and $v = s^{-1}t \bmod q$, and checking whether $\text{xcoord}([u]\boldsymbol{P} + [v]\boldsymbol{Q}) = t$. A legitimate signature indeed satisfies $[u]\boldsymbol{P} + [v]\boldsymbol{Q} = [s^{-1}(h + t \cdot x) \cdot h \bmod q]\boldsymbol{P} = [k]\boldsymbol{P}$.

2.2 Target Implementation

The attacks presented in this paper target an ECC-signature implementation that relies on a regular scalar multiplication algorithm, which is assumed to be binary in the sense that each loop iteration handles a single bit of the scalar. A prominent example of such a binary regular algorithm is the Montgomery ladder [Mon87] which is widely used for its security and efficiency features [JY03, IMT02, GJM+11].

The target implementation is also assumed to include common countermeasures against side-channel attacks such as as the classic scalar blinding [Cor99] and the Euclidean blinding [CJ03]:

Classic blinding scheme:

1. $r \xleftarrow{\$} [\![0, 2^\lambda - 1]\!]$
2. $a \leftarrow k + r \cdot q$
3. return $[a]\boldsymbol{P}$

Euclidean blinding scheme:

1. $r \xleftarrow{\$} [\![1, 2^\lambda - 1]\!]$
2. $a \leftarrow \lfloor k/r \rfloor;\ b \leftarrow k \bmod r$
3. return $[r]([a]\boldsymbol{P}) + [b]\boldsymbol{P}$

2.3 Leakage Model

The computation performed during one iteration of any regular binary scalar multiplication is deterministic with respect to the current scalar-bit b and the two points $\boldsymbol{P}_0$ and $\boldsymbol{P}_1$ in input of the iteration. The side-channel leakage produced in such an iteration can hence be modeled as a noisy function $\psi(b, \boldsymbol{P}_0, \boldsymbol{P}_1)$. In the following we shall assume that for randomized points $(\boldsymbol{P}_0, \boldsymbol{P}_1)$, the leakage $\psi(b, \boldsymbol{P}_0, \boldsymbol{P}_1)$ can be modeled by a multivariate Gaussian distribution:

$$\psi(b, \boldsymbol{P}_0, \boldsymbol{P}_1) \sim \mathcal{N}(m_b, \Sigma), \tag{1}$$

where m_0 and m_1 are T-dimensional mean leakage vectors and where Σ is a $T \times T$ covariance matrix. In what follows, we shall simply denote this leakage $\psi(b)$. The overall leakage of the scalar multiplication $[a]\boldsymbol{P}$ is hence modeled as $\big(\psi(a_{\ell_a-1}), \ldots, \psi(a_1), \psi(a_0)\big)$ where we further assume the mutual independence between the $\psi(a_i)$ (where the length ℓ_a of a depends on the randomization scheme).

2.4 Profiling Attack

We consider a profiling attacker that owns templates for the leakage of a scalar multiplication iteration w.r.t. the input bit. Based on these templates, the attacker can mount a maximum likelihood attack to recover each bit of the scalar with a given probability. More precisely, the considered attacker can

- measure the side-channel leakage of a scalar multiplication $[a]\boldsymbol{P}$,
- divide the measured traces into sub-traces $\psi(a_i)$,
- compare the measured leakage with templates for $\psi(0)$ and $\psi(1)$, and determine the probability that $a_i = 0$ or $a_i = 1$ given an observation $\psi(a_i)$.

In particular, we consider the ideal case where the attacker knows the exact distribution of $\psi(0)$ and $\psi(1)$, namely he knows the leakage parameters m_0, m_1 and Σ. Although this might be viewed as a strong assumption, efficient techniques exist to derive precise leakage templates in practice, especially in our context where the key-space is of size 2 ($b \in \{0, 1\}$). Even if model-error might lower the probability of correctly recovering a target bit in practice, it would not invalidate the principle of our attacks.

Considering the above leakage model, the probability that the bit a_i equals $b \in \{0, 1\}$ given a leakage sample $\psi(a_i) = x_i$ satisfies

$$p_b(x_i) := \Pr(a_i = b \mid \psi(a_i) = x_i) \propto \exp\left(-\frac{1}{2}(x_i - m_b)^{\mathrm{t}} \cdot \Sigma^{-1} \cdot (x_i - m_b)\right), \quad (2)$$

where $\propto$ means *is equal up to a constant factor*, such that $p_0(x_i) + p_1(x_i) = 1$. Let us define the *multivariate signal-to-noise ratio* (SNR) θ as

$$\theta = \Lambda \cdot (m_0 - m_1), \quad (3)$$

where Λ is the Cholesky decomposition matrix of Σ^{-1}, that is the upper triangular matrix satisfying $\Lambda^{\mathrm{t}} \cdot \Lambda = \Sigma^{-1}$. This decomposition always exists provided that Σ is full-rank (*i.e.* no coordinate variable of the multivariate Gaussian is the exact linear combination of the others) which can be assumed without loss of generality. We have the following result:

Proposition 1. *For every $x_i \in \mathbb{R}^{\mathrm{t}}$*

$$p_0(x_i) \propto \begin{cases} \exp\left(-\frac{1}{2}y_i^{\mathrm{t}}y_i\right) & \text{if } a_i = 0 \\ \exp\left(-\frac{1}{2}(y_i^{\mathrm{t}}y_i + 2\theta^{\mathrm{t}}y_i + \theta^{\mathrm{t}}\theta)\right) & \text{if } a_i = 1 \end{cases} \quad (4)$$

where $y_i = \Lambda \cdot (x_i - m_{a_i})$. Moreover if x_i follows a distribution $\mathcal{N}(m_{a_i}, \Sigma)$, then y_i follows a distribution $\mathcal{N}(\mathbf{0}, I_T)$, where I_T is the identity matrix of dimension T.

The above proposition shows that, for any T-dimensional leakage distribution, the outcome of the considered template attack only depends on the multivariate SNR θ. That is why, in Sect. 6 we provide attack experiments for several values of θ. In practice, one could evaluate the vulnerability of an implementation to our attack by constructing leakage templates for $\psi(0)$ and $\psi(1)$, deriving the corresponding multivariate SNR θ, and simulating probability scores based on Proposition 1.

3   Lattice Attack with Partially-Known Blinded Nonces

In this section, we recall the lattice attack by Howgrave-Graham and Smart [HS01] against ECDSA with partially known nonces, and we extend it to the case where the attacker has partial information on the blinded nonces. The attacker is assumed to collect $n+1$ signatures $\sigma_0, \sigma_1, \ldots, \sigma_n$, for which he knows some bits of the blinded nonces $a_0, a_1, \ldots, a_n$. These blinded nonces are defined as $a_i = k_i + r_i \cdot q$, where k_i is the original (non-blinded) nonce, and r_i is the λ-bit random used in the blinding of k_i. Additionally, we denote by t_i and s_i the two parts of the ECDSA signature $\sigma_i = (t_i, s_i)$ and by h_i the underlying hash value.

3.1   Attack Description

By definition, we have $s_i - a_i^{-1}(h_i + t_i x) \equiv 0 \bmod q$ for every signature σ_i. We can then eliminate the secret key x from the equations since we have

$$x \equiv \frac{a_i s_i - h_i}{t_i} \bmod q \quad \Longrightarrow \quad \frac{a_0 s_0 - h_0}{t_0} \equiv \frac{a_i s_i - h_i}{t_i} \bmod q \qquad (5)$$

for every $i \in \{1, \ldots, n\}$. The above can then be rewritten as

$$a_i + A_i a_0 + B_i \equiv 0 \bmod q \qquad (6)$$

where $A_i = -\frac{s_0 t_i}{s_i t_0} \bmod q$ and $B_i = \frac{h_0 t_i - h_i t_0}{t_0 s_i} \bmod q$.

The goal of the attack is to use the information we have on each a_i to derive a system of equations which can be reduced to a lattice *closest vector problem* (CVP). Let us assume we can obtain several blocks of consecutive bits so that a_i can be expressed as:

$$a_i = \sum_{j=1}^{N} x_{i,j} 2^{\kappa_{i,j}} + x_i' \qquad (7)$$

where x_i' is known, and $x_{i,1}, x_{i,2}, \ldots, x_{i,N}$ are the N unknown blocks. We will denote by $\mu_{i,j}$ the bit-length of each $x_{i,j}$ so that we have $0 \le x_{i,j} < 2^{\mu_{i,j}}$.

We can now rewrite (6) to obtain a system of linear equations in the $x_{i,j}$ as follows:

$$x_{i,1} + \sum_{j=2}^{N} \alpha_{i,j} x_{i,j} + \sum_{j=1}^{N} \beta_{i,j} x_{0,j} + \gamma_i = \eta_i q \qquad (8)$$

for some known coefficients $\alpha_{i,j}, \beta_{i,j}, \gamma_i \in [\![0, q-1]\!]$, and unknown integers $x_{i,j}$ and η_i. Let $C \in \mathcal{M}_{n, N(n+1)-n}(\mathbb{Z})$ be the matrix defined as

$$C = \begin{pmatrix} \alpha_{1,2} \cdots \alpha_{1,N} & & & & \beta_{1,1} \ \beta_{1,2} \cdots \beta_{1,N} \\ & \alpha_{2,2} \cdots \alpha_{2,N} & & & \beta_{2,1} \ \beta_{2,2} \cdots \beta_{2,N} \\ & & \ddots & & \vdots \quad \vdots \quad \ddots \quad \vdots \\ & & & \alpha_{n,2} \cdots \alpha_{n,N} & \beta_{n,1} \ \beta_{n,2} \cdots \beta_{n,N} \end{pmatrix} \qquad (9)$$

and let $\boldsymbol{x} \in \mathbb{Z}^{N(n+1)-n}$ and $\boldsymbol{\eta} \in \mathbb{Z}^n$ be the (column) vectors defined as

$$\boldsymbol{x} = (x_{1,2}, \ldots, x_{1,N} \mid x_{2,2}, \ldots, x_{2,N} \mid \cdots \mid x_{n,2}, \ldots, x_{n,N} \mid x_{0,1}, \ldots, x_{0,N})^{\mathrm{t}},$$
$$\boldsymbol{\eta} = (\eta_1, \eta_2, \ldots, \eta_n)^{\mathrm{t}}.$$

According to (7), we have

$$C \cdot \boldsymbol{x} - q \cdot \boldsymbol{\eta} = -(\gamma_1 + x_{1,1}, \gamma_2 + x_{2,1}, \ldots, \gamma_n + x_{n,1})^{\mathrm{t}} \qquad (10)$$

We consider the lattice $\mathcal{L}$ spanned by the columns of the following matrix:

$$M = \begin{pmatrix} C & -q \cdot I_n \\ -I_{N(n+1)-n} & 0 \end{pmatrix} \qquad (11)$$

where I_n and $I_{N(n+1)-n}$ denote the identity matrices of dimension n and $N(n+1) - n$ respectively. In particular, the vector $\boldsymbol{y}^t = -(\boldsymbol{x}^t | \boldsymbol{\eta}^t) \in \mathbb{Z}^{N(n+1)}$ yields the following lattice vector:

$$M \cdot \boldsymbol{y} = (\gamma_1 + x_{1,1}, \gamma_2 + x_{2,1}, \ldots, \gamma_n + x_{n,1} \mid \boldsymbol{x}^t)^t \qquad (12)$$

which might be close to the non-lattice vector

$$\boldsymbol{v} = (\gamma_1, \gamma_2, \ldots, \gamma_n, 0, 0, \ldots, 0)^t \qquad (13)$$

It is indeed easy to check that the Euclidean distance $\|M \cdot \boldsymbol{y} - \boldsymbol{v}\|$ satisfies

$$\|M \cdot \boldsymbol{y} - \boldsymbol{v}\|^2 = \sum_{\substack{0 \le i \le n \\ 1 \le j \le N}} x_{i,j}^2 \le \sum_{\substack{0 \le i \le n \\ 1 \le j \le N}} 2^{2\mu_{i,j}} \qquad (14)$$

If the distance is small enough, and if the lattice dimension $(n+1)N$ is not too high, one can find $M \cdot \boldsymbol{y}$ as the closest lattice vector to $\boldsymbol{v}$ and hence get the solution to the system. From the latter solution, one can recover the randomized nonces a_i (by (7)) which in turn yields the secret key x (by (5)).

Normalization. We consider unknown blocks $x_{i,j}$ that may be of different bit-lengths $\mu_{i,j}$. Therefore, we shall normalize the matrix M in order to have the same weight in all the coordinates of $M \cdot \boldsymbol{y}$, which shall lead to a better heuristic on the resolution of the lattice problem. To do so, let us define $\rho_{i,j} \in \mathbb{Z}$ as

$$\rho_{i,j} = \frac{1}{2^{\mu_{i,j}}}. \qquad (15)$$

We then consider the lattice $\mathcal{L}'$ spanned by the rows of the matrix $M' = D \cdot M$, where D denotes the following diagonal matrix:

$$D = \mathcal{D}\big((\rho_{i,1})_{i=1}^n \mid (\rho_{1,j})_{j=2}^N \mid (\rho_{n,j})_{j=2}^N \mid (\rho_{0,j})_{j=1}^N\big) \qquad (16)$$

where $\mathcal{D}$ is the function mapping a vector to the corresponding diagonal matrix.[1] The non-lattice vector $\boldsymbol{v}'$ is then defined as

$$\boldsymbol{v}' = D \cdot \boldsymbol{v} = (\rho_{1,1}\gamma_1, \rho_{2,1}\gamma_2, \ldots, \rho_{n,1}\gamma_n, 0, 0, \ldots, 0)^t$$

and the target closest lattice vector is $M' \cdot \boldsymbol{y}$. The distance between these two vectors then satisfies

$$\|M' \cdot \boldsymbol{y} - \boldsymbol{v}'\| = \|D \cdot (M \cdot \boldsymbol{y} - \boldsymbol{v})\| = \Big(\sum_{\substack{0 \le i \le n \\ 1 \le j \le N}} (\rho_{i,j} x_{i,j})^2 \Big)^{\frac{1}{2}} \le \sqrt{(n+1)N}. \qquad (17)$$

---

[1] In practice, we might take $\rho_{i,j} = 2^{\mu_{max} - \mu_{i,j}}$ where $\mu_{max} = \max_{i,j} \mu_{i,j}$ to work over the integers.

3.2   Attack Parameters

We make the classic heuristic assumption that the CVP can be efficiently solved as long as the dimension is not too high and the distance is upper-bounded by

$$\|M' \cdot \boldsymbol{y} - \boldsymbol{v}'\| \leq c_0 \sqrt{\dim(M')} \det(M')^{\frac{1}{\dim(M')}}, \qquad (18)$$

for some constant $c_0 = 1/\sqrt{2\pi e} \simeq 0.24197$ (see for instance [FGR13]). In our attacks, we will actually use the polynomial-time LLL [LLL82] algorithm in order to recover the lattice vector $M' \cdot \boldsymbol{y}$. This gives a slightly worse inequality in (18) and the CVP can be efficiently solved for smaller values c_0 that may actually depend on the dimension of the lattice.

Let us denote by $\mu_i = \sum_j \mu_{i,j}$ the number of unknown bits in a_i, and by $\mu = \sum_i \mu_i$ the total number of unknown bits. Let us also denote by $\delta_i = \ell + \lambda - \mu_i$ the number of known bits in a_i, and by $\delta = \sum_i \delta_i = (n+1)(\ell + \lambda) - \mu$ the total number of known bits. The dimension of the matrix M' is $(n+1)N$ and its determinant is $\det(M') = \det(M) \cdot \det(D) = q^n \cdot 2^{-\mu}$. By (17), the above inequality is satisfied whenever we have[2]

$$1 \leq c_0 (q^n \cdot 2^{-\mu})^{\frac{1}{N_b}} \iff \mu + c_1 N_b \leq n \cdot \log_2(q) < n\ell, \qquad (19)$$

where $N_b = (n+1)N$ denotes the total number of unknown blocks in the $n+1$ signatures, and where c_1 is a constant defined as $c_1 = -\log_2(c_0) > 0$. We can deduce a sufficient condition on the number of bits δ that must be known for the attack to succeed:

$$\delta \geq \ell + (n+1)\lambda + c_1 N_b \iff \sum_{i=0}^{n} (\delta_i - \lambda - c_1 N) \geq \ell. \qquad (20)$$

This means that in order to mount a lattice attack against ECDSA protected with classic randomization of the nonce, one must recover at least $(\lambda + c_1 N)$ bits of each blinded nonce plus some extra bits, where the total amount of extra bits must reach ℓ.

Varying number of Blocks. For the sake of clarity, we described the attack by assuming a constant number N of unknown blocks for all the signatures. Note that the attack works similarly with a varying number of blocks $N_0, N_1, \ldots, N_n$ and the above condition keep unchanged. To see this, simply consider the above description with $N = \max_i N_i$ and $\mu_{i,j} = 0$ for every $j > N_i$. Note however that for different block sizes, we have $N_b = \sum_{i=0}^{n} N_i$ instead of $N_b = (n+1)N$.

3.3   Experiments

The CVP can be solved using Babai's rounding algorithm [Bab86], but in practice one shall prefer the *embedding technique* from [GGH97]. This technique

---

[2] In [HS01], Howgrave-Graham and Smart overlooked the linear dependency in N_b in (19) as they assumed $c_0 = 1$. As we show in Sect. 3.3, this increase in the number of known bits is not an artifact of the technique but is actually necessary.

reduces the CVP to the *shortest vector problem* (SVP), by defining the *embedding lattice* $\mathcal{L}_{\text{emb}}$ spanned by the columns of

$$\begin{pmatrix} M' & v \\ \hline 0 \cdots 0 & -K \end{pmatrix} \tag{21}$$

where K is a constant taken to be of the same order as the greatest coefficient of M'. Informally, if $M' \cdot y \in \mathcal{L}$ is close to v, then $w = \left((M' \cdot y - v)^{\text{t}} \mid K \right)^{\text{t}}$ is a short vector of the embedding lattice $\mathcal{L}_{\text{emb}}$. By applying LLL to $\mathcal{L}_{\text{emb}}$, one can therefore recover w as the shortest vector in the reduced basis, from which $M' \cdot y$ is easily deduced.

In order to validate the constraint on the parameters (see (20)) and determine the actual value of c_1, we experienced the attack using the embedding technique with different set of parameters. Specifically, we used the ANSSI 256-bit elliptic curve ($\ell = 256$), different random sizes ($\lambda \in \{0, 16, 32, 64\}$), different numbers of signatures ($(n + 1) \in \{5, 10, 20\}$), as well as different numbers of blocks per signature ($N \in \{1, 2, 5, 10\}$). In each experiments, the blocks were randomly distributed in the blinded nonces (by randomly sampling the starting index among possible ones). Additionally, we considered that the number of bits per block could vary according to a standard deviation parameter σ as follows. The bit-length $\delta_{i,j}$ of each block is randomly sampled according to the distribution $\mathcal{N}(m, \sigma)$ with $m = \delta/N_b$, where we recall that $\delta = \sum_{i,j} \delta_{i,j}$ is the total number of known bits and $N_b = (n + 1)N$ is the total number of unknown blocks. Samples are then rounded to the nearest integer with rejection for samples lower than 1 (minimum number of bits per block). Finally, samples are randomly incremented or decremented until they sum to the desired value δ. In a nutshell, taking a standard deviation $\sigma = 0$ makes all the $\delta_{i,j}$'s equal to m. On the other hand, taking a standard deviation to $\sigma \approx m$ makes the $\delta_{i,j}$'s to vary over $[1, 2m]$ (with a few ones beyond $2m$). For our experiments, we took $\sigma = 0.1m$ (slight deviation), $\sigma = 0.5m$ (medium deviation), and $\sigma = m$ (strong deviation).

In each setting, the number of known bits is set to $\delta = \ell + (n + 1)\lambda + \tau$ for a varying margin τ that shall correspond to $c_1 N_b$. For each tested value of τ, we record the success rate of the attack over 100 trials. Table 1 summarizes the obtained ratio τ_{95}/N_b where τ_{95} is the margin required to obtain a 95% success rate. This ratio gives a good estimation of the practical value of c_1.

We observe that for the tested parameters, the experimental value of c_1 lies between 3 and 5 most of the time. We also observe that the size of the random (λ) and the deviation on the number of bits per known blocks have a small impact on the resulting c_1, whereas the number of signatures and the total number of blocks have a stronger impact.

Table 1. Ratio τ_{95}/N_b for various set of parameters.

	$(n+1) = 5$				$(n+1) = 10$				$(n+1) = 20$		
$N_b =$	5	10	25	50	10	20	50	100	20	40	100
$\lambda = 0$ (slight dev)	2.80	2.50	2.36	2.90	3.90	3.55	3.58	4.62	5.20	4.90	5.26
$\lambda = 0$ (medium dev)	3.60	2.60	2.56	2.90	4.10	3.30	3.52	3.57	4.85	4.42	4.51
$\lambda = 0$ (strong dev)	3.20	2.50	2.28	2.54	4.10	2.95	3.14	3.07	5.25	4.17	3.93
$\lambda = 16$ (slight dev)	3.40	2.80	2.44	2.90	3.80	3.35	3.62	4.75	5.05	4.95	5.5
$\lambda = 16$ (medium dev)	3.40	2.60	2.40	3.02	4.20	3.15	3.40	4.20	5.25	4.77	4.96
$\lambda = 16$ (strong dev)	3.60	2.50	2.36	2.66	3.70	3.05	3.24	3.29	5.25	4.67	4.28
$\lambda = 32$ (slight dev)	3.80	2.70	2.68	3.06	3.80	3.45	3.74	4.71	4.75	4.70	5.17
$\lambda = 32$ (medium dev)	3.40	2.60	2.60	2.68	3.90	3.10	3.60	4.07	4.95	4.50	5.12
$\lambda = 32$ (strong dev)	3.00	2.90	2.36	2.60	4.00	3.05	3.32	3.41	4.90	4.73	4.62
$\lambda = 64$ (slight dev)	3.00	2.90	2.52	2.98	3.80	3.35	3.44	4.72	4.70	4.77	5.24
$\lambda = 64$ (medium dev)	3.20	2.80	2.36	2.98	3.70	3.55	3.68	4.31	4.80	4.60	5.23
$\lambda = 64$ (strong dev)	3.20	2.80	2.44	2.72	3.50	3.40	3.68	3.78	5.45	4.30	4.75

4   Attacking Implementations with Classic Blinding

In this section we focus on attacking implementations of elliptic-curve signatures that leak side-channel information on the blinded nonce as described in Sect. 2. In this model, one performs a template attack on the randomized scalar multiplication and recovers a probability score $\Pr(a_{i,j} = 0)$ for every bit $a_{i,j}$ of every blinded nonce a_i. The goal is then to select a set of bits among all the recovered *noisy bits* that has the required properties for solving the associated lattice system and that has the highest possible probability of success (*i.e.* all the selected bits must have been correctly guessed based on the probability scores).

For the lattice construction, we shall use the most probable value $\hat{a}_{i,j}$ of each bit $a_{i,j}$, that is

$$\hat{a}_{i,j} = \underset{b \in \{0,1\}}{\operatorname{argmax}} \Pr(a_{i,j} = b), \tag{22}$$

and we shall denote by $p_{i,j}$ the probability that the bit $a_{i,j}$ is correctly guessed, that is

$$p_{i,j} = \Pr(\hat{a}_{i,j} = a_{i,j}) = \max_{b \in \{0,1\}} \Pr(a_{i,j} = b). \tag{23}$$

Let I denote the set of indices corresponding to the selected signatures $\{\sigma_i\}_{i \in I}$ and let J_i denote the set of indices corresponding to the selected guessed bits $\{\hat{a}_{i,j}\}_{i \in J_i}$ within a random nonce for every $i \in I$. Then we will use the bits $\{\hat{a}_{i,j} | i \in I, j \in J_i\}$ to construct the lattice. The CVP's algorithm will only succeed in recovering the right solution if all these bits are correctly guessed, namely if $\hat{a}_{i,j} = a_{i,j}$ for every $i \in I$ and $j \in J_i$. The probability p that these success events happen satisfies

$$p = \prod_{i \in I} p_i \text{ with } p_i = \prod_{j \in J_i} p_{i,j}, \tag{24}$$

where p_i is the probability that all the guessed bits within the random nonce a_i are correct. Our goal is hence to define the sets I, and J_i for every $i \in I$ such that the success probability is maximal and such that the prerequisites of the lattice attack are well met.

Let N_i denote the number of unknown blocks in the blinded nonce a_i, *i.e.* the number of non-adjacent blocks of indices $j \notin J_i$. The dimension Δ of the lattice satisfies $\Delta = \sum_{i \in I} N_i$ (see Sect. 4). Let Δ_{max} be the maximal dimension of a lattice for which the attack can be practical, *i.e.* the LLL-reduction can be done in a reasonable time.[3] The selected bits must then satisfy

$$\sum_{i \in I} N_i \leq \Delta_{max}. \tag{25}$$

Following Sect. 4, we denote by $\delta_i = |J_i|$ the number of recovered bits in the nonce a_i. We recall the necessary condition for the lattice reduction to work:

$$\sum_{i \in I} (\delta_i - \lambda - c_1 N_i) \geq \ell. \tag{26}$$

From the two above conditions, a requirement for the selected bits within a blinded nonce a_i is to satisfy:

$$\frac{\delta_i - \lambda}{N_i} > \frac{\ell}{\Delta_{max}} + c_1. \tag{27}$$

Otherwise, the contribution in terms of known bits is too small with respect to the increase of the lattice dimension. In other words, if the above condition is not satisfied then the maximal dimension Δ_{max} is reached before the number of necessary known bits.

The above condition on the pair (δ_i, N_i) is not sufficient in itself as it does not specify how to select the sets J_i maximizing the success probability. Our goal is now to define a sound criterion on each signature that will allow us to select the best sets J_i *i.e.* the set maximizing the success probability defined in (24). Maximizing this probability is equivalent to maximizing the log-success probability $\log(p) = \sum_{i \in I} \log(p_i)$, and maximizing the latter sum while satisfying (26) can be done by selecting the sets J_i with maximal ratio $\frac{\log p_i}{\delta_i - \lambda - c_1 N_i}$. We hence look for the set of indices J_i that maximizes the following value

$$\gamma_i = p_i^{\frac{1}{\delta_i - \lambda - c_1 N_i}}. \tag{28}$$

Let $f \colon N \mapsto \lceil \frac{\ell \cdot N}{\Delta_{max}} \rceil + c_1 N + \lambda$ be the function that define the minimum number of known bits from a number of unknown blocks for condition (27) to hold. For each signature σ_i and for each possible number of blocks $N \in [\![1, N_{max}]\!]$ we define:

$$J_i(N) = \operatorname*{argmax}_{\substack{J; |J| = f(N) \\ g(J) \leq N}} \Big(\prod_{j \in J_i} p_{i,j} \Big), \tag{29}$$

where the function $g(J)$ gives the number of unknown blocks in the set J.

---

[3] For our experiments, we took $\Delta_{max} = 256$ for which *SageMath* (using the fplll library) performs the reduction within 20 s on a 2.0 GHz Intel Xeon E5649 processor.

We then define

$$p_i(N) = \prod_{j \in J_i(N)} p_{i,j} \text{ and } \gamma_i(N) = p_i(N)^{\frac{1}{f(N) - \lambda - c_1 N}}. \tag{30}$$

Eventually, we shall set

$$N_i = \operatorname*{argmax}_N \gamma_i(N), \gamma_i = \gamma_i(N_i), J_i = J_i(N_i), \text{ and } \delta_i = f(N_i).$$

Remark 1. At this point, a natural idea is to take I as the set of indices with the highest values γ_i. However, this strategy is not reliable in practice. Indeed, it can occur that a bad guess $\hat{a}_{i,j} \neq a_{i,j}$ comes with a strong probability $p_{i,j}$ (*i.e.* one gets a strong confidence in the wrong choice). Since each γ_i aggregates many probabilities $p_{i,j}$, the occurrence of such an extreme event is only marginally correlated to the actual value of γ_i.

In view of the above remark, our approach in practice is to try several random combinations of signatures, namely to take I uniformly at random among the subsets of signatures (tightly) satisfying the constraint (26).

Evaluating Eq. (29). We now explain how to evaluate the function

$$N \mapsto \max_{\substack{J; |J| = f(N) \\ g(J) \leq N}} \left(\prod_{j \in J_i} p_{i,j} \right) \tag{31}$$

for some family of probabilities $p_{i,0}, p_{i,1}, \ldots, p_{i,\ell+\lambda-1}$ (getting the argmax is then straightforward). For this purpose, we define maxp as the function mapping a triplet (j, δ, N) to the above max but with the condition $J \subseteq [\![j, \ell + \lambda - 1]\!]$, $|J| = \delta$, and $g(J) \leq N$, so that the target function in (31) can be rewritten as:

$$N \mapsto \mathsf{maxp}(0, f(N), N) \tag{32}$$

The maxp function can then be evaluated by the following recurrence relation:

$$\mathsf{maxp}(j, \delta, N) = \max \left(\mathsf{block}(j, \delta, N), \mathsf{next}(j, \delta, N) \right) \tag{33}$$

where

$$\mathsf{block}(j, \delta, N) = \max_{w \in [\![1, \delta]\!]} \left(\prod_{k=j}^{j+w-1} p_{i,k} \right) \mathsf{maxp}(j + w + 1, \delta - w, N - 1), \tag{34}$$

$$\mathsf{next}(j, \delta, N) = \mathsf{maxp}(j + 1, \delta, N). \tag{35}$$

The term $\mathsf{block}(j, \delta, N)$ in the above max represents the case where the next w bits are added to the set J, while the term $\mathsf{next}(j, \delta, N)$ is for the case where the next bit is not taken in the set J.[4] The tail of the recursion is fourfold:

$$\mathsf{maxp}(j, \delta, N) = \begin{cases} 0 & \text{if } j + \delta > \ell + \lambda \\ \prod_{k=j}^{\ell+\lambda-1} p_{i,k} & \text{if } j + \delta = \ell + \lambda \\ 0 & \text{if } N = 0 \text{ and } \delta > 0 \\ 1 & \text{if } \delta = 0 \end{cases} \tag{36}$$

---

[4] A particular case occurs when $j = 0$ for which the recursive call to maxp in block does not decrement N since no unknown block appears before the index $j = 0$.

Remark 2. In order to get a higher success probability, we can also use exhaustive search to guess certain bits in a signature. In fact, between two blocks of bits selected by the block function, we can find a small block of bits with poor probability score which will discard the chance of selecting a larger unique block. To tackle this issue and improve the probability of success, we can use exhaustive search to recover these bits with probability 1 (independently of the underlying $p_{i,j}$ values). This improvement of our method is described in Appendix A.

5   Attacking Implementations with Euclidean Blinding

In this section we focus on attacking implementations protected with Euclidean blinding [CJ03], *i.e.* for which the nonces are decomposed as $k_i = a_i \cdot r_i + b_i$ for $i \in \{1, \ldots, n\}$ where r_i is picked uniformly at random over $[\![1, 2^\lambda - 1]\!]$, $a_i = \lceil \frac{k_i}{r_i} \rceil$ and $b_i = k_i \mod r_i$ (see Sect. 3). We suppose that the attacker can recover the probability score for every bit of a_i, r_i and b_i for $i \in \{1, \ldots, n\}$ with the same template attack as in the previous section.

One approach to mount an attack is to use the information we have on each nonce to derive a system of equations which we may solve using a lattice attack. The equations are of degree 2 and the number of involved monomials increases quadratically with the number of unknown blocks. A natural idea is to use the famous Coppersmith method [Cop96b, Cop96a] which is a family of lattice-based techniques to find small integer roots of polynomial equations. The method generates more non-linear equations by multiplication of several non-linear equations before the linearization step in order to improve attacks. However, in this case the structure of the variables is complex and even using polynomials of small degree leads to a lattice of very large dimension.

Our approach is different and practical: from the recovered noisy bits, one can compute probability scores for the individual bits of the nonces k_i themselves. If the probability that the individual bits in a_i, r_i and b_i for $i \in [\![1, n]\!]$ are all correctly guessed is equal to $1/2 + \varepsilon$, for some $\varepsilon > 0$, it can be checked that the obtained probabilities $\Pr(\hat{k}_{i,j} = k_{i,j})$ that the j-th bit of the nonce k_i is correctly guessed tends towards $1/2$ quickly as j grows towards the middle bit $\ell/2$.

To show this fact, we ran the following experiments:

- for several bias $\varepsilon \in \{1/4, 1/8, \ldots, 1/24\}$, we picked uniformly at random 10,000 nonces k_i and random r_i (for $\lambda = 16$) and we computed the corresponding a_i and b_i such that $a_i \cdot r_i + b_i = k_i$;
- we computed noisy versions of $(\tilde{a}_i, \tilde{r}_i, \tilde{b}_i)$ of (a_i, r_i, b_i) as if they were transmitted over a binary symmetric channel with crossover probability ε and we computed $\tilde{k}_i = \tilde{a}_i \cdot \tilde{r}_i + \tilde{b}_i$;
- for each bias and each individual bit, we computed the proportion of $\tilde{k}_{i,j}$ that matches $k_{i,j}$.

The estimated error probability $\Pr(\tilde{k}_{i,j} \neq k_{i,j})$ is plotted in Fig. 1 for the first 64 bits (in log-scale for the x-axis). These experiments show that the probabilities $\Pr(\hat{k}_{i,j} = k_{i,j})$ tends towards $1/2$ exponentially as j comes close to the middle

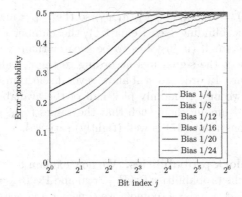

Fig. 1. Estimated error probability of individual nonce bits

bit $\ell/2$. For this reason, in the case of the Euclidean blinding, we will only focus on two particular blocks of k_i for each signature: the block of $\delta_{i,1}$ least significant bits (lsb) and the block of $\delta_{i,2}$ most significant bits (msb), for some $\delta_{i,1}, \delta_{i,2}$. In this model, the necessary condition for the lattice reduction to work (see (20)) becomes:

$$\sum_{i \in I} (\delta_i - c_1) \geq \ell \text{ where } \delta_i = \delta_{i,1} + \delta_{i,2}, \tag{37}$$

since $\lambda = 0$ and $N_i = 1$ (there is a single unknown block per nonce), implying:

$$\delta_i \geq \left\lceil \frac{\ell}{\Delta_{max}} \right\rceil + c_1 \tag{38}$$

as a sound condition on each signature. As seen in the previous sections, the CVP algorithm will only succeed in recovering the right solution if the two blocks of k_i are correctly guessed for each signature. In order to select the blocks and their respective sizes $\delta_{i,1}, \delta_{i,2}$, we use an approach similar to the one proposed in Sect. 5.

Let $B_{i,1}$ and $B_{i,2}$ denote the blocks of $\delta_{i,1}$ lsb and $\delta_{i,2}$ msb of k_i. The guessed blocks are defined as:

$$\hat{B}_{i,j} = \operatorname*{argmax}_{x \in \{0,1\}^{\delta_{i,j}}} \Pr(B_{i,j} = x) \tag{39}$$

for $j \in \{1, 2\}$. The block probabilities are then defined as

$$p_{i,j} = \Pr(B_{i,j} = \hat{B}_{i,j}) = \max_x \Pr(B_{i,j} = x)$$

for $j \in \{1, 2\}$ and the probability for one signature is $p_i = p_{i,1} \cdot p_{i,2}$. Then, we select $\delta_{i,1}$ and $\delta_{i,2}$ such that they maximize the value $\gamma_i = p_i^{\frac{1}{\delta_i - c_1}}$.

Clearly, γ_i depends on the sizes $\delta_{i,1}$ and $\delta_{i,2}$ of the blocks $B_{i,1}$ and $B_{i,2}$. We shall denote $\gamma_i(\delta_{i,1}, \delta_{i,2})$ to see γ_i as a function of these sizes. We then set $\delta_{i,1}$

and $\delta_{i,2}$ as $(\delta_{i,1}, \delta_{i,2}) = \mathrm{argmax}_{(\delta_1,\delta_2)} \gamma_i(\delta_1, \delta_2)$ so that $\gamma_i = \max_{(\delta_1,\delta_2)} \gamma_i(\delta_1, \delta_2)$. Unlike for the classic blinding (see Remark 1), the number of selected bits δ_i per signature can be small (it just has to be greater than c_1). Therefore, it is more relevant to select the signatures according to the γ_i value in the case of the Euclidean blinding. In order to still allow several selection trials, we suggest a hybrid approach: we first randomly pick half of the available signatures and then select a subset I among them such that the values $(\gamma_i)_{i \in I}$ are the highest, and such that the constraint (37) is well (tightly) satisfied.

Computation of block probabilities. We now explain how to evaluate (39), *i.e.* how to evaluate the probabilities $\Pr(B_{i,1} = x)$ and $\Pr(B_{i,2} = x)$. For the sake of simplicity, we drop the index i, namely we consider a nonce $k = a \cdot r + b$ for which we know some probability score $\Pr(a_j = 1)$, $\Pr(r_j = 1)$ and $\Pr(b_j = 1)$ for the bits of a, r, and b. Moreover, for some integer v, we shall denote $v_{[j_0 : j_1]} = \lfloor \frac{v}{2^{j_0}} \rfloor \bmod 2^{j_1 - j_0}$, *i.e.* the integer value composed of the j_0-th bit to the $(j_1 - 1)$-th bit of v. It is easy to check that the block $B_1 = k_{[0:\delta_1]}$ only depends on the δ_1 lsb of a, r and b. We can then compute the probability $\Pr(B_1 = x)$ as:

$$\Pr(B_1 = x) = \sum_{x,y,z} \chi_1(x, w, y, z) \prod_{j=0}^{\delta_1 - 1} \Pr(a_j = w_j) \Pr(r_j = y_j) \Pr(b_j = z_j) \quad (40)$$

with $\chi_1(x, w, y, z) = 1$ if $x = (w \cdot y + z)_{[0:\delta_1]}$ and $\chi(x, w, y, z) = 0$ otherwise.

For the case of the block $B_2 = k_{[\ell - \delta_2 : \ell]}$, we shall denote by a_h and r_h the δ msb of a and r respectively for some δ (*i.e.* $a_h = a_{[\ell - \lambda - \delta : \ell - \lambda]}$ and $r_h = r_{[\lambda - \delta : \lambda]}$). We then have

$$k_{[\ell - 2\delta : \ell]} = a_h r_h + c \quad \text{where} \quad c = \left\lfloor \frac{a \cdot r + b - a_h r_h 2^{\ell - 2\delta}}{2^{\ell - 2\delta}} \right\rfloor. \quad (41)$$

It can be checked that the carry c is lower than $2^{\delta + 1}$. Then if we take δ sufficiently greater than δ_2, we get $\tilde{B}_2 = B_2$ with high probability, where $\tilde{B}_2$ denotes the δ_2 msb of $a_h r_h$ (*i.e.* $\tilde{B}_2 = (a_h r_h)_{[2\delta - \delta_2 : 2\delta]}$). We then approximate

$$\Pr(B_2 = x) \approx \Pr(\tilde{B}_2 = x) = \sum_{x_0 = 0}^{2^{2\delta - \delta_1}} \Pr(a_h r_h = x_0 + 2^{2\delta - \delta_1} x), \quad (42)$$

where the probability mass function of $a_h r_h$ satisfies

$$\Pr(a_h r_h = x) \doteq \sum_{w,y} \chi_2(x, w, y) \prod_{j=0}^{\delta - 1} \Pr(a_{\ell - \lambda - \delta + j} = w_j) \Pr(r_{\lambda - \delta + j} = y_j) \quad (43)$$

where $\chi_2(x, w, y) = 1$ if $x = w \cdot y$ and $\chi_2(x, w, y) = 0$ otherwise.

The above approximation is sound as long as we take δ sufficiently greater than δ_2. Since we have $c \le 2^{\delta + 1}$, it can be checked that the approximation error is upper bounded by $2^{\delta_1 + 1 - \delta}$. We shall then take δ such that the latter approximation error does not impact the selection of the maximal probability $\Pr(B_2 = x)$.

6   Experimental Results

This section provides experimental results: we have implemented our attacks against ECDSA on the ANSSI 256-bit elliptic curve (*i.e.* $\ell = 256$) with the two considered blinding schemes for 3 different random sizes, specifically $\lambda \in \{16, 32, 64\}$. We considered a three-dimensional leakage model with unbalanced multivariate SNR $\theta = \alpha \cdot (0.5, 1, 2)$, where $\alpha \in \{1.5, 2.0\}$. For each setting, the used signatures and blinded nonces were randomly generated, and the corresponding likelihood scores were simulated from the multivariate SNR θ as shown in Proposition 1. We then applied our filtering methods to select the best blocks of bits from which we constructed the Howgrave-Graham and Smart lattice with the constant c_1 set to 4, and we checked whether the correct solution was well retrieved by the CVP algorithm. We experimented our attacks with n_{sig} available signatures and n_{tr} trials for the subset selection (and the underlying lattice reduction) for $(n_{sig}, n_{tr}) \in \{(10, 1), (20, 5), (20, 10), (100, 10), (100, 50), (100, 100)\}$.

The obtained success rates are reported in Table 2. The lattice reduction worked almost every time the selected bits were correctly guessed (we noticed only a few lattice failures in all the performed experiments). These results suggest that the classic blinding is more sensitive to our attack than the Euclidean blinding. However, one should note that the classic blinding is inefficient when the group order is sparse. We also observe that for the Euclidean blinding, the random size λ has a small impact on the resulting success rate, which is not surprising since it can be checked from Sect. 5 that this parameter is not expected to play a significant role.

Table 2. Success rate of our attacks.

(n_{sig}, n_{tr})		(10,1)	(20, 5)	(20, 10)	(100, 10)	(100, 50)	(100, 100)
Classic blinding							
$\alpha = 1.5$	$\lambda = 16$	13.5%	38.3%	54.0%	70.1%	99.0%	99.9%
	$\lambda = 32$	3.5%	13.6%	22.7%	27.8%	73.9%	91.9%
	$\lambda = 64$	0.2%	0.6%	1.2%	1.5%	6.2%	11.7%
$\alpha = 2$	$\lambda = 16$	91.2%	99.9%	100%	100%	100%	100%
	$\lambda = 32$	90.5%	99.5%				
	$\lambda = 64$	85.7%	99.3%				
Euclidean blinding							
$\alpha = 1.5$	$\lambda = 16$	0%	0%	0%	0%	0%	0%
	$\lambda = 32$						
	$\lambda = 64$						
$\alpha = 2$	$\lambda = 16$	0.7%	3.1%	5.8%	42.8%	76.8%	83.3%
	$\lambda = 32$	0.1%	0.4%	0.8%	41.1%	74.9%	82.6%
	$\lambda = 64$	0.1%	0.4%	1.0%	40.2%	75.0%	82.8%

Acknowledgments. The authors were supported in part by the French ANR JCJC ROMAnTIC project (ANR-12-JS02-0004).

A   Using Exhaustive Search

In order to increase the success probability of our attack, we can make use of exhaustive search to guess certain bits in a signature. Specifically, we can select m bits among all the signatures for which we shall try the 2^m possible values and apply the previously defined lattice attack. To select the guessed bits in a signature we follow the same approach as previously but with a new dimension for the parameters: the number of bits allowed to be exhaustively guessed. Specifically, we shall compute the optimal parameters γ_i, J_i and N_i for each possible $m_i \leq m_{max}$, where m_i is the number of bits exhaustively guessed in the i-th signature and m_{max} is the maximal value for m. We hence get a new constraint to the selection of the signatures, that is:

$$\sum_{i=0}^{n} m_i \leq m_{max}. \tag{44}$$

Let us introduce the function π_i defined for every $J \subseteq [\![0, \ell + \lambda - 1]\!]$ and $m \in \mathbb{N}$ as:

$$\pi_i(J, m) = \max_{\substack{J' \subseteq J \\ |J'| = |J| - m}} \prod_{j \in J'} p_{i,j}. \tag{45}$$

Namely, $\pi_i(J, m)$ is the maximal product of $|J| - m$ probabilities among $(p_{i,j})_{j \in J}$. This gives the probability of correctly guessing the bits $(a_{i,j})_{j \in J}$ while m among them are exhaustively guessed. We then define

$$J_i(m, N) = \operatorname*{argmax}_{\substack{J; |J| = f(N) \\ g(J) \leq N}} \pi_i(J, m) \tag{46}$$

and

$$p_i(m, N) = \pi_i(J_i(m, N), m), \gamma_i(m, N) = p_i(m, N)^{\frac{1}{f(N) - \lambda - c_1 N}}, \tag{47}$$

from which we set

$$\gamma_i(m) = \max_N \gamma_i(m, N), N_i(m) = \operatorname*{argmax}_N \gamma_i(m, N), \text{ and } J_i(m) = J_i(m, N_i(m)). \tag{48}$$

Eventually, we select I and $(J_i(m_i))_{i \in I}$ such that (44) and

$$\ell \leq \sum_{i \in I} |J_i(m_i)| - \lambda - c_1 N_i \tag{49}$$

are both satisfied.

Evaluating Eq. (46). We now explain how to evaluate the function

$$(m, N) \mapsto \max_{\substack{J; |J| = f(N) \\ g(J) \leq N}} \pi_i(J, m), \tag{50}$$

from which getting the argmax is then straightforward. For such a purpose, we extend the maxp function of Sect. 4 to deal with the m parameter *i.e.* the number of bits that can be exhaustively guessed. Here, maxp maps a quadruple (j, δ, N, m) to the above max with the condition $J \subseteq [\![j, \ell + \lambda - 1]\!]$, $|J| = \delta$, and $g(J) \le N$, so that the target function in (50) can be rewritten as:

$$(m, N) \mapsto \mathsf{maxp}(0, f(N), N, m) \tag{51}$$

The difference with the original method is that in a recursive call to the maxp function, one can spend v out of m bits to be exhaustively guessed. Specifically, the recurrence relation becomes:

$$\mathsf{maxp}(j, \delta, N, m) = \max(\mathsf{block}(j, \delta, N, m), \mathsf{next}(j, \delta, N, m)) \tag{52}$$

with

$$
\begin{aligned}
\mathsf{block}(j, \delta, N, m) &= \max_{\substack{w \in [\![1, \delta]\!] \\ v \in [\![0, m]\!]}} \pi_i([\![j, j + w - 1]\!], v) \\
&\qquad \times \mathsf{maxp}(j + w + 1, \delta - w, N - 1, m - v) \\
\mathsf{next}(j, \delta, N, m) &= \mathsf{maxp}(j + 1, \delta, N, m)
\end{aligned}
$$

The recursion tail is pretty similar to the original case:

$$
\mathsf{maxp}(j, \delta, N, m) = \begin{cases}
0 & \text{if } j + \delta > \ell + \lambda \\
\pi_i([\![j, \ell + \lambda]\!], m) & \text{if } j + \delta = \ell + \lambda - 1 \\
0 & \text{if } N = 0 \text{ and } \delta > 0 \\
1 & \text{if } \delta = 0
\end{cases} \tag{53}
$$

Only the second case changes, where the m left bits of exhaustive search are spend for the final block.

References

[Bab86] Babai, L.: On Lovász' lattice reduction and the nearest lattice point problem. Combinatorica **6**(1), 1–13 (1986)

[Ble00] Bleichenbacher, D.: On the generation of one-time keys in dl signature schemes. Presentation at IEEE P1363 Working Group meeting, Unpublished, November 2000

[BV15] Bauer, A., Vergnaud, D.: Practical key recovery for discrete-logarithm based authentication schemes from random nonce bits. In: Güneysu, T., Handschuh, H. (eds.) CHES 2015. LNCS, vol. 9293, pp. 287–306. Springer, Heidelberg (2015). doi:10.1007/978-3-662-48324-4_15

[CJ03] Ciet, M., Joye, M.: (Virtually) free randomization techniques for elliptic curve cryptography. In: Qing, S., Gollmann, D., Zhou, J. (eds.) ICICS 2003. LNCS, vol. 2836, pp. 348–359. Springer, Heidelberg (2003). doi:10.1007/978-3-540-39927-8_32

138   D. Goudarzi et al.

[Cop96a] Coppersmith, D.: Finding a small root of a bivariate integer equation; factoring with high bits known. In: Maurer, U. (ed.) EUROCRYPT 1996. LNCS, vol. 1070, pp. 178–189. Springer, Heidelberg (1996). doi:10.1007/3-540-68339-9_16

[Cop96b] Coppersmith, D.: Finding a small root of a univariate modular equation. In: Maurer, U. (ed.) EUROCRYPT 1996. LNCS, vol. 1070, pp. 155–165. Springer, Heidelberg (1996). doi:10.1007/3-540-68339-9_14

[Cor99] Coron, J.-S.: Resistance against differential power analysis for elliptic curve cryptosystems. In: Koç, Ç.K., Paar, C. (eds.) CHES 1999. LNCS, vol. 1717, pp. 292–302. Springer, Heidelberg (1999). doi:10.1007/3-540-48059-5_25

[CRR03] Chari, S., Rao, J.R., Rohatgi, P.: Template attacks. In: Kaliski, B.S., Koç, K., Paar, C. (eds.) CHES 2002. LNCS, vol. 2523, pp. 13–28. Springer, Heidelberg (2003). doi:10.1007/3-540-36400-5_3

[ElG84] ElGamal, T.: A public key cryptosystem and a signature scheme based on discrete logarithms. In: Blakley, G.R., Chaum, D. (eds.) CRYPTO 1984. LNCS, vol. 196, pp. 10–18. Springer, Heidelberg (1985). doi:10.1007/3-540-39568-7_2

[FGR13] Faugère, J.-C., Goyet, C., Renault, G.: Attacking (EC)DSA given only an implicit hint. In: Knudsen, L.R., Wu, H. (eds.) SAC 2012. LNCS, vol. 7707, pp. 252–274. Springer, Heidelberg (2013). doi:10.1007/978-3-642-35999-6_17

[GGH97] Goldreich, O., Goldwasser, S., Halevi, S.: Public-key cryptosystems from lattice reduction problems. In: Kaliski, B.S. (ed.) CRYPTO 1997. LNCS, vol. 1294, pp. 112–131. Springer, Heidelberg (1997). doi:10.1007/BFb0052231

[GJM+11] Goundar, R.R., Joye, M., Miyaji, A., Rivain, M., Venelli, A.: Scalar multiplication on Weierstraß elliptic curves from Co-Z arithmetic. J. Cryptogr. Eng. $\mathbf{1}(2)$, 161–176 (2011)

[HM09] Herbst, C., Medwed, M.: Using templates to attack masked montgomery ladder implementations of modular exponentiation. In: Chung, K.-I., Sohn, K., Yung, M. (eds.) WISA 2008. LNCS, vol. 5379, pp. 1–13. Springer, Heidelberg (2009). doi:10.1007/978-3-642-00306-6_1

[HMHW09] Hutter, M., Medwed, M., Hein, D., Wolkerstorfer, J.: Attacking ECDSA-enabled RFID devices. In: Abdalla, M., Pointcheval, D., Fouque, P.-A., Vergnaud, D. (eds.) ACNS 2009. LNCS, vol. 5536, pp. 519–534. Springer, Heidelberg (2009). doi:10.1007/978-3-642-01957-9_32

[HS01] Howgrave-Graham, N., Smart, N.P.: Lattice attacks on digital signature schemes. Des. Codes Cryptogr. $\mathbf{23}$(3), 283–290 (2001)

[IMT02] Izu, T., Möller, B., Takagi, T.: Improved elliptic curve multiplication methods resistant against side channel attacks. In: Menezes, A., Sarkar, P. (eds.) INDOCRYPT 2002. LNCS, vol. 2551, pp. 296–313. Springer, Heidelberg (2002). doi:10.1007/3-540-36231-2_24

[JY03] Joye, M., Yen, S.-M.: The montgomery powering ladder. In: Kaliski, B.S., Koç, K., Paar, C. (eds.) CHES 2002. LNCS, vol. 2523, pp. 291–302. Springer, Heidelberg (2003). doi:10.1007/3-540-36400-5_22

[KJJ99] Kocher, P., Jaffe, J., Jun, B.: Differential power analysis. In: Wiener, M. (ed.) CRYPTO 1999. LNCS, vol. 1666, pp. 388–397. Springer, Heidelberg (1999). doi:10.1007/3-540-48405-1_25

[Kob87] Koblitz, N.: Elliptic curve cryptosystems. Math. Comput. $\mathbf{48}$(177), 203–209 (1987)

[Koc96] Kocher, P.C.: Timing attacks on implementations of Diffie-Hellman, RSA, DSS, and other systems. In: Koblitz, N. (ed.) CRYPTO 1996. LNCS, vol. 1109, pp. 104–113. Springer, Heidelberg (1996). doi:10.1007/3-540-68697-5_9

[LLL82] Lenstra, A., Lenstra, H., Lovász, L.: Factoring polynomials with rational coefficients. Math. Ann. **261**, 515–534 (1982)

[LPS04] Leadbitter, P.J., Page, D., Smart, N.P.: Attacking DSA under a repeated bits assumption. In: Joye, M., Quisquater, J.-J. (eds.) CHES 2004. LNCS, vol. 3156, pp. 428–440. Springer, Heidelberg (2004). doi:10.1007/978-3-540-28632-5_31

[MHMP13] De Mulder, E., Hutter, M., Marson, M.E., Pearson, P.: Using Bleichenbacher's solution to the hidden number problem to attack nonce leaks in 384-bit ECDSA. In: Bertoni, G., Coron, J.-S. (eds.) CHES 2013. LNCS, vol. 8086, pp. 435–452. Springer, Heidelberg (2013). doi:10.1007/978-3-642-40349-1_25

[Mil86] Miller, V.S.: Use of elliptic curves in cryptography. In: Williams, H.C. (ed.) CRYPTO 1985. LNCS, vol. 218, pp. 417–426. Springer, Heidelberg (1986). doi:10.1007/3-540-39799-X_31

[MO09] Medwed, M., Oswald, E.: Template attacks on ECDSA. In: Chung, K.-I., Sohn, K., Yung, M. (eds.) WISA 2008. LNCS, vol. 5379, pp. 14–27. Springer, Heidelberg (2009). doi:10.1007/978-3-642-00306-6_2

[Mon87] Montgomery, P.L.: Speeding the Pollard and elliptic curve methods of factorization. Math. Comput. **48**, 243–264 (1987)

[Nat00] National Institute for Standards and Technology. FIPS PUB 186-2: Digital Signature Standard (DSS). National Institute for Standards and Technology, Gaithersburg (2000)

[NS02] Nguyen, P.Q., Shparlinski, I.: The insecurity of the digital signature algorithm with partially known nonces. J. Cryptol. **15**(3), 151–176 (2002)

[NS03] Nguyen, P.Q., Shparlinski, I.: The insecurity of the elliptic curve digital signature algorithm with partially known nonces. Des. Codes Cryptogr. **30**(2), 201–217 (2003)

[Sch91] Schnorr, C.-P.: Efficient signature generation by smart cards. J. Cryptol. **4**(3), 161–174 (1991)

Loop-Abort Faults on Lattice-Based Fiat-Shamir and Hash-and-Sign Signatures

Thomas Espitau[1], Pierre-Alain Fouque[2], Benoît Gérard[3],
and Mehdi Tibouchi[4(✉)]

[1] École normale supérieure de Cachan & Sorbonne Universités,
UPMC Univ Paris 06, LIP6, Paris, France
tespitau@ens-cachan.fr
[2] Institut Universitaire de France & IRISA & Université de Rennes I, Rennes, France
pierre-alain.fouque@univ-rennes1.fr
[3] DGA.MI and IRISA, Rennes, France
benoit.gerard@irisa.fr
[4] NTT Secure Platform Laboratories, Tokyo, Japan
tibouchi.mehdi@lab.ntt.co.jp

Abstract. Although postquantum cryptography is of growing practical concern, not many works have been devoted to implementation security issues related to postquantum schemes.

In this paper, we look in particular at fault attacks against implementations of lattice-based signature schemes, looking both at Fiat-Shamir type constructions (particularly BLISS, but also GLP, PASSSing and Ring-TESLA) and at hash-and-sign schemes (particularly the GPV-based scheme of Ducas–Prest–Lyubashevsky). These schemes include essentially all practical lattice-based signatures, and achieve the best efficiency to date in both software and hardware. We present several fault attacks against those schemes yielding a full key recovery with only a few or even a single faulty signature, and discuss possible countermeasures to protect against these attacks.

Keywords: Fault attacks · Digital signatures · Postquantum cryptography · Lattices · BLISS · GPV

1 Introduction

Lattice-based cryptography. Recent progress in quantum computation [7], the NSA advisory memorandum recommending the transition away from Suite B and to postquantum cryptography [1], as well as the announcement of the NIST standardization process for postquantum cryptography [6] all suggest that research on postquantum schemes, which is already plentiful but mostly focused on theoretical constructions and asymptotic security, should increasingly take into account real world implementation issues.

Among all postquantum directions, lattice-based cryptography occupies a position of particular interest, as it relies on well-studied problems and comes

© Springer International Publishing AG 2017
R. Avanzi and H. Heys (Eds.): SAC 2016, LNCS 10532, pp. 140–158, 2017.
https://doi.org/10.1007/978-3-319-69453-5_8

with uniquely strong security guarantees, such as worst-case to average-case reductions [35]. A number of works have also focused on improving the performance of lattice-based schemes, and actual implementation results suggest that properly optimized schemes may be competitive with, or even outperform, classical factoring- and discrete logarithm-based cryptography.

The literature on the underlying number-theoretic problems of lattice-based cryptography is extensive (even though concrete bit security is not nearly as well understood as for factoring and discrete logarithms; in addition, ring-based schemes have recently been subjected to new families of attacks that might eventually reduce their security, especially in the postquantum setting). On the other hand, there is currently a distinct lack of cryptanalytic results on the *physical* security of implementations of lattice-based schemes (or in fact, postquantum schemes in general! [39]). It is well-known that physical attacks, particularly against public-key schemes, are often simpler, easier to mount and more devastating than attacks targeting underlying hardness assumptions: it is often the case that a few bits of leakage or a few fault injections can reveal an entire secret key (the well-known attacks from [3,5] are typical examples). We therefore deem it important to investigate how fault attacks may be leveraged to recover secret keys in the lattice-based setting, particularly against signature schemes as signatures are probably the most likely primitive to be deployed in a setting where fault attacks are relevant, and have also received the most attention in terms of efficient implementations both in hardware and software.

Practical implementations of lattice-based signatures. Efficient signature schemes are typically proved secure in the random oracle model, and can be roughly divided in two families: the hash-and-sign family (which includes schemes like FDH and PSS), as well as signatures based on identification schemes, using the Fiat-Shamir heuristic or a variant thereof. Efficient lattice-based signatures can also be divided along those lines, as observed for example in the survey of practical lattice-based digital signature schemes presented by O'Neill and Güneysu at the NIST workshop on postquantum cryptography [23,24].

The Fiat-Shamir family is the most developed, with a number of schemes coming with concrete implementations in software, and occasionally in hardware as well. Most schemes in that family follow Lyubashevsky's "Fiat-Shamir with aborts" paradigm [26], which uses rejection sampling to ensure that the underlying identification scheme achieves honest-verifier zero-knowledge. Among lattice-based schemes, the exemplar in that family is Lyubashevsky's scheme from EUROCRYPT 2012 [27]. It is, however, of limited efficiency, and had to be optimized to yield practical implementations. This was first carried out by Güneysu et al., who described an optimized hardware implementation of it at CHES 2012 [20], and then to a larger extent by Ducas et al. in their scheme BLISS [9], which includes a number of theoretical improvements and is the top-performing lattice-based signature. It was also implemented in hardware by Pöppelmann et al. [36]. Other schemes in that family include Hoffstein et al.'s PASSSign [22], which incorporates ideas from NTRU, and Akleylek et al.'s Ring-TESLA [2], which boasts a tight security reduction.

On the hash-and-sign side, there were a number of early proposals with heuristic security (and no actual security proofs), particularly GGH [18] and NTRUSign [21], but despite several attempts to patch them[1] they turned out to be insecure. A principled, provable approach to designing lattice-based hash-and-sign signatures was first described by Gentry et al. in [16], based on discrete Gaussian sampling over lattices. The resulting scheme, GPV, is rather inefficient, even when using faster techniques for lattice Gaussian sampling [30]. However, Ducas et al. [11] later showed how it could be optimized and instantiated over NTRU lattices to achieve a relatively efficient scheme with particularly short signature size. The DLP scheme is somewhat slower than BLISS in software, but still a good contender for practical lattice-based signatures, and seemingly the only one in the hash-and-sign family.

Our contributions. In this work, we initiate the study of fault attacks against lattice-based signature schemes, and obtain attacks against all the practical schemes mentioned above.

As noted previously, early lattice-based signature schemes with heuristic security have been broken using standard attacks [15,17,32] but recent constructions including [9,11,16,26,27] are provably secure, and cryptanalysis therefore requires a more powerful attack model. In this work we consider fault attacks.

We present two attacks, both using a similar type of faults which allows the attacker to cause a loop inside the signature generation algorithm to abort early. Successful loop-abort faults have been described many times in the literature, including against DSA [31] and pairing computations [34], and in our attacks they can be used to recover information about the private signing key. The underlying mathematical techniques used to actually recover the key, however, are quite different in the two attacks.

Our first attack applies to the schemes in the Fiat-Shamir family: we describe it against BLISS [9,36], and show how it extends to GLP [20], PASSSign [22] and Ring-TESLA [2]. In that attack, we inject a fault in the loop that generates the random "commitment value" y of the sigma protocol associated with the Fiat-Shamir signature scheme. That commitment value is a random polynomial generated coefficient by coefficient, and an early loop abort causes it to have abnormally low degree, so that the protocol is no longer zero-knowledge. In fact, this will usually leak enough information that *a single faulty signature is enough to recover the entire signing key.* More specifically, we show that the faulty signature can be used to construct a point that is very close to a vector in a suitable integer lattice of moderate dimension, and such that the difference is essentially (a subset of) the signing key, which can thus be recovered using lattice reduction.

---

[1] There is a provably secure scheme due to Melchor et al. [29] that claims to "seal the leak on NTRUSign", but it actually turns the construction into a Fiat-Shamir type scheme, using rejection sampling à la Lyubashevsky.

Our second attack targets the GPV-based hash-and-sign signature scheme of Ducas et al. [11]. In that case, we consider early loop abort faults against the discrete Gaussian sampling in the secret trapdoor lattice used in signature generation. The early loop abort causes the signature to be a linear combination of the last few rows of the secret lattice. A few faulty signatures can then be used to recover the span of those rows, and using the special structure of the lattice, we can then use lattice reduction to find one of the rows up to sign, which is enough to completely reconstruct the secret key. In practice, if we can cause loop aborts after up to m iterations, we find that $m + 2$ *faulty signatures are enough for full key recovery* with high probability.

Both of our attacks are supported by extensive simulations in Sage [38], whose source code is provided in the full version of this paper [13].

We also take a close look at the concrete software and hardware implementations of the schemes above, and discuss the concrete feasibility of injecting the required loop-abort faults in practice. We find the attacks to be highly realistic. Finally, we discuss several possible countermeasures to protect against our attacks.

Related work. To the best of our knowledge, the first previous work on fault attacks against lattice-based signatures, and in particular the only one mentioned in the survey of Taha and Eisenbarth [39], is the fault analysis work of Kamal and Youssef on NTRUSign [25]. It is, however, of limited interest since NTRUSign is known to be broken [12,32]; it also suffers from a very low probability of success.

Much more recently, a relevant preprint has also been made available online by Bindel et al. [4] concurrently with this work. That paper proposes various fault attacks against the same Fiat-Shamir type schemes that we consider in this paper. Most of the attacks, however, are either in a contrived model (targeting key generation), or require unrealistically many faults and are arguably straightforward (bypassing rejection sampling in signature generation or size/-correctness checks in signature verification). One attack described in the paper can be seen as posing a serious threat, namely the one described in [4, Sect. IV-B], but it amounts to a weaker variant of our Fiat-Shamir attack, using simple linear algebra rather than lattice reduction. As a result, it requires several hundred faulty signatures, whereas our attack needs only one.

Another interesting concurrent work is the recent cache attack against BLISS of Bruinderink et al. [19]. It uses cache side-channels to extract information about the coefficients of the commitment polynomial **y**, and then lattice reduction to recover the signing key based on that side-channel information. In that sense, it is similar to our Fiat-Shamir attack. However, since the nature of the information to be exploited is quite different than in our setting, the mathematical techniques are also quite different. In particular, again, in contrast with our fault attack, that cache attack requires many signatures for a successful key recovery.

2   Description of the Lattice-Based Signature Schemes We Consider

Notation. For any integer q, we represent the ring $\mathbb{Z}_q$ by $[-q/2, q/2) \cap \mathbb{Z}$. Vectors are considered as column vectors and will be written in bold lower case letters and matrices with upper case letters. By default, we will use the ℓ_2 Euclidean norm, $\|\mathbf{v}\|_2 = (\sum_i v_i^2)^{1/2}$ and ℓ_∞-norm as $\|\mathbf{v}\|_\infty = \max_i |v_i|$.

The Gaussian distribution with standard deviation $\sigma \in \mathbb{R}$ and center $c \in \mathbb{R}$ at $x \in \mathbb{R}$, is defined by $\rho_{c,\sigma}(x) = \exp\left(\frac{-(x-c)^2}{2\sigma^2}\right)$ and more generally by $\rho_{\mathbf{c},\sigma}(\mathbf{x}) = \exp\left(\frac{-(\mathbf{x}-\mathbf{c})^2}{2\sigma^2}\right)$ and when $\mathbf{c} = \mathbf{0}$, by $\rho_\sigma(\mathbf{x})$. The discrete Gaussian distribution over $\mathbb{Z}$ centered at $\mathbf{0}$ is defined by $D_\sigma(x) = \rho_\sigma(x)/\rho_\sigma(\mathbb{Z})$ (or $D_{\mathbb{Z},\sigma}$) and more generally over $\mathbb{Z}^m$ by $D_\sigma^m(\mathbf{x}) = \rho_\sigma(\mathbf{x})/\rho_\sigma(\mathbb{Z}^m)$, where $\rho_\sigma(\mathbb{Z}^m) = \sum_{\mathbf{x} \in \mathbb{Z}^m} \rho_\sigma(\mathbf{x})$.

Description of BLISS. The BLISS signature scheme [9] is possibly the most efficient lattice-based signature scheme so far. It has been implemented in both software [10] and hardware [36], and boasts performance numbers comparable to classical factoring and discrete-logarithm based schemes. BLISS can be seen as a ring-based optimization of the earlier lattice-based scheme of Lyubashevsky [27], sharing the same "Fiat-Shamir with aborts" structure [26]. One can give a simplified description of the scheme as follows: the public key is an NTRU-like ratio of the form $\mathbf{a}_q = \mathbf{s}_2/\mathbf{s}_1 \bmod q$, where the signing key polynomials $\mathbf{s}_1, \mathbf{s}_2 \in \mathcal{R} = \mathbb{Z}[\mathbf{x}]/(\mathbf{x}^n + 1)$ are small and sparse. To sign a message μ, one first generates commitment values $\mathbf{y}_1, \mathbf{y}_2 \in \mathcal{R}$ with normally distributed coefficients, and then computes a hash $\mathbf{c}$ of the message μ together with $\mathbf{u} = -\mathbf{a}_q\mathbf{y}_1 + \mathbf{y}_2 \bmod q$. The signature is then the triple $(\mathbf{c}, \mathbf{z}_1, \mathbf{z}_2)$, with $\mathbf{z}_i = \mathbf{y}_i + \mathbf{s}_i\mathbf{c}$, and there is rejection sampling to ensure that the distribution of $\mathbf{z}_i$ is independent of the secret key. Verification is possible because $\mathbf{u} = -\mathbf{a}_q\mathbf{z}_1 + \mathbf{z}_2 \bmod q$. The real BLISS scheme, described in full in Fig. 1, includes several optimizations on top of the above description. In particular, to improve the repetition rate, it targets a bimodal Gaussian distribution for the $\mathbf{z}_i$'s, so there is a random sign flip in their definition. In addition, to reduce key size, the signature element $\mathbf{z}_2$ is actually transmitted in compressed form $\mathbf{z}_2^\dagger$, and accordingly the hash input includes only a compressed version of $\mathbf{u}$. These various optimizations are essentially irrelevant for our purposes.

Description of the GPV-based scheme of Ducas et al. The second signature scheme we consider is the one proposed by Ducas et al. at ASIACRYPT 2014 [11]. It is an optimization using NTRU lattices of the GPV hash-and-sign signature scheme of Gentry et al. [16], and has been implemented in software by Prest [37]. As in GPV, the signing key is a "good" basis of a certain lattice Λ (with short, almost orthogonal vectors), and the public key is a "bad" basis of the same lattice (with longer vectors and a large orthogonality defect). To sign a message μ, one simply hashes it to obtain a vector $\mathbf{c}$ in the ambient space of Λ, and uses the good, secret basis to sample $\mathbf{v} \in \Lambda$ according to a discrete Gaussian

---

1: **function** KEYGEN()
2:     sample $\mathbf{f}, \mathbf{g} \in \mathcal{R} = \mathbb{Z}[\mathbf{x}]/(\mathbf{x}^n + 1)$, uniformly with $\lceil \delta_1 n \rceil$ coefficients in $\{\pm 1\}$, $\lceil \delta_2 n \rceil$ coefficients in $\{\pm 2\}$ and other equal to zero
3:     $\mathbf{S} = (\mathbf{s}_1, \mathbf{s}_2)^T \leftarrow (\mathbf{f}, 2\mathbf{g} + 1)^T$
4:     **if** $N_\kappa(\mathbf{S}) \geq C^2 \cdot 5 \cdot (\lceil \delta_1 n \rceil + 4 \lceil \delta_2 n \rceil) \cdot \kappa$ **then restart**
5:     $\mathbf{a}_q = (2\mathbf{g} + 1)/\mathbf{f} \bmod q$ (restart if $\mathbf{f}$ is not invertible)
6:     **return** $(pk = \mathbf{a}_1, sk = \mathbf{S})$ where $\mathbf{a}_1 = 2\mathbf{a}_q \bmod 2q$
7: **end function**

1: **function** VERIFY($\mu, pk = \mathbf{a}_1, (\mathbf{z}_1, \mathbf{z}_2^\dagger, \mathbf{c})$)
2:     **if** $\|(\mathbf{z}_1, 2^d \cdot \mathbf{z}_2^\dagger)\|_2 > B_2$ **then reject**
3:     **if** $\|(\mathbf{z}_1, 2^d \cdot \mathbf{z}_2^\dagger)\|_\infty > B_\infty$ **then reject**
4:     **accept iff** $\mathbf{c} = H(\lfloor \zeta \cdot \mathbf{a}_1 \cdot \mathbf{z}_1 + \zeta \cdot q \cdot \mathbf{c} \rceil_d + \mathbf{z}_2^\dagger \bmod p, \mu)$
5: **end function**

1: **function** SIGN($\mu, pk = \mathbf{a}_1, sk = \mathbf{S}$)
2:     $\mathbf{y}_1 \leftarrow D_{\mathbb{Z}, \sigma}^n$, $\mathbf{y}_2 \leftarrow D_{\mathbb{Z}, \sigma}^n$
3:     $\mathbf{u} = \zeta \cdot \mathbf{a}_1 \cdot \mathbf{y}_1 + \mathbf{y}_2 \bmod 2q$
4:     $\mathbf{c} \leftarrow H(\lfloor \mathbf{u} \rceil_d \bmod p, \mu)$
5:     choose a random bit b
6:     $\mathbf{z}_1 \leftarrow \mathbf{y}_1 + (-1)^b \mathbf{s}_1 \mathbf{c}$
7:     $\mathbf{z}_2 \leftarrow \mathbf{y}_2 + (-1)^b \mathbf{s}_2 \mathbf{c}$
8:     rejection sampling: **restart** to step 2 except with probability $1/\big(M \exp(-\|\mathbf{Sc}\|/(2\sigma^2)) \cosh(\langle \mathbf{z}, \mathbf{Sc} \rangle/\sigma^2)\big)$
9:     $\mathbf{z}_2^\dagger \leftarrow (\lfloor \mathbf{u} \rceil_d - \lfloor \mathbf{u} - \mathbf{z}_2 \rceil_d) \bmod p$
10:    **return** $(\mathbf{z}_1, \mathbf{z}_2^\dagger, \mathbf{c})$
11: **end function**

Fig. 1. Description of the BLISS signature scheme. The random oracle H takes its values in the set of polynomials in $\mathcal{R}$ with $0/1$ coefficients and Hamming weight exactly κ, for some small constant κ. The value ζ is defined as $\zeta \cdot (q - 2) = 1 \bmod 2q$. The authors of [9] propose four different sets of parameters with security levels at least 128 bits. The interesting parameters for us are: $n = 512$, $q = 12289$, $\sigma \in \{215, 107, 250, 271\}$, $(\delta_1, \delta_2) \in \{(0.3, 0), (0.42, 0.03), (0.45, 0.06)\}$ and $\kappa \in \{23, 30, 39\}$. We refer to the original paper for other parameters and for the definition of notation like N_κ and $\lfloor \cdot \rceil_d$, as they are not relevant for our attack. The instruction in red (sampling of $\mathbf{y}_1$) is where we introduce our faults. (Color figure online)

distribution of small variance supported on Λ and centered at $\mathbf{c}$. That vector $\mathbf{v}$ is the signature; it is, in particular, a lattice point very close to $\mathbf{c}$. That property can be checked using the bad, public basis, but that basis is too large to sample such close vectors (this, combined with the fact that the discrete Gaussian leaks no information about the secret basis, is what makes it possible to prove security). The actual scheme of Ducas–Lyubashevsky–Prest, described in Fig. 2, uses a lattice of the same form as NTRU: $\Lambda = \{(\mathbf{y}, \mathbf{z}) \in \mathcal{R}^2 \mid \mathbf{y} + \mathbf{z} \cdot \mathbf{h} = 0\}$, where the public key $\mathbf{h}$ is again a ratio $\mathbf{g}/\mathbf{f} \bmod q$ of small, sparse polynomials in $\mathcal{R} = \mathbb{Z}[\mathbf{x}]/(\mathbf{x}^n + 1)$. The use of such a lattice yields a very compact representation of the keys, and makes it possible to compress the signature as well by publishing only the second component of the sampled vector $\mathbf{v}$. As a result, this hash-and-sign scheme is very space efficient (even more than BLISS). However, the use of lattice Gaussian sampling makes signature generation significantly slower than BLISS at similar security levels.

```
1: function KEYGEN(n, q)
2:     f ← D_{σ_0}^n, g ← D_{σ_0}^n              ▷
       σ_0 = 1.17√(q/2n)
3:     if ||(g, -f)||_2 > σ then restart         ▷
       σ = 1.17√q
4:     if ||(qf̄/(ff̄+gḡ), (q⁻g)/(ff̄+gḡ))||_2 > σ then
       restart
5:     using the extended Euclidean algo-
       rithm, compute ρ_f, ρ_g ∈ R and R_f, R_g ∈
       Z s.t. ρ_f · f = R_f mod x^n + 1 and ρ_g · g =
       R_g mod x^n + 1
6:     if gcd(R_f, R_g) ≠ 1 or gcd(R_f, q) ≠ 1
       then restart
7:     using the extended Euclidean algo-
       rithm, compute u, v ∈ Z s.t. u · R_f + v ·
       R_g = 1
8:     F ← qvρ_g, G ← -quρ_f
9:     repeat
10:        k ← ⌊(F·f̄+G·ḡ)/(ff̄+gḡ)⌉ ∈ R
11:        F ← F - k · f, G ← G - k · g
12:    until k=0
13:    h ← g · f^{-1} mod q
14:    B ← ( M_g  -M_f )  ∈ Z^{2n×2n}        ▷
           ( M_G  -M_F )
       short lattice basis
15:    return sk = B, pk = h
16: end function
```

```
1: function GAUSSIANSAMPLER(B, σ, c)  ▷
   we denote by b_i (resp. b̃_i) the rows of B
   (resp. of its Gram–Schmidt matrix B̃)
2:     v ← 0
3:     for i = 2n down to 1 do
4:         c' ← ⟨c, b̃_i⟩/||b̃_i||_2^2
5:         σ' ← σ/||b̃_i||_2
6:         r ← D_{Z,σ',c'}
7:         c ← c - rb_i and v ← v + rb_i
8:     end for
9:     return v   ▷ v sampled according to
       the lattice Gaussian distribution D_{Λ,σ,c}
10: end function

1: function SIGN(μ, sk = B)
2:     c ← H(μ) ∈ Z_q^n
3:     (y, z)        ←        (c, 0) -
       GAUSSIANSAMPLER(B, σ, (c, 0))   ▷ y, z
       are short and satisfy y + z · h = c mod q
4:     return z
5: end function

1: function VERIFY(μ, pk = h, z)
2:     accept iff ||z||_2 + ||H(μ) - z · h||_2 ≤
       σ√(2n)
3: end function
```

Fig. 2. Description of the GPV-based signature scheme of Ducas–Lyubashevsky–Prest. The random oracle H takes its values in $\mathbb{Z}_q^n$. We denote by $\mathbf{f} \mapsto \bar{\mathbf{f}}$ the conjugation involution of $\mathcal{R} = \mathbb{Z}[\mathbf{x}]/(\mathbf{x}^n + 1)$, i.e. for $\mathbf{f} = \sum_{i=0}^{n-1} f_i x^i$, $\bar{\mathbf{f}} = f_0 - \sum_{i=1}^{n-1} f_{n-i} x^i$. $\mathbf{M_a}$ represents the matrix of the multiplication by $\mathbf{a}$ in the polynomial basis of $\mathcal{R}$, which is anticirculant of dimension n. For 128 bits of security, the authors of [11] recommend the parameters $n = 256$ and $q \approx 2^{10}$. The constant 1.17 is an approximation of $\sqrt{e/2}$. The steps in red (main loop of the Gaussian sampler) is where we introduce our faults. (Color figure online)

3 Attack on Fiat-Shamir Type Lattice-Based Signatures

The first fault attack that we consider targets the lattice-based signature schemes of Fiat-Shamir type, and specifically the generation of the random "commitment" element in the underlying sigma protocols, which is denoted by $\mathbf{y}$ in our descriptions. That element consists of one or several polynomials generated coefficient by coefficient, and the idea of the attack is to introduce a fault in that random sampling to obtain a polynomial of abnormally small degree, in which case signatures will leak information about the private signing key. For simplicity's sake, we introduce the attack against BLISS in particular, but it works against the other Fiat-Shamir type schemes (GLP, PASSSign and Ring-TESLA) with almost no changes: see the full version of this paper [13] for details.

In BLISS, the commitment element actually consists of two polynomials $(\mathbf{y}_1, \mathbf{y}_2)$, and it suffices to attack $\mathbf{y}_1$. Intuitively, $\mathbf{y}_1$ should mask the secret key element $\mathbf{s}_1$ in the relation $\mathbf{z}_1 = \pm\mathbf{s}_1\mathbf{c} + \mathbf{y}_1$, and therefore modifying the distribution of $\mathbf{y}_1$ should cause some information about $\mathbf{s}$ to leak in signatures. The actual picture in the Fiat-Shamir with aborts paradigm is in fact slightly different (namely, rejection sampling ensures that the distribution of $\mathbf{z}_1$ is independent of $\mathbf{s}_1$, but only does so under the assumption that $\mathbf{y}_1$ follows the correct distribution), but the end result is the same: perturbing the generation of $\mathbf{y}_1$ should lead to secret key leakage.

Concretely speaking, in BLISS, $\mathbf{y}_1 \in \mathcal{R}_q$ is a ring element generated according to a discrete Gaussian distribution[2], and that generation is typically carried out coefficient by coefficient in the polynomial representation. Therefore, if we can use faults to cause an early termination of that generation process, we should obtain signatures in which the element $\mathbf{y}_1$ is actually a low-degree polynomial. If the degree is low enough, we will see that this reveals the whole secret key right away, from a single faulty signature!

Indeed, suppose that we can obtain a faulty signature obtained by forcing a termination of the loop for sampling $\mathbf{y}_1$ after the m-th iteration, with $m \ll n$. Then, the resulting polynomial $\mathbf{y}_1$ is of degree at most $m - 1$. As part of the faulty signature, we get the pair $(\mathbf{c}, \mathbf{z}_1)$ with $\mathbf{z}_1 = (-1)^b \mathbf{s}_1\mathbf{c} + \mathbf{y}_1$. Without loss of generality, we may assume that $b = 0$ (we will recover the whole secret key only up to sign, but in BLISS, $(\mathbf{s}_1, \mathbf{s}_2)$ and $(-\mathbf{s}_1, -\mathbf{s}_2)$ are clearly equivalent secret keys). Moreover, with high probability, $\mathbf{c}$ is invertible: if we heuristically assume that $\mathbf{c}$ behaves like a random element of the ring from that standpoint, we expect it to be the case with probability about $(1 - 1/q)^n$, which is over 95% for all proposed BLISS parameters. We thus get an equation of the form:

$$\mathbf{c}^{-1}\mathbf{z}_1 - \mathbf{s}_1 \equiv \mathbf{c}^{-1}\mathbf{y}_1 \equiv \sum_{i=0}^{m-1} y_{1,i}\mathbf{c}^{-1}\mathbf{x}^i \pmod{q} \tag{1}$$

Thus, the vector $\mathbf{v} = \mathbf{c}^{-1}\mathbf{z}_1$ is very close to the sublattice of $\mathbb{Z}^n$ generated by $\mathbf{w}_i = \mathbf{c}^{-1}\mathbf{x}^i \bmod q$ for $i = 0, \ldots, m-1$ and $q\mathbb{Z}^n$, and the difference should be $\mathbf{s}_1$.

The previous lattice is of full rank in $\mathbb{Z}^n$, so the dimension is too large to apply lattice reduction directly. However, the relation given by Eq. (1) also holds for all subsets of indices. More precisely, let I be a subset of $\{0, \ldots, n-1\}$ of cardinality ℓ, and $\varphi_I \colon \mathbb{Z}^n \to \mathbb{Z}^I$ be the projection $(u_i)_{0 \le i < n} \mapsto (u_i)_{i \in I}$. Then we also have that $\varphi_I(\mathbf{z}_1)$ is a close vector to the sublattice L_I of $\mathbb{Z}^I$ generated by $q\mathbb{Z}^I$ and the images under φ_I of the $\mathbf{w}_i$'s; and the difference should be $\varphi_I(\mathbf{s}_1)$.

Equivalently, using Babai's nearest plane approach to the closest vector problem, we hope to show that $(\varphi_I(\mathbf{s}_1), B)$, for a suitably chosen positive constant B, is the shortest vector in the sublattice L_I' of $\mathbb{Z}^I \times \mathbb{Z}$ generated by $(\varphi_I(\mathbf{v}), B)$ as well as the vectors $(\varphi_I(\mathbf{w}_i), 0)$ and $q\mathbb{Z}^I \times \{0\}$.

---

[2] In the other Fiat-Shamir schemes such as [20], the distribution of each coefficient is uniform in some interval rather than Gaussian, but this doesn't affect our attack strategy at all.

The volume of L'_I is given by:

$$\text{vol}(L'_I) = B \cdot \text{vol}(L_I) = B \cdot \frac{\text{vol}(q\mathbb{Z}^I)}{[L_I : q\mathbb{Z}^I]} = Bq^{\ell - r}$$

where r is the rank of the family $(\varphi_I(\mathbf{w}_0), \ldots, \varphi_I(\mathbf{w}_{m-1}))$ in $\mathbb{Z}_q^I$, which is at most m. Hence $\text{vol}(L'_I) \geq Bq^{\ell - m}$, and the Gaussian heuristic predicts that the shortest vector should be of norm:

$$\lambda_I \approx \sqrt{\frac{\ell + 1}{2\pi e}} \cdot \text{vol}(L'_I)^{1/(\ell+1)} \gtrsim \sqrt{\frac{\ell + 1}{2\pi e}} \cdot B^{1/(\ell+1)} q^{1 - (m+1)/(\ell+1)}.$$

Thus, we expect that $(\varphi_I(\mathbf{s}_1), B)$ will actually be the shortest vector of L'_I provided that its norm is significantly smaller than this bound λ_I. Now $\varphi_I(\mathbf{s}_1)$ has roughly $\delta_1 \ell$ entries equal to ± 1, $\delta_2 \ell$ entries equal to ± 2 and the rest are zeroes; therefore, the norm of $(\varphi_I(\mathbf{s}_1), B)$ is around $\sqrt{(\delta_1 + 4\delta_2)\ell + B^2}$. Let us choose $B = \lceil \sqrt{\delta_1 + 4\delta_2} \rceil$. The condition for $\mathbf{s}_1$ to be the shortest vector L_I can thus be written as:

$$\sqrt{(\delta_1 + 4\delta_2) \cdot (\ell + 1)} \ll \sqrt{\frac{\ell + 1}{2\pi e}} \cdot B^{1/(\ell+1)} q^{1 - (m+1)/(\ell+1)}$$

or equivalently:

$$\ell + 1 \gtrsim \frac{m + 1 + \frac{\log \sqrt{\delta_1 + 4\delta_2}}{\log q}}{1 - \frac{\log \sqrt{2\pi e(\delta_1 + 4\delta_2)}}{\log q}}. \tag{2}$$

The denominator of the right-hand side of (2) ranges from about 0.91 for the BLISS–I and BLISS–II parameter sets down to about 0.87 for BLISS–IV. In all cases, we thus expect to recover $\varphi_I(\mathbf{s}_1)$ if we can solve the shortest vector problem in a lattice of dimension slightly larger than m. This is quite feasible with the LLL algorithm for m up to about 50, and with BKZ for m up to 100 or so.

To complete the attack, it suffices to apply the above to a family of subsets I of $\{0, \ldots, n - 1\}$ covering the whole set of indices, which reveals the entire vector $\mathbf{s}_1$. The second component of the secret key is then obtained as $\mathbf{s}_2 = \mathbf{a}_1 \mathbf{s}_1/2 \bmod q$.

Simulations using our Sage implementation (see the full version of this paper [13]) confirm the theoretical estimates, and show that full key recovery can be achieved in practice in a time ranging from a few seconds to a few hours depending on m. Detailed experimental results are reported in Table 1.

Remark 1. A variant of that attack which is possibly slightly simpler consists in observing that $\varphi_I(\mathbf{s}_1)$ should be the shortest vector in the lattice generated by L_I and $\varphi_I(\mathbf{v})$. The bound on the lattice dimension becomes essentially the same as (2). The drawback of that approach, however, is that we obtain each $\varphi_I(\mathbf{s}_1)$ up to sign, and so one needs to use overlapping subsets I to ensure the consistency of those signs.

Table 1. Experimental success rate of the attack and average CPU time for key recovery for several values of m, the iteration after which the loop-abort fault is injected. We attack the BLISS–II parameter set $(n, q, \sigma, \delta_1, \delta_2, \kappa) = (512, 12289, 10, 0.3, 0, 23)$ from [9]. Since the choice of ℓ has no effect on the concrete fault injection (e.g. it does not affect the required number of faulty signatures, which is always 1), we did not attempt to optimize it very closely. The simulation was carried out using our Sage implementation (see the full version of this paper [13]) on a single core of an Intel Xeon E5-2697v3 workstation, using 100 trial runs for each value of m.

Fault after iteration number $m =$	2	5	10	20	40	60	80	100
Theoretical minimum dimension ℓ_{min}	3	6	11	22	44	66	88	110
Dimension ℓ in our experiment	3	6	12	24	50	80	110	150
Lattice reduction algorithm	LLL	LLL	LLL	LLL	BKZ–20	BKZ–25	BKZ–25	BKZ–25
Success probability (%)	100	99	100	100	100	100	100	98
Avg. CPU time to recover ℓ coeffs. (s)	0.002	0.005	0.022	0.23	7.3	119	941	33655
Avg. CPU time for full key recovery	0.5 s	0.5 s	1 s	5 s	80 s	14 min	80 min	38 h

Remark 2. Note that a single *faulty* signature is enough to recover the entire secret key with this attack, a successful key recovery may require several *fault injections*. This is due to rejection sampling: after a faulty y_1 is generated, the whole signature may be thrown away in the rejection step. On average, the fault attacker may thus need to inject the same number of faults as the repetition rate of the scheme, which is a small constant ranging from 1.6 to 7.4 depending on chosen parameters [9], and even smaller with the improved analysis of BLISS–B [8].

Remark 3. Finally, we note that in certain hardware settings, fault injection may yield a faulty value of y_1 in which all coefficients upwards of a certain degree bound are non zero but equal to a common constant (see the discussion in Sect. 5.3). Our attack adapts to that setting in a straightforward way: that simply means that y_1 is a linear combination of the x^i for small i and of the all-one vector $(1, \ldots, 1)$, so it suffices to add that vector to the set of lattice generators.

4 Attack on Hash-and-Sign Type Lattice-Based Signatures

Our second attack targets the practical hash-and-sign signature scheme of Ducas et al. [11], which is based on GPV-style lattice trapdoors. More precisely, the faults we consider are again early loop aborts, this time in the lattice-point Gaussian sampling routine used in signature generation.

4.1 Description of the Attack

The attack can be described as follows. A correctly generated signature element is of the form $\mathbf{z} = \mathbf{R} \cdot \mathbf{f} + \mathbf{r} \cdot \mathbf{F} \in \mathbb{Z}[\mathbf{x}]/(\mathbf{x}^n + 1)$, where the short polynomials $\mathbf{f}$ and $\mathbf{F}$ are components of the secret key, and $\mathbf{r}, \mathbf{R}$ are short random polynomials sampled in such a way that $\mathbf{z}$ follows a suitable Gaussian distribution. In fact, $\mathbf{r}, \mathbf{R}$ are generated coefficient by coefficient, in a single loop with $2n$ iterations, going from the top-degree coefficient of $\mathbf{r}$ down to the constant coefficient of $\mathbf{R}$.

Therefore, if we inject a fault aborting the loop after $m \leq n$ iterations (in the first half of the loop), the resulting signature simply has the form:

$$\mathbf{z} = r_0\mathbf{x}^{n-1}\mathbf{F} + r_1\mathbf{x}^{n-2}\mathbf{F} + \cdots + r_{m-1}\mathbf{x}^{n-m}\mathbf{F}.$$

Any such faulty signature is, in particular, in the lattice L of rank m generated by the vectors $\mathbf{x}^{n-i}\mathbf{F}$, $i = 1, \ldots, m$, in $\mathbb{Z}[\mathbf{x}]/(\mathbf{x}^n + 1)$.

Suppose then that we obtain several signatures $\mathbf{z}^{(1)}, \ldots, \mathbf{z}^{(\ell)}$ of the previous form. If ℓ is large enough (slightly more than m is sufficient; see Sect. 4.2 below for an analysis of success probability depending on ℓ), the corresponding vectors will then generate the lattice L. Assuming the lattice dimension is not too large, we should then be able to use lattice reduction to recover a shortest vector in L, which is expected to be one of the signed shifts $\pm\mathbf{x}^{n-i}\mathbf{F}$, $i = 1, \ldots, m$, since the polynomial $\mathbf{F}$ is constructed in a such a way as to make it quite short relative to the Gram–Schmidt norm of the ideal lattice it generates. Hence, we can recover $\mathbf{F}$ among a small set of at most $2m$ candidates.

And recovering $\mathbf{F}$ is actually sufficient to reconstruct the entire secret key $(\mathbf{f}, \mathbf{g}, \mathbf{F}, \mathbf{G})$, and hence completely break the scheme. This is due to the particular structure of the NTRU lattice. On the one hand, $\mathbf{G}$ is linked to $\mathbf{F}$ via the public key polynomial $\mathbf{h}$: $\mathbf{G} = \mathbf{F} \cdot \mathbf{h} \bmod q$, so we obtain it directly. On the other hand, the basis completion algorithm of Hoffstein et al. [21] allows to recover the pair $(\mathbf{f}, \mathbf{g})$ from $(\mathbf{F}, \mathbf{G})$ via the defining relation $\mathbf{f} \cdot \mathbf{G} - \mathbf{g} \cdot \mathbf{F} = q$. This is actually used in the opposite direction in the key generation algorithm of the scheme of Ducas et al. (i.e. they construct $(\mathbf{F}, \mathbf{G})$ from $(\mathbf{f}, \mathbf{g})$: see steps 5–12 of KEYGEN in Fig. 2), but applying [21, Theorem 1], the technique is easily seen to work in both ways.

Moreover, if we start from a polynomial of the form $\zeta\mathbf{F}$ where ζ is of the form $\pm\mathbf{x}^\alpha$, then applying the previous steps yields the quadruple $(\zeta\mathbf{f}, \zeta\mathbf{g}, \zeta\mathbf{F}, \zeta\mathbf{G})$, which is also a valid secret key equivalent to $(\mathbf{f}, \mathbf{g}, \mathbf{F}, \mathbf{G})$, in the sense that signing with either keys produces signatures with exactly the same distributions. Thus, we don't even need to carry out an exhaustive search on several possible values of $\mathbf{F}$ after the lattice reduction step: it suffices to use the first vector of the reduced basis directly.

4.2   How Many Faults Do We Need?

Let us analyze the probability of success of the attack depending on the iteration m at which the iteration is inserted and the number $\ell > m$ of faulty signatures $\mathbf{z}^{(i)}$ available. As we have seen, a sufficient condition for the attack to succeed (provided that our lattice reduction algorithm actually finds a shortest vector) is that the ℓ faulty signatures generate the rank-m lattice L defined above. This is not actually necessary (the attack works as soon as *one* of the shifts of $\mathbf{F}$ is in sub-lattice generated by the signatures, rather than all of them), but we will be content with a lower bound on the probability of success.

Now, that condition is equivalent to saying that the vectors $(r_0^{(i)}, \ldots, r_{m-1}^{(i)}) \in \mathbb{Z}^m$ (sampled according to the distribution given by the GPV algorithm) that define the faulty signatures:

$$\mathbf{z}^{(i)} = r_0^{(i)} \mathbf{x}^{n-1} \mathbf{F} + \cdots + r_{m-1}^{(i)} \mathbf{x}^{n-m} \mathbf{F}$$

generate the whole integer lattice $\mathbb{Z}^m$. But the probability that $\ell > m$ random vectors generate $\mathbb{Z}^m$ has been computed by Maze et al. [28] (see also [14]), and

Table 2. Experimental success probability of the attack and average CPU time for key recovery for several values of m, the iteration after which the loop-abort fault is injected. We consider the attack with $\ell = m + 1$ and $\ell = m + 2$ faulty signatures. The attacked parameters are $(n, q) = (256, 1021)$ as suggested in [11] for signatures. The simulation was carried out using our Sage implementation (see the full version of this paper [13]) on a single core of an Intel Xeon E5-2697v3 workstation, using 100 trial runs for each pair (ℓ, m).

Fault after iteration number $m =$	2	5	10	20	40	60	80	100
Lattice reduction algorithm	LLL	LLL	LLL	LLL	LLL	LLL	BKZ–20	BKZ–20
Success probability for $\ell = m + 1$ (%)	75	77	90	93	94	94	95	95
Avg. CPU time for $\ell = m + 1$ (s)	0.001	0.003	0.016	0.19	2.1	8.1	21.7	104
Success probability for $\ell = m + 2$ (%)	89	95	100	100	99	99	100	100
Avg. CPU time for $\ell = m + 2$ (s)	0.001	0.003	0.017	0.19	2.1	8.2	21.6	146

is asymptotically equal to $\prod_{k=\ell-m+1}^{\ell} \zeta(k)^{-1}$. In particular, if $\ell = m+d$ for some integer d, it is bounded below by:

$$p_d = \prod_{k=d+1}^{+\infty} \frac{1}{\zeta(k)}.$$

Thus, if we take $\ell = m + 1$ (resp. $\ell = m + 2$, $\ell = m + 3$), we expect the attack to succeed with probability at least $p_1 \approx 43\%$ (resp. $p_2 \approx 71\%$, $p_3 \approx 86\%$).

As shown in Table 2, this is well verified in practice (and the lower bound is in fact quite pessimistic). Moreover, the attack is quite fast even for relatively large values of m: only a couple of minutes for full key recovery for $m = 100$.

5   Implementation of the Faults

Once again, due to the obvious similarities between the four instances of the Fiat-Shamir family that we choose to attack, we only give details of the attack on the BLISS scheme. We also give details for the GPV scheme but they are essentially the same as the one for BLISS since the underlying fault introduced is strictly identical.

In this section we investigate how an attacker may obtain helpful faulty signatures for the proposed attacks. We base our discussion on two available implementations of BLISS signature, namely the software implementation from Ducas and Lepoint [10] and the FPGA implementation by Pöppelmann et al. [36], and on Prest's software implementation of the GPV-based scheme of Ducas et al. [37]. Notice that the discussion on the hardware implementation is also valid for the implementation of [20] since both share some common components and architecture that we exploit (for instance BRAM storage).

We emphasize the fact that those three implementations were not supposed to have any resilience with respect to fault attacks and were only developed as proofs of concept to illustrate the efficiency properties of the schemes. The point here is to show that the fault attacks presented in this paper are relevant based on the analysis of freely available and published implementations to put forward the need of dedicated protections against faults attacks (when attackers have such abilities).

5.1   Classical Fault Models

Faults during a computation may be induced by different means as a laser beam shot, electromagnetic injection, under-powering, glitches, etc. These faults are mainly characterized by their

- range: impacting a single bit or many bits (e.g. register or memory word);
- effect: typically target chunk is set to a chosen value, random value or all-zero/all-one value;
- persistence: a fault may only modify the target for a short period or it may be definitive.

Obviously, some fault models are close from being purely theoretical: it is very unlikely to be able to set a 32-bit register to `0xbad00dad` during precisely 2 cycles. Nevertheless many recent works have been published showing that some faults models that seemed overdone are actually obtained during lab experiments. One example is the work of Ordas et al. at CARDIS 2014 [33] showing that with finely tuned EM probes it is possible to induce a single-bit fault (bit-set or bit-reset).

In the next subsections we discuss which fault models[3] may lead to faulty signatures relevant with respect to the attacks presented in this paper. We did not investigate clock glitches or under-powering which induce violation of the setup time and which actual side-effects are implementation and compilation-dependent (with large ranges of possible parameters to test). Nevertheless, they may not be overseen in the evaluation of a chip since they may also lead to the generation of relevant faulty signatures.

5.2 Fault Attacks on Software Implementations

Polynomial y_1 can be generated using a loop over the n coefficients. This is, again, how the implementation in [10] is made: a loop is constructing polynomials y_1 and y_2 one coefficient at a time using a Gaussian sampler (function `Sign::signMessage`). The condition to perform the attack is rather few restrictive since we only require y_1 to have at most (roughly) a quarter of unknown coefficients. Such result can be obtain by going out the loop after a few iterations. A random fault on the loop counter or skipping the jump operation will lead to such result.

Notice here that it is less trivial here to decide whether a faulty signature will be helpful or not. Hopefully, the timing precision is much less important here since the attack will succeed even with 50 unknown coefficients out of 512. This means that the time-window for the fault to occur is composed of decades of loop iterations. Moreover, we may use side-channel analysis to detect the loop iteration pattern to trigger the fault injection. Such pattern is likely to be detected after much less than 50 iterations and thus it seems that the synchronization here will be relatively easy.

Similarly, the short random polynomials R and r used in the GPV scheme are generated in a single loop [37] ranging from leading coefficient of r to the constant term in R which allows to fault both polynomials using a single fault. Again, a random fault on the counter or skipping a jump makes it work and the time-window large according to the results shown in Table 2.

To conclude, these attacks seems to be a real threat since synchronization (which is a major difficulty when performing fault attacks) is eased by the loose condition on the number of known coefficients in faulted polynomials.

5.3 Fault Attacks on Hardware Implementations

Generation of polynomial y_1 requires n random coefficients. It is very unlikely that all these coefficients are obtained at the same time (n is too large) thus y_1

---

[3] We only focus on single fault attacks here.

generation will be sequential. This is the case in the implementation we took as example where the super memory is linked to the sampler through a 14-bit port. We may fault a flag or a state register to fool the control logic (here the bliss processor) and keep part of the BRAM cells to their initial state. If this initial state is known then we know all the corresponding coefficients and hopefully the number of unknown ones will be small enough for the attack to work. The large number of unknown coefficients handled by the attack again helps the attacker by providing a large time window for the fault to occur. The feasibility of the attack will mostly depend on the precise flag/state implementation and the knowledge of memory cells previous/initial value.

There is a second way of performing the fault injection here. The value of y_1 has to be stored somehow until the computation of z_1 (close to the end of the signature generation). In the example implementation a BRAM is used. We may fault BRAM access to fix some coefficients to a known value. A possible fault would be to set the `rstram` or `rstreg` signal to one (Xilinx's nomenclature). Indeed, when set to one, this will set the output latches (*resp.* register) of the RAM block to some fixed value `SRVAL` defined by the designer. We may notice two points to understand why this kind of fault enables the proposed attack.

(i) The value y_1 used to compute u will not be the faulted one but this has no impact on the attack.

(ii) If we do not know the default value for the output register, all coefficients are unknown but a big part of them are equal to the same unknown default value. In that case, the attack is still applicable by adding one generator to the constructed lattice: see Remark 3 in Sect. 3.

Again a large time window is given to the attacker due to sequential read induced by the size of y_1.

The BRAM storage of y_1 helps here the attacker since a single bit-set fault may have effects on many coefficients. The only difficulty seems to be able to perform a single-bit fault—which seems to be possible according to [33]—and the `rstram` signal localization[4].

6   Conclusion and Possible Countermeasures

We have shown that unprotected implementations of the lattice-based signature schemes that we considered are vulnerable to fault attacks, in fault models that our analysis suggests are quite realistic: the faulty signatures required by our attacks can be obtained on actual implementations. As a result, countermeasures should be added in applications where such a physical attacker is relevant to the threat model.

---

[4] Since y_1 is not directly outputted checking if the attack actually worked is a bit more tricky. Again side-channel collision analysis may help here. We may also notice that if the faulty y_1 is sparse (that is known coefficients have been set to zero) then the number of non-zero coefficients in the corresponding z_1 should be significantly smaller then for a z_1 corresponding to a dense y_1.

Simple countermeasures exist to thwart the single fault attacks proposed. There are simple, non-cryptographic countermeasures that consist in validating that the full loop have been correctly performed. This can be achieved for instance by adding a second loop counter and doing a consistency check after exiting the loop. Such a countermeasure is very cheap and we therefore recommend introducing it in all deployed implementations.

Nevertheless, it will only detect early-abort faults while an attacker may succeed in getting the same kind of faulty signature using another technique. For instance, we mentioned the possibility of faulting BRAM blocks so that they output a fixed value. For software implementations, the compiler may decide to put the coefficient in some RAM location which address could be faulted to point to another part of the memory leading in many coefficients having the same value. A single fault may also alter instruction cache leading to a nop operation instead of a load from memory and thus not updating the coefficient. We propose now other countermeasures that may deal with this issue for both types of signature schemes we considered.

We have described our attack on the Fiat-Shamir schemes in a setting where the attacker can obtain a commitment polynomial $\mathbf{y}$ of low degree, and it works more generally with a sparse $\mathbf{y}$, provided that the attackers *knows* where the non zero coefficients are located. If the locations are unknown, however, the attack does not work, so one possible countermeasure is to randomize the order of the loop generating $\mathbf{y}$. One should be careful that this may not protect against faults introduced after the very first few iterations, however: in the case of BLISS, for example, we have seen that we could easily attack polynomials $\mathbf{y}$ in which the non zero coefficients are located in the 20% lower degree coefficients, say; then, if a fault attacker can collect a few hundred faulty signatures with $\mathbf{y}$ of very low Hamming weight (say 3 or 4) at random positions, they have a good chance of finding one fault with all non zero coefficients in the lower 20%, and hence be able to attack.

Another possible approach for the Fiat-Shamir schemes is to check that the degree of the generated $\mathbf{y}$ is not too low. One cannot demand that all its coefficients are non zero, as this would skew the distribution and invalidate the security argument, but verifying that the top $\varepsilon \cdot n$ coefficients of $\mathbf{y}$ are not all zero for some small constant $\varepsilon > 0$, say $\varepsilon = 1/16$, would be a practical countermeasure that does not affect the security proof. Indeed, in the case of BLISS for example, the probability that all of these coefficients vanish is roughly $(1/\sigma\sqrt{2\pi})^{\varepsilon n}$, which is exponentially small. Thus, the resulting distribution of $\mathbf{y}$ after this check is statistically indistinguishable from the original distribution, and security is therefore preserved. Moreover, the lattice dimension required to mount our fault attack is then greater than $(1 - \varepsilon)n$, so it will not work. An additional advantage of that countermeasure is that it also adapts easily to thwart faults that cause all the top coefficients of $\mathbf{y}$ to be equal to some constant non-zero value.

Regarding the hash-and-sign signature of Ducas et al., one possible countermeasure is to simply check the validity of generated signatures. This will usually work due to the fact that a faulty signature generated from an early loop abort from the GAUSSIANSAMPLER algorithm is of significantly larger norm than a valid signature: a rough estimate of the norm after $m \leq n$ iterations is $\|\mathbf{F}\|_2 \sqrt{mq/12}$ (as $q/12$ is the variance of a uniform random variable in $\{-(q-1)/2, \ldots, (q-1)/2\}$), which is too large for correct verification even for very small values of m. An added benefit of that countermeasure is that even the correct signature generation algorithm has a very small but non zero probability of generating an invalid signature, so this countermeasure doubles up as a safeguard against those rare accidental failures.

Acknowledgments. We would like to thank Keita Xagawa and anonymous reviewers for useful comments on earlier versions of this paper.

References

1. Commercial national security algorithm suite and quantum computing FAQ. Technical report, National Security Agency, January 2016. https://www.iad.gov/iad/library/ia-guidance/ia-solutions-for-classified/algorithm-guidance/cnsa-suite-and-quantum-computing-faq.cfm
2. Akleylek, S., Bindel, N., Buchmann, J., Krämer, J., Marson, G.A.: An efficient lattice-based signature scheme with provably secure instantiation. In: Pointcheval, D., Nitaj, A., Rachidi, T. (eds.) AFRICACRYPT 2016. LNCS, vol. 9646, pp. 44–60. Springer, Cham (2016). doi:10.1007/978-3-319-31517-1_3
3. Biehl, I., Meyer, B., Müller, V.: Differential fault attacks on elliptic curve cryptosystems. In: Bellare, M. (ed.) CRYPTO 2000. LNCS, vol. 1880, pp. 131–146. Springer, Heidelberg (2000). doi:10.1007/3-540-44598-6_8
4. Bindel, N., Buchmann, J.A., Krämer, J.: Lattice-based signature schemes and their sensitivity to fault attacks. IACR Cryptology ePrint Archive 2016:415 (2016)
5. Boneh, D., DeMillo, R.A., Lipton, R.J.: On the importance of eliminating errors in cryptographic computations. J. Cryptol. **14**(2), 101–119 (2001)
6. Chen, L., Jordan, S., Liu, Y.-K., Moody, D., Peralta, R., Perlner, R., Smith-Tone, D.: Report on post-quantum cryptography. Technical report, National Institute of Standards and Technology, February 2016. http://csrc.nist.gov/publications/drafts/nistir-8105/nistir_8105_draft.pdf
7. Denchev, V.S., Boixo, S., Isakov, S.V., Ding, N., Babbush, R., Smelyanskiy, V., Martinis, J., Neven, H.: What is the Computational Value of Finite Range Tunneling? ArXiv e-prints, December 2015
8. Ducas, L.: Accelerating BLISS: the geometry of ternary polynomials. Cryptology ePrint Archive, Report 2014/874 (2014). http://eprint.iacr.org/
9. Ducas, L., Durmus, A., Lepoint, T., Lyubashevsky, V.: Lattice signatures and bimodal Gaussians. In: Canetti, R., Garay, J.A. (eds.) CRYPTO 2013. LNCS, vol. 8042, pp. 40–56. Springer, Heidelberg (2013). doi:10.1007/978-3-642-40041-4_3
10. Ducas, L., Lepoint, T.: A proof-of-concept implementation of BLISS. Available under the CeCILL License at http://bliss.di.ens.fr
11. Ducas, L., Lyubashevsky, V., Prest, T.: Efficient identity-based encryption over NTRU lattices. In: Sarkar, P., Iwata, T. (eds.) ASIACRYPT 2014. LNCS, vol. 8874, pp. 22–41. Springer, Heidelberg (2014). doi:10.1007/978-3-662-45608-8_2

12. Ducas, L., Nguyen, P.Q.: Learning a zonotope and more: cryptanalysis of NTRUSign countermeasures. In: Wang, X., Sako, K. (eds.) ASIACRYPT 2012. LNCS, vol. 7658, pp. 433–450. Springer, Heidelberg (2012). doi:10.1007/978-3-642-34961-4_27

13. Espitau, T., Fouque, P., Gérard, B., Tibouchi, M.: Loop-abort faults on lattice-based Fiat-Shamir and hash-and-sign signatures. IACR Cryptology ePrint Archive (2016). Full version of this paper

14. Fontein, F., Wocjan, P.: On the probability of generating a lattice. J. Symb. Comput. **64**, 3–15 (2014)

15. Gentry, C., Jonsson, J., Stern, J., Szydlo, M.: Cryptanalysis of the NTRU signature scheme (NSS) from Eurocrypt 2001. In: Boyd, C. (ed.) ASIACRYPT 2001. LNCS, vol. 2248, pp. 1–20. Springer, Heidelberg (2001). doi:10.1007/3-540-45682-1_1

16. Gentry, C., Peikert, C., Vaikuntanathan, V.: Trapdoors for hard lattices and new cryptographic constructions. In: Dwork, C. (ed.), STOC, pp. 197–206. ACM (2008)

17. Gentry, C., Szydlo, M.: Cryptanalysis of the revised NTRU signature scheme. In: Knudsen, L.R. (ed.) EUROCRYPT 2002. LNCS, vol. 2332, pp. 299–320. Springer, Heidelberg (2002). doi:10.1007/3-540-46035-7_20

18. Goldreich, O., Goldwasser, S., Halevi, S.: Public-key cryptosystems from lattice reduction problems. In: Kaliski, B.S. (ed.) CRYPTO 1997. LNCS, vol. 1294, pp. 112–131. Springer, Heidelberg (1997). doi:10.1007/BFb0052231

19. Bruinderink, L.G., Hülsing, A., Lange, T., Yarom, Y.: Flush, gauss, and reload - a cache attack on the BLISS lattice-based signature scheme. IACR Cryptology ePrint Archive 2016:300 (2016)

20. Güneysu, T., Lyubashevsky, V., Pöppelmann, T.: Practical lattice-based cryptography: a signature scheme for embedded systems. In: Prouff, E., Schaumont, P. (eds.) CHES 2012. LNCS, vol. 7428, pp. 530–547. Springer, Heidelberg (2012). doi:10.1007/978-3-642-33027-8_31

21. Hoffstein, J., Howgrave-Graham, N., Pipher, J., Silverman, J.H., Whyte, W.: NTRUSign: digital signatures using the NTRU lattice. In: Joye, M. (ed.) CT-RSA 2003. LNCS, vol. 2612, pp. 122–140. Springer, Heidelberg (2003). doi:10.1007/3-540-36563-X_9

22. Hoffstein, J., Pipher, J., Schanck, J.M., Silverman, J.H., Whyte, W.: Practical signatures from the partial fourier recovery problem. In: Boureanu, I., Owesarski, P., Vaudenay, S. (eds.) ACNS 2014. LNCS, vol. 8479, pp. 476–493. Springer, Cham (2014). doi:10.1007/978-3-319-07536-5_28

23. Howe, J., Pöppelmann, T., O'Neill, M., O'Sullivan, E., Güneysu, T.: Practical lattice-based digital signature schemes. ACM Trans. Embed. Comput. Syst. **14**(3), 41 (2015)

24. Howe, J., Pöppelmann, T., O'Neill, M., O'Sullivan, E., Güneysu, T., Lyubashevsky, V.: Practical lattice-based digital signature schemes. In: Slides of the Presentation at the NIST Workshop of Cybersecurity in a Post-Quantum World (2015). http://csrc.nist.gov/groups/ST/post-quantum-2015/presentations/session9-oneill-maire.pdf

25. Kamal, A.A., Youssef, A.M.: Fault analysis of the NTRUSign digital signature scheme. Cryptogr. Commun. **4**(2), 131–144 (2012)

26. Lyubashevsky, V.: Fiat-Shamir with aborts: applications to lattice and factoring-based signatures. In: Matsui, M. (ed.) ASIACRYPT 2009. LNCS, vol. 5912, pp. 598–616. Springer, Heidelberg (2009). doi:10.1007/978-3-642-10366-7_35

27. Lyubashevsky, V.: Lattice signatures without trapdoors. In: Pointcheval, D., Johansson, T. (eds.) EUROCRYPT 2012. LNCS, vol. 7237, pp. 738–755. Springer, Heidelberg (2012). doi:10.1007/978-3-642-29011-4_43

28. Maze, G., Rosenthal, J., Wagner, U.: Natural density of rectangular unimodular integer matrices. Linear Algebra Appl. **434**(5), 1319–1324 (2011)
29. Melchor, C.A., Boyen, X., Deneuville, J.-C., Gaborit, P.: Sealing the leak on classical NTRU signatures. In: Mosca, M. (ed.) PQCrypto 2014. LNCS, vol. 8772, pp. 1–21. Springer, Cham (2014). doi:10.1007/978-3-319-11659-4_1
30. Micciancio, D., Peikert, C.: Trapdoors for lattices: simpler, tighter, faster, smaller. In: Pointcheval, D., Johansson, T. (eds.) EUROCRYPT 2012. LNCS, vol. 7237, pp. 700–718. Springer, Heidelberg (2012). doi:10.1007/978-3-642-29011-4_41
31. Naccache, D., Nguyên, P.Q., Tunstall, M., Whelan, C.: Experimenting with faults, lattices and the DSA. In: Vaudenay, S. (ed.) PKC 2005. LNCS, vol. 3386, pp. 16–28. Springer, Heidelberg (2005). doi:10.1007/978-3-540-30580-4_3
32. Nguyen, P.Q., Regev, O.: Learning a parallelepiped: cryptanalysis of GGH and NTRU signatures. J. Cryptol. **22**(2), 139–160 (2009)
33. Ordas, S., Guillaume-Sage, L., Tobich, K., Dutertre, J.-M., Maurine, P.: Evidence of a larger EM-induced fault model. In: Joye, M., Moradi, A. (eds.) CARDIS 2014. LNCS, vol. 8968, pp. 245–259. Springer, Cham (2015). doi:10.1007/978-3-319-16763-3_15
34. Page, D., Vercauteren, F.: A fault attack on pairing-based cryptography. IEEE Trans. Comput. **55**(9), 1075–1080 (2006)
35. Peikert, C.: A decade of lattice cryptography. Cryptology ePrint Archive, Report 2015/939 (2015). http://eprint.iacr.org/
36. Pöppelmann, T., Ducas, L., Güneysu, T.: Enhanced lattice-based signatures on reconfigurable hardware. In: Batina, L., Robshaw, M. (eds.) CHES 2014. LNCS, vol. 8731, pp. 353–370. Springer, Heidelberg (2014). doi:10.1007/978-3-662-44709-3_20
37. Prest, T.: Implementation of the GPV-based scheme of Ducas et al. https://github.com/tprest/Lattice-IBE
38. Stein, W., et al.: Sage Mathematics Software (Version 7.0) (2016). http://www.sagemath.org
39. Taha, M., Eisenbarth, T.: Implementation attacks on post-quantum cryptographic schemes. In: Aleisa, E.A. (ed.) ICACC. IEEE Social Implications of Technology Society (2015)

Design and Implementation of Symmetric Cryptography

On the Construction of Hardware-Friendly 4 × 4 and 5 × 5 S-Boxes

Stjepan Picek[✉], Bohan Yang, Vladimir Rozic, and Nele Mentens

KU Leuven ESAT/COSIC and iMinds, Kasteelpark Arenberg 10,
3001 Leuven-Heverlee, Belgium
stjepan@computer.org

Abstract. With the emergence of the Internet of Things and lightweight cryptography, one can observe a gradual shift of interest in the design of block ciphers. Naturally, security is still of paramount importance, but one is willing to trade a part of that security in order to obtain higher speed and/or smaller implementation area. Accordingly, a common metric in many cipher proposals has been the gate count for realizing the cipher in hardware. On the other side, it is also important, especially for battery powered devices, to have a small energy consumption. That is why we can observe the following shift of research focus: from the analysis of the energy consumption of existing ciphers and their building blocks to the design of new ciphers and building blocks, specifically for low energy. Existing research results focusing on the energy consumption of symmetric ciphers, suggest that the S-box is the most expensive part in the majority of lightweight implementations. If we only consider purely combinatorial S-boxes, we can focus on reducing the power consumption of the S-box in order to minimize the energy consumption of the overall cipher. In this paper, we propose several methods to obtain 4 × 4 and 5 × 5 S-boxes that are either power or area efficient. Our results show that heuristics should be considered as a viable choice for the generation of S-boxes with good implementation properties.

1 Introduction

When designing a cryptographic cipher, security is the most important concern. Indeed, there exist a number of threats to be evaluated and it is advisable to build in a security margin large enough to withstand future attacks. By doing so, ciphers are usually large and often don't fit on constrained platforms like smart cards or microprocessors. Furthermore, even if they do fit, the speed of the execution renders the cipher often impractical for most use cases. This has led to the advent of lightweight cryptography. In lightweight cryptography, the security constraints are usually relaxed in order to make smaller and faster ciphers. Naturally, the security of such ciphers is still of prime importance, but the implementation properties are also taken into proper consideration. In battery-powered devices, considering only the area or speed is not enough, since the lifetime of the battery is determined by the energy efficiency of the device. In the examination

© Springer International Publishing AG 2017
R. Avanzi and H. Heys (Eds.): SAC 2016, LNCS 10532, pp. 161–179, 2017.
https://doi.org/10.1007/978-3-319-69453-5_9

of lightweight ciphers, we can observe that many are realized as Substitution Permutation Networks (SPNs) where the substitution part is done by one or more Substitution boxes (S-boxes). Recent results show that in a number of such ciphers, the S-box is the most power hungry building block [1]. Therefore, this work focuses on the design of S-boxes for lightweight cryptography. To be more precise, we experiment with S-boxes of size 4×4 and 5×5 that are implementation-friendly while remaining with good cryptographic properties.

We emphasize that our approach is to generate power efficient and/or small S-boxes using methods that are as simple (i.e. computationally easy) as possible. The size of the search space containing all 4×4 lookup tables is equal to $2^{2^4 \times 4} = 2^{64}$. Since we only consider bijections in the construction of S-boxes, the search space is reduced to 16! ($\approx 2^{44}$). Reducing the search space even further, considering only affine equivalent S-boxes of 16 optimal classes [2], would still result in a search space size larger than 2^{35}. Evaluating the power consumption of all S-boxes in the search space would consume too much time. This is even worse for 5×5 S-boxes. Therefore, we advocate the use of heuristics for optimizing S-boxes for power/area efficiency and we offer the experimental results that support our choice.

Considering the design choices for ciphers that use S-boxes, there are three main scenarios. The first one is to use different S-boxes in the encryption and decryption process, which is done in e.g. [3]. This is less efficient for area as well as energy optimized ciphers, since it requires the implementation of two different S-boxes. From a heuristics perspective, this scenario would increase the search space size, because it requires the optimization of the area and/or power consumption of the combination of both the encryption and the decryption S-boxes. Therefore, we do not consider this option in our experiments. The second scenario is to use a cipher in counter mode or in a sponge construction; in that case, the inverse of the S-box is not needed. This scenario is good for area and energy efficiency as well as for the minimization of the search space size. Finally, the third scenario is to use involutive S-boxes, for which the S-box is the same as its inverse. This is the approach used for instance in the Midori [4] and Noekeon ciphers [5].

The contributions of this paper are as follows:

1. In this work, we concentrate on the selection mechanism of S-boxes with power/area efficiency as a goal. As far as we know, we are the first to conduct such investigation. Since power/area efficiency plays an important role in lightweight ciphers, we also concentrate only on the S-box dimension usually found in such constructs, namely, we experiment with 4×4 and 5×5 S-box sizes.
2. We experiment with several different design methods and we identify the advantages of each of the methods. As a result we obtain a number of power/area efficient S-boxes of which the best one has a more than two times smaller power consumption and an almost two times smaller area than the PRESENT S-box. Naturally, our S-boxes also fulfill relevant cryptographic properties as discussed in [2].

3. Besides the main contribution, we also analyze the power/area efficiency of a number of S-boxes used in modern lightweight ciphers.

We emphasize that the main goal of this paper is to present a methodology on the construction of power/area efficient S-boxes, and not to concentrate on the specific results or technology. In all experiments, we use the NANGATE 45 open cell library.

The paper is organized as follows. Section 2 gives an overview of related work in terms of lightweight cipher design and power/area evaluation/optimization. In Sect. 3, we discuss basic notions about power and energy as well as the relevant cryptographic properties of S-boxes. Section 4 presents the methodology we use for obtaining power/area efficient S-boxes and discusses the results. Finally, in Sect. 5, we end with a conclusion and possible future work.

2 Related Work

There exist a number of research studies on lightweight ciphers, two of the most prominent ones being the PRESENT [6] and PRINCE [7] ciphers. In the rest of this paper, we will also concentrate on the comparison with those two ciphers/S-boxes. However, we mention several other ciphers that are SPN constructions for which we evaluate the energy consumption of the S-boxes. Those ciphers are RECTANGLE [3], Klein [8], Noekeon [5], and Luffa [9].

From the energy perspective of lightweight ciphers, Batina et al. give a comprehensive study of the area, power, and energy considerations in a number of lightweight ciphers [10]. In their paper, the authors also show that area is not always correlated with the power and energy consumption. This result further justifies our approach that to find a power efficient S-box, one needs to consider the power and not only the area. Knežević et al. analyze lightweight ciphers from the latency perspective and they discuss trade-offs between latency on the one hand and area, power, and energy on the other hand [11].

Kerckhof et al. present an evaluation of several lightweight ciphers with a focus on the energy cost [12]. Next, Banik et al. study the energy consumption of 9 lightweight ciphers as well as the AES cipher [1]. The authors also develop a model that predicts the optimal value r at which the r-round unrolled architecture should have the best energy efficiency. Banik et al. propose the energy efficient cipher Midori that uses involutive S-boxes that are extremely power efficient [4]. We note that the smaller version of the cipher, Midori64, has been broken and the authors recommend to use an S-box different from the one used in Midori128, since that leads to another invariant subspace attack [13].

Besides those results, there exist a number of papers considering heuristics to evolve S-boxes with good cryptographic properties [14,15], but we note that none of those papers consider the implementation properties of S-boxes.

3   Preliminaries

3.1   Power and Energy

The power consumption of a CMOS device is given by

$$P_{total} = P_{dynamic} + P_{static}. \tag{1}$$

The dynamic power consumption originates from the switching activity of the circuit, while the static power consumption is caused by subthreshold currents and gate leakage. The static power consumption is constant over time and does not depend on the clock frequency or the switching activity. In older technology nodes the dynamic power consumption was dominant in the total power consumption and the static power consumption was negligible. By moving to smaller transistor dimensions and thinner gate oxide layers, subthreshold currents and gate tunneling currents have increased causing higher leakage currents. Therefore, with smaller technology nodes, the relative contribution of the static leakage power consumption has increased.

The dynamic energy relates to the dynamic power consumption as follows:

$$P_{dynamic} = E_{dynamic} \cdot f_{CLK}, \tag{2}$$

where f_{CLK} is the clock frequency. The dynamic energy is given by

$$E_{dynamic} = \alpha \cdot C_{load} \cdot V_{dd}^2, \tag{3}$$

where α is the switching activity of the signal and C_{load} is the load capacitance.

3.2   Cryptographic Properties of S-Boxes

The inner product of vectors $\boldsymbol{a} = (a_0, a_1, \ldots, a_{n-1})$ and $\boldsymbol{b} = (b_0, b_1, \ldots, b_{n-1})$ is denoted as $\boldsymbol{a} \cdot \boldsymbol{b}$ and equals $\boldsymbol{a} \cdot \boldsymbol{b} = \oplus_{i=1}^n a_i b_i$. The addition modulo 2 is denoted as "$\oplus$". The Hamming weight HW of a vector $\boldsymbol{a}$, where $\boldsymbol{a} \in \mathbb{F}_2^n$, is the number of non-zero positions in the vector.

An (n, m)-function is a function from n bits to m bits. It is called **bijective** if it takes every value of $\mathbb{F}_2^m$ the same number of times, namely 2^{n-m} [16]. Balanced (n, n)-functions are permutations on $\mathbb{F}_2^n$.

The **nonlinearity** N_F of an (n, m)-function F is equal to the minimum nonlinearity of all non-zero linear combinations $\boldsymbol{b} \cdot F$, with $\boldsymbol{b} \neq 0$, of its coordinate functions f_i [16].

$$N_F = 2^{n-1} - \frac{1}{2} \max_{\substack{\boldsymbol{a} \in \mathbb{F}_2^n \\ \boldsymbol{v} \in \mathbb{F}_2^{m*}}} |W_F(\boldsymbol{a}, \boldsymbol{v})|. \tag{4}$$

$W_F(\boldsymbol{a}, \boldsymbol{v})$ is the Walsh-Hadamard transform of F:

$$W_F(\boldsymbol{a}, \boldsymbol{v}) = \sum_{\boldsymbol{x} \in \mathbb{F}_2^n} (-1)^{\boldsymbol{v} \cdot F(\boldsymbol{x}) \oplus \boldsymbol{a} \cdot \boldsymbol{x}}. \tag{5}$$

The nonlinearity N_F of any (n, n)-function F must satisfy the inequality [16]:

$$N_F \leq 2^{n-1} - 2^{\frac{n-1}{2}}. \tag{6}$$

An S-box F has **fixed points** if there exist x such that $x = F(x)$ [17]. Let F be a function from $\mathbb{F}_2^n$ into $\mathbb{F}_2^n$ and $a, b \in \mathbb{F}_2^n$. We denote:

$$D(a, b) = |\{x \in \mathbb{F}_2^n : F(x + a) + F(x) = b\}|. \tag{7}$$

The entry at the position (a, b) corresponds to the cardinality of $D(a, b)$ and is denoted as $\delta(a, b)$. The δ-uniformity δ_F is then defined as [18,19]:

$$\delta_F = \max_{a \neq 0, b} \delta(a, b). \tag{8}$$

To define the algebraic degree of an S-box, first we use the algebraic normal form (ANF) representation of a Boolean function f [20] represented by a polynomial in $\mathbb{F}_2[x_0, \ldots, x_{n-1}]/(x_0^2 - x_0, \ldots, x_{n-1}^2 - x_{n-1})$. ANF is a multivariate polynomial defined as:

$$f(x) = \oplus_{a \in \mathbb{F}_2^n} h(a) \cdot x^a, \tag{9}$$

where $h(a)$ is defined by the Möbius inversion principle

$$h(a) = \oplus_{x \preceq a} f(x), \text{ for any } a \in \mathbb{F}_2^n. \tag{10}$$

The algebraic degree deg_f of a Boolean function f is defined as the number of variables in the largest product term of the function's ANF having a non-zero coefficient [20]:

$$deg_f = max(HW(a) : h(a) = 1). \tag{11}$$

The **algebraic degree** deg_F of an S-box F is the maximum algebraic degree of all non-zero linear combinations of the coordinate functions (i.e. component functions) of F [16]:

$$deg_F = \max_{\substack{b \in \mathbb{F}_2^{m*} \\ HW(b)=1}} deg(b \cdot F). \tag{12}$$

In the case of equality in Eq. (6), such functions are called almost bent (AB) functions [16]. When a function is differentially 2-uniform, it is called almost perfect nonlinear (APN) function [16]. Every AB function is also APN, but the other direction does not hold. AB functions exist only in an odd number of variables, while APN functions also exist for an even number of variables. Furthermore, the maximal algebraic degree of AB functions equals $(n + 1)/2$ while for the inverse APN equals $n - 1$ [21].

Size 4 × 4. Leander and Poschmann define optimal 4-bit S-boxes as being bijective, with the minimal possible linearity (or, maximal possible nonlinearity) and with a minimal δ-uniformity value. For optimal S-boxes, both N_F and the δ-uniformity are equal to 4 [2].

Furthermore, Leander and Poschmann show that all optimal 4-bit S-boxes belong to 16 classes, i.e. all optimal S-boxes are affine equivalent to one of those classes [2]. For two S-boxes S_a and S_b to be equivalent, the following equation needs to hold:

$$S_a(x) = B(S_b(A(x) \oplus \boldsymbol{a})) \oplus \boldsymbol{b}, \qquad (13)$$

where A and B are invertible 4×4 matrices and $\boldsymbol{a}, \boldsymbol{b} \in \mathbb{F}_2^4$.

For PRESENT, S-boxes from the 16 classes mentioned in [2] are considered, but some lightweight ciphers use S-boxes with different cryptographic conditions. For instance, the authors of the PRINCE cipher impose several additional criteria on the S-box and therefore there are only 8 out of the 16 classes that are acceptable [7]. Alternatively, one can follow a different classification of S-boxes as for example given in [22].

Size 5×5. When considering 5×5 S-boxes, the cryptographic properties one can obtain differ with regards to the choice of the S-box. As a first example, we consider the Keccak S-box [23] for which both the nonlinearity and δ-uniformity are equal to 8. Note that those values are relatively far from the optimal ones. Furthermore, the algebraic degree of Keccak is low, and it actually equals the minimal possible algebraic degree for a nonlinear function. However, the Keccak S-box has an extremely efficient hardware implementation. The S-box used in Ascon [24] is an affine transformation of the Keccak S-box in order to remove the fixed points and to increase the branch number value [25]. On the other hand, the PRIMATEs S-box [26] is based on an almost bent permutation, which means it has a nonlinearity equal to 12 and a δ-uniformity equal to 2, while the algebraic degree is only 2.

4   Methodology and Results

4.1   Power Estimation

Before the optimization procedure, the working frequency is specified. To illustrate the methodology, we work with a clock frequency of 10 MHz. This is because the dynamic power and the cell leakage power have similar orders of magnitude for this frequency for the technology used in this paper. This enables us to optimize both shares of the power at the same time. Furthermore, for a fixed clock frequency and computation time, optimizing for energy is the same as optimizing for average power. We note that our methodology can be used for any other frequency. In this work, the power consumption of S-boxes is estimated by means of simulation.

In the first step of our simulation setup, an S-box is generated in the style of a lookup table (LUT). A Matlab (R2014b) script is used to generate the HDL description of the S-box (Verilog file *S-box.v*). For logic synthesis, we use a standard cell approach using the NANGATE 45 open cell library (PDKv1_3_v2010_12). Synopsys Design Compiler (I-2013.12) is used to produce the gate-level netlist and the delay file (*.sdf*). The standard method for estimating the power consumption using the Synopsys tool chain is based on the random switching activity of the internal nodes. While this approach may be suitable for first-order estimation it does not give realistic application-specific data. In order to obtain a more realistic estimation, one needs to use a real test-bench to approximate the switching activity for each gate. For this purpose, we have developed a test-bench that goes through all possible $n \times (n-1)$ input transitions of the S-box. Then, Modelsim SE PLUS 6.6d is used to simulate the wave file (.vcd) containing the switching activity of all nodes. This file is then converted to an activity file (.saif) using vcd2saif (D-2010.06-SP2). Finally, Design Compiler is used to estimate the power consumption. The obtained results are used as the fitness value for the optimization algorithm for both 4×4 and 5×5 S-box sizes. In Fig. 1, we depict our simulation setup in which the communication of our search strategies with the simulation part of the framework can be observed.

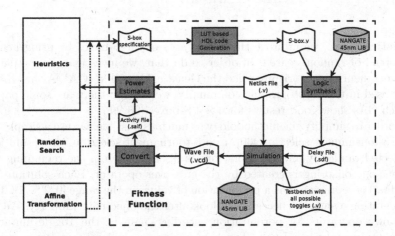

Fig. 1. Simulation setup for the generation/evaluation of S-boxes.

4.2   4 × 4 S-Boxes

The results for several commonly used 4×4 S-boxes are given in Table 1.

Random Search. As a first step in finding power/area optimized S-boxes, we run a simple random search to evaluate whether the optimization problem is trivial (disregarding the fact that randomly finding an optimal S-box is possible, but not trivial). We emphasize that this step serves only for comparison purposes. We create random S-boxes as permutations of values between 0 and $2^n - 1$ and check the results in terms of area and power. When evaluating only the optimal S-boxes, our results show that the power consumption is higher than 550 nW which makes this method quite inefficient when looking for power efficient S-boxes. In terms of area, the optimal S-boxes obtained through random search have an area larger than 20 GE.

Table 1. Reference 4×4 S-boxes

S-box	Dynamic power	Leakage power	Area
PRESENT [6]	470.2837 nW	430.608 nW	22.6667 GE
PRINCE [7]	256.176 nW	326.0947 nW	17 GE
Klein [8]	592.4351 nW	568.9604 nW	28.33 GE
Noekeon [5]	353.266 nW	383.658 nW	19.333 GE
Luffa [9]	413.1651 nW	457.9016 nW	23.667 GE
Rectangle [3]	535.5948 nW	473.317 nW	24 GE
Midori Sb_0 [4]	173.4473 nW	259.3096 nW	13.6667 GE
Midori Sb_1	283.6654 nW	297.5085 nW	15.333 GE
Piccolo [27]	334.1657 nW	342.5687 nW	17.333 GE

Heuristics. Here, we improve the power/area of the S-boxes by using heuristics instead of random search. In order to do that, we investigate a population based metaheuristic algorithm called the Genetic Algorithm (GA). Although not widely used in the cryptographic community, we observe there are some papers in which GAs show good results for 4×4 S-boxes [28, 29].

In order to simplify the methodology as much as possible, we use a simple GA with a 3-tournament selection [30]. In a 3-tournament selection, three solutions are selected randomly and the worst one is discarded. From the remaining two solutions one offspring is created by the crossover operator. Each solution (i.e. individual) is represented as a permutation of values in the range $[0, 2^n - 1]$. This representation avoids the necessity to look after the bijectivity property. We use well-known operators for permutation encoding, namely, the Toggle mutation and the Order crossover. In the Toggle mutation we randomly select two values and swap them. The Order crossover (OX) works by first randomly selecting two crossover points and copying everything between those two points from the first parent to the offspring. Then, starting from the second crossover point in the second parent, the unused numbers are copied in the order they appear in that parent [30]. The initial population is created uniformly at random and its size equals 100 individuals. We note that the computational complexity of the GA is negligible when compared with the evaluation cost, i.e. estimating the area or power consumption as further discussed later. As a stopping criterion, we use the number of evaluations without improvement, which is 30 generations in our case. Note that this algorithm also has the property of elitism, which means that the best solution will always remain intact in the population [30]. In order to better understand how GA works, we give a short pseudocode description:

```
 1: Input :  Parameters of the algorithm
 2: Output :  Optimal solution set
 3: t ← 0
 4: P(0) ← CreateInitialPopulation
 5: while TerminationCriterion do
 6:    t ← t + 1
 7:    P'(t) ← SelectMechanism (P(t − 1))
 8:    P(t) ← VariationOperators(P'(t))
 9: end while
10: Return OptimalSolutionSet(P)
```

In Fig. 2, we display one iteration of tournament selection, crossover and mutation (i.e., one generation of the GA). The numbers written next to the solutions represent the solutions' fitness values. We note that we expect that similar results could be obtained with other heuristic techniques, like local search for instance. However, we opted to work with GAs since they use a population of individuals, which allows us to generate a number of solutions before sending the data for evaluation. Since the evaluation part is the most expensive one, it makes sense to run the power and area estimation at once for a whole population, while with local search, every evaluation would consist of only one individual. For further details about GAs, we refer the readers to [30].

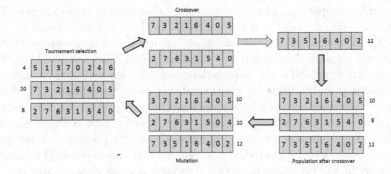

Fig. 2. One generation of the GA.

To evaluate the quality of each obtained S-box, we use a fitness function that consists of two parts. The first part checks the cryptographic properties of the S-box and only if all the criteria are met, it progresses to the second part where the power/area measurements are done. All S-boxes are ranked on the basis of their fitness where a higher value means an S-box is better. Therefore, since lower δ-uniformity is better, we subtract δ from a constant value. In summary, for the first part (cryptographic evaluation), the fitness function equals:

$$fitness = N_F + (2^m - \delta). \tag{14}$$

With this equation, we allow that our solutions have fixed points, but since we observe that the removal of fixed points can affect the power/area consumption, we also add that part to the fitness function. Since we work under the assumption that the smaller number of fixed points the better, we subtract the number of fixed points from the maximal possible number of fixed points:

$$fitness = N_F + (2^m - \delta) + (2^m - nr\_fixed\_points). \qquad (15)$$

We note that we experimented with more complex fitness functions where we added weights to each parameter, but here we present the simplest version of fitness function that yields good results. Such simple fitness function has advantages that it is more intuitive and there is no need to tune the weights in it. To state it differently, for size 4×4, this fitness function is more than sufficient to find solutions with maximal nonlinearity and minimal δ-uniformity (with or without fixed points). However, when working with size 5×5, there are no weight factors for the fitness function that reach values as obtained in e.g., AB functions. To improve the cryptographic properties, we believe one should use a completely different fitness function considering not only the nonlinearity value, but also all the values present in the Walsh-Hadamard spectrum. Naturally, this holds also for sizes larger than 5×5. We leave this research direction for future work.

In the second part of the evaluation, only those S-boxes that have the maximal nonlinearity and the minimal δ-uniformity are evaluated with regards to the power/area consumption. This means that all our solutions **must have** optimal cryptographic properties before the power/area estimation is performed. When evaluating power, we take into account both static power and dynamic power (i.e., we consider the sum of those two values). Naturally, this also means that the results could be somewhat different if only one power value is considered. Still, we believe our approach is the most general one, and we note that changing the fitness functions and consequently optimization process would be trivial. We also discuss the influence of the operators used on the obtained results. For instance, since the power consumption can change significantly with a single mutation operation, the question is how that influences the search process. It is not possible to give a definitive answer to this question, and for sure there will be a number of occasions in the evolution process where such a small change influences the fitness value significantly. However, from the other perspective, there will also be a number of building blocks (i.e., subsets of the solutions/permutations) that have a low power consumption and when combined also have a low power consumption. Because of that, the search process will eventually converge to better solutions, as evident from our results. In Table 2, we give results for the best evolved S-boxes, both for S-boxes with and without fixed points. Note that all S-boxes are optimal, so we do not add the cryptographic properties to the table. Furthermore, all mentioned S-boxes also have a maximal algebraic degree of 3. It is interesting to note that when optimizing with regards to the power consumption, we also found an S-box with smaller area than when optimizing for area (for the case without fixed points). A possible reason for such a result is that when considering power, there are more values one can obtain and therefore the search space is more fine grained. On average, our search process needed several hours to reach those solutions.

Table 2. Best evolved 4×4 S-boxes

Area results	
With fixed points	13, 1, 3, 11, 12, 2, 7, 10, 0, 5, 8, 9, 4, 6, 14, 15
Area:	14.33 GE
Without fixed points	7, 10, 11, 8, 5, 3, 1, 9, 6, 2 15, 0, 4, 14, 12, 13
Area:	13.33 GE
Power results	
With fixed points	3, 1, 2, 10, 14, 5, 7, 15, 4, 6, 0, 11, 13, 12, 8, 9
Dynamic power:	237.16 nW Leakage power: 297.52 nW Area: 14.67 GE
Without fixed points	13, 5, 10, 4, 7, 1, 2, 0, 14, 6, 8, 12, 15, 3, 9, 11
Dynamic power:	206.1 nW Leakage power: 240.73 nW Area: 12.67 GE

Involutive S-Boxes. The total number of involutions for an S-box of size $n \times n$ equals [31]:

$$\#Involution = \sum_{i=0}^{2^{n-1}} \frac{2^n!}{(2^{n-1}-i)!2^{2^{n-1}-i}}. \tag{16}$$

If we consider the 4×4 case, there are in total $462\,067\,736$ involutive S-boxes. This search space can be exhaustively searched if we consider only relevant cryptographic properties, but when power/area estimation is necessary, it still represents a search space too large to be efficiently exhausted. In order to conduct this search, we implemented a recursive swap algorithm that traverses all possible involutions with a defined number of fixed points. We tested more than $250\,000$ involutive S-boxes that are optimal (i.e., with the best possible nonlinearity and δ-uniformity values) and the best obtained result for area equals 13 GE. On the other hand, when considering power results, the best S-box has a dynamic power of 201.84 nW and a static leakage power of 271.48 nW. We note that when considering power results, we found two S-boxes with the same result and both of them are S-boxes with 4 fixed points. Finally, to put these results into perspective from the computational complexity point and with a conservative estimate of only 10 s per S-box power/area evaluation, we needed around 30 days of continuous computation to conduct this experiment.

Next, we concentrate only on involutive S-boxes with 4 and 6 fixed points that are optimal. There are in total $18\,918\,900$ involutive S-boxes with 4 fixed points, and $7\,567\,560$ involutive S-boxes with 6 fixed points. We opted to follow this line of research since for instance in Midori, both S-boxes have 4 fixed points, while 6 fixed points is the maximal number of fixed points we could find in 4×4 S-boxes that are optimal. Furthermore, we additionally prune the results in order to keep only those that have an algebraic degree equal to 3. For S-boxes with 4 fixed points, we investigate $30\,000$ optimal involutive S-boxes. The best result for area is 13 GE while the best result for power is an S-box with a dynamic power of 201.8418 nW and a leakage power of 255.1868 nW. For optimal involutive S-boxes with 6 fixed points, we evaluate 3 000 S-boxes. The best result for area is 15 GE and the best result for power is an S-box with a dynamic power of 223.3748 nW and a leakage power of 293.5608 nW. As can be seen, the best results are obtained for the search within optimal S-boxes with 4 fixed points. However, this search yields a somewhat larger (to be exact, 0.33 GE larger) S-box with a higher power consumption compared to the GA approach. As a future work, it would be interesting to run a heuristic search only within involutive S-boxes. However, we note that in that scenario, one would need to design custom heuristic initialization in order to seed the algorithm with only involutive S-boxes. Furthermore, in that scenario, custom-made crossover and mutation operators are also needed when only involutive S-boxes are produced. In order to give a better perspective to those results, we give the average results over 100 random involutions with 0 to 8 fixed points in Table 3. Note that those S-boxes are mostly not optimal.

Table 3. Random involutive 4×4 S-boxes

0 fixed points			
Dynamic power: 361.43 nW	Leakage Power: 388.53 nW	Area:	19.61 GE
2 fixed points			
Dynamic power: 383.69 nW	Leakage Power: 403.7 nW	Area:	20.55 GE
4 fixed points			
Dynamic power: 364.62 nW	Leakage Power: 390.43 nW	Area:	19.7367 GE
6 fixed points			
Dynamic power: 343.25 nW	Leakage Power: 380.74 nW	Area:	19.31 GE
8 fixed points			
Dynamic power: 328.83 nW	Leakage Power: 360.05 nW	Area:	18.2467 GE

4.3 5×5 S-Boxes

We omit the random search results since our experiments show that this problem is too difficult and the obtained results are far from power/area efficient. Furthermore, randomly created S-boxes also have cryptographic properties far from those observed in literature. In Table 4, we give the results for area and power for several S-boxes used in literature as well as for an "APN S-box", which is an S-box we generated with the multiplicative inverse function and irreducible polynomial $x^5 + x^4 + x^3 + x^2 + 1$ [19].

Heuristics. We use heuristics in the same way as in the 4×4 case. We note that we are unable to obtain AB 5×5 S-boxes with heuristics, but we are able to find S-boxes with cryptographic properties similar or somewhat better than those in the Keccak S-box. All the presented S-boxes have a nonlinearity equal to 8, δ-uniformity 6, and algebraic degree 4. The results are given in Table 5.

Table 4. Reference 5 × 5 S-boxes

S-box	Dynamic power	Leakage power	Area
Keccak [23]	318.9477 nW	299.5233 nW	17 GE
PRIMATEs [26]	676.173 nW	668.4548 nW	36 GE
APN S-box	1.3846 μW	1.1463 μW	57.33 GE
Ascon [24]	741.5331 nW	606.4438 nW	30.6667 GE
Icepole [32]	590.9029 nW	621.8677 nW	32.33 GE

Table 5. Best evolved 5 × 5 S-boxes

Area results	
With fixed points	10, 5, 2, 4, 29, 21, 17, 7, 15, 13, 24, 16, 26, 20, 11, 23, 31, 0, 19, 6, 25, 3, 1, 22, 30, 8, 28, 18, 27, 9, 14, 12
Area:	39.33 GE
Without fixed p.	14, 10, 28, 29, 1, 9, 0, 15, 4, 23, 20, 17, 24, 25, 16, 27, 8, 11, 12, 13, 31 22, 26, 2, 6, 30, 5, 1, 18, 7, 19, 3
Area:	38 GE
Power results	
With fixed points	24, 29, 12, 14, 8, 19, 4, 2, 25, 16, 13, 9, 10, 26, 5, 11, 21, 18, 22, 20, 7, 23, 6, 0, 17, 1, 30, 27, 3, 15, 28, 31
Dynamic Power:	801.8934 nW Leakage power: 777.7131 nW Area: 39.67 GE
Without fixed p.	13, 14, 22, 27, 24, 10, 0, 29, 4, 6, 30, 26, 9, 2, 1, 17, 3, 15, 19, 23, 11, 12, 7, 18, 16, 20, 31, 25, 8, 28, 5, 21
Dynamic Power:	734.7164 nW Leakage power: 754.2006 nW Area: 39.33 GE

Affine Transformations. Since we are unable to find 5 × 5 S-boxes that have cryptographic properties closer to the optimal values (either AB or APN), we use the fact that **affine transformations can change the power/area** of an S-box. Therefore, we aim to optimize the affine transformation in order to reduce the area/power.

Recall from Eq. (13) the matrices A and B need to be invertible in $\mathbb{F}_2$ and the number of such matrices equals:

$$GL = \prod_{i=0}^{n-1} (2^n - 2^i). \tag{17}$$

For $n = 4$ there are in total 20 160 invertible matrices. However, since there are two matrices and additionally two constants $a, b \in \mathbb{F}_2^n$, the total number of combinations is $\approx 2^{36}$. When calculating cryptographic properties, this number is within reach, but for implementation properties like power where the time necessary to calculate the results for a single 4 × 4 S-box is in the order of magnitude of 10 s, this task becomes impossible. Moreover, for the 4 × 4 size, there are 16 optimal classes, which means that we need to run such a search 16 times. For the 5×5 size, there are $9\,999\,360$ invertible matrices and therefore, the total number of combinations equals $\approx 2^{56}$.

Based on the aforesaid, we see that an exhaustive search is most often not a realistic option. Therefore, we need a faster way to obtain good results. To be able to do so, we again use the same genetic algorithms setting, only now the individuals are encoded as a set of genotypes of bitstring values. Each genotype represents one matrix or a constant as in Eq. (13). Each individual has four genotypes of which the first two represent matrices A and B and genotypes 3 and 4 represent constants a and b. For an easier visualization of the solutions, one can imagine genotypes A and B as row vectors of size n^2 where the transformation to a matrix is done by splitting the vector in n rows of size n.

The fitness function we aim to minimize is:

$$fitness = Power. \tag{18}$$

Since we know that affine transformations cannot change the cryptographic properties we consider here, we do not need to check them during the evolution. Here, we investigate 3 S-boxes: the Keccak S-box, the PRIMATEs S-box (AB), and our APN S-box. The results for the best obtained affine transformations with regards to power consumption are given in Table 6. Note that for the Keccak and PRIMATEs case our search did not reveal any better S-boxes that are affine equivalent to the original S-box. Still, our best S-box that is affine equivalent to Keccak has a significantly smaller area/power than for instance affine transformations of the Keccak S-box as used in the Ascon and ICEPOLE ciphers. We note that when we optimize the affine transformation for Keccak with the goal of improving the area, the best S-box we find has an area of 21 GE and is without fixed points (recall that Keccak has 2 fixed points) like those used in Ascon and ICEPOLE. For the "APN S-box" the result is quite improved. This shows it is not easy to find better S-boxes with respect to power efficiency, but also that the S-boxes in Keccak and PRIMATEs are also good candidates from the area/power perspective. Since "APN S-box" is an S-box we created with a randomly selected irreducible polynomial of degree 5, one could expect that the results for such S-box could be significantly improved and our analysis confirms that.

4.4 Discussion

When implementing an S-box, one can follow the encoder/decoder structure as presented in [33], but we note that that scheme is effective only on larger S-boxes, for example size 8×8. The extra cost to implement an encoder/decoder is cumbersome for 4×4 S-boxes both from the power and area perspective.

Therefore, we advocate here the usage of heuristics when generating S-boxes with good power/area properties. To put our solutions into an adequate perspective, we compare them with S-boxes that are used in a number of lightweight designs. As can be observed, the Midori S-boxes have the smallest power consumption as well as the smallest area when considering currently used S-boxes.

Table 6. Best evolved 5 × 5 S-boxes, affine transformations

Keccak	9, 24, 6, 23, 14, 30, 2, 18, 29, 13, 11, 27, 4, 21, 17, 0, 3, 26, 28, 5, 20, 12, 8, 16, 31, 7, 25, 1, 22, 15, 19, 10
Dynamic power:	488.6914 nW Leakage Power: 496.4189 nW Area: 26 GE
PRIMATEs	30, 22, 16, 31, 2, 18, 26, 13, 9, 21, 15, 20, 23, 19, 7, 4, 6, 8, 10, 3, 11, 29, 17, 0, 1, 27, 5, 24, 14, 12, 28, 25
Dynamic power:	751.4109 nW Leakage Power: 723.7496 nW Area: 37 GE
APN S-box	31, 16, 28, 27, 22, 24, 10, 30, 13, 17, 1, 25, 21, 29, 0, 2, 14, 5, 4, 8, 26, 15, 19, 23, 9, 12, 11, 18, 20, 3, 6, 7
Dynamic power:	913.5057 nW Leakage Power: 942.5685 nW Area: 48 GE

However, when compared with the evolved S-boxes in this paper, we see that our S-box without fixed points, has the smallest area (12.67 GE) and the smallest power consumption except for the Midori Sb_0 S-box where the difference is only 14 nW. As a matter of fact, our best evolved 4 × 4 S-box has a more than two times smaller power consumption than the PRESENT S-box while retaining the same nonlinearity and δ-uniformity values as the PRESENT S-box. However, we emphasize that our evolved S-boxes are not involutive.

On the other side, we observe that the involutive S-boxes that have the smallest power consumption (both in our work and other work) have 4 fixed points. Indeed, both Midori S-boxes (432.75 nW and 581.17 nW) as well as the two involutive S-boxes found in our search (473.32 nW) represent the best results for power consumption when considering involutive S-boxes. Therefore, in scenarios where involutive S-boxes with an as small as possible power consumption are necessary, it seems to be prudent first to conduct an exhaustive search within involutive S-boxes with 4 fixed points. Still, we emphasize that such a comparison in not completely fair since the PRESENT S-box has branch number equal to 3, while our S-boxes has branch number 2. Naturally, this is expected since we did not include the branch number property into our optimization process. Indeed, obtaining a high branch number in our current setting, would be more a matter of luck than the optimization process itself. However, we note a number of currently used S-boxes also have branch numbers equal to 2 (e.g., Klein, Noekeon, Rectangle, Prince). Finally, adding the branch number to the objective function is trivial, and we plan to explore that direction in future work. Moreover, we see that our evolved S-boxes have smaller area and power consumption than the S-box used in Piccolo, which is an S-box known to be extremely efficient from both area/power perspectives. However, we note that the Piccolo S-box is not intended to be implemented as a lookup table, which makes a fair comparison somewhat difficult.

For the 5 × 5 size, we observe that the problem is much more difficult, but we offer two heuristic techniques to improve the results; one based on the direct evolution of solutions, and another one that looks for the best affine transformation of a certain S-box. There, we managed to find an S-box that is affine equivalent to the Keccak S-box, with slightly worse area/power results, but without fixed points. However, that S-box has better area/power properties than those used in the Ascon and ICEPOLE ciphers.

A possible drawback of our approach is that S-boxes could be also implemented in other ways and not only as lookup tables. This is also the reason why we do not include results for S-boxes larger than 5×5, since those S-boxes are too big to be implemented as lookup tables in most realistic scenarios. With our approach, there is no guarantee that a certain power efficient S-box in lookup table style will remain power efficient when implemented using some other technique, but our results suggest that this is most often the case. Still, we believe our approach is as fair as possible since our technique can always serve as a strong indicator of S-box behavior. Implementing S-boxes in another way would make the search even more difficult (and computationally complex) since then we do not only look for S-boxes with good properties (i.e., the first level of the search) but also for different implementation methods (can be considered as the second level of search).

From the scalability perspective, our technique shows good behavior. Indeed, the same technique given for the 5×5 size (i.e., affine transformation-based search) works good for larger sizes. Still, those results are more difficult to interpret since such sizes usually necessitate different styles of the implementation of S-boxes. All our experiments were conducted on a PC that has an i7 4720HQ processor and 8 GB of RAM. For all relevant sizes ($4 \times 4, 5 \times 5$, and 8×8) the evaluation cost of the implementation properties is dominant. For instance, for size 4×4, to test all relevant cryptographic properties of a single S-box we need around 4 ms, for size 5×5 we need 8 ms, and for size 8×8, we need around 15 ms. Even for the smallest size of 4×4, evaluating the power consumption takes more than 10 s.

Naturally, these results should be taken with care. We do not suggest just to use our S-box and replace some of the existing ones with it. Indeed, doing that without a proper cryptanalytic analysis could be devastating for the security of the cipher. Rather, we suggest to use S-boxes we created in some new designs that specifically target low power consumption and area. Since we concentrated here only on the S-box part, we cannot give any cryptanalysis results since our S-boxes are not intended to replace existing S-boxes in modern ciphers. Therefore, we give relevant cryptographic properties that can be used as a comparison with other S-boxes. Furthermore, we focus here on a small set of cryptographic properties of S-boxes, but if other criteria need to be fulfilled, our heuristic approach can easily be adjusted. Finally, we tested our approach with one library using all possible input transitions of the $n \times n$ S-box to do the power estimation. Our method is easily transferable to other libraries and other ways of power estimation. We believe that such adaptability of our framework to different settings is the main advantage of our approach. Indeed, if a researcher needs to run experiments with different constraints, the running time of our approach coupled with good results makes a good choice.

5 Conclusions and Future Work

In this work, we focused on the power and area efficiency of S-boxes of small sizes, namely 4 × 4 and 5 × 5. First, we defined an objective experimental setting for testing the power/area efficiency and we conducted experiments based on several different approaches. The best results were obtained using the heuristic approach, for which our best S-box has a more than two times smaller power consumption than the PRESENT S-box. We emphasize that we do not recommend for instance to exchange the PRESENT S-box with this new one, but rather use the new S-box when designing new ciphers that are energy efficient. When further cryptographic constraints are imposed in the choice of an S-box, our heuristic approach can be readily adapted. We note that any automatic search strategy is only as good as the synthesis tool. Therefore, with a more powerful synthesis algorithm, our search strategy would also be more efficient.

As future work, we plan to investigate the possibility of finding one S-box that performs optimally when considering more cryptographic properties as well as both power and area over several implementation libraries and to use such an S-box in the design of a new cipher that is energy efficient.

Acknowledgments. This work has been supported in part by Croatian Science Foundation under the project IP-2014-09-4882. In addition, this work was supported in part by the Research Council KU Leuven (C16/15/058) and IOF project EDA-DSE (HB/13/020).

References

1. Banik, S., Bogdanov, A., Regazzoni, F.: Exploring energy efficiency of lightweight block ciphers. In: Dunkelman, O., Keliher, L. (eds.) SAC 2015. LNCS, vol. 9566, pp. 178–194. Springer, Cham (2016). doi:10.1007/978-3-319-31301-6_10
2. Leander, G., Poschmann, A.: On the classification of 4 bit S-boxes. In: Carlet, C., Sunar, B. (eds.) WAIFI 2007. LNCS, vol. 4547, pp. 159–176. Springer, Heidelberg (2007). doi:10.1007/978-3-540-73074-3_13
3. Zhang, W., Bao, Z., Lin, D., Rijmen, V., Yang, B., Verbauwhede, I.: RECTANGLE: a bit-slice ultra-lightweight block cipher suitable for multiple platforms. IACR Crypto. ePrint Arch. **2014**, 84 (2014)
4. Banik, S., Bogdanov, A., Isobe, T., Shibutani, K., Hiwatari, H., Akishita, T., Regazzoni, F.: Midori: a block cipher for low energy (extended version). Cryptology ePrint Archive, Report 2015/1142 (2015). http://eprint.iacr.org/
5. Daemen, J., Peeters, M., Assche, G.V., Rijmen, V.: Nessie proposal: the block cipher Noekeon. Nessie submission (2000). http://gro.noekeon.org/
6. Bogdanov, A., Knudsen, L.R., Leander, G., Paar, C., Poschmann, A., Robshaw, M.J.B., Seurin, Y., Vikkelsoe, C.: PRESENT: an ultra-lightweight block cipher. In: Paillier, P., Verbauwhede, I. (eds.) CHES 2007. LNCS, vol. 4727, pp. 450–466. Springer, Heidelberg (2007). doi:10.1007/978-3-540-74735-2_31
7. Borghoff, J., et al.: PRINCE – a low-latency block cipher for pervasive computing applications. In: Wang, X., Sako, K. (eds.) ASIACRYPT 2012. LNCS, vol. 7658, pp. 208–225. Springer, Heidelberg (2012). doi:10.1007/978-3-642-34961-4_14

8. Gong, Z., Nikova, S., Law, Y.W.: KLEIN: a new family of lightweight block ciphers. In: Juels, A., Paar, C. (eds.) RFIDSec 2011. LNCS, vol. 7055, pp. 1–18. Springer, Heidelberg (2012). doi:10.1007/978-3-642-25286-0_1

9. Canniere, C., Sato, H., Watanabe, D.: Hash function Luffa: specification 2.0.1. Submission to NIST (Round 2) (2009). http://www.sdl.hitachi.co.jp/crypto/luffa/

10. Batina, L., Das, A., Ege, B., Kavun, E.B., Mentens, N., Paar, C., Verbauwhede, I., Yalçın, T.: Dietary recommendations for lightweight block ciphers: power, energy and area analysis of recently developed architectures. In: Hutter, M., Schmidt, J.-M. (eds.) RFIDSec 2013. LNCS, vol. 8262, pp. 103–112. Springer, Heidelberg (2013). doi:10.1007/978-3-642-41332-2_7

11. Knežević, M., Nikov, V., Rombouts, P.: Low-latency encryption – is "lightweight = light + wait"? In: Prouff, E., Schaumont, P. (eds.) CHES 2012. LNCS, vol. 7428, pp. 426–446. Springer, Heidelberg (2012). doi:10.1007/978-3-642-33027-8_25

12. Kerckhof, S., Durvaux, F., Hocquet, C., Bol, D., Standaert, F.-X.: Towards green cryptography: a comparison of lightweight ciphers from the energy viewpoint. In: Prouff, E., Schaumont, P. (eds.) CHES 2012. LNCS, vol. 7428, pp. 390–407. Springer, Heidelberg (2012). doi:10.1007/978-3-642-33027-8_23

13. Guo, J., Jean, J., Nikolić, I., Qiao, K., Sasaki, Y., Sim, S.M.: Invariant subspace attack against full midori64. Cryptology ePrint Archive, Report 2015/1189 (2015). http://eprint.iacr.org/

14. Clark, J.A., Jacob, J.L., Stepney, S.: The design of S-boxes by simulated annealing. New Gener. Comput. **23**(3), 219–231 (2005)

15. Ivanov, G., Nikolov, N., Nikova, S.: Reversed genetic algorithms for generation of bijective s-boxes with good cryptographic properties. Crypt. Commun. **8**(2), 247–276 (2016)

16. Carlet, C.: Vectorial Boolean functions for cryptography. In: Crama, Y., Hammer, P.L. (eds.) Boolean Models and Methods in Mathematics, Computer Science, and Engineering, 1st edn, pp. 398–469. Cambridge University Press, New York (2010)

17. Daemen, J., Rijmen, V.: The Design of Rijndael. Springer-Verlag New York Inc., Secaucus (2002)

18. Biham, E., Shamir, A.: Differential cryptanalysis of DES-like cryptosystems. In: Menezes, A.J., Vanstone, S.A. (eds.) CRYPTO 1990. LNCS, vol. 537, pp. 2–21. Springer, Heidelberg (1991). doi:10.1007/3-540-38424-3_1

19. Nyberg, K.: Perfect nonlinear S-boxes. In: Davies, D.W. (ed.) EUROCRYPT 1991. LNCS, vol. 547, pp. 378–386. Springer, Heidelberg (1991). doi:10.1007/3-540-46416-6_32

20. Carlet, C.: Boolean functions for cryptography and error correcting codes. In: Crama, Y., Hammer, P.L. (eds.) Boolean Models and Methods in Mathematics, Computer Science, and Engineering, 1st edn, pp. 257–397. Cambridge University Press, New York (2010)

21. Budaghyan, L., Carlet, C., Pott, A.: New classes of almost bent and almost perfect nonlinear polynomials. IEEE Trans. Inf. Theory **52**(3), 1141–1152 (2006)

22. Zhang, W., Bao, Z., Rijmen, V., Liu, M.: A new classification of 4-bit optimal S-boxes and its application to PRESENT, RECTANGLE and SPONGENT. In: Leander, G. (ed.) FSE 2015. LNCS, vol. 9054, pp. 494–515. Springer, Heidelberg (2015). doi:10.1007/978-3-662-48116-5_24

23. Bertoni, G., Daemen, J., Peeters, M., Van Assche, G.: Keccak. In: Johansson, T., Nguyen, P.Q. (eds.) EUROCRYPT 2013. LNCS, vol. 7881, pp. 313–314. Springer, Heidelberg (2013). doi:10.1007/978-3-642-38348-9_19

24. Dobraunig, C., Maria Eichlseder, F.M., Schläffer, M.: Ascon (2014). CAESAR submission. http://ascon.iaik.tugraz.at/

25. Ullrich, M., De Cannière, C., Indesteege, S., Küçük, Ö., Mouha, N., Preneel, B.: Finding Optimal Bitsliced Implementations of 4 × 4-bit S-Boxes (2011)
26. Andreeva, E., Bilgin, B., Bogdanov, A., Luykx, A., Mendel, F., Mennink, B., Mouha, N., Wang, Q., Yasuda, K.: PRIMATEs v1 Submission to the CAESAR Competition (2014). http://competitions.cr.yp.to/round1/primatesv1.pdf
27. Shibutani, K., Isobe, T., Hiwatari, H., Mitsuda, A., Akishita, T., Shirai, T.: Piccolo: an ultra-lightweight blockcipher. In: Preneel, B., Takagi, T. (eds.) CHES 2011. LNCS, vol. 6917, pp. 342–357. Springer, Heidelberg (2011). doi:10.1007/978-3-642-23951-9_23
28. Picek, S., Ege, B., Papagiannopoulos, K., Batina, L., Jakobovic, D.: Optimality and beyond: the case of 4 × 4 S-boxes. In: IEEE International Symposium on Hardware-Oriented Security and Trust, HOST 2014, Arlington, VA, USA, 6–7 May 2014, pp. 80–83. IEEE Computer Society (2014)
29. Picek, S., Papagiannopoulos, K., Ege, B., Batina, L., Jakobovic, D.: Confused by confusion: systematic evaluation of DPA resistance of various S-boxes. In: Meier, W., Mukhopadhyay, D. (eds.) INDOCRYPT 2014. LNCS, vol. 8885, pp. 374–390. Springer, Cham (2014). doi:10.1007/978-3-319-13039-2_22
30. Eiben, A.E., Smith, J.E.: Introduction to Evolutionary Computing. Springer, Heidelberg, New York (2003). doi:10.1007/978-3-662-44874-8
31. Youssef, A., Tavares, S., Heys, H.: A new class of substitution-permutation networks. In: Proceedings of SAC 1996 - Workshop on Selected Areas in Cryptography, pp. 132–147 (1996)
32. Morawiecki, P., Gaj, K., Homsirikamol, E., Matusiewicz, K., Pieprzyk, J., Rogawski, M., Srebrny, M., Wójcik, M.: ICEPOLE: high-speed, hardware-oriented authenticated encryption. In: Batina, L., Robshaw, M. (eds.) CHES 2014. LNCS, vol. 8731, pp. 392–413. Springer, Heidelberg (2014). doi:10.1007/978-3-662-44709-3_22
33. Bertoni, G., Macchetti, M., Negri, L., Fragneto, P.: Power-efficient ASIC synthesis of cryptographic sboxes. In: Proceedings of the 14th ACM Great Lakes Symposium on VLSI, GLSVLSI 2004, pp. 277–281. ACM, New York (2004)

All the AES You Need on Cortex-M3 and M4

Peter Schwabe and Ko Stoffelen[✉]

Digital Security Group, Radboud University, Nijmegen, The Netherlands
peter@cryptojedi.org, k.stoffelen@cs.ru.nl

Abstract. This paper describes highly-optimized AES-$\{128, 192, 256\}$-CTR assembly implementations for the popular ARM Cortex-M3 and M4 embedded microprocessors. These implementations are about twice as fast as existing implementations. Additionally, we provide the fastest bitsliced constant-time and masked implementations of AES-128-CTR to protect against timing attacks, power analysis and other (first-order) side-channel attacks. All implementations, including an architecture-specific instruction scheduler and register allocator, which we use to minimize expensive loads, are released into the public domain.

Keywords: AES · Software implementation · ARM Cortex-M · Constant-time · Bitslicing · Masking

1 Introduction

AES was published as Rijndael in 1998 and standardized in FIPS PUB 197 in 2001. Highly optimized implementations have been written for most common architectures, ranging from 8-bit AVR microcontrollers to x86-64 and NVIDIA GPUs. See, for example, [4,17,24]. Implementing optimized AES on any of these architectures essentially requires to start from scratch to find out which implementation approach is going to be the most efficient. The past decades have seen a large shift toward ARM architectures and while we have seen efficient AES implementations for high-end processors used in modern smartphones [5] and for older microprocessors used in smart cards [1,6], there is little to choose from for modern low-end embedded devices and Internet of Things applications.

Sometimes an embedded device contains a coprocessor that can perform AES encryption in hardware, but such a coprocessor is not always available. It makes a device more expensive and it can increase the power consumption of a device. Simply compiling an existing implementation written in, for example, the C programming language, is unlikely to produce optimal performance. Even worse, embedded systems are typical targets for timing attacks, power analysis attacks,

This work was supported by the European Commission through the Horizon 2020 program under project number ICT-645622 (PQCRYPTO) and by Netherlands Organization for Scientific Research (NWO) through Veni 2013 project 13114. Permanent ID of this document: 9fc0b970660e40c264e50ca389dacd49. Date: October 19, 2016.

R. Avanzi and H. Heys (Eds.): SAC 2016, LNCS 10532, pp. 180–194, 2017.
https://doi.org/10.1007/978-3-319-69453-5_10

and other forms of side-channel attacks, so software for those devices typically needs to include adequate protection against such attacks.

We fill these gaps by providing highly optimized AES software implementations for two of the most popular modern microprocessors for constrained embedded devices, the ARM Cortex-M3 and the Cortex-M4. Our implementations of AES-{128, 192, 256}-CTR are more than twice as fast as existing implementations. We also provide a single-block AES-128 implementation, a constant-time AES-128-CTR implementation and a masked implementation that is secure against first-order power analysis attacks. All of them are the fastest of their kind. They are put into the public domain and available at https://github.com/Ko-/aes-armcortexm.

The results of this paper are not only interesting for "stand-alone" AES encryption. In the ongoing CAESAR competition for authenticated encryption schemes, 14 out of the 29 remaining second round candidates are based on AES or the AES round function. Our implementations will be helpful to speed up those candidates on embedded ARM microcontrollers.

Organization of the Paper. In Sect. 2, we will first discuss AES and give an outline of the different implementation approaches. We will also provide an overview of the target architecture and what features we can benefit from when optimizing software for speed. Section 3 then discusses our fastest AES implementations, based on the T-tables approach, while Sects. 4 and 5 consider our constant-time bitsliced and our masked implementation, respectively. We report performance benchmarks and provide a comparison to related work at the end of each of the Sects. 3–5.

2 Preliminaries

2.1 Implementing AES

AES is a substitution-permutation network that operates on 128-bit blocks. Key sizes of 128, 192, and 256 bits are supported. Depending on the key size, the network has 10, 12, or 14 rounds, respectively. The nonlinear substitution layer consists of the SubBytes step, where an 8-bit S-box is applied to each byte of the state. The linear permutation layer consists of ShiftRows and MixColumns, to provide diffusion. In the beginning, between all rounds, and at the end, the AddRoundKey step xors the state with round keys that are derived from the main key during a key schedule. MixColumns is omitted in the final round [12]. In software, there are four main implementation approaches:

Traditional: All steps are implemented "as is"; typically SubBytes is implemented through a 256-byte lookup table.

T-tables: SubBytes, ShiftRows, and MixColumns are combined in 4 1024-byte lookup tables. Each AES round then consists of 16 masks, 16 loads from the lookup tables and 4 loads from the round keys, and 16 XORs. This leads to very efficient implementations on platforms with a word size of at least 32 bits. At the cost of extra rotations, only 1 lookup table is required. This strategy was already suggested in the original Rijndael proposal [11]. Our fastest implementations in Sect. 3 are based on this approach.

Vector permute: The disadvantage of the T-tables approach is that key- and data-dependent lookups open the door for timing attacks on architectures with cache. See, for example, [3,25,33]. Another approach to implementing AES, which avoids such data-dependent lookups, uses vector-permute instructions [15]. However, such instructions are unavailable on our target platform, which is why we do not go into more detail on this strategy.

Bitslicing: Another approach that does not require lookup tables is bitslicing, originally introduced for DES by Biham [7]. The core idea is that data is split over multiple registers, but that other blocks are used to fill the registers. Multiple blocks can then be processed in parallel in a SIMD fashion. This approach is especially beneficial for architectures with large registers. For AES, the 128-bit state is usually bytesliced over 8 registers, as this allows for an efficient linear layer. Various papers describe bitsliced implementations of AES on Intel processors [19–21]; the most recent one by Käsper and Schwabe from 2009 is still the software speed-record holder [17]. Our implementations in Sects. 4 and 5 are also using bitslicing.

2.2 ARM Cortex-M

The Cortex-M is a family of 32-bit processors by ARM meant for use in embedded microcontrollers. They are designed to be cheap and to be energy efficient, while still being powerful enough to offer adequate performance in applications such as automotive systems, medical instruments, the Internet of Things, or other consumer products. As of 2015, over 10 billion of these processors have been shipped [27].

The Cortex-M3 was announced in 2004, while the Cortex-M4 is a more recent successor from 2010. Both microprocessors have 16 32-bit registers, of which three are reserved for program counter, stack pointer, and link register. The link pointer can be pushed to the stack to free another register. Both microprocessors support the ARMv7-M architecture and the Thumb-2 technology, but the Cortex-M4 supports additional instructions for digital signal processing, i.e., the ARMv7E-M architecture. However, we do not use these extensions.

Bitwise and arithmetic instructions take one cycle on these architectures, except for divisions or writes to the program counter. Branches, loads, and stores may take more cycles, which is why they can easily bottleneck the performance. A distinguishing feature of the ARM architecture is the availability of barrel-shifting registers. This means that we can do arithmetic on rotated or shifted registers, without any additional cost for the rotation or shift.

We used the STM32L100C and STM32F407VG development boards. The first comes with 256 KB of flash memory, 16 KB of RAM, and 4 KB of EEPROM. It can run a Cortex-M3 core at up to 32 MHz. The second is more powerful and has a 168 MHz Cortex-M4 core, 1024 KB of flash memory, 192 KB of RAM, and a true-random-number generator.

2.3   Accelerating Memory Access

Memory access can be expensive in terms of CPU cycles. Additionally, there are a lot of ways to introduce penalty cycles. Carefully optimized software therefore avoids as many potential delays as possible. Here we list a number of generic strategies related to memory access to reduce the cycle count of programs running on the Cortex-M3 and M4. A siginifcant portion of our speedups of AES stem from a combination of these strategies.

Flash. The instructions and tables are typically stored in flash memory. Accessing flash can introduce a number of wait states, depending on the relative clock frequency of the microprocessor core and the memory chip. For our development boards, the STM32L100C and STM32F407VG, STMicroelectronics describes in its documentation when it is possible to have zero wait states [29, p. 59, tbl. 13][30, p. 80, tbl. 10]. For example, on the STM32L100C, the CPU clock can only run at 16 MHz for a supply voltage of 3.3 V. To be able to compare the performance of implementations across different devices or boards, it is important to be in this scenario.

RAM. Something similar holds for accessing RAM, where the stack is stored. On the STM32F407VG, four different regions of RAM are available: SRAM1, SRAM2, SRAM3, and CCM. In our case it turned out to be faster to use the core coupled memory (CCM), as it uses the D-bus directly.

Alignment. The Cortex-M3 and M4 support Thumb-2 technology, which means that 16-bit and 32-bit encodings of instructions can freely be mixed. However, consider the case that a 16-bit instruction starts at a word-aligned address, followed by one or more 32-bit instructions. The 32-bit instructions are then no longer word-aligned, which may cause penalty cycles for the instruction fetcher, which fetches multiple instructions at a time. In this case, forcing the use of a 32-bit encoding for the first instruction by adding a .w suffix can improve the instruction alignment and reduce the cycle count. Our implementations take this into consideration. Penalty cycles may also be introduced when branching to addresses that are not word-aligned, when loading from memory at addresses that are not word-aligned or when not loading full words from memory. Implementers needs to take care of the alignment themselves. Our implementations carefully avoid these penalty cycles.

Pipelining Loads. Most str instructions take 1 cycle, because of the availability of a write buffer, but ldr instructions generally take at least 2 cycles. However, n ldr instructions can be pipelined together to be executed in $n + 1$ cycles if there are no address dependencies and the program counter remains untouched. An instruction such as ldm pipelines all of its loads together, but when it is followed by an ldr, those will not be pipelined together. For our implementations, we pipeline as many loads as possible.

Caches and Prefetch Buffers. The Cortex-M3 and M4 by themselves do not have any caches. However, caches can be added in embedded devices or development boards to boost the performance. For example, the STM32F407VG contains 64 128-bit lines of instruction cache memory and 8 128-bit lines of data cache memory [30, p. 90]. It also contains an instruction prefetch buffer to reduce the experienced number of wait states when a microprocessor running at a high clock frequency accesses flash memory to fetch 128 bits of instructions [30, p. 82]. The STM32L100C supports a similar prefetch buffer when 64-bit flash access is enabled [29, p. 59].

Data Location. When one wants to read data that is stored in the flash memory, one first needs to load the address of the data block before one can load the data itself. However, when data is located within 4096 bytes of the value of the program counter, the first load instruction can be replaced by an `adr` pseudo-instruction that is really an addition or subtraction of the program counter, which may save one cycle, depending on whether the load could be pipelined. It is therefore useful to store data close to where the data is being used.

3   Making AES Fast

Ever since Rijndael was standardized as AES, a lot of effort has been put into making fast and secure software implementations for a large range of platforms and architectures. Numerous optimization tricks have been suggested to improve the performance. For T-table-based implementations, the majority is summarized in [4]. In this section we discuss which strategies are useful to apply on the Cortex-M3 and M4.

Using the T-table-based approach, AES-128-CTR can typically be implemented in 720 instructions: 208 loads, 4 stores, 160 shifts, 176 masks, 168 xors and 4 others [4]. Thanks to ARM's barrel-shifting registers, we can do combined shifts and masks, saving 160 instructions. [4] also mentions *scaled-index loads* and *second-byte instructions*. A scaled-index load is the option to shift the offset of a load instruction for free, while a second-byte instruction allows for extracting the second byte of a register in one instruction. Both features are supported by our architecture, but as all shifts are already fully subsumed, these optimizations no longer yield any additional advantage.

Byte loads and *two-byte loads* could save another 8 instructions by not requiring an additional mask, but loads that are not word-aligned cause a penalty cycle, so for speed these optimizations are of little use. Other potential optimization strategies, such as combining *masks and inserts* or *loads and xors*, are not possible in a single instruction on these platforms. Being able to do *byte extraction via loads* allows to exchange arithmetic instructions for load instructions, but loads are either as fast or slower, so this strategy gives no advantage either.

With *round-key recomputation*, only one out of four round-key words is stored for all rounds except the first. During encryption, the other parts of the round keys can be recomputed on the fly, exchanging 30 loads for 30 xors. However, in our case the loads can be fully pipelined and the round keys from the previous round would not fit into registers anymore, so this would also not reduce the total number of cycles. *Round-key caching*, where all round keys are kept in registers when encrypting multiple blocks, would require even more registers.

Another technique called *padded registers* exists, where a 32-bit value is stored in a 64-bit register in such a way that combing shifts and masks can be done a bit more cleverly. However, our registers are too small to use anything like this.

However, *counter-mode caching* helps to save another 81 instructions in the main loop. In counter mode, for 256 consecutive blocks, only 1 byte of the input changes. This means that through the first and second AES round, computations that do not depend on this one byte can be cached and reused. Starting from the third round, everything will depend on all input bytes. While there is some additional overhead involved in storing and retrieving the cached values, this trick already leads to a speedup when only 2 blocks are processed.

3.1   Our Implementations

Our implementations of AES-128 encryption, AES-128-CTR, AES-192-CTR, and AES-256-CTR use one 1024-byte lookup table. The extra rotates that this would normally cause come for free thanks to ARM's barrel shifting registers. Using four tables would save another 40 1-cycle instructions in the key schedule, and 16 1-cycle instructions in the final round for encryption, but as there is typically little memory available on microcontrollers and the improvement in speed is only marginal, we decided that this trade-off was not worth it. AES-128 decryption needs two 1024-byte lookup tables. On the other hand, the 16 mask instructions in the final round are no longer required.

Key expansion is performed separately, as the round keys can be reused for multiple blocks. In our implementation of counter mode, there is a 32-bit counter and a 96-bit nonce. The reason is that then we do not have to deal with a carry from the counter and a conditional add for the second counter word, which gives another small speedup. We consider a 32-bit counter, providing a maximum stream length of $2^{32} \cdot 16 = 68719476736$ bytes, to be large enough in a typical microcontroller environment.

The performance of our speed-optimized implementations is summarized in Table 1. All results are averages over 10000 runs with random keys, inputs, and, if applicable, nonces. For encryption in counter mode, the number of cycles reflects the average number of cycles per block when processing 256 blocks, or 4096 bytes. Loops are fully unrolled, so the code size can be reduced drastically with only a small performance penalty. Note that data in ROM is typically shared by key expansion and encryption/decryption, so it has to be in memory only once. Under RAM usage, I/O refers to the amount of RAM that is required to store the input and output for the functions, e.g., $192 + 2m$ means that we require 4 bytes for the counter, 12 for the nonce, 176 for all round keys, m for our m-byte input, and m for the m-byte output. Again, I/O data is typically shared by key expansion and encryption/decryption and the same stack space can be reused for the encryption/decryption function call. It turns out that the same code runs in slightly fewer cycles on the Cortex-M3, which is most likely caused by the different way that instructions are fetched.

Table 1. Performance of unprotected AES

Algorithm	Speed (cycles)		ROM (bytes)		RAM (bytes)	
	M3	M4	Code	Data	I/O	Stack
AES-128 key expansion encryption	289.8	294.8	902	1024	176	32
AES-128 key expansion decryption	1180.0	1174.6	3714	2048	176	176
AES-128 single block encryption	659.4	661.7	2034	1024	$176 + 2m$	44
AES-128 single block decryption	642.5	648.3	1974	2048	$176 + 2m$	44
AES-128-CTR	546.3	554.4	2192	1024	$192 + 2m$	72
AES-192 key expansion	264.9	272.2	810	1024	240	32
AES-192-CTR	663.2	673.0	2576	1024	$224 + 2m$	72
AES-256 key expansion	364.8	371.8	1166	1024	240	32
AES-256-CTR	786.9	791.7	2960	1024	$256 + 2m$	72

3.2   Comparison to Existing Implementations

There are few publicly available AES implementations optimized for the Cortex-M3 and M4:

- In the SharkSSL crypto library, a speed of 1066.7 cycles per block is claimed for AES-128-ECB on the Cortex-M3 [28]. CTR mode is unavailable.
- A company called Cryptovia sells an implementation that does AES-128 on a single block in 1463 cycles [10], also on the Cortex-M3.
- The latest version of mbed TLS [26], formerly known as PolarSSL, contains a table-based AES-128-CTR implementation that takes 1247.4 cycles per block on the M3, while AES-128 key expansion takes 41545 cycles[1].
- NXP hosts the AN11241 AES library [23], but its implementation is very slow. AES-128-ECB runs in 4179.1 cycles per block on the M3, while the AES-128 key expansion takes 1089 cycles (See footnote 1).
- The fastest implementation currently listed by the FELICS benchmarking framework [13] encrypts a single block with AES-128 in 1816 cycles on a Cortex-M3. The fastest key scheduling takes 724 cycles[2].

We therefore claim that our CTR-mode implementations are about twice as fast as existing implementations. We also require fewer cycles than optimized implementations for older yet similar ARM architectures [1], even though in [1] heavy use is made of the fact that the full lookup tables fit in the data cache on a StrongARM-1110, which does not hold for our platforms.

3.3   Benchmarking with FELICS

The FELICS framework [13] has been proposed as an open system to benchmark the performance of implementations of lightweight cryptographic systems on

---

[1] We used `gcc -O3 -funroll-loops -fno-schedule-insns` with GCC 6.1.1 for these benchmarks, the best set of compiler flags we could find, based on all sets that are tried in the SUPERCOP benchmarking framework.

[2] `AES_128_128_V06` in scenario 0 with `-Os` and with `-O3`, respectively.

three different microprocessors, one of them being the ARM Cortex-M3. Cycle counts and memory usage are measured for three usage scenarios. Scenario 0 deals with single-block encryption, where the round keys are stored in RAM. In scenario 1, 128 bytes are encrypted in CBC mode. In scenario 2, 128 bits are encrypted in CTR mode.

This choice of scenarios means that our implementation needs to be adapted to fit in the framework. In particular, counter-mode caching can no longer be used and needs to be removed, which impacts the performance. Furthermore, the decryption algorithm and decryption key expansion are now required as well in scenarios 0 and 1. But most importantly, the FELICS framework does not set the number of wait states, which means that a load from memory will cost more than 2 cycles and that reported cycle counts are biased toward implementations with less load instructions. This greatly slows down the overall performance of our implementation.

The framework reports 1641 cycles for our encryption in scenario 0 and 578 cycles for our key schedule. Although this is still faster than currently listed results, the margin is smaller. This also holds for scenarios 1 and 2.

4   Protecting Against Timing Attacks

While the availability of caches allows for speedups on platforms with relatively slow memory, it also makes table-based AES implementations vulnerable to cache-timing attacks [3,18]. A popular technique for writing a constant-time AES implementation that is still reasonably fast, is by applying bitslicing. Of course, caches can be simply disabled when performing cryptographic operations, but this implementation also serves as a step toward the masked implementation.

Bitslicing is often explained as a technique where every bit of the state is stored in a separate register, such that we can do operations on the bits independently and such that we can process 32 blocks in parallel on 32-bit machines. However, in the case of AES this is not the fastest way to bitslice, as most operations are byte-oriented. Full bitslicing would also increase the amount of registers needed to store the state by a factor of 32. There are very few architectures that have enough registers to keep the bitsliced state in registers, so there would be a lot of overhead in storing and loading data to other types of memory.

Könighofer suggested in [19] to 'byteslice' and to process 4 blocks in parallel on an architecture with 64-bit registers. Käsper and Schwabe were able to process 8 blocks in parallel using 128-bit registers [17]. Unfortunately, the Cortex-M3 and M4 only have 32-bit registers, so we can only process 2 blocks in parallel while still retaining an efficient implementation of the linear layer.

4.1   Our Implementation

After key expansion, the round keys are stored in their bitsliced representation. To transform to bitsliced representation, we require 12 SWAPMOVE operations [22].

```
SWAPMOVE(a,b,n,m) {
  t = ((a ≫ n) ⊕ b) & m
  b = b ⊕ t
  a = a ⊕ (t ≪ n)
}
```

Due to ARM's barrel shifter, we can implement SWAPMOVE in just 4 1-cycle instructions, which gives a transformation overhead of 48 cycles.

```
eor t, b, a, lsl #n
and t, m
eor b, t
eor a, a, t, lsr #n
```

During encryption, the AES state is first transformed to bitsliced representation. AddRoundKey is then again just a matter of xoring the bitsliced round keys with the bitsliced state.

For SubBytes, a lot of research has been done on an efficient hardware implementation of the AES S-box [9]. These results are also very useful for bitsliced software implementations. Boyar and Peralta found a circuit with only 115 gates [8], which was later improved to 113: 32 AND gates, 77 XOR gates, and 4 XNOR gates. This is the smallest known implementation, which is why we used it as a basis for our implementation. However, with only 14 available registers, it is impossible to implement it directly in 113 instructions. We need more instructions to deal with storing values on the stack or with recomputation of values. We wrote an ad hoc combined instruction scheduler and register allocator that is tailored to our microprocessors.

Scheduling. Both instruction scheduling and register allocation are hard problems, as is the combined problem. Compilers usually implement a graph coloring algorithm and/or linear-scan allocation. They aim to schedule well on average, but do not necessarily generate the most efficient assembly for a specific part of code.

Existing compilers do not provide a lot of options to play with different scheduling and allocation strategies, which is why we decided to write an ARM-specific instruction scheduler and register allocator. This allows us to focus on ARM's three-operand instructions and to try several approaches. We aim to minimize the number of loads and stores and the usage of the stack. We first reschedule instructions to reduce the size of the active data set, by pushing instructions down based on their left-hand side and by pushing instructions up based on their right-hand side. Then we allocate registers in a greedy fashion, where we insert loads and stores when necessary and try to leave the output in registers. A more thorough overview of the tool is provided in [31], including a comparison against the compilers GCC, Clang, and the ARM Compiler.

Our tool is nondeterministic because of hash randomization in Python, so we try several scheduling strategies multiple times and only use the best result.

With our scheduler we are able to compute the AES S-box in 145 instructions: the 113 original operations, 16 loads and 16 stores. It is unknown whether this is optimal.

ShiftRows on a bitsliced state can be computed very efficiently on modern Intel CPUs using 8 SSSE3 byte-shuffling instructions [17]. However, something like this is unavailable on the Cortex-M3 and M4. We use the ubfx and uxtb bit-field instructions, together with eor on shifted registers, to compute ShiftRows in $8 \cdot 13 = 104$ 1-cycle instructions. The code below performs ShiftRows on r9, while r12 and r5 are used as temporary registers.

```
uxtb.w r12, r9
ubfx  r5,  r9, #14, #2
eor  r12,  r12, r5,  lsl #8
ubfx  r5,  r9, #8,  #6
eor  r12,  r12, r5,  lsl #10
ubfx  r5,  r9, #20, #4
eor  r12,  r12, r5,  lsl #16
ubfx  r5,  r9, #16, #4
eor  r12,  r12, r5,  lsl #20
ubfx  r5,  r9, #26, #6
eor  r12,  r12, r5,  lsl #24
ubfx  r5,  r9, #24, #2
eor  r9,  r12, r5,  lsl #30
```

In contrast, the barrel shifters allow us to compute MixColumns in just 27 eor instructions on shifted registers, which is even more efficient than in [17].

To update the counter for the next blocks, one can either store the bitsliced representation and operate on this, or one can use the original representation and transform this to bitsliced representation every two blocks. While the first may appear to be faster, we implemented both and it turned out that the latter is in fact more efficient. This is due to overhead caused by the limited way in which you can do conditional execution with IT-blocks on these microprocessors.

Table 2 contains performance benchmarks of our implementation. Again, speed is measured as the average number of cycles per block when encrypting 256 consecutive blocks, which explains the decimal for the encryption. The amount of cycles is exactly equal for all 10000 combinations of random nonces, keys, and inputs that we tried. We see a slowdown of roughly a factor 2.9 compared to our previous implementation. Note, however, that when one can disable the caches during the AES execution or when caches are not available at all, our previous faster implementations are also constant-time and should be favored. We verified that after disabling caches, the cycle counts are exactly equal for random combinations of inputs and keys. There is little related work that would make a fair comparison.

Table 2. Performance of constant-time AES

Algorithm	Speed (cycles)		ROM (bytes)		RAM (bytes)	
	M3	M4	Code	Data	I/O	Stack
AES-128 bitsliced key expansion	1027.8	1033.8	3434	1036	368	188
AES-128-CTR bitsliced constant-time	1616.6	1617.6	12120	12	$368 + 2m$	108

5 Protecting Against Side-Channel Attacks

Microprocessors are typical targets for side-channel attacks such as differential power analysis or differential electromagnetic analysis. A well-known counter-measure against first-order side-channel attacks that is used in practice is by Boolean masking, where a secret intermediate value a is split into two statistically independent shares, i.e., r_a and $\bar{a} = (a \oplus r_a)$, where r_a is called a random mask. Linear operations can be computed on both shares independently. After a linear operation, the shares can be xored together to unmask the result. Nonlinear operations are more difficult to mask securely. Trichina suggested the following provably secure method to mask $a \cdot b$ [32], where $\bar{a} = (a \oplus r_a)$, $\bar{b} = (b \oplus r_b)$, and r_a, r_b, r are random masks:

$$((\bar{a} \cdot \bar{b}) \oplus ((r_a \cdot \bar{b}) \oplus ((r_a \cdot r_b) \oplus r))) \oplus (r_b \cdot \bar{a}).$$

This means that every AND operation requires 4 AND operations, 4 XOR operations, and 1 load (of r) to mask.

We added first-order Boolean masking using Trichina gates to our constant-time bitsliced implementations to find out how much this additional security would cost on common microprocessors.

5.1 Our Implementation

To generate the masks, we need a source of randomness. The STM32F407VG contains a random number generator (RNG) that guarantees a new 32-bit random word every 40 periods of the RNG clock. In the case of AES, 8 random words are required to mask the input, as two blocks are processed in parallel, and 320 random words are required for a single encryption, as SubBytes contains 32 AND operations and is executed in all 10 rounds. While interleaving randomness generation and executing instructions can decrease the waiting time, the performance of the implementation will greatly depend on the performance of the RNG and the relative clock frequency between the core and the RNG.

All other operations are linear, so at least a factor of 2 slowdown can be expected there. However, because the size of the active data set doubles and will not fit in 14 registers anymore, a lot of overhead is created by additional loads and stores. Our scheduler manages to generate a securely masked bitsliced

SubBytes implementation in $2 \cdot 83 + 4 \cdot 32 = 294$ XORs, $4 \cdot 32 = 128$ ANDs, 99 stores and 167 loads, that are pipelined as much as possible. Once more, the speed is measured as the average number of cycles per block when encrypting 256 consecutive blocks. The cycle counts are precisely equal for all combinations of inputs, keys, and nonces.

Table 3. Performance of masked constant-time AES

Algorithm	Speed (cycles)		ROM (bytes)		RAM (bytes)	
	M3	M4	Code	Data	I/O	Stack
AES-128-CTR masked constant-time	N/A	7422.6	39916	12	$368 + 2m$	1588

The performance of the final implementation is summarized in Table 3. Note that of these 7422.6 cycles per block, 2132.5 are spent on generating random words and pushing them to the stack, while all the rest takes 5290.1 cycles per block. A faster RNG could significantly boost the total speed. Of the 1588 bytes on the stack, 1312 are taken by the 328 random words.

5.2 Comparison to Existing Implementations

Balasch et al. [2] showed at CHES 2015 that adding first-order Boolean masking with Trichina gates slows the implementation down by roughly a factor of 5 on the Cortex-A8. On the Cortex-M4, we see something similar compared to the unmasked bitsliced implementation, with a factor 4.6, although a faster RNG could reduce this to almost a factor of 3.5. Furthermore, we require less randomness because we based ourselves on the 113-gate SubBytes implementation.

Goudarzi and Rivain [14] investigated the performance of different approaches to higher-order masking based on the ISW masking scheme [16] by implementing masked versions of AES and PRESENT on the ARM7TDMI-S microprocessor, a somewhat older architecture from 2001 that is still widely deployed. For first-order masking, their fastest implementation takes 49329 cycles [14, tbl. 16, standard AES with parallel Kim-Hong-Lim S-box, 2 shares], which is a factor 5.6 more than ours, but that comparison is not entirely fair as we do not support higher-order masking. However, instruction timings appear to be similar between the two architectures.

6   Conclusion and Outlook

This paper presented various speed-optimized AES software implementations for multiple use case scenarios, including side-channel attack protection, for the ARM Cortex-M3 and M4. All of them are the fastest of their kind. Additionally, we provide an ARM-specific instruction scheduler and register allocator that is of independent interest to optimize other software for these platforms. All software is put into the public domain, which also may benefit the performance of (AES-based) CAESAR candidates on modern embedded microcontrollers.

We admit that the 'all the AES you need' claim in our tittle does not hold for use cases that need to protect against *higher-order* side-channel attacks. We plan to have an assembly generator for higher-order masked AES implementations, although one then may want to resort to masking schemes other than gate-level masking.

References

1. Atasu, K., Breveglieri, L., Macchetti, M.: Efficient AES implementations for ARM based platforms. In: Proceedings of the 2004 ACM Symposium on Applied Computing, pp. 841–845. ACM (2004)
2. Balasch, J., Gierlichs, B., Reparaz, O., Verbauwhede, I.: DPA, bitslicing and masking at 1 GHz. In: Güneysu, T., Handschuh, H. (eds.) CHES 2015. LNCS, vol. 9293, pp. 599–619. Springer, Heidelberg (2015). doi:10.1007/978-3-662-48324-4_30
3. Bernstein, D.J.: Cache-timing attacks on AES. https://cr.yp.to/antiforgery/cachetiming-20050414.pdf, 2005
4. Bernstein, D.J., Schwabe, P.: New AES software speed records. In: Chowdhury, D.R., Rijmen, V., Das, A. (eds.) INDOCRYPT 2008. LNCS, vol. 5365, pp. 322–336. Springer, Heidelberg (2008). doi:10.1007/978-3-540-89754-5_25
5. Bernstein, D.J., Schwabe, P.: NEON crypto. In: Prouff, E., Schaumont, P. (eds.) CHES 2012. LNCS, vol. 7428, pp. 320–339. Springer, Heidelberg (2012). doi:10.1007/978-3-642-33027-8_19
6. Bertoni, G., Breveglieri, L., Fragneto, P., Macchetti, M., Marchesin, S.: Efficient software implementation of AES on 32-bit platforms. In: Kaliski, B.S., Koç, K., Paar, C. (eds.) CHES 2002. LNCS, vol. 2523, pp. 159–171. Springer, Heidelberg (2003). doi:10.1007/3-540-36400-5_13
7. Biham, E.: A fast new DES implementation in software. In: Biham, E. (ed.) FSE 1997. LNCS, vol. 1267, pp. 260–272. Springer, Heidelberg (1997). doi:10.1007/BFb0052352
8. Boyar, J., Peralta, R.: A new combinational logic minimization technique with applications to cryptology. In: Festa, P. (ed.) SEA 2010. LNCS, vol. 6049, pp. 178–189. Springer, Heidelberg (2010). doi:10.1007/978-3-642-13193-6_16
9. Canright, D.: A very compact S-box for AES. In: Rao, J.R., Sunar, B. (eds.) CHES 2005. LNCS, vol. 3659, pp. 441–455. Springer, Heidelberg (2005). doi:10.1007/11545262_32
10. Cryptovia: AES algorithms for ARM CPU. http://www.cryptovia.com/ARM-AES.html
11. Daemen, J., Rijmen, V.: AES proposal: rijndael, version 2 (1999). http://csrc.nist.gov/archive/aes/rijndael/Rijndael-ammended.pdf
12. Daemen, J., Rijmen, V.: The Design of Rijndael: AES - The Advanced Encryption Standard. Springer, Heidelberg (2013). doi:10.1007/978-3-662-04722-4
13. Dinu, D., Corre, Y.L., Khovratovich, D., Perrin, L., Großschädl, J., Biryukov, A.: Triathlon of lightweight block ciphers for the Internet of Things. Cryptology ePrint Archive, Report 2015/209 (2015). http://eprint.iacr.org/2015/209/
14. Goudarzi, D., Rivain, M.: How fast can higher-order masking be in software? Cryptology ePrint Archive, Report 2016/264 (2016). http://eprint.iacr.org/2016/264/

15. Hamburg, M.: Accelerating AES with vector permute instructions. In: Clavier, C., Gaj, K. (eds.) CHES 2009. LNCS, vol. 5747, pp. 18–32. Springer, Heidelberg (2009). doi:10.1007/978-3-642-04138-9_2

16. Ishai, Y., Sahai, A., Wagner, D.: Private circuits: securing hardware against probing attacks. In: Boneh, D. (ed.) CRYPTO 2003. LNCS, vol. 2729, pp. 463–481. Springer, Heidelberg (2003). doi:10.1007/978-3-540-45146-4_27

17. Käsper, E., Schwabe, P.: Faster and timing-attack resistant AES-GCM. In: Clavier, C., Gaj, K. (eds.) CHES 2009. LNCS, vol. 5747, pp. 1–17. Springer, Heidelberg (2009). doi:10.1007/978-3-642-04138-9_1

18. Kocher, P.C.: Timing attacks on implementations of diffie-hellman, RSA, DSS, and other systems. In: Koblitz, N. (ed.) CRYPTO 1996. LNCS, vol. 1109, pp. 104–113. Springer, Heidelberg (1996). doi:10.1007/3-540-68697-5_9

19. Könighofer, R.: A fast and cache-timing resistant implementation of the AES. In: Malkin, T. (ed.) CT-RSA 2008. LNCS, vol. 4964, pp. 187–202. Springer, Heidelberg (2008). doi:10.1007/978-3-540-79263-5_12

20. Matsui, M.: How far can we go on the x64 processors? In: Robshaw, M. (ed.) FSE 2006. LNCS, vol. 4047, pp. 341–358. Springer, Heidelberg (2006). doi:10.1007/11799313_22

21. Matsui, M., Nakajima, J.: On the power of bitslice implementation on intel core2 processor. In: Paillier, P., Verbauwhede, I. (eds.) CHES 2007. LNCS, vol. 4727, pp. 121–134. Springer, Heidelberg (2007). doi:10.1007/978-3-540-74735-2_9

22. May, L., Penna, L., Clark, A.: An implementation of bitsliced DES on the pentium MMX^{TM} processor. In: Dawson, E.P., Clark, A., Boyd, C. (eds.) ACISP 2000. LNCS, vol. 1841, pp. 112–122. Springer, Heidelberg (2000). doi:10.1007/10718964_10

23. NXP Semiconductors N.V. AN11241: AES encryption and decryption software on LPC microcontrollers. https://www.lpcware.com/content/nxpfile/an11241-aes-encryption-and-decryption-software-lpc-microcontrollers

24. Osvik, D.A., Bos, J.W., Stefan, D., Canright, D.: Fast software AES encryption. In: Hong, S., Iwata, T. (eds.) FSE 2010. LNCS, vol. 6147, pp. 75–93. Springer, Heidelberg (2010). doi:10.1007/978-3-642-13858-4_5

25. Osvik, D.A., Shamir, A., Tromer, E.: Cache attacks and countermeasures: the case of AES. In: Pointcheval, D. (ed.) CT-RSA 2006. LNCS, vol. 3860, pp. 1–20. Springer, Heidelberg (2006). doi:10.1007/11605805_1

26. ARM Holdings plc: mbed TLS v2.3.0. https://tls.mbed.org/

27. ARM Holdings plc: ARM's Cortex-M and Cortex-R embedded processors (2015). http://www.arm.com/zh/files/event/2_2015_ARM_Embedded_Seminar_Ian_Johnson.pdf

28. RealTimeLogic: SharkSSL/RayCrypto v2.4 crypto library - benchmarks with ARM Cortex-M3. https://realtimelogic.com/products/sharkssl/Cortex-M3/

29. STMicroelectronics: RM0038 reference manual (2015). http://www2.st.com/content/ccc/resource/technical/document/reference_manual/cc/f9/93/b2/f0/82/42/57/CD00240193.pdf/files/CD00240193.pdf/jcr:content/translations/en.CD00240193.pdf

30. STMicroelectronics: RM0090 reference manual (2015). http://www2.st.com/content/ccc/resource/technical/document/reference_manual/3d/6d/5a/66/b4/99/40/d4/DM00031020.pdf/files/DM00031020.pdf/jcr:content/translations/en.DM00031020.pdf

31. Stoffelen, K.: Instruction scheduling and register allocation on ARM Cortex-M. In: Software Performance Enhancement for Encryption and Decryption, and Benchmarking - SPEED-B (2016). http://ccccspeed.win.tue.nl/papers/armscheduler-final.pdf
32. Trichina, E.: Combinational logic design for AES SubByte transformation on masked data. Cryptology ePrint Archive, Report 2003/236 (2003). http://eprint.iacr.org/2003/236/
33. Tromer, E., Osvik, D.A., Shamir, A.: Efficient cache attacks on AES, and countermeasures. J. Cryptol. **23**(1), 37–71 (2010). http://people.csail.mit.edu/tromer/papers/cache-joc-official.pdf

Efficient Symmetric Primitives

Hold Your Breath, PRIMATEs Are Lightweight

Danilo Šijačić[1]([⊠]), Andreas B. Kidmose[2], Bohan Yang[1], Subhadeep Banik[3],
Begül Bilgin[1], Andrey Bogdanov[2], and Ingrid Verbauwhede[1]

[1] ESAT/COSIC, KU Leuven and iMinds, Leuven, Belgium
{danilo.sijacic,bohan.yang,begul.bilgin,
ingrid.verbauwhede}@esat.kuleuven.be
[2] Technical University of Denmark, Kongens Lyngby, Denmark
s113242@student.dtu.dk, anbog@dtu.dk
[3] Temasek Labs, Nanyang Technological University, Singapore, Singapore
bsubhadeep@ntu.edu.sg

Abstract. This work provides the first hardware implementations of PRIMATEs family of authenticated encryption algorithms. PRIMATEs are designed to be lightweight in hardware, hence we focus on designs for constrained devices. We provide several serial implementations, smallest of which requires only 1.2 kGE. Additionally, we present a variety of threshold implementations that range from 4.7 kGE to 10.3 kGE.

The second part of this work presents a design of a lightweight PRIMATEs coprocessor. It is designed to conform versatile use of the core permutation, which allows implementation of the entire PRIMATEs family, with small differences in hardware. We implement HANUMAN-80 coprocessor, adapted for a 16-bit microcontroller from the Texas Instruments MSP430 family of microcontrollers. The entire HANUMAN-80 coprocessor is tested on a Spartan-6 (XC6SLX45) development board, where it occupies 72 slices (1.06% of available resources). ASIC synthesis yields a 2 kGE implementation using 90 nm library, achieves 33 kbits/sec throughput at 100 kHz operating frequency. It dissipates 0.53 μW of power on average, resulting in energy consumption of 15.60 pJ/bit.

Keywords: PRIMATEs · CAESAR · Authenticated encryption · Hardware implementation · Threshold implementation · Lightweight

1 Introduction

Motivation. Emerging Internet of Things (IOT) technologies require a swarm of lightweight devices scattered in our surroundings. Various sensors, actuators, or authenticators, have to provide reliable, uninterrupted service, while protecting users' privacy and data confidentiality through encryption, and data authenticity and integrity through authentication. Since adversaries may easily gain access to these devices, protection against physical attacks must be taken into account. Moreover, all of this has to be achieved at a very low price in terms of chip area, power, and energy consumption. While exact constraints vary between different kinds of these devices, we believe that passively powered Radio Frequency IDentification (RFID) tags—which are used for identification,

© Springer International Publishing AG 2017
R. Avanzi and H. Heys (Eds.): SAC 2016, LNCS 10532, pp. 197–216, 2017.
https://doi.org/10.1007/978-3-319-69453-5_11

access management and shipment tracking, handling payments; and are great assets in aiding medical treatment—present the worst-case in terms of area and power-budget limitations. Even though the notion of lightweightness seems subjective and application-bound, statements from industry [28] and research community [6,13,14] agree that the area footprint of the cryptographic algorithm must not exceed 2000 two-input NAND-gates equivalent (GE) in the selected library. Having at least 12 kbits/sec throughput at the operating frequency of 100 kHz is the only bound used by researchers [6,14] whereas industry requires having 1 bit/cycle [28]. Unfortunately, there are no widely accepted upper and lower bounds for low power and high throughput respectively, even though the discussions suggest that it is of interest [22,28]. Lastly, in terms of average power usage industry suggests between 1 and $10\,\mu W/MHz$, with peaks between 3 and $30\,\mu W$, respectively.

Many standardized cryptographic algorithms, such as the Advanced Encryption Standard (AES) [23] of which the smallest implementation requires 2400 GE, are unfit for the lightweight area of application especially when they are wrapped with a mode of operation to provide both encryption and authentication. This results in the increasing number of stream ciphers [20], block ciphers [6,14,17,19,30] and hash functions [5,12,18] together with the recent standardization of the lightweight block cipher PRESENT which can be implemented using 1000 GE [27].

Traditionally, the security goals of a system are achieved using generic compositions of cryptographic primitives providing only authentication or encryption often resulting in exploitable weaknesses [2,7]. This problem of authenticated encryption is formalized in [26], where a set of possible generic schemes named *authenticated-encryption with associated data* (AEAD) is discussed. Advancement in lightweight cipher design, and the formalization of the problem, followed by discussions in research community lead to the Competition for Authenticated Encryption: Security, Applicability, and Robustness (CAESAR) [1] which is running since 2014 with the goal of selecting a portfolio of AEAD ciphers. Currently 29 s-round candidates are being analyzed for security, software and hardware performance. Out of several candidates that claim to be lightweight, PRIMATEs [3] family grasped our attention for their claims of versatile usability, and efficiency in hardware.

Related Work. PRIMATEs [3] is a family of single-pass nonce-based AEAD schemes. All members of PRIMATEs are designed for constrained hardware. They differ slightly to provide trade-offs between security and performance. High level of granularity allows PRIMATEs to find application in a number of lightweight scenarios. Authors claim that PRIMATEs can efficiently be protected against Side-Channel Analysis (SCA), especially Differential Power Analysis (DPA) [25]. Namely, non-linear part of PRIMATEs has low algebraic degree, which results in efficient Threshold Implementations (TI) [24]. TI provides provable security against DPA for symmetric key algorithms (e.g., [11]).

Until now, only a reference software implementation of PRIMATEs is provided. The claims on efficiency in hardware, of unprotected and SCA-resistant implementations, are still to be examined.

Contribution. We challenge lightweightness and versatility claims by focusing on low-end designs. Namely, we design several lightweight architectures in order to analyze area, throughput and energy trade-offs. We discuss the overall performance, and how our implementations can be used in practice. Additionally, we provide a variety of TI to examine efficiency of SCA resistant implementations. We discuss the overall performance, and how our implementations can be used in practice. Furthermore, in order to accommodate practical lightweight scenarios, we present a PRIMATEs interface, designed to minimize area and latency overhead. Lastly, we design and implement a PRIMATEs coprocessor based on the aforementioned interface.

2 Preliminaries

We inherit the notation suggested by the PRIMATEs designers [3]. Namely, calligraphic, capital and small letters represent a set, an element of the set and the bit size of the element respectively, i.e. $\mathcal{X} := \{0,1\}^x$, $X \in \mathcal{X}$ and $|X| = x$. Let $X \in \{0,1\}^x$ and $Y \in \{0,1\}^y$, then $X||Y \in \{0,1\}^{x+y}$ represents the concatenation of X and Y.

2.1 AEAD Scheme

An AEAD scheme is defined by the three tuple $\Pi = (\mathcal{K}, \mathcal{E}, \mathcal{D})$ as follows:

Key Space. $\mathcal{K}$ is a non-empty set of k-bit strings, i.e. $\mathcal{K} := \{0,1\}^k$.

Encryption. $\mathcal{E}$ is a deterministic algorithm which returns a pair of strings $(C, T) = \mathcal{E}_K(N, A, M)$; where: the secret key $K \in \mathcal{K}$, the public nonce $N \in \mathcal{N} := \{0,1\}^n$, the public associated data $A \in \mathcal{A} := \{0,1\}^*$, the message $M \in \mathcal{M} := \{0,1\}^*$, the ciphertext $C \in \mathcal{C} := \{0,1\}^*$, and the tag $T \in \mathcal{T} := \{0,1\}^t$. The algorithm must work even if $|M| = 0$ and/or $|A| = 0$.

Decryption. $\mathcal{D}_K(N, A, C, T)$ is a deterministic algorithm that generates the pair (M, T'). $\mathcal{D}$ returns the value M to user if $T = T'$. Otherwise $\mathcal{D}$ returns a unique symbol $\perp$. It is possible to release the message even if the tags do not match if the AEAD scheme follows certain properties [4].

2.2 PRIMATEs

PRIMATEs is a family of three modes of operation named APE, HANUMAN and GIBBON [3] with sponge-like construction [8]. Namely, they rely on a permutation which operates on a binary state $B \in \{0,1\}^b$, comprised of the *rate* $R \in \{0,1\}^r$, and the *capacity* $C \in \{0,1\}^c$ (i.e. $B = R||C$).

Each mode of operation may provide two levels of security. The security level $s \in \{80, 120\}$ defines several parameters, as described in Table 1. The input block size is $r = 40$ bits independent of the mode or the security level.

PRIMATE. PRIMATEs family is based on a set of permutations, called PRI-MATE permutations. Depending on the security level s, two subsets are distinguished, PRIMATE-80 and PRIMATE-120. PRIMATE-80 (resp. PRIMATE-120) is based on P80 (P120), a 200-bit (resp. 280-bit) core permutation. Permutations are designed as substitution-permutation networks (SPNs).

In both cases states are divided into 5-bit *elements*, with big-endian encoding. Elements themselves are arranged into matrices with 5 (resp. 7) 8-element rows for PRIMATE-80 (resp. PRIMATE-120). The element in the ith row and j^{th} column of this matrix is denoted by $a_{i,j}$ where $i \in \{0, \ldots, 4\}$ (resp. $i \in \{0, \ldots, 6\}$) and $j \in \{0, \ldots, 7\}$. The first row $a_{0,*}$ in the state matrix contains the rate of the state, and will henceforth be referred to as the rate row. P80, and P120 are calculated using a sequence of four transformations described as follows:

1. *SubElements* (SE) is the only non-linear transformation. It consists of an element-wise permutation $X \rightarrow S(X) : \{0,1\}^5 \rightarrow \{0,1\}^5$ (S-Box) applied to each element of a state.
2. *ShiftRows* (SR) performs cyclical shifts of each row for a different number of elements. Row i is shifted left by $s_i = \{0, 1, 2, 4, 7\}$ in P80, or by $s_i = \{0, 1, 2, 3, 4, 5, 7\}$ in P120.
3. *MixColumns* (MC) operates on a state column at a time. It is a left-hand multiplication by a 5×5 (7×7) Maximum Distance Separable (MDS) matrix [3]. The matrices are chosen in a way that allows recursive calculation of a smaller matrix five (resp. seven) times.
4. *ConstantAddition* (CA) modifies a single state element $a_{1,1}$ by bitwise XOR-ing a 5-bit constant in each round.

Round constants are generated by a 5-bit Fibonacci LFSR [3]. Varying on the sequence of values sampled from this LFSR and the number of rounds, four permutations p_1, p_2, p_3, and p_4 are derived from the core permutation (either P80 or P120), as shown in Table 2.

PRIMATE Modes of Operation. All modes of operation are generic constructs, designed based on Sponge [8] methodology principles with slight differences in input output behavior, parameter size and used permutations. Table 1 gives an overview for the latter two differences. Please refer to [3] for details on the former difference. We only emphasize the fact that decryptions of HANUMAN and GIBBON do not require the inverse transformation of PRIMATE whereas APE does.

Table 1. PRIMATEs modes of operation.

PRIMATEs	APE-s	HANUMAN-s	GIBBON-s
k	$2s$	s	s
t	$2s$	s	s
n	s	s	s
PRIMATE	p_1	p_1, p_4	p_1, p_2, p_3

Table 2. PRIMATE permutations.

PRIMATE	p_1	p_2	p_3	p_4
# of rounds of P-s	12	6	6	12
Init. val. of the LFSR	1	24	30	24

3   Implementations of PRIMATE

Following the design rationale of the PRIMATEs family, we focus on hardware implementations for heavily constrained devices. We abstain from using power-saving techniques (e.g., clock gating). Instead we perform architectural optimizations that lead to reduced area and power consumption. Lastly, we strive towards the 12 kbit/sec throughput at the operating frequency of 100 kHz, discussed in Sect. 1, as the performance criterion.

As in all Sponge-based designs, the majority of implementation cost of PRIMATEs comes from core permutations. Therefore, we investigate several ways to serialize P80 and P120. Additionally, we provide round-based versions of both core permutations.

Lastly, in order to benefit from the granulated nature of the PRIMATE family we have fragmented our implementations into several hierarchical levels, thus creating a generic serial-implementation strategy independent of the permutation design. Therefore, we present the design of the control logic required for generation of p_1, p_2, p_3, and p_4 separately from the core permutations.

3.1   PRIMATE Permutations' Control

The 5-bit Fibonacci LFSR used for CA transformation is one of the lightweight features predicted by design. Firstly, depending on the selected p_i, rounds can be decoded from the values of the LFSR, thereby alleviating the need for additional counters. Secondly, p_i's defer from one another only by the sequence of Fibonacci constants. Therefore, for each of the modes of operations, regardless of the security level, control module MODECTRL is realized using simple hardware.

Permutation p_i starts by loading a corresponding constant of the Fibonacci sequence into the LFSR. After each round—underlying permutation core must provide a RNDDONE signal—LFSR progresses along its sequence. Output of the LFSR is used as a round constant RCON for the underlying permutation core. Lastly, small 5-bit decoders are used to detect rounds of interest (e.g., first, last) for the control of upper layers of logic.

3.2   Core Permutations P80 and P120

In combination with the control module from Sect. 3.1, any of the implementations from this section can be used to provide encryption and decryption of HANUMAN and GIBBON as well as APE encryption. Since APE requires inverse of p_1 we have abstained from implementing this functionality. Justification for this is twofold: from the area-performance-power perspective implementation of p_1 is negligibly different from its inverse; from the usability perspective it is likely that heavily constrained devices perform only encryption, while the decryption is performed in the backend. Lastly, note that these implementations include only the functionality of the core permutation, i.e. they allow computation of only one round. Depending on the way the core is used, different overhead will be introduced in the design (e.g., a feedback multiplexer). Nevertheless, we find this sort of results useful, for the various use-cases that may be anticipated for these architectures. More discussion on this topic is given in Sect. 5.

Round-Based Implementations of P80 and P120. Figure 1, where each line represents a 5-bit element value, depicts the round-based architecture of P80, called P80-1. S-Boxes, MDS matrix multiplications, and constant addition, are implemented as combinational networks. SR transformation can be realized by rewiring of rows, hence it is free in hardware. P120 version (P120-1) is obtained using the same design approach. Lastly, these implementations include a state-sized register—for fair comparison.

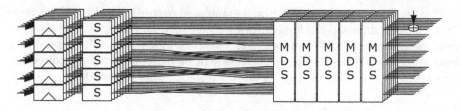

Fig. 1. Round-based architecture of the core permutation, P80-1.

Serial Implementations of P80 and P120. Due to the relatively large state, and combinatorial logic designed to be efficient in hardware, major cost of a serial implementation of P80 and P120 comes from the register file used to store the state. Consequently, we aim to minimize the number of multiplexed inputs to every bit of the State Register File (SRF), and to avoid additional registers in the design. Therefore, we abstain from using additional multiplexers for controlling data flow, and design the SRF as a column-wise FIFO register. Hence, all serial implementations of P80 feature a 25-bit data path, while P120 implementations use a 35-bit data path; which correspond to the number of bits in a column of the state matrix.

Lastly, each of the permutation cores requires a dedicated Finite State Machine (FSM) for controlling the data flow, in addition to the MODECTRL module used to iterate rounds of p_i.

P80-9 and P120-9. Figure 2 depicts a 9 clock cycle serial implementation (P80-9). Here, the SRF has only two modes of operation MC, and SR. When MC is active SRF is configured as a 25-bit FIFO register which feeds the data into the combinational network at its output. This mode is used for data input, as well. SR mode is always active during the first cycle of computation, during which it rewires the SRF to perform the SR transformation. P120 version (P120-9) is obtained using the same design approach.

P80-41 and P120-57. Based on the 9 clock cycle approach, we present another serial version by serializing the MC step. Namely, 5 matrix multiplications that are required for MC transformation of PRIMATE-80, are now performed in 5 clock

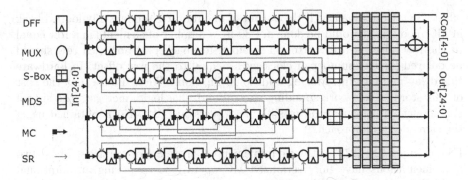

Fig. 2. Data path architecture of the P80-9 permutation core.

cycles using 5 times less hardware. The same concept applies for PRIMATE-120 with the difference that factor of 7 applies instead of 5 (since the state of PRIMATE-120 has 7 rows). Therefore instead of 8 additional cycles 5×8 (for PRIMATE-80), or 7×8 (for PRIMATE-120) are required to perform this computation. SR transformation remains performed in a single clock cycle as before.

P80-16 *and* P120-16. Figure 3 depicts a 16 clock cycle serial implementation (P80-16). Instead of serializing MC transformation, we serialize the SR transformation. Namely, SR is performed by shifting the position of the column-wise input by the number of shifts prescribed by the SR operation for each row of SRF. After 8 clock cycles of shifting in this manner, SRF is reconfigured to perform SE and MC for another 8 clock cycles, as done previously. This way a number of 2-to-1 multiplexers around the SRF is traded for additional latency of 7 clock cycles, resulting in minimal multiplexer overhead.

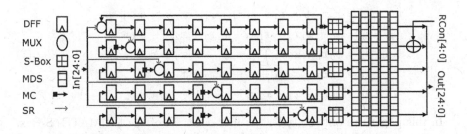

Fig. 3. Data path architecture of the P80-16 permutation core.

On the downside, this core can not preserve state between two consecutive rounds. Namely, in both configurations SRF is written in 8 clock cycles, using 2 different patterns: regular column-wise shift (MC), and columns-wise shift with offset (SR). Consequently, if the output of one round is sequentially looped back in MC mode (as it would be done in P80-9), a number of elements in a row equal to the row offset is overwritten. Therefore, feedback path for each row should be delayed for the number of clock cycles equal to the row offset. In hardware this delay maps to introducing additional flip-flops. We estimate that the cost of this storage, and corresponding control and glue logic, increases the size of this implementation beyond feasible. P120 version (P120-16) is obtained using the same design approach.

P80-95 *and* P120-127. Lastly, we implemented single S-Box versions. These two implementations have fully serialized MC operations (requiring an additional cycle to load new column), and require 7 clock cycles to perform SR operation. Since 5-bit PRIMATEs S-Boxes are small (30–40 GE), we believe that area savings are not worth the performance impact when it comes to unprotected implementations. Nevertheless, this approach may lead to major area savings with TI versions; since shared S-Boxes take 246 GE and 255 GE (see Sect. 4.1).

4   Threshold Implementations

Since the application of TI on affine functions is trivial [24], we mainly focus on the sharing of the S-box. Then, we briefly discuss the shared architectures.

4.1   The Shared S-Box

The PRIMATEs S-Box is an almost bent permutation with excellent linear and differential properties and is affine equivalent to the cubic power mapping in $GF(2^5)$. As can be seen from its algebraic normal form in Eq. (1) (with x_i and y_i correspond to input and output bits assuming 0 to be the index for MSB), the S-Box is quadratic which makes it suitable for efficient TI.

$$
\begin{aligned}
y_0 &= x_0x_2 + x_0x_3 + x_1x_4 + x_1 + x_2x_3 + x_2 + x_3 \\
y_1 &= x_0 + x_1x_2 + x_1x_3 + x_2x_3 + x_2x_4 + x_3 \\
y_2 &= x_0x_1 + x_0x_4 + x_0 + x_1 + x_2x_3 + x_2x_4 \\
y_3 &= x_0x_2 + x_0x_4 + x_0 + x_1x_2 + x_3x_4 \\
y_4 &= x_0x_3 + x_1 + x_2x_4 + x_4 + 1
\end{aligned}
\tag{1}
$$

In this paper, we only consider the first-order TI of the PRIMATEs S-Box. Since at least $d + 1$ shares are required to implement any function of degree d, we first implement the shared S-Box with 3 shares. We provide the component functions of y_0 for this version in Eq. (2) as an example where x_i^j refers to the j-th share of x_i.

$$y_0^1 = ((x_0^2 + x_0^3)(x_2^2 + x_2^3)) + ((x_0^2 + x_0^3)(x_3^2 + x_3^3)) +$$
$$((x_1^2 + x_1^3)(x_4^2 + x_4^3)) + ((x_2^2 + x_2^3)(x_3^2 + x_3^3)) + x_2^2 + x_3^2 + x_1^2$$
$$y_0^2 = (x_0^1 x_2^3 + x_0^3 x_2^1 + x_0^1 x_2^1) + (x_0^1 x_3^3 + x_0^3 x_3^1 + x_0^1 x_3^1) +$$
$$(x_1^1 x_4^3 + x_1^3 x_4^1 + x_1^1 x_4^1) + (x_2^1 x_3^3 + x_2^3 x_3^1 + x_2^1 x_3^1) + x_1^3 + x_2^3 + x_3^3 \qquad (2)$$
$$y_0^3 = (x_0^1 x_2^2 + x_0^2 x_2^1) + (x_0^1 x_3^2 + x_0^2 x_3^1) +$$
$$(x_1^1 x_4^2 + x_1^2 x_4^1) + (x_2^1 x_3^2 + x_2^2 x_3^1) + x_1^1 + x_2^1 + x_3^1$$

This particular sharing fails to satisfy the uniformity property of TI. Since we were not able to find a uniform 3-share TI with our limited computational resources, we re-mask the S-box output in order to attain provable security. Re-masking is performed similarly to [10] in order to reduce the fresh randomness requirement.

Additionally, we implement a 4-share uniform TI which is provided for y_0 in Eq. (3), as an example. Even though this implementation has bigger area compared to its 3-share counterpart, it does not require fresh randomness after the initial sharing. This may lead to significant savings once a random number generator is included in the design.

$$y_0^1 = ((x_0^2 + x_0^3 + x_0^4)(x_2^2 + x_2^3 + x_2^4)) + ((x_0^2 + x_0^3 + x_0^4)(x_3^2 + x_3^3 + x_3^4)) +$$
$$((x_1^2 + x_1^3 + x_1^4)(x_4^2 + x_4^3 + x_4^4)) + ((x_2^2 + x_2^3 + x_2^4)(x_3^2 + x_3^3 + x_3^4)) +$$
$$x_1^2 + x_2^2 + x_3^2$$
$$y_0^2 = ((x_0^1(x_2^3 + x_2^4)) + (x_2^1(x_0^3 + x_0^4)) + (x_0^1 x_2^1)) +$$
$$((x_0^1(x_3^3 + x_3^4)) + (x_3^1(x_0^3 + x_0^4)) + (x_0^1 x_3^1)) + ((x_1^1(x_4^3 + x_4^4)) + (x_4^1(x_1^3 + x_1^4)) +$$
$$(x_1^1 x_4^1)) + x_1^3 + ((x_2^1(x_3^3 + x_3^4)) + (x_3^1(x_2^3 + x_2^4)) + (x_2^1 x_3^1)) + x_2^3 + x_3^3$$
$$y_0^3 = ((x_0^1 x_2^2) + (x_0^2 x_2^1)) + ((x_0^1 x_3^2) + (x_0^2 x_3^1)) + ((x_1^1 x_4^2) + (x_1^2 x_4^1)) +$$
$$x_1^4 + ((x_2^1 x_3^2) + (x_2^2 x_3^1)) + x_2^4 + x_3^4$$
$$y_0^4 = x_1^1 + x_2^1 + x_3^1$$

$$(3)$$

4.2   Architectures

We implement a total of 6 threshold implementations of PRIMATEs. Firstly, implementations vary in number of shares. Secondly, we utilize different degrees of serialization to ensure tradeoffs between cost and performance. For convenience we name them P80-9³, P80-9⁴, P120-9⁴, P80-95³, P80-95⁴, P120-127⁴, where the superscript numbers indicate the number of shares.

Figure 4 depicts the datapath of P80-9³. For each of the S-Box shares a copy of the MC circuit, and an additional SRF, needs to be added to the design to maintain the masked state. The additional control logic required to implement the circuit is minimal. This is in contrast to the serialized threshold implementation of AES [23], where the S-Box needed to be implemented using pipelined stages. The structure of the Primates S-Box obviates the need for such pipelining.

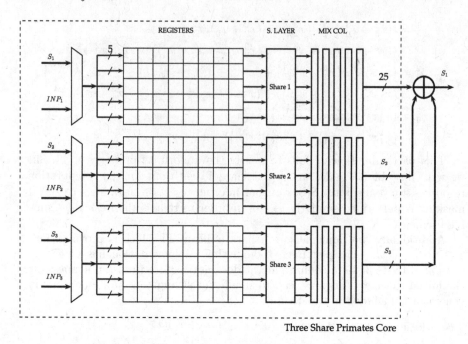

Fig. 4. Data path of the 3-share P80-9 permutation core.

5   Implementation Results

All implementations are synthesized from RTL code written in VHDL. We use Synopsys Design Compiler v2015.06 to synthesize each design. Furthermore, we use Synopsys PrimeTime v2015.06 with PX add-on to perform more accurate static timing analysis and switching activity based power estimation.

We provide synthesis results using 2 standard-cell libraries: Faraday UMC 90 nm, generic core in Low-K RVT process, 1.2 V power supply (UMC 90), and NangateOpenCellLibrary 45 nm, PDKv1_3_v2010_12 (NAN 45), in Table 3.

Along the maximum frequency and area we provide performance figures at the operating frequency of 100 kHz. These include throughput, implementation efficiency (throughput per unit of area), dynamic and static power consumption, and energy efficiency (energy per state bit in every round). We use these figures to benchmark P80 and P120 permutations, in order to identify the most suitable lightweight scenario for each of the permutation cores.

Firstly, we observe that the control logic MODECTRL discussed in Section utilizes negligible amount of resources—under 100 GE. All control modules have shorter critical paths than any of the permutation cores, therefore they can not pose as a computational bottleneck. Secondly, we observe that areas of P80 and P120 scale in a linear fashion with respect to the state-size. Therefore, we focus discussion on the P80, for simplicity.

Table 3. Post-synthesis hardware implementation results.

Design	Library	Max. Freq. [MHz]	Area [kGE]	@ 100 kHz				
				T'put [$\frac{Mbit}{s}$]	Impl. Eff'cy [$\frac{Mbit}{kGE\cdot s}$]	D. Pwr [μW]	S. Pwr [nW]	E. Eff'cy [$\frac{pJ}{bit}$]
ApeCtrl	UMC90	361.58	0.06	—	—	0.01	1.15	—
	NAN45	606.76	0.09	—	—	0.01	1.37	—
HanCtrl	UMC90	487.02	0.05	—	—	0.01	1.02	—
	NAN45	683.06	0.08	—	—	0.01	1.11	—
GibCtrl	UMC90	599.23	0.05	—	—	0.01	1.04	—
	NAN45	749.29	0.07	—	—	0.01	1.23	—
P80-1	UMC90	179.60	3.68	20.00	5.43	2.32	74.00	0.12
	NAN45	341.53	4.72		4.24	1.63	83.30	0.09
P80-9	UMC90	256.74	1.43	2.22	1.56	0.74	29.80	0.35
	NAN45	439.77	2.05		1.08	0.78	32.80	0.37
P80-16	UMC90	509.50	1.20	1.25	1.04	0.68	25.20	0.57
	NAN45	896.38	1.78		0.70	0.42	27.60	0.36
P80-41	UMC90	204.18	1.32	0.49	0.37	0.46	26.70	0.99
	NAN45	267.61	1.98		0.25	0.30	31.80	0.68
P120-1	UMC90	142.27	6.32	28.00	4.42	4.61	137.00	0.17
	NAN45	281.31	8.23		3.51	3.65	159.00	0.14
P120-9	UMC90	183.69	2.17	3.11	1.43	1.26	46.00	0.42
	NAN45	490.17	3.10		1.00	1.17	165.00	0.43
P120-16	UMC90	447.21	1.82	1.75	0.96	1.13	38.60	0.67
	NAN45	722.33	2.69		0.65	0.80	42.60	0.48
P120-57	UMC90	114.32	1.87	0.49	0.26	0.63	36.80	1.37
	NAN45	239.24	2.79		0.18	0.40	44.80	0.91
P80-9[3]	UMC90	162.60	5.18	2.22	0.428	0.81	1.04	0.36
	NAN45	251.25	7.20		0.308	0.49	65.90	0.25
P80-95[3]	UMC90	151.74	4.72	0.21	0.044	0.60	0.86	2.55
	NAN45	315.45	6.33		0.033	0.35	54.40	1.92
P80-9[4]	UMC90	133.15	6.15	2.22	0.360	0.87	1.07	0.39
	NAN45	249.33	9.24		0.240	0.53	86.20	0.28
P120-9[4]	UMC90	79.05	10.30	3.11	0.302	1.41	2.34	0.45
	NAN45	104.20	13.84		0.225	0.81	125.00	0.30
P80-95[4]	UMC90	181.49	6.19	0.21	0.033	0.78	1.12	3.71
	NAN45	298.50	8.31		0.025	0.50	70.40	2.71
P120-127[4]	UMC90	204.45	8.60	0.22	0.025	1.04	1.60	4.72
	NAN45	300.00	11.51		0.019	0.70	96.50	3.61

On the one hand, P80-1 is dominant in throughput and energy efficiency. Area costs of computing the entire round in parallel make P80-1 large for most resource-constrained devices (e.g., RFID tags). Therefore P80-1 is better suited for battery powered devices, and applications where high throughput and long battery life is of greater interest (e.g., wireless sensor nodes).

On the other hand, serial implementations seem very suited for constrained devices. P80-16 is the smallest implementation, which requires only 1.2 kGE. Unfortunately, this low resource cost comes at the requirement of storing value of the state between rounds externally. Hence, feasibility of this implementation strongly depends on the specifics of the application and resources of the platform that relies on PRIMATEs-80. Also, this implementation has maximum frequency considerably higher than the rest. Second largest implementation, P80-41 removes the problem of external storage requirement, and has the lowest power consumption. Decrease in power consumption is due to the decreased size of combinatorial logic used for MC transformation. Nevertheless, it is not followed by area decrease, since SRF of P80-41 is more costly. Namely, SRF size is increased by additional multiplexers[1] required for performing SR transformation in one clock cycle, as well as additional control that makes rest of the SRF idle while MC transformation is performed on a column. Still, the pitfall of this implementation is the heavily reduced throughput due to the high latency. Lastly, P80-9 is 50% (320%) more efficient than P80-16 (P80-41), at the price of 20% (8%) area, and 9% (61%) power, increase.

When it comes to TI, we see that due to the efficient design of the S-Box, there is no need for pipelining S-Boxes. Therefore, the circuit size is increased approximately linearly with respect to the number of shares.

6   Usability, Comparison, and Discussion

Implementation results presented in Sect. 5 serve the purpose of benchmarking the core permutation. Here we discuss how these results fit real-world applications. Figure 5 gives estimated encryption throughput of PRIMATEs based on different serial implementations, with respect to the size of authenticated data and plaintext in bytes; assuming 100 kHz operating frequency. Throughput is estimated based on the latency of encryption in all 3 modes of operation, APE, GIBBON, HANUMAN. Due to the fact that PRIMATEs may be used for applications that require encryption and (or) authentication of very short messages, we include the latency of initialization for each mode. Lastly, note that Fig. 5 does not take any interface overhead into account, other than assuming that input of data into state (e.g., initialization of key and nonce), and XOR-ing of data into state (e.g., for tag generation) introduces latency of one round of the core permutation.

Since GIBBON employs p_2, and p_3 which use only 6 rounds, it is asymptotically twice as fast as the other two modes, allowing throughput up to 70 kbits/sec for GIBBON using P80-9. Therefore it is preferred when performance takes precedence over slightly lowered security. Furthermore, APE is slightly

---

[1] Note that further area decrease SRF can be done by replacing flip-flop-multiplexer pairs with scan flip-flops. This has no practical significance as scan flip-flops are intended for test inputs in the during production.

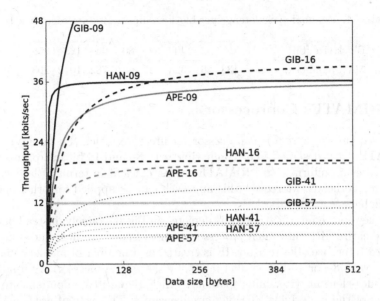

Fig. 5. Estimated encryption throughput at 100 kHz operating frequency.

slower than HANUMAN, due to the initial processing of the nonce (cf. [3]) and the highest level of security that follows. Taking the 12 kbits/sec throughput at 100 kHz operating frequency into account, serializing MC transformation in P80-41, and P120-57 versions makes them suitable for devices where their low power consumption outweighs low throughput. Namely, as p_1, and p_4 permutations which are effectively used for encryption (authentication) of each block require 12 rounds, this results in 492 clock cycles latency for P80-41 (684 for P120-57). On the other hand, 9 (108 cycles per block in APE, HANUMAN; 54 cycles per block in GIBBON), and 16 (144 cycles per block in APE, HANUMAN; 72 cycles per block in GIBBON) clock cycle version satisfies these requirements with a significant margin; and we deem them very usable for most constrained application even with significant interface overhead. Moreover, since it requires no external storage, we recommend P80-9, and P120-9 as the most sound choices.

Lastly, we look into some of the industrial standards, devised for lightweight devices (e.g., smartcards). For example, EPCGlobal Gen2 and ISO/IEC 18000-63 passive UHF RFID air interface standards discussed in [28] prescribes the following constraints: clock frequency (1.5–2.5 MHz) and response latency (39.06–187.50 µs). These constraints allow RFID devices between 58 and 468 clock cycles to respond. Considering this type of constraints, and the 12 kbits/sec at 100 kHz requirement, we believe that lightweight ciphers can be fairly compared based on the metric presented in Table 4. Assuming a fixed operating frequency, and the corresponding throughput, performance constraints of implementations can be compared solely based on the block size. We believe this is a practical usability metric, which can be easily used in conjecture with area, power, and energy constraints.

Table 4. Maximal cycle latency per block assuming 12 kbits/sec, 100 kHz.

Block size [bit]	16	32	40	64	80	96	128	256
Max. number of cycles	132	261	326	521	651	782	1042	2084

7 PRIMATEs Coprocessor

In this section we present a coprocessor architecture which can be used for all PRIMATEs. It is designed to be compatible with 8-, and 16-bit microprocessors; and features an interface to PRIMATEs cores, efficient in terms of latency and hardware overheads. Moreover, this approach can be applied for other sponge-based ciphers (e.g., [9,15]).

The key to the interface design is depicted in Fig. 6. Namely, instead of mapping the entire SRF into microprocessor memory space, we introduced a single row-sized InterFace (IF) register. IF is treated as a number of memory mapped registers with 8-, or 16-bit parallel input, by the microprocessor. Alternatively, IF provides element-wise shift capability; which allows it to communicate with each row of the underlying SRF via circular shifts. This way of accessing SRF (row-wise) has multiple benefits: IF allows data to be written to the permutation core in block-pipeline fashion, effectively introducing clock cycle latency overhead equal to the number of elements in a row; it provides translation from the microprocessor word to element-sized word without any precomputation; element-sized data path conforms FIFO construction of each row, hence it results in zero area overhead; allows implementation of all PRIMATEs. Namely, all steps required for PRIMATEs schemes (cf. [3]) can be divided into two groups: computationally expensive p_i permutations, and computationally feasible data flow operations (e.g., data parsing and writing to the core, etc.). Therefore, we believe that the best design strategy is to leave data flow operations to a microprocessor (or an upper level FSM), and dedicate coprocessor to performing p_i. All required read, write, and XOR operations can be achieved using three types of row interfaces depicted in Fig. 6.

Namely, RC encapsulates a row of the SRF with all of its logic, can only be written to, and introduces no hardware overhead. RB is a row which can also be read-written via circular shifts, where reading introduces overhead of 5-bit multiplexer entry at the output (5-bit AND2 is 6.25 GE in UMC 90), and slightly more control logic. RA allows the data from IF register to be circularly shifted as is (same as RB), or circularly XOR-ed to the value stored in the row, depending on the value of the XOR bit of the instruction (see Table 5). This approach requires no additional latency for the encryption of blocks, as they are XOR-ed to the rate as they are being written to SRF. Lastly, RA row can be read without changing its content by a circular XOR of 0^{40}. The cost of each 5-bit XOR-multiplexer required to support XOR is 23.75 GE in UMC 90.

In the particular case of PRIMATEs IF register is 40-bits wide, hence it can be mapped to 5 8-bit registers, or 3 16-bit registers. Additionally, an 8-bit Instruction Register (IR) is required for PRIMATEs instructions, which can be mapped to the remaining byte of one of the 3 16-bit registers.

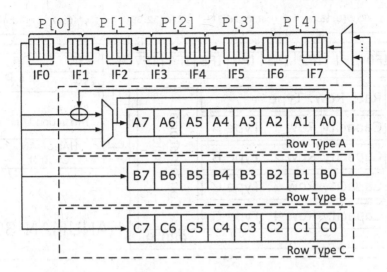

Fig. 6. Coprocessor interface and the XOR instruction support.

7.1   HANUMAN-80 Coprocessor

As an example, we design and implement a HANUMAN-80 coprocessor, based on the preferred P80-9 core. Subset of PRIMATEs micro-instructions, required for HANUMAN-80 encryption and decryption (cf. [3]), is given in Table 5. Top-level architecture, depicted in Fig. 7, is adapted to MSP430 microcontroller family [21].

Table 5. Instruction Set of the HANUMAN-80 coprocessor.

Mnemonic	Code	Description
RESET	0-------	Perform software reset
WAIT	1000-000	Put coprocessor in a idle state
P1	1----001	Perform p_1 permutation
P1S	1----101	Perform p_1 permutation with padding spill into capacity
P4	1----001	Perform p_4 permutation
RATEX	10011111	XOR in to rate
RATES	10010111	Shift in to rate
RDRATE	10011111	XOR in 0^{40} to rate; emulated rate read
CAP1S	10100111	Shift in to capacity row 1, R/W
CAP2S	10110111	Shift in to capacity row 2, R/W
CAP3S	11000111	Shift in to capacity row 3, W
CAP4S	11010111	Shift in to capacity row 4, W

Dash, "-", can be replaced by either zero or one.

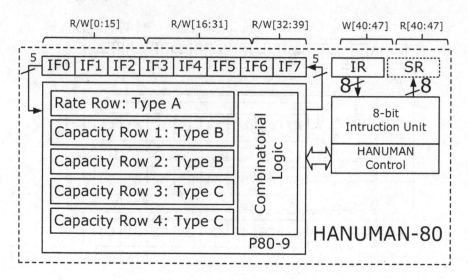

Fig. 7. HANUMAN-80 coprocessor architecture.

We use Spartan-6 FPGA (XC6SLX45-3CSG324) to implement and test our design, next to an OpenMSP430 implementation [16]. On this platform coprocessor fits in a total of 72 (1.06%) slices (206 FFs and 278 LUTs.) In ASIC, using UMC 90 standard-cell library from Sect. 5, the entire coprocessor requires 2 kGE. Note that HANUMAN-80 compliant P80-9 (with the data path arhictecture from Fig. 2) requires 1.69 kGE. Overhead of 0.26 kGE (18.68% larger than the raw P80-9 core of 1.43 kGE) includes all the glue logic; and entire control logic, including HANCTRL and the FSM of the coprocessor for fetching, decoding, and executing micro-instructions. Since each row has separate enable signal, area as well as power savings can easily be achieved by gating the clock instead of using flip-flops with enable. Furthermore overhead of 0.31 kGE is introduced for the 8-bit instruction unit and the 40-bit IF register, which enables circular access to SRF in a block-pipeline manner, allowing to almost negligible interface overhead. Alternatively, this register can be removed, and the SRF redesigned to be accessible to the microprocessor. This leads to area decrease; but also increases the latency, and makes it heavily dependent on the latency of the write cycle of the microprocessor, since pipeline feature is absent. Average dynamic power at the operating frequency of 100 kHz is 0.49 μW, while the 39.90 nW of power are dissipated statically; consuming 5.3 μW/MHz in total, which fits requirements of the industry [28]. Throughput estimated based on the 118 clock cycles (12×9 for p_i, 8 for circular shift, and 2 for instruction fetch and decode) latency per data block (asymptotically) is 33 kbits/sec. This is a valid assumption, since no additional storage is required for pre-computing and storing the initialization phase, while tag generation is simply the XOR operation. Under these assumptions, estimated energy consumption is 15.60 pJ/bit.

Implementation Comparison. Performing a fair evaluation of different candidates is a difficult task for several reasons. Firstly, it requires a common interface, equally suitable for all candidates. Secondly, broad area of use cases, ranging from RFID chips to high-end hardware accelerators, might not make use of a single interface objective enough. Thirdly, implementers present their results in different technological libraries and processes of each library; which makes area and power comparison more difficult. Consistently with the lightweight tone of this work, and assumed real-world limitations, we use Table 6 to benchmark several implementations of second-round CAESAR candidates, against the smallest implementation of AES [23]. Coherently to the discussion from Sect. 5, we use area, clock cycle latency(# TCLK), and block size as main comparison parameters. Additionally, we present how well does each implementation fit the constraints setting from Table 4 (lower percentage is better). Only two candidates are chosen at this time for the lack of lightweight ASIC implementations of others.

Table 6. Implementation comparison.

Design	Tech.	Area [kGE]	Block size [bit]	# TCLK	$\frac{\text{\# TCLK}}{\text{Table 4}} \cdot 100$ [%]
AES♣ [23]	UMC 180	2.4	128	226	21.69
HANUMAN-80	UMC 90	2.00	40	118	36.20
GIBBON-80	UMC 90	≈2.00◇	40	64	19.63
Minalpher [29]	NAN 45	2.81♣	256	304	14.59
Ascon-64 [15]	UMC 90	5.86	64	354	67.95
Ascon-x-low-area [15]	UMC 90	3.75	64	3072	589.63

♣UMCL18G212T3 based on a UMC 180 nm library. ◇GIBBON-80 coprocessor estimated area. ♣Authors state that no optimization is performed.

8   Conclusions and Future Work

Based on the hardware implementations of PRIMATEs family of authenticated ciphers, and adjacent discussion we find PRIMATEs to be very suitable for constrained devices. Namely, uninterfaced implementations of the permutation that lies in the heart of PRIMATEs takes only 60–72% of the 2 kGE lightweightness criteria. As shown by example of the HANUMAN-80 coprocessor, this leaves plenty of space for the implementation of interface and control logic. Furthermore, without any circuit-level optimizations (e.g., clock gating, power gating), or picking technology library for low-power application, our coprocessor fits the all of the commonly accepted criteria in practice; in terms of throughput, area, and average power consumption proposed in [28]. Additionally, presented variety of TI shows that securing PRIMATEs against first order DPA can be achieved using as little as 4.3 kGE.

Additionally, by looking at the PRIMATEs AEAD schemes in [3], and the design of our interface, we observe that by simply using different row interfaces depicted in Fig. 6 coprocessor can be turned into GIBBON-80, and APE-80. Similarly, by using P120-9 instead of P80-9 all 3 modes of operation can be satisfied for the increased security level, with minor changes in hardware, conforming the same architecture. Therefore, both security levels, for all modes of operation can be achieved on the same chip—or any reasonable combination tailored for the specific application—with very little hardware overhead.

Further evaluation of this family requires a tapeout of a versatile PRIMATEs chip, which would allow detailed assessment of SCA security. This study would also allow us to study how efficiently can different modes of operation and security levels coexist on a single chip. Furthermore, we plan to study TI of PRIMATEs in order to achieve same levels of security using less randomness and resources, as well as higher-order DPA security.

Acknowledgments. This work has started during a short-term research mission, COST Action IC1306: Cryptography for Secure Digital Interaction. In addition, this work is supported in part by the Research Council KU Leuven (C16/15/058), by the Flemish Government (G.00130.13N and FWO G.0876.14N), by the Flemish iMinds projects, by the Hercules Foundation (AKUL/11/19), and by the European Commission through the Horizon 2020 research and innovation programme under contract No H2020-ICT-2014-644371 WITDOM, H2020-ICT-2014-644209 HEAT, H2020-MSCA-ITN-2014-643161 ECRYPT-NET, and under grant agreement 644052 HECTOR. D. Šijačić is a beneficiary of a Marie Curie-Sklodowska research grant. B. Bilgin is a postdoctoral researcher of the Research Foundation—Flanders (FWO). B. Yang is supported by the Scholarship from China Scholarship Council (No. 201206210295).

References

1. Cryptographic Competitions. http://competitions.cr.yp.to/index.html
2. Al Fardan, N., Paterson, K.: Lucky thirteen: breaking the TLS and DTLS record protocols. In: IEEE Symposium on Security and Privacy (SP) 2013, pp. 526–540, May 2013
3. Andreeva, E., Bilgin, B., Bogdanov, A., Luykx, A., Mendel, F., Mennink, B., Mouha, N., Wang, Q., Yasuda, K.: PRIMATEs v1.02, Submission to the CAESAR Competition. http://primates.ae/wp-content/uploads/primatesv1.02.pdf
4. Andreeva, E., Bogdanov, A., Luykx, A., Mennink, B., Mouha, N., Yasuda, K.: How to securely release unverified plaintext in authenticated encryption. In: Sarkar, P., Iwata, T. (eds.) ASIACRYPT 2014. LNCS, vol. 8873, pp. 105–125. Springer, Heidelberg (2014). doi:10.1007/978-3-662-45611-8_6
5. Aumasson, J.-P., Henzen, L., Meier, W., Naya-Plasencia, M.: QUARK: a lightweight hash. In: Mangard, S., Standaert, F.-X. (eds.) CHES 2010. LNCS, vol. 6225, pp. 1–15. Springer, Heidelberg (2010). doi:10.1007/978-3-642-15031-9_1
6. Beaulieu, R., Shors, D., Smith, J., Treatman-Clark, S., Weeks, B., Wingers, L.: The SIMON and SPECK families of lightweight block ciphers. Cryptology ePrint Archive, Report 2013/404, 2013. http://eprint.iacr.org/
7. Bellare, M., Namprempre, C.: Authenticated encryption: relations among notions and analysis of the generic composition paradigm. IACR ePrint Archive 2000:25 (2000)

8. Bertoni, G., Daemen, J., Peeters, M., Van Assche, G.: Cryptographic sponge functions. http://sponge.noekeon.org/CSF-0.1.pdf
9. Bertoni, G., Daemen, J., Peeters, M., Van Assche, G., Keer, R.: Caesar submission: Ketje v1. http://competitions.cr.yp.to/round1/ketjev11.pdf
10. Bilgin, B., Daemen, J., Nikov, V., Nikova, S., Rijmen, V., Van Assche, G.: Efficient and first-order DPA resistant implementations of KECCAK. In: Francillon, A., Rohatgi, P. (eds.) CARDIS 2013. LNCS, vol. 8419, pp. 187–199. Springer, Cham (2014). doi:10.1007/978-3-319-08302-5_13
11. Bilgin, B., Gierlichs, B., Nikova, S., Nikov, V., Rijmen, V.: A more efficient AES threshold implementation. In: Pointcheval, D., Vergnaud, D. (eds.) AFRICACRYPT 2014. LNCS, vol. 8469, pp. 267–284. Springer, Cham (2014). doi:10.1007/978-3-319-06734-6_17
12. Bogdanov, A., Knežević, M., Leander, G., Toz, D., Varıcı, K., Verbauwhede, I.: SPONGENT: a lightweight hash function. In: Preneel, B., Takagi, T. (eds.) CHES 2011. LNCS, vol. 6917, pp. 312–325. Springer, Heidelberg (2011). doi:10.1007/978-3-642-23951-9_21
13. Bogdanov, A., Knudsen, L.R., Leander, G., Paar, C., Poschmann, A., Robshaw, M.J.B., Seurin, Y., Vikkelsoe, C.: PRESENT: an ultra-lightweight block cipher. In: Paillier, P., Verbauwhede, I. (eds.) CHES 2007. LNCS, vol. 4727, pp. 450–466. Springer, Heidelberg (2007). doi:10.1007/978-3-540-74735-2_31
14. De Cannière, C., Dunkelman, O., Knežević, M.: KATAN and KTANTAN—a family of small and efficient hardware-oriented block ciphers. In: Clavier, C., Gaj, K. (eds.) CHES 2009. LNCS, vol. 5747, pp. 272–288. Springer, Heidelberg (2009). doi:10.1007/978-3-642-04138-9_20
15. Dobraunig, C., Eichlseder, M., Mendel, F., Schläffer, M.: Ascon a family of authenticated encryption algorithms. http://ascon.iaik.tugraz.at/index.html
16. Girard, O.: openMSP430, rev. 1.15, 19 May 2015. http://opencores.org
17. Gong, Z., Nikova, S., Law, Y.W.: KLEIN: a new family of lightweight block ciphers. In: Juels, A., Paar, C. (eds.) RFIDSec 2011. LNCS, vol. 7055, pp. 1–18. Springer, Heidelberg (2012). doi:10.1007/978-3-642-25286-0_1
18. Guo, J., Peyrin, T., Poschmann, A.: The PHOTON family of lightweight hash functions. In: Rogaway, P. (ed.) CRYPTO 2011. LNCS, vol. 6841, pp. 222–239. Springer, Heidelberg (2011). doi:10.1007/978-3-642-22792-9_13
19. Guo, J., Peyrin, T., Poschmann, A., Robshaw, M.: The LED block cipher. In: Preneel, B., Takagi, T. (eds.) CHES 2011. LNCS, vol. 6917, pp. 326–341. Springer, Heidelberg (2011). doi:10.1007/978-3-642-23951-9_22
20. Hell, M., Johansson, T., Maximov, A., Meier, W.: The grain family of stream ciphers. In: Robshaw, M., Billet, O. (eds.) New Stream Cipher Designs. LNCS, vol. 4986, pp. 179–190. Springer, Heidelberg (2008). doi:10.1007/978-3-540-68351-3_14
21. Texas Instruments. Msp430x1xx family user's guide. http://www.ti.com/lit/ug/slau049f/slau049f.pdf
22. Knežević, M., Nikov, V., Rombouts, P.: Low-latency encryption – is "Lightweight = Light + Wait"? In: Prouff, E., Schaumont, P. (eds.) CHES 2012. LNCS, vol. 7428, pp. 426–446. Springer, Heidelberg (2012). doi:10.1007/978-3-642-33027-8_25
23. Moradi, A., Poschmann, A., Ling, S., Paar, C., Wang, H.: Pushing the limits: a very compact and a threshold implementation of AES. In: Paterson, K.G. (ed.) EUROCRYPT 2011. LNCS, vol. 6632, pp. 69–88. Springer, Heidelberg (2011). doi:10.1007/978-3-642-20465-4_6
24. Nikova, S., Rechberger, C., Rijmen, V.: Threshold implementations against side-channel attacks and glitches. In: Ning, P., Qing, S., Li, N. (eds.) ICICS 2006. LNCS, vol. 4307, pp. 529–545. Springer, Heidelberg (2006). doi:10.1007/11935308_38

25. Kocher, P., Jaffe, J., Jun, B.: Differential power analysis. In: Wiener, M. (ed.) CRYPTO 1999. LNCS, vol. 1666, pp. 388–397. Springer, Heidelberg (1999). doi:10. 1007/3-540-48405-1_25

26. Rogaway, P.: Authenticated-encryption with associated-data. In: Proceedings of the 9th ACM Conference on Computer and Communications Security, CCS 2002, pp. 98–107. ACM, New York (2002)

27. Rolfes, C., Poschmann, A., Leander, G., Paar, C.: Ultra-lightweight implementations for smart devices – security for 1000 gate equivalents. In: Grimaud, G., Standaert, F.-X. (eds.) CARDIS 2008. LNCS, vol. 5189, pp. 89–103. Springer, Heidelberg (2008). doi:10.1007/978-3-540-85893-5_7

28. Saarinen, M.-J.O., Engels, D.: A do-it-all-cipher for RFID: design requirements (extended abstract). IACR ePrint Archive (2012). http://eprint.iacr.org/

29. Sasaki, Y., Todo, Y., Aoki, K., Naito, Y., Sugaware, T., Murakami, Y., Matsui, M., Hirose, S.: Minalpher v1. https://competitions.cr.yp.to/round1/minalpherv1. pdf

30. Wu, W., Zhang, L.: LBlock: a lightweight block cipher. In: Lopez, J., Tsudik, G. (eds.) ACNS 2011. LNCS, vol. 6715, pp. 327–344. Springer, Heidelberg (2011). doi:10.1007/978-3-642-21554-4_19

Keymill: Side-Channel Resilient Key Generator, A New Concept for SCA-Security by Design

Mostafa Taha[1]([⊠]), Arash Reyhani-Masoleh[1], and Patrick Schaumont[2]

[1] Department of Electrical and Computer Engineering, Western University,
London, ON N6A 5B9, Canada
{mtaha9,areyhani}@uwo.ca
[2] Secure Embedded Systems, Center for Embedded Systems for Critical
Applications, Bradley Department of Electrical and Computer Engineering,
Virginia Tech, Blacksburg, VA 24061, USA
schaum@vt.edu

Abstract. In the crypto community, it is widely acknowledged that any cryptographic scheme that is built with no countermeasure against side-channel analysis (SCA) can be easily broken. In this paper, we challenge this intuition. We investigate a novel approach in the design of cryptographic primitives that promotes inherent security against side-channel analysis without using redundant circuits. We propose Keymill, a new keystream generator that is immune against SCA attacks. Security of the proposed scheme depends on mixing key bits in a special way that expands the size of any useful key hypothesis to the full entropy, which enables SCA-security that is equivalent to the brute force. Doing so, we do not propose a better SCA countermeasure, but rather a new one. The current solution focuses exclusively on side-channel analysis and works on top of any unprotected block cipher for mathematical security. The proposed primitive is generic and can turn any block cipher into a protected mode using only 775 equivalent NAND gates, which is almost half the area of the best countermeasure available in the literature.

1 Introduction

Side-Channel Analysis (SCA) is a major threat to the embedded implementation of cryptographic schemes. It is an implementation attack, where the adversary exploits side-channel outputs in order to recover information about secret values. Side-channel outputs include power consumption, electromagnetic radiation, execution time, and more. SCA targets the underlying implementation rather than the mathematical structure of the scheme. Its concept depends on predicting changes in the behavior of a crypto module using key hypotheses. Then, the hypothesis can be confirmed or rejected based on the actual behavior of the module. There are many variations in the details involved in applying this attack, but the overall concept remains the same.

Traditional countermeasures comes in three categories: Masking, Hiding, and Leakage Resiliency. Masking depends on blinding the internal operations using

© Springer International Publishing AG 2017
R. Avanzi and H. Heys (Eds.): SAC 2016, LNCS 10532, pp. 217–230, 2017.
https://doi.org/10.1007/978-3-319-69453-5_12

a random variable. The effect of randomness should be removed at the end of computation to retrieve the legitimate output. This countermeasure prevents correct prediction about the power consumption. Hiding depends on minimizing the signal-to-noise ratio in the leakage using a complement processing module, shuffling, dedicated noise generator or other means. This countermeasure prevents internal operations from affecting the power consumption. Leakage Resiliency depends on updating the secret value after every operation to prevent aggregating unbounded information against the same secret.

After much research in this field, it was acknowledged that protecting an already designed cryptographic algorithm can become very costly in terms of area and clock cycles. Hence, Medwed et al. [12] proposed using two different primitives: one to achieve security against side-channel analysis, while the other is used to protect the design against mathematical cryptanalysis, as shown in Fig. 1(a). They proposed modular multiplication between the key and a random number (function $g_k(r)$) that can be easily protected against SCA using masking and shuffling. The random number works as an Initialization Vector for block cipher modes, and should be sent to the other party. The output is a unique secret that can be used to encrypt plain data using any block cipher. Essentially, they proposed separation of duties while still depending on the common SCA countermeasure techniques. They acknowledged that any cryptographic primitive that is built with no sound SCA countermeasure can be easily broken.

In this paper, we challenge this intuition by proposing a new primitive that is secure against SCA attacks inherently by design without requiring any redundant circuit. We follow the separation guidelines of Medwed et al. [12] to better focus on side-channel properties. However, we propose a keystream generator that can encrypt plain data of any length, as shown in Fig. 1(b). The keystream generator, Keymill, depends on a special class of NLFSRs, augmented with some implementation aspects that are hardware specific. Security of the proposed scheme depends on mixing the key bits in a novel way so that no key hypothesis that is smaller than 128 bits can break the system. Our design can be implemented using 775 GEs in 130 nm CMOS technology.

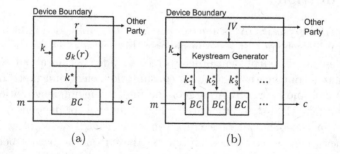

Fig. 1. (a) Domain separation proposed in [12]. (b) Generalized domain separation, as followed in this paper. BC denotes a block cipher.

The paper is organized as follows: Sect. 2 reviews some background about NLFSR and some insights about SCA that are mandatory for this research. Sect. 3 proposes a new definition about SCA-security and highlights the problem statement. Sect. 4 introduces some toy models that will be helpful in the analysis of the full system. Sect. 5 shows the proposed design, while its security analysis is highlighted in Sect. 6. The implementation cost and comparison with previous techniques are discussed in Sect. 7. We conclude the paper in Sect. 8.

2   Background

2.1   Nonlinear Feedback Shift Registers

An NLFSR is a common component in cryptographic stream ciphers. NLFSRs are known to be challenging targets for SCA [7], while having high performance at small implementation cost [15]. An NLFSR consists of n binary storage units called stages. In each cycle, the register is shifted by one bit, while the new value of the first bit is the output of the feedback function $f(S)$; where $f(S)$ is a non-linear function computed over the state of the register with mapping $f : \{0,1\}^n \rightarrow \{0,1\}$. The output of the NLFSR is the sequence that shows at the last stage. The period of an NLFSR is the length of the longest cyclic output that it can produce. The NLFSR that can generate the full period of $2^n - 1$ (excluding the zero state) is called a primitive NLFSR, where n is the length of the register.

2.2   Taxonomy of SCA

SCA depends on recovering information from leakage traces. The number of points that are involved in recovering complete information about any piece of the secret key determine the attack class.

For example, Simple Power Analysis (SPA) works by recovering information from a single point in the trace. Differential Power Analysis (DPA) works by combining information from a selected trace point across many different inputs. Higher Order DPA (HO-DPA) works by computing higher order moments before applying regular DPA attacks. Finally, a Multi-Variate DPA attack combines information from different trace points along the time at different input patterns.

2.3   SCA's Divide-and-Conquer

SCA works only for its ability to break complexity of the secret value. The typical trend in designing cryptographic algorithms is to mix the secret key in its original format at full entropy with the input data. We understand that, mathematically speaking, there is no reason to reduce entropy of the key before using it. However, this is exactly where the hardware fails, as the adversary can control (or monitor) the input data and observe the hardware's behavior as the input data interacts with the secret value. Usually, the input data width is equal to or larger than the width of the secret value, giving so much flexibility in the attacker's hands to isolate, test, and collect unbounded information about small parts of the secret key. This is known as the divide-and-conquer principle of SCA.

For example, in the AES encryption algorithm, 8-bits of the key can be isolated and recovered at the output of the SBox by controlling 8-bits of the input data. Similarly, 4-bits can be isolated by controlling 4-bits of the input in the PRESENT cipher. Also, in the typical implementation of RSA, singular key-bits can be isolated by monitoring changes in the entire input data. In these examples (and many others), increasing mathematical security of the algorithm does not affect its side-channel security, as a longer key length can be broken by recursively recovering smaller segments.

In this regard, our design has two features as detailed later. We shrink the input data width to only 1-bit, which is the smallest possible. Also, we reduce entropy of the key before interacting with the input data, which preserves its secrecy.

3   Design Goals

3.1   A New Definition: SCA-Security

In this paper, we propose a new definition for SCA-security.

SCA-security is the minimum size of key hypothesis (in bits) such that the leakage-model using the correct key correlates to the measured leakage significantly higher than the leakage-model using any other key.

Let L_s be the leakage-model using s bits of the secret key:

$$L_s = f(x, |k|_s),$$

where x is the known public data (IV), and $|k|_s$ represents s bits of the secret key. Let L_s^* represents the leakage using the correct secret key (k^*). Also, let M be the measured leakage using the same input data set.

The SCA-security can be defined as the minimum value of s so that:

$$\rho(L_s^*, M) >> \rho(L_s, M),$$

where ρ is the Pearson product-moment correlation coefficient.

Under this definition, SCA-security of the regular square-and-multiply algorithm of RSA is only 1-bit. SCA-security of AES is 8-bits, while that of PRESENT cipher is 4-bits.

The goal in this paper is to design a cryptographic primitive with SCA-security that equals its brute force.

3.2   Practical Applications

The AES encryption modes CBC, CFB, OFB, and CTR, and the authenticated encryption modes CCM, GCM and OCB [1,5] are equally vulnerable to SCA attacks as they use one fixed key k in every call to the underlying block cipher.

The direct application of our proposal as highlighted in Fig. 1(b) is to convert any of the aforementioned modes into SCA-secure. This is possible by using the secret key k along with the Initialization Vector (IV) as a seed for the random number generator. Each 128 bits of the pseudorandom output should be used only once to encrypt a message using any block cipher (denoted BC in the figure). In this case, the public input data that can be monitored (or controlled) by the adversary is the IV. In the following sections, we assume that the IV is 128 bits while the proposed keystream generator can be used with any other length.

4   Introductory Toy Models

A Simple Power Analysis against the internal state of a linear feedback shift register was proposed in [3]. They observed that the difference between power consumption following the Hamming Distance model of two consecutive clock cycles depends only on the edge bits, as the effect of internal bits will cancel out. This attack is not directly applied to the NLFSR case as proposed here. Zadeh and Heys [17] concluded that an SCA attack can reduce complexity of the secret key only if the adversary can detect changes within the underlying gates in the non-triggering edge of the clock cycle. This condition is very tricky and has never been tested through actual power traces. Hence, in the following toy models, we will focus on Differential Power Analysis.

4.1   Toy Model I: One 8-Bit NLFSR

Without loss of generality, we focus on a toy example of NLFSR that holds some security properties that are similar to the proposed structure. In this example, we study an 8-bit shift register shown in Fig. 2 where all the taps are connected to the feedback function. The structure is initialized with 8-bits of secret key, and the public data is added one-bit at a time by xoring with the feedback function.

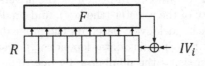

Fig. 2. Toy model I: one 8-bit NLFSR

We denote the state at clock i by S^i, with its internal bit number j as s_j, ($j \in [0 : 7]$). *In the first clock cycle*, the power leakage following the Hamming Distance power model is:

$$L = HD(S^0, S^1),$$
$$= HW(S^0 \oplus S^1),$$
$$= HW((s_0, s_1, s_2, ..., s_6, s_7) \oplus (s_1, s_2, s_3, ..., s_7, F(S^0) \oplus i_0)).$$

where HD is the Hamming Distance function, which is the number of bit-flips between its two inputs. HW is the Hamming Weight function, which is the number of set bits in the binary representation of its input. $F(S)$ is the feedback function. i_x is the input bit number x.

At a fixed secret value, the first terms are not data-dependent $((s_0, s_1, s_2, ..., s_6) \oplus (s_1, s_2, s_3, ..., s_7))$, hence their power consumption will not change by changing the input data and their effect will be canceled out by correlation. Hence, the data-dependent power leakage will be:

$$L = s_7 \oplus F(S^0) \oplus i_0,$$

where, the HW function was removed as its input is only one bit.

This equation shows a linear relationship between the measurable power consumption and one-bit of the input data, which does not reveal any information to the attacker. The reason is that the leakage will directly follow changes in the input ($L = 1$ at $i_0 = 1$) or the exact opposite ($L = 0$ at $i_0 = 1$) with equal probabilities of 50%, i.e. no advantage to the adversary.

In the second clock cycle, if we keep the register isolated with no other connected registers, the power leakage will depend on:

$$L = s_7 \oplus F(S^0) \oplus i_0 + F(S^0) \oplus i_0 \oplus F(S^1) \oplus i_1.$$

Here i_0 becomes part of $F(S^1)$, interacting non-linearly with other key bits to generate the output. This equation reveals information leakage that can be used by the adversary to break the system. Here, $F(S^1)$ compromises SCA-security of the system but the system is not completely broken yet. The reason is that the attacker can control only one bit-input of an 8-to-1 non-linear function. If function $F(S^1)$ is balanced over its input bits, the input data sequence will cause the output to flip in 50% of the cases, i.e. half of the secret space will be equally ranked first in the analysis, hiding the original secret.

The adversary can aggregate knowledge about the first two registers by addressing both $F(S^0)$ and $F(S^1)$ as highlighted in the equation above, to further reduce SCA-security of the system. In principle, the adversary makes a hypothesis over the initial state of the register (the key), and predicts the output of the feedback function in each clock cycle based on the input data. Then, he focuses on clock cycle number x to predict data-dependent power variations affecting all the taps $[1 : x]$. Increasing the number x by one reduces complexity of the unknown secret by at least one bit. The exact SCA-security reduction depends on nonlinearity of the feedback function. In this example, attacking clock cycle number 8 can uniquely determine the secret key.

To conclude, any regular NLFSR that is similar to *Model I* can be broken by SCA, regardless of complexity of the feedback function. Next, we will propose a novel modification in order to improve security of the structure.

4.2 Toy Model II: Two 8-Bit NLFSRs with Rotating Cross-Connect

In this model, we use two 8-bit registers, $R1$ and $R2$, each with its own function $F1(S1)$ and $F2(S2)$, as shown in Fig. 3. Here, $F1(S1)$ (or $F2(S2)$) is a nonlinear

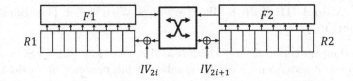

Fig. 3. Toy model II: two 8-bit NLFSRs with rotating cross-connect

function over the state bits of register $R1$ (or $R2$, respectively). Here, the feedback functions are connected to the register using a rotating cross-connect. In the odd clock cycles, the output of each function is normally connected back to its own register. In the even clock cycles, the output of function $F1(S1)$ is connected to $R2$, while $F2(S2)$ is connected to $R1$. Also, the system accepts two fresh IV bits per clock. Each bit is xored with the output of the non-linear functions before being stored in the first tap of the register.

Analysis of the system in the first clock cycle is equivalent to the previous model. Although the algorithmic noise here is higher, this noise alone cannot support sound security against SCA attacks.

In the second clock cycle, the data-dependent power leakage will be:

$$L = s1_7 \oplus F1(S1^0) \oplus i_0 + F1(S1^0) \oplus i_0 \oplus F2(S2^1) \oplus i_2$$
$$+ s2_7 \oplus F2(S2^0) \oplus i_1 + F2(S2^0) \oplus i_1 \oplus F1(S1^1) \oplus i_3.$$

where $s1, s2$ are the state bits of register $R1$ and $R2$, respectively. Similarly, $S1$ and $S2$ represent the state of registers.

The first part of this equation represents power variations in the first register $R1$, while the other part represents $R2$. The equation shows the effect of using two registers with a cross-connect. If the adversary predicts the initial state one register $S1$ (or $S2$), $F2(S2)$ (or $F1(S1)$ respectively) will act as a source of data-dependent noise. In this case, the adversary can still break the system, but only with a hypothesis over the entire secret space (16 bits in this particular example). To the best of our knowledge, this is the first cryptographic structure that can combine the effect of two non-linear functions, while being immune against the divide-and-conquer principle of SCA.

One may think that the adversary may try a specially crafted input sequence to focus on manipulating only one register. One possible choice is to switch between one random bit and one fixed bit (i_0 0 i_2 0 i_4 0...). However, this will not result in any better attack. Register $R2$ will still show data-dependent variations brought by the other feedback function ($F1(S1^1)$) as highlighted in the equation above.

To conclude, more than one non-linear register can be combined using a rotating cross-connect to increase the secret space of the structure.

4.3 Toy Model III: Two 8-Bit Registers with 4-Bit Feedback Function

In this section, we try to answer the interesting question: Should SCA-security depend on the register length or the number of bits involved in evaluating the feedback function? In the previous models, the two numbers were identical. In Model III, we keep the length of registers as 8-bits, but we use 4-to-1 nonlinear feedback functions. For instance, we assume that only the odd numbered taps are connected to the feedback function.

Here, the data-dependent power leakage in the second clock cycle will be:

$$L = s1_7 \oplus F1(S1^0_{odd}) \oplus i_0 + F1(S1^0_{odd}) \oplus i_0 \oplus F2(S2^1_{even}) \oplus i_2$$
$$+ s2_7 \oplus F2(S2^0_{odd}) \oplus i_1 + F2(S2^0_{odd}) \oplus i_1 \oplus F1(S1^1_{even}) \oplus i_3.$$

where S_{odd} represents the odd numbered taps, while S_{even} represents the even numbered taps.

The equation shows that the adversary needs a correct hypothesis over the entire register in order to correctly model the power consumption. This is true if knowledge about the value of some taps does not help in predicting the output of the feedback function in the next clock cycle. This is best achieved by connecting the feedback function over the odd taps of the register.

To conclude, we do not have to find a nonlinear feedback function that covers all the register taps. Interestingly, a feedback function that connects only half of the taps can provide the same level of SCA-security. Feedback functions with a smaller number of inputs will have a degraded level of SCA-security.

5 Keymill, The Proposed Design

Our first option for a full system with 128-bits of secret key is to combine 16 8-bit registers using AES SBox as a non-linear function. However, the size of the rotating cross-connect circuit will be significant. More importantly, the combined secret space will be less-optimal. One full rotation of the cross-connect will require 16 cycles, while the IV input will vanish in only 8 clock cycles (assuming 16-bits input per clock with IV of 128 bits).

Hence, we propose to use only four registers featuring 8 full rotations while accepting new IV bits, which will be done within $128/4 = 32$ clock cycles.

Unfortunately, there is no theory on how to construct NLFSRs with good cryptographic properties and long period. However, there are many constructions in the literature that have been carefully designed for cryptography with a guaranteed maximum period. Hence, we will not design a new NLFSR. Instead, we will focus on how to use one of the established NLFSRs in the proposed construction.

Achterbahn [8] is a cryptographic stream cipher that was designed as part of the eSTREAM competition. An innovative part of Achterbahn stream cipher is the design of 13 primitive NLFSRs of different sizes (21 bits to 33 bits). Although Achterbahn did not advance to the eSTREAM portfolio due to some limitations

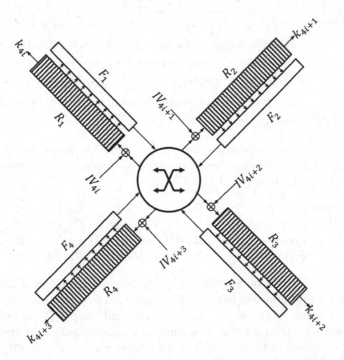

Fig. 4. Structure of Keymill, the proposed keystream generator

in the combining function, which we are not using here, the NLFSRs are still a valuable contribution. Next, we will use three of the Achterbahn NLFSRs to build the proposed scheme.

The proposed construction is composed of four NLFSRs, as shown in Fig. 4. Register $R1$ is a 31-bit register. Registers $R2$ and $R3$ are 32 bits each. Register $R4$ has 33 bits. We use feedback functions from Achterbahn stream cipher [8] (where they are named A_{10}, A_{11}, A_{12}) with the following equations:

$$
\begin{aligned}
F1(S) = \ &s_0 + s_2 + s_5 + s_6 + s_{15} + s_{17} + s_{18} + s_{20} + s_{25} + s_8 s_{18} \\
&+ s_8 s_{20} + s_{12} s_{21} + s_{14} s_{19} + s_{17} s_{21} + s_{20} s_{22} + s_4 s_{12} s_{22} + s_4 s_{19} s_{22} \\
&+ s_7 s_{20} s_{21} + s_8 s_{18} s_{22} + s_8 s_{20} s_{22} + s_{12} s_{19} s_{22} + s_{20} s_{21} s_{22} \\
&+ s_4 s_7 s_{12} s_{21} + s_4 s_7 s_{19} s_{21} + s_4 s_{12} s_{21} s_{22} + s_4 s_{19} s_{21} s_{22} \\
&+ s_7 s_8 s_{18} s_{21} + s_7 s_8 s_{20} s_{21} + s_7 s_{12} s_{19} s_{21} + s_8 s_{18} s_{21} s_{22} \\
&+ s_8 s_{20} s_{21} s_{22} + s_{12} s_{19} s_{21} s_{22}.
\end{aligned}
$$

$$
\begin{aligned}
F2(S) = F3(S) = \ &s_0 + s_3 + s_{17} + s_{22} + s_{28} + s_2 s_{13} + s_5 s_{19} + s_7 s_{19} + s_8 s_{12} \\
&+ s_8 s_{13} + s_{13} s_{15} + s_2 s_{12} s_{13} + s_7 s_8 s_{12} + s_7 s_8 s_{14} + s_8 s_{12} s_{13} \\
&+ s_2 s_7 s_{12} s_{13} + s_2 s_7 s_{13} s_{14} + s_4 s_{11} s_{12} s_{24} + s_7 s_8 s_{12} s_{13} \\
&+ s_7 s_8 s_{13} s_{14} + s_4 s_7 s_{11} s_{12} s_{24} + s_4 s_7 s_{11} s_{14} s_{24}.
\end{aligned}
$$

$$F4(S) = s_0 + s_2 + s_7 + s_9 + s_{10} + s_{15} + s_{23} + s_{25} + s_{30} + s_8 s_{15}$$
$$+ s_{12} s_{16} + s_{13} s_{15} + s_{13} s_{25} + s_1 s_8 s_{14} + s_1 s_8 s_{18} + s_8 s_{12} s_{16}$$
$$+ s_8 s_{14} s_{18} + s_8 s_{15} s_{16} + s_8 s_{15} s_{17} + s_{15} s_{17} s_{24} + s_1 s_8 s_{14} s_{17}$$
$$+ s_1 s_8 s_{17} s_{18} + s_1 s_{14} s_{17} s_{24} + s_1 s_{17} s_{18} s_{24} + s_8 s_{12} s_{16} s_{17}$$
$$+ s_8 s_{14} s_{17} s_{18} + s_8 s_{15} s_{16} s_{17} + s_{12} s_{16} s_{17} s_{24} + s_{14} s_{17} s_{18} s_{24}$$
$$+ s_{15} s_{16} s_{17} s_{24}.$$

Hence, the internal state of the structure is 128-bits, similar to the common length of AES' key. The feedback functions are mixed using rotating cross-connect as follows. Assuming that $i = [1, 5, 9, 13, ...]$,

Clock cycle i: F1 → R1, F2 → R2, F3 → R3, F4 → R4
Clock cycle $i + 1$: F1 → R2, F2 → R3, F3 → R4, F4 → R1
Clock cycle $i + 2$: F1 → R3, F2 → R4, F3 → R1, F4 → R2
Clock cycle $i + 3$: F1 → R4, F2 → R1, F3 → R2, F4 → R3

The structure starts by loading the secret key into the internal state. Then on each clock cycle, four bits from the IV, one for each register, are added to the feedback functions. After $128/4 = 32$ clock cycles, adding the IV should be completed. Then, the structure goes free running without inputs or outputs for 33 clock cycles, equivalent to the length of the longest register. From this point forward, the NLFSRs will generate 4 bits in each clock cycle, one from each register.

Every 128 bits of the output should be used only once to encrypt plain data with the AES block cipher.

6   Security Analysis

First of all, we selected 4 registers with a total number of taps equal to 128 bits in order to preserve the entropy of the secret key. Also, the output is not generated until after 33 clock cycles from the acceptance of the last IV bit. This number allows the last input bit to go through all the taps of the longest register, which allows the structure to distribute its effect over all the internal state. This lets each unique IV generate a unique bit stream.

Regarding SCA-security, the number of taps that are connected to functions $F1(S)$, $F2(S)$ and $F3(S)$ is 17 bits each. The number of connected taps to $F4(S)$ is 18 bits. The unique taps that are connected to each register are as follows:

$$F1(S) : 0, 2, 4, 5, 6, 7, 8, 12, 14, 15, 17, 18, 19, 20, 21, 22, 25$$
$$F2(S), F3(S) : 0, 2, 3, 4, 5, 7, 8, 11, 12, 13, 14, 15, 17, 19, 22, 24, 28$$
$$F4(S) : 0, 1, 2, 7, 8, 9, 10, 12, 13, 14, 15, 16, 17, 18, 23, 24, 25, 30$$

The number of connected taps in each register is slightly higher than half the register length, with a good distribution that is close to the distribution recommended in Model III (see Sect. 4.3). Hence, it is very reasonable to declare that the registers will have SCA-security equivalent to their length (31, 32, 32, and 34 bits).

Moreover, the feedback functions are mixed with a rotating cross-connect that is similar to Model II (see Sect. 4.2). Hence, SCA-security of the system will be equivalent to the aggregated length of the involved registers, which is $(31 + 32 + 32 + 34 = 128$ bits). Essentially, the adversary cannot make an accurate estimate about the data-dependent power changes in the structure unless he makes a correct hypothesis over the entire secret key.

To the best of our knowledge, this is the first cryptographic structure ever proposed that has an SCA-security that equals its brute force security.

6.1 Failure of Other NLFSRs

NLFSRs have long been used in the design of stream ciphers including, KeeLoq [6] and Grain [10] as notable examples. One of the design principles that we followed in this proposal is to limit the attacker's control on the internal behavior by shrinking the data-width of public data that is used in each operation. In this regard, KeeLoq's design features control the entire internal state, while the key bits are used one at a time [6]. Grain makes use of a non-linear function (called H) that involves four bits of the input data and one bit key [7]. Another key difference is that the output of the aforementioned non-linear functions are feedback to the register, which originally holds the input data. Hence, the previous state of the HD power model is known to enable an easy recovery of the new state.

6.2 Similarity to GGM Structures

GGM is a tree-based structure, named after its inventors [9], that is used to realize pseudorandom functions from any pseudorandom generator. It was reintroduced in many recent contributions as a structure that is capable of initializing leakage resilient primitives in an SCA-secure manner [2,4,16].

The GGM structure starts from the secret key and inserts the IV one bit at a time followed by a randomization step. The value of each bit determines the next branch in the tree. The randomization step is used to distribute the effect of the inserted bit over the entire internal state. Hence, the attacker will face a new secret at each step, which renders DPA attacks almost impossible. Randomization can be realized using block ciphers [13] or hashing functions [11].

The proposed Keymill is similar to the GGM in accepting one bit of the IV at each clock cycle, however the core concept for SCA-protection is different. GGM employs a *leaky* function, that is not secure against SCA attacks, to build an SCA-secure algorithm. Here, the randomization primitive used in the GGM (block cipher or hashing function) is still vulnerable to SCA attacks, while protection is achieved by preventing the adversary from aggregating information across different executions. On the contrary, SCA-security of the proposed Keymill depends on expanding the size of key hypotheses to the full size of the secret key. Hence, Keymill, as an isolated primitive, is inherently secure against SCA attacks without being part of any special algorithm.

6.3   Cautionary Notes

There are a couple of cautionary notes that come with this new protection mechanism.

Proposing a new masking or hiding countermeasure must be evaluated with actual power consumption traces. This is a typical requirement in order to ensure that the engineering defects (glitches and balanced routing) are resolved. On the contrary, we found it difficult to evaluate the SCA-security of our scheme on the same grounds. The reason is that our countermeasure depends on expanding the size of key hypothesis to 128 bits. Hence, we could not enumerate all the possible 2^{128} cases in order to measure feasibility of the proposed scheme. Rather, we built our security on mathematical modeling. In fact, this is in line with leakage resilient schemes that depend on updating the secret key after each run.

Another similarity with leakage resilient schemes is that the proposed countermeasure slightly changes the algorithm. Hence, the same algorithm needs to be applied in the two sides of communication even if one side is physically protected (server or so).

One last note is that high entropy of the input key is required in order to generate high-entropy keystream. In other words, if the input key is all zeros, the output bit stream will also be zeros. This limitation is inherited from the Achterbahn NLFSRs [8]. Here, we focused exclusively on SCA-security and we did not add any cryptographic part to break symmetry of the scheme, which can be a topic for future improvement.

7   Hardware Results

Using the hardware budget of the individual NLFSRs as discussed in [8], the hardware cost of the proposed structure at a low-Vt 1.5 V standard cell library targeting 130 nm CMOS technology is:

$$Area = 608 + 125.5 + 41.5 = 775 \; GE$$

The $(4.75 \times 128) = 608$ GE covers the internal state of the registers. The $(31.75 + 31 + 31 + 31.75) = 125.5$ GE covers the feedback functions $F1$, $F2$, $F3$, and $F4$ respectively. The 41.5 GE covers the rotating cross-connect. The rotating cross-connect can be implemented very efficiently using four 4-to-1 multiplexers $(4 \times 7.5 = 30$ GE$)$ and a 2-bit counter (11.5 GE).

A comparison between the hardware cost of the proposed scheme and that of the previous work is shown in Table 1. The results of [12] are taken at the first-order masked implementation, while the results of the minimum SP network are taken at the lightest implementation of [2]. The table shows superiority of the proposed scheme in terms of both area and clock cycles.

Also, the proposed scheme shows superior performance over typical masking schemes. The smallest threshold implementation of AES (to prevent leakage caused by glitches) requires 8,393 GE of area overhead [14].

Although there is no initialization required for threshold implementations, the initialization overhead of our scheme requires only 65 clock cycles. Then, one key is generated every 32 clock cycles.

Table 1. Comparison against similar schemes

Contribution	Area (GE)	Clock cycles
Modular Mul of [12]	7,300	562
Minimum SP network of [2]	5,302	61
The proposed Keymill	775	97

8 Conclusion

In this paper, we have proposed a new solution to SCA attacks. Our solution depends on a new cryptographic structure that expands the size of key hypothesis, and breaks the divide-and-conquer principle of SCA. Our structure is generic, lightweight and can turn any block cipher into an SCA-secured encryption mode.

Acknowledgment. This work was supported in part by the Natural Sciences and Engineering Research Council (NSERC) of Canada under the Discovery and Discovery Accelerate Supplement (DAS) Grants awarded to A. Reyhani-Masoleh.

References

1. Information technology, security techniques, authenticated encryption. In: ISO/IEC 19772:2009. Accessed 12 Mar 2013
2. Belaid, S., Santis, F.D., Heyszl, J., Mangard, S., Medwed, M., Schmidt, J.-M., Standaert, F.-X., Tillich, S.: Towards fresh re-keying with leakage-resilient PRFs: cipher design principles and analysis. Cryptology ePrint Archive, Report 2013/305 (2013)
3. Burman, S., Mukhopadhyay, D., Veezhinathan, K.: LFSR based stream ciphers are vulnerable to power attacks. In: Srinathan, K., Rangan, C.P., Yung, M. (eds.) INDOCRYPT 2007. LNCS, vol. 4859, pp. 384–392. Springer, Heidelberg (2007). doi:10.1007/978-3-540-77026-8_30
4. Dodis, Y., Pietrzak, K.: Leakage-resilient pseudorandom functions and side-channel attacks on feistel networks. In: Rabin, T. (ed.) CRYPTO 2010. LNCS, vol. 6223, pp. 21–40. Springer, Heidelberg (2010). doi:10.1007/978-3-642-14623-7_2
5. Dworkin, M.: NIST special publication 800-38A, recommendation for block cipher modes of operation: Methods and techniques (2001)
6. Eisenbarth, T., Kasper, T., Moradi, A., Paar, C., Salmasizadeh, M., Shalmani, M.T.M.: On the power of power analysis in the real world: a complete break of the KEELOQ code hopping scheme. In: Wagner, D. (ed.) CRYPTO 2008. LNCS, vol. 5157, pp. 203–220. Springer, Heidelberg (2008). doi:10.1007/978-3-540-85174-5_12
7. Fischer, W., Gammel, B.M., Kniffler, O., Velten, J.: Differential power analysis of stream ciphers. In: Abe, M. (ed.) CT-RSA 2007. LNCS, vol. 4377, pp. 257–270. Springer, Heidelberg (2006). doi:10.1007/11967668_17
8. Gammel, B. M., Göttfert, R., Kniffler, O.: Achterbahn-128/80. In: eSTREAM, ECRYPT Stream Cipher Project (2006)
9. Goldreich, O., Goldwasser, S., Micali, S.: How to construct random functions. J. ACM (JACM) **33**(4), 792–807 (1986)

10. Hell, M., Johansson, T., Meier, W.: Grain: a stream cipher for constrained environments. Int. J. Wirel. Mob. Comput. **2**(1), 86–93 (2007)
11. Kocher, P.: Complexity and the challenges of securing SoCs. In: 2011 48th ACM/EDAC/IEEE Design Automation Conference (DAC), pp. 328–331 (2011)
12. Medwed, M., Standaert, F.-X., Großschädl, J., Regazzoni, F.: Fresh re-keying: security against side-channel and fault attacks for low-cost devices. In: Bernstein, D.J., Lange, T. (eds.) AFRICACRYPT 2010. LNCS, vol. 6055, pp. 279–296. Springer, Heidelberg (2010). doi:10.1007/978-3-642-12678-9_17
13. Medwed, M., Standaert, F.-X., Joux, A.: Towards super-exponential side-channel security with efficient leakage-resilient PRFs. In: Prouff, E., Schaumont, P. (eds.) CHES 2012. LNCS, vol. 7428, pp. 193–212. Springer, Heidelberg (2012). doi:10.1007/978-3-642-33027-8_12
14. Moradi, A., Poschmann, A., Ling, S., Paar, C., Wang, H.: Pushing the limits: a very compact and a threshold implementation of AES. In: Paterson, K.G. (ed.) EUROCRYPT 2011. LNCS, vol. 6632, pp. 69–88. Springer, Heidelberg (2011). doi:10.1007/978-3-642-20465-4_6
15. Robshaw, M., Billet, O.: New Stream Cipher Designs: The Estream Finalists, vol. 4986. Springer, Heidelberg (2008)
16. Standaert, F.-X., Pereira, O., Yu, Y., Quisquater, J.-J., Yung, M., Oswald, E.: Leakage resilient cryptography in practice. In: Towards Hardware-Intrinsic Security, pp. 99–134 (2010)
17. Zadeh, A.A., Heys, H.M.: Simple power analysis applied to nonlinear feedback shift registers. Inf. Secur. IET **8**(3), 188–198 (2014)

Lightweight Fault Attack Resistance in Software Using Intra-instruction Redundancy

Conor Patrick[✉], Bilgiday Yuce, Nahid Farhady Ghalaty,
and Patrick Schaumont

Bradley Department of Electrical and Computer Engineering,
Virginia Tech, Blacksburg, USA
{conorpp,bilgiday,farhady,schaum}@vt.edu

Abstract. Fault attack countermeasures can be implemented by storing or computing sensitive data in redundant form, such that the faulty data can be detected and restored. We present a class of lightweight, portable software countermeasures for block ciphers. Our technique is based on redundant bit-slicing, and it is able to detect faults in the execution of a single instruction. In comparison to earlier techniques, we are able to intercept data faults as well as instruction sequence faults using a uniform technique. Our countermeasure thwarts precise bit-fault injections through pseudo-random shifts in the allocation of data bit-slices. We demonstrate our solution on a full AES design and confirm the claimed security protection through a detailed fault simulation for a 32-bit embedded processor. We also quantify the overhead of the proposed fault countermeasure, and find a minimal increase in footprint (14%), and a moderate performance overhead between 125% to 317%, depending on the desired level of fault-attack resistance.

Keywords: Fault attacks · Fault resistance · Intra-instruction redundancy · Bitslicing · Block ciphers

1 Introduction

The injection of faults in cryptographic software is a well-studied technique to extract cryptographic keys. Originally demonstrated against public-key cryptography [1], their scope has since been widened to the symmetric-key case. The current state of the art in differential fault analysis on the Advanced Encryption Standard can extract an AES-128 key with just two faults [2]. Therefore, for applications where fault injection is a relevant threat, it is crucial to detect the occurrence of even a single fault and respond appropriately. In this contribution, we study and develop countermeasure techniques that are applicable to software, and that does not need any special hardware. Software countermeasures against fault attacks are commonly developed using redundancy. However, all redundancy based techniques share a common weakness: they are ineffective against an adversary who can inject consistent faults in redundant sections of code or

© Springer International Publishing AG 2017
R. Avanzi and H. Heys (Eds.): SAC 2016, LNCS 10532, pp. 231–244, 2017.
https://doi.org/10.1007/978-3-319-69453-5_13

data. This is especially relevant for implementations that are time-redundant, since they only require the adversary to inject the same fault sequentially.

In this paper, we propose a technique that enables the software to exploit redundancy for fault attack protection *within* a single instruction; we propose the term *intra-instruction redundancy* to describe it. Our technique is cross-platform and requires that an algorithm be bit-sliced. We focus on block ciphers, as they all can be bit-sliced. To protect from computation faults, we allocate some bit-slices as redundant copies of the true payload data slices. A data fault can then be detected by the difference between a data slice and its redundant copy after the encryption of the block completes. To protect the computations from instruction faults such as instruction skip, we also allocate some of the bit-slices as check-slices which compute a known result. The *intra-instruction redundancy* countermeasure is thus obtained through the bit-sliced design of a cipher, with redundant data-slices to detect computation faults and check-slices to detect instruction faults. This basic mechanism is then further strengthened against targeted fault-injection as follows. First, we pipeline the bit-sliced computation such that each slice computes a different encryption round. Since a fault-injection adversary is typically interested in the last or penultimate round, and since there will be only a few bits in a word that contain such a round, the *pipelined intra-instruction redundancy* countermeasure reduces the attack surface considerably. Second, we also randomize the slice assignment after each encryption, such that a cipher round never remains on a single slice for more than a single encryption. We show that this final countermeasure, the *shuffled pipelined intra-instruction redundancy*, is very effective and requires an adversary who can control fault injection with single-cycle, bit-level targeting chosen bits. We are not aware of a fault injection mechanism that achieves this level of precision.

The contributions of the paper are as follows.

- We propose a software countermeasure based on redundant bit slicing. The bit slices are used for data redundancy as well as control redundancy. The latter is achieved by computing a known answer.
- The proposed countermeasure is generic and can still be used in combination with other software countermeasures such as infective countermeasures or side-channel resistant techniques based on masking.
- The security of the proposed countermeasure is quantitatively analyzed to establish estimated fault coverage. In addition, it is empirically tested using simulation for different fault models including instruction skip, random word, random byte and bit-precision faults.
- The bit-sliced design leads to a secure fault detection and fault handling approach that is purely computational, and that avoids comparison and decision making. This avoids a well-known single point-of-failure in redundancy-based countermeasures.
- We evaluate the overhead of the countermeasure over an unprotected, bit-sliced implementation of AES-128 that runs at 469.3 cycles/bytes, we show that the highest level of protection is achieved at 1957 cycles per byte, which protects against targeted, repeatable, multiple bit faults.

The rest of the paper is organized as follows. In the next section, we provide additional details on the fault models used in this work. In Sect. 3, we highlight the differences of previous software countermeasures with our proposed countermeasure. Section 4 is an up-close discussion of the design of our countermeasure; we elaborate on the bit-slice allocation strategy and on the integration of the protected design on embedded platforms. Section 5 estimates the fault coverage of the proposed countermeasures under different fault models. Section 6 presents the implementation overhead for a 32-bit embedded processor and empirically demonstrates the fault countermeasure operation using fault simulation. We conclude the paper in Sect. 7.

2   Fault Models

This section details the fault models that we used in this paper to evaluate our countermeasures. The fault model is the expected effect of the fault injection on a cryptosystem. The manipulated data may affect instruction opcodes as well as data, and we distinguish these two cases as instruction faults and computation faults.

Computation Faults: These faults cause errors in the data that is processed by a program. There is a trade-off between the accuracy by which an adversary can control the fault injection, and the required sophistication of a fault countermeasure that thwarts it. Therefore, we assume four computational fault models:

1. *Random Word*: The adversary can target a specific word in a program and change its value into a random value unknown to the adversary.
2. *Random Byte*: The adversary can target a specific word in a program and change a single byte of it into a random value unknown to the adversary.
3. *Random Bit*: The adversary can target a specific word in a program and change a single bit of it into a random value unknown to the adversary.
4. *Chosen Bit Pair*: The adversary can target a chosen bit pair of a specific word in a program, and change it into a random value unknown to the adversary.

Instruction Faults: This fault model assumes that an attacker can change the opcode of an instruction by fault injection. A very common model is the *Instruction Skip* fault model, which replaces the opcode of an instruction with a nop instruction. Using this model, an attacker can skip the execution of a specific instruction in the program.

3   Related Work

The two principal techniques for a fault countermeasure are detection-based and infection-based [3]. We start with detection-based countermeasures in software, as they are most similar to our proposal. A classic technique relies on time redundancy, such as duplicate encryption or encryption followed by decryption.

This allows the detection of faults by comparing the consistency of the redundant executions. These techniques, however, do not work well against an adversary who is able to inject consistent faults in the redundant copies, or against an adversary who directly targets the comparison. Several time redundant techniques have been proposed to make consistent fault injection more difficult. For example, instruction duplication and triplication [4] are used because it is assumed that back-to-back fault injection is harder than fault injections that are relatively far spaced apart. Duplication and triplication were found to incur 3.4 and 10.6 times performance overhead, respectively [4]. However, duplication and triplication are relatively easy to overcome with a modern fault injection setup. More sophisticated techniques are possible, but they are algorithm-specific. Examples are techniques based on invariant properties of a block cipher [5], or based on storing sensitive variables in a transformed format [6]. However, we are interested in a generic, algorithm independent technique.

Another category of detection-based countermeasures use information redundancy, which uses additional check variables or parity bits [4] to detect faults in the data. This was found to incur between 3.5–4.7 times performance overhead [4]. A recent proposal observed that Wave Dynamic Differential Logic (WDDL), which represents data in complementary format, is able to detect computation faults [7] but no performance metrics are provided. While these information-based countermeasures are generic and easy to apply to a broad class of algorithms, they are unable to detect low-level instruction-level faults in the underlying processor when implemented in software.

The second major class of countermeasures uses infection. The idea is that injected faults will also destroy the invariant properties of the fault. This effectively eliminates the possibility of differential fault analysis. However, for every infective countermeasure proposed so far, a corresponding attack has been demonstrated [8,9].

Most related works show that they have good fault coverage but it's under a narrow fault model. A common fault model is to assume that attackers can only inject one fault at a time. But if an attacker can inject more than one fault or affect multiple instructions with one fault, the fault coverage is likely to plummet.

In this paper, we propose detection-based countermeasures against fault attacks in software that are based on intra-instruction redundancy. We go beyond redundant encoding of information by also including the ability to detect instruction faults as well as computation faults. We show that these countermeasures can protect against a variety of realistic fault models. To the best of our knowledge, this is the first work that provides comprehensive coverage against processor-level fault attacks.

4   Proposed Software Countermeasures for Fault Attacks

We will explain the motivation and main idea of our countermeasures. We will then explain how they can be implemented. Finally, we will provide some discussion on the performance and footprint impact.

The countermeasures are based on bit-slicing. As we will explain further, bit-slicing allows for a program to dynamically select different data flows to be present in a processor word. This is attractive for a fault attack countermeasure because it presents a new way to leverage redundancy. Each data word can be split amongst regular data streams and redundant data streams, allowing redundancy to be present spatially in all instructions without actually having to re-execute anything. Because the redundancy is interleaved with the data in every instruction, we call this Intra-Instruction Redundancy (IIR). By never separating the data from the redundancy in the processor word, we use pure spatial redundancy rather than commonly used time-based redundancy, which is vulnerable to repeated fault injections [10].

In this work we consider two ways to detect faults using redundancy. First, if you are computing data where the result is unknown, you can only detect a fault by recomputing the data an additional time to compare the results. Second, if the result is already known before computation, you can store a read only copy of the result and only need to execute on the data once to reproduce the result and check that it is the same.

Our countermeasure scheme relies on making comparisons at the end of encryption rounds. Because the comparisons are a very small part of the code, we assume we can cheaply duplicate them enough such that an adversary may not reasonably skip all of them using faults. In the advent of a fault detection, a random cipher text is output and the program may either restart encryption or enact a different, application-specific policy.

4.1 Bit-Slicing Without Fault Attack Protection

Bit-slicing is a technique used commonly in block ciphers and embedded systems to fully utilize the word length of a processor for all operations, potentially increasing the total throughput. Bit-slicing avoids data-dependent memory lookups and because of that, data-dependent cache effects. It involves decomposing all components into boolean operations and orienting the data such that one bit can be computed at a time per instruction. If one bit is computed at a time, then a 32 bit processor word can be filled with 32 different blocks, computing all blocks simultaneously.

A prerequisite for bit-slicing is to transpose the layout of input blocks, as shown in Fig. 1. At the top, a traditional layout of blocks is depicted. There are 32 blocks, each composed of 4, 32 bit data words. All of them must be transposed. In the transposed layout, each word contains one bit from every block. In this format, each bit from 32 different blocks can be computed simultaneously for any instruction. A *slice* refers to a bit location in all words that together make up one block.

4.2 Intra-instruction Redundancy

In traditional bit-sliced implementations, each slice is allocated to operate on a different input block for maximum throughput (Fig. 1). Instead, we separate

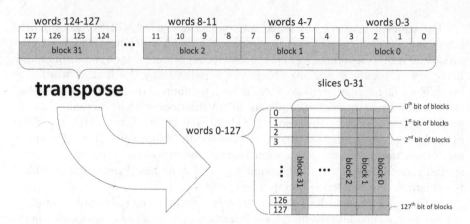

Fig. 1. Transpose of 32 blocks to fit bitwise into 128 32-bit words.

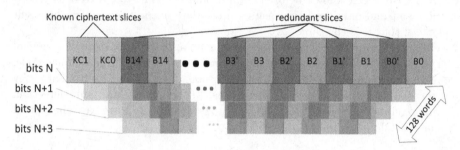

Fig. 2. Bit-slicing with intra-instruction redundancy using 15 data (B), 15 redundant (B'), and 2 known ciphertext (KC) slices. Each KC slice is aligned with its corresponding round key slices in other words.

slices into three categories: data slices (B), redundant slices (B'), and Known Ciphertext (KC) slices for fault detection (Fig. 2). Data slices and redundant slices operate on the same input plaintext, and thus, they produce the same ciphertext if no fault occurs. If a fault occurs during their execution, then it will be detected when results are compared at the end of encryption.

However, if both B and B' experience the same fault, then both of them will have the same faulty ciphertext and a fault cannot be detected. For example, this would always be the case for instruction skips. To address this issue, we include KC slices in addition to data and redundant slices. Instead of encrypting the input plaintexts with the run-time secret key, KC slices encrypt internally stored plaintexts with a different key, each of which are decided at design time. Therefore, the correct ciphertexts corresponding to these internally stored plaintexts are known by the software designer beforehand. If no fault is injected into the execution of a KC slice, it will produce a run-time ciphertext that is equal to the known, design-time ciphertext. In case of a computation or instruction fault, the run-time ciphertext will be different than the design-time ciphertext. Therefore, the run-time and design-time ciphertexts of the KC slices are compared at the end of encryption for fault detection. Because the round keys for the data slices and round keys for the KC slices can be intermixed at the slice level, we can execute KC slices together with the data and redundant slices. Figure 2 shows the slice allocation used in this work, which includes 15 data slices ($B0$–14), 15 redundant slices ($B'0$–14), and 2 KC slices ($KC0$–1). All slices are split across 128 words for a 128 bit block size.

A set of known plaintext-ciphertext pairs is included in the program from which KC slices can be selected from randomly for an encryption. This is because each KC slice only has a 50% chance of detecting an instruction fault. If only a couple of them are used, then there will likely be parts of the block cipher where instruction faults do not affect the KC slices. By selecting from a larger set of ciphertext-plaintext pairs, we significantly reduce the chance of an adversary finding such parts of the program. Each plaintext-ciphertext pair will be the size of two blocks of the cipher.

An adversary can bypass this countermeasure by injecting two bit faults that are next to each other in the processor word. The two bit fault has to align with any of the B slices and the corresponding B' slice. Then both will produce the same faulty ciphertext, going undetected.

4.3   Pipelined Intra-instruction Redundancy

For an adversary to carry out a fault analysis attack, he must inject a fault into a target round of the block cipher [11,12]. It is not enough to cause an undetected fault in the wrong round, as the faulty ciphertext will not be useful in analysis. Previously, all data and redundant slice pairs in the target word operate on the same round. Therefore, an adversary can target any combinations of these pairs to bypass IIR. Here we will explain how we can make the rounds spatial by making them correspond to slices within each word, instead of different words executed at different times. This makes fault injection harder as the faults will have to target specific bit locations.

Because block cipher rounds differ only in the round key used, we can make different bits correspond to different rounds by aligning slices with different round keys. Doing this means blocks will be computed in a pipelined fashion as shown in Fig. 3, which shows ten rounds. The round keys are doubled and interleaved with the known ciphertext key beforehand to align with the pipeline. Each block is transposed one at a time rather than 32 at a time. For every iteration, 3 slices are shifted into the 128 word state (1 data, 1 redundant, and 1 KC). Initially shifted in is $B0$. Running for one iteration will compute round one of $B0$. Applying another shift aligns $B0$ for round 2 and shifts in $B1$ for round 1. This eliminates the need to have a set of plaintext-ciphertext pairs as it will be okay to have one pair. One pair will effectively make 10 different KC slices amongst the 10 rounds.

In this pipeline, because each set of 3 bits corresponds to a different round, any two bit fault will not suffice to undo the countermeasure. There is now only one valid bit location to successfully inject a 2 bit fault. For example, to fault

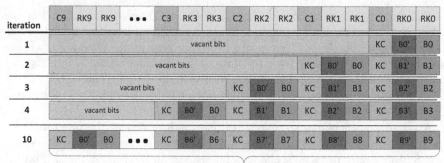

Fig. 3. Pipelined bit-sliced layout for 32 bit processor. $RK0$–9 are ten different round keys. $B0$–9 are different input blocks and $B'09$ are their redundant copies. KC is a known ciphertext slice. $C0$–9 are the round keys used to produce the known ciphertext.

round 9, a 2 bit fault must be injected at bit location 27. It is non-trivial for an adversary to inject a fault that is in a target bit location and consists of two adjacent bits.

Astute readers will point out that the last round in some block ciphers differs in more than just the round key. For example, in AES, the last round does not have the mix-columns step. Some additional masking can done to remove the effect of particular steps on any round(s). To be able to pipeline rounds that differ in steps, we add the following computation to each operation in the special step.

$$B = (BS \ \& \ RM) \ | \ (B \ \& \ {\sim}RM)$$

Where B is the block going through a particular step, BS is the computed result after the step, and RM is a mask representing the rounds that use the step. By doing this, the step will be applied to only the rounds that use it and leave the other round(s) the same.

4.4   Shuffled, Pipelined Intra-instruction Redundancy

For our final countermeasure stage, we assume a highly skilled adversary who can inject multiple bit faults into target bit locations. In our case, we need to protect from a targeted 2-bit fault.

For each plaintext, we can effectively apply a random rotation to all of the slices and their corresponding round keys. The randomness is from an initial secret number that is continually XOR'd with generated ciphertext. We can reasonably assume that the adversary will not be able to predict the random rotation. Despite the adversary being able to inject known bit location faults, he will not know which bit corresponds to what round, making the attack much more difficult.

To support random shifts, we support dynamic allocation of each slice in the processor word, rather than statically defining which bits correspond to each round. The transpose step will have to support transposing into and out of any target bit location, rather than always shifting into bit location 0 and shifting out of bit location 29.

4.5   Secure, Comparison-Free Fault Handling

Rather than checking the memory using excessively duplicated comparisons, fault handling can be done in a purely computational approach. This approach is ideal because an application no longer needs to have a secure response to a fault injection.

If either a block, B, or its respective redundant slice, B', contain an error, we would expect the XOR of them $B \oplus B'$ to be nonzero. Whereas a non-faulty operation would always produce zero. Building upon this, we can make a method such that when a fault is injected, only a random number is output, foiling any attempt of fault analysis.

After encryption, we can compute the following mask.

MASK = (−(B ⊕ B') >> 128)

If B and B' are the same, then $B \oplus B'$ will be zero and the signed shift will move in all zeros. If $B \oplus B'$ is nonzero, then the signed shift will move in all ones. We can easily extend this mask to check a KC slice as well for instruction faults using our known ciphertext KC'.

MASK = (−(B ⊕ B') >> 128) | (−(KC ⊕ KC') >> 128)

As in our pipelined countermeasure, a redundant slice, data slice, and KC slice can be shifted out every iteration to compute the mask. We can then use this mask to protect our ciphertext block before it is output.

OUTPUT = (MASK & R) | (∼MASK & B);

By doing this, only our random number R will be output when a fault is detected. Otherwise, the correct ciphertext B will be output. Because these computations are not covered by intra-instruction redundancy, they would have to be duplicated using traditional approaches to protect from instruction faults. They are a small part of the code, so they can easily be duplicated without significantly increasing the footprint size. Computation faults need not be protected from as they would either cause $B \oplus B'$ or $KC \oplus KC'$ to be nonzero or just flip bits in the already computed ciphertext.

5   Security Analysis of the Proposed Countermeasures

In this section, we provide a security analysis for the proposed countermeasures in Sect. 4 against the fault models defined in Sect. 2.

Similar to Guo et al. [13], we use the Fault Coverage (FC) to quantify the security level of countermeasures. For a given countermeasure c and fault model f, we compute the fault coverage using Eq. 1. In our computations, we assume

Table 1. Theoretical security analysis of the proposed countermeasures

Countermeasure	Computation fault models				Instruction fault models
	Random word	Random byte	Random bit	Chosen bit pair	Instruction skip
Unprotected AES	0%	0%	0%	0%	0%
IIR-AES	≈ 100%	94.90%	100%	51.61%	75%
Pipelined IIR-AES	≈ 100%	99.90%	100%	96.77%	99.90%
Shuffled pipelined IIR-AES	≈ 100%	99.90%	100%	96.77%	99.90%

that the adversary aims at injecting a computation or instruction skip fault into the execution of a target round.

$$(FC)_c^f = 1 - \frac{F_{undetected}}{F_{total}} \tag{1}$$

In Eq. 1, F_{total} is the total number of faults covered by the fault model f. $F_{undetected}$ is the number of faults that affect the execution of the target round r, but cannot be detected by the given countermeasure c. More capable adversaries can increase the $F_{undetected}$, and reduce the F_{total} by accurately tuning fault injection parameters. We list our FC computations in Table 1.

5.1 Security Analysis of Unprotected AES

In the unprotected, bit-sliced AES implementation, any computation or instruction fault during the execution of the target round r will be useful for the adversary. As there is no detection mechanism for this implementation, F_{total} and $F_{undetected}$ will be equal to each other. As a result, fault coverage will be 0 in any case. The detailed explanations for each fault model are as follows.

In the *Random Word* fault model, the adversary has no control on the number of faulty bits. The adversary can only create random faults in the target word (32-bit). For each fault injection, the difference between the correct word and the corresponding faulty word can have $(2^{32} - 1)$ different values. Therefore, F_{total} and $F_{undetected}$ are $(2^{32} - 1)$.

In the *Random Byte* fault model, an adversary can tune the fault injection to randomly affect a single byte of the 32-bit data. This adversary can inject a fault into one of the four bytes of the data. Each fault injection can create $(2^8 - 1)$ different faults in a byte. As a result, F_{total} and $F_{undetected}$ are $4 \times (2^8 - 1)$.

In the *Random Bit* fault model, the fault injection can be tuned to affect single bit of the target word. Therefore, F_{total} and $F_{undetected}$ are 32.

In the *Chosen Bit Pair* fault model, the adversary can inject faults into two chosen, adjacent bits of the target word. Therefore, F_{total} and $F_{undetected}$ are 31.

5.2 Security Analysis of IIR-AES

To thwart this countermeasure, the adversary needs to create the same effect on the data slices and their corresponding redundant slices, without affecting any KC slice. Affecting any combination of 15 data and redundant slice pairs will create undetected faults. A *Random Word* fault can achieve this in $\sum_{i=1}^{15} \binom{15}{i} -$ $\binom{15}{0} = (2^{15} - 1)$ different ways. Therefore, $F_{undetected}$ is $(2^{15} - 1)$ and the FC (using Eq. 1) is 99.9992% ($\approx 100\%$).

This countermeasure has three data and redundant slice pairs in the most significant byte of the target word, while it has four pairs in each of the remaining bytes (Fig. 2). Thus, for *Random Byte* fault model, the $F_{undetected}$ is $(2^3 - 1) + 3 \times (2^4 - 1) = 52$, and the FC is 94.90%.

As a *Random Bit* fault can manipulate only a single KC slice, data slice, or redundant slice, the $F_{undetected}$ is 0, and the FC is 100%.

As a *Chosen Bit Pair* fault can target a specific pair of data and redundant slices, the $F_{undetected}$ is 15, and the FC is 51.61%.

An *Instruction skip* fault will have the same effect on a data slice and its corresponding redundant slice. Thus, data and redundant slice pairs cannot detect an instruction skip. Each KC slice has a 50% chance of detecting an instruction skip. As we have 2 KC slices, the fault coverage is $1 - \frac{1}{2^2} = 75\%$.

5.3   Security Analysis of Pipelined IIR-AES

In this countermeasure, the 32-bit word consists of 10 KC, 10 redundant, 10 data, and 2 spare slices (Fig. 3). Each data and redundant slice pair apply a different round of AES on a different block. As there is only one data and redundant slice pair running the target round r, the only way to obtain a useful and undetected computation fault is by targeting this pair of slices.

For *Random Word* and *Random Byte* faults, the $F_{undetected}$ is equal to 1. Therefore the corresponding fault coverage for Random Word and Random Byte faults are $\approx 100\%$ and 99.90%, respectively.

As no *Random Bit* fault can bypass this countermeasure, the $F_{undetected}$ is 0, the fault coverage is 100%.

A *Chosen Bit Pair* fault can manipulate the data and redundant slice pair that computes the target round r. The $F_{undetected}$ is 1 and the fault coverage is 96.77%.

This countermeasure significantly increases the fault coverage against *instruction skip* attacks as we use 10 constant slices. The only undetected instruction skip fault is the one that does not affect any of the constant bits. Therefore, the fault coverage against instruction skip is $1 - \frac{1}{2^{10}} = 99.90\%$.

5.4   Security Analysis of Shuffled Pipelined IIR-AES

This countermeasure improves the security of the previous countermeasure by dynamically allocating the positions of the slices within a word. In this work, the slices are rotated by a random number after each encryption. In this scheme, we have 32 different allocations. This reduces the chance of an attacker to inject a useful and undetected fault 32 times because attacker's chance of guessing the position of the target round is 1/32. In addition, this countermeasure significantly reduces the chance of an attacker from repeating the same fault on successive encryptions.

6   Results

In this section we will cover our performance, footprint, and experimental fault coverage. We verified our results in a simulation of a 32 bit SPARC processor called LEON3. We used Aeroflex Gaisler's LEON3 CPU simulator, TSIM. To inject faults and determine the coverage, we wrote a wrapper program[1] for Gaisler's TSIM simulator. The wrapper enabled us to use TSIM commands to inject faults into any instruction, memory location, or register during the execution of the code.

6.1   Performance and Footprint

Our performance and footprint results are presented in Table 2. We wrote a bit-sliced implementation of AES in C and benchmarked it without any fault attack countermeasures added to it. We made three forks of our reference AES implementation and added each stage of our countermeasure to them[2]. Performance was measured by running AES in CTR mode and on a large input size. Footprint was calculated by measuring the compiled program size. Overheads were calculated by dividing by the corresponding reference bit-sliced AES metric.

Using Shuffled Pipelined IIR-AES will be about four times as slow as the reference implementation. Considering it can protect against side channels from the most dangerous of fault attacks, it is a good compromise.

The original AES metric is slow compared to other works because it is an unoptimized implementation. Other works have been able to get bit-sliced AES implementations down to about 20 cycles/byte on 32 bit ARM [14]. We believe the performance overhead for adding IIR would scale with the reference performance.

Table 2. Performance and footprint of multilevel countermeasure. Unprotected AES is the reference bit-sliced implementation with no added countermeasure.

	Performance	Footprint
Unprotected AES	469.3 cycles/byte	5576 bytes
IIR-AES	1055.9 cycles/byte	6357 bytes
Overhead IIR-AES	2.25	1.14
Pipelined IIR-AES	1942.9 cycles/byte	5688 bytes
Overhead pipelined IIR-AES	4.14	1.02
Shuffled pipelined IIR-AES	1957 cycles/byte	6134 bytes
Overhead shuffled pipelined IIR-AES	4.17	1.10

---

[1] The wrapper program may be accessed on Github: https://github.com/Secure-Embedded-Systems/tsim-fault.

[2] Our implementations can be accessed on Github: https://github.com/Secure-Embedded-Systems/fault-resistant-aes.

Table 3. Experimental fault coverage averages for different fault injections. Every register or instruction in the S-box stage in the last round was targeted one at a time per run for fault injection.

Countermeasure	Computation fault models				Instruction fault models
	Random word	Random byte	Random bit	Chosen bit pair	Instruction skip
Unprotected AES	0.0%	0.0%	0.0%	0.0%	0.0%
IIR-AES	99.98%	91.45%	93.66%	53.96%	80.56%
Pipelined IIR-AES	100.0%	100.00%	100.0%	98.51%	98.6%
Shuffled pipelined IIR-AES	100.0%	99.99%	100.0%	98.86%	98.6%

6.2 Experimental Fault Coverage

We ran fault simulations that emulated our considered adversaries. We injected faults for every register or instruction used in the S-box step of the last round. For data faults, our simulation would enumerate each register, injecting 1 fault, then letting the program run to completion to check the resulting ciphertext. For instruction skips, each instruction is similarly enumerated for skipping and checking the resulting ciphertext.

The S-box step has 144 instructions, consisting of 18 memory operations and 126 computational operations. Of the 144 instructions, 404 operands were registers. Each fault injection simulation was repeated 50 times and averaged together. For each data fault simulation, 20,200 faults were injected. For each instruction skip simulation, 7,200 faults were injected.

Table 3 shows the average fault coverages for each countermeasure. Most of the experiments match closely with our theoretical fault coverages. IIR-AES has slightly lower fault coverage then theorized for random bit and byte faults because memory addresses stored in a register can change to a different but valid location, resulting in a control fault. Because of this, the theoretical fault coverage for data faults will be slightly averaged with the control fault coverage. Random bit and byte coverage is slightly lower than expected and chosen bit pair is slightly higher than expected for IIR-AES.

Instruction skip coverage in IIR-AES is 5.56% higher then expected, which is likely just specific to the S-box and key constants we used.

7 Conclusion

We have introduced a set of novel and state of the art methods for detecting faults in block ciphers. We use only software and introduce intra-instruction redundancy. We can protect from well timed, repeatable faults. By adding pipelining, we make our block cipher rounds spatial and much harder to target. And by finally applying random rotations, we make it even more difficult to fault the target round more than once. We show that the performance overhead of our countermeasure is acceptable and scales depending on the desired security level. Our program size overhead is considerably lightweight. We theoretically show why our countermeasure meets the requirements for different fault models. We support our theoretical claims using experimental simulation results based on Gaisler's LEON3 simulator.

Acknowledgments. This research was supported in part by the National Science Foundation Grant 1441710, Semiconductor Research Corporation Task 2552.001, and Nation Science Foundation CyberCorps Program.

References

1. Boneh, D., DeMillo, R.A., Lipton, R.J.: On the Importance of checking cryptographic protocols for faults. In: Fumy, W. (ed.) EUROCRYPT 1997. LNCS, vol. 1233, pp. 37–51. Springer, Heidelberg (1997). doi:10.1007/3-540-69053-0_4
2. Ali, S., Hyay, D.M., Tunstall, M.: Differential fault analysis of AES: towards reaching its limits. J. Cryptogr. Eng. **3**, 73–97 (2013)
3. Lomné, V., Roche, T., Thillard, A.: On the need of randomness in fault attack countermeasures - application to AES. In: 2012 Workshop on Fault Diagnosis and Tolerance in Cryptography, Leuven, Belgium, 9 September 2012, pp. 85–94 (2012)
4. Barenghi, A., Breveglieri, L., Koren, I., Pelosi, G., Regazzoni, F.: Countermeasures against fault attacks on software implemented AES: effectiveness and cost. In: Proceedings of the 5th Workshop on Embedded Systems Security, WESS 2010, Scottsdale, AZ, USA, 24 October 2010, pp. 7:1–7:10 (2010)
5. Guo, X., Karri, R.: Invariance-based concurrent error detection for advanced encryption standard. In: The 49th Annual Design Automation Conference 2012, DAC 2012, San Francisco, CA, USA, 3–7 June 2012, pp. 573–578 (2012)
6. Patranabis, S., Chakraborty, A., Mukhopadhyay, D., Chakrabarti, P.P.: Using state space encoding to counter biased fault attacks on AES countermeasures. IACR Cryptology ePrint Arch. **2015**, 806 (2015)
7. Breier, J., Jap, D., Bhasin, S.: The other side of the coin: Analyzing software encoding schemes against fault injection attacks. In: IEEE International Symposium on Hardware Oriented Security and Trust, HOST 2016, Washington, DC, USA, 2016 (2016)
8. Battistello, A., Giraud, C.: Fault analysis of infective AES computations. In: 2013 Workshop on Fault Diagnosis and Tolerance in Cryptography, Los Alamitos, CA, USA, 20 August 2013, pp. 101–107 (2013)
9. Battistello, A., Giraud, C.: Lost in translation: Fault analysis of infective security proofs. In: 2015 Workshop on Fault Diagnosis and Tolerance in Cryptography, FDTC 2015, Saint Malo, France, 13 September 2015, pp. 45–53 (2015)
10. Endo, S., Homma, N., Hayashi, Y., Takahashi, J., Fuji, H., Aoki, T.: A multiple-fault injection attack by adaptive timing control under black-box conditions and a countermeasure. In: Prouff, E. (ed.) COSADE 2014. LNCS, vol. 8622, pp. 214–228. Springer, Cham (2014). doi:10.1007/978-3-319-10175-0_15
11. Ghalaty, N.F., Yuce, B., Taha, M., Schaumont, P.: Differential fault intensity analysis. In: Fault Diagnosis and Tolerance in Cryptography (FDTC), Workshop on. 2014, pp. 49–58 (2014)
12. Tunstall, M., Mukhopadhyay, D.: Differential fault analysis of the advanced encryption standard using a single fault. Cryptology ePrint Archive, Report 2009/575 (2009). http://eprint.iacr.org/
13. Guo, X., Mukhopadhyay, D., Karri, R.: Provably secure concurrent error detection against differential fault analysis. Cryptology ePrint Archive, Report 2012/552 (2012). http://eprint.iacr.org/
14. Atasu, K., Breveglieri, L., Macchetti, M.: Efficient aes implementations for arm based platforms. In: Proceedings of the 2004 ACM Symposium on Applied Computing. SAC 2004, New York, pp. 841–845. ACM (2004)

Cryptanalysis of Symmetric Primitives

New Second Preimage Attacks on Dithered Hash Functions with Low Memory Complexity

Muhammad Barham[1], Orr Dunkelman[1(✉)], Stefan Lucks[2], and Marc Stevens[3]

[1] Computer Science Department, University of Haifa, Haifa, Israel
muhammad.barham@gmail.com, orrd@cs.haifa.ac.il
[2] Bauhaus-Universität Weimar, Weimar, Germany
stefan.lucks@uni-weimar.de
[3] Centrum Wiskunde & Informatica, Amsterdam, The Netherlands
marc@marc-stevens.nl

Abstract. Dithered hash functions were proposed by Rivest as a method to mitigate second preimage attacks on Merkle-Damgård hash functions. Despite that, second preimage attacks against dithered hash functions were proposed by Andreeva et al. One issue with these second preimage attacks is their huge memory requirement in the precomputation and the online phases. In this paper, we present new second preimage attacks on the dithered Merkle-Damgård construction. These attacks consume significantly less memory in the online phase (with a negligible increase in the online time complexity) than previous attacks. For example, in the case of MD5 with the Keränen sequence, we reduce the memory complexity from about 2^{51} blocks to about $2^{26.7}$ blocks (about 545 MB). We also present an essentially memoryless variant of Andreeva et al. attack. In case of MD5-Keränen or SHA1-Keränen, the offline and online memory complexity is $2^{15.2}$ message blocks (about 188–235 KB), at the expense of increasing the offline time complexity.

1 Introduction

Cryptographic hash functions have many information security applications, notably in digital signatures and message authentication codes (MACs). The need for hash functions renders its security as one of the important topics in the design analysis of cryptographic primitives.

Designing hash function usually consists of two parts:

- Designing a compression function (or a secure permutation, in the case of sponge functions [4]).
- Designing the mode of iteration (also called domain extension).

The first and second authors were supported in part by the Israeli Science Foundation through grant No. 827/12.

© Springer International Publishing AG 2017
R. Avanzi and H. Heys (Eds.): SAC 2016, LNCS 10532, pp. 247–263, 2017.
https://doi.org/10.1007/978-3-319-69453-5_14

These two parts complement one another, and to create a secure hash function, a secure compression function and a secure mode of iteration are needed.

The most common and used mode of iteration is Merkle-Damgård [6,14]. Though believed to be secure, the Merkle-Damgård construction was found vulnerable to different multi-collision and second preimage attacks [1–3,7–11]. One of the alternatives that was suggested to replace Merkle-Damgård and to increase Merkle-Damgård security, is the dithered Merkle-Damgård [18]. Dithered Merkle-Damgård was designed by Rivest to overcome the *Expandable Message* attack of [11]. The main idea of dithered Merkle-Damgård is to add a third input (derived from some sequence) to the compression function to perturb the hashing process. However, dithered Merkle-Damgård was found vulnerable to two second preimage attacks by Andreeva et al. [1,3]. While taking less than 2^n time to find a second preimage, these attacks consume a huge amount of memory.

1.1   Related Work

Andreeva et al. described in [1,3] two second preimage attacks on dithered Merkle-Damgård. The first attack, the "adapted Kelsey-Kohno", uses a diamond structure (similarly to our attack). Assume that the dithering sequence z (over alphabet $\mathcal{A}$) is used, the compression function $f : \{0,1\}^n \times \{0,1\}^b \times \mathcal{A} \to \{0,1\}^n$ and that the target message is 2^k blocks. The online time complexity of the first attack is[1] $\frac{2^{n-k}}{Freq_z(w_{mc}^{\ell+1})} + 2^{n-\ell}$, and the offline and the online memory complexities are $2^{n/2+\ell/2+1/2}$ and $2^{\ell+1}$, respectively.

The second attack, the "Kite Generator", requires $2 \cdot |\mathcal{A}| \cdot 2^n$ time complexity in the offline phase, and $max(2^k, 2^{(n-k)/2})$ time complexity in the online phase. Its memory complexity is $|\mathcal{A}| \cdot 2^{n-k}$ in both the offline and the online phases.

To summarize both attacks require a tremendous amount of memory in the online phase. In this paper we reduce the memory complexity (of the online and offline phases) of second preimage attacks on dithered.

1.2   Our Results

This paper describes novel second preimage attacks on dithered Merkle-Damgård hash function with very low memory complexities. We first explore attacks that have low online memory complexity with almost no increase in the time complexities compared with the attacks of [1,3]. The online and offline memory complexities of the basic attack are $|\mathcal{A}| \cdot Fact_z(\ell) \cdot (\ell+1)$ and $Fact_z(\ell) \cdot 2^{\ell+1} + 2^{n/2+\ell/2+1/2}$, respectively. For example, the memory time complexity of the attack, in case of MD5-Keränen for $\ell = 50$ is about $2^{26.7}$ blocks (about 507 MB).

Then, we introduce ideas and optimizations of the attack that reduce the offline time and memory complexities. Lastly, we use these ideas to present

---

[1] $Freq_z(w)$ is the frequency of the word w in the sequence z, $w_{mc}^{\ell+1}$ is the most common word of length $\ell + 1$ in the sequence z.

an essentially memoryless attack, again without increasing the online time complexity. The online and the offline memory complexity of the attack is $(\ell + 1) \cdot Fact_z(\ell + 1)$ blocks. However, for this reduced memory attack, the offline time complexity is increased.

In Table 1, we compare the complexities of second preimage attacks on dithered Merkle-Damgård. In Table 2, we compare the complexities of the second preimage attacks on real hash functions with concrete parameters.[2]

Table 1. Comparison of the second preimage attacks on dithered hash functions.

	Time Complexity		Memory Complexity (blocks)	
	Offline	Online	Offline	Online
Adapted Kelsey-Kohno [1,3]	$2^{n/2+\ell/2+2}$	$\frac{2^{n-k}}{Freq_z(w_{mc}^{\ell+1})} + 2^{n-\ell}$	$2^{n/2+\ell/2+1/2}$ $+2^{\ell+1}$	$2^{\ell+1}$
Kite Generator [1,3]	$2 \cdot |\mathcal{A}| \cdot 2^n$	$max(2^k, 2^{(n-k)/2})$	$2 \cdot |\mathcal{A}| \cdot 2^{n-k}$	$2 \cdot |\mathcal{A}| \cdot 2^{n-k}$
Basic Attack (Section 4)	$Fact_z(\ell) \cdot 2^{n/2+\ell/2+2}$ $+|\mathcal{A}| \cdot Fact_z(\ell)^2 \cdot 2^{n-\ell}$	$\frac{2^{n-k}}{Freq_z(w_{mc}^{\ell+1})}$	$Fact_z(\ell) \cdot 2^{\ell+1}$ $+2^{n/2+\ell/2+1/2}$	$|\mathcal{A}| \cdot (\ell + 1) \cdot Fact_z(\ell)^2$
Time Optimization I (Section 5)	$Fact_z(\ell) \cdot 2^{n/2+\ell/2+2}$ $+|\mathcal{A}| \cdot \left(\binom{Fact_z(\ell)}{2} + \ell \right) \cdot 2^{n-\ell}$	$\frac{2^{n-k}}{Freq_z(w_{mc}^{\ell+1})}$	$Fact_z(\ell) \cdot 2^{\ell+1}$ $+2^{n/2+\ell/2+1/2}$	$|\mathcal{A}| \cdot (\ell + 1) \cdot Fact_z(\ell)^2$
Time Optimization II (Section 5)	$Fact_z(\ell) \cdot 2^{n/2+\ell/2+2}$ $+|\mathcal{A}| \cdot Fact_z(\ell) \cdot 2^{n-\ell}$	$\frac{2^{n-k}}{Freq_z(w_{mc}^{\ell+1})}$	$Fact_z(\ell) \cdot 2^{\ell+1}$ $+2^{n/2+\ell/2+1/2}$	$|\mathcal{A}| \cdot (\ell + 1) \cdot Fact_z(\ell)^2$
Time Optimization III (Section 5)	$Fact_z(\ell) \cdot 2^{n/2+\ell/2+2}$ $+Fact_z(\ell + 1) \cdot 2^{n-\ell}$	$\frac{2^{n-k}}{Freq_z(w_{mc}^{\ell+1})}$	$Fact_z(\ell) \cdot 2^{\ell+1}$ $+2^{n/2+\ell/2+1/2}$	$(\ell + 1) \cdot Fact_z(\ell + 1)$
Memory Optimization (Section 6)	$2 \cdot Fact_z(\ell) \cdot 2^{n/2+\ell/2+2}$ $+Fact_z(\ell + 1) \cdot 2^{n-\ell}$	$\frac{2^{n-k}}{Freq_z(w_{mc}^{\ell+1})}$	$2^{n/2+\ell/2+1/2}$ $+2^{\ell+1}$	$(\ell + 1) \cdot Fact_z(\ell + 1)$
The Memoryless Attack (Section 7)	$2 \cdot Fact_z(\ell) \cdot 2^{n/2+\ell}$ $+Fact_z(\ell + 1) \cdot 2^{n-\ell}$	$\frac{2^{n-k}}{Freq_z(w_{mc}^{\ell+1})}$	$(\ell + 1) \cdot Fact_z(\ell + 1)$	$(\ell + 1) \cdot Fact_z(\ell + 1)$

1.3 Organization of the Paper

We introduce some terminology, describe the Merkle-Damgård construction, and the dithered Merkle-Damgård construction in Sect. 2. We describe the previous attacks in Sect. 3. We then present our new basic attack (which has comparable time complexity to the attack of [1,3]) in Sect. 4. We show optimizations and improvements for the offline time complexity of this attack in Sect. 5. In Sect. 6, we show offline memory optimizations. We then show an essentially memoryless attack on dithered hash functions in Sect. 7. Finally, we conclude the paper in Sect. 8.

---

[2] In Appendix A we discuss a compact representation of message blocks both in the generation of the diamond structure and in the online phase. The results reported in Table 2 assume these compact representations.

Table 2. Comparison of the second preimage attacks on dithered hash functions which uses the Keränen sequence. ℓ was chosen as an optimal value for "adapted Kelsey-Kohno". The analysis in [1,3] about dithering sequence, showed that for Keränen sequence and $\ell = 50$ holds $Fact_{\text{Keränen}}(\ell) \leq 732$, $\frac{1}{Freq_{\text{Keränen}}(w_{mc}^{50})} \leq 340$ and $\frac{1}{Freq_{\text{Keränen}}(w_{mc}^{110})} \leq 1020$.

Function (n,k,ℓ)	-		MD5-Keränen $(128,55,50)$	SHA1-Keränen $(160,55,50)$	SHA256-Keränen $(256,55,50)$	SHA512-Keränen $(512,118,110)$
Adapted Kelsey-Kohno	Time	Offline	2^{91}	2^{107}	2^{155}	2^{313}
		Online	$2^{81.5}$	$2^{113.5}$	$2^{209.5}$	$2^{404.3}$
	Memory	Offline	$2^{89.5}$	$2^{105.5}$	$2^{153.5}$	$2^{311.5}$
		Online	2^{51}	2^{51}	2^{51}	2^{111}
			11259 TB	15763 TB	15763 TB	37530 TB
Kite Generator	Time	Offline	2^{131}	2^{163}	2^{259}	2^{515}
		Online	2^{55}	2^{55}	$2^{100.5}$	2^{197}
	Memory	Offline	2^{76}	2^{108}	2^{204}	2^{397}
		Online	2^{76}	2^{108}	2^{204}	2^{397}
			$3.778 \cdot 10^{11}$ TB	$2 \cdot 10^{21}$ TB	$1.6 \cdot 10^{50}$ TB	$4.1 \cdot 10^{108}$ TB
Basic Attack (Section 4)	Time	Offline	$2^{100.9}$	2^{131}	2^{227}	$2^{424.6}$
		Online	$2^{81.4}$	$2^{113.4}$	$2^{209.4}$	2^{404}
	Memory	Offline	$2^{89.5}$	$2^{105.5}$	$2^{153.5}$	$2^{311.5}$
		Online	$2^{26.7}$	$2^{26.7}$	$2^{26.7}$	$2^{29.4}$
			519.8 MB	727.8 MB	727.8 MB	9.237 GB
Optimization III (Section 5)	Time	Offline	$2^{100.5}$	$2^{119.5}$	$2^{215.5}$	$2^{412.3}$
		Online	$2^{81.4}$	$2^{113.4}$	$2^{209.4}$	2^{404}
	Memory	Offline	$2^{89.5}$	$2^{105.5}$	$2^{153.5}$	$2^{311.5}$
		Online	$2^{15.2}$	$2^{15.2}$	$2^{15.2}$	$2^{17.1}$
			188.2 KB	263.5 KB	263.5 KB	1.976 GB
Offline Memory Optimization (Section 6)	Time	Offline	$2^{101.5}$	$2^{119.5}$	$2^{215.5}$	$2^{412.3}$
		Online	$2^{81.4}$	$2^{113.4}$	$2^{209.4}$	2^{404}
	Memory	Offline	2^{51}	2^{51}	2^{51}	2^{111}
		Online	$2^{15.2}$	$2^{15.2}$	$2^{15.2}$	$2^{17.1}$
			188.2 KB	263.5 KB	263.5 KB	1.976 GB
Memoryless Attack (Section 7)	Time	Offline	$2^{124.5}$	$2^{140.5}$	$2^{215.5}$	$2^{412.3}$
		Online	$2^{81.4}$	$2^{113.4}$	$2^{209.4}$	2^{404}
	Memory	Offline	$2^{15.2}$	$2^{15.2}$	$2^{15.2}$	$2^{17.1}$
		Online	$2^{15.2}$	$2^{15.2}$	$2^{15.2}$	$2^{17.1}$
			301.1 KB	376.4 KB	602.2 KB	4.495 MB

2   Background and Notations

2.1   General Notations

- $\{0,1\}^n$ — all the strings over '0' and '1' of length n.
- $\{0,1\}^*$ — all the strings of finite length.
- m — a message $m \in \{0,1\}^*$.
- $|m|_b$ — the length of m in b-bit block units.
- $\mathcal{A}$ — a finite alphabet.
- $w[i]$ — the ith letter of w.
- w_{mc}^ℓ — the most common word (or factor) of length ℓ in a sequence z.
- $Freq_z(w)$ is the frequency of the word w over the sequence z.

- $Fact_z(\ell)$ — the *sequence's complexity* of a sequence z, given an integer ℓ, as the number of different factors in z of length ℓ.
- P_{w_h} — given a binary tree w, the path P_{w_h} is the sequence of edges which connects the leaf h to the root of w.

2.2 Merkle-Damgård

Let $f : \{0,1\}^n \times \{0,1\}^b \to \{0,1\}^n$ be a compression function, then the Merkle-Damgård hash function $H^f : \{0,1\}^* \to \{0,1\}^n$ is:

- $m_1, m_2, \ldots, m_L \leftarrow pad_{MD}(m)$.
- $h_0 = IV$.
- For $i = 1$ to L, compute $h_i = f(h_{i-1}, m_i)$.
- $H^f(m) \triangleq h_L$.

where $pad_{MD}(m)$ is the conventional Merkle-Damgård padding function, also called Merkle-Damgård Strengthening: Given a message m, it pads a single '1' to the end of the message m also up to $b-1$ zeros, and an embedding the original length of the message at the end, such that the length of the padded message will be a multiple of b.

2.3 Dithering Sequence

To overcome the attack of [11], which is based on *Expandable Message*, Rivest suggested to add a third input (dithered symbol) to the compression function derived from an infinite sequence [18]. Rivest proposed to use one of two sequences:

- Keränen sequence.[3]
- His concrete proposal (a combination of the Keränen sequence and a counter) [18].

Let $f : \{0,1\}^n \times \{0,1\}^b \times \mathcal{A} \to \{0,1\}^n$ be a compression function that accepts an n-bit chaining value, b-bit message block, and $\mathcal{A}$ dither symbol taken from the sequence z. The Dithered Merkle-Damgård hash function $H^f : \{0,1\}^* \to \{0,1\}^n$ is:

- $m_1, m_2, \ldots, m_L \leftarrow pad_{MD}(m)$.
- $h_0 = IV$.
- For $i = 1$ to L, compute $h_i = f(h_{i-1}, m_i, z[i])$.
- $H^f(m) \triangleq h_L$.

In [2], second preimage attacks were shown on dithered hash functions, and the conclusion was that the more *complex* the sequence is, the more secure the dithered hash function is against second preimage attacks.

---

[3] In 1992, Keränen showed in [12] an infinite abelian square-free sequence over a four letter alphabet (Hereafter called the Keränen sequence).

2.4 Diamond Structure

The diamond structure was introduced in [10]. It was used in the attacks of [2,10], and also in the second preimage attack on dithered Merkle-Damgård in [1,3]. A diamond structure is a tree T of depth ℓ, where the 2^ℓ leafs are the possible chaining values, denoted by $D_T = \{h_i^\ell\}$. The nodes in the tree are labeled by digest values and the edges are labeled by message blocks. The adversary builds the diamond structure starting from the 2^ℓ leafs, she tries to map the 2^ℓ leafs to $2^{\ell-1}$ digest values (to the next level in the structure). She does so by generating about $2^{n/2+1/2-\ell/2}$ message blocks from each leaf, then she detects collisions in the generated values. She repeats the process until she reaches the root (with adjusted number of message blocks from each node in each level). The expected time complexity of building a diamond structure is $2^{n/2+\ell/2+2}$.

The diamond structure has the interesting property that there is a path of message blocks from any chaining value leaf h_i^ℓ to the the digest value h_T (the root). See an example in Fig. 1.

The diamond structure was introduced at first to attack classic Merkle-Damgård hash functions [10]. But it can be easily adapted for dithered Merkle-Damgård, by labeling the tree edges with a dither symbol $\alpha \in \mathcal{A}$ as well.

We say that a diamond structure "uses" a sequence w' when all the edges between level i and level $i + 1$ in the structure are labeled in addition to the message block also with $w[i]$. We denote the diamond structure T that uses the sequence w' by $T_{w'}$.

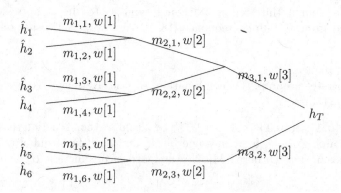

Fig. 1. A diamond structure that uses the dithering sequence w.

3 Previous Attacks on Merkle-Damgård Hash Functions

3.1 Dean's Attack

Dean showed in [7] a second preimage attack on Merkle-Damgård hash functions. The idea is to generate an *expandable message* using a fixed point of the compression function, to connect to the targeted message. Consider a message

$M = m_1 m_2 \ldots m_L$, an initial value IV, and let $H(M) = h$. Denote the intermediate values of processing the message M by $h_1, h_2, \ldots, h_L = h$, i.e., $f(h_{i-1}, m_i) = h_i$ (where $h_0 = IV$).

At the first step of the attack, the adversary generates $2^{n/2}$ random block messages, denoted by $m_1^r, m_2^r, \ldots, m_{2^{n/2}}^r$. Then, she computes $\mathcal{X}_1 = \{f(IV, m_j^r) | \forall j \in \{1, 2, \ldots, 2^{n/2}\}\}$. She then generates $2^{n/2}$ random fixed points of the compression function,[4] denoted by $\mathcal{X}_2 = \{(h_k^f, m_k^c) | f(h_k^f, m_k^f) = h_k^f\}$. Due to the birthday paradox, with non-negligible probability there is m_j^r such that $f(IV, m_j^r) = h_k^f$ (which also means that $f(IV, m_j^r) = f(h_k^f, m_k^f)$). Now, she tries to connect h_k^f to the message, so she generates $2^n/L$ random message blocks, denoted by m_z, $1 \le z \le 2^n/L$. With a non–negligible probability, there is a h_i, such that $f(h_k^f, m_z) = h_i$. At this stage the adversary can output the message $M' = m_j' \underbrace{m_k^f \ldots m_k^f}_{i-2 \text{ times}} m_z m_{i+1} \ldots m_L$ as a second preimage for M. Note that $|M'| = |M|$, which means that after processing the messages, the Merkle-Damgård Strengthening has a similar affect on the digest value in both messages.

The time complexity of the attack is $2^{n/2+1} + 2^n/L$ compression function calls.[5]

3.2   Kelsey and Schneier's Expandable Messages Attack

Kelsey and Schneier showed in [11] a second preimage attack on Merkle-Damgård hash functions. They presented a new technique to build expandable messages without any assumption about the compression function (unlike in Dean's attack), the new technique is based on Joux's multi-collision technique [9], producing multiple messages of varying lengths, with the same digest value. The time complexity of the attack is about $k \cdot 2^{n/2+1} + 2^{n-k+1}$, for a 2^k-block length message.

3.3   Adapted Kelsey-Kohno

One of the second preimage attacks presented in [1,3], is against Merkle-Damgård hash functions. The attack depends heavily on diamond structures which were introduced in [10]. The adversary generates a diamond structure of depth ℓ and tries to connect the diamond structure to the message by a connecting message block. After a successful attempt, she generates a prefix P which connects the IV to the diamond structure of appropriate length. The second preimage message is the concatenation of the prefix P, the path in the diamond structure (which connects the prefix to the root) and the remaining of the

---

[4] The attack is efficient when it is "easy" to find fixed points of the compression function. For example, in Davis-Meyer compression functions.

[5] We note that the complexity is for the case where finding a fixed point is trivial, i.e., takes one compression function call.

original message blocks (what come after the connecting message block). The complexity of the attack is $2^{n/2+\ell/2+2} + 2^{n-k} + 2^{n-\ell}$.

The attack can also be adapted to the dithered Merkle-Damgård hash functions, which we refer to as "Adapted Kelsey-Kohno". Most of the attack steps are similar to the original attack. The adversary generates a diamond structure of depth ℓ. The diamond structure uses the first ℓ symbols of the most common factor of length $\ell + 1$ of the dithering sequence z. The last symbol of the most common factor of length $\ell + 1$ is used to connect the diamond structure's root to the message in appropriate location. The number of possible points to connect the diamond structure is equal to $\frac{1}{Freq_z(w_{mc}^{\ell+1})}$. So, the time complexity of the attack $2^{n/2+\ell/2+2} + \frac{2^{n-k}}{Freq_z(w_{mc}^{\ell+1})} + 2^{n-\ell}$, and the online memory complexity is $2^{\ell+1}$.

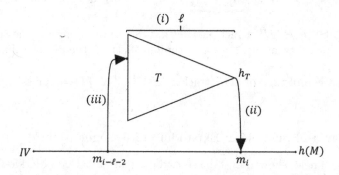

Fig. 2. Illustration of Andreeva' et al.'s attack: (i) Build the diamond structure T (ii) Connect the root to the message (iii) Connect the message to the leafs of T.

3.4 Kite Generator and More Second Preimage Attacks

The "Kite Generator" was introduced in [1,3]. It is a labeled directed graph of 2^{n-k} vertices. Every vertex labeled by a chaining value (including the IV) and every edge is labeled by a message block and a symbol from the alphabet of the dithering sequence. For any symbol in the dithering alphabet (i.e., $\alpha \in \mathcal{A}$) there are two edges labeled by the symbol α. The result is that every vertex has $2 \cdot |\mathcal{A}|$ outgoing edges. The structure is highly connected, that is to say, there is an exponential number of paths for any dithering sequence that starts from a single vertex. The time complexity of building such a structure is $2 \cdot |\mathcal{A}| \cdot 2^n$ and it requires $|A| \cdot 2^{n-k}$ memory.

For a given message m of 2^k blocks and 2^k intermediate digest values, there is a non-negligible probability that there is an intermediate digest value which is also a label of a vertex in the kite generator structure. Denote this value by h_i. The adversary picks a path in the kite generator starting from the IV of length $i - (n - k)$ with the dithering sequence $z[0 \ldots (i - (n - k))]$. Then, from the last chaining value of the generated path, build a binary tree of depth $(n - k)/2$

that uses the dithering sequence $z[(i - (n - k) + 1) \ldots (i - (n - k)/2)]$ (this is achieved by traversing all possible routes in the graph corresponding the required dithering sequence). In the last step, she builds a binary tree from h_i of depth $(n - k)/2$ that uses the dithering sequence $z[(i - (n - k)/2 + 1) \ldots i]^R$. With a non-negligible probability there is a collision in the leafs of the two trees. The second preimage is the concatenation the the generated path, the path which connects the two roots of the trees, and the remaining blocks of the original message (the message blocks which come after h_i). The online time and memory complexities of the attack are $max(2^{n/2}, 2^{n-k})$ and $|\mathcal{A}| \cdot 2^{n-k}$, respectively.

4   A New Second Preimage Attack on Dithered Merkle-Damgård

We now present a new second preimage attack on dithered Merkle-Damgård. The new attack has a slightly longer precomputation time, but in exchange, the online memory complexity of the attack is significantly reduced to practical levels.

Similarly to the attacks of [1,3], the attack consists of two phases: the precomputation (the offline) and the online phase. In the offline phase, we generate $Fact_z(\ell)$ diamond structures, every structure with a unique factor of the sequence z of length ℓ. Then, we connect every diamond structure to *all* diamond structures (including itself). We then purge unnecessary paths from the memory. These purged structures are then used in the online phase to find a second preimage by connecting one of the purged diamond structures to the message, and starting from the IV traversing through the purged diamond structures to reach the connecting point. The total amount of memory that is needed for keeping the purged diamond structures is significantly smaller than the amount of memory needed for storing a full diamond structure.

4.1   Adapting Diamond Structure to Dithered Merkle-Damgård

We now give the details of the attack:

- **Offline phase:**
  1. Build $Fact_z(\ell)$ diamond structures of depth ℓ each (denoted by $\{T_i | 1 \leq i \leq Fact_z(\ell)\}$), where every T_i uses a different factor of z of length ℓ. Every diamond structure T_i has a digest value h_{T_i}. Note that the IV is a leaf in all the generated diamond structures.
  2. Connect every diamond structure T_i to every diamond structure T_j (T_i may be T_j) with all $|\mathcal{A}|$ possible dithering symbols. Namely, for every pair of T_i, T_j and any dithering symbol α, find $m_{i \to j}^{\alpha}$ such that $f(h_{T_i}, m_{i \to j}^{\alpha}, \alpha)$ is a leaf of T_j.
  3. Prune (reduce) the diamond structures by removing all unnecessary nodes and edges that do not belong to any path that connects two roots h_{T_i} and h_{T_j}. Formally, let $\mathcal{G}_i = \{P_{j,i} | \exists T_j \text{ such that } P_{j,i} \text{ is a path from a } h_{T_j}$

to h_{T_i}}. Then, for all $T_{i'}$, remove the nodes $\{n' \in T_{i'} | n' \notin P_{j',i'}$ for any $P_{j',i'} \in \mathcal{G}_i\}$.

Between two diamond structures, there are $|\mathcal{A}|$ such paths. So overall, keep $|\mathcal{A}| \cdot Fact_z^2(\ell)$ paths (of length $\ell + 1$ each, as we also store the connecting message block $m_{j \to i}^\alpha$).

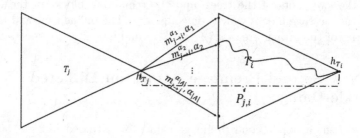

Fig. 3. Connecting diamond structure T_i to diamond structure T_j

Complexity Analysis: Constructing $Fact_z(\ell)$ diamond structures takes $Fact_z(\ell) \cdot 2^{n/2+\ell/2+2}$ compression function calls. Connecting one diamond structure to another takes $2^{n-\ell}$ compression function calls for a given dither sequence. Therefore, connecting all the diamond structures to each other takes $|\mathcal{A}| \cdot Fact_z(\ell)^2 \cdot 2^{n-\ell}$ time. Finally, pruning the diamond structures takes $|\mathcal{A}| \cdot Fact_z^2(\ell) \cdot (\ell + 1)$ time and memory.

Therefore, the overall time complexity of the offline phase is $Fact_z(\ell) \cdot 2^{n/2+\ell/2+2} + |\mathcal{A}| \cdot Fact_z(\ell)^2 \cdot 2^{n-\ell}$ compression function calls, and it passes $|\mathcal{A}| \cdot Fact_z(\ell)^2 \cdot (\ell + 1)$ memory blocks to the online phase.

- **Online phase:** In the online phase, given a message m, such that $|m| = 2^k$ blocks. Let $w' = w_r \alpha$ ($|w'| = \ell + 1$) be the most common factor in positions $0, \ell + 1, 2(\ell + 1), \dots$ (positions that are multiple of $\ell + 1$) of the sequence z. Let $Range$ be $\{i \in \mathcal{N} | i \leq 2^k \wedge z[i - (\ell + 1)] \dots z[i] = w'\}$, namely, $Range$ is the set of all indexes of chaining values which were produced of hashing of any consecutive $\ell + 1$ blocks with w_r, perform:
  - Find a connecting block B_r such that $f(h_{T_{w_r}}, B_r, \alpha) = h_{i_0}$ for $i_0 \in Range$.
  - Traverse the structures and find a path from IV to B_r while preserving the dithering sequence order.
  - The second preimage is generated by concatenating the path that was found in the previous step, with the rest of the original m from the block after the connecting point till the end.

The complexity of the online phase is $\frac{2^{n-k}}{Freq_z(w')} + 2^k$, which is essentially the same as the adapted Kelsey-Schneier's attack [1,3] (we note that the connection "into" the diamond structure is eliminated).

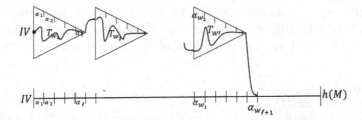

Fig. 4. Illustration of the attack.

4.2 Generalization

The previous attack worked with the most common factor of length $\ell + 1$ in positions $0, \ell + 1, 2 \cdot (\ell + 1), \ldots$ (i.e., a multiple of $\ell + 1$). Traversing the diamond structures gives always a path whose length is a multiple of $\ell + 1$. Therefore, it is limited to work with the most common factor in specific positions, which may not be the most common factor of the whole sequence.

To overcome this issue, we generate from the IV a chain of length ℓ. We pick at random a message $m' = m'_1 m'_2 \ldots m'_\ell$ and evaluate $h'_1 = h(IV, m'_1, z[1]), h'_2 = h(IV, m'_1 m'_2, z[1, 2]), h'_3 = h(IV, m'_1 m'_2 m'_3, z[1, 2, 3])$. etc. until h'_ℓ. We then, use h'_1 as one of the leafs in $T_{z[1\ldots]}$, h'_2 as one of the leafs in $T_{z[2\ldots]}$, etc.

In the online phase, let w' be the most common word of length $\ell + 1$. We connect $T_{w'}$ to the message in position t. Let $d = t \mod (\ell + 1)$, and use $m'_1 m'_2 \ldots m'_d$ to connect the IV to the diamond structure $T_{z[d..(d+\ell)]}$, then traverse from $T_{z[d..(d+\ell)]}$ to $T_{w'}$.

Complexity Analysis: The additional complexity of the new algorithm is ℓ compression function calls (generating $h'_1, h'_2, \ldots, h'_\ell$). This amount of complexity is negligible, and does not affect the offline time complexity. However, it does improve the online time complexity, as we now use the global most common factor in the sequence, which may allow for more connecting points, instead of a most common factor in specific positions. Therefore, the new complexity of the online phase is $\frac{2^{n-k}}{Freq_z(w_{mc}^{\ell+1})} + 2^k$.

5 Optimizations and Improvements

In this section, we present several optimizations and improvements. Some ideas reduce the offline time complexity, while other ideas reduce the offline memory complexity. All these improvements *do not* increase the online time and memory complexities.

5.1 Reducing Offline Time Complexity

Optimization I - Use the Diamond Structure Roots as Leafs. A major factor in the offline phase is connection of the diamond structures to each other.

We now present a simple way to reduce the time to connect the diamond structures by about half: After generating a structure, let its root and all other previous structures' roots be part of the 2^ℓ leafs of the following generated structures. Namely,

- Set $D_1 = \{IV\}$.
- For $i = 1$ to $Fact_z(\ell)$:
  - Generate the diamond structure T_i, such that D_i is a subset of the leafs of T_i, i.e., make sure that $D_i \subseteq D_{T_i}$.
  - Set $D_{i+1} = D_i \cup \{h_{T_i}\}$.

As the IV is a leaf in all the diamond structures, one can start the exploration of the dither sequence factor from it, independently of the first factor. Similarly, when generating a new diamond structure, if all the roots of the previously generated diamond structures are leafs in the new diamond structure, then they are already connected to it, and there is no need in connecting the previous diamond structures to the new one.

After the generation of the diamond structures, every pair of different diamond structures is already connected in one direction, but needs to connect in the other direction. This reduces the connection time from $|\mathcal{A}| \cdot Fact_z^2(\ell) \cdot 2^{n-\ell}$ to $|\mathcal{A}| \cdot \left(\binom{Fact_z(\ell)}{2} + \ell\right) \cdot 2^{n-\ell}$, and the total offline time complexity to $Fact_z(\ell) \cdot 2^{n/2+\ell/2+2} + |\mathcal{A}| \cdot \left(\binom{Fact_z(\ell)}{2} + \ell\right) \cdot 2^{n-\ell}$.

Optimization II - All the Diamond Structures Have the Same Leafs. Another simple optimization which reduces the offline time complexity by a factor of $Fact_z(\ell)$, is to let the leafs of all the diamond structures be the same. In other words, choose 2^ℓ random values, and let those values be the leafs of all the diamond structures. This way, when connecting a diamond structure to another, it connects the diamond structure to all the other diamond structures. This reduces the diamond structures connection time complexity to $|\mathcal{A}| \cdot Fact_z(\ell) \cdot 2^{n-\ell}$, and the total offline time complexity to $Fact_z(\ell) \cdot 2^{n/2+\ell/2+2} + |\mathcal{A}| \cdot Fact_z(\ell) \cdot 2^{n-\ell}$.

Optimization III - Validate the Connections Between the Diamond Structures. Another simple observation that reduces the time complexity of the offline phase is that not all the factors of length ℓ (or any length greater than 1) are sequential. Meaning, if a factor x of length ℓ in sequence z, does not appear before another factor y of length ℓ, then there is no need to connect T_x to T_y. In fact, the number of the needed connections is $Fact_z(\ell+1)$ (the number of the factors of length $\ell+1$). So the overall diamond structures connection time complexity is $Fact_z(\ell+1) \cdot 2^{n-\ell}$.

Note that as a result of this improvement (together with optimization II), the online memory complexity is also reduced to $Fact_z(\ell+1) \cdot (\ell+1)$.

5.2   Treating (Almost)-Regular Sequences

Our attacks were designed for any possible sequence z over a dithering alphabet $\mathcal{A}$. However, some sequences, such as the Keränen show some regular behavior. Namely, the Keränen sequence is built by taking a sequence over a 4-letter alphabet $\{a, b, c, d\}$ and replacing each character by its own 85-character sequence.[6] We now show how to use the "regularity" of this sequence in reducing the complexities of attacking the sequence.

The basic idea is that if $\ell \leq 84$, the most frequent factor necessarily starts at the beginning of one of the 85-character chunks. Moreover, the sequence itself is divided into four such chunks, and thus, one can build in advance only four diamond structures (of depth $\ell \leq 84$) and connect each of them to the others by paths of length $85 - \ell$. We note that the online time complexity is not affected by this change (as one of the most frequent factors starts at the first character of the chunk), whereas the offline time complexity is reduced to $4 \cdot 2^{n/2+\ell+2} + 12 \cdot 2^{n-\ell}$ (compared with $732 \cdot 2^{n/2+\ell/2+2} + 2928 \cdot 2^{n-\ell}$ for the best general attack).

The idea also reduces the memory complexities (both offline and online). For example, the attack of Sect. 5.1 takes $(\ell + 1) \cdot Fact_z(\ell + 1)$ which are $732 \cdot (\ell + 1)$ blocks of memory (for $\ell < 85$), or merely $12 \cdot 85 = 1020$ blocks of memory when the regularity of the sequence is used.

For $84 < \ell < 169$, the attack spans over two dither chunks. The offline time complexity is thus $12 \cdot 2^{n/2+\ell+2} + 132 \cdot 2^{n-\ell}$, and the memory complexity is $36 \cdot 170 = 6120$ blocks of memory.

6   Memory Optimizations

In this section, we discuss how to reduce the increased memory complexity in the offline phase in exchange for additional offline computations.

6.1   Reducing Memory in the Offline Phase

The offline memory complexity could be reduced from storing $Fact_z(\ell)$ diamond structures to only one diamond structure. The basic idea is to generate only one diamond structure at a time, and reconstruct it in the preprocessing when needed. The improvement is based on optimization methods II and III, where all diamond structures share leafs. Below we describe the algorithm and then explain it:

- Let D be the leafs of the diamond structures with cardinality of 2^ℓ.
- For $i = 1$ to $Fact_z(\ell)$:
  - Generate the diamond structure T_i.
  - Find and save $(h_{T_i}, \mathcal{M}_i)$, where $\mathcal{M}_i = \{m^\alpha_{h_{T_i} \to D} | \forall \alpha \in A, h(h_{T_i}, m^\alpha_{h_{T_i} \to D}, \alpha) \in D$ and $w_i\alpha$ is a factor of z $\}$.
  - Delete T_i.

---

[6] We refer the interested reader to [12] for the full specification of the sequence.

– For $i = 1$ to $Fact_z(\ell)$:
  • Regenerate the diamond structure T_i.
  • Compute and save $\mathcal{G}_i$ (recall that $\mathcal{G}_i$ are the paths from the leafs connected to the root from other diamonds).
  • Delete T_i.

After the offline phase, pass the paths ($\mathcal{G}_i$'s) to the online phase.

We first note that when generating a diamond structure T_i, the leafs of other diamond structure T_j should be predictable to allow the connection from T_i to T_j. By fixing the leafs of all the diamond structures to be the same, we can overcome this obstacle, and we also reduce the time complexity, because connecting T_i to T_j is also connecting T_i to all other diamond structures under the same dithering symbol.

To enable regenerating the same diamond structure twice independently, we can use a fixed pseudo-random sequence (e.g., by seeding some PRNG) to determine the message blocks used (and their order) along with any randomness needed for other decisions.

This reduces the offline memory complexity to $Fact_z(\ell+1) \cdot 2^{n-\ell}$. In exchange, the diamond structure generation time complexity is increased to $2 \cdot Fact_z(\ell) \cdot 2^{n/2+1/2+\ell/2}$, and the total offline time complexity to $2 \cdot Fact_z(\ell) \cdot 2^{n/2+\ell/2+2} + Fact_z(\ell+1) \cdot 2^{n-\ell}$.

6.2   Time-Memory Trade-Off

It is possible to balance the offline memory complexity with the offline time complexity. One could store in memory x diamond structures that are computed only once. The offline memory and time complexities are $x \cdot 2^{\ell+1}$ and $(2 \cdot Fact_z(\ell) - x) \cdot 2^{n/2+\ell/2+2} + Fact_z(\ell+1) \cdot 2^{n-\ell}$, respectively.

7   Memoryless Diamond Structure Generation

We now show that it is possible to essentially eliminate the memory used in the offline phase: By building the diamond structure as a Merkle hash tree [15] (i.e., deciding in advance which leaf collides with which leaf), and using memoryless collision search [13,16,17], we can reduce the offline memory completely to the online complexity.

Each diamond structure has $2^{\ell+1} - 1$ collisions, each can be found in time $\mathcal{O}(2^{n/2})$ without additional memory, allowing for a memoryless diamond structure generation in time $2^{n/2+\ell+1}$. Obviously, the randomness in the generation needs to be replaced with a pseudo-random sequence (the same as in Sect. 6.1). The total offline time complexity of the attack is $2 \cdot Fact_z(\ell) \cdot 2^{n/2+\ell} + Fact_z(\ell+1) \cdot 2^{n-\ell}$. The online time complexity does not change, and remains at $(\ell+1) \cdot Fact_z(\ell+1)$. The total (offline and online) memory complexity is $(\ell+1) \cdot Fact_z(\ell+1)$.

8   Summary

In this work we present a series of second preimage attacks on dithered Merkle-Damgård hash functions. The proposed attacks have the same online time complexity as the previous works of [1,3], but enjoy a significantly reduced memory complexity.

The first set of attacks concentrate on reducing the memory complexity in the online phase (while maintaining, or slightly increasing, the offline memory complexity). This set is motivated by the fact that an adversary may be willing to spend some extra memory (or time) in the offline phase, so the online phase could use a smaller amount of memory (which may be more suitable for FPGA-based cryptanalytic efforts). We believe that this line of research (reducing online memory complexity, possibly at the expense of an increased offline complexities), would open up a new way to look at cryptanalytic problems.

The last attack we present offers an essentially memoryless attack on dithered hash functions which is still considerably better than generic attacks. To conclude, it seems that any dithered hash function should use as complex sequences (namely, with as many different factors) as possible.

Acknowledgements. The authors would like to thank the anonymous referees for their constructive comments that have improved the results of the paper. In addition, the interaction of the authors during the Dagstuhl seminar on symmetric cryptography in January 2016, have contributed significantly to improving the results.

A   Compact Representations of Message Blocks in the Considered Attacks

We now turn our attention to a small constant improvement in the memory consumption of both the generation and the online storage of the diamond structures (similar ideas can be applied to the kite generator, though we do not discuss these in detail). Recall the generation of a diamond structure: 2^ℓ chaining values are chosen (or given). For each such chaining value, we compute $n/2-\ell/2+1/2$ calls to the compression function using different message blocks. To find collisions among the $2^{n/2+\ell/2+1/2}$ chaining values one can use several data structures, where the easiest one is a hash table (indexed by the chaining value). Such a hash table can store all the (chaining value, message block) pairs, and allow for an easy and efficient detection of collisions.

The main question is what is the entry size that needs to be stored in such a table. The trivial solution requires n bits for the chaining value and b bits for the message block, i.e., a total of $n+b$ bits (e.g., 640 for MD5). One can immediately note that as there are only 2^ℓ different chaining values, it is possible to assign to each chaining value an index of ℓ bits, and store only the index. Another trivial improvement is to note that one can use the same message blocks for all chaining values (or determine given the chaining values the message blocks in a pseudo-random manner), and thus, one needs to store only $n/2 - \ell/2 + 1/2$ bits for describing the message block in use. Hence, one can easily use a simpler and more compact representation of $n/2 + \ell/2 + 1/2$ bits (i.e., 90 bits in our attacks on MD5).

We devise an even more compact representation, which is based on storing only the chaining value in the hash table. Then, once a collision is found, one can try all message blocks sequentially to recover the message blocks that led to the collision. While this doubles the computational effort of the generation of the diamond structure, by storing a few additional bits of the message block along the chaining value, is sufficient to make this approach quite computationally efficient. Hence, one can use $\ell + t$ bits, where even a small t (of 3–4 bits) can ensure the reconstruction does not affect the time complexity by much.

We note that when $\ell > n/2 - \ell/2$ i.e., when $3 \cdot \ell > n$, it is possible to store in the table the message blocks themselves, and then in the reconstruction try all possible message blocks. The resulting representation in this case is $n/2 - \ell/2 + t$ bits.

Finally, we briefly discuss the online data structure. For that structure one needs to store only the message block that connects the current chaining value to the next one. Hence, in the online phase, the memory block size is $n/2 - \ell/2$.

References

1. Andreeva, E., Bouillaguet, C., Dunkelman, O., Fouque, P., Hoch, J.J., Kelsey, J., Shamir, A., Zimmer, S.: New second-preimage attacks on hash functions. J. Cryptol. **29**(4), 657–696 (2016)
2. Andreeva, E., Bouillaguet, C., Dunkelman, O., Kelsey, J.: Herding, second preimage and trojan message attacks beyond merkle-damgård. In: Jacobson, M.J., Rijmen, V., Safavi-Naini, R. (eds.) SAC 2009. LNCS, vol. 5867, pp. 393–414. Springer, Heidelberg (2009). doi:10.1007/978-3-642-05445-7_25
3. Andreeva, E., Bouillaguet, C., Fouque, P.-A., Hoch, J.J., Kelsey, J., Shamir, A., Zimmer, S.: Second preimage attacks on dithered hash functions. In: Smart, N. (ed.) EUROCRYPT 2008. LNCS, vol. 4965, pp. 270–288. Springer, Heidelberg (2008). doi:10.1007/978-3-540-78967-3_16
4. Bertoni, G., Daemen, J., Peeters, M., Van Assche, G.: On the indifferentiability of the sponge construction. In: Smart, N. (ed.) EUROCRYPT 2008. LNCS, vol. 4965, pp. 181–197. Springer, Heidelberg (2008). doi:10.1007/978-3-540-78967-3_11
5. Brassard, G. (ed.): CRYPTO 1989. LNCS, vol. 435. Springer, New York (1990)
6. Damgård, I.B.: A design principle for hash functions. In: Brassard [5], pp. 416–427 (1990)
7. Dean, R.D.: Formal aspects of mobile code security. Ph.D. thesis, princeton university (1999)
8. Hoch, J.J., Shamir, A.: Breaking the ICE – finding multicollisions in iterated concatenated and expanded (ICE) hash functions. In: Robshaw, M. (ed.) FSE 2006. LNCS, vol. 4047, pp. 179–194. Springer, Heidelberg (2006). doi:10.1007/11799313_12
9. Joux, A.: Multicollisions in iterated hash functions. Application to cascaded constructions. In: Franklin, M. (ed.) CRYPTO 2004. LNCS, vol. 3152, pp. 306–316. Springer, Heidelberg (2004). doi:10.1007/978-3-540-28628-8_19

10. Kelsey, J., Kohno, T.: Herding hash functions and the nostradamus attack. In: Vaudenay, S. (ed.) EUROCRYPT 2006. LNCS, vol. 4004, pp. 183–200. Springer, Heidelberg (2006). doi:10.1007/11761679_12
11. Kelsey, J., Schneier, B.: Second preimages on n-bit hash functions for much less than 2^n work. In: Cramer, R. (ed.) EUROCRYPT 2005. LNCS, vol. 3494, pp. 474–490. Springer, Heidelberg (2005). doi:10.1007/11426639_28
12. Keränen, V.: Abelian squares are avoidable on 4 letters. In: Kuich, W. (ed.) ICALP 1992. LNCS, vol. 623, pp. 41–52. Springer, Heidelberg (1992). doi:10.1007/3-540-55719-9_62
13. Knuth, D.E.: The Art of Computer Programming: Seminumerical Algorithms, vol. 2. Addison-Wesley, Boston (1969)
14. Merkle, R.C.: A certified digital signature. In: Brassard, G. (ed.) CRYPTO 1989. LNCS, vol. 435, pp. 218–238. Springer, New York (1990). doi:10.1007/0-387-34805-0_21
15. Merkle, R.C.: One Way Hash Functions and DES. In: Brassard [5], pp. 428–446 (1990)
16. Nivasch, G.: Cycle detection using a stack. Inf. Process. Lett. **90**(3), 135–140 (2004)
17. van Oorschot, P.C., Wiener, M.J.: Parallel collision search with cryptanalytic applications. J. Cryptol. **12**(1), 1–28 (1999)
18. Rivest, R.L.: Abelian Square-Free Dithering for Iterated Hash Functions. In: Presented at ECrypt Hash Function Workshop, 21 June 2005, Cracow, and at the Cryptographic Hash workshop, 1 November 2005, Gaithersburg, Maryland, August 2005

New Differential Bounds and Division Property of LILLIPUT: Block Cipher with Extended Generalized Feistel Network

Yu Sasaki[✉] and Yosuke Todo[✉]

NTT Secure Platform Laboratories,
3-9-11 Midori-cho, Musashino-shi, Tokyo 180-8585, Japan
{sasaki.yu,todo.yosuke}@lab.ntt.co.jp

Abstract. This paper provides security analysis of lightweight block cipher LILLIPUT, which is an instantiation of extended generalized Feistel network (EGFN) developed by Berger *et al.* at SAC 2013. Its round function updates a part of the state only linearly, which yields several security concerns. The first important discovery is that the lower bounds of the number of active S-boxes provided by the designers are incorrect. Then the new bounds are derived by using mixed integer linear programming (MILP), which shows an interesting fact that the actual bounds are better than the designers originally expected. Another contribution is the best third-party cryptanalysis. Owing to its unique computation structure, the designers expected that EGFN efficiently enhances security against integral cryptanalysis. However, the security is not enhanced as the designers expect. In fact, division property, which is a new method to find integral distinguishers, finds a 13-round distinguisher which improves the previous distinguisher by 4 rounds. The new distinguisher is further extended to a 17-round key recovery attack which improves the previous best attack by 3 rounds.

Keywords: Block-cipher · LILLIPUT · Extended generalized Feistel network · Mixed integer linear programming · Division property

1 Introduction

Lightweight cryptography is one of the most actively discussed topics in the current symmetric-key community. A huge number of designs have been proposed especially for the last decade. Here, we omit the list of all the lightweight primitives. Readers may refer to [1] for such a list. An important challenge that is common for most of those designs is achieving good security without significantly sacrificing efficiency.

One of the major approaches to design lightweight cipher is using Feistel network or generalized Feistel network (GFN), which has a property that its transformation is basically involutive thus the overhead to implement decryption circuit is minimized. Meanwhile, diffusion speed of the standard Feistel network

© Springer International Publishing AG 2017
R. Avanzi and H. Heys (Eds.): SAC 2016, LNCS 10532, pp. 264–283, 2017.
https://doi.org/10.1007/978-3-319-69453-5_15

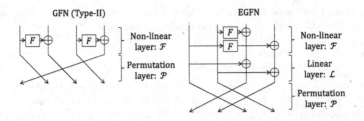

Fig. 1. Comparison of GFN (Left) and EGFN (Right) with four branches.

is often much slower than other design approaches. To overcome this drawback, several researches have developed new ideas. Suzaki and Minematsu pointed out that security of GFN can be enhanced by replacing the way of mixing branches [2]. This is called *block-shuffle* and TWINE [3] was designed based on this idea. Zhang and Wu used modified Feistel network to design LBlock [4], which turned out to be the same network as one in TWINE [3]. The latest approach, which is a main focus in this paper, is *extended GFN (EGFN)* proposed by Berger *et al.* [5], in which an additional linear diffusion layer is inserted between the application to F-function and branch network. The comparison of GFN and EGFN is depicted in Fig. 1. In many designs, the non-linear layer is the most expensive, thus the linear layer leads to better diffusion speed with a small extra cost.

Berger *et al.* [5] specified two concrete examples of EGFN with security analysis. Unfortunately, mistakes in the security analysis were pointed out by Zhang and Wu [6] and very effective differential trails were constructed for those original choices of EGFN. To fix this drawback, Berger *et al.* combined block-shuffle [2] with EGFN, and proposed a new cipher preventing the attack by Zhang and Wu. The cipher was named LILLIPUT [7].

LILLIPUT is a lightweight block cipher, supporting 64-bit block and 80-bit key. LILLIPUT is a 16-branch EGFN with block-shuffle, in which the size of each branch is 4 bits (nibble) and the non-linear function is an application of a 4-bit S-box. Those parameter sizes are the same as TWINE and LBlock. The number of rounds is 30, which is 2 rounds less than TWINE and LBlock. This shows that the additional linear layer of LILLIPUT allows to ensure its security with a smaller number of rounds than TWINE and LBlock. The designers of LILLIPUT provided several security analysis, including minimal number of active S-boxes for every round, impossible differential attack, integral attack, differential/linear cryptanalysis, related-key attacks and chosen-key attacks. Regarding differential cryptanalysis, the minimal number of active S-boxes is listed in Table 1. Other single-key attacks are summarized in Table 2.

Our Contributions. In this paper, we show that the linear layer of EGFN and LILLIPUT yields several security concerns to be carefully discussed.

We first study differential cryptanalysis. We show that the linear layer makes the evaluation of truncated differential very complicated. The linear layer allows differences to go through the round function without going through S-box.

Table 1. Lowerbounds of number of active S-boxes for each round. NW and BW represent nibble-wise model and bit-wise model, respectively.

Approach	Rounds																Tightness
	1	2	3	4	5	6	7	8	9	10	11	12	13	14	15	16	
Branching [7]	0	1	2	3	5	9	12	14	15	17	21	24	26	28	29	31	Claimed as tight
MILP (NW, basic)	0	1	2	3	5	9	12	14	15	17	19	22	25	27	29	31	Not tight
MILP (NW, advanced)	0	1	2	3	5	9	12	14	15	17	19	23	25	28	30	32	Not tight
MILP (BW)	0	1	2	3	5	9	12	15	17	19	22	?	?	?	?	?	Tight

Table 2. Key recovery attacks in the single-key model against LILLIPUT. Related-key attack and chosen-key attacks reach 23 rounds, which are not included in this table.

Approaches	Distinguisher	Key recovery	Data	Time	Ref
Integral	9 rounds	13 rounds	2^{62}	2^{72}	[7]
Impossible differential	8 rounds	14 rounds	2^{63}	2^{77}	[7]
Division property	13 rounds	17 rounds	2^{63}	2^{77}	**Ours**

This implies that attackers need to trace the impact of linearly diffused difference over many rounds. This is quite opposite for SPN-based ciphers, say AES, in which difference in all cells is randomly updated in every round. To illustrate this fact, an example of contradicting truncated differential searched by a simple search is shown in Fig. 3. We search for the lower bounds of the number of active S-boxes with MILP. The results show that the lower bounds provided by the designers are incorrect. This is the reason why our bounds are sometimes larger and sometimes smaller than the original bounds. Then, we derive new bounds with MILP in two approaches; nibble-wise and bitwise models. The former can evaluate many rounds while the derived bounds are loose. The latter can derive tight bounds while its expensive search cost restricts the search range up to 11 rounds. The results are shown in Table 1. Interestingly, our results show that LILLIPUT is more secure than the designers have expected, e.g. the designers reported that the best characteristic could reach 16 rounds while we prove this is impossible.

We next study integral cryptanalysis. The designers evaluated the security in [5,7], where the propagation characteristic of the integral property [8] was used to search for the integral distinguisher. They showed that EGFN and LILLIPUT have higher security than GFN with block-shuffle. Actually, while TWINE and LBlock allow 15-round integral distinguisher, LILLIPUT only allows the 9-round integral distinguisher. It implies that the linear layer enhances security against the integral cryptanalysis by $6(= 15 - 9)$ rounds. On the other hand, the linear layer does not increase the algebraic degree. Hence by constructing the integral characteristic by estimating the algebraic degree, which is often called the higher order differential cryptanalysis, the attack may be improved drastically. The division property is a new method to find integral distinguisher, which is a generalization of the integral property and can exploit low algebraic degree in the same time [9]. Thus security contribution of the linear layer can be evaluated more accurately with the the division property. As a result, we show that the division property finds a 13-round integral distinguisher, and it implies that the security is not enhanced as the designers expected. Moreover, the new distinguisher leads the attack against 17-round LILLIPUT (see Table 2), which is the current best attack against LILLIPUT.

Paper Outline. Related work and specification are introduced in Sect. 2. High-level overview of the properties we discuss on EGFN and LILLIPUT is given in Sect. 3. In Sect. 4 we search for new bounds of number of active S-boxes using MILP. In Sect. 5, we improve the previous best attack with division property. Finally, we conclude this paper in Sect. 6.

2   Related Work

2.1   Extended Generalized Feistel Network (EGFN)

Previous GFN has two computation layers per round; one is applying non-linear functions to some of branches and xoring the results to other branches (non-linear layer $\mathcal{F}$), and the other is permuting branches (permutation layer $\mathcal{P}$), which is often designed as a simple cyclic shift of branches. EGFN [5] adds a new diffusion layer (linear layer $\mathcal{L}$). In many designs, the non-linear layer $\mathcal{F}$ is the most expensive part, thus the linear layer $\mathcal{L}$ helps to increase the diffusion speed with a small additional cost. Berger *et al.* showed two concrete choices of $\mathcal{F}$ and $\mathcal{L}$ when the number of branches is 8 and 16 along with some security analysis. It is notable that the permutation $\mathcal{P}$ was assumed to be a simple swap of the left half and right half of the state.

Zhang and Wu [6] pointed out that the security evaluation in [5] was wrong and presented efficient differential characteristics against concrete examples in [5]. The attack relies on the choice of $\mathcal{P}$, which is a simple swap of branches.

2.2   LILLIPUT Specification

LILLIPUT [7] was designed by Berger et al. in 2015. So as to prevent the attack by Zhang and Wu [6], the designers adopted block-shuffle network [2] proposed by Suzaki *et al.* on top of EGFN so as to achieve even faster diffusion.

The block size and the key size of LILLIPUT are 64 bits and 80 bits, respectively. Its round function consists of 16 branches of size 4 bits. 64-bit plaintext is first loaded to sixteen 4-bit array $X_{15}, X_{14} \ldots, X_0$. Then, the round function consisting of three layers $\mathcal{F}, \mathcal{L}$, and $\mathcal{P}$ is iterated 30 times. The permutation layer $\mathcal{P}$ is omitted in the last round for involution reasons. An illustration of the round function is shown in Fig. 2.

The key schedule first expands the 80-bit key to 32-bit round keys for round $j, j = 0, \ldots, 29$ dented by RK^j. Because we do not analyze the key schedule, we omit its description.

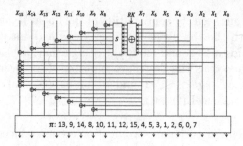

Fig. 2. Round function of LILLIPUT.

Table 3. S-box.

x	0	1	2	3	4	5	6	7	8	9	A	B	C	D	E	F
$S(x)$	4	8	7	1	9	3	2	E	0	B	6	F	A	5	D	C

Table 4. Nibble permutation.

x	0	1	2	3	4	5	6	7	8	9	10	11	12	13	14	15
$\pi(x)$	13	9	14	8	10	11	12	15	4	5	3	1	2	6	0	7

Non-linear Layer $\mathcal{F}$. At first, the state input and round key are xored. Then, a 4-bit S-box is applied to each of eight nibbles in the right half of the state, and the results are xored to the left half of the state. Let RK_i^j and X_i^j be the i-th nibble of the j-th round key RK^j and j-th round state X^j, respectively. Then, the nonlinear layer can be defined as $X_{8+i}^j \leftarrow X_{8+i}^j \oplus S(X_{7-i}^j \oplus RK_i^j)$, $i = 0, 1, \ldots, 7$, where $S(\cdot)$ is a 4-bit to 4-bit S-box defined in Table 3.

Linear Layer $\mathcal{L}$. The idea in $\mathcal{L}$ is, along with diffusion by $\mathcal{F}$, having X_7^j propagate to all nibbles in the left half of the state and having X_{15}^j be propagated from all nibbles from the right half of the state. $\mathcal{L}$ is defined as follows.

$$X_{15}^j \leftarrow X_{15}^j \oplus X_7^j \oplus X_6^j \oplus X_5^j \oplus X_4^j \oplus X_3^j \oplus X_2^j \oplus X_1^j,$$
$$X_{15-i}^j \leftarrow X_{15-i}^j \oplus X_7^j \text{ for } i = 1, 2, \ldots, 6.$$

Permutation Layer $\mathcal{P}$. Nibble positions are permuted with permutation π defined in Table 4. The designers chose π to achieve the highest number of active S-boxes after 18, 19 and 20 rounds.

3   Difficulties of Analyzing LILLIPUT Round Function

In Sect. 4, we will show that the lower bounds of the number of active S-boxes provided by the authors are wrong. However, this is not because of careless mistakes. In Sect. 5, we will present a current best attack against LILLIPUT using division property. Before explaining details, in this section, we extract overview of the essential difficulties of analyzing EGFN and LILLIPUT with respect to differential cryptanalysis and division property.

Differential Cryptanalysis. Evaluating security of EGFN and LILLIPUT against differential cryptanalysis is quite difficult owing to their unique computation structure, $\mathcal{L}$. The previous truncated differential search, both dedicated search or more structural approach such as wide trail strategy in AES [10], yields a correct result only if the cipher can be assumed to be Markov cipher [11] with respect to truncated differential. Namely, the probability to achieve a truncated differential in round $i+1$ needs to be determined only depending on a truncated differential in round i (or possibly in any fixed round before round $i+1$).

A main obstacle for EGFN and LILLIPUT is that this assumption does not hold after a few rounds because of the linear layer $\mathcal{L}$. Let us discuss the LILLIPUT round function (Fig. 2).

- For some round j, X_{15}^j easily gets active thanks to $\mathcal{L}$, then X_{15}^j moves to X_7^{j+1} after $\mathcal{P}$.
- X_7^{j+1} duplicates *an identical difference* to X_9^{j+1} to X_{14}^{j+1}, and those will propagate to subsequent rounds.
- In a truncated differential, we only remember active/inactive of each nibble, thus we lose information that those differences are identical, which with high probability causes contradiction after a few rounds. (In Markov cipher, difference in round $j+2$ or later rounds should not depend on difference in round j.)

An example of contradicting truncated differential is shown in Figs. 3 and 4. The differential is 3 middle rounds of 16-round differential evaluated by the basic nibble-wise MILP model, which will be explained later. Figure 3 shows that the truncated differential is valid under the assumption that difference of all nibbles are reset to be a random difference in every round. Meanwhile, Fig. 4 traces the impact of linear diffusion. It shows that the difference of $x_{14}^{i+2}, x_{13}^{i+2}$, and x_9^{i+2} are the same as the one in x_7^i, which are denoted by Δ in Fig. 4. Here, we denote the difference of x_7^{i+2} by α, Then, the difference of the 9th, 13th, and 14th branches after the linear layer in round $i+2$ are denoted by $\Delta \oplus \alpha$. It is unknown if $\Delta \oplus \alpha$ is 0 or not, however, differences in those three branches must be identical. As one can see, Fig. 4 assumes that the 13th and 14th branches are inactive while the 9th branch is active. Thus this differential is contradicted.

Even with contradiction, it is still possible to provide lower bounds. However, the derived bounds are not tight as the linear layer $\mathcal{L}$, a source of contradiction, diffuses many truncated differential at once. Alternative approach is simulating differential propagation bit-by-bit precisely instead of truncated differential. However, this approach requires a very expensive search cost, and simulating all rounds is infeasible. All in all, evaluating security of EGFN and LILLIPUT against differential cryptanalysis is challenging work.

Integral Cryptanalysis. The designers of EGFN and LILLIPUT already showed the security against the integral cryptanalysis in [5,7], and the propagation characteristic of the integral property [8] was used to search for the integral distinguisher. When a d-round EGFN reaches the full diffusion, the integral

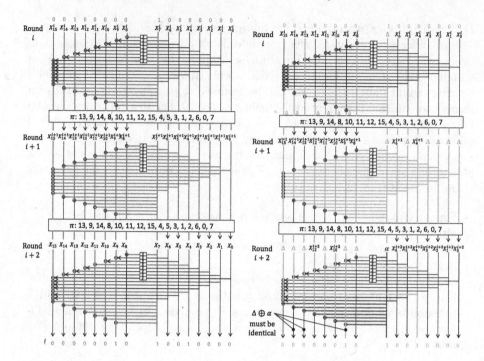

Fig. 3. Valid if differential is reset in every round.

Fig. 4. Contradiction if linear propagation is considered.

distinguisher of the EGFN covers at most $2d + 2$ rounds. Moreover, LILLIPUT, which is a specific block cipher based EGFN with $d = 4$, has the 9-round integral distinguisher. Compared with 15-round integral distinguishers of TWINE and LBlock, it implies that the linear layer enhances the security against the integral cryptanalysis by $6(= 15 - 9)$ rounds. On the other hand, if we construct the integral distinguisher by estimating the algebraic degree, which is often called the higher order differential cryptanalysis, the security is not likely to dramatically improve because the linear layer does not increase the algebraic degree.

The division property is a new method to find integral distinguishers, and it is the generalization of the integral property so that can exploit the algebraic degree in the same time. Therefore, we can more accurately evaluate the contribution of the linear layer by using the division property. In Sect. 5, we will show a new integral distinguisher with the division property, and it covers 13 rounds, which is beyond $2d + 2 = 10$. Very recently, Zhang and Wu showed that TWINE and LBlock have 16-round integral distinguishers by using the division property [12]. Therefore, the true contribution by the linear layer is $3(= 16 - 13)$ rounds. Moreover, this 13-round integral distinguisher leads to a 17-round attack, which is a current best attack against LILLIPUT.

4 New Differential Bounds

We search for lower bounds of number of active S-boxes of Lilliput with MILP. Section 4.1 explains background of MILP based search. Section 4.2 explains nibble-wise search and proves the 16-round truncated differential shown by the designers are incorrect. Section 4.3 explains bit-wise search, which proves better bounds than the evaluation by the designers up to 11 rounds.

4.1 Background of Mixed Integer Linear Programming (MILP)

An MILP-based search was proposed by Mouha et al. [13]. The approach has two stages; (1) describing valid active byte/nibble/bit propagation patterns with a system of linear inequalities, and (2) solving the system with an MILP solver. Cryptographer's task is for (1) to efficiently describe active byte/nibble/bit patterns. Regarding stage (2), many softwares are available, some are license-free and other are in commerce. In this research, we used Gurobi Optimizer [14] for stage (2). Hereafter we explain stage (1).

The following discussion focuses on nibble-oriented ciphers. The goal is counting the number of active S-boxes, thus truncated differential is analyzed. Each nibble in each round is represented by a binary variable x_i meaning that the nibble is active when $x_i = 1$ and inactive when $x_i = 0$. Then, we specify an object to be optimized, called *objective function*. Our goal is finding a minimal number of active S-boxes, thus if S-box is applied to all nibbles, the objective function is "minimize $\sum_i x_i$." The main task is giving *constraint inequalities* to specify valid differential propagations with linear inequalities.

Inequations to Describe XOR by Mouha et al. Suppose that the nibble corresponding to x_3 is computed by other two nibbles corresponding to x_1 and x_2, i.e. $x_1 \oplus x_2 = x_3$. Mouha et al. describe all possible differential patterns by introduced a dummy binary variable d as follows.

$$x_1 + x_2 + x_3 - 2d \geq 0,$$
$$x_1 - d \leq 0,$$
$$x_2 - d \leq 0,$$
$$x_3 - d \leq 0.$$

Bit-Wise Model by Sun et al. Several nibble-oriented ciphers cannot be evaluated with the approach by Mouha et al. An example is PRESENT, in which 4 bits output from a 4-bit S-box will be input to different S-box in the next round. Thus, it is necessary to look inside the S-box. Sun et al. proposed MILP-based search in a bit-wise model to simulate such a case [15], in which each binary variable x_i represents active/inactive of each bit. This approach is more advantageous for versatility, while it loses efficiency (the number of evaluated rounds is less).

A notable technique in [15] is to rule out impossible differential patterns from a feasible region of MILP. Recall the XOR case explained above, $x_3 = x_2 \oplus x_1$. We need to rule out $(x_1, x_2, x_3) = (1,0,0), (0,1,0), (0,0,1)$ and Sun et al. showed each impossible pattern can be ruled out with 1 inequality. For example, $(x_1, x_2, x_3) = (1,0,0)$ is ruled out with $-x_1 + x_2 + x_3 \geq 0$. Indeed, any other value of (x_1, x_2, x_3) satisfy this inequality, and thus only $(x_1, x_2, x_3) = (1,0,0)$ is ruled out. $(0,1,0)$ and $(0,0,1)$ can be ruled out similarly.

4.2   Nibble-Wise Search

We first explain a basic method which assumes that the difference of each active nibble is reset to a random difference in every round. This assumption is clearly incorrect for the real specification because the linear layer $\mathcal{L}$ diffuses difference only linearly (difference in round j uniquely determines difference in round $j+1$ in $\mathcal{L}$). Hence, the derived lower bounds are loose. We then show that equivalently transforming the cipher's description helps us to improve the model that can derive tighter lower bounds.

Constructing Basic Model. We assign a binary variable to each nibble in every round. Thus we use $16r$ variables for r rounds; $x_0, \ldots, x_{15}$ for round 1, $x_{16}, \ldots, x_{31}$ for round 2, and so on.

As for the objective function, our goal is minimizing the number of active S-boxes, thus we minimize the sum of x_i in the right half of the state, i.e. "minimize $\sum_r \sum_{j=0}^{7} x_{16r+j}$."

Constraint inequalities can be derived round-by-round. For simplicity, we explain constraints between $x_0, \ldots, x_{15}$ and $x_{16}, \ldots, x_{31}$, which are depicted in Fig. 5. The other rounds can be modeled just by replacing indices. S-box and key addition do not impact to truncated differential, thus we omit them in Fig. 5. First, we list variables before the permutation layer, which are $\pi^{-1}(x_{16}, \ldots, x_{31})$. Here the right most one, x_{29}, can be represented by $x_{\pi(0)+16}$. Similarly, the other 15 variables can be represented by $x_{\pi(1)+16}, x_{\pi(2)+16}, \ldots, x_{\pi(15)+16}$. This representation is useful to systematically construct MILP models. We then derive constraint inequalities between $x_0, x_1, \ldots x_{15}$ and $x_{\pi(0)+16}, x_{\pi(1)+16}, \ldots, x_{\pi(15)+16}$ dividing them into four types.

Type 1: Right half of the state is not updated. Constraints are $x_{\pi(i)+16} = x_i$ for $i = 0, 1, \ldots, 7$.

Type 2: $x_{\pi(8)+16}, x_8$ and x_7 must be a valid xor, i.e. $(x_{\pi(8)+16}, x_8, x_7) = (1,0,0), (0,1,0), (0,0,1)$ are impossible. We rule out those three patterns with the following three inequalities;

$$-x_{\pi(8)+16} + x_8 + x_7 \geq 0,$$
$$x_{\pi(8)+16} - x_8 + x_7 \geq 0,$$
$$x_{\pi(8)+16} + x_8 - x_7 \geq 0.$$

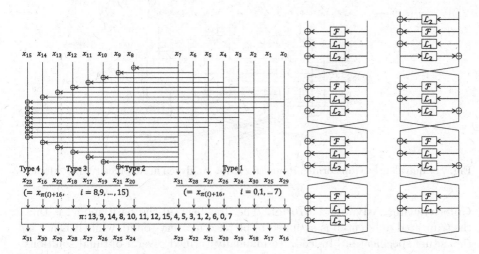

Fig. 5. Nibble based MILP model for LILLIPUT.   **Fig. 6.** Equivalent descriptions.

Type 3: For $j = 9, 10, \ldots, 14$, $x_{\pi(j)+16}, x_{8+j}, x_{7-j}, x_7$ must be a valid xor. We rule out $(x_{\pi(j)+16}, x_{8+j}, x_{7-j}, x_7) = (1, 0, 0, 0), (0, 1, 0, 0), (0, 0, 1, 0), (0, 0, 0, 1)$. Similarly to Type 2, this can be done with four inequalities for each j.

Type 4: $x_{\pi(15)+16}$ and other 9 input variables must be a valid xor. Similarly to Type 2 and Type 3, differential propagation is impossible if and only if exactly one variable is active. There are ten impossible patterns, and these are ruled out with ten inequalities.

In total, we use $8 + 3 + (6 * 4) + 10 = 45$ inequalities per round, thus $45r$ for r rounds. In addition we use 1 inequalities $\sum_{j=0}^{15} x_i > 0$ to ensure at least one nibble is active in plaintext.

Results of Basic Model. Execution time is reasonably short. The system for 16 rounds was solved in a few minutes by a standard PC. The results are shown in Table 1. At first glance, the derived bounds are worse than the designers' evaluation. However this is not right. The designers claimed that the best 16-round characteristic activates 31 S-boxes [7, Sect. 7].

> we provide here the best truncated differential and linear masks we found for 16 rounds of LILLIPUT with 31 active S-boxes · The best truncated differential path is given by an input of the form $(\alpha_0, 0, \alpha_0, \alpha_0, \alpha_0, \alpha_0, \alpha_0, \alpha_1, \alpha_0, 0, 0, 0, 0, 0, 0, 0,)$ that gives after 16 rounds an output of the form $(0, 0, 0, 0, 0, 0, 0, 0, 0, 0, 0, 0, 0, 0, 0, \beta, 0) \cdots$

We tested their input and output differential masks. We obtained that the (loose) lower bound for those masks is 34 for 16 rounds, thus their claim is wrong.[1]

---

[1] We communicated to the designers and asked to provide more details, in particular differential masks for every round. The designers have not provide us the details.

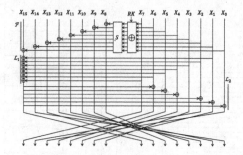

Fig. 7. Equivalently Transformed Round Function Analyzed in Advanced MILP Model.

Constructing Advanced Model. The drawback in the basic model is that the truncated differential is assumed to be reset in every round, while it is not in the actual specification. Indeed, we manually verified several optimal solutions returned by a solver, but they always include contradiction. Namely, the bound is not tight (though 30 rounds seem sufficient to resist differential cryptanalysis). Let us analyze more details. We divide the linear layer $\mathcal{L}$ into two layers $\mathcal{L}_1$ and $\mathcal{L}_2$, in which $\mathcal{L}_1$ is the diffusion from $X_1, X_2, \ldots, X_7$ to X_{15} and $\mathcal{L}_2$ is the diffusion from X_7 to $X_9, X_{10}, \ldots, X_{15}$ defined below (illustrated in Fig. 7).

$$\mathcal{L}_1 : X_{15} \leftarrow X_{15} \oplus X_7 \oplus X_6 \oplus X_5 \oplus X_4 \oplus X_3 \oplus X_2 \oplus X_1,$$
$$\mathcal{L}_2 : X_{15-i} \leftarrow X_{15-i} \oplus X_7 \text{ for } i = 1, 2, \ldots, 6.$$

Our observation is that the impact of linear diffusion with $\mathcal{L}_1$ and $\mathcal{L}_2$ never interact within one round. X_{15} is (easily) activated through $\mathcal{L}_1$, and this moves to x_7 after $\mathcal{P}$, and in the next round, x_7 diffuses with $\mathcal{L}_2$. In the basic MILP model, the above combination effect via $\mathcal{L}_1$ and $\mathcal{L}_2$ over two rounds cannot be captured due to the difference reset in every round.

Our improving idea is moving the position of the linear layer $\mathcal{L}_2$ so that the cancellation through $\mathcal{L}_1$ and $\mathcal{L}_2$ can be simulated within one round. In details, we move $\mathcal{L}_2$ for round i (diffusion from X_7 in round i) to round $i-1$ (diffusion from $\pi^{-1}(X_7) = X_{15}$ in round $i-1$). The converted computation structure is shown in the right-half of Fig. 6. Note that the original $\mathcal{L}_2$ in the first round can be regarded as a preprocessing and $\mathcal{L}_2$ in the last round is removed.

Results of Advanced Model. Execution time of the advanced model is almost the same as the basic one. The results are shown in Table 1. Compared to the basic model, the lower bounds are improved when the number of rounds is 12, 14, 15 and 16. Compared to the designers' original expectation, the lower bounds are improved, meaning that LILLIPUT is more secure than it was expected. In particular, proving 32 active S-boxes for 16 rounds is important owing to the 64-bit block size and the maximum differential probability of the S-box, 2^{-2}.

Even with the advanced model, contradiction via $\mathcal{L}$ over 3 rounds cannot be simulated, thus the bounds are not tight. This motivates us to generate tight bounds in the next section.

4.3   Bit-Wise Search

The bit-wise model traces active/inactive of each bit. The main advantage is that the cancellation by the xor operation, which is the main cause of the contradiction in the nibble-wise model, can be simulated precisely and solving the system becomes equivalent to finding the best characteristic. Meanwhile, S-box is not bit-wise thus cannot be ignored as the nibble-wise model, which requires a large number of constraint inequalities to describe valid differential propagations.

Variables in One Round. We assign a binary variable x_i bit by bit. To reduce a total number of variables, we introduce new variables only for updated 32 bits (right half of the state) in every round. Besides, active/inactive of each bit changes through S-box, thus we introduce a binary variable y_i to describe active/inactive of each bit of S-box output.

Permutation π needs to be adjusted to be bitwise, π_{bw}. The conversion is straightforward, thus we omit it.

Number of Active S-Boxes in Bitwise Model. We need to convert active-bit information into active-nibble one to count the number of active S-boxes. Here, we introduce a dummy binary variable, n. Suppose that n_{4i} is a nibble whose corresponding 4 input bits are $x_i, x_{i+1}, x_{i+2}, x_{i+3}$. We set constrains so that n_{4i} becomes 1 when at least one of $x_i, \ldots, x_{i+3}$ are active and $n_{4i} = 0$ if all of $x_i, \ldots, x_{i+3}$ are inactive. This can be done by borrowing the idea of simulating XOR by Mouha et al. [13], and we set the following five inequalities;

$$x_i + x_{i+1} + x_{i+2} + x_{i+3} - n_{4i} \geq 0,$$
$$n_{4i} - x_i \geq 0,$$
$$n_{4i} - x_{i+1} \geq 0,$$
$$n_{4i} - x_{i+2} \geq 0,$$
$$n_{4i} - x_{i+3} \geq 0.$$

If all of $x_i, x_{i+1}, x_{i+2}, x_{i+3}$ are inactive ($= 0$), n_{4i} becomes 0. If at least one of $x_i, x_{i+1}, x_{i+2}, x_{i+3}$ is active ($= 1$), n_{4i} becomes 1. Thus, n_{4i} represents active/inactive of the S-box.

Each round computes 8 S-boxes. The objective function for r rounds is "minimize $\sum_{i=0}^{8r-1} n_i$."

Constraints for S-Box. We first generate differential distribution table (DDT). DDT consists of 150 zero entries (impossible propagations). With the approach by Sun et al. [15], we can rule out each impossible propagation with one inequality.

For example, $x_{i+3}\|x_{i+2}\|x_{i+1}\|x_i = 0010$ and $y_{i+3}\|y_{i+2}\|y_{i+1}\|y_i = 0011$ is an impossible propagation and this can be ruled out by

$$x_{i+3} + x_{i+2} - x_{i+1} + x_i + y_{i+3} + y_{i+2} - y_{i+1} - y_i \geq -2.$$

All the impossible differential propagations can be ruled out with at most 150 inequalities. Sun *et al.* showed that several impossible propagations may be ruled out with 1 inequality.

For example, $x_{i+3}\|x_{i+2}\|x_{i+1}\|x_i\|y_{i+3}\|y_{i+2}\|y_{i+1}\|y_i = *00**101$ is impossible for any choice of $* \in \{0, 1\}$. Those 8 patterns are ruled out by

$$x_{i+2} + x_{i+1} - y_{i+2} + y_{i+1} - y_i \geq -1.$$

We exhaustively searched for such compact representations. The number of total constraint inequalities should be minimized. We followed the approach by Sun et al. [16] using the greedy algorithm to choose constraint inequalities. In the end, we rule out all 150 impossible differential patterns with 46 inequalities.

Constraints Other Than S-Box. Update on 28 bits, from bit positions 32 to 59, is rather simple. If the computation is the 2-input xor, e.g. $a \oplus b = c$, the number of impossible propagations is 4; $(a, b, c) = (1, 0, 0), (0, 1, 0), (0, 0, 1), (1, 1, 1)$, which can be ruled out with 4 inequalities. Note that differently from nibble-wise search, $(a, b, c) = (1, 1, 1)$ is impossible in the bitwise computation. Similarly, for 3-input xor, e.g. $a \oplus b \oplus c = d$, there are 8 impossible patterns, and we can rule them out with 8 inequalities.

The last 4 bits are updated with 9-input xor, thus the number of impossible propagations is $2^9 = 512$ per bit. Using 512 inequalities is too expensive. Here, we focus on the property that the sum of involved variables must be even. We introduce an *integer* dummy variable e, where $e \in \{0, 1, 2, 3, 4\}$. Let 9 input bits be $y_0, x_4, x_8, \ldots, x_{28}$ and 1 output bit be $x_{\pi_{bw}(60)}$. We set the following constraint;

$$y_0 + x_4 + x_8 + x_{12} + x_{16} + x_{20} + x_{24} + x_{28} - x_{\pi_{bw}(60)} = 2e.$$

Result of Bitwise Model. Owing to the expensive computational cost, the machine performance is an important factor for this research. We executed Gurobi Optimizer with Xeon Processor E5-2699 (18 cores) in 128 GB RAM. The results are shown in Table 1. It provides the best bound from 8 rounds and we confirmed the tightness. Namely the optimal solutions can be used for attacks. The running time for 8, 9, 10, and 11 rounds is 2746 s, 5512 s, 53099 s ($\approx$14 h), and about 1 week, respectively. Because of the complicated algorithm of the MILP solver, it is difficult to predict the running time for more rounds.

5   Attacks Based on Division Property

5.1   Background of Division Property

The division property proposed in [9] is a new method to find integral distinguishers. This section briefly shows the definition and propagation rules to understand this paper. Please refer to [9] for details.

The division property of a multiset is evaluated by using the bit product function defined as follows. Let $\pi_{\boldsymbol{u}} : (\mathbb{F}_2^n)^m \to \mathbb{F}_2$ be a bit product function for any $\boldsymbol{u} \in (\mathbb{F}_2^n)^m$. Let $\boldsymbol{x} \in (\mathbb{F}_2^n)^m$ be the input, and $\pi_{\boldsymbol{u}}(\boldsymbol{x})$ is defined as

$$\pi_{\boldsymbol{u}}(\boldsymbol{x}) := \prod_{i=1}^m \left(\prod_{j=1}^n x_i[j]^{u_i[j]} \right).$$

Notice that $x_i[j]^1 = x_i[j]$ and $x_i[j]^0 = 1$.

Definition 1 (Division Property [9]). *Let* $\mathbb{X}$ *be a multiset whose elements take a value of* $(\mathbb{F}_2^n)^m$. *When the multiset* $\mathbb{X}$ *has the division property* $\mathcal{D}_{\mathbb{K}}^{n^m}$, *where* $\mathbb{K}$ *denotes a set of* m-*dimensional vectors whose elements take a value between* 0 *and* n, *it fulfills the following conditions:*

$$\bigoplus_{\boldsymbol{x} \in \mathbb{X}} \pi_{\boldsymbol{u}}(\boldsymbol{x}) = \begin{cases} unknown & \text{if there are } \boldsymbol{k} \in \mathbb{K} \text{ s.t. } W(\boldsymbol{u}) \succeq \boldsymbol{k}, \\ 0 & otherwise, \end{cases}$$

where $W(\boldsymbol{u}) = (w(u_m), \dots, w(u_1)) \in \mathbb{Z}^m$ *and* $w(u_j) = \sum_{i=1}^n u_j[i]$. *Moreover,* $\boldsymbol{k} \succeq \boldsymbol{k}'$ *denotes* $k_i \geq k_i'$ *for all* $i \in \{1, 2, \dots, m\}$.

If there are $\boldsymbol{k} \in \mathbb{K}$ and $\boldsymbol{k}' \in \mathbb{K}$ satisfying $\boldsymbol{k} \succeq \boldsymbol{k}'$ in the division property $\mathcal{D}_{\mathbb{K}}^{n^m}$, $\boldsymbol{k}$ can be removed from $\mathbb{K}$ because the vector $\boldsymbol{k}$ is redundant. Let $\mathbb{X}$ be the set of texts encrypted by r rounds, and $e_i \in \mathbb{Z}^m$ denotes an unit vector whose ith element is one and the others are zero. Assuming that $\mathbb{X}$ fulfills the division property $\mathcal{D}_{\mathbb{K}}^{n^m}$ and e_i does not belong to $\mathbb{K}$, the cipher has the r-round integral distinguisher, where the ith element is balanced.

We summarize propagation rules that we use in this paper as follows.

Rule 1 (Substitution). Let F be a function that consists of m S-boxes, where the bit length and the algebraic degree of S-boxes is n bits and d, respectively. The input and the output take a value of $(\mathbb{F}_2^n)^m$ and $\mathbb{X}$ and $\mathbb{Y}$ denote the input multiset and the output multiset, respectively. Assuming that the multiset $\mathbb{X}$ has the division property $\mathcal{D}_{\mathbb{K}}^{n^m}$, the multiset $\mathbb{Y}$ has the division property $\mathcal{D}_{\mathbb{K}'}^{n^m}$, where $\mathbb{K}'$ is calculated as follows: First, $\mathbb{K}'$ is initialized to ϕ. Then, for all $\boldsymbol{k} \in \mathbb{K}$,

$$\mathbb{K}' = \mathbb{K}' \cup \left[\left\lceil \frac{k_1}{d} \right\rceil, \left\lceil \frac{k_2}{d} \right\rceil, \dots, \left\lceil \frac{k_m}{d} \right\rceil \right],$$

is calculated. Here, when the ith S-box is bijective and $k_i = n$, the ith element of the propagated property becomes n not $\lceil n/d \rceil$.

Rule 2 (Copy). Let F be a copy function, where the input x takes a value of $\mathbb{F}_2^n$ and the output is calculated as $(y_1, y_2) = (x, x)$. Let $\mathbb{X}$ and $\mathbb{Y}$ be the input multiset and output multiset, respectively. Assuming that the multiset $\mathbb{X}$ has the division property $\mathcal{D}_k^n$, the multiset $\mathbb{Y}$ has the division property $\mathcal{D}_{\mathbb{K}'}^{n,n}$, where $\mathbb{K}'$ is calculated as follows: First, $\mathbb{K}'$ is initialized to ϕ. Then, for all i ($0 \leq i \leq k$),

$$\mathbb{K}' = \mathbb{K}' \cup [k - i, i],$$

is calculated.

Rule 3 (Compression by XOR). Let F be a function compressed by an XOR, where the input (x_1, x_2) takes a value of $(\mathbb{F}_2^n \times \mathbb{F}_2^n)$ and the output is calculated as $y = x_1 \oplus x_2$. Let $\mathbb{X}$ and $\mathbb{Y}$ be the input multiset and output multiset, respectively. Assuming that the multiset $\mathbb{X}$ has the division property $\mathcal{D}_\mathbb{K}^{n,n}$, the division property of the multiset $\mathbb{Y}$ is $\mathcal{D}_{k'}^n$ as

$$k' = \min_{[k_1, k_2] \in \mathbb{K}} \{k_1 + k_2\}.$$

Here, if the minimum value of k' is larger than n, the propagation characteristic of the division property is aborted. Namely, a value of $\oplus_{y \in \mathbb{Y}} \pi_v(y)$ is 0 for all $v \in \mathbb{F}_2^n$.

These propagation rules are proven in [9, 17].

5.2   Integral Distinguisher on LILLIPUT

The state of LILLIPUT is represented as sixteen 4-bit values, and the use of the division property $\mathcal{D}_\mathbb{K}^{4^{16}}$ is appropriate. Let $|\mathbb{K}|$ be the number of elements in $\mathbb{K}$, and the upper bound of $|\mathbb{K}|$ is $5^{16} \approx 2^{37.15}$. Since we can reduce $|\mathbb{K}|$ by removing redundant vectors in general, we can practically evaluate the propagation characteristic of $\mathcal{D}_\mathbb{K}^{4^{16}}$.

Propagation Characteristic. The round function of EGFN consists of three layers: the non-linear layer, the linear layer, and the permutation layer. In the non-linear layer of EGFN, the core operation is

$$x_i = x_i \oplus F(x_j)$$

for appropriate i and j. We only focus on the case that F is permutation because the most important instantiation LILLIPUT uses a bijective S-box. Let $\mathcal{D}_k^4$ and $\mathcal{D}_{k'}^4$ be the input and output division property for the S-box, respectively. As the algebraic degree of F is at most three, it holds

$$k' = D_S(k) = \begin{cases} 4 & \text{if } k = 4, \\ 1 & \text{if } k = 1, 2, 3, \\ 0 & \text{if } k = 0. \end{cases}$$

Assuming $\mathcal{D}_{(k_i, k_j)}^{4^2}$ be the input division property of the Feistel structure, the output division property $\mathcal{D}_\mathbb{K}^{4^2}$ is

$$\mathbb{K} = \{(k_i + D_S(x), k_j - x) \mid 0 \leq x \leq k_j, D_S(x) \leq 4 - k_i\}.$$

The propagation characteristic for the non-linear layer is shown in nonLinear of Algorithm 1.

Algorithm 1. Propagation from $\mathcal{D}_{\mathbb{K}}^{4^{16}}$ for the round function of Lilliput

1: **procedure** nonLinear($\mathbb{K}, i, j$)	1: **procedure** linear($\mathbb{K}, i, j$)
2: $\mathbb{K}' \Leftarrow \phi$	2: $\mathbb{K}' \Leftarrow \phi$
3: **for all** $k \in \mathbb{K}$ **do**	3: **for all** $k \in \mathbb{K}$ **do**
4: $k' \Leftarrow k$	4: $k' \Leftarrow k$
5: **for** $x = 0$ to k_j **do**	5: **for** $x = 0$ to k_j **do**
6: $k_i' \Leftarrow k_i + D_S(x)$	6: $k_j' \Leftarrow k_j - x$
7: $k_j' \Leftarrow k_j - x$	7: $k_i' \Leftarrow k_i + x$
8: **if** $k_i' \leq 4$ **then**	8: **if** $k_i' \leq 4$ **then**
9: $\mathbb{K}' \Leftarrow \mathbb{K}' \cup \{k'\}$	9: $\mathbb{K}' \Leftarrow \mathbb{K}' \cup \{k'\}$
10: **end if**	10: **end if**
11: **end for**	11: **end for**
12: **end for**	12: **end for**
13: remove redundant vectors from $\mathbb{K}'$	13: remove redundant vectors from $\mathbb{K}'$
14: return $\mathbb{K}'$	14: return $\mathbb{K}'$
15: **end procedure**	15: **end procedure**

The linear layer of EGFN consists of the iteration of XORs as

$$x_i = x_i \oplus x_j$$

for appropriate i and j. Therefore, assuming $\mathcal{D}_{(k_i, k_j)}^{4^2}$ be the input division property of the Feistel structure, the output division property $\mathcal{D}_{\mathbb{K}}^{4^2}$ is

$$\mathbb{K} = \{(k_i + x, k_j - x) \mid 0 \leq x \leq \min\{k_j, 4 - k_i\}\}.$$

The propagation characteristic for the linear layer is shown in linear of Algorithm 1.

About the permutation layer, the propagation characteristic is the only modification of the corresponding index. The entire algorithm to evaluate the propagation characteristic of the round function is shown in roundFunction of Algorithm 2.

New Integral Distinguisher. As the number of exploiting chosen plaintexts increases, the integral distinguisher can analyze more rounds in general. Therefore, we evaluate all integral distinguishers with 2^{63} chosen plaintexts where only one bit in the right half is constant. Note that these distinguishers are always better than distinguishers whose only one bit in the left half is constant. We choose one 4-bit value from X_0 to X_7, and we prepare chosen plaintexts such that any one bit in the chosen value is constant and the others are active.

We implemented Algorithm 2 and searched non-trivial integral distinguishers. Let $\mathcal{D}_k^{4^{16}}$ be the plaintext division property. When we choose one-bit constant from X_p, we use k as

$$k_i = \begin{cases} 4 & \text{if } i \neq p \\ 3 & \text{if } i = p \end{cases}$$

Algorithm 2. Propagation from $\mathcal{D}_{\mathbb{K}}^{4^{16}}$ for the round function of LILLIPUT

```
1: procedure roundFunction(𝕂)
2:     for all (i,j) ∈ {(8,7),(9,6),(10,5),(11,4),(12,3),(13,2),(14,1),(15,0)} do
3:         𝕂 = nonLinear(𝕂,i,j)
4:     end for
5:     for all (i,j) ∈ {(15,1),(15,2),(15,3),(15,4),(15,5),(15,6),(15,7)} do
6:         𝕂 = linear(𝕂,i,j)
7:     end for
8:     for all (i,j) ∈ {(14,7),(13,7),(12,7),(11,7),(10,7),(9,7)} do
9:         𝕂 = linear(𝕂,i,j)
10:    end for
11:    𝕂' ⇐ φ
12:    for all k ∈ 𝕂 do
13:        for i = 0 to 16 do
14:            k'_{π(i)} ⇐ k_i
15:        end for
16:        𝕂' ⇐ 𝕂' ∪ {k'}
17:    end for
18:    return 𝕂'
19: end procedure
```

Table 5. Propagation from $\mathcal{D}_{\{[4,4,\ldots,4,3]\}}^{4^{16}}$

#rounds	0	1	2	3	4	5	6 ⋆
$\|\mathbb{K}\|$	1	1	3	14	377	33948	5513237
$\max_w(\mathbb{K})$	63	63	63	63	63	55	≤ 57
$\min_w(\mathbb{K})$	63	63	61	59	55	19	35
#rounds	7 ⋆	8 ⋆	9 ⋆	10	11	12	13
$\|\mathbb{K}\|$	266813452	70804820	1385951	16960	572	52	16
$\max_w(\mathbb{K})$	≤ 51	≤ 43	≤ 25	13	6	4	2
$\min_w(\mathbb{K})$	22	9	6	3	2	1	1

In rounds labeled ⋆, the set $\mathbb{K}$ includes redundant vectors.

for $i \in \{0, 1, \ldots, 16\}$. We coded our algorithm with C++, and we executed it in Xeon Processor E5-2699 (18 cores) in 128 GB RAM. As a result, our algorithm found 13-round integral distinguishers for $p = 0$ and $p = 6$. For other p, our algorithm found 12-round integral distinguishers.

When $p = 0$ i.e., $\boldsymbol{k} = [4,4,4,4,4,4,4,4,4,4,4,4,4,4,4,3]$, we find a 13-round integral distinguisher, and the position X_9^{13} is balanced. Table 5 shows the propagation characteristic, where $\min_w(\mathbb{K})$ and $\max_w(\mathbb{K})$ are calculated as

$$\min_w(\mathbb{K}) = \min_{\boldsymbol{k} \in \mathbb{K}} \left\{ \sum_{i=1}^{16} k_i \right\}, \quad \max_w(\mathbb{K}) = \max_{\boldsymbol{k} \in \mathbb{K}} \left\{ \sum_{i=1}^{16} k_i \right\}.$$

Round 0 denotes the division property of the plaintext set, and we perfectly remove redundant vectors except for 6, 7, 8, and 9 rounds.

5.3 Key Recovery

Let X_i^j be the j-round nibble value in X_i, where the plaintext is represented as $(X_{15}^0, \ldots, X_0^0)$. Moreover, let Y_i^j be the output of the S-box as $Y_i^j = S(X_i^j \oplus RK_i^j)$. We prepare 2^{63} chosen plaintexts such that any one bit of X_0^0 is constant and the other 63 bits are active. Then, it holds $\bigoplus X_9^{13} = 0$, and we can attack 17-round LILLIPUT by using the 13-round integral distinguisher. In our attack, let $c = (c_{15}, \ldots, c_0)$ be the ciphertext, where the linear layer of the last round is removed. Note that the last round of LILLIPUT has the linear layer but this c is equivalent with the ciphertext of 17-round LILLIPUT because the linear layer is public.

Since LILLIPUT has many XORs in the round function, the procedure of the key recovery is very complicating. For simplicity, we use the following strategy. We first decompose four rounds of LILLIPUT into five subfunctions denoted by $f_{13}, f_{14}, f_{15}, f_{16}$, and L. Here the output of f_i is the XOR of Y^i involved in X_9^{13}, and the output of L is the linear part to compute X_9^{13} from ciphertext. Then

$$X_9^{13} = f_{13}(c, \mathcal{K}_{13}) \oplus f_{14}(c, \mathcal{K}_{14}) \oplus f_{15}(c, \mathcal{K}_{15}) \oplus f_{16}(c, \mathcal{K}_{16}) \oplus L(c),$$

where $\mathcal{K}_i$ is the set of round keys involved in f_i. The bit sizes of $\mathcal{K}_{13}, \mathcal{K}_{14}, \mathcal{K}_{15}$, and $\mathcal{K}_{16}$ are 44, 16, 48, and 28 bits, respectively. Then,

$$f_{13}(c, \mathcal{K}_{13}) = Y_6^{13},$$
$$f_{14}(c, \mathcal{K}_{14}) = Y_0^{14},$$
$$f_{15}(c, \mathcal{K}_{15}) = Y_0^{15} \oplus Y_1^{15} \oplus Y_3^{15} \oplus Y_5^{15} \oplus Y_6^{15} \oplus Y_7^{15},$$
$$f_{16}(c, \mathcal{K}_{16}) = Y_0^{15} \oplus Y_1^{15} \oplus Y_3^{15} \oplus Y_4^{15} \oplus Y_5^{15} \oplus Y_6^{15} \oplus Y_7^{15}.$$

We compute the sum of $f_i(c, \mathcal{K}_i)$ by guessing $\mathcal{K}_i$ independently of i. Then, we compute keys satisfying

$$\bigoplus_{X^0} f_{13}(c, \mathcal{K}_{13}) \oplus f_{14}(c, \mathcal{K}_{14}) \oplus f_{16}(c, \mathcal{K}_{16}) = \bigoplus_{X^0} f_{15}(c, \mathcal{K}_{15}) \oplus L(c) \quad (1)$$

Note that we do not need to guess round keys to compute the sum of $L(c)$. Note that $\mathcal{K}_{13} \cup \mathcal{K}_{14} \cup \mathcal{K}_{15} \cup \mathcal{K}_{16}$ is 72 bits, and the probability that Eq. (1) holds randomly is 2^{-4}. Therefore, we reduce the space of key candidates from 2^{72} to 2^{68}. Finally, we recover the correct key by additionally guessing the remaining 8 bits. It is enough to determine the correct key by using two known plaintexts. Thus, the total time complexity is $2^{76} \times 2 = 2^{77}$.

Note that the time complexity that we evaluate whether Eq. (1) holds or not is less than 2^{61} and it is negligible because of [18, 19]. Due to the limited space, we omit the detailed procedure.

6   Concluding Remarks

In this paper, we showed security evaluation of LILLIPUT. The linear layer $\mathcal{L}$, which is the main feature introduced by EGFN, gives several security concerns to be carefully discussed. By using MILP, we proved that the lower bounds of number of active S-boxes provided by the designers were incorrect. Then, we derived new bounds in two approaches; nibble-wise and bitwise models. Interestingly, it turned out that security of LILLIPUT is better than the original expectation. Further improving the lower bounds and deriving tight bounds for more rounds will be interesting future research directions. Meanwhile, we showed that the security enhance by the linear layer $\mathcal{L}$, which applies many xors without increasing S-box, is not so strong against division property, and improved the previous best key recovery attacks by three rounds. EGFN is a relatively new design approach. We believe that this paper leads to better understanding of EGFN.

References

1. Biryukov, A., Johann Großschädl, Y.L.C.: CryptoLUX, Lightweight Cryptography (2015). https://www.cryptolux.org/index.php/Lightweight_Cryptography
2. Suzaki, T., Minematsu, K.: Improving the generalized Feistel. In: Hong, S., Iwata, T. (eds.) FSE 2010. LNCS, vol. 6147, pp. 19–39. Springer, Heidelberg (2010). doi:10.1007/978-3-642-13858-4_2
3. Suzaki, T., Minematsu, K., Morioka, S., Kobayashi, E.: TWINE: a lightweight block cipher for multiple platforms. In: Knudsen, L.R., Wu, H. (eds.) SAC 2012. LNCS, vol. 7707, pp. 339–354. Springer, Heidelberg (2013). doi:10.1007/978-3-642-35999-6_22
4. Wu, W., Zhang, L.: LBlock: a lightweight block cipher. In: Lopez, J., Tsudik, G. (eds.) ACNS 2011. LNCS, vol. 6715, pp. 327–344. Springer, Heidelberg (2011). doi:10.1007/978-3-642-21554-4_19
5. Berger, T.P., Minier, M., Thomas, G.: Extended generalized Feistel networks using matrix representation. In: Lange, T., Lauter, K., Lisoněk, P. (eds.) SAC 2013. LNCS, vol. 8282, pp. 289–305. Springer, Heidelberg (2014). doi:10.1007/978-3-662-43414-7_15
6. Zhang, L., Wu, W.: Differential analysis of the extended generalized Feistel networks. Inf. Process. Lett. 114(12), 723–727 (2014)
7. Berger, T.P., Francq, J., Minier, M., Thomas, G.: Extended generalized Feistel networks using matrix representation to propose a new lightweight block cipher: LILLIPUT. IEEE Trans. Comput. 65, 2074–2089 (2015)
8. Knudsen, L., Wagner, D.: Integral cryptanalysis. In: Daemen, J., Rijmen, V. (eds.) FSE 2002. LNCS, vol. 2365, pp. 112–127. Springer, Heidelberg (2002). doi:10.1007/3-540-45661-9_9
9. Todo, Y.: Structural evaluation by generalized integral property. In: Oswald, E., Fischlin, M. (eds.) EUROCRYPT 2015. LNCS, vol. 9056, pp. 287–314. Springer, Heidelberg (2015). doi:10.1007/978-3-662-46800-5_12
10. Daemen, J., Rijmen, V.: The Design of Rijndeal: AES - The Advanced Encryption Standard (AES). Springer, Heidelberg (2002). doi:10.1007/978-3-662-04722-4
11. Lai, X., Massey, J.L., Murphy, S.: Markov ciphers and differential cryptanalysis. In: Davies, D.W. (ed.) EUROCRYPT 1991. LNCS, vol. 547, pp. 17–38. Springer, Heidelberg (1991). doi:10.1007/3-540-46416-6_2
12. Zhang, H., Wu, W.: Structural evaluation for generalized Feistel structures and applications to LBlock and TWINE. In: Biryukov, A., Goyal, V. (eds.) INDOCRYPT 2015. LNCS, vol. 9462, pp. 218–237. Springer, Cham (2015). doi:10.1007/978-3-319-26617-6_12

13. Mouha, N., Wang, Q., Gu, D., Preneel, B.: Differential and linear cryptanalysis using mixed-integer linear programming. In: Wu, C.-K., Yung, M., Lin, D. (eds.) Inscrypt 2011. LNCS, vol. 7537, pp. 57–76. Springer, Heidelberg (2012). doi:10. 1007/978-3-642-34704-7_5

14. Gurobi Optimization Inc.: Gurobi optimizer 6.5 (2015). Official webpage http:// www.gurobi.com/

15. Sun, S., Hu, L., Wang, P., Qiao, K., Ma, X., Song, L.: Automatic security evaluation and (related-key) differential characteristic search: application to SIMON, PRESENT, LBlock, DES(L) and other bit-oriented block ciphers. In: Sarkar, P., Iwata, T. (eds.) ASIACRYPT 2014. LNCS, vol. 8873, pp. 158–178. Springer, Heidelberg (2014). doi:10.1007/978-3-662-45611-8_9

16. Sun, S., Hu, L., Wang, M., Wang, P., Qiao, K., Ma, X., Shi, D., Song, L.: Automatic enumeration of (related-key) differential and linear characteristics with predefined properties and its applications. IACR Cryptol. ePrint Arch. **2014**, 747 (2014)

17. Todo, Y.: Integral cryptanalysis on full MISTY1. In: Gennaro, R., Robshaw, M. (eds.) CRYPTO 2015. LNCS, vol. 9215, pp. 413–432. Springer, Heidelberg (2015). doi:10.1007/978-3-662-47989-6_20

18. Ferguson, N., Kelsey, J., Lucks, S., Schneier, B., Stay, M., Wagner, D., Whiting, D.: Improved cryptanalysis of Rijndael. In: Goos, G., Hartmanis, J., van Leeuwen, J., Schneier, B. (eds.) FSE 2000. LNCS, vol. 1978, pp. 213–230. Springer, Heidelberg (2001). doi:10.1007/3-540-44706-7_15

19. Sasaki, Y., Wang, L.: Meet-in-the-middle technique for integral attacks against Feistel ciphers. In: Knudsen, L.R., Wu, H. (eds.) SAC 2012. LNCS, vol. 7707, pp. 234–251. Springer, Heidelberg (2013). doi:10.1007/978-3-642-35999-6_16

Cryptanalysis of Simpira v1

Christoph Dobraunig, Maria Eichlseder$^{(\boxtimes)}$, and Florian Mendel

Graz University of Technology, Graz, Austria
`maria.eichlseder@iaik.tugraz.at`

Abstract. Simpira v1 is a recently proposed family of permutations, based on the AES round function. The design includes recommendations for using the Simpira permutations in block ciphers, hash functions, or authenticated ciphers. The designers' security analysis is based on computer-aided bounds for the minimum number of active S-boxes. We show that the underlying assumptions of independence, and thus the derived bounds, are incorrect. For family member Simpira-4, we provide differential trails with only 40 (instead of 75) active S-boxes for the recommended 15 rounds. Based on these trails, we propose full-round collision attacks on the proposed Simpira-4 Davies-Meyer hash construction, with complexity $2^{82.62}$ for the recommended full 15 rounds and a truncated 256-bit hash value, and complexity $2^{110.16}$ for 16 rounds and the full 512-bit hash value. These attacks violate the designers' security claims that there are no structural distinguishers with complexity below 2^{128}.

Keywords: Simpira · Permutation-based cryptography · Cryptanalysis · Hash functions · Collisions

1 Introduction

The Advanced Encryption Standard AES and its underlying wide-trail design strategy are among the most popular building blocks for new symmetric designs. There are several good reasons for this. New AES-like designs profit both from the insights in efficient implementations and from the extensive cryptanalysis and well-understood security bounds of AES. In particular, if new designs not only reuse the general design ideas, but the AES block cipher itself or its round function, then Intel's AES-NI instruction set can provide high software performance on modern CPUs. However, while block ciphers are a versatile building block for other cryptographic primitives, the fixed block size of AES of 128 bits implies a certain limitation. Modern designs often require larger states for efficiency or security. Examples include permutation-based cryptography (hash functions, authenticated encryption, etc.), wide-block encryption, security beyond 2^{64} inputs without resorting to beyond-birthday-security schemes, and more.

These considerations have motivated the design of numerous cryptographic algorithms based on the AES round function. Notable recent examples of dedicated designs include several authenticated encryption algorithms with excellent

© Springer International Publishing AG 2017
R. Avanzi and H. Heys (Eds.): SAC 2016, LNCS 10532, pp. 284–298, 2017.
https://doi.org/10.1007/978-3-319-69453-5_16

software performance, such as the CAESAR round-2 candidates AEGIS [15] and Tiaoxin [12], but also more specialized primitives like the Haraka hash function for short inputs [9]. Very recently, Jean and Nikolić [5] analyzed a more general family of AES-round-based building blocks that generalizes several of the previous dedicated designs. However, except for the last work, these dedicated designs target only specific state sizes, and do not offer scalable, easily reusable building blocks for other cryptographic applications.

Simpira is a recently proposed family of permutations designed by Gueron and Mouha [2] that aims to fill this gap. The design goal is to provide very efficient permutations for arbitrarily large input sizes of $b \cdot 128$ bits, $b \in \mathbb{N}^+$, while taking advantage of the Intel AES-NI instruction set for optimized software implementations. To achieve these goals, Simpira plugs the AES round function into a generalized Feistel construction. Additionally, the designers provide computer-aided bounds for the minimum number of active S-boxes, and argue that these bounds provide security against a wide range of attack vectors. To showcase the versatility of the Simpira permutations, the designers propose a number of application scenarios, including Even-Mansour block cipher constructions, or a keyless Davies-Meyer variant for hash functions with limited-length inputs.

Our Contribution. We analyze members of the original Simpira v1 family [2]. We show that the underlying assumptions of independence, and thus the derived bounds on the minimum number of active S-boxes, are incorrect. We focus our analysis on family member Simpira-4 with its 512-bit state, but similar observations also apply to other family members with larger state sizes. For Simpira-4, we provide differential trails with only 40 (instead of 75) active S-boxes for the recommended 15 rounds. Based on these trails, we propose collision attacks on the proposed Simpira-4 Davies-Meyer hash construction. For 16 rounds of the permutation, we obtain collisions for the full 512-bit hash output with complexity $2^{110.16}$. We also adapt the attack to the originally recommended 15 rounds, providing second-order collisions and truncated collisions. We consider several truncation variants, and obtain, among others, collisions on truncated 384-bit output with complexity $2^{110.16}$, or collisions on the 256-bit output with complexity $2^{82.62}$ – the details depend on the implemented truncation variant. These attacks violate the designers' security claims that there are no structural distinguishers below 2^{128}.

Related Work. Rønjom [14] independently analyzed Simpira v1, and identified invariant subspaces for any even number of rounds of Simpira-4. Both attacks on Simpira v1 exploit properties of the underlying Type-1.x Generalized Feistel Structure by Yanagihara and Iwata [16] and the sparse, structured round constants. In response to Rønjom's and our attacks, Gueron and Mouha proposed a new version of the design, Simpira v2 [3], published at ASIACRYPT 2016. Simpira v2 replaces both the Feistel construction and the round constant schedule. In the remaining document, Simpira always refers to Simpira v1.

Simpira is not the first AES-round-based design with problematic round constants. Other examples include the analysis of the hash function Haraka [9] by Jean [4], the analysis of the withdrawn CAESAR round-1 candidate PAES [17] by Jean et al. [6,7], or the analysis of SHAvite-3 [1] by Peyrin [13]. In all three cases, the structure of the round constants failed to break the symmetry properties of the unkeyed AES round function. However, our attack exploits different properties, in particular the incomplete diffusion of differences in the structured round constants.

Outline. We first describe the Simpira family of permutations in Sect. 2. We then propose our attacks in Sect. 3, beginning with an iterative truncated differential trail with fewer S-boxes than expected in Sect. 3.1. In Sect. 3.2, we select the bitwise differences of our truncated trail to obtain an 8-round differential trail with probability $2^{-110.16}$. Based on this trail, we propose a collision attack on the 16-round Simpira-4 hash construction in Sect. 3.3. Finally, in Sect. 3.4, we adapt our attack to the recommended 15-round design.

2   Description of Simpira

Simpira is a family of permutations designed by Gueron and Mouha [2]. By using the AES round function in a generalized Feistel construction, it can be adapted to any input size of $b \cdot 128$ bits, $b \in \mathbb{N}^+$. We refer to Simpira family members as Simpira-b.

2.1 F-Function

The Feistel update function $F = F_{c,b}$ applies two rounds of AES, where the Simpira family member b and the round counter c define the round constants. Like for AES, the 128-bit intermediate state of F is represented as a 4×4-matrix of bytes, labelled $s_0, \ldots, s_{15}$:

$$S = \begin{array}{|c|c|c|c|} \hline s_0 & s_4 & s_8 & s_{12} \\ \hline s_1 & s_5 & s_9 & s_{13} \\ \hline s_2 & s_6 & s_{10} & s_{14} \\ \hline s_3 & s_7 & s_{11} & s_{15} \\ \hline \end{array} .$$

We also refer to the value at byte position s_i in state S as $S[i]$.

The operations SubBytes, ShiftRows, and MixColumns are defined identically to AES, whereas AddConstant adds counters that define an invocation counter and the value b:

- SubBytes (SB): Applies the 8-bit AES S-box $\mathcal{S}$ to each of the 16 state bytes.
- ShiftRows (SR): Rotates row i of the state, $0 \le i \le 3$, by i bytes to the left.
- MixColumns (MC): Multiplies each byte column of the state by the MDS-matrix M over $\mathbb{K} = \mathbb{F}_2[\alpha]/(\alpha^8 + \alpha^4 + \alpha^3 + \alpha + 1)$,

$$M = \begin{pmatrix} \alpha & \alpha+1 & 1 & 1 \\ 1 & \alpha & \alpha+1 & 1 \\ 1 & 1 & \alpha & \alpha+1 \\ \alpha+1 & 1 & 1 & \alpha \end{pmatrix} = \begin{pmatrix} 02 & 03 & 01 & 01 \\ 01 & 02 & 03 & 01 \\ 01 & 01 & 02 & 03 \\ 03 & 01 & 01 & 02 \end{pmatrix}$$

- AddConstant (AC): In the cth invocation of F for Simpira-b, xors the following round constant $C_{c,b}$ to the state:

$$C_{c,b} = \begin{array}{|c|c|c|c|} \hline c_0 & b_0 & 0 & 0 \\ \hline c_1 & b_1 & 0 & 0 \\ \hline c_2 & b_2 & 0 & 0 \\ \hline c_3 & b_3 & 0 & 0 \\ \hline \end{array} .$$

In the remaining paper, we focus on Simpira-4, so $b_0 = 04$ and $b_1 = b_2 = b_3 = 00$. Also, since the number of invocations of F is limited to 30 in Simpira-4, $c_1 = c_2 = c_3 = 00$. This constant is only added in the first of the two AES rounds of F, while the second round adds 0.

To refer to intermediate states of F for an input S, we use the following notation:

$$S \overset{\text{SB}}{\longmapsto} S^{\text{SB1}} \overset{\text{SR}}{\longmapsto} S^{\text{SR1}} \overset{\text{MC}}{\longmapsto} S^{\text{MC1}} \overset{\text{AC}}{\longmapsto} S^{\text{AC}} \overset{\text{SB}}{\longmapsto} S^{\text{SB2}} \overset{\text{SR}}{\longmapsto} S^{\text{SR2}} \overset{\text{MC}}{\longmapsto} S^{\text{MC2}} = F(S) .$$

2.2 Round Function and Permutation

The permutation Simpira-b keeps a state of $b \cdot 128$ bits. The generalized Feistel round function for $b \geq 4$, where $b \neq 6, 8$, is illustrated in Fig. 1. The final output of Simpira-b for $b \geq 4$, $b \neq 6, 8$, is the state after $6b - 9$ such rounds. Note that if the number of rounds is not a multiple of b, the state words are output in a permuted order to allow for more efficient implementations.

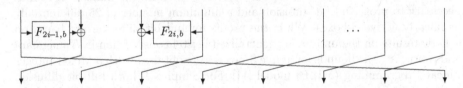

Fig. 1. Round function for round i of Simpira-b for $b \geq 4$, $b \neq 6, 8$.

In case of Simpira-4, we denote the 4 state words before round $i \geq 1$ by $S_i^A, S_i^B, S_i^C, S_i^D$, so the state update rule corresponds to

$$S_{i+1}^A = F_{2i-1,4}(S_i^A) \oplus S_i^B ,$$
$$S_{i+1}^B = F_{2i,4}(S_i^D) \oplus S_i^C ,$$
$$S_{i+1}^C = S_i^D ,$$
$$S_{i+1}^D = S_i^A .$$

The recommended number of rounds for Simpira-4 is 15, with output words $(S_{16}^B, S_{16}^C, S_{16}^D, S_{16}^A)$.

2.3   Permutation-Based Hashing

Simpira's designers identify several application areas for the Simpira permutation, such as block ciphers via an Even-Mansour construction. One particular suggested application is permutation-based hashing for short inputs, where "short" means the state size of any Simpira variant. The proposal is to use a single-block, keyless Davies-Meyer-like construction with a feed-forward, and compute the hash $h(x)$ of x as

$$h(x) = \text{Simpira-}b(x) \oplus x.$$

This approach provides an efficient construction for hashing inputs of limited length, which is required by many applications, such as Lamport signatures [10].

3   Collision Attacks on Simpira-4 Hash

In this section, we show that the number of rounds recommended by the designers is not sufficient to obtain a secure permutation. In particular, we provide collisions for full-round Simpira-4 when used in the hash mode suggested by the designers. While our analysis is focused primarily on Simpira-4, the basic observations also apply to the larger Simpira variants with the same construction approach, that is, Simpira-b with $b \geq 4$, $b \neq 6, 8$.

3.1   Differential Trail with 40 Active S-Boxes over 15 Rounds

The analysis performed by Simpira's designers [2] relies on two basic bounds: full bit diffusion, and minimum number of active S-boxes. The recommended number of rounds for each variant is selected as 3 times the number of rounds necessary to prove full bit diffusion and a minimum number of 25 differentially or linearly active S-boxes. While the proofs for full bit diffusion are based on generic results on the underlying generalized Feistel construction by Yanagihara and Iwata [16], the bounds for active S-boxes were obtained with a Mixed-Integer Linear Programming (MILP) model [11]. For Simpira-4, both full bit diffusion and at least 25 active S-boxes are claimed to be provided by 5 rounds of the round function. For the full number of 15 rounds, this method would imply at least 75 active S-boxes.

The bound is derived under the assumption that all F-function inputs are processed independently. For example, if the F-functions were indeed independent, the 4-round differential trail illustrated in Fig. 2 would contain 20 independently active S-boxes. Since the trail is iterative, and adds 5 active S-boxes per round, this trail also demonstrates the tightness of the 15-round bound.

Of course, in an unkeyed primitive like a permutation or a hash function, the S-boxes are not really independent, since there are no random, independent round keys. Nevertheless, it is usually a reasonable assumption that the differential probabilities behave as if the values were actually independently random.

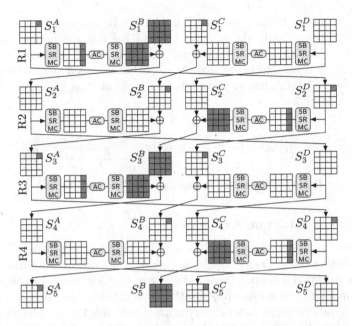

Fig. 2. Iterative 4-round trail for Simpira-4 with 10 independently active S-boxes.

We thus count S-boxes as independently active when it can reasonably be expected that their multiplied differential probabilities give a good estimate for the overall differential probability of the trail.

However, for all instances of Simpira-b with $b \geq 4$, $b \neq 6, 8$, this independence is violated by the generalized Feistel construction, and the particular definition of F. Consider, for example, the inputs to the active F-functions in rounds 1 and 2, S_1^A and S_2^D. The input values to the two F-functions are identical. Recall the definition of $F = F_{c,b}$, in our case $F_{1,4}$ and $F_{3,4}$. The only difference between $F_{1,4}$ and $F_{3,4}$ is the round-constant addition at the end of the first AES round. This means that the inputs and outputs of the S-boxes of the first AES round must be identical, i.e., $S_1^{A,\mathsf{MC1}} = S_2^{D,\mathsf{MC1}}$. The round constant only differs in state byte s_0, so this means the S-box transitions in the second AES round will also be identical except in s_0. In fact, the outputs $S_1^{A,\mathsf{MC2}}$ of $F_{1,4}$ an $S_2^{D,\mathsf{MC2}}$ of $F_{3,4}$ will have identical values except for the first column.

Considering the 4-round trail of Fig. 2, this means that the entire output difference of $F_{3,4}$ will be identical to that of $F_{1,4}$ with probability 1, as illustrated in Fig. 3. Note that s_0 is not active in the second AES round, and the differential behaviour of MixColumns is independent of the actual values of s_0. Consequently, if we fix all full-state differences to the same bitwise difference pattern, all single-byte differences to the same difference pattern, and all columnwise differences to the same difference pattern, the actual cost of the iterative trail of Fig. 2 is equivalent to only 5 active S-boxes per 2 rounds, or 40 S-boxes overall for the recommended 15 rounds, which is about half as many as suggested by the

MILP-based bound. In fact, the MILP model can be adapted to take this into account by counting only the activity of the left-hand F-functions, and only S-box s_0 for the right-hand F-functions, except in the first round. With this modification, it is easy to prove that 40 active S-boxes is a tight bound for 25 rounds. The minimum number of rounds to achieve at least 25 active S-boxes is then 9, instead of 5.

Fig. 3. Trail for the F-function with 5 active S-boxes.

3.2   Collision Attack on 8 Rounds

We now want to use this iterative differential trail of Fig. 2 to find collisions for the permutation-based hash construction suggested for Simpira permutations. Recall that in this short-input Davies-Meyer construction, the $b \cdot 128$-bit message is used as input to the Simpira permutation, and finally added as a feed-forward to the permutation output to produce the untruncated $b \cdot 128$-bit hash value. Our trail is incidentally very well suited to produce collisions for this feed-forward construction. Observe that if we fix all state differences to the same patterns as discussed in Sect. 3.1, the feed-forward will cancel out the message difference with probability 1 for any number of rounds that is a multiple of $b = 4$.

To optimize the complexity of the collision attack, we need to fix the bitwise difference patterns suitably. Recall that the AES S-box has maximum differential probability $\frac{4}{256} = 2^{-6}$. For each nonzero input difference, there is exactly one output difference with this probability (and vice versa), while the other probabilities are either $\frac{2}{256} = 2^{-7}$ or 0. We can easily choose difference patterns so that all S-box transitions have this optimal probability, at least for uniformly random round constants. For example, if we fix the one-byte input difference to 75, the trail illustrated in Fig. 4 satisfies our requirements. The probability of the differential for the F-function is then at least 2^{-30}. Overall, the probability of such an 8-round trail is at least $2^{-30 \cdot 4} = 2^{-120}$, and the resulting complexity for finding the 512-bit collision is at most 2^{120}.

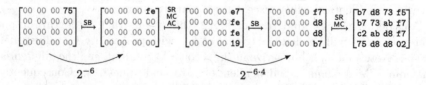

Fig. 4. Trail for the F-function with probability 2^{-30}

Note that we are actually not interested in the probability of the trail within the F-function, but just in the input-output differential from the fixed 1-byte difference to the fixed 16-byte difference. The probability of this differential is typically higher than that of the trail, since several different trails can contribute to the same differential. In the case of 2-round AES, Keliher and Sui [8] proved that for a random round constant, the probability of the differential in Fig. 4 is actually $2^{-30} + 74 \cdot 2^{-35} \approx 2^{-28.272}$.

If we consider additionally that the round constant is not random, but in our case fixed to $(00, 00, 00, 00)^\top$ for the relevant state bytes, the transition probabilities can increase even further. For example, the differential in Fig. 5 is satisfied with probability $22 \cdot 2^{-32} \approx 2^{-27.54}$. With this differential, the probability of the 8-round trail is increased to $2^{4 \times 27.54} = 2^{-110.16}$.

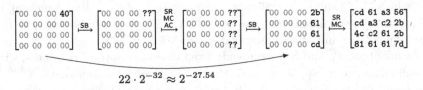

$$22 \cdot 2^{-32} \approx 2^{-27.54}$$

Fig. 5. Differential for F-function with probability $2^{-27.54}$

3.3   Collision Attack on 16 Rounds

Since the permutation involves no round keys, we can try to satisfy the conditions for some active F-functions with message modification. We will try to find messages (or rather, initial structures for intermediate Simpira states) such that the conditions for several rounds are satisfied "for free" with probability 1, and append the previous 8-round trails of Sect. 3.2 to be satisfied probabilistically. We first propose a simple initial structure covering 6 rounds, and then improve it to satisfy all conditions over 8 rounds, thus extending the previous 8-round trail to a 16-round trail with the same probability.

Initial Structure for 6 Rounds. It is sufficient to set the 4 bytes x_1, x_6, x_{11}, x_{12} of a state S_i^A to a suitable assignment in order to follow the trail for this F-function deterministically. We will refer to these 4 bytes as the diagonal in the following, and to a valid assignment as a valid diagonal. We can reuse one precomputed valid diagonal for all necessary diagonals.

We want to fix the values of the diagonals in S_1^A, S_3^A, and S_5^A to the valid diagonal. Observe that $S_1^A = S_3^C$, and $S_3^A = S_5^C$. Thus, by fixing the diagonals of S_5^A and S_5^C, we have already satisfied 2 F-trails. The remaining $12 + 16 + 12$ bytes of S_5^A, S_5^B, S_5^C can be filled arbitrarily, which will immediately determine the value of S_3^D and thus $S_3^{D,\mathrm{MC2}}$. If we now set the diagonal of S_3^C to the valid diagonal, and fill its remaining 12 bytes with arbitrary values, we completely determine S_5^D via S_4^B and S_4^A, and thus complete the state after 4 rounds. By varying the 52 arbitrary byte values, we can obtain the necessary $2^{110.16}$ candidates to satisfy the 8-round trail. The approach is illustrated in rounds 1–6 of Fig. 6, where ⊠ and ⊠ mark the 52 arbitrary bytes.

Improved Initial Structure for 8 Rounds by Matching Diagonals. With some additional effort, we can find initial structures that also satisfy the F-trail in round 7. We will again initialize the values of $S_5^A, S_5^B, S_5^C, S_3^C$ as in the previous 6-round initial structure. However, we can use the $12 + 12$ arbitrary bytes of S_5^A and S_5^C to obtain a valid diagonal in S_7^A. This will provide us with a 16-round collision attack with the same computational complexity as the 8-round trail in Sect. 3.2.

Our goal is to obtain a match between the diagonals of $S_5^{D,MC2}$ and $S_6^{A,MC2}$, as illustrated in Fig. 6. If these two diagonals sum to zero, the diagonal of S_7^A will take the exact same value as that of S_5^C, which is the valid diagonal. For this purpose, we want to initialize part of the initial structure to generate random values in $S_6^{A,MC2}$, and independently a different part of the initial structure, to independently get random values in $S_5^{D,MC2}$. Then, any match between the two corresponds to an initial structure that satisfies 4 F-trails.

Assume that S_3^C and S_5^B are already fixed to some arbitrary constants, with the valid diagonal in S_3^C. We first use the free bytes of S_5^A to randomize $S_6^{A,MC2}$. Any complete assignment of S_5^A will directly determine $S_6^{A,MC2}$ via $S_5^{A,MC2}$ and S_6^A. We can assume the values are distributed reasonably close to uniformly random, since the values are processed by 4 AES rounds, and only 4 input bytes are fixed.

Independently, we can vary the 12 bytes of S_5^C to randomize the diagonal of $S_5^{D,MC2}$. To see the independence of the values in S_5^A, consider the diagonal of $S_4^{A,MC2}$. Its values will always be identical to that of $S_5^{D,MC2}$, except for the first column, which is influenced by the round constant and will be considered separately in a moment. Since the diagonals of S_5^A and S_3^C are fixed and predetermined, these values can further be traced back right to $S_3^{D,MC2}$. Thus, knowing the diagonal of $S_3^{D,MC2}$ is equivalent to knowing the target diagonal of $S_5^{D,MC2}$, except for 1 byte in s_1. This equivalent diagonal is derived easily from S_5^C, again by 4 AES rounds via $S_4^D, S_4^{D,MC2}, S_4^C$.

Evaluating the Missing Match Byte s_1 of $S_5^{D,MC2}$. Now we still need to account for the missing byte s_1. Fortunately, with some minor modifications of our guessing strategy, this value can also be computed directly from $S_3^{D,MC2}$. Instead of varying all 12 arbitrary bytes of S_5^A to produce our matching candidates, we will keep the first column (bytes s_0, s_2, s_3) fixed. In fact, for simplicity, we will set them to the exact same values as the first column of S_3^C:

$$S_5^A[0,\dots,3] = S_3^C[0,\dots,3].$$

This implies that the values of the first column and diagonal (bytes $s_0, \dots, s_3, s_6,$ s_{11}, s_{12}) must be identical between $S_3^{D,MC2}$ and $S_4^{A,MC2}$. By partially inverting the last few steps of F, we can also easily verify that this means that

$$S_3^{D,AC}[0] = S_4^{A,AC}[0].$$

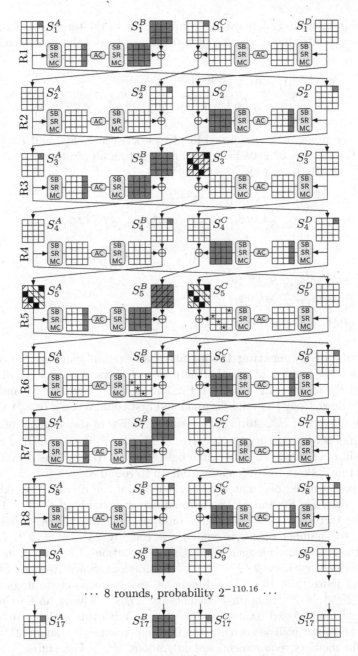

Fig. 6. 16-round collision attacks on Simpira-4 hash using 8-round initial structure. ■·fixed difference, ■ valid diagonal, ▨ arbitrary bytes, ▩ matching inputs, ⊞ match

To determine our target·value s_1 in $S_5^{D,\text{MC2}}$, consider a differential view of the intermediate variables in the computations $F(S_4^A)$ and $F(S_5^D)$. The input values are identical, but a difference in s_0 is introduced by AddConstant. We are interested in how this difference ΔS^{AC} propagates to the target byte in ΔS^{MC2}. Since we only introduced a single-byte difference before the final MixColumns, we get

$$\Delta S^{MC2}[1] = 01 \cdot \Delta S^{SB2}[0]$$
$$= \mathcal{S}\left(S_4^{A,AC}[0]\right) \oplus \mathcal{S}\left(S_4^{A,AC}[0] \oplus \Delta S^{AC}[0]\right).$$

By using the previously established identities between $F(S_4^A)$ and $F(S_3^D)$, and observing $\Delta S^{AC}[0] = 07 \oplus 0A = 0D$, we finally obtain our target match bytes in $S_5^{D,MC2}$ directly from $F(S_3^D)$:

$$S_5^{D,MC2}[1] = S_4^{A,MC2}[1] \oplus \Delta S^{MC2}[1]$$
$$= S_4^{A,MC2}[1] \oplus \mathcal{S}\left(S_4^{A,AC}[0]\right) \oplus \mathcal{S}\left(S_4^{A,AC}[0] \oplus 0D\right)$$
$$= S_3^{D,MC2}[1] \oplus \mathcal{S}\left(S_3^{D,AC}[0]\right) \oplus \mathcal{S}\left(S_3^{D,AC}[0] \oplus 0D\right),$$
$$S_5^{D,MC2}[6] = S_3^{D,MC2}[6],$$
$$S_5^{D,MC2}[11] = S_3^{D,MC2}[11],$$
$$S_5^{D,MC2}[12] = S_3^{D,MC2}[12].$$

Complexity of Generating Initial Structures. Summarizing, we can now generate a large number of initial structures as follows. First, fix the diagonals in S_3^C and S_5^C to any valid diagonal. Fix all remaining bytes of S_3^C and S_5^B to arbitrary values. Copy the valid diagonal and first column of S_3^C to S_5^A. Vary the remaining 9 bytes of S_5^A, storing the resulting values of the diagonal of $S_6^{A,MC2}$ in a list. Independently vary the 12 bytes of S_5^C, derive the diagonal of $S_5^{D,MC2}$, and store it in a second list. Any match between the two lists gives a valid initial structure that follows the differential trail up to round 8.

If we only wanted one match on the 4 bytes of the diagonal, we could try 2^{16} values each for S_5^A and S_5^C, and would expect roughly $2^{2 \cdot 16 - 32} = 1$ match due to the birthday effect. However, consider using 2^{32} values each instead. The expected number of 4-byte matches is roughly $2^{2 \cdot 32 - 32} = 2^{32}$. Now we evaluate the complexity for generating these 2^{32} solutions. Computing the match bytes requires to evaluate $2 \cdot 2 \cdot 2^{32} = 2^{34}$ F-functions. Since 16-round Simpira-4 evaluates more than $16 = 2^4$ F-functions, this corresponds to a complexity of about $2^{32-4} = 2^{30}$ Simpira-4 evaluations. Thus, we were able to produce solutions with amortized complexity less than 1. With this initial structure, we obtain a 16-round collision with computational complexity about $2^{4 \times 27.54} = 2^{110.16}$. The memory requirements are only about $2^{32} \cdot 2$ AES states.

3.4   Collision Attack on 15 Rounds with Truncation

In Sect. 3.3, we actually attacked more than the recommended number of 15 rounds for the Simpira-4. In the following, we discuss the applicability of the analysis to the original 15-round design.

Permutation Distinguisher. Clearly, the 16-round trail of Fig. 6 also immediately leads to a 15-round permutation distinguisher. With a computational complexity of $2^{110.16}$, we can find pairs of inputs with a fixed input difference such that the permutation outputs collide in 62 of 64 bytes, or actually in 510 of 512 bits, since we use the 1-byte differences of Fig. 5. This property implies, for example, second-order collisions for the hash construction with complexity $2 \cdot 2^{110.16}$, whereas the generic complexity bound is at least about $2^{512/4} = 2^{128}$. This distinguisher violates the security claims for Simpira-4.

Furthermore, if we impose no constraints on the active F-function in round 15 by allowing arbitrary constraints in $S_{15}^{A,MC2}$ and thus in S_{16}^A, we still get a collision on at least 46 of 64 bytes, or in at least 382 of 512 bits, with a fixed input difference. Then, only the 3 active F-functions in rounds 9, 11, and 13 need to be satisfied probabilistically. The probability for this trail is $2^{-3\times27.54} = 2^{-82.62}$.

Truncated Collisions. The trail no longer automatically leads to full-state collisions for the hash construction, since the 2 active state words we get after an odd number of rounds cannot cancel all 3 active state words at the input. However, we can consider truncated versions of the hash construction. Since the permutation-based Simpira-4 hash construction claims only 128-bit security, but the state size is 512 bits, Simpira's designers comment that "truncation of the output of Simpira may be required [...] to match the intended application". An obvious choice would be to truncate the state to 256 bits, so that the security claim matches the generic bound. The details and complexity of the collision attack then vary depending on the implementation of this truncation. Below, we consider 3 natural choices for truncation.

Truncation Variant 1: Left/right Half. The most intuitive choice is to simply truncate to the right (or left) half of the final state. Consider the rightmost 256 bits. With the previous 16-round trail of Figs. 6 and 7a, the permutation of the output words means that this conveniently corresponds to a hash output of

$$(S_1^C \oplus S_{16}^D,\ S_1^D \oplus S_{16}^A) = \left(\;\boxplus \oplus \boxplus,\ \boxplus \oplus \boxplus\;\right) = \left(\;\boxplus,\ \boxplus\;\right).$$

In fact, we can extend this to collisions up to the rightmost 384 bits if we just shift our iterative trail down by 1 round, as illustrated in Fig. 7b. The probabilistic part of the trail is then moved to rounds 1 (input S_1^D) and rounds 10, 12, and 14 (inputs S^A). For the same complexity of $2^{110.16}$, we get a 384-bit hash collision of the output

$$(S_1^B \oplus S_{16}^C,\ S_1^C \oplus S_{16}^D,\ S_1^D \oplus S_{16}^A).$$

Truncation Variant 2: Every Second Word. Assume the truncation function selects every second word, that is, the 256-bit hash output is

$$(S_1^A \oplus S_{16}^B,\ S_1^C \oplus S_{16}^D).$$

Then, we can even take advantage of the improved permutation distinguisher with complexity $2^{82.62}$, as in Fig. 7c.

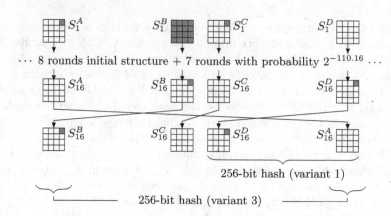

(a) Truncation variants 1 and 3: 256-bit collisions with complexity $2^{110.16}$

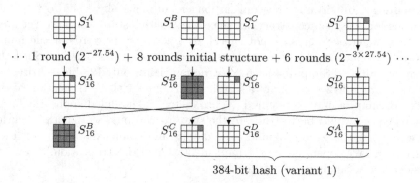

(b) Truncation variant 1: 384-bit collisions with complexity $2^{110.16}$

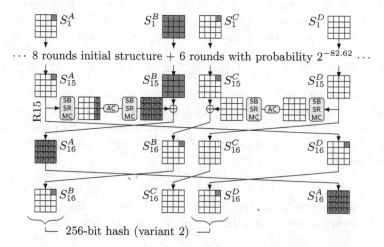

(c) Truncation variant 2: 256-bit collisions with complexity $2^{82.62}$

Fig. 7. Collisions for truncated 15-round Simpira-4 hash.

Truncation Variant 3: Updated Words. In the previous truncation variants, we took advantage of the fact that the output of one of the last round's two F-functions was truncated. Consequently, another good candidate for a truncation function is to select exactly the words that depend on the last round's F-outputs, S_{16}^A and S_{16}^B, so the hash output is

$$(S_1^A \oplus S_{16}^B, \ S_1^D \oplus S_{16}^A).$$

Nevertheless, the trail of Fig. 7a still provides hash collisions with complexity $2^{110.16}$.

4 Conclusion

In this paper, we analyzed the permutations Simpira-b, $b \geq 4$, $b \neq 6, 8$, of the Simpira v1 family, with a focus on Simpira-4. Due to properties of the underlying Type-1.x Generalized Feistel Structure and the sparse round constants, the computer-aided bounds given by the designers for the minimum number of active S-boxes are invalid. The count includes many pairs of S-boxes whose inputs are not independent, in particular, many actually share the exact same inputs. Based on differential trails that exploit this property, we propose full-round collision attacks on the proposed Simpira-4 Davies-Meyer hash construction, with complexities down to $2^{82.62}$ for the recommended full 15 rounds and the truncated 256-bit hash value, depending on the truncation rule, and complexity $2^{110.16}$ for 16 rounds and the full 512-bit hash value.

The attacks exploit Generalized Feistel Structures which apply multiple F-functions to a Feistel branch without xoring other F-outputs in between, as would be the case in a standard Feistel construction. While it is not clear whether this property could be exploited in general for independent F, it certainly becomes a problem when the F-functions differ only by using different, sparse round constants. In Simpira v1, this is the case for all family members $b \geq 4$, $b \neq 6, 8$. The consequence is that two branches of the state will be updated with two closely related F-outputs.

To address the problems described in this paper and by Rønjom [14], Gueron and Mouha subsequently tweaked their design [3]. The new Simpira v2 fixes the issue by replacing both the Feistel construction, to ensure disjoint F-inputs, and the round constants with denser values.

 Acknowledgments. We thank the Simpira designers Shay Gueron and Nicky Mouha for verifying our results and providing useful suggestions. The research leading to these results has received funding from the European Union's Horizon 2020 research and innovation programme under grant agreement No 644052 (HECTOR) and from the Austrian Science Fund (project P26494-N15).

References

1. Biham, E., Dunkelman, O.: The SHAvite-3 hash function. Submission to NIST (2009). http://www.cs.technion.ac.il/orrd/SHAvite-3/Spec.15.09.09.pdf
2. Gueron, S., Mouha, N.: Simpira: a family of efficient permutations using the AES round function. Cryptology ePrint Archive, Report 2016/122 (2016). http://eprint.iacr.org/2016/122/20160214:005409
3. Gueron, S., Mouha, N.: Simpira v2: a family of efficient permutations using the AES round function. In: Cheon, J.H., Takagi, T. (eds.) ASIACRYPT 2016. LNCS, vol. 10031, pp. 95–125. Springer, Heidelberg (2016). doi:10.1007/978-3-662-53887-6_4
4. Jean, J.: Cryptanalysis of Haraka. Cryptology ePrint Archive, Report 2016/396 (2016). http://ia.cr/2016/396
5. Jean, J., Nikolić, I.: Efficient design strategies based on the AES round function. In: Peyrin, T. (ed.) FSE 2016. LNCS, vol. 9783, pp. 334–353. Springer, Heidelberg (2016). doi:10.1007/978-3-662-52993-5_17
6. Jean, J., Nikolić, I., Sasaki, Y., Wang, L.: Practical cryptanalysis of PAES. In: Joux, A., Youssef, A. (eds.) SAC 2014. LNCS, vol. 8781, pp. 228–242. Springer, Cham (2014). doi:10.1007/978-3-319-13051-4_14
7. Jean, J., Nikolić, I., Sasaki, Y., Wang, L.: Practical forgeries and distinguishers against PAES. IEICE Trans. 99(A(1)), 39–48 (2016)
8. Keliher, L., Sui, J.: Exact maximum expected differential and linear probability for two-round advanced encryption standard. IET IFS 1(2), 53–57 (2007)
9. Kölbl, S., Lauridsen, M.M., Mendel, F., Rechberger, C.: Haraka - efficient short-input hashing for post-quantum applications. Cryptology ePrint Archive, Report 2016/098 (2016). http://ia.cr/2016/098
10. Lamport, L.: Constructing digital signatures from a one-way function. Technical Report SRI-CSL-98, SRI International Computer Science Laboratory (1979)
11. Mouha, N., Wang, Q., Gu, D., Preneel, B.: Differential and linear cryptanalysis using mixed-integer linear programming. In: Wu, C.K., Yung, M., Lin, D. (eds.) Inscrypt 2011. LNCS, vol. 7537, pp. 57–76. Springer, Heidelberg (2012). doi:10.1007/978-3-642-34704-7_5
12. Nikolić, I.: Tiaoxin v2. Submission to the CAESAR competition (2015). http://competitions.cr.yp.to/round2/tiaoxinv2.pdf
13. Peyrin, T.: Chosen-salt, chosen-counter, pseudo-collision for the compression function of SHAvite-3. NIST mailing list (2009). http://ehash.iaik.tugraz.at/uploads/e/ea/Peyrin-SHAvite-3.txt
14. Rønjom, S.: Invariant subspaces in Simpira. Cryptology ePrint Archive, Report 2016/248 (2016). http://ia.cr/2016/248
15. Wu, H., Preneel, B.: AEGIS v1: Submission to the CAESAR competition (2014). http://competitions.cr.yp.to/round1/aegisv1.pdf
16. Yanagihara, S., Iwata, T.: Type 1.x generalized feistel structures. IEICE Trans. 97(A(4)), 952–963 (2014)
17. Ye, D., Wang, P., Hu, L., Wang, L., Xie, Y., Sun, S., Wang, P.: PAES v1. Submission to the CAESAR competition (2014). http://competitions.cr.yp.to/round1/paesv1.pdf

An Efficient Affine Equivalence Algorithm for Multiple S-Boxes and a Structured Affine Layer

Jung Hee Cheon[1], Hyunsook Hong[1], Joohee Lee[1($\boxtimes$)], and Jooyoung Lee[2]

[1] Seoul National University (SNU), Seoul, Republic of Korea
{jhcheon,hongsuk07,skfro6360}@snu.ac.kr
[2] KAIST, Daejeon, Republic of Korea
hicalf@gmail.com

Abstract. An affine equivalence problem is to find affine mappings A and B such that $F = B \circ S \circ A$ for given two permutations F and S, which was first studied by Biryukov et al. Their algorithm for solving an affine equivalence problem is quite efficient and has been used in the cryptanalytic toolbox for many cryptographic schemes. Recently, Baek et al. presented a specialized affine equivalence algorithm (SAEA), which solves an affine equivalence problem in the case that S is a concatenation of several smaller S-boxes. The SAEA is more efficient than the affine equivalence algorithm for special cases, but its complexity mainly depends on the entire input size of F.

In this paper, we revisit the affine equivalence problem for a special ASA structure with multiple S-boxes and a structured input affine layer. We show that the work factor in SAEA can be reduced if the input affine layer in ASA has a certain structure. Moreover, the complexity of our algorithm mainly depends on the input size of smaller S-boxes, and not on the entire input size of F. We also present a new attack algorithm on the white-box AES implementation proposed by Baek et al. The cryptanalysis efficiently extracts the secret key from the implementation with a complexity of 2^{33}, where the claimed security level is 2^{110}.

Keywords: Affine equivalence algorithm · ASA structure · Multiple S-boxes · Structured affine mapping · White-box implementation

1 Introduction

In 1997, Even and Mansour [9] showed that for the independent n-bit keys K and K' and the random permutation P, the block cipher $E_{(K,K')}(x) = P(x \oplus K) \oplus K'$ is secure against an adversary with up to $\mathcal{O}(2^{n/2})$ queries. This block cipher, often referred to as the Even-Mansour cipher, is regarded as a minimal block cipher construction [8]. The three-layer scheme $E_{(A,B)}(x) = B \circ S \circ A(x)$ for which S is a substitution layer and B and A are secret affine mappings is a generalization of the Even-Mansour cipher, say ASA structure or three-layer scheme ASA. The problem of finding the affine layers for a given three-layer scheme ASA with a known S can be seen as the *affine equivalence problem*, which was introduced in [4].

© Springer International Publishing AG 2017
R. Avanzi and H. Heys (Eds.): SAC 2016, LNCS 10532, pp. 299–316, 2017.
https://doi.org/10.1007/978-3-319-69453-5_17

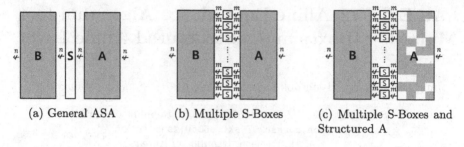

(a) General ASA   (b) Multiple S-Boxes   (c) Multiple S-Boxes and Structured A

Fig. 1. Variants of the ASA structure

More precisely, the *affine equivalence problem* is to find the affine mappings A and B satisfying $F = B \circ S \circ A$ for two given permutations F and S of n bits, if they exist, as in Fig. 1(a). Biryukov et al. [4] proposed an algorithm, which solves the affine equivalence problem with a complexity of $O(n^3 2^{2n})$. Their algorithm is quite efficient and has been used as a cryptanalytic tool [11–16] for many cryptographic schemes. A variant of this problem appears in the white-box implementations, where the middle layer S consists of a concatenation of several m-bit S-boxes as in Fig. 1(b). Baek et al. [1] presented a specialized affine equivalence algorithm (SAEA), which solves the affine equivalence problem in this case. They showed that an ASA structure with multiple S-boxes requires $O\left(\min\left\{(n^{m+4}/m)2^{2m}, (n^4/m)2^{3m} + n\log n \cdot 2^{n/2}\right\}\right)$ steps to recover the secret affine mappings under the previous attacks.

In this paper, we propose an efficient attack algorithm for the special ASA structure with multiple S-boxes and a *structured* input affine layer. Especially, we consider a variant of the affine equivalence problem depicted as in Fig. 1(c) where S is a concatenation of m-bit S-boxes for $m = n/s$ and A is an $s \times s$ block matrix with $m \times m$ matrix entries which are zeros in at least one position of each row except one. Our algorithm has a complexity that mainly depends on the size of the smaller S-boxes, and not the entire input/output size of F. Furthermore, the main factor of the complexity of our algorithm related to n drops from n^{m+3} to n^3 compared to SAEA. In Table 1, we precisely compare our affine equivalence algorithm to previous results [1,4].

Table 1. Comparison to previous affine equivalence algorithms

Algorithm	Complexity (dominant part)
Naive approach	$n^3 2^{n^2+n}$
Affine equivalence algorithm [4]	$n^3 2^{2n}$
SAEA [1]	$\min\left\{\dfrac{n}{m} \cdot n^{m+3} 2^{2m}, \dfrac{n}{m} \cdot n^3 2^{3m} + n\log n \cdot 2^{n/2}\right\}$
Our algorithm	$5\left(\dfrac{n}{m}\log\dfrac{n}{m}\right)n^3 + 5n^2 2^m + nm^2 2^{2m}$

m: the input size of smaller S-boxes in the S-layer of ASA sturcture (in [4], $m = n$)
n: an entire input/output size of the instance functions
The "naive approach" is to check if $B = F \circ A^{-1} \circ S^{-1}$ is affine and invertible for all As

Application to White-Box Implementations. A white-box implementation aims to obfuscate the secret key inside a cryptographic algorithm itself [6]. It is a way of implementing a cryptographic algorithm with a specialized attack model, thereby protecting the secret keys even in the situation that the adversary has a full access to the implementation of the cryptosystem and full control over its execution platform.

Given n-bit block ciphers as in [2,7], a naive approach to hide the secret key in such situations is to provide an input/output table of the original cipher with the secret key. However, this is not a practical solution since it is too heavy, e.g. It needs about 2^{102} GB for $n = 128$. To reduce storage requirements, the most popular approach is to decompose a cipher into round functions and split each round function as a sum of small tables [1,5,6,10,18]. Since the secret key can be easily exposed from the input/output behaviors of the round function, the table representations of round functions need to be obfuscated by secret encoding functions.

To obfuscate the secret key efficiently, the composition of an affine layer and a substitution layer with tiny S-boxes was usually considered as a secret encoding (SA as an output encoding and AS as an input encoding). Baek et al. [1] showed that composing the substitution layers of tiny S-boxes to the input/output encodings does not help to improve the security of the white-box implementations. Hence, the secret encodings would be reduced up to affine layers so that encoded round functions may have the ASA structure. One approach to split the table of ASA structure into smaller ones is to use an affine map whose linear part is a block diagonal matrix of $m \times m$ blocks as an input A layer, where m is the size of S-boxes. In this case, we can express the three layer scheme ASA as a sum of 2^m-by-n tables. However, this type of construction allows the block-wise attacks with the affine equivalence algorithm in [4], which results in a low complexity depending on the block size.

Recently, Baek et al. [1] proposed a white-box AES implementation (referred to as the BCH implementation) that uses the special input affine encoding with sparse non-zero $m \times m$ blocks which is depicted in Fig. 2. They made a point of trade-off between the above approach and a naive approach (to store an entire input/output table) to hide the secret key into the ASA structure and suggested a method for constructing the look-up tables of the encoded round functions with this special input affine encodings. The encoded round function in their implementation can be expressed as a sum of 2^{2m}-by-n tables instead of the 2^n-by-n table in the naive approach.

By the way, the affine input encodings in the BCH implementation exactly have a structure that we define. Applying our attack algorithm, we can efficiently extract the secret round key in the implementation with a complexity of 2^{33} for the case that the input size of the encoded round function is 256 bits, where the claimed security level is 2^{110}. We provide the attack complexities for the other parameters in the BCH implementation in Table 2. In future works, our attack algorithm for the special ASA would be a useful attack tool for white-box implementations.

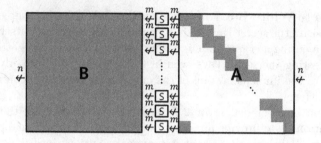

Fig. 2. The special structure lying in the input A layers of the BCH implementation

Table 2. The security of the BCH implementation, where n is the block size of encoded round function

n	Claimed security level in [1]	New security level
128	75 bits	32 bits
256	110 bits	33 bits
384	117 bits	34 bits

Outline of the Paper: In Sect. 2, we give some preliminaries used in this paper. Our attack for the special ASA structure is presented in Sect. 3. We give a cryptanalysis of the BCH implementation in Sect. 4. Finally, we conclude the paper in Sect. 5.

2 Preliminaries

2.1 Structured Matrix

Fix parameters n, m, s such that $n = s \cdot m$ (throughout this paper), and we will consider an n-bit ASA scheme

$$F = B \circ S \circ A$$

such that the inner S-box S is given as a concatenation of s S-boxes of m-bit input/output size. We will also give a certain condition on the linear part L of A: when L is viewed as an $s \times s$ block matrix of $m \times m$ blocks, each row contains some zero entries except one row. The motivation of this particular structure is that such a scheme allows an efficient white-box implementation based on table look-ups. The block-wise density of a matrix can be represented by its block representing matrix, as defined as follows.

Definition 1 (Block Representing Matrix). *Let n, m, s be integers such that $n = s \cdot m$, and let L be an $n \times n$ matrix that is represented by a block matrix as follows.*

$$L = \begin{bmatrix} L_{1,1} & L_{1,2} & \cdots & L_{1,s} \\ L_{2,1} & L_{2,2} & \cdots & L_{2,s} \\ \vdots & \vdots & \ddots & \vdots \\ L_{s,1} & L_{s,2} & \cdots & L_{s,s} \end{bmatrix}$$

where $L_{i,j}$ is an $m \times m$ matrix for every i and j. Then the block representing matrix of L, denoted by B_L, is defined as a binary $s \times s$ matrix where the (i,j)-entry is 0 if $L_{i,j}$ is the zero matrix and 1 otherwise.

Definition 2 (Structured Matrix). Let n, m, s be integers such that $n = s \cdot m$. A matrix L is called structured with respect to the block length m if L is invertible and the rows of its block representing matrix B_L are pairwise distinct.

Example 1. The MixColumn step of AES-128 can be represented by a 128×128 matrix, say MC. When it is partitioned into 8×8 blocks, its 16×16 block representing matrix becomes

$$\mathsf{B}_{\mathsf{MC}} = \left[\begin{array}{cccc|cccc|cccc|cccc} 1&0&0&0&0&0&0&0&0&0&0&0&0&0&0&0 \\ 0&0&0&0&0&1&0&0&0&0&0&0&0&0&0&0 \\ 0&0&0&0&0&0&0&0&0&0&1&0&0&0&0&0 \\ 0&0&0&0&0&0&0&0&0&0&0&0&0&0&0&1 \\ \hline 0&0&0&0&1&0&0&0&0&0&0&0&0&0&0&0 \\ 0&0&0&0&0&0&0&0&0&1&0&0&0&0&0&0 \\ 0&0&0&0&0&0&0&0&0&0&0&0&0&0&1&0 \\ 0&0&0&1&0&0&0&0&0&0&0&0&0&0&0&0 \\ \hline 0&0&0&0&0&0&0&0&1&0&0&0&0&0&0&0 \\ 0&0&0&0&0&0&0&0&0&0&0&0&0&1&0&0 \\ 0&0&1&0&0&0&0&0&0&0&0&0&0&0&0&0 \\ 0&0&0&0&0&0&0&1&0&0&0&0&0&0&0&0 \\ \hline 0&0&0&0&0&0&0&0&0&0&0&0&1&0&0&0 \\ 0&1&0&0&0&0&0&0&0&0&0&0&0&0&0&0 \\ 0&0&0&0&0&0&1&0&0&0&0&0&0&0&0&0 \\ 0&0&0&0&0&0&0&0&0&0&0&1&0&0&0&0 \end{array} \right].$$

Since MC is invertible over $\mathbb{F}_2$ and any two rows of above matrix B_{MC} are pairwise distinct, MC is structured.

An affine mapping A that maps n bits to n bits can be decomposed into a linear part L and a constant translation C as follows:

$$A(x) = L \cdot x + C$$

where L is an $n \times n$ matrix and C is an $n \times 1$ matrix over $\mathbb{F}_2$. We will say A is *structured* with respect to the block size m if the linear part L is structured with respect to the block size m.

2.2   Notation

We would set our notation used in Sects. 3 and 4. Throughout this paper, we set our target as a three-layer scheme $F = B \circ S \circ A$ of n bits which consists of a substitution and affine transformations. Our attack considers the case that the S layer contains s invertible S-boxes $S_1, S_2, \cdots, S_s$ of m bits, the output affine layer B is invertible, and the input affine layer A is structured. For the affine mappings A and B, we use the notation L and M to represent the linear part of A and B, and C and D to represent to the constant part of A and B, respectively. *i.e.,* The affine functions A and B are represented as follows:

$$A(x) = L \cdot x + C \text{ and } B(x) = M \cdot x + D$$

We consider the linear part L of A to be partitioned into s^2 $m \times m$ blocks. The (i, j)-th block matrix of size $m \times m$ is denoted by $L_{i,j}$. *i.e.,*

$$L = \begin{bmatrix} L_{1,1} & L_{1,2} & \cdots & L_{1,s} \\ L_{2,1} & L_{2,2} & \cdots & L_{2,s} \\ \vdots & \vdots & \ddots & \vdots \\ L_{s,1} & L_{s,2} & \cdots & L_{s,s} \end{bmatrix}$$

The linear part M of B can be partitioned into s vertical strips of size $n \times m$. We denote the i-th strip by M_i so that

$$M = \begin{bmatrix} M_1 \big| M_2 \big| \cdots \big| M_s \end{bmatrix}$$

For an arbitrary rectangular matrix N, we use a notation $\mathrm{col}(N)$ to represent the column space of N, namely a subspace of $\mathbb{F}_2^n$ spanned by the columns of N. We write the operation '+' to denote the bitwise XOR operation. We define $\oplus_K$ as the map $\oplus(x) = x + K$. Using this notation, we represent the key additions in a block cipher. We also split the n-bit string x into s m-bit blocks and write it as $x = (x_1, \cdots, x_s)$.

2.3   Our Problem Related to the Affine Equivalence Problem

We will formulate a problem, namely *specialized affine equivalence problem*. It can be regarded as a special variant of the affine equivalence problem. So, we first present the problem definition of the affine equivalence problem defined in [4] and then our problem related to the affine equivalence problem.

Given two permutations F and S, we say that F and S are *affine equivalent* if there exist invertible affine mappings A and B such that $F = B \circ S \circ A$. The *affine equivalence problem* is to find such affine mappings if they exist, by making a certain number of oracle queries to F and S.

We also take an attacker who can make oracle queries to F into account. The goal of this attacker might be to recover the affine layers with the knowledge of the three-layer scheme structure and input/output tables of m-bit S-boxes.

Definition 3 (Specialized   Affine   Equivalence   Problem). *Consider a three-layer invertible ASA scheme $F = B \circ S \circ A$ of n-bit for which S is a concatenation of m-bit S-boxes and A is structured with respect to the block size* *m. We assume that the s m-bit S-boxes are given as input/output tables, and the block representing matrix of A with respect to the block length m is known. By making a certain number of oracle queries to F, we want to recover affine mappings A' and B' which are equivalent to A and B in the sense that:*

- *F = B' ∘ S ∘ A'*
- *The block representing matrices of A and A' with respect to the block length m are the same.*

We can erase the assumption that m-bit S-boxes are given as tables. Then, we need to allow the oracle queries to S and store $sm2^m$ bits of input/output pairs of S-boxes in our algorithm in Sect. 3. We added an assumption that the block representing matrix of A with respect to the block length m is known since we can easily retrieve it with input/output behaviors of F in a practical scheme or it would be contained in an algorithm of a practical scheme, e.g. BCH implementation [1].

2.4 Useful Lemmas

In this subsection, we introduce useful lemmas which are used in our cryptanalysis.

Affine Equivalence Algorithm. Biryukov et al. [4] proposed an *affine equivalence algorithm* that efficiently solves the affine equivalence problem compared to the exhaustive search for A and B. The following lemma summarizes their result in terms of the complexity of the algorithm.

Lemma 1. *Let S_1 and S_2 be m-bit permutations. If S_1 and S_2 are affine equivalent, one can find all the pairs of affine mappings A and B such that $S_2 = B \circ S_1 \circ A$ in time $O(m^3 2^{2m})$.*

Rank of a Random Matrix over $\mathbb{F}_2$. The following lemma presented by Wan [17] tells us the property of random binary matrices.

Lemma 2. *Let n, k, r be integers such that $1 \leq r \leq \min(n, k)$. The probability that a random $n \times k$ binary matrix has rank r over $\mathbb{F}_2$ is*

$$P(n, k, r) = \frac{1}{2^{(n-r)(k-r)}} \cdot \prod_{i=0}^{r-1} \frac{(1 - 2^{i-k})(1 - 2^{i-n})}{(1 - 2^{i-r})}.$$

By Lemma 2, the simulation result shows that the probability that a random $n \times k$ binary matrix has rank $r \geq k - 5$ is greater than or equal to 0.99 for $n \leq 1000$.

Affine Self-equivalences in Rijndael. The affine equivalence problem can have many equivalent solutions. For a permutation $\hat{S}$, if there exists nontrivial affine mappings a, b such that $\hat{S} = b \circ \hat{S} \circ a$, then we say that (a, b) is a self-equivalence of $\hat{S}$. The following lemma proposed by Biryukov et al. [4] tells us the number of affine self-equivalence of the S-box used in Rijndael [7].

Lemma 3. *The S-box $\hat{S}$ used in Rijndael has 2040 affine self-equivalences. In other words, there exists 2040 pairs of affine mappings (a, b) such that $\hat{S} = b \circ \hat{S} \circ a$.*

Intersection of Subspaces. For given two subspaces of $\mathbb{F}_2^n$, a complexity for computing an intersection of these two subspaces is less than $5n^3$ and is more precisely presented as follows.

Lemma 4. *For $0 < m_1 < m_2 < n$, suppose that V and W are subspaces of $\mathbb{F}_2^n$ of dimensions m_1 and m_2, respectively. For given bases of V and W, we can compute a basis for a subspace*

$$V \cap W$$

over $\mathbb{F}_2$ in a complexity of $n(2m_1^2 + 2m_1 m_2 + m_2^2)$.

Proof. To calculate an intersection, consider the basis matrices $\bar{V}$ and $\bar{W}$ for V and W, respectively. Since

$$\bar{V} \cdot x = \bar{W} \cdot y \quad \Longleftrightarrow \quad [\bar{V} | \bar{W}] \cdot \begin{bmatrix} x \\ -y \end{bmatrix} = \mathbf{0},$$

we need to find the null space of $[\bar{V} | \bar{W}]$ with a Gaussian elimination in $n(m_1 + m_2)^2$ steps and then multiply $\bar{V}$ to the x's to obtain a basis for $V \cap W$ in less than nm_1^2 steps. $\qquad\square$

3   Cryptanalysis of the ASA Structure with a Structured Affine Layer

In this section, we present an efficient algorithm solving the specialized affine equivalence problem defined in Definition 3. To avoid an abuse of notation, we first describe an instance of our algorithm for the specific cases which can be directly applied to the BCH implementation and then present a theorem for the general cases.

For an ASA structure $F = B \circ S \circ A$ whose notation is defined in Sect. 2.2, we would specify a class of L by defining its block representing matrix B_L with respect to block length m as follows.

$$(\mathsf{B}_L)_{i,j} = \begin{cases} 1, & \text{if } 1 \leq i \leq s - \beta + 1 \text{ and } i \leq j \leq i + \beta - 1 \\ 1, & \text{if } s - \beta + 1 < i \leq s \text{ and } 1 \leq j \leq i + \beta - s - 1, \\ 0, & \text{otherwise} \end{cases}$$

for some positive integer $\beta < \left\lfloor \dfrac{s}{2} \right\rfloor$.

In other words, the $s \times s$ block representing matrix B_L of L would be depicted as:

$$
\mathsf{B}_L = \begin{bmatrix}
1 & 1 & \cdots & 1 & & & & \\
& 1 & 1 & \cdots & 1 & & & \\
& & 1 & 1 & \cdots & 1 & & \\
& & & \ddots & \ddots & \ddots & \ddots & \\
& & & & 1 & 1 & \cdots & 1 \\
1 & & & & & 1 & \cdots & 1 \\
\vdots & \ddots & & & & & \ddots & \vdots \\
1 & \cdots & 1 & & & & & 1
\end{bmatrix}, \tag{1}
$$

where each row and column contains β nonzero entries. Note that all the rows of B_L are distinct so that L is structured.

Summary of Our Approach. Our cryptanalysis is divided into two phases. Before we start to describe our attack, we summarize our cryptanalysis as below.

Phase 1. We first find the column spaces $\mathrm{col}(M_i)$ for all $1 \leq i \leq s$. Then, we can recover the linear part of output affine layer B up to a block diagonal matrix of block size m. Though we cannot obtain the exact M, it is an essential step to reduce output sizes of F from n to m.

Phase 2. From the phase 1, we can split F into $\tilde{F}_i$ for $1 \leq i \leq s$ which are the ASA structures from βm bits to m bits, respectively. We transform $\tilde{F}_i$ into an invertible ASA structure on m bits reducing the input sizes from βm to m. Then, the affine equivalence algorithm can be applied to the invertible ASA structure on m bits.

3.1 Decomposing the Linear Part of B

The first phase of our attack is to recover the linear part of B upto a block diagonal matrix. For each index $1 \leq i \leq s$, we will choose a certain number of pairs of plaintexts (P_1, P_2) having a difference only in the i-th m-bit block. Namely, when we write

$$P_1 = (x_1, x_2, \cdots, x_i, \cdots, x_s)$$
$$P_2 = (y_1, y_2, \cdots, y_i, \cdots, y_s)$$

for m-bit blocks x_j and y_j, $j = 1, \ldots, s$, we have $x_j = y_j$ for every $j \neq i$, but $x_i \neq y_i$. For any of such pairs (P_1, P_2), $S \circ A(P_1)$ and $S \circ A(P_2)$ will have non-zero differences exactly in β blocks since each column of B_L contains β 1's and S is defined as a concatenation of m-bit S-boxes. Specifically, we have

$$S \circ A(P_1) + S \circ A(P_2) = (\Delta_1, \cdots, \Delta_s),$$

where $\Delta_{i-\beta+1}, \cdots, \Delta_i$ are all non-zero blocks and the others are all zero blocks (cyclically indexed modulo s). So the positions of non-zero blocks are cyclically shifted as the index i increases. Since

$$F(P_1) + F(P_2) = B \circ S \circ A(P_1) + B \circ S \circ A(P_2) = M \cdot (S \circ A(P_1) + S \circ A(P_2))$$

$F(P_1) + F(P_2)$ would be always a linear combination of the βm columns from $M_{i-\beta+1}$ to M_i, namely

$$F(P_1) + F(P_2) \in \mathsf{col}(M_{i-\beta+1}|M_{i-\beta+2}|\cdots|M_i).$$

In order to find the column space $\mathsf{col}(M_{i-\beta+1}|M_{i-\beta+2}|\cdots|M_i)$, we set P_1+P_2 to have nonzero entries exactly in β blocks and compute $F(P_1) + F(P_2)$ for random P_1's in $\{0,1\}^n$ to collect βm linearly independent vectors over $\mathbb{F}_2$. Note that the probability that a random $n \times (\beta m+5)$ binary matrix has rank $r \geq \beta m$ is greater than 0.99 when $n \leq 1000$ by Lemma 2. Hence, from $\beta m + 5$ vectors of the form $F(P_1) + F(P_2)$, we can find the basis of this column space with a high probability(≥ 0.99) via the Gaussian elimination which takes $n(\beta m +5)^2$ time. Since M is invertible over $\mathbb{F}_2$ and $\beta < \left\lfloor \dfrac{s}{2} \right\rfloor$, we have

$$\mathsf{col}(M_i) = \mathsf{col}(M_{i-\beta+1}|M_{i-\beta+2}|\cdots|M_i) \cap \mathsf{col}(M_i|M_{i+1}|\cdots|M_{i+\beta-1}).$$

Therefore we can compute a basis of $\mathsf{col}(M_i)$ in $5n(\beta m)^2$ time by Lemma 4. Overall, this phase requires $sn[(\beta m+5)^2+5(\beta m)^2]$ time complexity and $2s(\beta m+5)$ chosen plaintexts.

Now, we obtained the basis of each space $\mathsf{col}(M_i)$ for $1 \leq i \leq s$. Let $\tilde{M}_i \in \mathbb{F}_2^{n \times m}$ denote the matrix whose columns are the basis of $\mathsf{col}(M_i)$. Then each column of M_i can be represented by a linear combination of the columns of $\tilde{M}_i$ with certain unknown coefficients. So we have a decomposition as follows.

$$M = \tilde{M} \cdot U$$

where

$$\tilde{M} = \left[\tilde{M}_1 \middle| \tilde{M}_2 \middle| \cdots \middle| \tilde{M}_s\right] \quad \text{and} \quad U = \begin{bmatrix} U_1 & 0 & 0 & \cdots & 0 \\ 0 & U_2 & 0 & \cdots & 0 \\ 0 & 0 & U_3 & \cdots & 0 \\ \vdots & \vdots & \vdots & \ddots & \vdots \\ 0 & 0 & 0 & \cdots & U_s \end{bmatrix}$$

for some (unknown) $m \times m$ invertible matrices $U_1, \ldots, U_s$.

3.2   Recovering A and B

The second phase is to split the entire structure F on n bits into smaller ASA structures on m bits, and then apply the affine equivalence algorithm given in Lemma 1 to each of the smaller structures.

Let $\tilde{F}$ be a map defined by $\tilde{F}(X) = \tilde{M}^{-1} \cdot F(X)$ for every $X \in \mathbb{F}_2^n$. When $\tilde{F}$ is splitted into m-bit blocks as

$$\tilde{F} = (\tilde{F}_1, \cdots, \tilde{F}_s),$$

it is easily shown that each $\tilde{F}_i$, $i = 1, \ldots, s$, depends only on βm bits of an n-bit input X: precisely we can write

$$\tilde{F}_i(X) = U_i\left(S_i\left([L_{i,i}|L_{i,i+1}|\cdots|L_{i,i+\beta-1}] \cdot X' + C_i'\right)\right) + D_i'$$

where S_i is an m-bit S-box in S layer, X' denotes the βm bits of X from the i-th m-bit block to $(i + \beta - 1)$-th m-bit block, and C'_i and D'_i are the i-th m-bit block of C and $\tilde{M}^{-1} \cdot D$, respectively. In this way, we can view $\tilde{F}_i$ as an ASA structure based on a single m-bit S-box that takes as input βm bits and outputs m bits.

The first step of this phase is to fix $(\beta - 1)m$ bits of inputs X' for each $\tilde{F}_i$ and then apply the affine equivalence algorithm of Lemma 1 to the resulting m-bit to m-bit ASA structure. Since the affine map A is invertible, the $m \times \beta m$ matrix

$$[L_{i,i}|L_{i,i+1}|\cdots|L_{i,i+\beta-1}]$$

has full row rank$(= m)$ over $\mathbb{F}_2$, and hence the column rank m. In order to find the positions of m linearly independent columns from this unknown matrix, we fix a set of m positions of X', and then evaluate $\tilde{F}_i$ for all the possible 2^m values on this set of positions with the other positions fixed as zero. If all the possible outputs of $\mathbb{F}_2^m$ are obtained from this evaluation, then the columns corresponding to these m positions would be linearly independent.

The probability that we choose m linearly independent columns from βm columns is $(1 - \frac{1}{2}) \cdot (1 - \frac{1}{2^2}) \cdots (1 - \frac{1}{2^m}) > 0.288$ for the random full rank $m \times \beta m$ matrix. So, we would iterate the procedures to guess m positions of X' and check if all the possible outputs come out for about 5 times in average. It takes n^3 time to compute $\tilde{M}^{-1}$ and for each iteration, $nm2^m$ time to perform a matrix multiplication and $m2^m$ time to sort 2^m instances, with 2^m chosen plaintexts needed. Since five iterations would be held for each $1 \leq i \leq s$, it takes totally $n^3 + 5s(nm2^m + m2^m) = n^3 + 5(n^2 + n)2^m$ steps with $5s2^m$ chosen plaintexts to find the positions of m linearly independent columns for all $1 \leq i \leq s$.

After this step, by fixing the other $(\beta - 1)m$ positions of X' as zero, we obtain an invertible m-bit ASA structure. By applying the affine equivalence algorithm of Lemma 1 to this small construction which takes $m^3 2^{2m}$ time, we can recover the affine layers of $\tilde{F}_i$ for every $i = 1, \ldots, s$, and hence F. More precisely, after running the affine equivalence algorithms, we achieve U_i, C'_i, D'_i and the m linearly independent columns of $[L_{i,i}|L_{i,i+1}|\cdots|L_{i,i+\beta-1}]$. We recover the affine maps A and B from this information as follows. We first recover B multiplying $\tilde{M}$ to the affine map $U \cdot X + (D'_1, \cdots, D'_s)$ in time n^3, and compute B^{-1} in time n^3. Then the unknown $(\beta - 1)m$ columns of $[L_{i,i}|L_{i,i+1}|\cdots|L_{i,i+\beta-1}]$ remain for each i. The j-th unknown column of this matrix is obtained by

$$S_i^{-1}(i\text{-th } m\text{-bit block of } (B^{-1} \cdot F(e_j))) + C'_i,$$

where e_j is the j-th coordinate vector in $\mathbb{F}_2^n$. To calculate all of them for $1 \leq i \leq s$, we need to compute $B^{-1} \cdot F(e_j)$ for all j, which takes $n \cdot (n^2)$ time with n chosen plaintexts. Now, we can obtain the whole matrix $[L_{i,i}|L_{i,i+1}|\cdots|L_{i,i+\beta-1}]$ for each i, and finally achieve A.

The overall work factor of the second phase is $4n^3 + 5(n^2 + n)2^m + nm^2 2^{2m}$ with $s(5 \cdot 2^m + m)$ chosen plaintexts.

We can conclude the overall work factor in our attack including the first and second phases would be calculated as

$$sn[(\beta m + 5)^2 + 5(\beta m)^2] + 4n^3 + 5(n^2 + n)2^m + nm^2 2^{2m}$$
$$\approx 6\beta^2 n^2 m + 4n^3 + 5n^2 2^m + nm^2 2^{2m},$$

with about $s(2\beta m + 5 \cdot 2^m + m + 10)$ chosen plaintexts.

Example 2. For $n = 128$, $m = 8$ and $\beta = 3$, the time complexity of our attack would grow up to 2^{29}. For $n = 256$, $m = 8$ and $\beta = 2$, the complexity would be less than 2^{31}. In these examples, the complexity of our attack algorithm is dominated by the term $nm^2 2^{2m}$.

3.3   Generalizations

In Sects. 3.1 and 3.2, we cryptanalyze the three-layer scheme ASA with specific input affine layers. We would provide an upper bound for the complexity of the attack algorithm for ASA with structured input affine layers.

Theorem 1. *Consider a three-layer scheme ASA, $F = B \circ S \circ A$ on n bits for which A is a structured affine mapping with respect to block length m and S is a concatenation of m-bit S-boxes. One can solve the specialized affine equivalence problem for F in time*

$$5 \cdot \left(\frac{n}{m} \cdot \log_2 \frac{n}{m} \right) \cdot n^3 + 5 \cdot n^2 \cdot 2^m + n \cdot m^2 \cdot 2^{2m}$$

with $\frac{n}{m}(2n + 5 \cdot 2^m + m + 10)$ chosen plaintexts.

Proof. The proof of theorem follows the attack scenario of Sects. 3.1 and 3.2. Since the attack procedure in the second phase is appliable to the general cases with no changes in time complexity, it suffices to show the following claim related to the first phase (with the same notations as in Sects. 3.1 and 3.2).

Claim. Let col_i be the column space obtained by picking plaintexts with no differentials except the i-th block in *Phase 1* (e.g. In our example in Sect. 3.1, $\mathrm{col}_i = \mathrm{col}(M_{i-\beta+1} | M_{i-\beta+2} | \cdots | M_i)$ for $1 \leq i \leq s$). Given col_i for $1 \leq i \leq s$, performing less than $s(\log_2 s + 1)$ operations of intersections of subspaces in $\mathbb{F}_2^n$,[1] we can achieve bases for $\mathrm{col}(M_i)$ for $1 \leq i \leq s$ over $\mathbb{F}_2$, respectively.

Proof of Claim (Sketch). Note that since L is invertible, every column of B_L is not a zero vector. The following algorithm terminates in $\log_2 s$ iterations and outputs $\mathrm{col}(M_i)$ for some single strip M_i.

– Let l be an index in $\{1, \cdots, s\}$. Set the initial values $v \leftarrow$ (the l-th column of B_L) and $\mathrm{col} \leftarrow \mathrm{col}_l$. We iterate the followings while $k > 1$.

---

[1] Each operation of subspaces takes less than $5n^3$ steps by Lemma 4.

- $k \leftarrow$ (hamming weight of v).
- Let $\{i_1 < \cdots < i_k\}$ be the set of indices in which components of v are nonzeros.
- For the i_1-th row and i_2-th row of B_L, find j such that the i_1-th component of the j-th column of B_L is different from the i_2-th component of the j-th column (such j exists since L is structured).
- Set w as the j-th column of B_L.
  * If w has more than $\lfloor k/2 \rfloor$ nonzero overlapped components with v, then
    $$v \leftarrow v + (v \wedge w)$$ where "$\wedge$" indicates componentwise multiplication and compute $\mathsf{col} \leftarrow \mathsf{col} \cap \mathsf{col}_j^{\perp}$ where $\mathsf{col}_j^{\perp} \in \mathbb{F}_2^n$ is an orthogonal space of col_j.
  * Otherwise, set $v \leftarrow v \wedge w$ and compute $\mathsf{col} \leftarrow \mathsf{col} \cap \mathsf{col}_j$.
- Output v and col.

Remark 1. The algorithm outputs v whose components are all zeros except one. Suppose that the output $v \in \mathbb{F}_2^s$ has all zero entries except the i-th entry. Then, we can observe that the output col is equal to $\mathsf{col}(M_i)$. In other words, v indicates the index of the strip of which column space is obtained from the above algorithm.

Note that this algorithm does not guarantee to output distinct column spaces. So, to find distinct column spaces, we remove the indices i's from the initial $\{i_1, i_2, \cdots, i_k\}$, check if the set remains nonempty (if it is empty, then choose another l and repeat), and then replace the initial col with an intersection of col and the spaces $\mathsf{col}(M_i)^{\perp}$'s to run the algorithm again. Totally, we could output $\mathsf{col}(M_i)$ for $1 \le i \le s$ with $\log_2 s + (s-1)(\log_2 s + 1)$ operations of subspaces in $\mathbb{F}_2^n$. Though the above algorithm is not optimized for a particular A, it provides an approximate upper bound of complexity of finding $\mathsf{col}(M_i)$'s for the structured A with our strategies in general. $\qquad\qquad\square$

4   Application to the White-Box AES Implementation

To see the background of the BCH implementation, let us take a glance at the historical aspects briefly. In the first white-box implementations presented by Chow et al. [6], the composition of a linear map and a nonlinear permutation with multiple S-boxes is used as an encoding. The linear map in their encoding contains a block diagonal matrix in which block provides a linear mixing bijection. However, the implementation is vulnerable to the Billet et al. attack [3]. Since then, Xiao and Lai proposed a white-box AES implementation with linear mappings as encodings [18]. They expected their implementation would resist the Billet et al. attack, using the linear encodings of block diagonal matrices whose block size is twice of the size of S-boxes. But the implementation was also broken by Mulder et al. attack [14] using linear equivalence algorithm in [4].

Recently, Baek et al. [1] showed that the substitution layers of the encodings in the previous constructions do not help to improve the security of the white-box implementations and the linear parts of the affine input encodings should not be split into the block diagonal matrices of small blocks to resist their attack toolbox. Hence, they constructed the special input encoding in which linear part can not be split, called *sparse unsplit encoding*. They presented their white-box AES implementation using the sparse unsplit encodings in [1], which was claimed to be secure against all known attacks including their attack toolbox.

However, the special structure of their sparse unsplit encodings threw new light on the cryptanalysis for us. We will explain our attack against the BCH implementation in this section. We can efficiently extract the round key of the implementation for all rounds except the first round, in 2^{33} time with 2^{14} chosen plaintexts for $n = 256$. This attack can also be applied for other parameters. The attack complexities for other parameters are presented in Table 2.

4.1 The BCH Implementation

The strategy of the BCH implementation is to obfuscate several parallel AES round functions at the same time using the special input encoding and to decompose the encoded round function into table lookups with small inputs so that their composition is equivalent to the encoded round function. Especially, the structured affine mapping with respect to the block length 8 was used as an input encoding in the BCH implementation.[2]

Let an input encoding $\widehat{A}^{(r)}$ be a structured affine mapping on n bits with respect to block length 8 of the form in Eq. 1 for $\beta = 2$. The r-th encoded round function $F^{(r)}$ of AES-128 in the BCH implementation is of the form:

$$F^{(r)} = \widehat{B}^{(r)} \circ \underbrace{(\hat{S}, \cdots, \hat{S})}_{\# \text{ of S-boxes}=s} \circ \underbrace{\oplus_{(K^{(r)}, \cdots, K^{(r)})}}_{\# \text{ of round Key}=n/128} \circ \widehat{A}^{(r)},$$

where $\hat{S}$ is the S-box on 8 bits used in Rijndael, $K^{(r)}$ is the r-th round key of 128 bits in AES-128, and the output encoding $\widehat{B}^{(r)}$ is an affine map defined as $\widehat{B}^{(r)} = (\widehat{A}^{(r+1)})^{-1} \circ (MC \circ SR, \cdots, MC \circ SR)$ for $r < 10$, where MC and SR are the functions of MixColumn and ShiftRow steps in AES-128, respectively. Then, the encoded round function $F^{(r)}$ in the BCH implementation has ASA structure on n bits with $n = 8s$, where the S layer is a concatenation of s S-boxes on 8 bits and the input affine layer contains structured input affine mapping.

4.2 Cryptanalysis of the BCH Implementation

In our notations of Eq. (1), the input encoding of the BCH implementation is the case of $\beta = 2$ and $m = 8$. Hence, our cryptanalysis can be directly applied to the BCH implementation, setting $m = 8$. The encoded round function of the BCH Implementation is of the form in Sects. 3.1 and 3.2 for $\beta = 2$. For each round, we can solve the specialized affine equivalence problem for $F^{(r)}$ in

$$6\beta^2 n^2 m + 4n^3 + 5n^2 2^m + nm^2 2^{2m}$$

time with $s(2\beta m + 5 \cdot 2^m + m + 10)$ chosen plaintexts.

---

[2] In [1], they called the input encodings used in the BCH implementation as the *sparse unsplit* affine mapping.

We would regard $\widehat{B}^{(r)}$ as B, and $\oplus_{(K^{(r)},\cdots,K^{(r)})} \circ \widehat{A}^{(r)}$ as A, according to the notations in Sects. 3.1 and 3.2. For example, to find the image space of M_1, we would start with the plaintexts P_1, P_2, P_3 and P_4 such that P_1 and P_2 have the same values except the first 8-bit blocks, and P_3 and P_4 have same values except the second 8-bit blocks. From such plaintexts, we can find the column spaces as follows:

$\mathsf{col}(M_1|M_s) = \{F(P_1) + F(P_2) \mid P_1, P_2 \in \{0,1\}^n \text{ with } P_1 + P_2 = (*, 0, \cdots, 0) \in \{0,1\}^{8 \cdot s}\},$

$\mathsf{col}(M_1|M_2) = \{F(P_3) + F(P_4) \mid P_3, P_4 \in \{0,1\}^n \text{ } with \text{ } P_3 + P_4 = (0, *, 0, \cdots, 0) \in \{0,1\}^{8 \cdot s}\}$

The column space $\mathsf{col}(M_1)$ is obtained by computing an intersection of $\mathsf{col}(M_1|M_s)$ and $\mathsf{col}(M_1|M_2)$. The work factor of the first phase in Sect. 3.1 is $sn[(2m+5)^2 + 5(2m)^2] \approx 2^{24}$ for $n = 256$.

In the second phase, for example, we know $\tilde{M}_1$ such that $M_1 = \tilde{M}_1 \cdot U_1$ for some (unknown) 8×8 matrix U_1. So, we have the function

$$\tilde{F}_1 = U_1 \circ \hat{S} \circ ((L_{1,1}|L_{1,2}) \cdot X' + C_1') + D_1',$$

where $\hat{S}$ is the 8-bit S-box in Rijndael, X' consists the first and second 8-bit blocks of an n-bit input X, and C_1' and D_1' are the first 8-bit blocks of C and $\tilde{M}^{-1} \cdot D$. To transform $\tilde{F}_1 : \mathbb{F}_2^{16} \to \mathbb{F}_2^8$ into an invertible map $\hat{F}_1$, we search for the set of eight indices $\{i_1, \cdots, i_8\}$ such that the output values of $\tilde{F}_1$ restricting j-th bits to be zeros for all $j \in \{1, \cdots, 16\} \backslash \{i_1, \cdots, i_8\}$ covers all 2^8 possible values. After then, applying Lemma 1 for $\hat{F}_1$ and $\hat{S}$, we can obtain U_1, C_1', D_1' and the eight columns of $[L_{1,1}|L_{1,2}]$. Each unknown column of $[L_{1,1}|L_{1,2}]$ can be recovered by computing $\hat{S}^{-1}(\text{the first 8 bits of } B^{-1} \cdot F(e_j))) + C_1'$ for $j \in \{1, \cdots, 16\} \backslash \{i_1, \cdots, i_8\}$. The overall complexity of the second phase is $4n^3 + 5(n^2 + n)2^m + nm^2 2^{2m} \lesssim 2^{31}$ for $n = 256$.

Hence, we can recover a pair A and B, a solution for the specialized affine equivalence problem in 2^{31} time for $n = 256$.

Extracting the Round Keys. Now, our goal is to extract the round key bits except for the first round. Note that it suffices to have the adjacent two round keys to extract the full 128-bit AES key.

Following the above strategies, we have possibly many candidates of $\widehat{B}^{(r)}$ and $\oplus_{(K^{(r+1)},\cdots,K^{(r+1)})} \circ \widehat{A}^{(r+1)}$ on consecutive rounds. However, just one representative of the solutions, say $B^{(r)}$ and $A^{(r+1)}$, would be used to recover the exact $\widehat{B}^{(r)}$ and $\oplus_{(K^{(r+1)},\cdots,K^{(r+1)})} \circ \widehat{A}^{(r+1)}$ and extract the $(r+1)$-th round key bits, with the set of self-equivalences of $\hat{S}$.

We know that the exact pair of $\oplus_{(K^{(r+1)},\cdots,K^{(r+1)})} \circ \widehat{A}^{(r+1)}$ and $\widehat{B}^{(r)}$ differs from the obtained pair $A^{(r+1)}$ and $B^{(r)}$ by the $2s$ pairs of affine self-equivalences of the S-box $\hat{S}$. Recall that

$$\widehat{A}^{(r+1)} \circ \widehat{B}^{(r)} = (\mathsf{MC} \circ \mathsf{SR}, \cdots, \mathsf{MC} \circ \mathsf{SR}).$$

Hence, to find the exact pair of $\oplus_{(K^{(r+1)},\cdots,K^{(r+1)})} \circ \widehat{A}^{(r+1)}$ and $\widehat{B}^{(r)}$, it suffices to find the set of self-equivalences of $\hat{S}$,

$$\{(a_1, b_1), \cdots, (a_s, b_s), (a_1', b_1') \cdots, (a_s', b_s')\}$$

such that

$$(\ell_{a_1}, \cdots, \ell_{a_s}) \circ L^{(r+1)} \circ M^{(r)} \circ (\ell_{b_1'}, \cdots, \ell_{b_s'}) = (\mathsf{MC} \circ \mathsf{SR}, \cdots, \mathsf{MC} \circ \mathsf{SR}),$$

where ℓ_{a_i}, ℓ_{b_j} are the linear part of the small affine maps a_i and b_j on 8 bits, respectively. So, we do the followings.

- Searching for all self-equivalences, we would find self-equivalences (a_1, b_1) and (a_1', b_1') of $\hat{S}$ such that

$$\ell_{a_1} \cdot [(1,1)\text{-th block of } L^{(r+1)} \cdot M^{(r)}] \cdot \ell_{b_1'}$$

is equal to the corresponding $(1,1)$-th block of the matrix $(\mathsf{MC} \circ \mathsf{SR}, \cdots, \mathsf{MC} \circ \mathsf{SR})$.
- If we find the right pairs (a_1, b_1) and (a_1', b_1'), then fix b_1' and then search for all self-equivalences to find (a_j, b_j) such that

$$\ell_{a_j} \cdot [(j,1)\text{-th block of } L^{(r+1)} \cdot M^{(r)}] \cdot \ell_{b_1'}$$

is equal to the corresponding $(j,1)$-th block of the matrix $(\mathsf{MC} \circ \mathsf{SR}, \cdots, \mathsf{MC} \circ \mathsf{SR})$ for all $1 \le j \le s$.
- Samely, fix a_1 and then search for all self-equivalences of $\hat{S}$ to find (a_j', b_j') such that

$$\ell_{a_1} \cdot [(1,j)\text{-th block of } L^{(r+1)} \cdot M^{(r)}] \cdot \ell_{b_j'}$$

is equal to the corresponding $(1,j)$-th block of the matrix $(\mathsf{MC} \circ \mathsf{SR}, \cdots, \mathsf{MC} \circ \mathsf{SR})$ for all $1 \le j \le s$.
- Now we have the set of $\{(a_1, b_1), \cdots, (a_s, b_s), (a_1', b_1') \cdots, (a_s', b_s')\}$ so that we can obtain

$$(a_1, \cdots, a_s) \circ A^{(r+1)} \circ B^{(r)} \circ (b_1', \cdots, b_s') = \oplus_{(K^{(r+1)}, \cdots, K^{(r+1)})} \circ \widehat{A}^{(r+1)} \circ \widehat{B}^{(r)}.$$

Since the number of self-equivalences of $\hat{S}$ is about 2^{11} by Lemma 3, the work factor to find the exact pair of $\oplus_{(K^{(r+1)}, \cdots, K^{(r+1)})} \circ \widehat{A}^{(r+1)}$ and $\widehat{B}^{(r)}$ is $[(2^{11})^2 + 2 \cdot (s-1) \cdot 2^{11}] \cdot (2 \cdot m^3) + 2 \cdot n^3 \approx 2^{32}$ for $n = 256$.

Now, we know the exact affine maps $\oplus_{(K^{(r+1)}, \cdots, K^{(r+1)})} \circ \widehat{A}^{(r+1)}$ and $\widehat{B}^{(r)}$. We can achieve the round key bits $K^{(r+1)}$ from

$$(\oplus_{(K^{(r+1)}, \cdots, K^{(r+1)})} \circ \widehat{A}^{(r+1)}) \circ \widehat{B}^{(r)} = \oplus_{(K^{(r+1)}, \cdots, K^{(r+1)})} \circ (\mathsf{MC} \circ \mathsf{SR}, \cdots, \mathsf{MC} \circ \mathsf{SR}),$$

in time complexity n^2. In fact, $(K^{(r+1)}, \cdots, K^{(r+1)})$ is the sum of the constant of $\widehat{A}^{(r+1)}$ and $L^{(r+1)} \times$ (the constant of $\widehat{B}^{(r)}$).

Thus, the total work factor of our attack for the BCH implementation to extract the round key is less than 2^{33} for $n = 256$. The complexity of our attack is stable for other parameters as in Table 2, since it mainly depends on the input size of S-boxes.

5   Conclusion

In this paper, we suggested an optimized algorithm to solve the affine equivalence problem in the case that the middle S layer is a concatenation of S-boxes and the input affine layer is structured. For the three-layer scheme $F = B \circ S \circ A$ satisfying our problem setting, one can find the secret affine layers via oracle queries to F (as black boxes) with our algorithm in low complexity. Our algorithm is more efficient than previous algorithms such as the affine equivalence algorithm [4] and SAEA [1].

The structured affine map could induce an efficient white-box implementation. In the BCH implementation [1], the structured affine mapping was used as an input encoding to resist known attacks. Baek et al. expected that their implementation is secure against a cryptanalysis using SAEA. In this paper, we showed that the overall work factor of SAEA can be significantly reduced. As a result, our cryptanalysis on the BCH implementation efficiently extracted the round key with low complexity, 2^{32}, 2^{33}, and 2^{34} for $n = 128, 256$, and 384, respectively.

Acknowledgements. This work was supported by Samsung Research Funding Center of Samsung Electronics under Project Number SRFC-TB1403-03. The authors would like to thank the anonymous reviewers of SAC 2016 for their helpful comments.

References

1. Baek, C.H., Cheon, J.H., Hong, H.: White-box AES implementation revisited. J. Commun. Netw. **18**(3), 273–287 (2016)
2. Biham, E., Anderson, R., Knudsen, L.: Serpent: a new block cipher proposal. In: Vaudenay, S. (ed.) FSE 1998. LNCS, vol. 1372, pp. 222–238. Springer, Heidelberg (1998). doi:10.1007/3-540-69710-1_15
3. Billet, O., Gilbert, H., Ech-Chatbi, C.: Cryptanalysis of a white box AES implementation. In: Handschuh, H., Hasan, M.A. (eds.) SAC 2004. LNCS, vol. 3357, pp. 227–240. Springer, Heidelberg (2004). doi:10.1007/978-3-540-30564-4_16
4. Biryukov, A., De Cannière, C., Braeken, A., Preneel, B.: A toolbox for cryptanalysis: linear and affine equivalence algorithms. In: Biham, E. (ed.) EUROCRYPT 2003. LNCS, vol. 2656, pp. 33–50. Springer, Heidelberg (2003). doi:10.1007/3-540-39200-9_3
5. Chow, S., Eisen, P., Johnson, H., van Oorschot, P.C.: A white-box DES implementation for DRM applications. In: Feigenbaum, J. (ed.) DRM 2002. LNCS, vol. 2696, pp. 1–15. Springer, Heidelberg (2003). doi:10.1007/978-3-540-44993-5_1
6. Chow, S., Eisen, P., Johnson, H., Van Oorschot, P.C.: White-box cryptography and an AES implementation. In: Nyberg, K., Heys, H. (eds.) SAC 2002. LNCS, vol. 2595, pp. 250–270. Springer, Heidelberg (2003). doi:10.1007/3-540-36492-7_17
7. Daemen, J., Rijmen, V.: AES Proposal: Rijndael (1999)
8. Dunkelman, O., Keller, N., Shamir, A.: Minimalism in cryptography: the Even-Mansour scheme revisited. In: Pointcheval, D., Johansson, T. (eds.) EUROCRYPT 2012. LNCS, vol. 7237, pp. 336–354. Springer, Heidelberg (2012). doi:10.1007/978-3-642-29011-4_21

9. Even, S., Mansour, Y.: A construction of a cipher from a single pseudorandom permutation. J. Cryptol. **10**(3), 151–161 (1997)
10. Karroumi, M.: Protecting white-box AES with dual ciphers. In: Rhee, K.-H., Nyang, D.H. (eds.) ICISC 2010. LNCS, vol. 6829, pp. 278–291. Springer, Heidelberg (2011). doi:10.1007/978-3-642-24209-0_19
11. Leander, G., Poschmann, A.: On the classification of 4 bit S-boxes. In: Carlet, C., Sunar, B. (eds.) WAIFI 2007. LNCS, vol. 4547, pp. 159–176. Springer, Heidelberg (2007). doi:10.1007/978-3-540-73074-3_13
12. Liu, F., Ji, W., Hu, L., Ding, J., Lv, S., Pyshkin, A., Weinmann, R.-P.: Analysis of the SMS4 block cipher. In: Pieprzyk, J., Ghodosi, H., Dawson, E. (eds.) ACISP 2007. LNCS, vol. 4586, pp. 158–170. Springer, Heidelberg (2007). doi:10.1007/978-3-540-73458-1_13
13. Michiels, W., Gorissen, P., Hollmann, H.D.L.: Cryptanalysis of a generic class of white-box implementations. In: Avanzi, R.M., Keliher, L., Sica, F. (eds.) SAC 2008. LNCS, vol. 5381, pp. 414–428. Springer, Heidelberg (2009). doi:10.1007/978-3-642-04159-4_27
14. De Mulder, Y., Roelse, P., Preneel, B.: Cryptanalysis of the Xiao – Lai white-box AES implementation. In: Knudsen, L.R., Wu, H. (eds.) SAC 2012. LNCS, vol. 7707, pp. 34–49. Springer, Heidelberg (2013). doi:10.1007/978-3-642-35999-6_3
15. Özbudak, F., Sınak, A., Yayla, O.: On verification of restricted extended affine equivalence of vectorial boolean functions. In: Koç, Ç.K., Mesnager, S., Savaş, E. (eds.) WAIFI 2014. LNCS, vol. 9061, pp. 137–154. Springer, Cham (2015). doi:10.1007/978-3-319-16277-5_8
16. Saarinen, M.-J.O.: Cryptographic analysis of all 4 × 4-bit S-boxes. In: Miri, A., Vaudenay, S. (eds.) SAC 2011. LNCS, vol. 7118, pp. 118–133. Springer, Heidelberg (2012). doi:10.1007/978-3-642-28496-0_7
17. Wan, Z.: Geometry of Classical Groups over Finite Fields, 2nd edn. Science Press, Beijing (2006)
18. Xiao, Y., Lai, X.: A secure implementation of white-box AES. In: Computer Science and its Applications - CSA 2009, pp. 1–6. IEEE (2009)

Estimating the Cost of Generic Quantum Pre-image Attacks on SHA-2 and SHA-3

Matthew Amy[1,4], Olivia Di Matteo[2,4], Vlad Gheorghiu[3,4($\boxtimes$)],
Michele Mosca[3,4,5,6], Alex Parent[2,4], and John Schanck[3,4]

[1] David R. Cheriton School of Computer Science, University of Waterloo,
Waterloo, Canada
[2] Department of Physics and Astronomy, University of Waterloo, Waterloo, Canada
[3] Department of Combinatorics and Optimization, University of Waterloo,
Waterloo, Canada
vgheorgh@gmail.com
[4] Institute for Quantum Computing, University of Waterloo, Waterloo, Canada
[5] Perimeter Institute for Theoretical Physics, Waterloo, Canada
[6] Canadian Institute for Advanced Research, Toronto, Canada

Abstract. We investigate the cost of Grover's quantum search algorithm when used in the context of pre-image attacks on the SHA-2 and SHA-3 families of hash functions. Our cost model assumes that the attack is run on a surface code based fault-tolerant quantum computer. Our estimates rely on a time-area metric that costs the number of logical qubits times the depth of the circuit in units of surface code cycles. As a surface code cycle involves a significant classical processing stage, our cost estimates allow for crude, but direct, comparisons of classical and quantum algorithms.

We exhibit a circuit for a pre-image attack on SHA-256 that is approximately $2^{153.8}$ surface code cycles deep and requires approximately $2^{12.6}$ logical qubits. This yields an overall cost of $2^{166.4}$ logical-qubit-cycles. Likewise we exhibit a SHA3-256 circuit that is approximately $2^{146.5}$ surface code cycles deep and requires approximately 2^{20} logical qubits for a total cost of, again, $2^{166.5}$ logical-qubit-cycles. Both attacks require on the order of 2^{128} queries in a quantum black-box model, hence our results suggest that executing these attacks may be as much as 275 billion times more expensive than one would expect from the simple query analysis.

Keywords: Post-quantum cryptography · Hash functions · Pre-image attacks · Symmetric cryptographic primitives

1 Introduction

Two quantum algorithms threaten to dramatically reduce the security of currently deployed cryptosystems: Shor's algorithm solves the abelian hidden subgroup problem in polynomial time [1,2], and Grover's algorithm provides a quadratic improvement in the number of queries needed to solve black-box search problems [3–5].

© Springer International Publishing AG 2017
R. Avanzi and H. Heys (Eds.): SAC 2016, LNCS 10532, pp. 317–337, 2017.
https://doi.org/10.1007/978-3-319-69453-5_18

Efficient quantum algorithms for integer factorization, finite field discrete logarithms, and elliptic curve discrete logarithms can all be constructed by reduction to the abelian hidden subgroup problem. As such, cryptosystems based on these problems can not be considered secure in a post-quantum environment. Diffie-Hellman key exchange, RSA encryption, and RSA signatures will all need to be replaced before quantum computers are available. Some standards bodies have already begun discussions about transitioning to new public key cryptographic primitives [6,7].

The situation is less dire for hash functions and symmetric ciphers. In a pre-quantum setting, a cryptographic primitive that relies on the hardness of inverting a one-way function is said to offer k-bit security if inverting the function is expected to take $N = 2^k$ evaluations of the function. An exhaustive search that is expected to take $O(N)$ queries with classical hardware can be performed with $\Theta(\sqrt{N})$ queries using Grover's algorithm on quantum hardware. Hence, Grover's algorithm could be said to reduce the bit-security of such primitives by half; one might say that a 128-bit pre-quantum primitive offers only 64-bit security in a post-quantum setting.

A conservative defense against quantum search is to double the security parameter (e.g. the key length of a cipher, or the output length of a hash function). However, this does not mean that the true cost of Grover's algorithm should be ignored. A cryptanalyst may want to know the cost of an attack even if it is clearly infeasible, and users of cryptosystems may want to know the minimal security parameter that provides "adequate protection" in the sense of [8–10].

In the context of pre-image search on a hash function, the cost of a pre-quantum attack is given as a number of invocations of the hash function. If one assumes that quantum queries have the same cost as classical queries, then the query model provides a reasonable comparison between quantum and classical search. However, realistic designs for large quantum computers call this assumption into question.

The main difficulty is that the coherence time of physical qubits is finite. Noise in the physical system will eventually corrupt the state of any long computation. If the physical error rate can be suppressed below some threshold, then *logical qubits* with arbitrarily long coherence times can be created using quantum error correcting codes. Preserving the state of a logical qubit is an active process that requires periodic evaluation of an error detection and correction routine. This is true even if no logical gates are performed on the logical qubit. Hence the classical processing required to evaluate a quantum circuit will grow in proportion to both the depth of the circuit and the number of logical qubits on which it acts.

We suggest that a cost model that facilitates direct comparisons of classical and quantum algorithms should take the classical computation required for quantum error correction into consideration. Clearly such estimates will be architecture dependent, and advances in quantum computing could invalidate architectural assumptions.

To better understand the impact of costing quantum error correction, we present an estimate of the cost of pre-image attacks on SHA-2 and SHA-3 assuming a quantum architecture based on the surface code with a logical Clifford+T gate set. We execute the following procedure for each hash function. First, we implement the function as a reversible circuit[1] over the Clifford+T gate set. We use a quantum circuit optimization tool, "T-par" [11], to minimize the circuit's T-count and T-depth[2]. With the optimized circuit in hand we estimate the additional overhead of fault tolerant computation. In particular, we estimate the size of the circuits needed to produce the ancillary states that are consumed by T-gates.

Grassl et al. presented a logical-layer quantum circuit for applying Grover's algorithm to AES key recovery [12]. Separately, Fowler et al. have estimated the physical resources required to implement Shor's factoring algorithm on a surface code based quantum computer [13]. Our resource estimates combine elements of both of these analyses. We focus on the number of logical qubits in the fault-tolerant circuit and the overall depth of the circuit in units of surface code cycles. While our cost model ties us to a particular quantum architecture, we segment our analysis into several layers so that the impact of a different assumptions at any particular level can be readily evaluated. We illustrate our method schematically in Fig. 2.

The structure of this article reflects our workflow. In Sect. 2 we state the problem of pre-image search using Grover's algorithm. Section 3 introduces our framework for computing costs, and Sect. 4 applies these principles to compute the intrinsic cost of performing Grover search. Sections 5 and 6 detail our procedure for generating reversible circuits for SHA-256 and SHA3-256 respectively. In Sect. 7 we embed these reversible implementations into a surface code, and estimate the required physical resources. We summarize our results and propose avenues of future research in Sect. 8.

2 Pre-image Search via Grover's Algorithm

Let $f : \{0,1\}^k \rightarrow \{0,1\}^k$ be an efficiently function. For a fixed $y \in \{0,1\}^k$, the value x such that $f(x) = y$ is called a *pre-image* of y. In the worst case, the only way to compute a pre-image of y is to systematically search the space of all inputs to f. A function that must be searched in this way is known as a *one-way function*. A one-way function that is bijective is a *one-way permutation*[3].

Given a one-way permutation f, one might ask for the most cost effective way of computing pre-images. With a classical computer one must query f on the

---

[1] Reversibility is necessary for the hash function to be useful as a subroutine in Grover search.

[2] The logical T gate is significantly more expensive than Clifford group gates on the surface code.

[3] A hash function that has been restricted to length k inputs is expected to behave roughly like a one-way permutation. The degree to which it fails to be injective should not significantly affect the expected probability of success for Grover's algorithm.

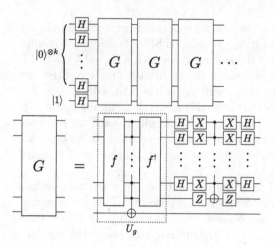

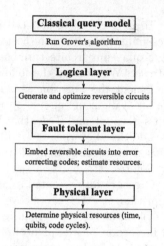

Fig. 1. Grover searching with an oracle for $f : \{0,1\}^k \to \{0,1\}^k$.

Fig. 2. Analyzing Grover's algorithm.

order of 2^k times before finding a pre-image. By contrast, a quantum computer can perform the same search with $2^{k/2}$ queries to f by using Grover's algorithm [3]. Of course, counting only the queries to f neglects the potentially significant overhead involved in executing f on a quantum computer.

Figure 1 gives a high-level description of Grover's algorithm. The algorithm makes $\lfloor \frac{\pi}{4} 2^{k/2} \rfloor$ calls to G, the *Grover iteration*. The Grover iteration has two subroutines. The first, U_g, implements the predicate $g : \{0,1\}^k \to \{0,1\}$ that maps x to 1 if and only if $f(x) = y$. Each call to U_g involves two calls to a reversible implementation of f and one call to a comparison circuit that checks whether $f(x) = y$.

The second subroutine in G implements the transformation $2|0\rangle\langle 0| - I$ and is called the *diffusion operator*. The diffusion operator is responsible for amplifying the probability that a measurement of the output register would yield x such that $f(x) = y$. As it involves only single-qubit gates and a one k-fold controlled-NOT, the cost of the diffusion operator is expected to be small compared with that of U_g.

3   A Cost Metric for Quantum Computation

Without significant future effort, the classical processing will almost certainly limit the speed of any quantum computer, particularly one with intrinsically fast quantum gates.

Fowler–Whiteside–Hollenberg [14]

The majority of the overhead for quantum computation, under realistic assumptions about quantum computing architectures, comes from error detection and correction. There are a number of error correction methods in the literature, however the most promising, from the perspective of experimental realizability, is the surface code [15].

The surface code allows for the detection and correction of errors on a two-dimensional array of nearest-neighbor coupled physical qubits. A distance d surface code encodes a single logical qubit into an $n \times n$ array of physical qubits ($n = 2d - 1$). A classical error detection algorithm must be run at regular intervals in order to track the propagation of physical qubit errors and, ultimately, to prevent logical errors. Every surface code *cycle* involves some number of one- and two-qubit physical quantum gates, physical qubit measurements, and classical processing to detect and correct errors.

The need for classical processing allows us to make a partial comparison between the cost of classical and quantum algorithms for any classical cost metric. The fact that quantum system engineers consider classical processing to be a bottleneck for quantum computation [14] suggests that an analysis of the classical processing may serve as a good proxy for an analysis of the cost of quantum computation itself.

Performing this analysis requires that we make a number of assumptions about how quantum computers will be built, not least of which is the assumption that quantum computers will require error correcting codes, and that the surface code will be the code of choice.

Assumption 1. *The resources required for any large quantum computation are well approximated by the resources required for that computation on a surface code based quantum computer.*

Fowler et al. [16] give an algorithm for the classical processing required by the surface code. A timing analysis of this algorithm was given in [14], and a parallel variant was presented in [17]. Under a number of physically motivated assumptions, the algorithm of [17] runs in constant time per round of error detection. It assumes a quantum computer architecture consisting of an $L \times L$ grid of logical qubits overlaid by a constant density mesh of classical computing units. More specifically, the proposed design involves one ASIC (application-specific integrated circuit) for each block of $C_a \times C_a$ physical qubits. These ASICs are capable of nearest-neighbor communication, and the number of rounds of communication between neighbors is bounded with respect to the error model. The number of ASICs scales linearly with the number of logical qubits, but the constant C_a, and the amount of computation each ASIC performs per time step, is independent of the number of logical qubits.

Each logical qubit is a square grid of $n \times n$ physical qubits where n depends on the length of the computation and the required level of error suppression. We are able to estimate n directly (Sect. 7). Following [14] we will assume that $C_a = n$. The number of classical computing units we estimate is therefore equal to the number of logical qubits in the circuit. Note that assuming $C_a = n$ introduces a dependence between C_a and the length of the computation, but we will ignore this detail. Since error correction must be performed on the time scale of hundreds of nanoseconds (200 ns in [15]), we do not expect it to be practical to make C_a much larger than n. Furthermore, while n depends on the length of the computation it will always lie in a fairly narrow range. A value of $n < 100$ is sufficient even for the extremely long computations we consider. The comparatively short modular exponentiation computations in [15] require $n > 31$. As long as it is not practical to take C_a much larger than 100, the assumption that $C_a = n$ will introduce only a small error in our analysis.

Assumption 2. *The classical error correction routine for the surface code on an $L \times L$ grid of logical qubits requires an $L \times L$ mesh of classical processors (i.e. $C_a = n$).*

The algorithm that each ASIC performs is non-trivial and estimating its exact runtime depends on the physical qubit error model. In [14] evidence was presented that the error correction algorithm requires $O(C_a^2)$ operations, on average, under a reasonable error model. This work considered a single qubit in isolation, and some additional overhead would be incurred by communication between ASICs. A heuristic argument is given in [17] that the communication overhead is also independent of L, i.e. that the radius of communication for each processor depends on the noise model but not on the number of logical qubits in the circuit.

Assumption 3. *Each ASIC performs a constant number of operations per surface code cycle.*

Finally we (arbitrarily) peg the cost of a surface code cycle to the cost of a hash function invocation. If we assume, as in [15], that a surface code cycle time on the order of 100 ns is achievable, then we are assuming that each logical qubit is equipped with an ASIC capable of performing several million hashes per second. This would be on the very low end of what is commercially available for Bitcoin mining today [18], however the ASICs used for Bitcoin have very large circuit footprints. One could alternatively justify this assumption by noting that typical hash functions require ≈ 10 cycles per byte on commercial desktop CPUs [19]. This translates to approximately ≈ 1000 cycles per hash function invocation. Since commercial CPUs operate at around 4 GHz, this again translates to a few million hashes per second.

Assumption 4. *The temporal cost of one surface code cycle is equal to the temporal cost of one hash function invocation.*

Combining Assumptions 1, 2, and 4 we arrive at the following metric for comparing the costs of classical and quantum computations.

Cost Metric 1. *The cost of a quantum computation involving ℓ logical qubits for a duration of σ surface code cycles is equal to the cost of classically evaluating a hash function $\ell \cdot \sigma$ times. Equivalently we will say that one logical qubit cycle is equivalent to one hash function invocation.*

We will use the term "cost" to refer either to logical qubit cycles or to hash function invocations.

4   Intrinsic Cost of Grover Search

Suppose there is polynomial overhead per Grover iteration, i.e. $\Theta(2^{k/2})$ Grover iterations cost $\approx k^v 2^{k/2}$ logical qubit cycles for some real v independent of k. Then an adversary who is willing to execute an algorithm of cost 2^C can use Grover's algorithm to search a space of k bits provided that

$$k/2 + v \log_2(k) \leq C. \tag{1}$$

We define the *overhead* of the circuit as v and the *advantage* of the circuit as k/C. Note that if we view k as a function of v and C then for any fixed v we have $\lim_{C \to \infty} k(v, C)/C = 2$, i.e. asymptotically, Grover's algorithm provides a quadratic advantage over classical search. However, here we are interested in non-asymptotic advantages.

When costing error correction, we must have $v \geq 1$ purely from the space required to represent the input. However, we should not expect the temporal cost to be independent of k. Even if the temporal cost is dominated by the k-fold controlled-NOT gate, the Clifford+T depth of the circuit will be at least $\log_2(k)$ [20]. Hence, $v \geq 1.375$ for $k \leq 256$. This still neglects some spatial overhead required for magic state distillation, but $v = 1.375$ may be used to derive strict upper bounds, in our cost model, for the advantage of Grover search.

In practice the overhead will be much greater. The AES-256 circuit from [12] has depth 130929 and requires 1336 logical qubits. This yields overhead of $v \approx 3.423$ from the reversible layer alone.

Substituting $z = \frac{k \ln 2}{2v}$, the case of equality in Eq. 1 is

$$ze^z = \frac{2^{C/v} \ln 2}{2v} \implies k(v, C) = \frac{2v}{\ln(2)} \cdot W\left(\frac{2^{C/v} \ln 2}{2v}\right) \tag{2}$$

where W is the Lambert W-function. Table 4 in Appendix A gives the advantage of quantum search as a function of its cost C and overhead v; k is computed using Eq. 2.

5   Reversible Implementation of a SHA-256 Oracle

The Secure Hash Algorithm 2 (SHA-2) [21] is a family of collision resistant cryptographic hash functions. There are a total of six functions in the SHA-2 family: SHA-224, SHA-256, SHA-384, SHA-512, SHA-512/224 and SHA-512/256. There are currently no known classical pre-image attacks against any of the SHA-2 algorithms which are faster then brute force. We will focus on SHA-256, a commonly used variant, and will assume a message size of one block (512 bits).

First the message block is stretched using Algorithm 2 and the result is stored in **W**. The internal state is then initialized using a set of constants. The round function is then run 64 times, each run using a single entry of **W** to modify the internal state. The round function for SHA-256 is shown in Algorithm 1.

Algorithm 1. SHA-256. All variables are 32-bit words.

1: **for** i=0 to 63 **do**
2:      $\Sigma_1 \leftarrow (\mathbf{E} \ggg 6) \oplus (\mathbf{E} \ggg 11) \oplus (\mathbf{E} \ggg 25)$
3:      $\mathbf{Ch} \leftarrow (\mathbf{E} \wedge \mathbf{F}) \oplus (\neg\mathbf{E} \wedge \mathbf{G})$
4:      $\mathbf{t}_1 \leftarrow \mathbf{H} + \Sigma_1 + \mathbf{Ch} + \mathbf{K}[i] + \mathbf{W}[i]$
5:      $\Sigma_0 \leftarrow (\mathbf{A} \ggg 2) \oplus (\mathbf{A} \ggg 13) \oplus (\mathbf{A} \ggg 22)$
6:      $\mathbf{Maj} \leftarrow (\mathbf{A} \wedge \mathbf{B}) \oplus (\mathbf{A} \wedge \mathbf{C}) \oplus (\mathbf{B} \wedge \mathbf{C})$
7:      $\mathbf{t}_2 \leftarrow \Sigma_0 + \mathbf{Maj}$
8:      $\mathbf{H} \leftarrow \mathbf{G}$
9:      $\mathbf{G} \leftarrow \mathbf{F}$
10:     $\mathbf{F} \leftarrow \mathbf{E}$
11:     $\mathbf{E} \leftarrow \mathbf{D} + \mathbf{t}_1$
12:     $\mathbf{D} \leftarrow \mathbf{C}$
13:     $\mathbf{C} \leftarrow \mathbf{B}$
14:     $\mathbf{B} \leftarrow \mathbf{A}$
15:     $\mathbf{A} \leftarrow \mathbf{t}_1 + \mathbf{t}_2$
16: **end for**

Algorithm 2. SHA-256 Stretch. All variables are 32-bit words.

1: **for** $i = 16$ to 63 **do**
2:      $\sigma_0 \leftarrow (\mathbf{W}_{i-15} \ggg 7) \oplus (\mathbf{W}_{i-15} \ggg 18) \oplus (\mathbf{W}_{i-15} \gg 3)$
3:      $\sigma_1 \leftarrow (\mathbf{W}_{i-2} \ggg 17) \oplus (\mathbf{W}_{i-2} \ggg 19) \oplus (\mathbf{W}_{i-2} \gg 10)$
4:      $w[i] \leftarrow \mathbf{W}_{i-16} + \sigma_0 + \mathbf{W}_{i-7} + \sigma_1$
5: **end for**

5.1   Reversible Implementation

Our implementation of the SHA-256 algorithm as a reversible circuit is similar to the one presented in [22] (with the addition of the stretching function). Each round can be performed fully reversibly (with access to the input) so no additional space is accumulated as rounds are performed. The in-place adders shown in the circuit are described in [23]. The adders perform the function $(a, b, 0) \mapsto (a, a + b, 0)$ where the 0 is a single ancilla bit used by the adder. Since the Σ blocks use only rotate and XOR operations, they are constructed using CNOT gates exclusively.

Maj is the bitwise majority function. The majority function is computed using a CNOT gate and two Toffoli gates as show in Fig. 4.

The Ch function is $ab \oplus \neg ac$ which can be rewritten as $a(b \oplus c) \oplus c$. This requires a single Toffoli gate as shown in Fig. 5.

There are a few options for constructing the round circuit. For example if space is available some of the additions can be performed in parallel, and the cleanup of the Σ, Ch, and Maj functions can be neglected if it is desirable to exchange space for a lower gate count. We select the round implementation shown in Fig. 3.

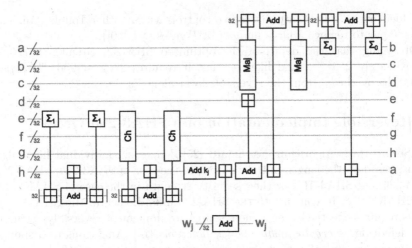

Fig. 3. SHA-256 round.

Fig. 4. Majority circuit implementation. The $a \oplus b$ line will be returned to b when the inverse circuit is applied.

Fig. 5. Ch circuit implementation. This circuit is applied bitwise to input of each Ch block.

5.2 Quantum Implementation

For the quantum implementation we converted the Toffoli-CNOT-NOT circuit (Fig. 3) discussed above into a Clifford+T circuit. To expand the Toffoli gates we used the T-depth 3 Toffoli reported in [24]. T-par was then used to optimize a single round. The results are shown in Table 1. Due to the construction of

Table 1. T-par optimization results for a single round of SHA-256, one iteration of the stretch algorithm and full SHA-256. Note that 64 iterations of the round function and 48 iterations of the stretch function are needed. The stretch function does not contribute to overall depth since it can be performed in parallel with the rounds function. No X gates are used so an X column is not included. The circuit uses 2402 total logical qubits.

	$T/T^\dagger$	$P/P^\dagger$	Z	H	CNOT	T-Depth	Depth
Round	5278	0	0	1508	6800	2262	8262
Round (Opt.)	3020	931	96	1192	63501	1100	12980
Stretch	1329	0	0	372	2064	558	2331
Stretch (Opt.)	744	279	0	372	3021	372	2907
SHA-256	401584	0	0	114368	534272	171552	528768
SHA-256 (Opt.)	228992	72976	6144	94144	4209072	70400	830720

the adders every Toffoli gate shares two controls with another Toffoli gate. This allows T-par to remove a large number of T-gates (see [20]).

Observing that the depth of the optimized SHA-256 circuit, 830720, is approximately $256^{2.458}$, and likewise that it requires $2402 \approx 256^{1.404}$ logical qubits, the overhead, from the reversible layer alone, is $v \approx 3.862$.

6   Reversible Implementation of a SHA3-256 Oracle

The Secure Hash Algorithm 3 standard [25] defines six individual hash algorithms, based on the length of their output in the case of SHA3-224, SHA3-256, SHA3-384 and SHA3-512, or their security strength in the case of SHAKE-128 and SHAKE-256. In contrast to the SHA-2 standard, each of the SHA-3 algorithms requires effectively the same resources to implement reversibly, owing to their definition as *cryptographic sponge functions* [26]. Analogous to a sponge, the sponge construction first pads the input to a multiple of the given rate constant then absorbs the padded message in chunks, applying a permutation to the state after each chunk, before "squeezing" out a hash value of desired length. Each of the SHA-3 algorithms use the same underlying permutation, but vary the chunk size, padding and output lengths.

The full SHA-3 algorithm is given in pseudocode in Algorithm 3. Each instance results from the sponge construction with permutation KECCAK-$p[1600, 24]$ described below, padding function pad10*1(x, m) which produces a length $-m$ mod x string of the form (as a regular expression) 10*1, and rate $1600 - 2k$. The algorithm first pads the input message M with the string 0110*1 to a total length some multiple of $1600 - 2k$. It then splits up this string into length $1600 - 2k$ segments and absorbs each of these segments into the current hash value S then applies the KECCAK-$p[1600, 24]$ permutation. Finally the hash value is truncated to a length k string.

The SHAKE algorithms are obtained by padding the input M with a string of the form 111110*1, but otherwise proceed identically to SHA-3.

Assuming the pre-image has length k, the padded message P has length exactly $1600 - 2k$ and hence $n = 1$ for every value of k, so Algorithm 3 reduces to one application of KECCAK-$p[1600, 24]$.

The KECCAK Permutation. The permutation underlying the sponge construction in each SHA-3 variant is an instance of a family of functions, denoted

---

Algorithm 3. SHA3-$k(M)$.

1: $P \leftarrow M01(\text{pad}10^*1(1600 - 2k, |M|))$
2: Divide P into length $1600 - 2k$ strings $P_1, P_2, \ldots, P_n$
3: $S \leftarrow 0^{1600}$
4: **for** i=1 **to** n **do**
5:         $S \leftarrow$ KECCAK-$p[1600, 24](S \oplus (P_i 0^{2k}))$
6: **end for**
7: **return** $S[0, c - 1]$

$$\theta : A'[x][y][z] \qquad \leftarrow A[x][y][z] \oplus \left(\bigoplus_{y' \in \mathbb{Z}_4} A[x-1][y'][z] \oplus A[x+1][y'][z-1] \right) \quad (3)$$

$$\rho : A'[x][y][z] \qquad \leftarrow A[x][y][z+c(x,y)] \qquad\qquad\qquad\qquad\qquad (4)$$

$$\pi : A'[y][2x+3y][z] \quad \leftarrow A[x][y][z] \qquad\qquad\qquad\qquad\qquad\qquad (5)$$

$$\chi : A'[x][y][z] \qquad \leftarrow A[x][y][z] \oplus A[x+2][y][z] \oplus A[x+1][y][z])A[x+2][y][z] \quad (6)$$

$$\iota_i : A'[x][y][z] \qquad \leftarrow A[x][y][z] \oplus RC(i)[x][y][z] \qquad\qquad\qquad\qquad (7)$$

Fig. 6. The component functions of R_i

KECCAK-$p[b, r]$. The KECCAK permutation accepts a 5 by 5 array of *lanes*, bit-strings of length $w = 2^l$ for some l where $b = 25w$, and performs r rounds of an invertible operation on this array. In particular, round i is defined, for $12 + 2l - r$ up to $12 + 2l - 1$, as $R_i = \iota_i \circ \chi \circ \pi \circ \rho \circ \theta$, where the component functions are described in Fig. 7. Note that array indices are taken mod 5 and A, A' denote the input and output arrays, respectively. The rotation array c and round constants $RC(i)$ are pre-computed values.

The KECCAK-$p[b, r]$ permutation itself is defined as the composition of all r rounds, indexed from $12 + 2l - r$ to $12 + 2l - 1$. While any parameters could potentially be used to define a hash function, only KECCAK-$p[1600, 24]$ is used in the SHA-3 standard. Note that the lane size w in this case is 64 bits (Fig. 6).

6.1   Reversible Implementation

Given the large size of the input register for the instance used in SHA3-256 (1600 bits), we sought a space-efficient implementation as opposed to a more straightforward implementation using Bennett's method [27] which would add an extra 1600 bits *per round*, to a total of 38400 bits. While this space usage could be reduced by using *pebble games* [28], the number of iterations of KECCAK-p would drastically increase. Instead, we chose to perform each round *in place* by utilizing the fact that each component function $(\theta, \rho, \pi, \chi, \iota_i)$ is invertible. The resulting circuit requires only a single temporary register the size of the input, which is returned to the all-zero state at the end of each round.

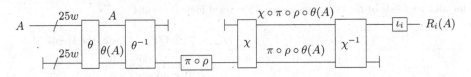

Fig. 7. Reversible circuit implementation for round i of KECCAK-p.

Figure 7 shows our circuit layout for a given round of KECCAK-$p[b, r]$. We compute $\theta(A)$ into the ancilla register by a straightforward implementation of (3), as binary addition ($\oplus$) is implemented reversibly by the CNOT gate. The implementation of $\theta^{-1} : |\psi\rangle|\theta(A)\rangle \mapsto |\psi \oplus A\rangle|\theta(A)\rangle$ is much less obvious – we adapted our implementation from the C++ library KECCAK tools [29] with minor modifications to remove temporary registers. To reduce the number of unnecessary gates, we perform the ρ and π operations "in software" rather than physically swapping bits. The χ and χ^{-1} operations are again straightforward implementations of (6) and the inverse operation from KECCAK tools, respectively, using Toffoli gates to implement the binary multiplications. Finally the addition of the round constant (ι_i) is a sequence of at most 5 NOT gates, precomputed for each of the 24 individual rounds.

As a function of the lane width w, θ comprises $275w$ CNOT gates. The inverse of θ is more difficult to assign a formula to, as it depends on some precomputed constants – in particular, $\theta-1$ is implemented using $125w \cdot j$ CNOT gates, where j is 170 for $b = 1600$. As ρ and π are implemented simply by re-indexing, they have no logical cost. We implement χ using $50w$ additions and $25w$ multiplications, giving $50w$ CNOT gates and $25w$ Toffoli gates in 5 parallel stages. Finally χ^{-1} requires $25w$ CNOT gates to copy the output back into the initial register, then $60w$ CNOT and $30w$ Toffoli gates in 6 parallel stages. As the cost of ι_i is dependent on the round, we don't give its per-round resources.

The final circuit comprises 3200 qubits, 85 NOT gates, 33269760 CNOT gates and 84480 Toffoli gates. Additionally, the Toffoli gates are arranged in 264 parallel stages.

6.2   Quantum Implementation

As with the Clifford+T implementation of SHA-256, we used the T-depth 3 Toffoli reported in [24] to expand each Toffoli gate. Since the χ (and χ^{-1}) transformation is the only non-linear operation of KECCAK-$p[1600, 24]$, we applied T-par just to the χ/χ^{-1} subcircuit to optimize T-count and depth. We used the formally verified reversible circuit compiler REVERC [30] to generate a machine readable initial circuit for χ/χ^{-1} – while the REVERC compiler performs some space optimization [31], the straightforward manner in which we implemented the circuit meant the compiled circuit coincided exactly with our analysis above. The optimized results are reported in Table 2. Note that each algorithm in the

Table 2. Clifford+T resource counts for the KECCAK-$p[1600, 24]$ components, as well as for the full oracle implementation of SHA3-256. ι gives the combined resource counts for all 24 rounds of ι_i. The circuit uses 3200 total logical qubits.

	X	$P/P^\dagger$	$T/T^\dagger$	H	CNOT	T-depth	Depth
θ	0	0	0	0	17600	0	275
θ^{-1}	0	0	0	0	1360000	0	25
χ	0	0	11200	3200	14400	15	55
χ^{-1}	0	0	13440	3840	18880	18	66
ι	85	0	0	0	0	0	24
SHA3-256	85	0	591360	168960	33269760	792	10128
SHA3-256 (Opt.)	85	46080	499200	168960	34260480	432	11040

SHA-3 family corresponds to one application of KECCAK-$p[1600, 24]$ for our purposes, so the resources are identical for any output size.

As an illustration of the overhead for SHA3-256, our reversible SHA3-256 circuit, having depth $11040 \approx 256^{1.679}$ and a logical qubit count of $3200 \approx 256^{1.455}$ yields $v \approx 3.134$ at the reversible layer.

7 Fault-Tolerant Cost

The T gate is the most expensive in terms of the resources needed for implementing a circuit fault-tolerantly in a surface code. Most known schemes implement the T gate using an auxiliary resource called a *magic state*. The latter is usually prepared in a faulty manner, and purified to the desired fidelity via a procedure called *magic state distillation*. Fault-tolerant magic state *distilleries* (circuits for performing magic state distillation) require a substantial number of logical qubits. In this section we estimate the additional resources required by distilleries in the particular case of SHA-256 and SHA3-256.

Let T_U^c denote the T-count of a circuit U (i.e., total number of logical T gates), and let T_U^d be the T-depth of the circuit. We denote by $T_U^w = T_U^c/T_U^d$ the T-width of the circuit (i.e., the number of logical T gates that can be done in parallel on average for each layer of depth). Each T gate requires one logical magic state of the form

$$|A_L\rangle := \frac{|0_L\rangle + e^{i\pi/4}|1_L\rangle}{\sqrt{2}} \tag{8}$$

for its implementation. For the entirety of U to run successfully, the magic states $|A_L\rangle$ have to be produced with an error rate no larger than $p_{out} = 1/T_U^c$.

The magic state distillation procedure is based on the following scheme. The procedure starts with a physical magic state prepared with some failure probability p_{in}. This faulty state is then *injected* into an error correcting code, and then by performing a suitable distillation procedure on the output carrier qubits of the encoded state a magic state with a smaller failure probability is distilled. If this failure probability is still larger than the desired p_{out}, the scheme uses another layer of distillation, i.e. concatenates the first layer of distillation with a second layer of distillation, and so forth. The failure probability thus decreases exponentially.

In our case, we use the Reed-Muller 15-to-1 distillation scheme introduced in [32]. Given a state injection error rate p_{in}, the output error rate after a layer of distillation can be made arbitrarily close to the ideal $p_{dist} = 35p_{in}^3$ provided we ignore the logical errors that may appear during the distillation procedure (those can be ignored if the distillation code uses logical qubits with high enough distance). As pointed out in [33] logical errors do not need to be fully eliminated. We also assume that the physical error rate per gate in the surface code, p_g, is approximately 10 times smaller than p_{in}, i.e. $p_g = p_{in}/10$, as during the state injection approximately 10 gates have to perform without a fault before error protection is available (see [13] for more details).

Algorithm 4. Estimating the required number of rounds of magic state distillation and the corresponding distances of the concatenated codes

---

1: **Input:** $\varepsilon, p_{in}, p_{out}, p_g(= p_{in}/10)$
2: $d \leftarrow$ empty list $[]$
3: $p \leftarrow p_{out}$
4: $i \leftarrow 0$
5: **repeat**
6:      $i \leftarrow i + 1$
7:      $p_i \leftarrow p$
8:      Find minimum d_i such that $192 d_i (100 p_g)^{\frac{d_i+1}{2}} < \frac{\varepsilon p_i}{1+\varepsilon}$
9:      $p \leftarrow \sqrt[3]{p_i/(35(1+\varepsilon))}$
10:     d.append(d_i)
11: **until** $p > p_{in}$
12: **Output:** $d = [d_1, \ldots, d_i]$

We define ε so that εp_{dist} represents the amount of logical error introduced, so $p_{out} = (1 + \varepsilon)p_{dist}$. In the balanced case $\varepsilon = 1$ the logical circuit introduces the same amount of errors as distillation eliminates. Algorithm 4 [33] summarizes the procedure for estimating the number of rounds of state distillation needed to achieve a given output error rate, as well as the required minimum code distances at each round. Note that d_1 represents the distance of the surface code used in the top layer of distillation (where by top we mean the initial copy of the Reed-Muller circuit), d_2 the distance of the surface code used in the next layer, and so forth.

7.1   SHA-256

The T-count of our SHA-256 circuit is $T^c_{\text{SHA-256}} = 228992$ (see Table 1), and the T-count of the k-fold controlled-NOT is $T^c_{\text{CNOT-k}} = 32k - 84$ [12]. With $k = 256$, we have $T^c_{\text{CNOT-256}} = 8108$ and the total T-count of the SHA-256 oracle U_g (of Fig. 1) is

$$T^c_{U_g} = 2T^c_{\text{SHA-256}} + T^c_{\text{CNOT-256}} = 2 \times 228992 + 8108 = 466092. \quad (9)$$

The diffusion operator consists of Clifford gates and a $(k-1)$-fold controlled-NOT, hence its T-count is $T^c_{\text{CNOT-255}} = 8076$. The T-count of one Grover iteration G is therefore

$$T^c_G = T^c_{U_g} + T^c_{\text{CNOT-255}} = 466092 + 8076 = 474168, \quad (10)$$

and the T-count for the full Grover algorithm (let us call it GA) is

$$T^c_{GA} = \lfloor \pi/4 \times 2^{128} \rfloor \times 474168 \approx 1.27 \times 10^{44}. \quad (11)$$

For this T-count the output error rate for state distillation should be no greater than $p_{out} = 1/T^c_{GA} \approx 7.89 \times 10^{-45}$. Assuming a magic state injection error rate

$p_{in} = 10^{-4}$, a per-gate error rate $p_g = 10^{-5}$, and choosing $\varepsilon = 1$, Algorithm 4 suggests 3 layers of distillation, with distances $d_1 = 33, d_2 = 13$ and $d_3 = 7$.

The bottom layer of distillation occupies the largest footprint in the surface code. Three layers of distillation consume $N_{dist} = 16 \times 15 \times 15 = 3600$ input states in the process of generating a single $|A_L\rangle$ state. These input states are encoded on a distance $d_3 = 7$ code that uses $2.5 \times 1.25 \times d_3^2 \approx 154$ physical qubits per logical qubit. The total footprint of the distillation circuit is then $N_{dist} \times 154 \approx 5.54 \times 10^5$ physical qubits. The round of distillation is completed in $10d_3 = 70$ surface code cycles.

The middle layer of distillation requires a $d_2 = 13$ surface code, for which a logical qubit takes $2.5 \times 1.25 \times d_2^2 \approx 529$ physical qubits. The total number of physical qubits required in the second round is therefore $16 \times 15 \times 529 \approx 1.27 \times 10^5$ physical qubits, with the round of distillation completed in $10d_2 = 130$ surface code cycles.

The top layer of state distillation requires a $d_1 = 33$ surface code, for which a logical qubit takes $2.5 \times 1.25 \times d_1^2 \approx 3404$ physical qubits. The total number of physical qubits required in the top layer is therefore $16 \times 3404 = 54464$ physical qubits, with the round of distillation completed in $10d_1 = 330$ surface code cycles.

Note that the physical qubits required in the bottom layer of state distillation can be reused in the middle and top layers. Therefore the total number of physical qubits required for successfully distilling one purified $|A_L\rangle$ state is $n_{dist} = 5.54 \times 10^5$. The concatenated distillation scheme is performed in $\sigma_{dist} = 70 + 130 + 330 = 530$ surface code cycles. Since the middle layer of distillation has smaller footprint than the bottom layer, distillation can potentially be pipelined to produce $\phi = (5.54 \times 10^5)/(1.27 \times 10^5) \approx 4$ magic states in parallel. Assuming, as in [13], a $t_{sc} = 200\,\text{ns}$ time for a surface code cycle, a magic state distillery can therefore produce $4 |A_L\rangle$ states every $\sigma_{dist} \times t_{sc} \approx 106\,\mu\text{s}$. Generating the requisite $T_{GA}^c = 1.27 \times 10^{44}$ magic states with a single distillery would take approximately $t_{dist} = 3.37 \times 10^{39}\,\text{s} \approx 1.06 \times 10^{32}$ years.

We now compute the distance required to embed the entire algorithm in a surface code. The number of Clifford gates in one iteration of G is roughly 8.76×10^6, so the full attack circuit performs around 2.34×10^{45} Clifford gates. The overall error rate of the circuit should therefore be less than 4.27×10^{-46}. To compute the required distance, we seek the smallest d that satisfies the inequality [14]

$$\left(\frac{p_{in}}{0.0125}\right)^{\frac{d+1}{2}} < 4.27 \times 10^{-46}, \tag{12}$$

and find this to be $d_{\text{SHA-256}} = 43$. The total number of physical qubits in the Grover portion of the circuit is then $2402 \times (2.5 \times 1.25 \times 43^2) = 1.39 \times 10^7$.

We can further estimate the number of cycles required to run the entire algorithm, σ_{GA}. Consider a single iteration of G from Fig. 1. The T-count is $T_{GA}^c = 1.27 \times 10^{44}$ and the T-depth is $T_{GA}^d = 4.79 \times 10^{43}$ for one iteration of SHA-256, yielding $T_G^w = T_{GA}^c/T_{GA}^d \approx 3$.

Our SHA-256 circuit has $N_{\text{SHA-256}} = 2402$ logical qubits. Between sequential T gates on any one qubit we will perform some number of Clifford operations. These are mostly CNOT gates, which take 2 surface code cycles, and Hadamard gates, which take a number of surface code cycles equal to the code distance [13].

Assuming the 8.76×10^6 Clifford gates in one Grover iteration are uniformly distributed among the 2402 logical qubits, then we expect to perform $8.76 \times 10^6/(2402 \times T_G^d) \approx 0.026$ Clifford operations per qubit per layer of depth. As a magic state distillery produces 4 magic states per 530 surface code cycles, we can perform a single layer of T depth every 530 surface code cycles. We thus need only a single distillery, $\Phi = 1$. On average about 2% of the Cliffords are Hadamards, and the remaining 98% are CNOTs. This implies that the expected number of surface code cycles required to implement the 0.025 average number of Clifford gates in a given layer of T depth is $2\% \times 0.025 \times 43 + 98\% \times 0.025 \times 2 = 0.071$. As this is significantly lower than 1, we conclude that performing the T gates comprises the largest part of the implementation, while the qubits performing the Clifford gates are idle most of the time. In conclusion, the total number of cycles is determined solely by magic state production, i.e.

$$\sigma_{GA} = \lfloor \pi/4 \times 2^{128} \rfloor \times 530 \times (2T_{\text{SHA-256}}^d) \approx 2^{153.8}.$$

As discussed in Sect. 3, the total cost of a quantum attack against SHA-256 equals the product of the total number of logical qubits (including the ones used for magic state distillation) and the number of code cycles, which in our case results in

$$(N_{\text{SHA-256}} + \Phi N_{dist})\sigma_{GA} = (2402 + 1 \times 3600) \times 2^{153.8} \approx 2^{166.4},$$

corresponding to an overhead factor of $v = (166.4 - 128)/\log_2(256) = 4.8$.

7.2   SHA3-256

We perform a similar analysis for SHA3-256. We have $T_{U_g}^c = 2 \times 499200 + 32 \times 256 - 84 = 1006508$, and $T_G^c = 1006508 + 32 \times 255 - 84 = 1014584$, and thus the full Grover algorithm takes T-count $T_{GA}^c = \lfloor \pi/4 \times 2^{128} \rfloor \times 1014584 \approx 2.71 \times 10^{44}$. If we choose, like in the case of SHA-256, $p_{in} = 10^{-4}, p_g = 10^{-5}$, and $\varepsilon = 1$, Algorithm 4 yields 3 layers of distillation with distances $d_1 = 33, d_2 = 13$, and $d_3 = 7$; these are identical to those of SHA-256. Thus, the distillation code requires takes 3600 logical qubits (and 5.54×10^5 physical qubits), and in 530 cycles is able to produce roughly 4 magic states.

We compute the distance required to embed the entire algorithm in a surface code. The total number of Cliffords in one iteration of G is roughly 6.90×10^7, so the total number will be around 1.84×10^{46} operations. We thus need the overall error rate to be less than 5.43×10^{-47}, which by Eq. 12 yields a distance $d_{\text{SHA3-256}} = 44$. The number of physical qubits is then 1.94×10^7.

Consider a single iteration of G from Fig. 1. $T_G^c = 1014584$ and $T_{\text{SHA3-256}}^d = 432$, which yields $T_G^w = 1014584/(2 \times 432) = 1175$. Above we figured we can compute 4 magic states in 530 code cycles. Then, to compute 1175 magic states in the same number of cycles we will need roughly $\Phi = 294$ distillation factories working in parallel to keep up. This will increase the number of physical qubits required for state distillation to 1.63×10^8. If we assume $t_{sc} = 200$ ns cycle time, generation of the full set of magic states will take 2.28×10^{37} s, or about $t_{dist} = 7.23 \times 10^{29}$ years.

Our SHA3-256 circuit uses $N_{\text{SHA3-256}} = 3200$ logical qubits. Assuming the 6.90×10^7 Clifford gates per Grover iteration are uniformly distributed among the qubits, and between the 864 sequential T gates, we must be able to implement $6.90 \times 10^7/(3200 \times 864) \approx 25$ Clifford operations per qubit per layer of T-depth. As the ratio of CNOTs to Hadamards is roughly 202 to 1, i.e. 99.5% of the Cliffords are CNOTs and only 0.5% are Hadamards, the expected number of surface code cycles required to implement the average of 25 Clifford gates in a given layer of T depth is $25 \times (0.005 \times 44 + 0.995 \times 2) \approx 55$. We have used just enough ancilla factories to implement a single layer of T-depth in 530 cycles, meaning that once again the limiting step in implementing this circuit is the production of magic states. Hence, we can compute the total number of surface code cycles required to implement SHA3-256 using just the T-depth:

$$\sigma_{GA} = \lfloor \pi/4 \times 2^{128} \rfloor \times 530 \times (2T^d_{\text{SHA3-256}}) \approx 1.22 \times 10^{44} \approx 2^{146.5}.$$

The total cost of a quantum attack against SHA3-256 is then

$$(N_{\text{SHA3-256}} + \Phi N_{dist})\sigma_{GA} = (3200 + 294 \times 3600) \times 2^{146.5} \approx 2^{166.5},$$

or an overhead of $v = (166.5 - 128)/\log_2(256) = 4.81$ (Table 3).

Table 3. Fault-tolerant resource counts for Grover search of SHA-256 and SHA3-256.

		SHA-256	SHA3-256
Grover	T-count	1.27×10^{44}	2.71×10^{44}
	T-depth	3.76×10^{43}	2.31×10^{41}
	Logical qubits	2402	3200
	Surface code distance	43	44
	Physical qubits	1.39×10^7	1.94×10^7
Distilleries	Logical qubits per distillery	3600	3600
	Number of distilleries	1	294
	Surface code distances	$\{33, 13, 7\}$	$\{33, 13, 7\}$
	Physical qubits	5.54×10^5	1.63×10^8
Total	Logical qubits	$2^{12.6}$	2^{20}
	Surface code cycles	$2^{153.8}$	$2^{146.5}$
	Total cost	$2^{166.4}$	$2^{166.5}$

8   Conclusions and Open Questions

We estimated the cost of a quantum pre-image attack on SHA-256 and SHA3-256 cryptographic hash functions via Grover's quantum searching algorithm. We constructed reversible implementations of both SHA-256 and SHA3-256 cryptographic hash functions, for which we optimized their corresponding T-count and depth. We then estimated the required physical resources needed to run a brute force Grover search on a fault-tolerant surface code based architecture.

We showed that attacking SHA-256 requires approximately $2^{153.8}$ surface code cycles and that attacking SHA3-256 requires approximately $2^{146.5}$ surface code cycles. For both SHA-256 and SHA3-256 we found that the total cost when including the classical processing increases to approximately 2^{166} basic operations.

Our estimates are by no means a lower bound, as they are based on a series of assumptions. First, we optimized our T-count by optimizing each component of the SHA oracle individually, which of course is not optimal. Dedicated optimization schemes may achieve better results. Second, we considered a surface code fault-tolerant implementation, as such a scheme looks the most promising at present. However it may be the case that other quantum error correcting schemes perform better. Finally, we considered an optimistic per-gate error rate of about 10^{-5}, which is the limit of current quantum hardware. This number will probably be improved in the future. Improving any of the issues listed above will certainly result in a better estimate and a lower number of operations, however the decrease in the number of bits of security will likely be limited.

Acknowledgments. We acknowledge support from NSERC and CIFAR. IQC and PI are supported in part by the Government of Canada and the Province of Ontario.

A   Tables

(See Tables 4 and 5).

Table 4. The advantage, k/C, of a quantum pre-image search that can be performed for cost $2^C = k^a 2^{k/2}$. The entries less than 1 correspond to a regime where quantum search is strictly worse than classical search.

$\frac{k(a,C)}{C}$	C							
	16	32	48	64	80	96	112	128
a \ 0	2.00	2.00	2.00	2.00	2.00	2.00	2.00	2.00
1	1.38	1.63	1.73	1.78	1.81	1.84	1.86	1.88
2	1.00	1.31	1.48	1.58	1.64	1.69	1.72	1.75
3	0.69	1.03	1.25	1.39	1.48	1.54	1.60	1.63
4	0.44	0.81	1.04	1.20	1.33	1.41	1.47	1.52
5	0.38	0.63	0.88	1.05	1.18	1.27	1.35	1.41

Table 5. The k for which the classical and quantum search costs are equal after accounting for the k^a overhead for quantum search.

a	1	2	3	4	5
k	5	16	30	44	59

B Parallel quantum search

Classical search is easily parallelized by distributing the 2^k bitstrings among 2^t processors. Each processor fixes the first t bits of its input to a unique string and sequentially evaluates every setting of the remaining $k - t$ bits. Since our cost metric counts only the number of invocations of g, the cost of parallel classical search is 2^k for all t. If one is more concerned with time (i.e. the number of sequential invocations) than with area, or vice versa, it may be more useful to report the cost as (T, A). Or, in this case, $(2^{k-t}, 2^t)$.

Quantum computation has a different time/area trade-off curve. In particular, parallel quantum strategies have strictly greater cost than sequential quantum search. Consider sequential quantum search with cost $C(1) = (C_T, C_A) = (k^a 2^{k/2}, k^b)$. Parallelizing this algorithm across 2^t quantum processors reduces the temporal cost per processor by a factor of $2^{t/2}$ and increases the area by a factor of 2^t. Fixing t bits of the input does not change the overhead of the Grover iteration, so the cost for parallel quantum search on 2^t processors is $C(2^t) = (2^{-t/2} C_T, 2^t C_A) = (k^a 2^{(k-t)/2}, k^b 2^t)$.

References

1. Shor, P.W.: Polynomial-time algorithms for prime factorization and discrete logarithms on a quantum computer. SIAM J. Comput. **26**(5), 1484–1509 (1997). http://link.aip.org/link/?SMJ/26/1484/1
2. Boneh, D., Lipton, R.J.: Quantum cryptanalysis of hidden linear functions. In: Coppersmith, D. (ed.) CRYPTO 1995. LNCS, vol. 963, pp. 424–437. Springer, Heidelberg (1995). doi:10.1007/3-540-44750-4_34
3. Grover, L.K.: Quantum mechanics helps in searching for a needle in a haystack. Phys. Rev. Lett. **79**, 325–328 (1997). http://link.aps.org/doi/10.1103/PhysRev Lett.79.325
4. Boyer, M., Brassard, G., Høyer, P., Tapp, A.: Tight bounds on quantum searching. Fortschritte der Physik **46**(4–5), 493–505 (1998). http://dx.doi.org/10.1002/ (SICI)1521--3978(199806)46:4/5⟨493::AID-PROP493⟩3.0.CO;2-P
5. Gilles, B., Peter, H., Michele, M., Alain, T.: Quantum amplitude amplification and estimation. Quantum Comput. Quantum Inf. **305**, 53–74 (2002). e-print arXiv:quant-ph/0005055. Lomonaco Jr., S.J. (ed.) AMS Contemporary Mathematics
6. U.S. National Security Agency: NSA Suite B Cryptography - NSA/CSS. NSA. https://www.nsa.gov/ia/programs/suiteb_cryptography/
7. Chen, L., Jordan, S., Liu, Y.K., Moody, D., Peralta, R., Perlner, R., Smith-Tone, D.: Report on post-quantum cryptography. National Institute of Standards and Technology Internal Report 8105, February 2016

8. Lenstra, A.K.: Key lengths. In: Handbook of Information Security. Wiley (2004)
9. Lenstra, A.K., Verheul, E.R.: Selecting cryptographic key sizes. J. Cryptol. **14**(4), 255–293 (2001)
10. Blaze, M., Diffie, W., Rivest, R., Schneier, B., Shimomura, T., Thompson, E., Weiner, M.: Minimal key lengths for symmetric ciphers to provide adequate commercial security. Technical report, An ad hoc group of cryptographers and computer scientists (1996)
11. Amy, M., Maslov, D., Mosca, M.: Polynomial-time t-depth optimization of Clifford+T circuits via matroid partitioning. IEEE Trans. Comput.-Aided Des. Integr. Circuits Syst. **33**(10), 1476–1489 (2014)
12. Grassl, M., Langenberg, B., Roetteler, M., Steinwandt, R.: Applying Grover's algorithm to AES: quantum resource estimates, e-print arXiv:1512.04965 [quant-ph]
13. Fowler, A.G., Mariantoni, M., Martinis, J.M., Cleland, A.N.: Surface codes: towards practical large-scale quantum computation. Phys. Rev. A **86**, 032324 (2012). http://link.aps.org/doi/10.1103/PhysRevA.86.032324
14. Fowler, A.G., Whiteside, A.C., Hollenberg, L.C.L.: Towards practical classical processing for the surface code: timing analysis. Phys. Rev. A **86**(4), 042313 (2012). http://link.aps.org/doi/10.1103/PhysRevA.86.042313
15. Fowler, A.G., Mariantoni, M., Martinis, J.M., Cleland, A.N.: Surface codes: towards practical large-scale quantum computation. Phys. Rev. A **86**(3), 032324 (2012). http://link.aps.org/doi/10.1103/PhysRevA.86.032324
16. Fowler, A.G., Whiteside, A.C., Hollenberg, L.C.L.: Towards practical classical processing for the surface code. Phys. Rev. Lett. **108**(18), 180501 (2012). http://link.aps.org/doi/10.1103/PhysRevLett.108.180501
17. Fowler, A.G.: Minimum weight perfect matching of fault-tolerant topological quantum error correction in average $O(1)$ parallel time. arXiv:1307.1740 [quant-ph], July 2013
18. Mining hardware comparison. Bitcoin Wiki, September 2015. https://en.bitcoin.it/wiki/Mining_hardware_comparison. Accessed 30 Mar 2016
19. Bernstein, D.J., Lange, T. (eds.): eBACS: ECRYPT Benchmarking of Cryptographic Systems. http://bench.cr.yp.to. Accessed 30 Mar 2016
20. Selinger, P.: Quantum circuits of T-depth one. Phys. Rev. A, **87**, 042302 (2013). http://link.aps.org/doi/10.1103/PhysRevA.87.042302
21. NIST: Federal information processing standards publication 180–2 (2002). See also the Wikipedia entry http://en.wikipedia.org/wiki/SHA-2
22. Parent, A., Roetteler, M., Svore, K.M.: Reversible circuit compilation with space constraints. arXiv preprint arXiv:1510.00377 (2015)
23. Cuccaro, S.A., Draper, T.G., Kutin, S.A., Moulton, D.P.: A new quantum ripple-carry addition circuit. arXiv preprint arXiv:quant-ph/0410184 (2004)
24. Amy, M., Maslov, D., Mosca, M., Roetteler, M.: A meet-in-the-middle algorithm for fast synthesis of depth-optimal quantum circuits. IEEE Trans. Comput.-Aided Des. Integr. Circuits Syst. **32**(6), 818–830 (2013)
25. NIST: Federal information processing standards publication 202 (2015). See also the Wikipedia entry http://en.wikipedia.org/wiki/SHA-3
26. Bertoni, G., Daemen, J., Peeters, M., Assche, G.V.: Sponge functions. In: Ecrypt Hash Workshop 2007, May 2007
27. Bennett, C.H.: Logical reversibility of computation. IBM J. Res. Dev. **17**, 525–532 (1973)
28. Bennett, C.H.: Time/space trade-offs for reversible computation. SIAM J. Comput. **18**, 766–776 (1989)

29. Bertoni, G., Daemen, J., Peeters, M., Assche, G.V.: KECCAKTOOLS software, April 2012. http://keccak.noekeon.org/
30. Amy, M., Parent, A., Roetteler, M.: REVERC software, September 2016. https://github.com/msr-quarc/ReVerC
31. Amy, M., Roetteler, M., Svore, K.M.: Verified compilation of space-efficient reversible circuits. arXiv preprint arXiv:1603.01635 (2016)
32. Bravyi, S., Kitaev, A.: Universal quantum computation with ideal Clifford gates and noisy Ancillas. Phys. Rev. A **71**, 022316 (2005). http://link.aps.org/doi/10.1103/PhysRevA.71.022316
33. Fowler, A.G., Devitt, S.J., Jones, C.: Surface code implementation of block code state distillation. Scientific Reports 3, 1939 EP - (2013). http://dx.doi.org/10.1038/srep.01939

MACs and PRNGs

Output Masking of Tweakable Even-Mansour Can Be Eliminated for Message Authentication Code

Shoichi Hirose[1], Yusuke Naito[2(✉)], and Takeshi Sugawara[2]

[1] University of Fukui, Fukui, Japan
hrs_shch@u-fukui.ac.jp
[2] Mitsubishi Electric Corporation, Kamakura, Kanagawa, Japan
Naito.Yusuke@ce.MitsubishiElectric.co.jp,
Sugawara.Takeshi@bp.MitsubishiElectric.co.jp

Abstract. In this paper we consider the simplest possible construction of PMAC from a permutation. PMAC-type schemes have been usually constructed from a tweakable blockcipher (TBC). Regarding TBCs, there have been research directions from (1) to (2) and from (1) to (3) described as follows. Here, $E_{K'} : \{0,1\}^n \to \{0,1\}^n$ is a blockcipher with a key K', $P : \{0,1\}^n \to \{0,1\}^n$ is a permutation, h_K is a hash function of a uniform and almost XOR universal family from some tweak space $\mathcal{TW}$ to $\{0,1\}^n$, $tw \in \mathcal{TW}$ is a tweak, and $x \in \{0,1\}^n$ is an input to a TBC.

(1) Liskov *et al.* proposed a blockcipher-based TBC defined as $(tw,x) \mapsto h_K(tw) \oplus E_{K'}(h_K(tw) \oplus x)$. They proved that this scheme is a secure tweakable SPRP (Strong Pseudo-Random Permutation) up to the birthday bound, assuming $E_{K'}$ is a secure SPRP.

(2) Kurosawa eliminated $E_{K'}$ from Liskov *et al.*'s TBC, where it is replaced with a permutation. This scheme is called Tweakable Even-Mansour (TEM), defined as $(tw,x) \mapsto h_K(tw) \oplus P(h_K(tw) \oplus x)$. He proved that TEM is a secure tweakable SPRP up to the birthday bound, assuming P is a public random permutation to which everyone can access. Therefore, one can construct a permutation-based PMAC by incorporating TEM with PMAC.

(3) Rogaway eliminated the output masking. The resultant scheme is called XE, defined as $(tw,x) \mapsto E_{K'}(h_K(tw) \oplus x)$. He proved that XE is a secure tweakable PRP (Pseudo-Random Permutation) up to the birthday bound, assuming $E_{K'}$ is a secure PRP. Indeed the XE-style constructions have been employed in almost all blockcipher-based PMAC-type schemes.

From these research directions, it is quite natural to consider the scheme defined as $(tw,x) \mapsto P(h_K(tw) \oplus x)$. We call the scheme XP (Xor-Permutation). From TEM to XP, the output masking is eliminated, and from XE to XP, the keyed blockcipher is eliminated, where it is replaced with a permutation. However, XP is not a secure tweakable (S)PRP, since the offset $h_K(tw)$ can be obtained by inverting the underlying permutation from the output of XP. The next question is to find a secure TBC-based cryptographic scheme, incorporating XP with it instead of a TBC. We prove that incorporating XP with PMAC and truncating some

© Springer International Publishing AG 2017
R. Avanzi and H. Heys (Eds.): SAC 2016, LNCS 10532, pp. 341–359, 2017.
https://doi.org/10.1007/978-3-319-69453-5_19

bits of XP at the last block of PMAC, the resultant scheme becomes a secure pseudorandom function up to the birthday bound, assuming P is a public random permutation.

Keywords: Tweakable Even-Mansour · XE · Output masking · PMAC · PRF · Coefficient H technique

1   Introduction

Simplification of Blockcipher Construction. Designing a cryptographic scheme with minimal components is a main theme in cryptographic research over the last thirty years. Even and Mansour [8,9] addressed this problem with respect to blockcipher design in 1991. They were motivated by DESX proposed by Rivest in 1984. DESX was designed to protect DES against exhaustive search attacks by XORing two independent prewhitening and postwhitening keys to the plaintext and ciphertext, respectively. The Even-Mansour (EM) scheme used such whitening keys but eliminated the keyed blockcipher, where it is replaced with a public random permutation. The constructions of DESX and EM are shown in Fig. 1, where $E_{K'} : \{0,1\}^n \to \{0,1\}^n$ is a blockcipher with a key K', $P : \{0,1\}^n \to \{0,1\}^n$ is a permutation, x is the input, and y is the output (Note that hereafter, we use these notations).

Dunkelman *et al.* [7] considered the minimal construction for EM. They showed that the two-key EM is not minimal in the sense that it can be further simplified into a single-key variant, i.e., $K_1 = K_2$, which has exactly the same provable security.

Tweakable Blockcipher Design. The same research direction has been done in the area of tweakable blockcipher (TBC) design. TBCs are a generalization of traditional blockciphers, which have been formalized by Liskov *et al.* [14,15]. A TBC takes, in addition to the usual inputs (message and key), an extra input for performing rekeying efficiently. This input is called tweak.

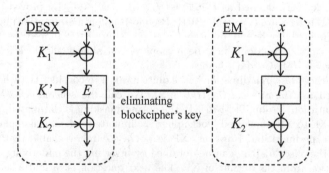

Fig. 1. Blockciphers: DESX and EM

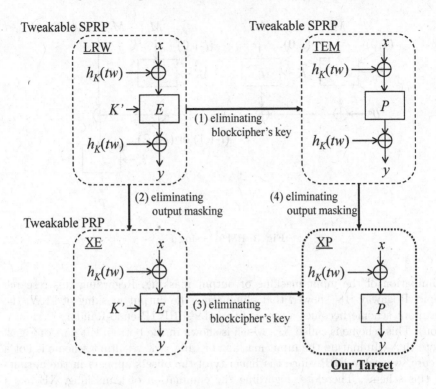

Fig. 2. Tweakable blockciphers: LRW, TEM, XE and our target: XP

Liskov *et al.* [14,15] proposed the so-called LRW that is based on a block-cipher and a uniform and almost XOR-universal (AXU) family of functions $\{h_K\}_{K \in \mathcal{K}_h}$ indexed by key set $\mathcal{K}_h$ from tweak set $\mathcal{TW}$ to $\{0,1\}^n$. In this scheme, the underlying blockcipher is sandwiched between two maskings of offset $h_K(tw)$. This construction is shown in the top left part of Fig. 2, where $tw \in \mathcal{TW}$ is a tweak and (K, K') is a key. They proved that LRW is a secure tweakable SPRP (Strong Pseudo-Random Permutation) up to the birthday bound, i.e., $2^{n/2}$ adversarial queries, assuming $E_{K'}$ is a secure SPRP.

Similar to the research direction from DESX to EM, Kurosawa [12,13] elim-inated the keyed blockcipher, where it is replaced with a permutation. This scheme is called tweakable Even-Mansour (TEM) [6], which is shown in the top right part of Fig. 2. He proved that TEM is a secure tweakable SPRP up to the birthday bound, assuming the underlying permutation is a public random per-mutation [12,13]. The research direction from LRW to TEM is shown in Fig. 2 cited as (1).

Eliminating Output Masking. In LRW, there are three components, the input masking, the output masking, and the keyed blockcipher. Therefore, besides the elimination of the keyed blockcipher, it is natural to consider the

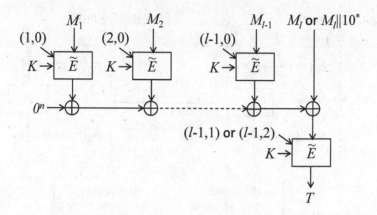

Fig. 3. PMAC using $\widetilde{E}$

elimination of the input masking or output masking. Regarding this research topic, Rogaway [18] showed that eliminating the output masking of LRW, the resultant scheme becomes a secure tweakable PRP (Pseudo-Random Permutation). This scheme is called XE, which is shown in the bottom left part of Fig. 2. Note that eliminating the input masking of LRW, the resultant scheme is not a secure tweakable PRP, since the linearity of the offsets appears in the outputs of this scheme. Therefore, regarding the elimination of a masking, XE has a minimal construction with tweakable PRP-security. The research direction from LRW to XE is shown in Fig. 2 cited as (2).

Main Question. From Fig. 2, it is quite natural to consider the research directions (3) and (4), both of which attain at the scheme shown in the bottom right part. The direction (3) eliminates the keyed blockcipher, where it is replaced with a permutation, and the direction (4) eliminates the output masking of TEM. We call the target scheme XP (Xor-Permutation). However XP is not a secure tweakable (S)PRP, since the offset can be obtained by inverting the underlying permutation from the output of XP. Therefore the next question is naturally arisen: *can we securely incorporate XP with a cryptographic scheme?*

Our Result. In this paper we consider PMAC [3,18] that is a TBC-based message authentication code (MAC) and is a main application of XE. Indeed almost all blockcipher-based PMAC-type schemes such as [3,18,21] use the XE-type schemes. The PMAC construction is shown in Fig. 3, where the tweak space is defined as $\mathcal{TW} := \mathbb{N} \times \{0,1,2\}$, $\widetilde{E}_K : \mathcal{TW} \times \{0,1\}^n \to \{0,1\}^n$ is a TBC having a key K, $M_1, M_2, \ldots, M_l$ are message blocks with $|M_i| = n$ $(i = 1, \ldots, l-1)$ and $|M_l| \leq n$, and $(1,0),(2,0),\ldots,(l-1,0),(l-1,1),(l-1,2) \in \mathcal{TW}$ are tweaks. In this construction, if $|M_l| = n$, then the tweak $(l-1,1)$ is used, and else if $|M_l| < n$, then 1 and zero strings are appended to M_l and the tweak $(l-1,2)$ is used.

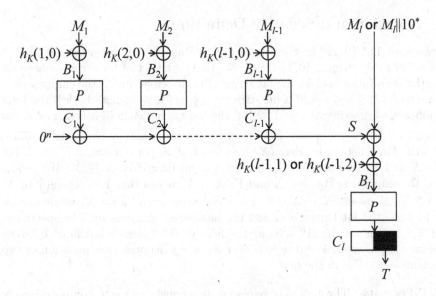

Fig. 4. PMAC using XP

Note that incorporating XP with PMAC, the resultant scheme does not become a secure PRF (Pseudo-Random Function), since the offset of XP at the last block can be obtained by inverting the underlying permutation from the output. In order to avoid this attack, we consider PMAC with output truncation. The resultant construction is shown in Fig. 4. We prove that truncating $n - t$ bits (i.e., the tag length is t bit), it is a secure PRF up to $\min\{2^{n-t}/t, 2^{n/2}\}$ permutation calls by adversarial queries, assuming the underlying permutation is a public random permutation. As a result, setting $t = n/2 - \log_2(n/2)$, it becomes a secure PRF up to the birthday bound.

In addition to the theoretical result, we discuss practical benefit of the scheme. It should be noted that the advantage of XP over TEM would be similar to that of XE over LRW, however, the advantage has not been thoroughly discussed so far. In the previous work, Rogaway mentioned that XE is slightly more efficient than LRW because some XOR instructions/gates can be reduced [18]. In this paper, we show that the benefit can be even more significant. That is because the elimination of the output masking relaxes data dependency and enables further optimization. In particular, an architectural optimization enabled by the relaxed data dependency is discussed in detail for hardware implementation.

Organization. We start by giving notations and security definitions in Sect. 2. In Sect. 3, we give the description of PMAC with XP, and the PRF-security bound. In Sect. 4, we give the security proof. Finally, we discuss the benefit of PMAC with XP over PMAC with TEM with respect to hardware implementation, and describe a future work from this paper in Sect. 5.

2   Notation and Security Definition

Notation. Let $\{0,1\}^*$ be the set of all bit strings, for an integer $n \geq 0$, $\{0,1\}^n$ the set of n-bit strings, $\{0,1\}^{\leq n} := \cup_{i=0}^{n}\{0,1\}^i$ the set of bit strings whose bit lengths are n bit or less, 0^n the bit string of n-bit zeroes, and λ the empty string. For integers $0 \leq i \leq n$ and a bit string $x \in \{0,1\}^n$, we denote by $[x]_i$ the least significant i-bit string of x and by $[x]^i$ the most significant i-bit string of x. For a finite set X, $x \xleftarrow{\$} X$ means that an element is randomly drawn from X and is set to x. For a set X, let $\mathsf{Perm}(X)$ be the set of all permutations: $X \to X$. For sets X and Y, let $\mathsf{Func}(X,Y)$ be the set of all functions: $X \to Y$. We denote by $\emptyset$ the empty set. For sets X and Y, $X \leftarrow Y$ means that Y is assigned to X, and $X \xleftarrow{\cup} Y$ means $X \leftarrow X \cup Y$. For a bit string x and a set X, we denote by $|x|$ and $|X|$ the bit length of x and the number of elements in X, respectively. Let $\mathbb{F}_{2^n}$ be the set $\{0,1\}^n$ seen as the field with 2^n elements defined by some irreducible polynomial of degree n over $\mathbb{F}_2$. $a \otimes b$ denotes multiplication of two elements $a, b \in \mathbb{F}_{2^n}$ in the field.

PRF-Security. Through this paper, a distinguisher **D** is a computationally unbounded algorithm. It is given query access to one or more oracles $\mathcal{O}$, denoted by $\mathbf{D}^{\mathcal{O}}$. Its complexity is solely measured by the number of queries made to its oracles. Let $t \geq 0$ be an integer, $\mathcal{K}$ a key set, and $\{\mathcal{F}_K^P\}_{K \in \mathcal{K}}$ a family of functions from $\{0,1\}^*$ to $\{0,1\}^t$ indexed by $\mathcal{K}$ and based on a permutation $P \in \mathsf{Perm}(\{0,1\}^n)$ for an integer $n > 0$. The security proof will be done in the ideal model, regarding the underlying permutation as a random permutation $\mathcal{P} \xleftarrow{\$} \mathsf{Perm}(\{0,1\}^n)$. We denote by $\mathcal{P}^{-1}$ its inverse.

The PRF-security of $\mathcal{F}$ is defined in terms of indistinguishability between the real world and the ideal world. In the real world, **D** has query access to $\mathcal{F}_K^{\mathcal{P}}$, $\mathcal{P}$, and $\mathcal{P}^{-1}$ for $K \xleftarrow{\$} \mathcal{K}$ and $\mathcal{P} \xleftarrow{\$} \mathsf{Perm}(\{0,1\}^n)$. In the ideal world, it has query access to a random function $\mathcal{R}$, $\mathcal{P}$, and $\mathcal{P}^{-1}$ for $\mathcal{R} \xleftarrow{\$} \mathsf{Func}(\{0,1\}^*, \{0,1\}^t)$ and $\mathcal{P} \xleftarrow{\$} \mathsf{Perm}(\{0,1\}^n)$. After interacting with oracles, **D** outputs $y \in \{0,1\}$. This event is denoted by $\mathbf{D} \Rightarrow y$. We define the advantage function as

$$\mathbf{Adv}_{\mathcal{F}}^{\mathsf{prf}}(\mathbf{D}) = \Pr[\mathbf{D}^{\mathcal{F}_K^{\mathcal{P}}, \mathcal{P}, \mathcal{P}^{-1}} \Rightarrow 1] - \Pr[\mathbf{D}^{\mathcal{R}, \mathcal{P}, \mathcal{P}^{-1}} \Rightarrow 1].$$

We call queries to $\mathcal{F}_K^{\mathcal{P}}/\mathcal{R}$ "online queries" and queries to $(\mathcal{P}, \mathcal{P}^{-1})$ "offline queries." Through this paper, without loss of generality, we assume that **D** is deterministic, and makes no repeated query which includes offline queries such that once **D** obtains (X, Y) such that $Y = \mathcal{P}(X)$, it does not ask X nor Y as an offline query.

3   PMAC with XP and the PRF-Security

In this section, first we give the description of PMAC using XP. This construction is denoted by PMAC_XP. Secondly, we define a uniform and almost XOR-universal (AXU) family of hash functions whose properties will be used in the security proof of PMAC_XP. Thirdly, we give the PRF-security bound of PMAC_XP.

3.1   PMAC_XP

Fix integers $n \geq 1$ and $p \geq 0$. Let $TW := \mathbb{Z}_p \times \{0,1,2\}$ be the set of tweaks, and $\mathcal{K}$ the set of keys. Let $\mathcal{H} = \{h_K\}_{K \in \mathcal{K}}$ be a family of functions from TW to $\{0,1\}^n$ indexed by $\mathcal{K}$. By PMAC_XP_K^P, we simply denote the function with PMAC_XP, which uses a permutation $P \in \text{Perm}(\{0,1\}^n)$ as the underlying permutation and a key $K \in \mathcal{K}$. For a message $M \in \{0,1\}^{\leq n \times (p+1)}$, the response $\text{PMAC\_XP}_K^P(M) = T$ is defined as follows. Here, $M \| 10^*$ means that first 1 is appended to M, and if the bit length of $M \| 1$ is not a multiple of n bits, then a sequence of the minimum number of zeros is appended to $M \| 1$ so that it becomes a multiple of n bits.

1. If $|M| \mod n = 0$ and $M \neq \lambda$ then $M' \leftarrow M$; Else $M' \leftarrow M \| 10^*$
2. Partition M' into n-bit blocks $M_1, \ldots, M_l$
3. $S \leftarrow 0^n$; For $i = 1, \ldots, l-1$ do $B_i \leftarrow M_i \oplus h_K(i, 0)$; $C_i \leftarrow P(B_i)$; $S \leftarrow S \oplus C_i$
4. If $|M| \mod n = 0$ and $M \neq \lambda$ then $B_l \leftarrow S \oplus M_l \oplus h_K(l-1, 1)$; $C_l \leftarrow P(B_l)$;
   Else $B_l \leftarrow S \oplus M_l \oplus h_K(l-1, 2)$; $C_l \leftarrow P(B_l)$
5. $T \leftarrow [C_l]_t$; Return T

3.2   Uniform AXU Hash Function Family

We will need the following property of the family of functions $\mathcal{H}$.

Definition 1. *Let* $\mathcal{H} = \{h_K\}_{K \in \mathcal{K}}$ *be a family of functions from (some set)* TW *to* $\{0,1\}^n$ *indexed by a set of keys* $\mathcal{K}$. $\mathcal{H}$ *is said to be uniform if for any* $tw \in TW$ *and* $y \in \{0,1\}^n$,

$$\Pr[K \xleftarrow{\$} \mathcal{K} : h_K(tw) = y] = 2^{-n}.$$

$\mathcal{H}$ *is said to be* ε-*almost XOR-universal* (ε-*AXU*) *if for all distinct* $tw, tw' \in TW$ *and all* $y \in \{0,1\}^n$,

$$\Pr[K \xleftarrow{\$} \mathcal{K} : h_K(tw) \oplus h_K(tw') = y] \leq \varepsilon.$$

$\mathcal{H}$ *is simply said to be XOR-universal* (*XU*) *if it is* 2^{-n}-*AXU*.

Example 1. Let $\mathcal{K} := \mathbb{F}_{2^n}$. For any integer $\ell \geq 1$, we define a family of functions $\mathcal{H} = \{h_K\}_{K \in \mathcal{K}}$ from $(\mathbb{F}_{2^n})^\ell$ to $\mathbb{F}_{2^n}$ as $h_K(X_1, \ldots, X_\ell) = \sum_{i=1}^{\ell} K^i \otimes X_i$. Then $\mathcal{H}$ is $\ell \cdot 2^{-n}$-AXU [20]. Note, however, that $\mathcal{H}$ is not uniform since the tweak with $(X_1, \ldots, X_\ell) = (0, \ldots, 0)$ is always mapped to 0 independently of the key. This can be handled by forbidding the all-zero input, in which case the family is not exactly uniform, but rather $\ell \cdot 2^{-n}$-almost uniform, i.e., for $\forall (X_1, \ldots, X_\ell) \in (\mathbb{F}_{2^n})^\ell \backslash \{(0, \ldots, 0)\}$ and $y \in \{0,1\}^n$, $\Pr[K \xleftarrow{\$} \mathcal{K} : h_K(X_1, \ldots, X_\ell) = y] \leq \ell \cdot 2^{-n}$.

Example 2. Rogaway [18] proposed a powering-up method that offers a uniform AXU function family, e.g., $TW := \{1, \ldots, 2^{n/2}\} \times \{0, \ldots, 10\} \times \{0, \ldots, 10\}$, $\mathcal{K} := \mathbb{F}_{2^n}$, and the family of functions $\mathcal{H} = \{h_K\}_{K \in \mathcal{K}}$ is defined as $h_K(i, j, r) :=$

$2^i 3^j 7^r \otimes K$.[1] The multiplications by 2, 3 and 7 can be calculated by XOR and shift operations. Using this method, the offsets of PMAC_XP can be efficiently calculated, e.g., $2 \otimes K, 2^2 \otimes K, 2^3 \otimes K, \ldots, 2^{l-1} \otimes K, 2^{l-1}3 \otimes K$, etc.

Example 3. Several methods for efficiently implementing the uniform AXU function family have been proposed such as Gray-code-based schemes [3,11] and LFSR-based schemes [4,10].

3.3   The PRF-Security of PMAC_XP

The PRF-security bound of PMAC_XP is given in the following, where the underlying permutation is modeled as a random permutation. The proof will be provided in the next section.

Theorem 1. *Let $\mathcal{H}$ be a uniform ε-AXU family of functions from $\mathcal{TW}$ to $\{0,1\}^n$. Let $\mathbf{D}$ be a distinguisher which makes Q offline queries and q online queries. Let σ be the total number of the blocks in q online queries, namely, $\sigma = \sum_{i=1}^{q} l_i$, where l_i is the number of the blocks l at the i-th online query. Then, we have*

$$\mathbf{Adv}^{\mathrm{prf}}_{\mathrm{PMAC\_XP}}(\mathbf{D}) \leq 0.5\sigma^2\varepsilon + \frac{0.5q^2 + 0.5\sigma^2 + 2\sigma Q}{2^n} + \frac{tQ}{2^{n-t}} + \left(\frac{8eqQ}{2^n}\right)^{1/2},$$

where $e = 2.71828\cdots$ is Napier's constant.

Theorem 1 can be interpreted as implying that setting $t = n/2$, PMAC_XP becomes a secure PRF as long as σ and Q do not exceed roughly $2^{n/2}$ and $2^{n/2}/n$, respectively, and setting $t \leq n/2 - \log_2(n/2)$, it becomes a secure PRF as long as both of σ and Q do not exceed roughly $2^{n/2}$, assuming $\varepsilon = 2^{-n}$.

Remark 1. The requirement for a secure MAC is unforgeability under chosen-message attacks, i.e., in the PMAC_XP case, for a key $K \xleftarrow{\$} \mathcal{K}$ and a random permutation $\mathcal{P} \xleftarrow{\$} \mathrm{Perm}(\{0,1\}^n)$, an attacker $\mathcal{A}$, given adaptive access to $\mathrm{PMAC\_XP}^{\mathcal{P}}_K$, cannot output a valid pair (M,T) such that $\mathrm{PMAC\_XP}^{\mathcal{P}}_K(M) = T$ and M was not a query to $\mathrm{PMAC\_XP}^{\mathcal{P}}_K$. We note that if $\mathbf{Adv}^{\mathrm{prf}}_{\mathrm{PMAC\_XP}}(\mathbf{D}) \leq \epsilon$ for any distinguisher $\mathbf{D}$ making $q+q_V$ online queries, then no attacker making q queries to $\mathrm{PMAC\_XP}^{\mathcal{P}}_K$ can output such a valid pair (M,T) within q_V attempts, except with probability at most $\epsilon+q_V/2^t$. Combining Theorem 1 with this fact and setting $t = n/2$, PMAC_XP is secure in the sense of unforgeability as long as q_V, σ and Q do not exceed roughly $2^{n/2}$, $2^{n/2}$ and $2^{n/2}/n$, respectively, and setting $t = n/2 - \log_2(n/2)$, PMAC_XP is secure in the sense of unforgeability as long as q_V, σ and Q do not exceed roughly $2^{n/2}/n$, $2^{n/2}$, and $2^{n/2}$, respectively.

---

[1] The original method by Rogaway is based on a blockcipher $E_{K'} : \{0,1\}^n \rightarrow \{0,1\}^n$, where K is defined as $K := E_{K'}(0^n)$.

4   Proof of Theorem 1

We give the PRF-security bound of $\text{PMAC\_XP}_K^{\mathcal{P}}$ via three games denoted by Game 1, Game 2, and Game 3. For $i \in \{1, 2, 3\}$, let $G_i := (L_i, \mathcal{P}, \mathcal{P}^{-1})$ be oracles to which $\mathbf{D}$ has query access in Game i. Note that in each game, $\mathcal{P}$ is independently drawn as $\mathcal{P} \xleftarrow{\$} \text{Perm}(\{0, 1\}^n)$. Let $L_1 := \text{PMAC\_XP}_K^{\mathcal{P}}$ and $L_3 := \mathcal{R}$. L_2 will be defined in Subsect. 4.1. Then,

$$\mathbf{Adv}_{\text{PMAC\_XP}}^{\text{prf}}(\mathbf{D}) = \sum_{i=1}^{2} \left(\Pr[\mathbf{D}^{G_i} \Rightarrow 1] - \Pr[\mathbf{D}^{G_{i+1}} \Rightarrow 1] \right). \tag{1}$$

Hereafter, we upper-bound $\Pr[\mathbf{D}^{G_i} \Rightarrow 1] - \Pr[\mathbf{D}^{G_{i+1}} \Rightarrow 1]$ for $i \in \{1, 2\}$. In this evaluation, we use the following notations. For $\alpha \in \{1, \dots, Q\}$, we denote the α-th offline query by X^α, resp. Y^α, and the response by Y^α, resp. X^α, where $Y^\alpha = \mathcal{P}(X^\alpha)$, resp. $X^\alpha = \mathcal{P}^{-1}(Y^\alpha)$. For $\alpha \in \{1, \dots, q\}$, we denote the α-th online query by M^α and the response by T^α. We also use superscripts for internal values defined by online queries except for their block length l, e.g., B_1^1, C_1^1, S_1^1, etc. For $\alpha \in \{1, \dots, q\}$, we denote the block length l at the α-th online query by l_α.

4.1   Upper-Bound of $\Pr[\mathbf{D}^{G_1} \Rightarrow 1] - \Pr[\mathbf{D}^{G_2} \Rightarrow 1]$

We start by defining L_2. Let $\mathcal{G} \xleftarrow{\$} \text{Func}(\mathcal{TW} \times \{0, 1\}^n, \{0, 1\}^n)$ be a random function (Note that $\mathcal{TW} = \mathbb{Z}_p \times \{0, 1, 2\}$). For an online query $M \in \{0, 1\}^{\leq n \times (p+1)}$, the response $L_2(M) = T$ is defined as follows.

1. If $|M| \mod n = 0$ and $M \neq \lambda$ then $M' \leftarrow M$; Else $M' \leftarrow M \| 10^*$
2. Partition M' into n-bit blocks $M_1, \dots, M_l$
3. $S \leftarrow 0^n$; For $i = 1, \dots, l-1$ do $C_i \leftarrow \mathcal{G}((i, 0), M_i)$; $S \leftarrow S \oplus C_i$
4. If $|M| \mod n = 0$ and $M \neq \lambda$ then $C_l \leftarrow \mathcal{G}((l-1, 1), S \oplus M_l)$; Else $C_l \leftarrow \mathcal{G}((l-1, 2), S \oplus M_l)$
5. $T \leftarrow [C_l]_t$; Return T

Independently of the above procedure, a key is defined as $K \xleftarrow{\$} \mathcal{K}$ before $\mathbf{D}$ makes the first query. In addition, at the α-th online query for $\alpha \in \{1, \dots, q\}$, B_i^α for $i \in \{1, \dots, l_\alpha - 1\}$ is defined as $B_i^\alpha := M_i^\alpha \oplus h_K(i, 0)$, and $B_{l_\alpha}^\alpha$ is defined as $B_{l_\alpha}^\alpha := S^\alpha \oplus M_l^\alpha \oplus h_K(l_\alpha - 1, 1)$ if $|M^\alpha| \mod n = 0$ and $M^\alpha \neq \lambda$; $B_{l_\alpha}^\alpha := S^\alpha \oplus M_{l_\alpha}^\alpha \oplus h_K(l_\alpha - 1, 2)$ otherwise. These values are defined after $\mathbf{D}$ ends all queries. Note that these values do not affect the procedure of L_2 but are used in the following proof.

Transcript

Since $\mathbf{D}$ is deterministic, its output is determined by the transcript, which is a list of values obtained by its queries. Let T_1 be the transcript in Game 1 obtained by sampling $K \xleftarrow{\$} \mathcal{K}$ and $\mathcal{P} \xleftarrow{\$} \mathsf{Perm}(\{0,1\}^n)$. Let T_2 be the transcript in Game 2 obtained by sampling $K \xleftarrow{\$} \mathcal{K}$, $\mathcal{P} \xleftarrow{\$} \mathsf{Perm}(\{0,1\}^n)$ and $\mathcal{G} \xleftarrow{\$} \mathsf{Func}(\mathcal{TW} \times \{0,1\}^n, \{0,1\}^n)$. We call a transcript τ *valid* if an interaction with their oracles could render this transcript, namely, $\Pr[\mathsf{T}_i = \tau] > 0$ for $i \in \{1, 2\}$. Then $\Pr[\mathbf{D}^{G_1} \Rightarrow 1] - \Pr[\mathbf{D}^{G_2} \Rightarrow 1]$ is upper bounded by the statistical distance of transcripts, i.e.,

$$\Pr[\mathbf{D}^{G_1} \Rightarrow 1] - \Pr[\mathbf{D}^{G_2} \Rightarrow 1] \leq \mathsf{SD}(\mathsf{T}_1, \mathsf{T}_2) = \frac{1}{2} \sum_\tau |\Pr[\mathsf{T}_1 = \tau] - \Pr[\mathsf{T}_2 = \tau]|,$$

where the sum is over all valid transcripts.

Regarding $\mathbf{D}$'s transcript, it obtains the following sets of query-response pairs after queries: $\tau_L := \{(M^1, T^1), \ldots, (M^q, T^q)\}$ the set of query-response pairs defined by online queries; $\tau_P := \{(X^1, Y^1), \ldots, (X^Q, Y^Q)\}$ the set of query-response pairs defined by offline queries. In addition to these sets, we define a set $\tau_{i,j}$ for $(i, j) \in \mathcal{TW}$, which keeps all pairs for (B_i, C_i) defined by using the tweak (i, j). Formally, $\tau_{i,j} := \cup_{\alpha=1}^q \{(B_{i,j}^\alpha, C_{i,j}^\alpha)\}$, where $\{B_{i,j}^\alpha, C_{i,j}^\alpha\} := \{(B_i^\alpha, C_i^\alpha)\}$ if $tw_i^\alpha = (i, j)$, and $\{B_{i,j}^\alpha, C_{i,j}^\alpha\} := \emptyset$ otherwise, where for $\alpha \in \{1, \ldots, q\}$ and $i \in \{1, \ldots, l_\alpha\}$, let tw_i^α denotes the tweak used at the i-th block of the α-th online query, i.e., if $i \neq l_\alpha$, then $tw_i^\alpha := (i, 0)$; if $i = l_\alpha \wedge |M^\alpha| \mod n = 0 \wedge M \neq \lambda$, then $tw_{l_\alpha}^\alpha := (l_\alpha -, 1)$; if $i = l_\alpha \wedge (|M^\alpha| \mod n \neq 0 \vee M = \lambda)$, then $tw_{l_\alpha}^\alpha := (l_\alpha - 1, 2)$. This proof permits $\mathbf{D}$ to obtain these sets and a secret key K after $\mathbf{D}$'s interaction but before it outputs a result. Let $\tau_{\mathrm{prim}} := \cup_{(i,j)\in\mathcal{TW}} \tau_{i,j}$. Consequently, $\mathbf{D}$'s transcript is summarized as $\tau := \{\tau_L, \tau_P, \tau_{\mathrm{prim}}, K\}$.

Coefficient H Technique

We upper-bound the statistical distance by using the coefficient H technique [5, 17], in which valid transcripts are partitioned into good transcripts $\mathcal{T}_{\mathrm{good}}$ and bad transcripts $\mathcal{T}_{\mathrm{bad}}$, and then the following lemma holds.

Lemma 1 (Coefficient H Technique). *Let* $0 \leq \delta \leq 1$ *be such that for all* $\tau \in \mathcal{T}_{\mathrm{good}}$, $\frac{\Pr[\mathsf{T}_1=\tau]}{\Pr[\mathsf{T}_2=\tau]} \geq 1 - \delta$. *Then,* $\mathsf{SD}(\mathsf{T}_1, \mathsf{T}_2) \leq \delta + \Pr[\mathsf{T}_2 \in \mathcal{T}_{\mathrm{bad}}]$.

The proof of the lemma is given in [5]. Hence, we can upper-bound $\Pr[\mathbf{D}^{G_1} \Rightarrow 1] - \Pr[\mathbf{D}^{G_2} \Rightarrow 1]$ by defining good and bad transcripts and by evaluating δ and $\Pr[\mathsf{T}_2 \in \mathcal{T}_{\mathrm{bad}}]$.

Good and Bad Transcripts

In order to define $\mathcal{T}_{\mathrm{good}}$ and $\mathcal{T}_{\mathrm{bad}}$, we need to recall the difference between Game 1 and Game 2. In Game 1, the i-th output block at the α-th query is defined as $\mathcal{P}(h_K(tw_i^\alpha) \oplus M_i^\alpha)$ $(i \neq l_\alpha)$; $\mathcal{P}(h_K(tw_i^\alpha) \oplus M_i^\alpha \oplus S^\alpha)$ $(i = l_\alpha)$. On the other hand, in Game 2, it is defined as $\mathcal{G}(tw_i^\alpha, M_i^\alpha)$ $(i \neq l_\alpha)$; $\mathcal{G}(tw_i^\alpha, M_i^\alpha \oplus S^\alpha)$ $(i = l_\alpha)$, which implies that in Game 2, (1) the output block is defined independently of all offline queries, since $\mathcal{G}$ is defined independently of $\mathcal{P}$, and (2) the output block is also defined independently of the other blocks with distinct inputs. Therefore, if Game 1 and Game 2 are indistinguishable, these independences should also hold in Game 1. Thus we consider four conditions $\mathsf{hit}_{BB}, \mathsf{hit}_{CC}, \mathsf{hit}_{BX}$, and hit_{CY}. hit_{BB} and hit_{CC} come from the independence (2), where hit_{BB} considers an input collision by online queries (collision in B-values) and hit_{CC} considers an output collision by online queries (collision in C-values). hit_{BX} and hit_{CY} come from the independence (1), where hit_{BX} considers an input collision between online and offline queries (collision between B-values and X-values) and hit_{CY} considers an output collision between online and offline queries (collision between C-values and Y-values). Formally, these conditions are defined as follows.

$$\mathsf{hit}_{BB} \Leftrightarrow \exists \alpha, \beta \in \{1, \dots, q\}, i \in \{1, \dots, l_\alpha\}, j \in \{1, \dots, l_\beta\}$$
$$\text{s.t. } B_i^\alpha = B_j^\beta \wedge tw_i^\alpha \neq tw_j^\beta$$
$$\mathsf{hit}_{CC} \Leftrightarrow \exists \alpha, \beta \in \{1, \dots, q\}, i \in \{1, \dots, l_\alpha\}, j \in \{1, \dots, l_\beta\}$$
$$\text{s.t. } C_i^\alpha = C_j^\beta \wedge (tw_i^\alpha, B_i^\alpha) \neq (tw_j^\beta, B_j^\beta)$$
$$\mathsf{hit}_{BX} \Leftrightarrow \exists \alpha \in \{1, \dots, q\}, i \in \{1, \dots, l_\alpha\}, \beta \in \{1, \dots, Q\} \text{ s.t. } B_i^\alpha = X^\beta$$
$$\mathsf{hit}_{CY} \Leftrightarrow \exists \alpha \in \{1, \dots, q\}, i \in \{1, \dots, l_\alpha\}, \beta \in \{1, \dots, Q\} \text{ s.t. } C_i^\alpha = Y^\beta$$

We define $\mathcal{T}_{\mathsf{bad}}$ by the set of transcripts which satisfy one of the above conditions, and $\mathcal{T}_{\mathsf{good}}$ by the set of transcripts which do not satisfy any of the above conditions.

Upper-Bound of $\Pr[\mathsf{T}_2 \in \mathcal{T}_{\mathsf{bad}}]$

We first note that the following inequation holds.

$$\Pr[\mathsf{T}_2 \in \mathcal{T}_{\mathsf{bad}}] \leq \Pr[\mathsf{hit}_{BB} \vee \mathsf{hit}_{CC} \vee \mathsf{hit}_{BX} \vee \mathsf{hit}_{CY}]$$
$$\leq \Pr[\mathsf{hit}_{BB}] + \Pr[\mathsf{hit}_{CC}] + \Pr[\mathsf{hit}_{BX}] + \Pr[\mathsf{hit}_{CY}]. \qquad (2)$$

Hereafter, we upper bound $\Pr[\mathsf{hit}_{BB}]$, $\Pr[\mathsf{hit}_{CC}]$, $\Pr[\mathsf{hit}_{BX}]$, and $\Pr[\mathsf{hit}_{CY}]$. Note that these events are considered within Game 2, and L_2 is independent of K.

Upper-Bound of $\Pr[\mathsf{hit}_{BB}]$. First we fix $\alpha, \beta \in \{1, \dots, q\}, i \in \{1, \dots, l_\alpha\}, j \in \{1, \dots, l_\beta\}$ such that $tw_i^\alpha \neq tw_j^\beta$, and evaluate the probability that hit_{BB} is satisfied due to B_i^α and B_j^β, that is, $B_i^\alpha = B_j^\beta$. Here, B_i^α is of the form $h_K(tw_i^\alpha) \oplus D_i^\alpha$, and B_j^β is of the form $h_K(tw_j^\beta) \oplus D_j^\beta$, where for $\gamma \in \{\alpha, \beta\}$, $D_i^\gamma := M_i^\gamma$ for $i \in \{1, \dots, l_\gamma - 1\}$ and $D_{l_\gamma}^\gamma := M_{l_\gamma}^\gamma \oplus S^\gamma$. Thus,

$$B_i^\alpha = B_j^\beta \Leftrightarrow h_K(tw_i^\alpha) \oplus D_i^\alpha = h_K(tw_j^\beta) \oplus D_j^\beta$$
$$\Leftrightarrow h_K(tw_i^\alpha) \oplus h_K(tw_j^\beta) = D_i^\alpha \oplus D_j^\beta$$

By the ε-AXU property of h, the probability that the above equation holds is at most ε. Finally, we have $\Pr[\mathsf{hit}_{BB}] \leq \binom{\sigma}{2} \times \varepsilon \leq 0.5\sigma^2\varepsilon$.

Upper-Bound of $\Pr[\text{hit}_{CC}]$. First we fix $\alpha, \beta \in \{1, \ldots, q\}, i \in \{1, \ldots, l_\alpha\}, j \in \{1, \ldots, l_\beta\}$ such that $(tw_i^\alpha, B_i^\alpha) \neq (tw_j^\beta, B_j^\beta)$, and evaluate the probability that hit_{CC} is satisfied due to C_i^α and C_j^β, that is, $C_i^\alpha = C_j^\beta$. By $(tw_i^\alpha, B_i^\alpha) \neq (tw_j^\beta, B_j^\beta)$, $(tw_i^\alpha, M_i^\alpha) \neq (tw_j^\beta, M_j^\beta)$ holds, and thereby, C_i^α and C_j^β are independently drawn. As a result, the probability that $C_i^\alpha = C_j^\beta$ is at most $1/2^n$. Finally, we have $\Pr[\text{hit}_{CC}] \leq \binom{\sigma}{2} \times \frac{1}{2^n} \leq \frac{0.5\sigma^2}{2^n}$.

Upper-Bound of $\Pr[\text{hit}_{BX}]$. First we fix $\alpha \in \{1, \ldots, q\}, i \in \{1, \ldots, l_\alpha\}$ and $\beta \in \{1, \ldots, Q\}$, and evaluate the probability that hit_{BX} is satisfied due to B_i^α and X^β, that is, $B_i^\alpha = X^\beta$. Here, B_i^α is of the form $h_K(tw_i^\alpha) \oplus D_i^\alpha$, where $D_i^\alpha := M_i^\alpha$ with $i \in \{1, \ldots, l_\gamma - 1\}$ and $D_{l_\alpha}^\alpha := M_{l_\alpha}^\alpha \oplus S^\alpha$. Thus,

$$B_i^\alpha = X^\beta \Leftrightarrow h_K(tw_i^\alpha) \oplus D_i^\alpha = X^\beta$$
$$\Leftrightarrow h_K(tw_i^\alpha) = D_i^\alpha \oplus X^\beta$$

By the property of uniformity of h, the probability that the above equation holds is at most $1/2^n$. Finally, we have $\Pr[\text{hit}_{BX}] \leq \frac{\sigma Q}{2^n}$.

Upper-Bound of $\Pr[\text{hit}_{CY}]$. Let ρ be any threshold, and $\mathcal{C}_{\text{last}} := \left\{ C_{l_\alpha}^\alpha : (\alpha \in \{1, \ldots, q\}) \wedge (\forall \beta \in \{1, \ldots, \alpha - 1\} : (tw_{l_\alpha}^\alpha, S^\alpha \oplus M_{l_\alpha}^\alpha) \neq (tw_{l_\beta}^\beta, S^\beta \oplus M_{l_\beta}^\beta)) \right\}$ the set of outputs of $\mathcal{G}$ at the last block with distinct inputs. Thus, all elements in $\mathcal{C}_{\text{last}}$ are independently drawn. Then we define the following condition.

$$\text{mcoll}(\rho) \Leftrightarrow \exists C^{(1)}, C^{(2)}, \ldots, C^{(\rho)} \in \mathcal{C}_{\text{last}} \text{ s.t. } [C^{(1)}]_t = [C^{(2)}]_t = \ldots = [C^{(\rho)}]_t$$

Then we have

$$\Pr[\text{hit}_{CY}] \leq \Pr[\text{mcoll}(\rho)] + \Pr[\text{hit}_{CY} | \neg \text{mcoll}(\rho)].$$

Hereafter, we evaluate the probabilities $\Pr[\text{mcoll}(\rho)]$ and $\Pr[\text{hit}_{CY} | \neg \text{mcoll}(\rho)]$.

- We evaluate $\Pr[\text{mcoll}(\rho)]$. Fixing $C \in \mathcal{C}_{\text{last}}$ and $C' \in \{0,1\}^t$, since $[C]_t$ is randomly drawn from $\{0,1\}^t$, the probability that $[C]_t = C'$ holds is at most $1/2^t$. Since all elements in $\mathcal{C}_{\text{last}}$ are independently drawn and $|\mathcal{C}_{\text{last}}| \leq q$, we have

$$\Pr[\text{mcoll}(\rho)] \leq 2^t \cdot \binom{q}{\rho} \cdot \left(\frac{1}{2^t}\right)^\rho \leq 2^t \cdot \left(\frac{eq}{\rho 2^t}\right)^\rho,$$

  using Stirling's approximation ($x! \geq (x/e)^x$ for any x).
- We evaluate $\Pr[\text{hit}_{CY} | \neg \text{mcoll}(\rho)]$. We assume that $\text{mcoll}(\rho)$ is not satisfied. First we fix $\beta \in \{1, \ldots, Q\}$, and evaluate the probability that hit_{CY} is satisfied due to Y^β, that is, $\exists \alpha \in \{1, \ldots, q\}, i \in \{1, \ldots, l_\alpha\}$ s.t. $C_i^\alpha = Y^\beta$.
  - We consider the case where $\exists \alpha \in \{1, \ldots, q\}, i \in \{1, \ldots, l_\alpha - 1\}$ s.t. $C_i^\alpha = Y^\beta$. Since C_i^α is randomly drawn from $\{0,1\}^n$, the probability that hit_{CY} is satisfied in this case is at most $\sigma/2^n$.

- Next we consider the case where $\exists \alpha \in \{1, \ldots, q\}$ s.t. $C_{l_\alpha}^\alpha = Y^\beta$. By $\neg \mathrm{mcoll}(\rho)$, the number of outputs at the last block whose inputs are distinct and whose last t bits equal $[Y^\beta]_t$ is at most ρ. Thus the probability that hit_{CY} is satisfied in this case is at most $\rho/2^{n-t}$.

We thus have

$$\Pr[\mathrm{hit}_{CY}|\neg \mathrm{mcoll}(\rho)] \leq \sum_{\beta=1}^{Q} \left(\frac{\rho}{2^{n-t}} + \frac{\sigma}{2^n} \right) = \frac{\rho Q}{2^{n-t}} + \frac{\sigma Q}{2^n}.$$

Finally, we have

$$\Pr[\mathrm{hit}_{CY}] \leq \frac{\rho Q}{2^{n-t}} + \frac{\sigma Q}{2^n} + 2^t \left(\frac{eq}{\rho 2^t} \right)^\rho.$$

and then putting $\rho = \max \left\{ t, \left(\frac{2eq2^{n-t}}{Q2^t} \right)^{1/2} \right\}$ gives

$$\Pr[\mathrm{hit}_{CY}] \leq \max \left\{ t, \left(\frac{2eq2^{n-t}}{Q2^t} \right)^{1/2} \right\} \times \frac{Q}{2^{n-t}} + \frac{\sigma Q}{2^n}$$

$$+ 2^t \left(\frac{eq}{\max \left\{ t, \left(\frac{2eq2^{n-t}}{Q2^t} \right)^{1/2} \right\} 2^t} \right)^{\max \left\{ t, \left(\frac{2eq2^{n-t}}{Q2^t} \right)^{1/2} \right\}}$$

$$\leq \frac{tQ}{2^{n-t}} + \left(\frac{2eqQ}{2^n} \right)^{1/2} + \frac{\sigma Q}{2^n} + 2^t \left(\frac{eq}{\left(\frac{2eq2^{n-t}}{Q2^t} \right)^{1/2} 2^t} \right)^t$$

$$\leq \frac{tQ}{2^{n-t}} + \left(\frac{2eqQ}{2^n} \right)^{1/2} + \frac{\sigma Q}{2^n} + \left(\frac{2eqQ}{2^n} \right)^{t/2}$$

$$\leq \frac{tQ}{2^{n-t}} + \frac{\sigma Q}{2^n} + \left(\frac{8eqQ}{2^n} \right)^{1/2}.$$

Upper-Bound of $\Pr[\mathsf{T}_2 \in \mathcal{T}_{\mathrm{bad}}]$. Finally, we have

$$\Pr[\mathsf{T}_2 \in \mathcal{T}_{\mathrm{bad}}] \leq 0.5\sigma^2 \varepsilon + \frac{0.5\sigma^2 + 2\sigma Q}{2^n} + \frac{tQ}{2^{n-t}} + \left(\frac{8eqQ}{2^n} \right)^{1/2}.$$

Upper-Bound of δ

Let $\tau \in \mathcal{T}_{\mathrm{good}}$. Let all_i be the set of all oracles in Game i for $i = 1, 2$. Let $\mathrm{comp}_i(\tau)$ be the set of oracles compatible with τ in Game i for $i = 1, 2$. Then

$$\Pr[\mathsf{T}_1 = \tau] = \frac{|\mathrm{comp}_1(\tau)|}{|\mathrm{all}_1|} \quad \text{and} \quad \Pr[\mathsf{T}_2 = \tau] = \frac{|\mathrm{comp}_2(\tau)|}{|\mathrm{all}_2|}.$$

Hereafter, we evaluate $|\text{all}_1|$, $|\text{all}_2|$, $|\text{comp}_1(\tau)|$ and $|\text{comp}_2(\tau)|$. In this evaluation, we use the following notations: $N_{tw} := |\mathcal{TW}|$, $N_K := |\mathcal{K}|$, $\gamma_{i,j} := |\tau_{i,j}|$ for $(i,j) \in \mathcal{TW}$, $\gamma_{\mathcal{P}} := |\tau_{\mathcal{P}}|$, and $\gamma := \gamma_{\mathcal{P}} + \sum_{(i,j)\in\mathcal{TW}} \gamma_{i,j}$.

Firstly, we evaluate $|\text{all}_1|$. By $K \in \mathcal{K}$ and $\mathcal{P} \in \text{Perm}(\{0,1\}^n)$, $|\text{all}_1| = N_K \cdot 2^n!$.

Secondly, we evaluate $|\text{all}_2|$. By $K \in \mathcal{K}$, $\mathcal{P} \in \text{Perm}(\{0,1\}^n)$, and $\mathcal{G} \in \text{Func}(\mathcal{TW} \times \{0,1\}^n, \{0,1\}^n)$, $|\text{all}_2| = N_K \cdot 2^n! \cdot (2^n)^{N_{tw} \cdot 2^n}$.

Thirdly, we evaluate $|\text{comp}_1(\tau)|$. $\tau_{i,j}$'s with $(i,j) \in \mathcal{TW}$ and $\tau_{\mathcal{P}}$ are defined so that they do not overlap each other. In this case, the number of input-output pairs of $\mathcal{P}$ defined by online and offline queries is γ, and thereby $|\text{comp}_1(\tau)| = (2^n - \gamma)!$.

Fourthly, we evaluate $|\text{comp}_2(\tau)|$. In this case, the number of input-output pairs of $\mathcal{P}$ defined by online queries is $\gamma_{\mathcal{P}}$, the number of input-output pairs of $\mathcal{G}$ with tweak (i,j) defined by offline queries is $\gamma_{i,j}$, and thereby

$$|\text{comp}_2(\tau)| = (2^n - \gamma_{\mathcal{P}})! \cdot \prod_{(i,j)\in\mathcal{TW}} (2^n)^{2^n - \gamma_{i,j}} = (2^n - \gamma_{\mathcal{P}})! \cdot (2^n)^{N_{tw} \cdot 2^n - \gamma + \gamma_{\mathcal{P}}}.$$

Finally, we have

$$\frac{\Pr[T_1 = \tau]}{\Pr[T_2 = \tau]} = \frac{|\text{comp}_1(\tau)|}{|\text{all}_1|} \times \frac{|\text{all}_2|}{|\text{comp}_2(\tau)|} = \frac{(2^n - \gamma)!}{N_K \cdot 2^n!} \times \frac{N_K \cdot 2^n! \cdot (2^n)^{N_{tw} \cdot 2^n}}{(2^n - \gamma_{\mathcal{P}})! \cdot (2^n)^{N_{tw} \cdot 2^n - \gamma + \gamma_{\mathcal{P}}}}$$

$$= \frac{(2^n)^{\gamma} \cdot (2^n - \gamma)!}{(2^n)^{\gamma_{\mathcal{P}}} \cdot (2^n - \gamma_{\mathcal{P}})!} \geq 1,$$

and thereby $\delta = 0$.

Upper-Bound of $\Pr[\mathbf{D}^{G_1} \Rightarrow 1] - \Pr[\mathbf{D}^{G_2} \Rightarrow 1]$

We apply the above results to Lemma 1, and thereby

$$\Pr[\mathbf{D}^{G_1} \Rightarrow 1] - \Pr[\mathbf{D}^{G_2} \Rightarrow 1] \leq 0.5\sigma^2\varepsilon + \frac{0.5\sigma^2 + 2\sigma Q}{2^n} + \frac{tQ}{2^{n-t}} + \left(\frac{8eqQ}{2^n}\right)^{1/2}. \tag{3}$$

4.2   Upper-Bound of $\Pr[\mathbf{D}^{G_2} \Rightarrow 1] - \Pr[\mathbf{D}^{G_3} \Rightarrow 1]$

First we prove the following lemma.

Lemma 2. G_2 and G_3 are indistinguishable unless the following condition holds in Game 2.

$$\text{coll} \Leftrightarrow \exists \alpha, \beta \in \{1, \dots, q\} \text{ s.t. } \alpha \neq \beta \wedge tw_{l_\alpha}^\alpha = tw_{l_\beta}^\beta \wedge M_{l_\alpha}^\alpha \oplus S^\alpha = M_{l_\beta}^\beta \oplus S^\beta.$$

Proof. We assume that coll does not hold. Then for any $\alpha, \beta \in \{1, \dots, q\}$ with $\alpha \neq \beta$, $(tw_{l_\alpha}^\alpha, M_{l_\alpha}^\alpha \oplus S^\alpha) \neq (tw_{l_\beta}^\beta, M_{l_\beta}^\beta \oplus S^\beta)$ holds, where for $\gamma \in \{\alpha, \beta\}$, $(tw_{l_\gamma}^\gamma, M_{l_\gamma}^\gamma \oplus S^\gamma)$ is the input to $\mathcal{G}$ at the last block of the γ-th online query. Hence, the outputs $C_{l_\alpha}^\alpha$ and $C_{l_\beta}^\beta$ are independently and randomly drawn from $\{0,1\}^n$. As a result, all outputs of L_2: $T^1, \dots, T^q$ are independently and randomly drawn from $\{0,1\}^n$, and thereby G_2 and G_3 are indistinguishable. $\square$

By the above lemma, $\Pr[\mathbf{D}^{G_2} \Rightarrow 1|\neg\mathsf{coll}] = \Pr[\mathbf{D}^{G_3} \Rightarrow 1]$ holds, and thereby

$$\Pr[\mathbf{D}^{G_2} \Rightarrow 1] - \Pr[\mathbf{D}^{G_3} \Rightarrow 1] \leq \Pr[\mathsf{coll}].$$

The detail for deriving the upper-bound is given in Appendix A. Hereafter, we upper bound $\Pr[\mathsf{coll}]$.

First we fix $\alpha, \beta \in \{1, \ldots, q\}$ such that $\alpha \neq \beta \wedge tw_{l_\alpha}^\alpha = tw_{l_\beta}^\beta$, and upper bound the probability that $M_{l_\alpha}^\alpha \oplus S^\alpha = M_{l_\beta}^\beta \oplus S^\beta$ holds. Note that

$$M_{l_\alpha}^\alpha \oplus S^\alpha = M_{l_\beta}^\beta \oplus S^\beta \Leftrightarrow M_{l_\alpha}^\alpha \oplus \left(\bigoplus_{i=1}^{l_\alpha - 1} C_i^\alpha\right) = M_{l_\beta}^\beta \oplus \left(\bigoplus_{i=1}^{l_\beta - 1} C_i^\beta\right)$$

$$\Leftrightarrow M_{l_\alpha}^\alpha \oplus M_{l_\beta}^\beta = \left(\bigoplus_{i=1}^{l_\alpha - 1} C_i^\alpha\right) \oplus \left(\bigoplus_{i=1}^{l_\beta - 1\cdot} C_i^\beta\right). \quad (4)$$

Let $\mathsf{twM}^{\alpha,\beta} := \cup_{\gamma \in \{\alpha,\beta\}} \cup_{i=1}^{l_\gamma - 1} \{(tw_i^\gamma, M_i^\gamma)\}$ be the set of the inputs to $\mathcal{G}$ at the α-th and β-th online queries except for the last blocks (thus $tw_i^\gamma = (i, 0)$), and $\mathsf{C}^{\alpha,\beta} := \cup_{\gamma \in \{\alpha,\beta\}} \cup_{i=1}^{l_\gamma - 1} \{C_i^\gamma\}$ the set of the corresponding outputs of $\mathcal{G}$.

- If $M_{l_\alpha}^\alpha = M_{l_\beta}^\beta$, then since $\mathbf{D}$ makes no repeated query, $M_1^\alpha \| \cdots \| M_{l_\alpha - 1}^\alpha \neq M_1^\beta \| \cdots \| M_{l_\beta - 1}^\beta$ holds. Note that $l_\alpha = l_\beta$ by $tw_{l_\alpha}^\alpha = tw_{l_\beta}^\beta$. Then there exist $\gamma \in \{\alpha, \beta\}, i \in \{1, \ldots, l_\gamma - 1\}$ such that $(tw_i^\gamma, M_i^\gamma) \notin \mathsf{twM}^{\alpha,\beta}\backslash\{(tw_i^\gamma, M_i^\gamma)\}$. Therefore, C_i^γ is drawn independently of $\mathsf{C}^{\alpha,\beta}\backslash\{C_i^\gamma\}$. Hence, the probability that the equation of (4) holds is at most $1/2^n$.
- If $M_{l_\alpha}^\alpha \neq M_{l_\beta}^\beta$, then in order to satisfy the equation of (4), $S^\alpha \neq S^\beta$ should hold. $S^\alpha \neq S^\beta$ implies that there exists $\gamma \in \{\alpha, \beta\}, i \in \{1, \ldots, l_\gamma - 1\}$ such that $C_i^\gamma \notin \mathsf{C}^{\alpha,\beta}\backslash\{C_i^\gamma\}$, namely, C_i is drawn independently of $\mathsf{C}^{\alpha,\beta}\backslash\{C_i^\gamma\}$. Hence, the probability that the equation of (4) holds is at most $1/2^n$.

By the above analysis, we have

$$\Pr[\mathbf{D}^{G_2} \Rightarrow 1] - \Pr[\mathbf{D}^{G_3} \Rightarrow 1] \leq \Pr[\mathsf{coll}] \leq \binom{q}{2} \times \frac{1}{2^n} \leq \frac{0.5q^2}{2^n}. \quad (5)$$

4.3 Upper-Bound of $\mathbf{Adv}_{\mathsf{PMAC\_XP}}^{\mathsf{prf}}(D)$

Finally, putting upper-bounds (3) and (5) into (1) gives

$$\mathbf{Adv}_{\mathsf{PMAC\_XP}}^{\mathsf{prf}}(D) \leq 0.5\sigma^2\varepsilon + \frac{0.5q^2 + 0.5\sigma^2 + 2\sigma Q}{2^n} + \frac{tQ}{2^{n-t}} + \left(\frac{8eqQ}{2^n}\right)^{1/2}.$$

5   Discussion

5.1   Benefit in Hardware Implementation

In this section, we discuss benefits of PMAC with XP over the previous permutation-based PMAC, PMAC with TEM, with respect to hardware implementation. In summary, there are two main advantages. Firstly and apparently, some XOR gates can be reduced. Secondly, architectural optimization is enabled because data dependency is relaxed.

The reduction of XOR gates is discussed. Two common architectures shown in Fig. 5 are considered. Figure 5(a) and (b) are ones for the TEM- and XP-based schemes, respectively. Both are based on a reference circuit found in the specification document of Minalpher [19]. Note that offsets are assumed to be serially updated for each permutation call (e.g., $2 \otimes K, 2^2 \otimes K, 2^3 \otimes K, \ldots$ in a field for $K \in \{0,1\}^n$) in the component labeled "offset update". If a single XOR gate is approximated by 2 [GE], then XOR gates corresponding to $2N$ [GE] are reduced by the XP-based scheme in which N is the datapath width. In addition, some accompanying gates can be reduced. In case of Minalpher, the permutation can be called without any masking and thus there are accompanying AND gates for disabling the XORs (see Fig. 5). The AND gates can also be reduced in the XP-based scheme.

Secondly, and more importantly, data dependency is relaxed by eliminating the output masking. The architectures in Fig. 5 are considered again. Data dependency is discussed using concurrency diagrams shown in Fig. 6. In the diagrams, horizontal axes represent time and squares represent that the resource is occupied. In the TEM-based scheme in Fig. 6(a), the offset should be maintained until the end of permutation and thus "offset update" should be suspended while permutation is being executed. Similarly, permutation should be suspended while the offset is being updated. In the XP-based scheme in Fig. 6(b), on the other hand, permutation and "offset update" can be processed simultaneously because the data dependency is relaxed by eliminating the output masking. The property brings advantages both in throughput and circuit area: (i) throughput is

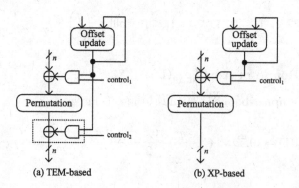

(a) TEM-based   (b) XP-based

Fig. 5. Common circuit architectures for (a) TEM-based and (b) XP-based schemes

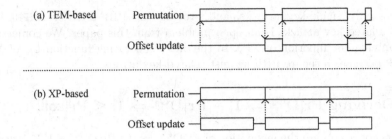

Fig. 6. Concurrency diagram: occupancy of resources in circuits for (a) TEM-based and (b) XP-based schemes

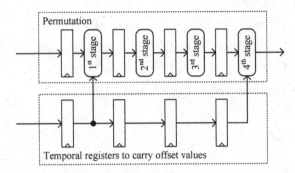

Fig. 7. A 4-stage pipeline architecture and pipeline registers for delaying tweak values

improved because the idling period is removed and (ii) a smaller implementation can be used for "offset update" because that is no longer a bottleneck for speed.

An alternative way to tackle the data dependency issue in the TEM-based scheme is to use a temporal register that stores the offset until the end of permutation. In that case, the XP-based scheme is advantageous in the sense that the temporal register can be removed. The reduction is effective because register is relatively expensive (i.e., a 1-bit register is approximated by 5–7 [GE]). The capability to reduce temporal register is more important in a pipelined implementation. Figure 7 shows a 4-stage pipelined implementation for the TEM-based scheme. In order to carry offset values to the last stage, multiple of temporal registers (i.e. pipeline registers) are needed. The registers can be simply eliminated in the XP-based scheme.

5.2   Open Problem

Recently, Mennink [16] discussed the tweakable SPRP-security (Strong Pseudo-Random Permutation security) of TEM against related-key attacks. He defined a family of functions calculating an offset from a tweak. He call the TEM construction with this function XPX. He showed sufficient conditions for functions to become secure tweakable SPRPs against related-key attacks within the framework of Bellare and Kohno [2] and Albrecht *et al.* [1]. Note that our result considers only single-key attacks, and proving the PRF-security of PMAC_XP against related-key attacks is an open problem from this paper. We conjecture that applying the function of XPX to the offset generating function h_K of XP, PMAC_XP becomes a secure PRF against related-key attacks.

A Deriving $\Pr[\mathbf{D}^{G_2} \Rightarrow 1] - \Pr[\mathbf{D}^{G_3} \Rightarrow 1] \leq \Pr[\mathsf{coll}]$

We show how to obtain the inequation of $\Pr[\mathbf{D}^{G_2} \Rightarrow 1] - \Pr[\mathbf{D}^{G_3} \Rightarrow 1] \leq \Pr[\mathsf{coll}]$, assuming $\Pr[\mathbf{D}^{G_2} \Rightarrow 1|\neg\mathsf{coll}] = \Pr[\mathbf{D}^{G_3} \Rightarrow 1]$.

$$\Pr[\mathbf{D}^{G_2} \Rightarrow 1] - \Pr[\mathbf{D}^{G_3} \Rightarrow 1]$$
$$= \Pr[\mathbf{D}^{G_2} \Rightarrow 1 \wedge \mathsf{coll}] + \Pr[\mathbf{D}^{G_2} \Rightarrow 1 \wedge \neg\mathsf{coll}] - \Pr[\mathbf{D}^{G_3} \Rightarrow 1]$$
$$= \Pr[\mathbf{D}^{G_2} \Rightarrow 1|\mathsf{coll}] \cdot \Pr[\mathsf{coll}] + \Pr[\mathbf{D}^{G_2} \Rightarrow 1|\neg\mathsf{coll}] \cdot \Pr[\neg\mathsf{coll}] - \Pr[\mathbf{D}^{G_3} \Rightarrow 1]$$
$$= \Pr[\mathbf{D}^{G_2} \Rightarrow 1|\mathsf{coll}] \cdot \Pr[\mathsf{coll}] + \Pr[\mathbf{D}^{G_3} \Rightarrow 1] \cdot (\Pr[\neg\mathsf{coll}] - 1)$$
$$= \Pr[\mathbf{D}^{G_2} \Rightarrow 1|\mathsf{coll}] \cdot \Pr[\mathsf{coll}] - \Pr[\mathbf{D}^{G_3} \Rightarrow 1] \cdot \Pr[\mathsf{coll}]$$
$$\leq \Pr[\mathsf{coll}].$$

References

1. Albrecht, M.R., Farshim, P., Paterson, K.G., Watson, G.J.: On cipher-dependent related-key attacks in the ideal-cipher model. In: Joux, A. (ed.) FSE 2011. LNCS, vol. 6733, pp. 128–145. Springer, Heidelberg (2011). doi:10.1007/978-3-642-21702-9_8
2. Bellare, M., Kohno, T.: A theoretical treatment of related-key attacks: RKA-PRPs, RKA-PRFs, and applications. In: Biham, E. (ed.) EUROCRYPT 2003. LNCS, vol. 2656, pp. 491–506. Springer, Heidelberg (2003). doi:10.1007/3-540-39200-9_31
3. Black, J., Rogaway, P.: A block-cipher mode of operation for parallelizable message authentication. In: Knudsen, L.R. (ed.) EUROCRYPT 2002. LNCS, vol. 2332, pp. 384–397. Springer, Heidelberg (2002). doi:10.1007/3-540-46035-7_25
4. Chakraborty, D., Sarkar, P.: A general construction of tweakable block ciphers and different modes of operations. IEEE Trans. Inf. Theory **54**(5), 1991–2006 (2008)
5. Chen, S., Steinberger, J.: Tight security bounds for key-alternating ciphers. In: Nguyen, P.Q., Oswald, E. (eds.) EUROCRYPT 2014. LNCS, vol. 8441, pp. 327–350. Springer, Heidelberg (2014). doi:10.1007/978-3-642-55220-5_19
6. Cogliati, B., Lampe, R., Seurin, Y.: Tweaking even-mansour ciphers. In: Gennaro, R., Robshaw, M. (eds.) CRYPTO 2015. LNCS, vol. 9215, pp. 189–208. Springer, Heidelberg (2015). doi:10.1007/978-3-662-47989-6_9
7. Dunkelman, O., Keller, N., Shamir, A.: Minimalism in cryptography: the even-mansour scheme revisited. In: Pointcheval, D., Johansson, T. (eds.) EUROCRYPT 2012. LNCS, vol. 7237, pp. 336–354. Springer, Heidelberg (2012). doi:10.1007/978-3-642-29011-4_21
8. Even, S., Mansour, Y.: A construction of a cipher from a single pseudorandom permutation. In: Imai, H., Rivest, R.L., Matsumoto, T. (eds.) ASIACRYPT 1991. LNCS, vol. 739, pp. 210–224. Springer, Heidelberg (1993). doi:10.1007/3-540-57332-1_17

9. Even, S., Mansour, Y.: A construction of a cipher from a single pseudorandom permutation. J. Cryptol. **10**(3), 151–162 (1997)
10. Granger, R., Jovanovic, P., Mennink, B., Neves, S.: Improved masking for tweakable blockciphers with applications to authenticated encryption. In: Fischlin, M., Coron, J.-S. (eds.) EUROCRYPT 2016. LNCS, vol. 9665, pp. 263–293. Springer, Heidelberg (2016). doi:10.1007/978-3-662-49890-3_11
11. Krovetz, T., Rogaway, P.: The software performance of authenticated-encryption modes. In: Joux, A. (ed.) FSE 2011. LNCS, vol. 6733, pp. 306–327. Springer, Heidelberg (2011). doi:10.1007/978-3-642-21702-9_18
12. Kurosawa, K.: Power of a public random permutation and its application to authenticated-encryption. IACR Cryptology ePrint Archive, 2002:127 (2002)
13. Kurosawa, K.: Power of a public random permutation and its application to authenticated encryption. IEEE Trans. Inf. Theory **56**(10), 5366–5374 (2010)
14. Liskov, M., Rivest, R.L., Wagner, D.: Tweakable block ciphers. In: Yung, M. (ed.) CRYPTO 2002. LNCS, vol. 2442, pp. 31–46. Springer, Heidelberg (2002). doi:10.1007/3-540-45708-9_3
15. Liskov, M., Rivest, R.L., Wagner, D.: Tweakable block ciphers. J. Cryptol. **24**(3), 588–613 (2011)
16. Mennink, B.: XPX: generalized tweakable even-mansour with improved security guarantees. IACR Cryptology ePrint Archive, 2015:476 (2015)
17. Patarin, J.: The "Coefficients H" technique. In: Avanzi, R.M., Keliher, L., Sica, F. (eds.) SAC 2008. LNCS, vol. 5381, pp. 328–345. Springer, Heidelberg (2009). doi:10.1007/978-3-642-04159-4_21
18. Rogaway, P.: Efficient instantiations of tweakable blockciphers and refinements to modes OCB and PMAC. In: Lee, P.J. (ed.) ASIACRYPT 2004. LNCS, vol. 3329, pp. 16–31. Springer, Heidelberg (2004). doi:10.1007/978-3-540-30539-2_2
19. Sasaki, Y., Todo, Y., Aoki, K., Naito, Y., Sugawara, T., Murakami, Y., Matsui, M., Hirose, S.: Minalpher v1.1. CAESAR Round 2 submission (2015)
20. Shoup, V.: On fast and provably secure message authentication based on universal hashing. In: Koblitz, N. (ed.) CRYPTO 1996. LNCS, vol. 1109, pp. 313–328. Springer, Heidelberg (1996). doi:10.1007/3-540-68697-5_24
21. Yasuda, K.: A new variant of PMAC: beyond the birthday bound. In: Rogaway, P. (ed.) CRYPTO 2011. LNCS, vol. 6841, pp. 596–609. Springer, Heidelberg (2011). doi:10.1007/978-3-642-22792-9_34

Improved Algebraic MACs and Practical Keyed-Verification Anonymous Credentials

Amira Barki[1,2], Solenn Brunet[1,3(✉)], Nicolas Desmoulins[1], and Jacques Traoré[1]

[1] Orange Labs, Caen, France
solenn.brunet@orange.com
[2] Sorbonne Universités, Université de Technologie de Compiègne (UTC), CNRS, UMR 7253 Heudiasyc, Compiègne, France
[3] Université de Rennes 1, Rennes, France

Abstract. Until quite recently, anonymous credentials systems were based on public key primitives. A new approach, that relies on algebraic Message Authentication Codes (MACs) in prime-order groups, has recently been introduced by Chase *et al.* at CCS 2014. They proposed two anonymous credentials systems referred to as *"Keyed-Verification Anonymous Credentials (KVAC)"* as they require the verifier to know the issuer secret key. Unfortunately, both systems presentation proof, for n unrevealed attributes, is of complexity $O(n)$ in the number of group elements. In this paper, we propose a new KVAC system that provides multi-show unlinkability of credentials and is of complexity $O(1)$ in the number of group elements while being almost as efficient as Microsoft's U-Prove anonymous credentials system (which does not ensure multi-show unlinkability) and many times faster than IBM's Idemix. Our credentials are constructed based on a new algebraic MAC scheme which is of independent interest. Through slight modifications on the verifier side, our KVAC system, which is proven secure in the random oracle model, can be easily turned into a public-key credentials system. By implementing it on a standard NFC SIM card, we show its efficiency and suitability for real-world use cases and constrained devices. In particular, a credential presentation, with 3 attributes, can be performed in only 88 ms.

Keywords: MAC · Anonymous credentials · Attributes · Multi-show unlinkability · Java Card SIM card

1 Introduction

Introduced by Chaum [16], anonymous credentials systems allow users to obtain a credential from an issuer and then, later, prove possession of this credential, in an unlinkable way, without revealing any additional information. This primitive has attracted a lot of interest as it complies with data minimization principles that consist in preventing the disclosure of irrelevant and unnecessary information. Typically, an anonymous credentials system is expected to enable users to

© Springer International Publishing AG 2017
R. Avanzi and H. Heys (Eds.): SAC 2016, LNCS 10532, pp. 360–380, 2017.
https://doi.org/10.1007/978-3-319-69453-5_20

reveal a subset of the attributes associated to their credentials while keeping the remaining ones hidden. For instance, a service provider only needs to know that a user is legitimate (*i.e.* he is authorized to access the service) without yet being able to collect personal information such as address, date of birth, etc.

Potential applications of anonymous credentials systems are numerous, including e-cash [21], public transport and electronic toll (for authentication purposes). In such applications, the system efficiency is an important requirement especially as it is usually deployed on constrained environments like smart cards.

Furthermore, it is desirable that an anonymous credentials system provides multi-show unlinkability. That is, one can prove possession of the same credential several times in an unlinkable manner. However, when it is intended for eCash applications, credentials should be one-show to prevent double spending of coins.

Related Work. One of the most prevalent anonymous credentials systems is Microsoft's U-Prove [23,24] which is based on a blind signature scheme due to Brands [6]. It is quite efficient, as it works in prime-order groups, and supports the selective disclosure of attributes. Nevertheless, U-Prove does not provide multi-show unlinkability unless the user uses a different credential at each proof of possession. Besides, to date, its security has not been formally proven.

A slightly less efficient anonymous attribute-based credentials system has been proposed by Baldimsti and Lysyanskaya [3]. Their proposal, which relies on an extension of Abe's blind signature scheme [1], is proven secure in the Random Oracle Model (ROM) under the DDH assumption. Recently, Fuchsbauer *et al.* [19] introduced another anonymous credentials system that is proven secure in the standard model. However, similarly to U-Prove, both systems are one-show (*i.e.* credential presentations are linkable if a credential is used more than once).

IBM's Identity Mixer, commonly known as Idemix [22], is built on Camenisch-Lysyanskaya (CL) signature scheme [10,11]. Unlike previously reviewed credentials systems, Idemix credentials provide multi-show unlinkability but at the cost of a less efficient proof of possession. Indeed, the used CL signatures are based on the Strong RSA assumption [4]. This implies large RSA parameters which make Idemix unsuitable for constrained devices. Despite this, Vullers and Alpár focused in [27] on the implementation of Idemix on MULTOS smart cards. Using a 1024-bit modulus, their implementation enables the presentation of a credential with three attributes, one of which is undisclosed, in 1 s. Moreover, de la Piedra *et al.* [25] addressed smart cards limited Random Access Memory (RAM) issues by proposing a RAM-efficient implementation of Idemix. Thereby, smart cards can support Idemix credentials with more than 5 attributes. Unfortunately, even with these implementation improvements, timing results far exceed the time constraints of some use cases, which limits the use of Idemix in practice.

Camenisch and Lysyanskaya introduced in [12] an efficient signature scheme defined in bilinear groups and used it to construct an anonymous credentials system. Shortly afterwards, Akagi *et al.* [2] provided a more effective Boneh Boyen-based anonymous credentials system. Recently, Camenisch *et al.* [9] proposed a Universally Composable (UC) secure anonymous credentials system that provides multi-show unlinkability and whose presentation proof is of constant size. Nevertheless, these three proposals require the prover to compute pairings and/or perform computations in $\mathbb{G}_2$. Thus, they cannot be implemented on SIM cards as the latter cannot handle such heavy computations.

Recently, Chase *et al.* [15] have opted for the use of symmetric key primitives, instead of digital signatures, so as to achieve better performances. More precisely, they used algebraic Message Authentication Codes (MACs), that relies on group operations rather than block ciphers or hash functions, as the main building block of their credentials system. Their two proposals, denoted $\mathsf{MAC_{GGM}}$ and $\mathsf{MAC_{DDH}}$, assume that the issuer of the credential and the verifier share a secret key: In such a setting, the anonymous credentials system is referred to as *Keyed-Verification Anonymous Credentials* (KVAC). Unfortunately, their presentation proofs, for n unrevealed attributes, are of complexity $O(n)$ in the number of group elements. Moreover, when credential blind issuance is required, their KVAC systems do not provide perfect anonymity as they rely on ElGamal encryption to hide attributes.

As pointed out in [15], one can switch between the use of public-key and keyed-verification anonymous credentials which are more efficient. For that, whenever interacting with a new entity, the user proves the possession of a publicly verifiable credential (such as a driving license anonymous credential issued by a government on a set of attributes) and gets back a keyed-verification credential on the same attributes without disclosing them. Thus, during subsequent interactions with that entity, the user will use the keyed-verification credential for better efficiency.

Contributions. In this paper, we aim to design an anonymous credentials system that provides multi-show unlinkability while being both efficient and suitable for resource constrained environments like SIM cards (that cannot handle pairing computations). To this end, following Chase *et al.* approach [15], we first build a new algebraic MAC scheme that relies on a pairing-free variant of the Boneh Boyen signature scheme. We prove the security of our proposal, which is of independent interest, under the $q-\mathsf{SDH}$ assumption. Then, we use it to construct a practical Keyed-Verification Anonymous Credentials (KVAC) system whose presentation proof is of complexity $O(1)$ in the number of group elements and linear in the number of scalars. Our KVAC system is proven secure in the ROM under classical assumptions. Furthermore, it can be easily turned into an efficient publicly verifiable anonymous credentials system through the use of pairings *solely* on the verifier side. To show its efficiency and suitability for constrained environment, we implemented our system on a standard NFC SIM card. The proof of possession of a credential on three attributes, with one unrevealed, takes just 88 ms. This confirms its suitability for real world applications.

Organization. The paper is structured as follows. Section 2 introduces our main notation and necessary building blocks. Then, Sect. 3 presents a novel algebraic MAC scheme based on a pairing-free variant of the Boneh Boyen signature scheme. Next, Sect. 4 describes our keyed-verification anonymous credentials system as well as the way it can be turned into a traditional public-key anonymous credentials system. Finally, Sect. 5 provides efficiency and complexity evaluations as well as implementation benchmarks of our KVAC system.

2 Preliminaries

2.1 Classical Tools

Notation. To state that x is chosen uniformly at random from the set X, we use one of the two following notations $x \xleftarrow{R} X$ or $x \in_R X$. In addition, $\overrightarrow{m}$ and $\{g_i\}_{i=1}^l$ respectively denote the vector $(m_1, \ldots, m_n)$ and the set $\{g_1, g_2, \ldots, g_l\}$.

Zero-Knowledge Proof of Knowledge. Zero-Knowledge Proofs of Knowledge (ZKPKs) allow a prover $\mathcal{P}$ to convince a verifier $\mathcal{V}$ that he knows some secrets verifying a given statement without revealing anything else about them. Following the usual notation introduced by Camenisch and Stadler [13], they are denoted by $\pi = \text{PoK}\{\alpha, \beta : statements\, about\, \alpha, \beta\}$ where Greek letters correspond to the knowledge of $\mathcal{P}$.

A ZKPK should satisfy three properties, namely (1) *completeness* (*i.e.* a valid prover should be able to convince an honest verifier with overwhelming probability), (2) *soundness* (*i.e.* a malicious prover should be rejected with overwhelming probability), (3) *zero-knowledge* (*i.e.* the proof reveals no information about the secret(s)).

In addition to classical ZKPKs (such as a proof of knowledge of a discrete logarithm [26], a proof of knowledge of a representation [8], or a proof of equality of discrete logarithms [17]), our KVAC system relies on a ZKPK that a committed value is non-zero. Such a proof has been introduced by Brands [7].

Indeed, a prover $\mathcal{P}$ may sometimes have to convince the verifier $\mathcal{V}$ that the value x committed in $C = g^x h^w$ is non-zero, where g and h are two random generators (*i.e.* the discrete logarithm of g in the base h is unknown). To do so, $\mathcal{P}$ has to prove the knowledge of the representation of g in the bases C and h. That is, $\mathcal{P}$ has to build a ZKPK π defined as $\pi = \{\alpha, \beta, \gamma, \delta : C = g^\alpha h^\beta \wedge g = C^\gamma h^\delta\}$.

Computational Hardness Assumptions. The security of our MAC scheme and KVAC system relies on a set of computational hardness assumptions. In what follows, $\mathbb{G}$ denotes a cyclic group of prime order p.

Discrete Logarithm (DL) *Assumption.* The Discrete Logarithm assumption states that, given a generator $g \in_R \mathbb{G}$ and an element $y \in_R \mathbb{G}$, it is hard to find the integer $x \in \mathbb{Z}_p$ such that $y = g^x$.

Decisional Diffie-Hellman (DDH) *Assumption.* The Decisional Diffie-Hellman assumption states that, given a generator $g \in_R \mathbb{G}$, two elements $g^a, g^b \in_R \mathbb{G}$ and a candidate $X \in \mathbb{G}$, it is hard to decide whether $X = g^{ab}$ or not. This is equivalent to decide, given g, h, g^a, g^b, whether $a = b$ or not.

q-Strong Diffie-Hellman $(q - \mathsf{SDH})$ *Assumption.* The q-Strong Diffie-Hellman assumption holds in $\mathbb{G}$ if, given a generator $g \in_R \mathbb{G}$ and $(g^y, g^{y^2}, \dots, g^{y^q}) \in \mathbb{G}^q$ as input, it is hard to output a pair $(x, g^{\frac{1}{y+x}}) \in \mathbb{Z}_p^* \times \mathbb{G}$.

This assumption is believed to be hard even in gap-DDH groups, *i.e.* groups in which there is an efficient test to determine, with probability 1, on input (g, h, g^x, h^y) if $x = y \bmod p$ or not. Moreover, it has been proven in [20] that the hardness of the $q - \mathsf{SDH}$ assumption in gap-DDH groups implies the hardness of the *gap* $q - \mathsf{SDH} - \mathsf{III}$ assumption defined as follows[1].

*Gap q-Strong Diffie-Hellman-*III *(gap* $q - \mathsf{SDH} - \mathsf{III})$ *Assumption.* The q-Strong Diffie-Hellman-III assumption states that, given $(g, h, g^y) \in \mathbb{G}^3$ and q distinct triples $(x_i, m_i, (g^{m_i} h)^{\frac{1}{y+x_i}}) \in \mathbb{Z}_p^2 \times \mathbb{G}$ and having access to a DDH oracle (which indicates whether a given quadruple $(g, h, g^x, h^y) \in \mathbb{G}^4$ is a DH quadruple or not), it is hard to output a *new* triple $(x, m, (g^m h)^{\frac{1}{y+x}})$ where $(x, m) \in \mathbb{Z}_p^2$.

2.2 Message Authentication Codes (MACs)

A Message Authentication Code (MAC) is an authentication tag computed using a secret key that is shared between the issuer and the verifier. More formally, a MAC scheme consists of the following four algorithms:

$\mathsf{Setup}(1^k)$ creates the public parameters pp, given a security parameter k.

$\mathsf{KeyGen}(pp)$ generates the secret key sk that is shared between the issuer and the verifier.

$\mathsf{MAC}(pp, sk, m)$ takes as input a message m and a secret key sk. It outputs a MAC, also known as a tag and denoted by τ, on the message m.

$\mathsf{Verify}(pp, sk, m, \tau)$ is a deterministic algorithm which outputs either 1 or 0 depending on the validity of the MAC τ with respect to the message m and the secret key sk.

UF-CMVA Security. Usually, a probabilistic MAC scheme is considered secure if it is *unforgeable under chosen message and verification attack* (UF-CMVA). In other words, the adversary $\mathcal{A}$ can query two oracles: $\mathcal{O}\mathsf{MAC}$ and $\mathcal{O}\mathsf{Verify}$. $\mathcal{O}\mathsf{MAC}$ provides him with a valid MAC on any message of his choice whereas $\mathcal{O}\mathsf{Verify}$ enables him to check the validity of any (message, MAC) pair. Such an adversary should not be able to compute a pair (m', τ') where τ' is a valid MAC on the message m' that has not already been queried to the $\mathcal{O}\mathsf{MAC}$ oracle. A yet stronger security notion for probabilistic MACs, denoted sUF-CMVA, exists. In such a variant, the adversary wins even if m' has already been queried to the $\mathcal{O}\mathsf{MAC}$ oracle, provided that the oracle did not output the pair (m', τ'). More formally, Fig. 1 details the sUF-CMVA experiment $\mathsf{Exp}_{\mathcal{A}}^{\mathrm{sUF\text{-}CMVA}}(1^k)$ between a challenger $\mathcal{C}$ and an adversary $\mathcal{A}$. The adversary's success probability, denoted by $\mathsf{Adv}_{\mathcal{A}}^{\mathrm{sUF\text{-}CMVA}}(1^k)$, is defined as $\Pr[\mathsf{Exp}_{\mathcal{A}}^{\mathrm{sUF\text{-}CMVA}}(1^k) = 1]$.

---

[1] For this reason, we will sometimes simply refer to the *gap* $q - \mathsf{SDH} - \mathsf{III}$ assumption as the $q - \mathsf{SDH}$ assumption.

$\text{Exp}_{\mathcal{A}}^{\text{sUF-CMVA}}(1^k)$

1. $pp \leftarrow \text{Setup}(1^k)$
2. $sk \leftarrow \text{KeyGen}(pp)$
3. $(m', \tau') \leftarrow \mathcal{A}^{\mathcal{O}\text{MAC}, \mathcal{O}\text{Verify}}(pp)$
4. If (m', τ') was obtained following a call to the $\mathcal{O}$MAC oracle, then return 0.
5. Return $\text{Verify}(pp, sk, m', \tau')$

Fig. 1. sUF-CMVA security

3 An Algebraic MAC Scheme Based on Boneh-Boyen Signatures

Based on a *pairing-free* variant [14] of the Boneh-Boyen signature scheme [5], we design a new algebraic MAC scheme. In this section, we detail our construction which can be applied to both a single message as well as a block of messages.

3.1 MAC$_{\text{BB}}$

Our algebraic MAC scheme for a single message m, referred to as MAC$_{\text{BB}}$, is defined as follows:

Setup(1^k) creates the system public parameters $pp = (\mathbb{G}, p, h, g_0, g_1, g)$ where $\mathbb{G}$ is a cyclic group of prime order p, a k-bit prime, and h, g_0, g_1, g are four random generators of $\mathbb{G}$.

KeyGen(pp) selects a random value $y \in_R \mathbb{Z}_p$ as the issuer's private key and *optionally* computes the corresponding public key $Y = g_0^y$.

MAC(m, y) picks two random values $r, s \in_R \mathbb{Z}_p$ and computes $A = (g_1^m g^s h)^{\frac{1}{y+r}}$. The MAC on the message m consists of the triple (A, r, s).

Verify(m, A, r, s, y) checks the validity of the MAC (A, r, s) with respect to the message m. The MAC is valid only if $(g_1^m g^s h)^{\frac{1}{y+r}} = A$.

Theorem 1. *Our* MAC$_{\text{BB}}$ *scheme is* sUF-CMVA *secure under the gap* $q - \text{SDH} - \text{III}$ *assumption*[2].

3.2 MAC$_{\text{BB}}^n$

Our algebraic MAC scheme can be generalized to support a block of n messages $(m_1, \ldots, m_n)$. This extension is referred to as MAC$_{\text{BB}}^n$ and works as follows:

Setup(1^k) creates the system public parameters $pp = (\mathbb{G}, p, g_1, g_2, \ldots, g_n, h, g_0, g)$ where $\mathbb{G}$ is a cyclic group of prime order p, a k-bit prime, and $h, g, g_0, g_1, \ldots, g_n$ are random generators of $\mathbb{G}$.

---

[2] The proof is detailed in Appendix A.1.

KeyGen(pp) selects a random value $y \in_R \mathbb{Z}_p$ as the issuer's private key and *optionally* computes the corresponding public key $Y = g_0^y$.

MAC($\overrightarrow{m}, y$) takes as input a block of n messages $\overrightarrow{m} = (m_1, \ldots, m_n)$ and computes $A = (g_1^{m_1} g_2^{m_2} \ldots g_n^{m_n} g^s h)^{\frac{1}{y+r}}$ where $r, s \in_R \mathbb{Z}_p$. The MAC on $\overrightarrow{m}$ consists of the triple (A, r, s).

Verify($\overrightarrow{m}, A, r, s, y$) checks the validity of the MAC with respect to the block of messages $\overrightarrow{m}$. The MAC is valid only if $(g_1^{m_1} g_2^{m_2} \ldots g_n^{m_n} g^s h)^{\frac{1}{y+r}} = A$.

Theorem 2. *Our* MAC$_{BB}^n$ *scheme is* sUF-CMVA *secure under the assumption that* MAC$_{BB}$ *is sUF-CMVA*[3].

One particular feature of our algebraic MAC scheme is that anyone can verify the validity of a given MAC by himself (*i.e.* without neither knowing the private key y nor querying the Verify algorithm). Indeed, a MAC on $\overrightarrow{m} = (m_1, \ldots, m_n)$ consists of the triple (A, r, s) such that $A = (g_1^{m_1} g_2^{m_2} \ldots g_n^{m_n} g^s h)^{\frac{1}{y+r}}$. This implies that $A^{y+r} = g_1^{m_1} g_2^{m_2} \ldots g_n^{m_n} g^s h$ and hence, $B = g_1^{m_1} g_2^{m_2} \ldots g_n^{m_n} g^s h \cdot A^{-r} = A^y$. Therefore, if the issuer of the MAC also provides a ZKPK defined as

$$\pi = \text{PoK}\{\gamma : B = A^\gamma \wedge Y = g_0^\gamma\},$$

then its receiver will be convinced that the MAC is valid.

Furthermore, unlike both algebraic MAC schemes due to Chase *et al.* [15], the issuer does not have to hold as many private keys as messages but rather a sole private key regardless of the number of messages.

4   A Keyed-Verification Anonymous Credentials System Based on MAC$_{BB}^n$

In this section, we first define Keyed-Verification Anonymous Credentials (KVAC) systems as well as their requirements. Next, we detail our new KVAC system that is built upon our MAC$_{BB}^n$ scheme.

4.1   Overview on KVAC Systems

A keyed-verification anonymous credentials system is defined through the following algorithms which involve three entities: a user $\mathcal{U}$, an issuer $\mathcal{I}$ and a verifier $\mathcal{V}$.

Setup(1^k) creates the system public parameters pp, given a security parameter k.

CredKeyGen(pp) generates the issuer's private key sk, which is shared with $\mathcal{V}$, and computes the corresponding public key pk.

BlindIssue($\mathcal{U}(\overrightarrow{m}, s), \mathcal{I}(sk)$) is an interactive protocol between a user $\mathcal{U}$ who wants to get an anonymous credential on a set of attributes $\overrightarrow{m} = (m_1, \ldots, m_n)$ and a secret value s, without revealing them, and the issuer $\mathcal{I}$ who holds the private key sk. If the protocol does not abort, the user gets a credential σ.

---

[3] The proof is detailed in Appendix A.2.

$\text{Show}(\mathcal{U}(s, \sigma, \overrightarrow{m}, \phi), \mathcal{V}(sk, \phi))$ is an interactive protocol between $\mathcal{U}$, who wants to prove that he holds a valid credential on attributes $\overrightarrow{m}$ satisfying a given set of statements ϕ, and $\mathcal{V}$, holding the private key sk, whose goal is to check that it is actually true.

Security Requirements. In addition to the usual *correctness* property, a KVAC system should satisfy four security properties, namely *unforgeability*, *anonymity*, *blind issuance* and *key-parameter consistency*. Roughly speaking, they are defined as follows (formal definitions are provided in [15]):

- *Unforgeability:* it should be infeasible for an adversary to generate a valid ZKPK that convinces a verifier that he holds a credential satisfying a given statement, or a set of statements, when it is not actually true;
- *Anonymity:* the presentation proof produced during the protocol Show reveals nothing else aside from the statement ϕ being proven;
- *Blind issuance:* BlindIssue is a secure two-party protocol for generating credentials on the user's attributes;
- *Key-parameter consistency*: an adversary should not be able to find two secret keys that correspond to the same issuer's public key.

4.2 Our Keyed-Verification Anonymous Credentials System

Based on the designed $\mathsf{MAC}^n_{\mathsf{BB}}$ scheme, we construct a KVAC system involving a user $\mathcal{U}$, an issuer $\mathcal{I}$ and a verifier $\mathcal{V}$. Our KVAC system consists of the following four phases. The two main phases (BlindIssue and Show) are depicted in Fig. 2.

Setup. Generate the public parameters $pp = (\mathbb{G}, p, g_1, g_2, \ldots, g_n, g, h, g_0, f)$ where $\mathbb{G}$ is a cyclic group of prime order p, a k-bit prime, and $(h, g, g_0, \{g_i\}_{i=1}^n, f)$ are random generators of $\mathbb{G}$ where DDH is hard. For $i \in \{1, \ldots, n\}$, g_i is associated with a specific type of attributes (*e.g.* age, gender, etc.). This allows us to differentiate attributes and avoid any ambiguity. Note that, in the sequel, all computations on exponents are computed modulo p (*i.e.* mod p).

Key Generation. Choose a random value $y \in_R \mathbb{Z}_p$ as the issuer's private key and compute the corresponding public key $Y = g_0^y$. Each user $\mathcal{U}$ is also provided with a private key sk_u and the associated public key pk_u which may be used to authenticate the user during the issuance of his credentials.

Blind Issuance. To issue a credential on the attributes $(m_1, \ldots, m_n)$, the issuer and the user (who has already been authenticated) engage in the following protocol. First, the user $\mathcal{U}$ builds a commitment $C_m = g_1^{m_1} \ldots g_n^{m_n} g^s$ on his attributes, where $s \in_R \mathbb{Z}_p^*$. Then, he sends it to the issuer $\mathcal{I}$ along with a ZKPK π_1 defined as $\pi_1 = \text{PoK}\{\alpha_1, \ldots, \alpha_{n+1} : C_m = g_1^{\alpha_1} g_2^{\alpha_2} \ldots g_n^{\alpha_n} g^{\alpha_{n+1}}\}$. If the proof is valid, $\mathcal{I}$ randomly picks $r, s' \in_R \mathbb{Z}_p$ and computes $A = (C_m \cdot g^{s'} \cdot h)^{\frac{1}{y+r}}$ which corresponds to a $\mathsf{MAC}^n_{\mathsf{BB}}$ on $(m_1, \ldots, m_n)$. He may also build a ZKPK π_2 ensuring that the credential is well-formed. Such a proof is defined as $\pi_2 = \text{PoK}\{\gamma : B = A^\gamma \wedge Y = g_0^\gamma\}$

where $B = C_m \cdot g^{s'} \cdot h \cdot A^{-r} = A^y$. Then, he provides $\mathcal{U}$ with the triple (A, r, s') along with the proof π_2. Upon receiving them, $\mathcal{U}$ first verifies the validity of π_2, then computes $\tilde{C}_m = C_m \, g^s h$ as well as $s_u = s + s'$, which is a secret value only known to $\mathcal{U}$. Finally, he sets his anonymous credential σ as $\sigma = (A, r, s_u, \tilde{C}_m)$.

Note that in case where $\mathcal{U}$ does not mind revealing his attributes (or a subset of them), he just sends them without using any commitment (respectively, only commits to the attributes that he does not want to reveal).

Public Input: pp, pk_u, Y	
(1) Issuance of a credential (BlindIssue)	
User $\mathcal{U}$	**Issuer $\mathcal{I}$**
Private Input: $sk_u \in_R \mathbb{Z}_p^*$	Private Input: $y \in_R \mathbb{Z}_p^*$
$\vec{m} = (m_1, \ldots, m_n)$	
Choose $s \xleftarrow{R} \mathbb{Z}_p^*$	
Compute $C_m \leftarrow g_1^{m_1} \cdots g_n^{m_n} g^s$	
Build	
$\pi_1 = \text{PoK}\{\alpha_1, \cdots, \alpha_{n+1} : C_m = g^{\alpha_{n+1}} \prod_{i=1}^n g_i^{\alpha_i}\} \xrightarrow{C_m, \pi_1}$ Check π_1 and Choose $r, s' \xleftarrow{R} \mathbb{Z}_p^*$	
	Compute $A \leftarrow (C_m \cdot g^{s'} \cdot h)^{\frac{1}{y+r}}$
Check π_2	$\xleftarrow{A, r, s', \pi_2}$ Build $\pi_2 = \text{PoK}\{\gamma : Y = g_0^\gamma \wedge$
Compute $\tilde{C}_m \leftarrow C_m \cdot g^{s'} \cdot h$ and $s_u \leftarrow s + s'$	$A^\gamma = C_m \cdot g^{s'} h \cdot A^{-r}\}$
$\sigma \leftarrow (A, r, s_u, \tilde{C}_m)$	

(2) Proving Knowledge of a Credential (Show)	
User $\mathcal{U}$	**Verifier $\mathcal{V}$**
Private Input: $(A, r, s_u, \tilde{C}_m), \vec{m}$	Private Input: y
Choose $l, t \xleftarrow{R} \mathbb{Z}_p^*$	
Compute $B_0 \leftarrow A^l$, $E \leftarrow C^{\frac{1}{t}} \cdot f^t$	
$C \leftarrow \tilde{C}_m^l \cdot B_0^{-r}$	
Build $\pi_3 = \text{PoK}\{\alpha, \beta, \lambda, \delta_1, \delta_2, \ldots, \delta_{n+1}, \gamma, \theta : \xrightarrow{B_0, C, E, \pi_3}$ Compute $C' \leftarrow B_0^y$	
$E \cdot h^{-1} = g_1^{\delta_1} g_2^{\delta_2} \cdots g_n^{\delta_n} g^{\delta_{n+1}} B_0^\lambda f^\beta \wedge E = C^\alpha f^\beta$ Check if $C' \overset{?}{=} C$	
$\wedge C = E^\theta f^\gamma\}$ Check π_3	

Fig. 2. Our keyed-verification anonymous credentials system

Credential Presentation. To anonymously prove that he holds a credential on the attributes $(m_1, \ldots, m_n)$, the user engages in an interactive protocol with the verifier $\mathcal{V}$. First, he randomly selects $l, t \in_R \mathbb{Z}_p^*$ and computes $B_0 = A^l$, a randomized version of his credential. He also computes $C = \tilde{C}_m^l B_0^{-r}$ as well as $E = C^{\frac{1}{t}} f^t$.

Note that by definition, $A^{y+r} = C_m \, g^{s'} h = g_1^{m_1} g_2^{m_2} \cdots g_n^{m_n} g^{s_u} h$. Thus, we have $(A^l)^{y+r} = g_1^{lm_1} g_2^{lm_2} \cdots g_n^{lm_n} g^{ls_u} h^l$. Hence, C is simply equal to $A^{ly} = B_0^y$.

$\mathcal{U}$ also builds a ZKPK π_3 to prove that he really holds a valid credential (*i.e.* he knows the associated attributes/secrets and the value committed in E is different from zero). π_3 is defined as $\pi_3 = \text{PoK}\{\alpha, \beta, \lambda, \delta_1, \ldots, \delta_{n+1}, \gamma, \theta : E = C^\alpha f^\beta \wedge E \cdot h^{-1} = g_1^{\delta_1} \cdots g_n^{\delta_n} g^{\delta_{n+1}} \cdot B_0^\lambda \cdot f^\beta \wedge C = E^\theta f^\gamma\}$. Once the required values have been computed, $\mathcal{U}$ provides $\mathcal{V}$ with B_0, C and E along with π_3[4].

---

[4] π_3 is detailed in Appendix C.

Upon their receipt, $\mathcal{V}$ first computes $C' = B_0^y$, then verifies that $C = C'$. If so, he checks that π_3 is valid. $\mathcal{V}$ is convinced that $\mathcal{U}$ really holds a valid credential on attributes $(m_1, \ldots, m_n)$ if, and only if, both checks succeed.

Theorem 3. *Our KVAC system is* unforgeable *under the assumption that* MAC_{BB}^n *is sUF-CMVA, perfectly anonymous and ensures* blind issuance *as well as* key-parameter consistency *in the Random Oracle Model*[5].

4.3 From Keyed-Verification to Public Key Anonymous Credentials

In this section, we explain how to turn our KVAC system into a public key anonymous credentials system. Thereby, a user would be able to prove possession of a credential to any entity (*i.e.* even if the issuer's private key is unknown).

For that, our system should be defined in bilinear groups. Let us first recall that bilinear groups are a set of three cyclic groups $\mathbb{G}_1$, $\mathbb{G}_2$ and $\mathbb{G}_T$ of prime order p along with a bilinear map $e : \mathbb{G}_1 \times \mathbb{G}_2 \to \mathbb{G}_T$ satisfying the following properties:

- For all $g \in \mathbb{G}_1, \tilde{g} \in \mathbb{G}_2$ and $a, b \in \mathbb{Z}_p, e(g^a, \tilde{g}^b) = e(g, \tilde{g})^{a.b}$;
- For $g \neq 1_{\mathbb{G}_1}$ and $\tilde{g} \neq 1_{\mathbb{G}_2}$, $e(g, \tilde{g}) \neq 1_{\mathbb{G}_T}$;
- e is efficiently computable.

In such a case, the system public parameters are defined as $pp = (\mathbb{G}_1, \mathbb{G}_2, \mathbb{G}_T, p, e, g_1, \ldots, g_n, g, h, g_0, f, \tilde{g}_0)$ where $\mathbb{G}_1$, $\mathbb{G}_2$ and $\mathbb{G}_T$ are three cyclic groups of prime order p, $(h, g, g_0, \{g_i\}_{i=1}^n, f)$ are random generators of $\mathbb{G}_1$ and $\tilde{g}_0$ is a random generator of $\mathbb{G}_2$. The other phases are updated as follows.

Key Generation. The issuer publishes a second public key $W = \tilde{g}_0^y$ associated with his private key y.

Blind Issuance. This phase does not require any changes.

Credential Presentation. As the verifier $\mathcal{V}$ does not hold the private key y, some changes are required on his side. More precisely, he must compute two pairings $e(C, \tilde{g}_0)$ and $e(B_0, W)$. $\mathcal{V}$ is convinced that the user really holds a valid credential on $(m_1, \ldots, m_n)$ only if $e(C, \tilde{g}_0) = e(B_0, W)$ and π_3 is valid.

5 Efficiency Comparison and Performance Assessment

We first compare the efficiency of our KVAC system to that of the main existing anonymous credentials schemes (*i.e.* U-Prove, Idemix, Bilinear CL, MAC_{GGM} and MAC_{DDH}) both in terms of credential size and computational cost related to the creation of a presentation proof since it is the most time-critical phase. Next, we focus on the complexity, in the number of group elements, of KVAC systems presentation proofs. Finally, we provide timing results of the implementation of our *Credential presentation* protocol on a standard NFC SIM card.

---

[5] Proofs are detailed in Appendix B.

Presentation Proof Computational Cost. We compare in Table 1 the estimated cost of creating a presentation proof in terms of total number of multi-exponentiations. We use the same notation as [15] where l-exp denotes the computation of the product of l powers and $l - exp(b_1, \ldots, b_l)$ corresponds to the computation of the product of l powers with exponents of $b_1, \ldots, b_l$ bits (for Idemix). The number of multi-exponentiations depends on three parameters: n, r and c which respectively denote the number of attributes in a credential, the number of revealed attributes and the number of attributes kept secret.

Table 1 shows that our KVAC system is competitive with U-Prove (which does not provide multi-show unlinkability) and $\mathsf{MAC_{GGM}}$ (which requires the verifier to know the issuer's private key and thus does not allow public verifiability). When most of the attributes are not disclosed, our proposal outperforms $\mathsf{MAC_{GGM}}$.

Table 1. Comparison of credential sizes (for s *unlinkable* shows) and presentation proof generation cost (for a credential on n attributes, c of which are not disclosed). Note that all schemes use a 256-bit elliptic curve group, except Idemix which uses a 2048-bit modulus.

Schemes	Credential size (in bits)	Number of exponentiations
U-Prove [23,24]	$1024s$	$2c$ 2-exp and 1 $(n - r + 1)$-exp
Idemix [22]	5369	1 1-exp(2048), c 2-exp(256, 2046), c 2-exp(592, 2385) and 1 $(n - r + 2)$-exp (456, 3060, 592, …, 592)
Bilinear CL [12]	$512n + 768$	$(3 + n)$ 1-exp, $2c$ 2-exp and $3 + n$ pairings
$\mathsf{MAC_{GGM}}$ [15]	512	3 1-exp, $2(n - r)$ 2-exp and 1 $(n - r + 1)$-exp
$\mathsf{MAC_{DDH}}$ [15]	1024	6 1-exp, $2(n - r + 1)$ 2-exp and 2 $(n - r + 1)$-exp
$\mathsf{MAC_{BB}^n}$	**1024**	**1 1-exp, 4 2-exp and 1 $(n - r + 3)$-exp**

Complexity in the Number of Group Elements. As it only requires a multi-commitment to all undisclosed attributes, our presentation proof is of complexity $O(1)$ in the number of group elements. This makes our KVAC system more efficient than Chase *et al.* systems (*i.e.* $\mathsf{MAC_{GGM}}$ and $\mathsf{MAC_{DDH}}$ [15]) whose presentation proof is of complexity $O(c)$. Indeed, both of their proposals presentation proof needs c commitments (one for each unreavealed attribute).

Implementation Results. Table 2 gives timing results of the implementation of our Show protocol on a Javacard 2.2.2 SIM card, Global Platform 2.2 compliant, embedded in a Samsung galaxy S3 NFC smartphone. Compared to the javacard specifications, the only particularity of our card is some additional API provided by the card manufacturer enabling operations in modular and elliptic curve arithmetic. To be able to handle asymmetric cryptography on elliptic curves, the used card is equipped with a cryptoprocessor. This makes it more powerful than most cards. It is, however, worth to emphasize that such SIM cards are already widely deployed by some phone carriers to provide NFC based services.

The implementation uses a 256-bit prime "pairing friendly" Barreto-Naehrig elliptic curve. In our implementation, the protocol is split into two parts: an *off-line* part that can be run in advance by the card (during which all the values necessary for an execution of the Show protocol in the worst case scenario, i.e. no revealed attributes, are computed) and an *on-line* part that needs to be performed on-line as it depends on the verifier's challenge. Indeed, in our implementation, the proof π_3 is made non-interactive: the verifier sends to the prover a challenge Ch which is included in the computation of the hash value c. Timings are given for $n = 3$, $r = 2$ and $c = 1$.

Table 2. Timings in ms ((min-max) average) of the implementation of the protocol Show

Off-line part (card) Battery-On: (1352–1392) 1378 ms			
On-line part			
Presentation proof (card)		*Proof verification (PC)*	
Battery-On	**Battery-Off**	y **known**	y **unknown**
(81–86) 83 ms	(123–124) 123.4 ms	(3–14) 5 ms	(5–17) 10 ms
Total On-line part			
Battery-On		**Battery-Off**	
y **known**	y **unknown**	y **known**	y **unknown**
(84–100) 88 ms	(86–103) 93 ms	(126–137) 128 ms	(128–141) 133 ms

The *presentation proof* by the card actually refers to the total time, from the applet selection to the proof reception, including the sending of the challenge by the verifier, but excluding the proof verification. Communication between the SIM card in the smartphone and the PC (Intel Xeon CPU 3.70 GHz), acting as the Verifier, was done in NFC using a standard PC/SC reader (an Omnikey 5321). "Battery-Off" denotes a powered-off phone either by the user, or because its battery is flat. In such a situation, as stated by NFC standards, NFC-access to the SIM card is still possible, but with degraded performances. Off-line computations are assumed to be automatically launched by the smartphone (battery-On) after a presentation proof, in anticipation for the next one. It is noteworthy that all computations are entirely done by the card: the smartphone is only used to trigger the Show protocol and to power the card. On-line computations refer to computations of R_i values and the hash c involved in the proof π_3 (see Appendix C), and can be potentially carried out even by a battery-Off phone. On average, the On-line part of the presentation proof is very fast even when the phone is powered-off. Actually, data exchange is the most time-consuming task.

6   Conclusion

In this paper, our contribution is twofold. First, we proposed a new algebraic MAC scheme that relies on a pairing-free variant of the Boneh Boyen signature scheme. Then, based on it, we designed a keyed-verification anonymous credentials (KVAC) system whose presentation proof is efficient both in terms of presentation cost and complexity (in the number of group elements). Our KVAC system provides multi-show unlinkability and requires the issuer to hold a single private key regardless of the number of attributes. Through slight modifications (solely on the verifier side), our KVAC system can be easily turned into a quite efficient public key anonymous credentials system. Thereby, it can also be used even if the verifier does not hold the issuer's private key. Finally, implementation results confirm its efficiency and suitability for delay sensitive applications, even when implemented on a standard NFC SIM card.

A   MAC Security

A.1   Security Proof of $\mathsf{MAC_{BB}}$ (Theorem 1)

Let $\mathcal{A}$ be an adversary who breaks the sUF-CMVA security of our $\mathsf{MAC_{BB}}$ scheme with non-negligible probability. Using $\mathcal{A}$, we construct a reduction $\mathcal{B}$ against the $q - \mathsf{SDH}$ assumption in gap-DDH groups (which implies the $gap\ q - \mathsf{SDH} - \mathsf{III}$ assumption). $\mathcal{A}$ can ask for tags on any message of his choice and receives the corresponding tags (A_i, r_i, s_i) for $i \in \{1, \ldots, q\}$ where q denotes the number of requests to the $\mathcal{O}\mathsf{MAC}$ oracle. Eventually, $\mathcal{A}$ outputs his forgery (A, r, s) for the message m. We distinguish two types of forgeries:

- **Type-1 Forger:** an adversary that outputs a valid tag (A, r, s) on m such that $(A, r) \neq (A_i, r_i)$ for all $i \in \{1, \ldots, q\}$.
- **Type-2 Forger:** an adversary that outputs a valid tag (A, r, s) on m such that $(A, r) = (A_j, r_j)$ for some $j \in \{1, \ldots, q\}$ and $(m, s) \neq (m_j, s_j)$.

We show that, regardless of their type, both adversaries can be used to break the gap $q - \mathsf{SDH}$ assumption. However, the reduction works differently for each type of forger. Consequently, $\mathcal{B}$ initially chooses a random bit $c_{mode} \in \{1, 2\}$ which indicates its guess for the type of forgery that $\mathcal{A}$ will output.

   • If $c_{mode} = 1$: $\mathcal{B}$ receives on input from its $q - \mathsf{SDH}$ challenger, denoted by $\mathcal{C}$, the public parameters (g_0, g_1, h) and the public key $Y = g_0^y$ as well as q random, and distinct, triples (A_i, r_i, m_i) such that $A_i = (g_1^{m_i} h)^{\frac{1}{r_i + y}}$ for $i \in \{1, \ldots, q\}$. As it is against the $gap\ q - \mathsf{SDH} - \mathsf{III}$ assumption, $\mathcal{B}$ has access to a DDH oracle, denoted by $\mathcal{O}\mathsf{DDH}$, that decides whether a given quadruple (g, h, g^x, h^y) is a valid Diffie-Hellman quadruple (*i.e.* whether $x \overset{?}{=} y \bmod p$) or not. $\mathcal{B}$ also randomly chooses $v \in_R \mathbb{Z}_p$ and computes $g = g_1^v$. Thereby, it can provide $\mathcal{A}$ with the public parameters (g_0, g_1, h, g, Y) and answer his requests as follows:

- $\mathcal{O}$MAC requests: given m as input, $\mathcal{B}$ first computes s_i such that $m + vs_i = m_i$. Then, it provides $\mathcal{A}$ with the triple (A_i, r_i, s_i) which is a valid MAC on m (i.e. $A_i^{y+r_i} = g_1^m g^{s_i} h$). The simulation of this oracle is perfect.
- $\mathcal{O}$Verify requests: given a quadruple (A, r, s, m), $\mathcal{B}$ first verifies that $A \neq 1$ and computes $B = A^{-r} g_1^m g^s h$. Then, it provides the quadruple (g_0, A, Y, B) as input to the $\mathcal{O}$DDH oracle so as to know if it is valid or not. $\mathcal{B}$ forwards the oracle's answer to $\mathcal{A}$, thus perfectly simulating $\mathcal{O}$Verify.

Eventually, after q queries to $\mathcal{O}$MAC and q_v queries to $\mathcal{O}$Verify, $\mathcal{A}$ outputs his forgery (A, r, s) on m such that it breaks the sUF-CMVA security of our $\mathsf{MAC_{BB}}$ scheme. Using these values, $\mathcal{B}$ computes $\tilde{m} = m + sv$ and outputs his forgery $(A, r, \tilde{m})$ thus breaking the $q - \mathsf{SDH}$ assumption with the same advantage as $\mathcal{A}$.

• If $c_{mode} = 2$: A Type-2 adversary $\mathcal{A}$ is rather used, as a subroutine, to construct a reduction $\mathcal{B}$ against the DL problem. In such a case, $\mathcal{B}$ receives on input from its DL challenger, denoted by $\mathcal{C}$, the challenge $(g_1, H = g_1^v)$. Its goal is to find the value v. For this purpose, it first randomly chooses $(y, g_0, h) \in_R \mathbb{Z}_p \times \mathbb{G}^2$ and computes $Y = g_0^y$. $\mathcal{B}$ also sets g as $g = H$. Thereby, it can provide $\mathcal{A}$ with the public parameters (g_1, g_0, h, g, Y) and answer his requests as follows:

- $\mathcal{O}$MAC requests: as it holds y, $\mathcal{B}$ can generate a valid MAC (A, r, s) on any queried message m. To do so, it computes $A = (g_1^m g^s h)^{\frac{1}{y+r}}$ where $r, s \in \mathbb{Z}_p^*$.
- $\mathcal{O}$Verify requests: given a quadruple (A, r, s, m), $\mathcal{B}$ computes $\tilde{A} = (g_1^m g^s h)^{\frac{1}{y+r}}$. To check its validity, and answer $\mathcal{A}$'s query, $\mathcal{B}$ verifies whether $\tilde{A} \overset{?}{=} A$.

Eventually, after q queries to $\mathcal{O}$MAC and q_v queries to $\mathcal{O}$Verify, $\mathcal{A}$ outputs his forgery (A, r, s) on m such that it breaks the sUF-CMVA security of our $\mathsf{MAC_{BB}}$ scheme. By assumption, (A, r) is equal to one of the (A_j, r_j) pairs output by the $\mathcal{O}$MAC oracle following $\mathcal{A}$'s request for some $j \in \{1, \ldots, q\}$. Since $(A, r) = (A_j, r_j)$, then $A^{y+r_j} = g_1^{m_j} g^{s_j} h = A^{y+r} = g_1^m g^s h$ and so $g_1^{m_j} g^{s_j} = g_1^m g^s$. We therefore necessarily have $s_j \neq s$, otherwise this would imply that $m = m_j$ (contradicting the fact that we have supposed $(m, s) \neq (m_j, s_j)$). Thereby, $g = (g_1)^{\frac{m - m_j}{s_j - s}}$. Using the values (m, m_j, s, s_j), $\mathcal{B}$ can recover v, hence breaking the DL problem. If $\mathcal{B}$ can break the DL problem, then it can break the $q - \mathsf{SDH}$ problem (by finding the discrete logarithm y of g^y in the base g).

$\mathcal{B}$ can guess which type of forgery a particular adversary $\mathcal{A}$ will output with probability $1/2$. So, $\mathcal{B}$ can break the gap $q - \mathsf{SDH}$ problem with probability $\varepsilon/2$ where ε is the probability that $\mathcal{A}$ breaks the sUF-CMVA security of our $\mathsf{MAC_{BB}}$ scheme. Therefore, under the gap $q - \mathsf{SDH}$ assumption, our $\mathsf{MAC_{BB}}$ scheme is sUF-CMVA secure.

A.2 Security Proof of $\mathsf{MAC_{BB}^n}$ (Theorem 2)

Let $\mathcal{A}$ be an adversary who breaks the unforgeability of our $\mathsf{MAC_{BB}^n}$ with non-negligible probability. Using $\mathcal{A}$, we construct an algorithm $\mathcal{B}$ against the unforgeability of $\mathsf{MAC_{BB}}$. $\mathcal{A}$ can ask for tags on blocks of messages $\vec{m_1} = (m_1^1, \ldots, m_n^1)$,

$\vec{m_2} = (m_1^2, \ldots, m_n^2), \ldots, \vec{m_q} = (m_1^q, \ldots, m_n^q)$ and receives the corresponding tags (A_i, r_i, s_i) for $i \in \{1, \ldots, q\}$. Eventually, $\mathcal{A}$ outputs his forgery (A, r, s) for the block of messages $\vec{m} = (m_1, \ldots, m_n)$. We differentiate two types of forgers:

- **Type-1 Forger:** an adversary that outputs a forgery where $(A, r, s) \neq (A_i, r_i, s_i)$ for $i \in \{1, \ldots, q\}$.
- **Type-2 Forger:** an adversary that outputs a forgery where $(A, r, s) = (A_i, r_i, s_i)$ for some $i \in \{1, \ldots, q\}$ and $(m_1', \ldots, m_n') \neq (m_1^i, \ldots, m_n^i)$.

We show that any forger can be used to forge $\mathsf{MAC_{BB}}$ tags. The reduction works differently for each forger type. Therefore, $\mathcal{B}$ initially chooses a random bit $c_{mode} \in \{1, 2\}$ that indicates its guess for the type of forgery that $\mathcal{A}$ will emulate.

• If $c_{mode} = 1$: $\mathcal{B}$ receives on input from its $\mathsf{MAC_{BB}}$ challenger, denoted by $\mathcal{C}$, the public parameters (g_0, g_1, g, h) as well as the public key $Y = g_0^y$. Then, $\mathcal{B}$ constructs the public parameters for $\mathcal{A}$ as follows: for $i \in \{2, \ldots, n\}$, $\mathcal{B}$ chooses $\alpha_i \in_R \mathbb{Z}_p^*$ and computes $g_i = g_1^{\alpha_i}$. The parameters g_0, g_1, g, h and Y are the same as those sent by $\mathcal{C}$. $\mathcal{B}$ can answer $\mathcal{A}$'s requests as follows:

- $\mathcal{O}$Verify requests: when $\mathcal{A}$ sends a verify request to $\mathcal{B}$ on (A, r, s) and a block of messages $(m_1, \ldots, m_n)$, $\mathcal{B}$ computes $M = m_1 + \alpha_2 m_2 + \ldots + \alpha_n m_n$. Then, it queries its $\mathsf{MAC_{BB}}$ $\mathtt{Verify}$ oracle on (A, r, s, M) and outputs the oracle's answer to $\mathcal{A}$.
- $\mathcal{O}$MAC requests: when $\mathcal{A}$ sends a tag request to $\mathcal{B}$ on the block of messages $(m_1, \ldots, m_n)$, $\mathcal{B}$ asks the $\mathsf{MAC_{BB}}$ oracle on $M = m_1 + \alpha_2 m_2 + \ldots + \alpha_n m_n$. Thus, $\mathcal{B}$ obtains the tag (A_i, r_i, s_i). It sends back (A_i, r_i, s_i) to $\mathcal{A}$ which is a valid $\mathsf{MAC_{BB}^n}$ tag on $(m_1, \ldots, m_n)$.

Eventually, $\mathcal{A}$ outputs his forgery (A, r, s) on the block of messages $(m_1, \ldots, m_n)$. Using these values, $\mathcal{B}$ directly outputs its $\mathsf{MAC_{BB}}$ forgery (A, r, s) on $M' = m_1 + \alpha_2 m_2 + \ldots + \alpha_n m_n$. Therefore, $\mathcal{B}$ breaks the unforgeability of $\mathsf{MAC_{BB}}$ with the same advantage as $\mathcal{A}$.

• If $c_{mode} = 2$: In this case, $\mathcal{A}$ is rather used as a subroutine to construct a reduction $\mathcal{B}$ against the DL problem. $\mathcal{B}$ receives as input from its DL challenger, denoted by $\mathcal{C}$, the challenge $(g, H = g^v)$. The goal of $\mathcal{B}$ consists in finding v. For that purpose, it first randomly chooses $(y, g_0, h) \in_R \mathbb{Z}_p \times \mathbb{G}^2$ and computes $Y = g_0^y$. Then, it chooses $I \in \{1, \ldots, n\}$ and $(n - 1)$ random values $\alpha_i \in \mathbb{Z}_p^*$. It computes, for $i \neq I, g_i = g^{\alpha_i}$ and defines $g_I = H$. $\mathcal{B}$ can answer $\mathcal{A}$'s requests as follows:

- $\mathcal{O}$Verify requests: when $\mathcal{A}$ sends a verify request to $\mathcal{B}$ on (A, r, s) and a block of messages $\vec{m} = (m_1, \ldots, m_n)$, $\mathcal{B}$ computes $\tilde{A} = (g_1^{m_1} \ldots g_n^{m_n} g^s \cdot h)^{\frac{1}{y+r}}$. It can thus check the validity of the quadruple $(A, r, s, \vec{m})$ by verifying whether $\tilde{A} \stackrel{?}{=} A$;
- $\mathcal{O}$MAC requests: as it holds y, $\mathcal{B}$ can generate a valid MAC (A, r, s) on any queried block of messages $(m_1, \ldots, m_n)$. Indeed, it chooses $r, s \in_R \mathbb{Z}_p^*$ and computes $A = (g_1^{m_1} \ldots g_n^{m_n} g^s \cdot h)^{\frac{1}{y+r}}$.

Eventually, $\mathcal{A}$ outputs his forgery (A, r, s) on a block of messages $\vec{m} = (m_1, \ldots, m_n)$. By assumption, (A, r, s) is equal to one of $\mathcal{A}$'s requests, let say (A_i, r_i, s_i), but it is a forgery on a new block of messages. Therefore, $(m_1^i, \ldots, m_n^i) \neq (m_1, \ldots, m_n)$ (one can easily show that there is at least one difference in the two blocks of messages). So with probability $\frac{1}{n}, m_I^i \neq m_I$. Thus, since $(A, r, s) = (A_i, r_i, s_i)$, we have $g_1^{m_1} g_2^{m_2} \ldots g_n^{m_n} = g_1^{m_1^i} g_2^{m_2^i} \ldots g_n^{m_n^i}$. So, the discrete logarithm v of $H = g_I$ in the base g is equal to: $v = \sum_{j \neq I}^n \alpha_j \frac{(m_j - m_j^i)}{m_I^i - m_I}$. Therefore, $\mathcal{B}$ can find v with probability $\frac{\varepsilon}{n}$ where ε is the probability that $\mathcal{A}$ breaks the unforgeability of $\mathsf{MAC_{BB}^n}$. If $\mathcal{B}$ can break the DL problem then, it can break the $\mathsf{MAC_{BB}}$ scheme (by finding the discrete logarithm of Y in the base g_0).

We can guess which of the two forgers a particular adversary $\mathcal{A}$ is with probability $1/2$. So, assuming the most pessimistic scenario (case 2), $\mathcal{B}$ can break the unforgeability of $\mathsf{MAC_{BB}}$ with probability $\varepsilon/2n$.

B   Security Proofs of Theorem 3

Relying on the KVAC security model provided in [15], we focus in this appendix on the security proofs of our KVAC system. Owing to the lack of space, we only detail the proofs of unforgeability and anonymity.

Unforgeability. Here, we prove unforgeability when $\mathcal{A}$ is given credentials generated by the $\mathsf{BlindIssue}$ protocol. We have shown (see Theorem 2) that $\mathsf{MAC_{BB}^n}$ is unforgeable under the gap $q - \mathsf{SDH}$ assumption.

Suppose there exists an adversary $\mathcal{A}$ who can break the unforgeability property of our anonymous credentials system. We will show that $\mathcal{A}$ can be used to construct an algorithm $\mathcal{B}$ that breaks the unforgeability of $\mathsf{MAC_{BB}^n}$. $\mathcal{B}$ receives $pp = (\mathbb{G}, p, g_1, \ldots, g_n, g, h, g_0)$ from its $\mathsf{MAC_{BB}^n}$ challenger along with Y, the issuer's public key. It sends pp and Y to $\mathcal{A}$ and answers his requests as follows:

- When $\mathcal{A}$ queries the $\mathcal{O}\mathsf{BlindIssue}$ oracle: $\mathcal{A}$ sends C_m and gives a proof π_1. If π_1 is invalid, $\mathcal{B}$ returns $\perp$. Otherwise, $\mathcal{B}$ runs the proof of knowledge extractor to extract $\{m_i\}_{i=1}^n$ and s. $\mathcal{B}$ then queries its $\mathsf{MAC_{BB}^n}$ oracle on $\{m_i\}_{i=1}^n$ which returns a tag (A, r, s_u) to $\mathcal{B}$. Finally, $\mathcal{B}$ simulates the corresponding proof[6] π_2 and forwards the tag $(A, r, s_u - s)$ along with π_2 to $\mathcal{A}$.
- When $\mathcal{A}$ queries the $\mathcal{O}\mathsf{ShowVerify}$ oracle: $\mathcal{A}$ sends B_0, C, E along with a proof π_3. If the proof π_3 is invalid, $\mathcal{B}$ returns $\perp$. Otherwise, $\mathcal{B}$ runs the proof of knowledge extractor to extract $\alpha, \beta, \lambda, \delta_1, \delta_2, \ldots, \delta_{n+1}, \gamma$ and θ. If $\alpha = 0$, $\mathcal{B}$ returns 0 to $\mathcal{A}$. Otherwise, $\mathcal{B}$ computes $A = B_0^\alpha$, $r = -\frac{\lambda}{\alpha}$ and $s = \delta_{n+1}$. Finally, it queries its Verify oracle with $((\delta_1, \delta_2, \ldots, \delta_n), (A, r, s))$ as input and returns the result to $\mathcal{A}$.

In the final Show protocol, $\mathcal{B}$ again extracts $\alpha, \beta, \lambda, \delta_1, \delta_2, \ldots, \delta_{n+1}, \gamma, \theta$ and outputs $((\delta_1, \delta_2, \ldots, \delta_n), (B_0^\alpha, -\frac{\lambda}{\alpha}, \delta_{n+1}))$ as its forgery.

---

[6] Such a proof can be easily simulated in the ROM, using standard techniques.

First, note that $\mathcal{B}$'s response to $\mathcal{O}$BlindIssue queries are identical to the ones of the honest $\mathcal{O}$BlindIssue algorithm. Then, we argue that its response to ShowVerify queries are also, with overwhelming probability, identical to the output of a real ShowVerify algorithm. To see this, note that the proof of knowledge property guarantees that the extractor succeeds in producing a valid witness with all but negligible probability. Furthermore, if the extractor gives valid $(\alpha, \beta, \lambda, \delta_1, \delta_2, \ldots, \delta_{n+1})$, we have from $E = C^\alpha f^\beta = g_1^{\delta_1} \ldots g_n^{\delta_n} g^{\delta_{n+1}} \cdot h \cdot B_0^\lambda \cdot f^\beta$ that

$$C^\alpha = g_1^{\delta_1} \ldots g_n^{\delta_n} g^{\delta_{n+1}} \cdot h \cdot B_0^\lambda \implies C^\alpha B_0^{-\lambda} = g_1^{\delta_1} \ldots g_n^{\delta_n} g^{\delta_{n+1}} \cdot h$$

If the $\mathsf{MAC}_{\mathsf{BB}}^n$ Verify oracle outputs 1 on input $((\delta_1, \delta_2, \ldots, \delta_n), (B_0^\alpha, -\frac{\lambda}{\alpha}, \delta_{n+1}))$, this implies that

$$
\begin{aligned}
(B_0^\alpha)^{y-\frac{\lambda}{\alpha}} &= g_1^{\delta_1} \ldots g_n^{\delta_n} g^{\delta_{n+1}} \cdot h \\
\Leftrightarrow (B_0^\alpha)^y \cdot B_0^{-\lambda} &= g_1^{\delta_1} \ldots g_n^{\delta_n} g^{\delta_{n+1}} \cdot h \\
\Leftrightarrow (B_0^\alpha)^y \cdot B_0^{-\lambda} &= C^\alpha B_0^{-\lambda} \\
\Leftrightarrow (B_0^\alpha)^y &= C^\alpha \\
\Leftrightarrow B_0^y &= C
\end{aligned}
$$

Note that α is necessarily different from 0, otherwise $B_0^\alpha = 1$ and would have been rejected by the $\mathsf{MAC}_{\mathsf{BB}}^n$ Verify oracle.

Thus, the honest verifier algorithm accepts, if and only if, $(B_0^\alpha, -\frac{\lambda}{\alpha}, \delta_{n+1})$ would be accepted by the $\mathsf{MAC}_{\mathsf{BB}}^n$ Verify algorithm for message $(\delta_1, \ldots, \delta_n)$. Similarly, we can argue that $\mathcal{B}$ can extract a valid MAC from the final Show protocol whenever $\alpha \neq 0$ and ShowVerify would have output 1. Thus, if $\mathcal{A}$ can cause ShowVerify to accept for some statement ϕ that is not satisfied by any of the attributes sets queried to $\mathcal{O}$BlindIssue, then $\mathcal{B}$ can extract a new message $(\delta_1, \ldots, \delta_n)$ and a valid tag for that message.

Anonymity. Suppose the user is trying to prove that he has a credential for attributes satisfying some statement ϕ. We want to show that there exists an algorithm SimShow that, for the adversary $\mathcal{A}$, is indistinguishable from Show but that only takes as input the statement ϕ and the secret key sk.

Let $\phi \in \Phi$ and $(m_1, \ldots, m_n) \in \mathcal{U}$ be such that $\phi(m_1, \ldots, m_n) = 1$. Let pp be the system public parameters, Y the issuer's public key and σ be such that CredVerify$(sk, \sigma, (m_1, \ldots, m_n)) = 1$. So, σ consists of a quadruple $(A, r, s_u, \tilde{C}_m) \in \mathbb{G} \times \mathbb{Z}_p \times \mathbb{Z}_p \times \mathbb{G}$ satisfying $A^{y+r} = g_1^{m_1} \ldots g_n^{m_n} g^{s_u} h$.

SimShow(sk, ϕ) behaves as follows: it chooses a random value $l' \in_R \mathbb{Z}_p^*$ as well as a random generator $E \in_R \mathbb{G}$. It then computes $B_0 = g_0^{l'}$ and $C = Y^{l'}$. It runs $\mathcal{A}$ with the values (B_0, C, E) as the first message, simulates the proof of knowledge π_3 and outputs whatever $\mathcal{A}$ outputs at the end of the proof.

Let us first show that the values B_0, C and E are distributed identically to those produced by Show. Note that since $A \neq 1$, there exists $x \in \mathbb{Z}_p$ such that $A = g_0^x$. For a random value $l \in_R \mathbb{Z}_p$, $B_0 = A^l = g_0^{lx} = g_0^{l'}$ for $l' = lx$. Therefore, we also have $C = A^{ly} = Y^{l'}$. Moreover, there exists t such that $E = C^{1/l}f^t$. Then, the values computed by SimShow are identical to those that the normal Show protocol would have produced. Owing to the zero-knowledge property of the proof of knowledge, we conclude that the resulting view is indistinguishable from that produced by the adversary interacting with Show.

C   ZKPK π_3 - Proof of Possession of a Credential

We describe an instantiation of our presentation protocol using non-interactive Schnorr-like proofs. As in [15], our protocol does not include proofs of any additional predicates ϕ, but outputs a commitment H on the attributes which may be used as input to further proof protocols.

Hereinafter, we detail the ZKPK $\pi_3 = \text{PoK}\{\alpha, \beta, \lambda, \delta_1, \ldots, \delta_{n+1}, \gamma, \theta : E = C^\alpha f^\beta \wedge H = g_1^{\delta_1} \ldots g_n^{\delta_n} g^{\delta_{n+1}} B_0^\lambda f^\beta \wedge C = E^\theta f^\gamma\}$ where $E = C^{1/l}f^t$, $H = E \cdot h^{-1} = g_1^{m_1} g_2^{m_2} \ldots g_n^{m_n} g^{s_u} B_0^{-r/l} f^t$ and $C = E^l f^{-tl}$.

Prover	Verifier
Private Input: $\vec{m} = (m_1, \ldots, m_n)$, l, t	
s_u and r	
Choose $a_1, a_2, \ldots, a_{n+6} \xleftarrow{R} \mathbb{Z}_q^*$	
Compute $t_1 \leftarrow C^{a_1} f^{a_2}$	
$t_2 \leftarrow g_1^{a_3} g_2^{a_4} \ldots g_n^{a_{n+2}} g^{a_{n+3}} B_0^{a_{n+4}} f^{a_2}$	
$t_3 \leftarrow E^{a_{n+5}} f^{a_{n+6}}$	
Compute $c = \mathcal{H}(Ch, t_1, t_2, t_3)$ $\quad\xleftarrow{\quad Ch \quad}$	**Choose** $Ch \in_R \mathbb{Z}_p^*$
Compute $R_1 \leftarrow a_1 + c/l$, $R_2 \leftarrow a_2 + ct$ $\xrightarrow{c, R_1, \ldots, R_{n+6}}$	**Compute** $t_1' = C^{R_1} f^{R_2} E^{-c}$
for $i \in \{1, \ldots, n\}$, $R_{i+2} \leftarrow a_{i+2} + cm_i$	$t_2' = g_1^{R_3} \ldots g_n^{R_{n+2}} g^{R_{n+3}} B_0^{R_{n+4}} f^{R_2} H^{-c}$
$R_{n+3} \leftarrow a_{n+3} + cs_u$, $R_{n+4} \leftarrow a_{n+4} - \frac{cr}{l}$	$t_3' = E^{R_{n+5}} f^{R_{n+6}} C^{-c}$
$R_{n+5} \leftarrow a_{n+5} + cl$, $R_{n+6} \leftarrow a_{n+6} - ctl$	**Check if** $c = \mathcal{H}(Ch, t_1', t_2', t_3')$

Proof. Let us prove that, when $C = B_0^y$, π_3 is a ZKPK of a MAC_{BB}^n (A, r, s_u) on a block of messages $(m_1, \ldots, m_n)$. The *completeness* of the protocol follows by inspection. The *soundness* follows from the extraction property of the underlying proof of knowledge[7]. In particular, the extraction property implies that for any prover $\mathcal{P}^*$ that convinces $\mathcal{V}$ with probability ε, there exists an extractor which interacts with $\mathcal{P}^*$ and outputs $(\alpha, \beta, \lambda, \delta_1, \ldots, \delta_{n+1}, \gamma, \theta)$ with probability $poly(\varepsilon)$. Moreover, if we assume that the extractor inputs consists of two transcripts *i.e.* $(\mathbb{G}, g, h, f, B_0, C, E, c, \tilde{c}, R_1, \ldots, R_{n+6}, \tilde{R}_1, \tilde{R}_2, \ldots, \tilde{R}_{n+6})$, the witness can be obtained by computing $\alpha = \frac{R_1 - \tilde{R}_1}{c - \tilde{c}}$; $\beta = \frac{R_2 - \tilde{R}_2}{c - \tilde{c}}$; $\delta_i = \frac{R_{i+2} - \tilde{R}_{i+2}}{c - \tilde{c}}$, $\forall i \in \{1, \ldots, n\}$; $\delta_{n+1} = \frac{R_{n+3} - \tilde{R}_{n+3}}{c - \tilde{c}}$; $\lambda = \frac{R_{n+4} - \tilde{R}_{n+4}}{c - \tilde{c}}$; $\theta = \frac{R_{n+5} - \tilde{R}_{n+5}}{c - \tilde{c}}$, $\gamma = \frac{R_{n+6} - \tilde{R}_{n+6}}{c - \tilde{c}}$; (all the computations are done mod p). The extractor succeeds when $(c - \tilde{c})$ is invertible in $\mathbb{Z}_p$. We know that $H = E \cdot h^{-1} = g_1^{\delta_1} \ldots g_n^{\delta_n} g^{\delta_{n+1}} B_0^\lambda f^\beta$

---

[7] For concurrent security, we could use the Dåmgard protocol [18] which converts any Σ protocol into a three-round interactive ZKPK secure under concurrent composition.

so $E = g_1^{\delta_1} \dots g_n^{\delta_n} g^{\delta_{n+1}} B_0^\lambda f^\beta h$. We also know that $E = C^\alpha f^\beta$ so $C^\alpha f^\beta = g_1^{\delta_1} \dots g_n^{\delta_n} g^{\delta_{n+1}} B_0^\lambda f^\beta h$ and then

$$C^\alpha = g_1^{\delta_1} \dots g_n^{\delta_n} g^{\delta_{n+1}} B_0^\lambda h. \tag{1}$$

Since $C = B_0^y$, we have $B_0^{\alpha y} = g_1^{\delta_1} \dots g_n^{\delta_n} g^{\delta_{n+1}} B_0^\lambda h$ and

$$B_0^{\alpha y - \lambda} = g_1^{\delta_1} \dots g_n^{\delta_n} g^{\delta_{n+1}} h. \tag{2}$$

If $\alpha \neq 0$, (2) implies that

$$(B_0^\alpha)^{y - \frac{\lambda}{\alpha}} = g_1^{\delta_1} \dots g_n^{\delta_n} g^{\delta_{n+1}} h. \tag{3}$$

Let $A = B_0^\alpha$, $r = -\frac{\lambda}{\alpha}$, $s_u = \delta_{n+1}$ and $m_i = \delta_i$ for $i \in \{1, \dots, n\}$.

If $\alpha \neq 0$, (3) implies that the prover knows a valid $\mathsf{MAC}_{\mathsf{BB}}^n$, (A, r, s_u) on a block of messages $(m_1, \dots, m_n)$. Note that $y - \frac{\lambda}{\alpha} \neq 0$, otherwise this would imply that the prover knows y which would be equal to $\frac{\lambda}{\alpha}$.

Let us now prove that $\alpha \neq 0$. We know that

$$C = E^\theta f^\gamma = (C^\alpha f^\beta)^\theta f^\gamma = C^{\alpha\theta} f^{\beta\theta + \gamma} \implies 1 = C^{\alpha\theta - 1} f^{\beta\theta + \gamma}. \tag{4}$$

- If the prover does not know the discrete logarithm of C in the base f, this implies that it only knows one representation $(0,0)$ of 1 in the base (C, f) [8]. Therefore, $\alpha\theta = 1$ which implies that $\alpha \neq 0$.

- Suppose now that the prover knows the discrete logarithm χ of C in the base f (*i.e.* $C = f^\chi$) and that $\alpha = 0$. Since $C = B_0^y$, we have $B_0^y = f^\chi$ and then $B_0 = f^{\frac{\chi}{y}}$ (since $Y = g_0^y \neq 1$, this implies that $y \neq 0 \mod p$). From (1) and since α is supposed to be equal to 0, we have that $h = g_1^{-\delta_1} g_2^{-\delta_2} \dots g_n^{-\delta_n} g^{-\delta_{n+1}} f^{-\lambda \frac{\chi}{y}}$.

So, the issuer could use the prover as a subroutine to compute a representation of h in the base $(g_1, g_2, \dots, g, f)$. As $(g_1, g_2, \dots, g, f)$ are random generators of $\mathbb{G}$, this is impossible under the DL assumption [8]. Therefore, this means that either $\mathcal{P}^*$ does not know the discrete logarithm of C in the base f or $\alpha \neq 0$. Both cases imply that $\alpha \neq 0$. We therefore conclude that $\alpha \neq 0$ and so the prover knows a valid $\mathsf{MAC}_{\mathsf{BB}}^n$ (A, r, s_u) on a block of messages $(m_1, \dots, m_n)$.

Finally, to prove (honest-verifier) *zero-knowledge*, we construct a simulator Sim that will simulate all interactions with any (honest verifier) $\mathcal{V}^*$.

1. Sim randomly chooses $l' \in_R \mathbb{Z}_p^*$ and a random generator $E \in_R \mathbb{G}$ and then computes $B_0 = g^{l'}$ and $C = Y^{l'}$.
2. Sim randomly chooses $c, R_1, \dots, R_{n+6} \in_R \mathbb{Z}_p^*$ and computes $t_1 = C^{R_1} f^{R_2} E^{-c}$, $t_2 = g_1^{R_3} \dots g_n^{R_{n+2}} g^{R_{n+3}} B_0^{R_{n+4}} f^{R_2} H^{-c}$ and $t_3 = E^{R_{n+5}} f^{R_{n+6}} C^{-c}$.
3. Sim outputs $S = \{B_0, C, E, c, R_1, R_2, \dots, R_{n+6}\}$.

Since $\mathbb{G}$ is a prime-order group, then the blinding is perfect in the first step. Indeed, there exists $x \in \mathbb{Z}_p$ such that for a valid $\mathsf{MAC}_{\mathsf{BB}}^n$ (A, r, s_u) on $(m_1, \dots, m_n)$: $A = g_0^x$.

For a random value $l \in \mathbb{Z}_p^*$, we therefore have $B_0 = A^l = g_0^{lx} = g_0^{l'}$ for $l' = lx$. This also implies that $C = A^{ly} = Y^{l'}$. Moreover, there exists t such that $E = C^{\frac{1}{t}} f^t$. Therefore S and $\mathcal{V}^*$'s view of the protocol are statistically indistinguishable.

References

1. Abe, M.: A secure three-move blind signature scheme for polynomially many signatures. In: Pfitzmann, B. (ed.) EUROCRYPT 2001. LNCS, vol. 2045, pp. 136–151. Springer, Heidelberg (2001). doi:10.1007/3-540-44987-6_9
2. Akagi, N., Manabe, Y., Okamoto, T.: An efficient anonymous credential system. In: Tsudik, G. (ed.) FC 2008. LNCS, vol. 5143, pp. 272–286. Springer, Heidelberg (2008). doi:10.1007/978-3-540-85230-8_25
3. Baldimtsi, F., Lysyanskaya, A.: Anonymous credentials light. In: CCS 2013, pp. 1087–1098. ACM (2013)
4. Barić, N., Pfitzmann, B.: Collision-free accumulators and fail-stop signature schemes without trees. In: Fumy, W. (ed.) EUROCRYPT 1997. LNCS, vol. 1233, pp. 480–494. Springer, Heidelberg (1997). doi:10.1007/3-540-69053-0_33
5. Boneh, D., Boyen, X.: Short signatures without random oracles. In: Cachin, C., Camenisch, J.L. (eds.) EUROCRYPT 2004. LNCS, vol. 3027, pp. 56–73. Springer, Heidelberg (2004). doi:10.1007/978-3-540-24676-3_4
6. Brands, S.: Untraceable off-line cash in wallet with observers. In: Stinson, D.R. (ed.) CRYPTO 1993. LNCS, vol. 773, pp. 302–318. Springer, Heidelberg (1994). doi:10.1007/3-540-48329-2_26
7. Brands, S.: Rapid demonstration of linear relations connected by boolean operators. In: Fumy, W. (ed.) EUROCRYPT 1997. LNCS, vol. 1233, pp. 318–333. Springer, Heidelberg (1997). doi:10.1007/3-540-69053-0_22
8. Brands, S.A.: An efficient off-line electronic cash system based on the representation problem. Technical report CS-R9323. CWI, Amsterdam (1993)
9. Camenisch, J., Dubovitskaya, M., Haralambiev, K., Kohlweiss, M.: Composable and modular anonymous credentials: definitions and practical constructions. In: Iwata, T., Cheon, J.H. (eds.) ASIACRYPT 2015. LNCS, vol. 9453, pp. 262–288. Springer, Heidelberg (2015). doi:10.1007/978-3-662-48800-3_11
10. Camenisch, J., Lysyanskaya, A.: An efficient system for non-transferable anonymous credentials with optional anonymity revocation. In: Pfitzmann, B. (ed.) EUROCRYPT 2001. LNCS, vol. 2045, pp. 93–118. Springer, Heidelberg (2001). doi:10.1007/3-540-44987-6_7
11. Camenisch, J., Lysyanskaya, A.: A signature scheme with efficient protocols. In: Cimato, S., Persiano, G., Galdi, C. (eds.) SCN 2002. LNCS, vol. 2576, pp. 268–289. Springer, Heidelberg (2003). doi:10.1007/3-540-36413-7_20
12. Camenisch, J., Lysyanskaya, A.: Signature schemes and anonymous credentials from bilinear maps. In: Franklin, M. (ed.) CRYPTO 2004. LNCS, vol. 3152, pp. 56–72. Springer, Heidelberg (2004). doi:10.1007/978-3-540-28628-8_4
13. Camenisch, J., Stadler, M.: Proof systems for general statements about discrete logarithms. Technical report (1997)
14. Canard, S., Coisel, I., Jambert, A., Traoré, J.: New results for the practical use of range proofs. In: Katsikas, S., Agudo, I. (eds.) EuroPKI 2013. LNCS, vol. 8341, pp. 47–64. Springer, Heidelberg (2014). doi:10.1007/978-3-642-53997-8_4
15. Chase, M., Meiklejohn, S., Zaverucha, G.: Algebraic MACs and keyed-verification anonymous credentials. In: CCS 2014, pp. 1205–1216. ACM (2014)
16. Chaum, D.: Security without identification: transaction systems to make big brother obsolete. Commun. ACM **28**(10), 1030–1044 (1985)
17. Chaum, D., Pedersen, T.P.: Wallet databases with observers. In: Brickell, E.F. (ed.) CRYPTO 1992. LNCS, vol. 740, pp. 89–105. Springer, Heidelberg (1993). doi:10.1007/3-540-48071-4_7

18. Damgård, I.: Efficient concurrent zero-knowledge in the auxiliary string model. In: Preneel, B. (ed.) EUROCRYPT 2000. LNCS, vol. 1807, pp. 418–430. Springer, Heidelberg (2000). doi:10.1007/3-540-45539-6_30

19. Fuchsbauer, G., Hanser, C., Slamanig, D.: Practical round-optimal blind signatures in the standard model. In: Gennaro, R., Robshaw, M. (eds.) CRYPTO 2015. LNCS, vol. 9216, pp. 233–253. Springer, Heidelberg (2015). doi:10.1007/978-3-662-48000-7_12

20. Fuchsbauer, G., Pointcheval, D., Vergnaud, D.: Transferable constant-size fair e-cash. Cryptology ePrint Archive, Report 2009/146 (2009). http://eprint.iacr.org/

21. Hinterwälder, G., Zenger, C.T., Baldimtsi, F., Lysyanskaya, A., Paar, C., Burleson, W.P.: Efficient e-cash in practice: NFC-based payments for public transportation systems. In: Cristofaro, E., Wright, M. (eds.) PETS 2013. LNCS, vol. 7981, pp. 40–59. Springer, Heidelberg (2013). doi:10.1007/978-3-642-39077-7_3

22. IBM: Specification of the identity mixer cryptographic library (revised version 2.3.0). IBM Research Report RZ 3730 (2010). http://domino.research.ibm.com/library/cyberdig.nsf/1e4115aea78b6e7c85256b360066f0d4/eeb54ff3b91c1d648525759b004fbbb1?OpenDocument

23. Paquin, C.: U-Prove Technology Overview V1.1 (Revision 2). In: Microsoft Technical report (2013). http://research.microsoft.com/pubs/166980/U-ProveTechnologyOverviewV1.1Revision2.pdf

24. Paquin, C., Zaverucha, G.: U-Prove cryptographic specification V1.1 (Revision 3). In: Microsoft Technical report (2013). http://research.microsoft.com/pubs/166969/U-ProveCryptographicSpecificationV1.1.pdf

25. de la Piedra, A., Hoepman, J.-H., Vullers, P.: Towards a full-featured implementation of attribute based credentials on smart cards. In: Gritzalis, D., Kiayias, A., Askoxylakis, I. (eds.) CANS 2014. LNCS, vol. 8813, pp. 270–289. Springer, Cham (2014). doi:10.1007/978-3-319-12280-9_18

26. Schnorr, C.P.: Efficient signature generation by smart cards. J. Cryptol. 4(3), 161–174 (1991)

27. Vullers, P., Alpár, G.: Efficient selective disclosure on smart cards using idemix. In: Fischer-Hübner, S., de Leeuw, E., Mitchell, C. (eds.) IDMAN 2013. IAICT, vol. 396, pp. 53–67. Springer, Heidelberg (2013). doi:10.1007/978-3-642-37282-7_5

A Robust and Sponge-Like PRNG with Improved Efficiency

Daniel Hutchinson[✉]

Royal Holloway, University of London, London, UK
dojh342@gmail.com

Abstract. Ever since Keccak won the SHA3 competition, sponge-based constructions are being suggested for many different applications, including pseudo-random number generators (PRNGs). Sponges are very desirable, being well studied, increasingly efficient to implement and simplistic in their design. The initial construction of a sponge-based PRNG (Bertoni et al. CHES 2010) based its security on the well known sponge indifferentiability proof in the random permutation model and provided no forward security.

Since then, another improved sponge-based PRNG has been put forward by Gaži and Tessaro (Eurocrypt 2016) who point out the necessity for a public seed to prevent an adversarial sampler from gaining non-negligible advantage. The authors further update the security model of Dodis et al. (CCS 2013) to accommodate a public random permutation, modelled in the ideal cipher model, and how this affects the notions of security.

In this paper we introduce Reverie, an improved and practical, sponge-like pseudo-random number generator together with a formal security analysis in the PRNG with input security model of Dodis et al. with the modifications of the Gaži and Tessaro paper.

We prove that Reverie is *robust* when used with a public random permutation; robustness is the strongest notion of security in the chosen security model. Robustness is proved by establishing two weaker notions of security, preserving and recovering security, which together, can be shown to imply the robustness result. The proofs utilise the H-coefficient technique that has found recent popularity in this area; providing a very useful tool for proving the generator meets the necessary security notions.

Keywords: Sponge · Pseudo-random number generator (PRNG) · Patarin's H-coefficient technique · Robustness · Keccak · SHA-3 · Ideal permutation model

1 Introduction

Randomness is an essential ingredient in almost every area of cryptography; yet in the literature, randomness is often sampled uniformly at random with little thought on how quickly this amount of "good" randomness can be generated in

© Springer International Publishing AG 2017
R. Avanzi and H. Heys (Eds.): SAC 2016, LNCS 10532, pp. 381–398, 2017.
https://doi.org/10.1007/978-3-319-69453-5_21

practice. The need for high quality randomness delivered quickly has spawned work on the various key aspects of a PRNG, such as the ability to produce randomness at a fast and reliable rate, and protection against adversaries who may be able to compromise parts of the generator's state or the environment in which it draws entropy. In practice, many generators in active use have not received valid security analysis, and, on the opposite side of the fence, many designs are created in a theoretical setting without the full scope of desirable properties for a PRNG in mind and as result, are impractical for active use.

Sponges. The sponge design is very simple and yet very powerful; it benefits from a large amount of analysis due to the success of Keccak [6] in the SHA3 competition in 2012. The design requires an n-bit state with a rate r and capacity c such that $n = r + c$; the r bits of the state s are known as the *outer* state, written $\bar{s}$ while the c bits are known as the *inner* state $\hat{s}$. The design initialises with an initial state of the zero state, and a random permutation π. The sponge has two algorithms; Absorb and Squeeze.

Previous constructions. The sponge-based PRNG construction first suggested by Bertoni et al. in [8] utilises a random permutation and relies on the sponge indifferentiability proof of [7] for security. This analysis, though useful, does not consider security in terms of a security model for PRNGs. More recently, work by Gaži and Tessaro has improved upon this design and security claims, but still requires multiple additional calls to the permutation to ensure forward security, along with several additional strings to give a seeded design.

Ideal permutation model. We prove all of our security claims in the ideal permutation model where π is a public, random permutation picked at the beginning of any game. Any party has access to the permutation and may make forward and backward queries. We denote by $\mathcal{A}^\pi$ an adversary with oracle access to $\pi \xleftarrow{\$} \mathcal{P}_n$ with $\mathcal{P}_n$ being the space of all permutations on n-bit strings. We say that $\mathcal{A}^\pi$ is a q_π-query adversary if it makes at most q_π queries π.

PRNG security models. The development of security models for PRNGs has been slow due to a complex combination of security goals and the difficulty in accurately capturing the environment both the PRNG and the associated adversary are working in. Security models for PRNGs include work by Barak and Halevi [2], from 2005, a brilliantly simple model that introduced a very strong notion of Robustness.

This model was later improved upon in successive work by Dodis et al. [10], which initially aims to address the situation where a PRNG accumulates entropy at a slow rate, and is at risk of "prematurely" being called before enough entropy has been gathered.

The model was then further improved in [11], which introduced the idea of a scheduler, inspired by the design of the Fortuna PRNG [12] which aimed at a design to improve the recovery time of a compromised PRNG. We will not be considering a scheduler in this paper and will keep to the definitions of [10];

however, the idea of a scheduler is an interesting prospect in terms of possibly replacing the need for seed described below.

Seedless design. More recently, work done concurrently to the first draft of this paper, by Gaži and Tessaro [13], concentrates on the importance of a "seeded" design when using a public ideal permutation. The authors argue that a publicly available permutation allows an adversary to generate PRNG inputs *dependent* on the permutation. These "bad" distributions can output high entropy inputs but result in a predictable bit of the state, and thus result in a non-negligible advantage for the adversary. The authors of this work ensure their implementation is seeded by requiring a small number ($s = 2$ or 3) of r-bit strings that are used as additional inputs to prevent this attack.

We note that this adds to the initial entropy requirements of the PRNG, which can already be one of the most restrictive and problematic situations for a PRNG. Another addition is the need for a counter to be kept; this is absorbed into the refresh procedure but should be an additional part of the state. Fortunately due to work in [15], this would not affect security. Alternatively, this could merely be an identifier of the system on which the PRNG is implemented, along with the current time of the system clock, which could be hashed to provide the seed, though in the security games the seed is chosen uniformly at random. Our design is aimed at being practical and efficient; in a practical scenario the distribution sampler or entropy accumulation mechanism is not so easily influenced and discovering these "bad" distributions is very difficult when good, studied entropy sources are used.

We include the option of a seed so that robustness can be achieved, but we question the necessity of the seed in a practical scenario; this can be likened to many PRNGs made for practical use having the option of a "personalisation" string [16], but note that this is often not used or even implemented. In practical implementations the PRNG does not have direct access to a noise source, but rather an entropy source that has been studied and provides a minimum entropy estimate, along with post processing and health checking [3,4].

Notation. In this paper we denote by s_i the ith n−bit state of a generator. In the context of sponges we work with an n−bit state s_i which is split into an inner state of c-bits, denoted by $\widehat{s}_i$. The rest of the state is called the outer state, of r-bits and is denoted $\bar{s}_i$. Thus, the state can be given as $s_i = (\bar{s}_i \| \widehat{s}_i)$ where $\|$ is the usual concatenation of strings. The construction defined in this paper utilises a public, random permutation π from the set $\mathcal{P}_n$ of all permutations on n-bits. We use $x \xleftarrow{\$} X$ to denote an element x of a set X chosen uniformly at random. We denote by I_i the ith r-bit input string, used to refresh the state of a generator. We denote by r_i the ith output of a generator. These counters are in fact dependent on the state counter, so rather than the i-th output, we refer to the output associated with state i.

Contributions. We put forward an improved sponge-like PRNG design which we prove is robust in the updated security model. The recent work by Gaži and

Tessaro updated the security of the sponge-based PRNG design of Bertoni et al. but did not seek to improve the design of the next procedure. We improve the design of the next function to ensure our design is more efficient, making a single call to the permutation π, compared with $1 + t$ calls; resulting in a design better suited for practical application, especially those that restrict the number of calls to π. Since the p.forget procedure of the previous generator calls the permutation $1 + t$ times, with zeroing, it presents the problem of increased collisions in the state, something that is avoided by our design and thus our bound is mainly limited by the collision factor associated with the refresh procedure. This potentially makes our generator comparatively more secure when first initialised on a random initial state and before any refreshes have been made. Below are the two main components of the new design Reverie.

- Reverie.refresh$^{\pi}(s_i, \mathrm{I}, \mathsf{seed}, j) = \pi((\bar{s}_i \oplus \mathrm{I} \oplus \mathsf{seed}_j) \| \hat{s}_i) = s_{i+1}, j = j + 1 \mod s$,
- Reverie.next$^{\pi}(s_i) = (\pi(s_i) \oplus (0^r \| \hat{s}_i), \bar{s}_i) = (s_{i+1}, r_{i+1})$.

The security notion of interest in this paper is the strongest security notion, "robustness" which, informally, refers to an adversary working in time t, with access to a distribution sampler $\mathcal{D}$ that outputs refresh material used to update the state of the generator.

The adversary is allowed up to $q_\mathcal{D}$ outputs from the distribution sampler $\mathcal{D}$, these strings are required to have a minimum entropy when being used to refresh the generator from a compromised state. The adversary also has access to two algorithms get-next and next-ror which give the adversary output from the generator or random. The adversary is allowed up to q_R queries between these two algorithms.

Lastly, the adversary has up to q_S queries to set-state and get-state which give the adversary the current state of the generator and in the case of the former, allow it to set the state. In addition, the generator is said to be "uncompromised" if the current state has minimum entropy $\geq \gamma^*$ for some value γ^*. We say a generator is $((t, q_\mathcal{D}, q_R, q_S), \gamma^*, \epsilon)$ robust where ϵ is the maximum advantage of any adversary playing the robustness game.

The design can be seen in Fig. 2 for further clarity. Although this design departs slightly from the sponge design, it can still be captured by the more generalised structure of the parazoa as defined in [1], and, given access to the underlying permutation function, easily implemented.

Organisation. This paper is organised into preliminaries in Sects. 2 and 3, followed by the description of the new generator in Sect. 4, the security analysis of the generator in Sect. 5 and finally a discussion of results in Sect. 6.

2   Preliminary Definitions

This section aims to provide a background on all the necessities of pseudo-random number generators (PRNG), the ideal permutation model, along with an introduction to Patarin's H-coefficient technique.

2.1 Probabilities and Further Notation

Definition 1. *The statistical distance between two discrete random variables* X *and* Y *over the set* $\mathcal{X}$ *is denoted*

$$\mathsf{SD}(X, Y) = \frac{1}{2} \sum_{x \in \mathcal{X}} |\Pr[X = x] - \Pr[Y = x]|.$$

Definition 2. *The minimum entropy of a random variable* X *is defined as* $H_\infty(X) = \min_{x \xleftarrow{\$} X}\{-\log(\Pr[X = x])\}.$

Definition 3. *For the purposes used in this paper, a source* S^π *is defined as an input-less randomised oracle which makes queries to* π *and outputs a string. The range of the source is denoted* $[S]$ *and is the set of all values the source outputs with positive probability, taken over the choice of* π *and the internal randomness of* S.

We use the usual game-based formalism from [5]; for a game G, $\mathsf{G}(\mathcal{A}) \Rightarrow 1$ denotes the event that an adversary $\mathcal{A}$ playing the game G, results in the game outputting 1, while $\mathsf{G}(\mathcal{A}) \to 1$ denotes the event that the $\mathcal{A}$ playing the game G outputs 1.

2.2 PRGs and PRNGs

In this document a PRG will refer to a pseudo-random number generator *without* input, while PRNG will refer to a pseudo-random number generator *with* input and in the form described in Definition 4.

Definition 4 (PRNG from [10]). *A PRNG with input is a triple of algorithms* $G = (setup, refresh, next)$ *and a triple* $(n, \ell, p) \in \mathbb{N}^3$ *where:* n *is the state length,* ℓ *is the output length,* p *is the input length of* G *and*

- setup: *is a probabilistic algorithm that outputs some public parameters* seed *for the generator.*
- refresh: *is a deterministic algorithm that, given* seed, *a state* $s_i \in \{0, 1\}^n$ *and an input* $\mathrm{I} \in \{0, 1\}^p$, *outputs a new state* $s_{i+1} := \mathsf{refresh}(s_i, \mathrm{I}, \mathsf{seed})$
- next: *is a deterministic algorithm that, given* seed *and a state* $s_i \in \{0, 1\}^n$, *outputs a pair* $(s_{i+1}, r_{i+1}) = \mathsf{next}(\mathsf{seed}, s_i)$, *where* s_{i+1} *is the new state and* $r_{i+1} \in \{0, 1\}^\ell$ *is the output. We write* $\mathsf{next}(s_i)$ *and omit* seed *for clarity.*

Definition 5 (Originally of [10] but as amended in [13]). *A* Q-*distribution sampler is a randomised stateful oracle algorithm* $\mathcal{D}$ *which operates as follows:*

- It takes a state σ_i, with initial state $\sigma_0 = \perp$.
- $\mathcal{D}^\pi(\sigma_i)$ outputs a tuple $(\sigma_i, \mathcal{S}_i, \gamma_i, z_i)$, where
  - σ_i is the new state of $\mathcal{D}^\pi$.
  - $\mathcal{S}_i$ is a source with range $[\mathcal{S}_i] \subseteq \{0, 1\}_i^\ell$ for some $\ell_i \geq 1$.
  - γ_i is an entropy estimation for $\mathcal{S}_i$ which will be discussed further below.
  - z_i is the leakage and/or auxiliary information about $\mathcal{S}_i$.
- When run $q_\mathcal{D}$ times, the number of queries to the permutation π made by $\mathcal{D}^\pi$ and $\mathcal{S}_1, \ldots, \mathcal{S}_{q_\mathcal{D}}$ is at most $Q(q_\mathcal{D})$.

For simplicity, $(\sigma_i, I_i, \gamma_i, r_i) \xleftarrow{\$} \mathcal{D}^\pi(\sigma_{i-1})$ is written as the overall process of running $\mathcal{D}$ and the generated source $\mathcal{S}_i$. Next, we note the requirement for some restriction on distribution samplers, namely we require the following:

Definition 6. *A distribution sampler $\mathcal{D}$ as defined above in Definition 5 is $(q_\mathcal{D}, q_\pi)$-legitimate, if, for every adversary $\mathcal{A}$ making q_π queries, every $i^* \in [q_\mathcal{D}]$, and for any possible values $(I_j)_{j \neq i^*}, (\gamma_1, z_1), \ldots, (\gamma_{q_\mathcal{D}}, z_{q_\mathcal{D}}), V_\mathcal{A}, Q_\mathcal{D}$ potentially output by the game $\mathsf{GLEG}_{q_\mathcal{D}, i^*}(\mathcal{A}, \mathcal{D})$ with positive probability,*

$$\Pr\left[I_{i^*} = x \mid (I_j)_{j \neq i^*}, (\gamma_1, z_1), \ldots, (\gamma_{q_\mathcal{D}}, z_{q_\mathcal{D}}), V_\mathcal{A}, Q_\mathcal{D}\right] \leq 2^{-\gamma_{i^*}},$$

for all $x \in \{0,1\}^{\ell_{i^*}}$, where the probability is conditioned on these particular values being output by the game. The game $\mathsf{GLEG}_{q_\mathcal{D}, i^*}(\mathcal{A}, \mathcal{D})$, is defined in full in [13, Definition 3, p. 10] and presented in Appendix A, but informally, the challenger samples a permutation π, $\mathcal{D}^\pi$ is run $q_\mathcal{D}$ times and the adversary $\mathcal{A}$ is run on all of the output from $\mathcal{D}^\pi$, apart from that of $\mathcal{S}_i$ and its associated queries. $V_\mathcal{A}$ is the adversary's final output, while $Q_\mathcal{D}$ is the input-output pairs of permutation queries made by $\mathcal{D}$.

2.3 The Ideal Permutation Model (IPM)

An implementation of a sponge-based PRNG would involve a publicly available permutation; hence, our analysis is done in the ideal permutation model. Formally, each party has oracle access to a public, random permutation $\pi \xleftarrow{\$} \mathcal{P}_n$, chosen by the challenger at the beginning of a game. The permutation can be queried as both π and π^{-1} but for simplicity, we write that an algorithm or entity, such as an adversary $\mathcal{A}$, has access to π by $\mathcal{A}^\pi$. We make use of the following, which denotes the advantage of an adversary $\mathcal{A}$ with oracle access to π in distinguishing between the distributions D_0, D_1 that also have access to π:

$$\mathsf{Adv}_\mathcal{A}^{\mathsf{dist}}(D_0, D_1) = \left| \Pr\left[X \xleftarrow{\$} D_0^\pi : \mathcal{A}^\pi(X) \Rightarrow 1\right] - \Pr\left[X \xleftarrow{\$} D_1^\pi : \mathcal{A}^\pi(X) \Rightarrow 1\right] \right|,$$

with $\mathcal{A}$ being called a q_π-query adversary if it asks at most q_π queries to π.

2.4 Patarin's H-Coefficient Technique

This section gives a brief introduction to Patarin's H-coefficient technique with a focus on functionality. Influenced by [9] and initially defined in [14], the H-coefficient technique is applied by splitting the "transcripts" of a game into two or more distinct sets; calculating the probability of the real or ideal world outputting transcripts in a particular set yields a close bound for the statistical distance of the real and ideal world.

A high level overview is that of a q-query information theoretic adversary $\mathcal{A}$ which can be assumed to be deterministic, making no redundant queries without loss of generality, interacting with an oracle ω representing either the real world or ideal world. The interaction $\mathcal{A}$ has with this oracle ω is represented in a transcript τ which includes a list of queries and their answers given by ω.

Let ω be an oracle that serves as the way the adversary $\mathcal{A}$ interacts with the challenger in the chosen world. Let Ω_X refer to the probability space of all real world oracles with the uniform probability distribution, and similarly Ω_Y is the probability space of all ideal world oracles again with the uniform distribution.

Let $\mathcal{T}$ be the set of all transcripts, with $\tau \in \mathcal{T}$ an individual transcript that describes, in full, the interactions and final output between the adversary $\mathcal{A}$ and the oracle she interacts with.

Further, the random variables X and Y are defined over the probability spaces respectively, where $X(\omega) = \tau$ refers to running $\mathcal{A}$ on oracle ω for $\omega \in \Omega_X$, which in turn produces the transcript τ.

For simplicity we will only consider two sets; good and bad transcripts, which are denoted $\mathcal{T}_G$ and $\mathcal{T}_B$ respectively. Defining this split is integral to the proof since the H-coefficient technique allows bounding the statistical distance of the random variables X and Y in the following way: suppose $\exists \epsilon \in [0, 1]$, such that $\forall \tau \in \mathcal{T}_G$, with $\Pr[Y = \tau] > 0$,

$$\frac{\Pr[X = \tau]}{\Pr[Y = \tau]} \geq 1 - \epsilon.$$

Finally,

Theorem 1 (H-coefficient). *Let* $X, Y, \mathcal{T}_G, \mathcal{T}_B, \tau, \epsilon$ *be as above, then,*

$$\mathsf{SD}(X, Y) \leq \epsilon + \Pr[Y \in \mathcal{T}_B].$$

3 Security Notions

This section defines the notion of robustness originally from [10], but augmented as in [13] to allow for the publicly available random permutation. Robustness is the strongest security notion of the security model. We also include definitions of two weaker notions of security; preserving and recovering security, which together imply that a PRNG fulfils the requirements of robustness.

As per the definitions of [10], a minimal "fresh" entropy in the PRNG system when security should be expected. Minimising γ^* corresponds to a stronger security guarantee.

An adversary is modelled using a pair $(\mathcal{A}, \mathcal{D})$ where $\mathcal{A}$ is the actual q_π-query adversary and $\mathcal{D}$ is a $(q_\mathcal{D}, q_\pi)$-legitimate distribution sampler. The adversary $\mathcal{A}$'s goal is to determine a challenge bit b picked during the initialise procedure, this procedure also returns seed to the adversary.

Definition 7. *A PRNG with input G, is called $((q_\pi, q_\mathcal{D}, q_R, q_S), \gamma^*, \epsilon_{\mathsf{rob}})$-robust* (ROB$_G^{\gamma^*}$) *if for any adversary $\mathcal{A}$ making at most q_π queries to $\pi^\pm$, making at most $q_\mathcal{D}$ calls to $\mathcal{D}$-refresh, q_R calls to Next-ror/Get-next and q_S calls to Get-state/Set-state and any legitimate distribution sampler $\mathcal{D}$, the advantage of any adversary in the robustness game is at most ϵ_{rob} which is defined below.*

The adversary $\mathcal{A}$ has access to a subset of the following oracles, dependent on the security game that it's playing; the full set is available in ROB$_G^{\gamma^*}(\mathcal{A}, \mathcal{D})$.

We say that an adversarial pair $(\mathcal{A}, \mathcal{D})$ playing the robustness game as described below in Sect. 3 for a PRNG G have advantage

$$\mathsf{Adv}_G^{\gamma^*-\mathsf{ROB}}(\mathcal{A}, \mathcal{D}) := \left| 2\Pr\left[\mathsf{ROB}_G^{\gamma^*}(\mathcal{A}, \mathcal{D}) \Rightarrow 1\right] - 1 \right| \leq \epsilon_{\mathsf{rob}}.$$

Next, we define two further security notions: preserving security and recovering security. If a PRNG satisfies both these notions, then by Theorem 1 of [10] (with updated version from [13]) the generator in question satisfies the robustness security notion under the corresponding parameters. Next we define preserving and recovering security (Fig. 1).

Proc. Initialise	**Proc. Next-ror**	**Proc. Get-state**

$\pi \xleftarrow{\$} \mathcal{P}_n$
 seed $\xleftarrow{\$}$ setup$^\pi$
 $s_0 \xleftarrow{\$} \{0,1\}^n$
 $\sigma \leftarrow \perp$
 corrupt $\leftarrow$ false
 $e \leftarrow n$
 $b \xleftarrow{\$} \{0,1\}$
 return seed
	$(s_{i+1}, r_0) \leftarrow \mathsf{next}^\pi(s_i, \mathsf{seed})$
	 $r_1 \xleftarrow{\$} \{0,1\}^l$
	 if corrupt $=$ true **then**
	 $\quad e \leftarrow 0$
	 $\quad$ **return** r_0
	 else
	 $\quad$ **return** r_b
	 end if
		$e \leftarrow 0$
		 corrupt $\leftarrow$ true
		 return s_i
		
		 Proc. Set-state(s^*)
		
		 $e \leftarrow 0$
		 corrupt $\leftarrow$ true
		 $s_i \leftarrow s^*$

Proc. Finalise(b^*)	**Proc. Get-next**	**Proc. $\mathcal{D}$-refresh**

if $b = b^*$ **then**
 $\quad$ **return** 1
 else
 $\quad$ **return** 0
 end if
	$(s_{i+1}, r_i) \leftarrow \mathsf{next}^\pi(s_i, \mathsf{seed})$
	 if corrupt $=$ true **then**
	 $\quad e \leftarrow 0$
	 end if
	 return r_i
		$(\sigma, \mathrm{I}, \gamma, z) \xleftarrow{\$} \mathcal{D}^\pi(\sigma)$
		 $s_{i+1} \leftarrow \mathsf{refresh}^\pi(s_i, \mathrm{I}, \mathsf{seed})$
		 $e \leftarrow e + \gamma$
		 if $e \geq \gamma^*$ **then**
		 $\quad$ corrupt $\leftarrow$ false
		 end if
		 return (γ, z)

Proc. $\pi(x)$	**Proc. $\pi^{-1}(x)$**
return $\pi(x)$	**return** $\pi^{-1}(x)$

Fig. 1. $\mathsf{ROB}_G^{\gamma^*}(\mathcal{A}, \mathcal{D})$

3.1 Preserving Security

Informally, preserving security states that if the state of a generator starts uncompromised, is refreshed using compromised input, then the next output and resulting state are still indistinguishable from random.

Definition 8. *A PRNG with input is said to have $(q_\pi, \epsilon_{\mathsf{pres}})$-preserving security if the advantage of any adversary $\mathcal{A}$ making at most q_π queries to $\pi^\pm$ in the following game is at most ϵ_{pres}, where the advantage is defined to be*

$$\mathsf{Adv}_G^{\mathsf{PRES}}(\mathcal{A}) := \left| 2\Pr\left[\mathsf{PRES}_G(\mathcal{A}) \Rightarrow 1\right] - 1 \right| \leq \epsilon_{\mathsf{pres}}.$$

$\mathsf{PRES}_G(\mathcal{A})$

$\pi \overset{\$}{\leftarrow} \mathcal{P}_n, \mathsf{seed} \overset{\$}{\leftarrow} \mathsf{setup}^\pi(), b \overset{\$}{\leftarrow} \{0,1\}, s_0 \overset{\$}{\leftarrow} \{0,1\}^n$

$(\mathrm{I}_1, \ldots, \mathrm{I}_d) \leftarrow \mathcal{A}^\pi(\mathsf{seed})$

for $j = 1, \ldots, d$ **do**

    $s_j \leftarrow \mathsf{refresh}^\pi(s_{j-1}, \mathrm{I}_j, \mathsf{seed})$ **if** $b = 0$ **then** $(S, T) \leftarrow \mathsf{next}^\pi(s_d, \mathsf{seed})$

    else $(S, T) \overset{\$}{\leftarrow} \{0,1\}^n \times \{0,1\}^r$

    $b^* \leftarrow \mathcal{A}^\pi((S, T))$

    return $b == b^*$

3.2 Recovering Security

Informally, recovering security implies that if a PRNG is compromised, inserting enough random entropy to refresh the internal state will ensure that the next output and state will be indistinguishable from random.

Definition 9. *A PRNG with input has* $(q_\pi, q_\mathcal{D}, \gamma^*, \epsilon_{\mathsf{rec}})$-*recovering security if the advantage of any adversary* $\mathcal{A}$ *making at most* $q_\mathcal{D}$ *queries to* $\pi^\pm$ *and distribution sampler* $\mathcal{D}$, *making at most* $Q(q_\mathcal{D})$ *queries to* $\pi^\pm$, *in the following game with* $\gamma^* > 0$ *is at most* ϵ_{rec} *where advantage is defined as*

$$\mathsf{Adv}_G^{(\gamma^*, q_\mathcal{D})-\mathsf{rec}}(\mathcal{A}, \mathcal{D}) := \left| 2\Pr\left[\mathsf{REC}_G^{(\gamma^*, q_\mathcal{D})} \Rightarrow 1\right] - 1 \right| \leq \epsilon_{\mathsf{rec}}.$$

$\mathsf{REC}_G^{(\gamma^*, q_\pi)}(\mathcal{A}, \mathcal{D})$ **Oracle get-refresh()**

$\pi \overset{\$}{\leftarrow} \mathcal{P}_n, \mathsf{seed} \overset{\$}{\leftarrow} \mathsf{setup}^\pi(), b \overset{\$}{\leftarrow} \{0,1\}, \sigma_0 \leftarrow \perp$ $k \leftarrow k + 1$

for $k = 1, \ldots, q_\mathcal{D}$ **do**   **return** I_k

  $(\sigma_k, \mathrm{I}_k, \gamma_k, z_k) \leftarrow \mathcal{D}^\pi(\sigma_{k-1})$

$k \leftarrow 0$

$(s_0, d) \leftarrow \mathcal{A}^{\pi, \mathsf{get\text{-}refresh}()}(\gamma_1, \ldots, \gamma_{q_\mathcal{D}}, z_1, \ldots, z_{q_\mathcal{D}}, \mathsf{seed})$

if $k + d > q_\mathcal{D}$ **then return** $\perp$

else

if $\displaystyle\sum_{j=k+1}^{k+d} \gamma_j < \gamma^*$ **then return** $\perp$

else

for $j = 1, \ldots, d$ **do**

  $s_j \leftarrow \mathsf{refresh}^\pi(s_{j-1}, \mathrm{I}_{k+j}, \mathsf{seed})$

if $b = 0$ **then** $(S, T) \leftarrow \mathsf{next}^\pi(\mathsf{seed}, s_d)$

else $(S, T) \overset{\$}{\leftarrow} \{0,1\}^n \times \{0,1\}^r$

$b^* \leftarrow \mathcal{A}^\pi((S, T), \mathrm{I}_{k+d+1}, \ldots, \mathrm{I}_{q_\mathcal{D}})$

return $b == b^*$

---

4   Improved Construction

The following algorithms describe Reverie, a sponge-like PRNG with forward security that does not require additional calls to the underlying public permutation. Let $s, r, c \geq 1$ and $c := n - r, \ell = p = r$, together with $\pi \xleftarrow{\$} \mathcal{P}_n$, then $\mathsf{Rev}^\pi_{s,n,r} := (\mathsf{Reverie.setup}^\pi, \mathsf{Reverie.refresh}^\pi, \mathsf{Reverie.next}^\pi)$ for:

Proc. .setup$^\pi()$	**Proc.** .refresh$^\pi(s_i, I, \mathsf{seed})$	**Proc.** .next$^\pi(s_i, \mathsf{seed})$
for $i = 0, \ldots, s-1$ **do** $\mathsf{seed}_i \xleftarrow{\$} \{0,1\}^r$ **end for** $\mathsf{seed} \leftarrow$ $(\mathsf{seed}_0, \ldots, \mathsf{seed}_{s-1})$ $j \leftarrow 1$	$s_{i+1} \leftarrow \pi((\bar{s}_i \oplus I \oplus \mathsf{seed}_j) \| \widehat{s}_i)$ $j \leftarrow j + 1 \mod s$ **return** s_{i+1}	$r_{i+1} \leftarrow \bar{s}_i$ $s_{i+1} \leftarrow$ $(\pi(s_i) \oplus (0^r \| \widehat{s}_i))$ **return** (s_{i+1}, r_{i+1})

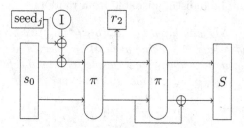

Fig. 2. Reverie.

5   Security of Reverie

This section consists of the security proofs of Reverie; the approach is to analyse the security of the next function, and then focus on the preserving and recovering security games, making use of the previous result.

Theorem 2. *For Reverie* $= \mathsf{Rev}^\pi_{s,n,r}$ *as defined above, let* $\gamma^* > 0$, *let* $\mathcal{D}$ *be a* $(q_\mathcal{D}, q_\pi)$-*legitimate distribution sampler, let* $\bar{q}_\pi := q_\pi + Q(q_\mathcal{D})$ *and* $\widehat{q} := \bar{q}_\pi + q_R + q_\mathcal{D}d$. *Then* $\mathsf{Rev}^\pi_{s,n,r}$ *is* $((q_\pi, q_\mathcal{D}, q_R, q_S), \gamma^*, \epsilon_{\mathsf{rob}})$-*robust, for* ϵ_{rob} *as below:*

$$\mathsf{Adv}^{\gamma^*-\mathsf{rob}}_{\mathsf{Rev}^\pi_{s,n,r}}(\mathcal{A}, \mathcal{D}) \leq q_R \cdot \left(\frac{\bar{q}_\pi + 1}{2^{\gamma^*}} + \frac{Q(q_\mathcal{D})}{2^{sr}} + \frac{7(\widehat{q}^2 + 1) + 29\widehat{q}}{2^{c-1}} \right.$$
$$\left. + \frac{(2d^2 + 3)\widehat{q} + d(3d + 2d)}{2^n} \right).$$

Proof. The theorem is the result of the preserving and recovering security bounds in Lemmas 2 and 3 respectively, combined by [13, Theorem 4]. □

Lemma 1 (Security of the *next* function). *Let U_x is the uniform distribution over x-bit strings, let next be as defined in Sect. 4, let $s_0 \xleftarrow{\$} \{0,1\}^n$, then for any q_π-query adversary $\mathcal{A}$,*

$$\epsilon_{\text{next}}(q_\pi) := \mathsf{Adv}_{\mathcal{A}}^{\text{dist}}(\mathsf{next}(U_n), (U_n, U_r)) \leq \left(2 - \frac{1}{2^r}\right)\frac{q_\pi}{2^{c-1}} + \frac{3q_\pi}{2^{c-1}}$$

$$= \left(5 - \frac{1}{2^r}\right)\frac{q_\pi}{2^{c-1}}.$$

Proof outline. Distinguishing between $\mathsf{next}(s_0)$ and random output $(S,T) \xleftarrow{\$} \{0,1\}^n \times \{0,1\}^r$ naively, it seems like the adversary's only option is to guess the inner state of the secret initial state, by either a direct forward query to π or by an indirect guess that would reveal a candidate for this inner state through a query to π^{-1}.

The proof, given in Appendix B proves that this is in fact the optimal strategy. Since there are two parts to the challenge, the logical approach is to split the proof into first proving that one part of the challenge can be replaced with random, before approaching the remaining part of the challenge.

We note that unlike [13], the next function requires a uniformly random state; the difference is made up for in a game jump in the proof, but allows us to avoid an additional call to π, as is required in [13]. This step can be reinstated at the cost of a single additional call to π.

5.1 Preserving Security

Now that we have this tool, we can prove the following:

Lemma 2. *Given Reverie as defined in Sect. 4, and with ϵ_{next} as above, then for every q_π-query adversary $\mathcal{A}$ playing the preserving security game defined in Definition 8 with d adversarial refresh inputs, we have*

$$\mathsf{Adv}_{\mathcal{A}}^{\text{pres}}(Reverie[\pi]) \leq \epsilon_{\text{next}}(q_\pi) + \frac{q_\pi' + d}{2^n} + \frac{(d+1)(2q_\pi' + d)}{2^n}$$

$$\leq \frac{5q_\pi}{2^{c-1}} + \frac{(2d+3)q_\pi + d(d+2)}{2^n}.$$

Proof outline. The proof relies on proving that for a random secret initial state s_0, the resulting state s_d will look random and thus, by our previous analysis of the next function, the challenge output will also be random. The complete proof can be found in the full version of the paper.

5.2 Recovering Security

Thanks to the impressive result of [13] the proof of recovering security can be expressed as an adaptation of their result; using the sponge as an extractor, and the security of the next function. To formalise this:

Lemma 3. *Let $q_\pi, \bar{q}_\pi := q_\pi + Q(q_\mathcal{D}), r, s, c$ be as in Sect. 4. Let $\epsilon_{ext}(q_\pi, q_\mathcal{D})$ be as described in [13, Sect. 5.3] and similarly let $\epsilon_{next}(\bar{q}_\pi)$ be the bound as in Lemma 1 as a function of $\bar{q}_\pi$; both with n, r, c as previously described. Given Reverie, also as in Sect. 4, $\gamma^* > 0, q_\mathcal{D} \geq 0, \mathcal{A}$, a q_π-query adversary against recovering security, and $\mathcal{D}$, a $(q_\mathcal{D}, q_\pi)$-legitimate distribution sampler as defined in Definition 5. Then,*

$$\mathsf{Adv}_{\mathsf{Rev}_{s,n,r}^\pi}^{(\gamma^*,q_\pi)-\mathsf{rec}}(\mathcal{A}, \mathcal{D}) \leq \epsilon_{ext}(q_\pi + 1, q_\mathcal{D}) + 2\epsilon_{next}(\bar{q}_\pi) + \frac{q_\pi}{2^{n-1}}$$

$$\leq \frac{\bar{q}_\pi + 1}{2^{\gamma^*}} + \frac{Q(q_\mathcal{D})}{2^{sr}} + \frac{7(\bar{q}_\pi^2 + 1) + 24\bar{q}_\pi}{2^{c-1}} + \frac{(\bar{q}_\pi + 1)d + d^2 + q_\pi - 2\bar{q}_\pi}{2^{n-1}}.$$

Proof outline. The strategy of the proof is to use the extractor properties of the sponge to replace the resulting state with a random state; following this the output of next will be random by the arguments of Lemma 1. The complete proof can be found in the full version of the paper.

6   Conclusion

We have presented an updated construction, Reverie, for a sponge-like PRNG. The construction incorporates an effective and efficient forward-security mechanism and we have provided proofs of both preserving and recovering security in the chosen security model. Our design makes a single call to the permutation on every invocation of Reverie.next, while the comparable generators make $1 + t$ calls. Our design choice ensures the underlying permutation is called far fewer times. Thus, the loss of security from collisions is reduced when compared to the relevant bounds of other designs.

The main limiting factor of the bound relates to the recovering security bound; and more precisely the extraction bound. This begs the question: can this bound be improved? This is briefly discussed in [13] in the present setting, but we would also like to consider other, possibly similar mechanisms that may present a better security bound; for instance, would a full state refresh yield a better bound? A full state refresh however, enables in practise an adversary to more easily affect or even set the state of the generator.

A $\mathsf{GLEG}_{q_\mathcal{D},i^*}(\mathcal{A}, \mathcal{D})$

Below the full game $\mathsf{GLEG}_{q_\mathcal{D},i^*}(\mathcal{A}, \mathcal{D})$ is given, as in [13, Definition 3, p. 10] and following on from Definition 6:

Let $\mathcal{D}$ be a distribution sampler, $\mathcal{A}$ an adversary and fix an $i^* \in [q_\mathcal{D}]$. Let $Q_\mathcal{D}$ be the set of all input-output pairs of permutation queries made by $\mathcal{D}$ and by all S_j for $j \in [q_\mathcal{D}]/\{i^*\}$.

$$\boxed{\begin{aligned}
&\textbf{Game GLEG}_{q_\pi, i^*}(\mathcal{A}, \mathcal{D}) \\[4pt]
\hline
&\pi \xleftarrow{\$} \mathcal{P}_n \\
&\textbf{for } j = 1, \ldots, q_\mathcal{D} \textbf{ do} \\
&\quad (\sigma_i, S_i, \gamma_i, z_i) \xleftarrow{\$} \mathcal{D}^\pi \\
&\quad I_i \xleftarrow{\$} S_i^\pi \\
&\textbf{endfor} \\
&V_\mathcal{A} \xleftarrow{\$} \mathcal{A}((\gamma_j, z_j)_{j \in [q_\mathcal{D}]}, (I_j)_{j \in [q_\mathcal{D}]/i^*}) \\
&\textbf{return } ((I_1, \gamma_1, z_1), \ldots, (I_{q_\mathcal{D}}, \gamma_{q_\mathcal{D}}, z_{q_\mathcal{D}}), V_\mathcal{A}, Q_\mathcal{D})
\end{aligned}}$$

Then $\mathcal{D}$ is said to be a $(q_\mathcal{D}, q_\pi)$-legitimate distribution sampler if for every adversary $\mathcal{A}$ making q_π queries and every $i^* \in [q_\mathcal{D}]$, all possible values of $(I_j)_{j \in [q_\mathcal{D}]/(i^*)}, (\gamma_1, z_1), \ldots, (\gamma_{q_\mathcal{D}}, z_{q_\mathcal{D}}), V_\mathcal{A}, Q_\mathcal{D}$ potentially output by the above game with positive probability,

$$\Pr\left[I_{i^*} = x \mid (I_j)_{j \neq i^*}, (\gamma_1, z_1), \ldots, (\gamma_{q_\mathcal{D}}, z_{q_\mathcal{D}}), V_\mathcal{A}, Q_\mathcal{D}\right] \leq 2^{-\gamma_{i^*}},$$

for all $x \in \{0, 1\}$.

B   Proof of next Security

Proof. Lemma 1

Algorithm 1 $\mathsf{next}_0^\pi(s_0)$	**Algorithm 2** $\mathsf{next}_1^\pi(s_0)$	**Algorithm 3** $\mathsf{next}_2(s_0)$
$s_0 \xleftarrow{\$} \{0,1\}^n$	$s_0 \xleftarrow{\$} \{0,1\}^n$	$s_0 \xleftarrow{\$} \{0,1\}^n$
$T \leftarrow \bar{s}$	$T \xleftarrow{\$} \{0,1\}^r$	$T \xleftarrow{\$} \{0,1\}^r$
$t \leftarrow \pi(s_0)$	$t \leftarrow \pi(s_0)$	$S \xleftarrow{\$} \{0,1\}^n$
$S \leftarrow t \oplus (0^r \| \hat{s}_0)$	$S \leftarrow t \oplus (0^r \| \hat{s}_0)$	

These algorithms are set up so that on input $s_0 \xleftarrow{\$} \{0,1\}^n$, next_0 is precisely the next function on input s_0 while next_2 has the same distribution as (U_n, U_r). next_1^π will be used as a hybrid game. Thus, by the triangle inequality,

$$\begin{aligned}
\mathsf{Adv}_\mathcal{A}^{\mathsf{dist}}(\mathsf{next}(s_0), (U_n, U_r)) &\leq \mathsf{Adv}_\mathcal{A}^{\mathsf{dist}}(\mathsf{next}_0^\pi(s_0), \mathsf{next}_1^\pi(s_0)) \\
&\quad + \mathsf{Adv}_\mathcal{A}^{\mathsf{dist}}(\mathsf{next}_1^\pi(s_0), \mathsf{next}_2(s_0)).
\end{aligned}$$

What follows is to prove the bound using the H-coefficient technique. As described in Sect. 2.4, we assume that $\mathcal{A}$ is deterministic and makes q_π non-repeating queries to the permutation π, denoted as

$$\tau_\mathcal{A} := (x_1, y_1, z_1), \ldots, (x_{q_\pi}, y_{q_\pi}, z_{q_\pi})$$

where $\forall i \in [1, \ldots, q_\pi]$,

$$y_i = \pi(x_i),$$
$$z_i = y_i \oplus (0^r \| \widehat{x}_i).$$

In addition to the challenge, the adversary in this distinguishing game is also given several other pieces of information at the end of the game, after all queries to π have been made, but before the adversary must output her decision. Formally, $\mathcal{A}$ is given $\widehat{s}_0$ and $t' := (\overline{S} \| (\widehat{s}_0 \oplus \widehat{S}))$ which it can compute for itself but is given for clarity. This completes the definition of a transcript for these experiments,

$$\tau := ((x_1, y_1, z_1), \ldots, (x_{q_\pi}, y_{q_\pi}, z_{q_\pi}), \widehat{s}_0, t', (S, T)).$$

We say a transcript τ is compatible with $\mathsf{next}_0^\pi(s_0)$ if it can be output in the experiment where $\mathcal{A}$ receives $\mathsf{next}_0^\pi(s_0)$. Since $\mathsf{next}_1^\pi(s_0)$ and $\mathsf{next}_2(s_0)$ differ only by replacing real output with random, it's clear that if a transcript is compatible with $\mathsf{next}_0^\pi(s_0)$ then it is compatible with $\mathsf{next}_1^\pi(s_0)$ and $\mathsf{next}_2(s_0)$.

What follows is bounding the probability of different transcripts from each experiment.

Lemma 4. *For the experiments* $\mathsf{next}_0^\pi(s_0)$, $\mathsf{next}_1^\pi(s_0)$ *as described above,*

$$\mathsf{Adv}_{\mathcal{A}}^{dist}(\mathsf{next}_0^\pi(s_0), \mathsf{next}_1^\pi(s_0)) \le \left(2 - \frac{1}{2^r}\right)\frac{q_\pi}{2^{c-1}} + 0 = \left(2 - \frac{1}{2^r}\right)\frac{q_\pi}{2^{c-1}}.$$

Proof. First we define the bad transcripts for this pair of experiments:

Definition 10 (Bad transcripts $\mathcal{T}_B$ for ($\mathsf{next}_0^\pi(s_0)\mathsf{next}_1^\pi(s_0)$)). *A compatible transcript as above, is called a bad transcript if any of the following occur:*

$$\text{State Collision (SC): } \exists j \in [q_\pi] \text{ such that } x_j = (T \| \widehat{s}_0),$$
$$\text{Image Collision (IC): } \exists j \in [q_\pi] \text{ such that } y_j = t',$$

The set of bad transcripts is denoted $\mathcal{T}_B$.

Let X_0, Y_0 be the random variables outputting transcripts that describe when $\mathcal{A}$ interacts with $\mathsf{next}_0^\pi(s_0)$ and $\mathsf{next}_1^\pi(s_0)$ respectively.

Lemma 5. *For an adversary making no more than* $q_\pi \le 2^{c-1}$ *queries to an oracle in the experiment* $\mathsf{next}_1(s_0)$,

$$\Pr[Y_0 \in \mathcal{T}_B] \le \left(2 - \frac{1}{2^r}\right)\frac{q_\pi}{2^{c-1}}.$$

Proof. Note that if $Y_0 \in \mathcal{T}_B$ then SC $\vee$ IC must occur.

$$\Pr[Y_0 \in \mathcal{T}_B] \le \Pr[\text{SC}] + \Pr[\text{IC} \mid \neg\text{SC}],$$

The first probability is relatively easy to bound,

$$\Pr[\mathrm{SC}] \le \frac{q_\pi}{2^{c-1}}. \tag{B.1}$$

Since the adversary is given T at the start of the game and s_0 is uniformly distributed over all the 2^c n-bit strings with outer bits equal to T, and recalling that $q_\pi \le 2^{c-1}$, the probability that $\mathcal{A}$'s i-th query is of the form $((T\|\widehat{s}_0), y_i, z_i)$ is $\frac{1}{2^c - i + 1}$. More formally, let $\Pr[win_i] := \Pr[x_i = (T\|\widehat{s}_0)]$, then

$$\Pr[win] \le \sum_{i=1}^{q_\pi} \Pr[win_i] = \sum_{i=1}^{q_\pi} \frac{1}{2^c - i + 1}$$

$$\le \sum_{i=1}^{q_\pi} \frac{1}{2^c - 2^{c-1}} = \frac{q_\pi}{2^{c-1}}.$$

The second, since SC has not occurred, must be where the adversary is interacting with $\mathsf{next}_1^\pi(s_0)$ where T was chosen uniformly at random from r-bit strings, and as such, was not used to produce S. There is the situation that the randomly chosen T matches the real value of $\widehat{s}_0$ which is reflected in the factor of $\left(1 - \frac{1}{2^r}\right)$.

The second probability is similar, in that the adversary has knowledge of $\bar{S}$, with $(\bar{S}\|(\widehat{s}_0 \oplus \widehat{S}))$ uniformly distributed over all the 2^c n-bit strings with outer bits equal to $\bar{S}$. It is also assumed that a SC has not occurred, meaning nothing beyond $\widehat{s}_0$ is known about s_0, then similarly to above,

$$\Pr[\mathrm{IC} \mid \neg\mathrm{SC}] \le \left(1 - \frac{1}{2^r}\right)\frac{q_\pi}{2^{c-1}}. \tag{B.2}$$

Equation (B.2), together with Eq. (B.1) complete the lemma.

Lemma 6. *For all compatible transcripts* $\tau \in \mathcal{T}_G$,

$$\Pr[X_0 = \tau] = \Pr[Y_0 = \tau].$$

Proof. For all $\tau \in \mathcal{T}_G$ (and for $\pi \xleftarrow{\$} \mathcal{P}_n$),

$$\Pr[X_0 = \tau] =$$

$$\Pr\left[\forall i \in [q_\pi], \pi(x_i) = y_i\right] \cdot \Pr\left[\pi(s_0) = \left(\bar{S}\|(\widehat{s}_0 \oplus \widehat{S})\right) \mid \neg\mathrm{SC} \vee \neg\mathrm{IC}\right]$$

$$= \frac{1}{2^r} 2^r \frac{(2^n - q_\pi - 1)!}{2^n!} = \Pr[Y_1 = \tau].$$

Putting Lemmas 5 and 6 together yields the result.

Next, we prove the following:

Lemma 7. *For the experiments* $\mathsf{next}_1^\pi(s_0), \mathsf{next}_2(s_0)$ *as described above and by Theorem 1,*

$$\mathsf{Adv}_{\mathcal{A}}^{dist}(\mathsf{next}_1^\pi(s_0), \mathsf{next}_2(s_0)) \le \frac{3q_\pi}{2^{c-1}} + 0 = \frac{3q_\pi}{2^{c-1}}.$$

Proof. This time, the transcript is slightly different, in that the adversary is given the entire s_0 at the end of her queries to π, so

$$\tau := ((x_1, y_1, z_1), \ldots, (x_{q_\pi}, y_{q_\pi}, z_{q_\pi}), s_0, t', (S, T)).$$

Comparing the distributions of these two experiments yields one more bad event, along with a modified state collision and unchanged image collision:

Definition 11 (Bad transcripts $\mathcal{T}_B$ for $(\mathbf{next}_1^\pi(s_0), \mathbf{next}_2(s_0))$). *A compatible transcript as above, is called a bad transcript if any of the following occur:*

$$\text{State Collision (SC): } \exists j \in [q_\pi] \text{ such that } x_j = s_0,$$
$$\text{Image Collision (IC): } \exists j \in [q_\pi] \text{ such that } y_j = t',$$
$$\text{Inversion (IN): } \exists j \in [q_\pi] \text{ such that } z_j = S.$$

The set of bad transcripts is denoted $\mathcal{T}_B$.

Let X_1, Y_1 be the random variables outputting transcripts that describe when $\mathcal{A}$ interacts with $\mathbf{next}_1^\pi(s_0)$ and $\mathbf{next}_2(s_0)$ respectively.

Lemma 8. *For an adversary making no more than $q_\pi \leq 2^{c-1}$ queries to an oracle in the experiment* $\mathbf{next}_2(s_0)$,

$$\Pr[Y_1 \in \mathcal{T}_B] \leq \frac{q_\pi}{2^{n-1}} + \left(2 - \frac{1}{2^r}\right)\frac{q_\pi}{2^{c-1}} = \frac{2q_\pi}{2^{c-1}}.$$

Proof. Note that if $Y_1 \in \mathcal{T}_B$ then SC $\vee$ IC $\vee$ IN must occur.

$$\Pr[Y_1 \in \mathcal{T}_B] \leq \Pr[\text{SC}] + \Pr[\text{IC} \mid \text{SC}] + \Pr[\text{IN} \mid \neg\text{SC} \wedge \neg\text{IC}],$$

The first probability is similar to before, but this time the adversary knows that $\bar{S}$ (with high probability) was not queried to π to produce the challenge. This results in the following:

$$\Pr[\text{SC}] \leq \frac{q_\pi}{2^{n-1}}.$$

The second probability is similar to the case where an IC occurs in a transcript in either $\mathbf{next}_0^\rho(s_0)$ or $\mathbf{next}_1^\pi(s_0)$. Once again since $\widehat{s_0}$ is uniformly distributed over $\{0,1\}^c$, the probability that any of the adversary's queries $(x_i) = y_i$ or $\pi^{-1}(y_i) = x_i$ is such that $y_i = (\bar{S} \| \widehat{S} \oplus \widehat{s_0})$ is at most $\frac{1}{2^{c-i+1}}$ resulting in the bound $\frac{q_\pi}{2^{c-1}}$. It is also assumed that a SC has not occurred, meaning nothing beyond $\widehat{s_0}$ is known about s_0. Thus,

$$\Pr[\text{IC} \mid \neg\text{SC}] \leq \left(1 - \frac{1}{2^r}\right)\frac{q_\pi}{2^{c-1}}.$$

Lastly, if neither a SC or IC has occurred, the probability of an IN can be expressed as

$$\Pr\left[\pi^{-1}(\bar{S} \| \widehat{y_i}) = \left(\widehat{x_i} \| (\widehat{y_i} \oplus \widehat{S})\right)\right],$$

which again is bounded by $\frac{q_\pi}{2^{c-1}}$ and together with the other events, yields the desired bound.

Lemma 9. *For all compatible transcripts* $\tau \in \mathcal{T}_G$,

$$\Pr[X_1 = \tau] = \Pr[Y_1 \doteq \tau].$$

For all $\tau \in \mathcal{T}_G$ (and for $\pi \xleftarrow{\$} \mathcal{P}_n$),

Proof.

$$\Pr[X_1 = \tau] = \Pr[\forall i \in [q_\pi], \pi(x_i) = y_i] \cdot \Pr\left[\pi(s_0) = \left(\overline{S} \| (\hat{s}_0 \oplus \hat{S})\right) \mid \neg\mathrm{SC} \vee \neg\mathrm{IC} \vee \neg\mathrm{IN}\right]$$

$$= \frac{(2^n - q_\pi - 1)!}{2^n!} = \frac{(2^n - q_\pi)!}{2^n} \cdot \frac{1}{2^n - q_\pi} = \Pr[Y_1 = \tau].$$

Putting Lemmas 8 and 9 together yields the result.

Finally, these two lemmas complete the proof of the security of next. □

References

1. Andreeva, E., Mennink, B., Preneel, B.: The parazoa family: generalizing the sponge hash functions. Cryptology ePrint Archive, Report 2011/028 (2011). http://eprint.iacr.org/2011/028
2. Barak, B., Halevi, S.: A model and architecture for pseudo-random generation with applications to /dev/random. In: Atluri, V., Meadows, C., Juels, A. (eds.) ACM CCS 05, Alexandria, Virginia, USA, 7–11 November, pp. 203–212. ACM Press (2005)
3. Barker, E., Kelsey, J.: Recommendation for random number generation using deterministic random bit generators, sp800-90a, April 2016. http://csrc.nist.gov/publications/drafts/800-90/sp800_90c_second_draft.pdf
4. Barker, E., Kelsey, J.: Recommendation for the entropy sources used for random bit generation, January 2016. http://csrc.nist.gov/publications/drafts/800-90/sp800-90b_second_draft.pdf
5. Bellare, M., Rogaway, P.: The security of triple encryption and a framework for code-based game-playing proofs. In: Vaudenay, S. (ed.) EUROCRYPT 2006. LNCS, vol. 4004, pp. 409–426. Springer, Heidelberg (2006). doi:10.1007/11761679_25
6. Bertoni, G., Daemen, J., Peeters, M., Van Assche, G.: The KECCAK reference, January 2011. http://keccak.noekeon.org/
7. Bertoni, G., Daemen, J., Peeters, M., Van Assche, G.: On the indifferentiability of the sponge construction. In: Smart, N. (ed.) EUROCRYPT 2008. LNCS, vol. 4965, pp. 181–197. Springer, Heidelberg (2008). doi:10.1007/978-3-540-78967-3_11
8. Bertoni, G., Daemen, J., Peeters, M., Van Assche, G.: Sponge-based pseudo-random number generators. In: Mangard, S., Standaert, F.-X. (eds.) CHES 2010. LNCS, vol. 6225, pp. 33–47. Springer, Heidelberg (2010). doi:10.1007/978-3-642-15031-9_3
9. Chen, S., Steinberger, J.: Tight security bounds for key-alternating ciphers. In: Nguyen, P.Q., Oswald, E. (eds.) EUROCRYPT 2014. LNCS, vol. 8441, pp. 327–350. Springer, Heidelberg (2014). doi:10.1007/978-3-642-55220-5_19
10. Dodis, Y., Pointcheval, D., Ruhault, S., Vergnaud, D., Wichs, D.: Security analysis of pseudo-random number generators with input: /dev/random is not robust. In: Sadeghi, A.-R., Gligor, V.D., Yung, M. (eds.) ACM CCS 13, Berlin, Germany, 4–8 November, pp. 647–658. ACM Press (2013)

11. Dodis, Y., Shamir, A., Stephens-Davidowitz, N., Wichs, D.: How to eat your entropy and have it too – optimal recovery strategies for compromised RNGs. In: Garay, J.A., Gennaro, R. (eds.) CRYPTO 2014. LNCS, vol. 8617, pp. 37–54. Springer, Heidelberg (2014). doi:10.1007/978-3-662-44381-1_3
12. Ferguson, N., Schneier, B.: Practical Cryptography. Wiley, Hoboken (2003)
13. Gaži, P., Tessaro, S.: Provably robust sponge-based PRNGs and KDFs. In: Fischlin, M., Coron, J.-S. (eds.) EUROCRYPT 2016. LNCS, vol. 9665, pp. 87–116. Springer, Heidelberg (2016). doi:10.1007/978-3-662-49890-3_4
14. Patarin, J.: The "Coefficients H" technique. In: Avanzi, R.M., Keliher, L., Sica, F. (eds.) SAC 2008. LNCS, vol. 5381, pp. 328–345. Springer, Heidelberg (2009). doi:10.1007/978-3-642-04159-4_21
15. Shrimpton, T., Terashima, R.S.: A provable-security analysis of intel's secure key RNG. In: Oswald, E., Fischlin, M. (eds.) EUROCRYPT 2015. LNCS, vol. 9056, pp. 77–100. Springer, Heidelberg (2015). doi:10.1007/978-3-662-46800-5_4
16. Turan, M.S., Barker, E., Kelsey, J., McKay, K.A., Baish, M.L., Boyle, M.: Recommendation for random number generation using deterministic random bit generators, sp800-90a, June 2015. http://nvlpubs.nist.gov/nistpubs/SpecialPublications/NIST.SP.800-90Ar1.pdf

Lattice-Based Cryptography

Fixed-Point Arithmetic in SHE Schemes

Anamaria Costache[1]([✉]), Nigel P. Smart[1], Srinivas Vivek[1],
and Adrian Waller[2]

[1] University of Bristol, Bristol, UK
anamaria.costache@bristol.ac.uk
[2] Thales UK Research and Technology, Reading, UK

Abstract. The purpose of this paper is to investigate fixed-point arithmetic in ring-based Somewhat Homomorphic Encryption (SHE) schemes. We provide three main contributions: firstly, we investigate the representation of fixed-point numbers. We analyse the two representations from Dowlin et al., representing a fixed-point number as a large integer (encoded as a scaled polynomial) versus a polynomial-based fractional representation. We show that these two are, in fact, isomorphic by presenting an explicit isomorphism between the two that enables us to map the parameters from one representation to another. Secondly, given a computation and a bound on the fixed-point numbers used as inputs and scalars within the computation, we achieve a way of producing lower bounds on the plaintext modulus p and the degree of the ring d needed to support complex homomorphic operations. Finally, as an application of these bounds, we investigate homomorphic image processing.

1 Introduction

The efficiency of Somewhat Homomorphic Encryption (SHE) schemes has improved dramatically in the seven years since their discovery by Gentry in 2009 [7]. The main effort in research now is to obtain practical schemes for a given class of interesting functions; since practical Fully Homomorphic Encryption seems out of reach using existing techniques.

When proposing to use SHE schemes in an application a key issue is how to map the data types of the application to the supported data types of the SHE scheme. Most theoretical treatments consider SHE schemes which work over bits, and the application is assumed to be the evaluation of some binary circuit. In practice this is likely to be very costly, and so some authors have considered other scenarios in which the computations are performed over arithmetic circuits or polynomial rings [6,8,11].

At their heart almost all SHE schemes make use of a plaintext space R_p, which is the reduction modulo p of a polynomial ring over the integers R. We shall refer to p as the plaintext modulus, which is often selected to be a prime. The ring is frequently selected to be the ring of integers of a cyclotomic number field; i.e.

$$R = \mathbb{Z}[X]/\Phi_m(X).$$

© Springer International Publishing AG 2017
R. Avanzi and H. Heys (Eds.): SAC 2016, LNCS 10532, pp. 401–422, 2017.
https://doi.org/10.1007/978-3-319-69453-5_22

In considering an application one has a number of factors to balance; first the SHE multiplicative depth of the functions which can be evaluated; secondly the plaintext modulus p and thirdly the security level required. These all imply bounds on the degree of the ring one is using; and hence the efficiency of the application[1]. Of importance in what follows is that an SHE scheme has a maximum multiplicative depth limiting what can be evaluated. In practice this consists of a number of levels, where each ciphertext is associated to a specific level. Multiplication of ciphertexts at levels i_0 and i_1 results in a ciphertext at level $\max(i_0, i_1) + 1$; whereas scalar multiplication is equivalent to the addition of roughly half a level. Once the maximum level is obtained, no further homomorphic operations are possible.

The first obvious method is to move away from binary circuits is to consider plaintext moduli other than $p = 2$, and hence to evaluate arithmetic circuits. Indeed the first application of SHE schemes to obtain an efficiency improvement upon other technologies did precisely this; for example the use of large plaintext moduli p in the SPDZ protocol [5]. However, using arithmetic circuits is also limited. For example, suppose one wished to perform integer arithmetic. In that case, naively increasing p to a large enough value to cope with the largest integer the application could obtain would impose considerable performance penalties.

One can think of using a large plaintext modulus p as using a plaintext space which is long and thin. Some authors have tried to balance the choice of p and the degree d of the ring R to obtain more efficient representation of integers, akin to a more short and fat plaintext space [11]. A problem overlooked by many authors is how to select p and d to enable such a plaintext representation of integer valued payloads; and in particular to bound p and d as a complex homomorphic operation is performed. This is the first problem we consider in this paper. Given a computation on integers, and a bound on the input integers, we are able to produce lower bounds on p and d needed to support such a homomorphic calculation. Our main general technical contribution is to derive such lower bounds on p and d.

Given an ability to process plaintext messages consisting of large integers the next task is to process fixed-point numbers. A number of authors have considered methodologies for this, most notably Dowlin et al. [6]. Dowlin et al. present two efficient methods to represent fixed-point numbers. In the first they encode a fixed point number as a scaled integer (which they then encode as a polynomial), whilst in the second they utilize a fractional representation (also based on polynomials). The advantage of the former method is that it is easier to analyse and it can be applied for any polynomial plaintext ring R_p. However, it also requires complex bookkeeping of the homomorphic ciphertexts during a calculation to ensure that the fixed-point numbers are correctly scaled. The fractional representation avoids such bookkeeping, but it appears harder to analyse so as to derive parameters which will support the homomorphic operations. Further, it requires R to be selected to be a cyclotomic ring $Z[X]/\Phi_m(X)$, where m is

---

[1] In this paper we will ignore issues such as SIMD operations obtained by selecting p and m in a special manner, see [4,8,12] for details.

a power of two. We show that the two representations are in fact isomorphic, when used with the same power of two cyclotomic ring; we present a concrete isomorphism between the two underlying rings and hence are able to map our parameters from the first representation to the second.

As a way of illustrating the use of our bounds, in Appendix A we analyse a relatively complex but useful fixed-point algorithm namely the Fast Fourier Transform (FFT). This is needed to perform applications such as homomorphic image processing. When examining fixed point algorithms for addition and multiplication it is be immediately seen that one needs to consider the homomorphic levels which a given calculation will consume. However, additionally, one must also consider how much the fixed point calculation increases the demands on the plaintext space, with repeated scalar multiplication being particularly costly. This is particularly interesting for the FFT algorithm, since at its heart it is a linear operation performed in a recursive manner (with an FFT of size n reduced to two FFTs of size $n/2$). This recursion decreases the *number* of scalar multiplications needed, but increases the *depth* of the scalar multiplications needed. The naive Fourier Transform is also a linear operation, but it consists of only scalar multiplications of depth one. Thus one has a trade off between reducing the number of operations against the required depth. In spite of the independent usefulness of computing the FFT homomorphically, we underline that this is just a minor application of our bounds, given as a purely illustrative example.

Thus in Appendix A we consider the homomorphic evaluation of an FFT operation in a standard image processing pipeline. We examine the resulting homomorphic algorithms, given bounds on the plaintext spaces derived from our earlier analysis, and present runtimes obtained from an implementation using the HElib library [9]. Whilst we are not able to process large images in the encrypted domain, one notes that processing of tiny (32×32 pixel) images have found application in some domains, e.g. [13]. In addition, even when processing large images, they are often divided into smaller patches during the processing pipeline.

2   Integer Arithmetic

We first consider the simpler case of integer arithmetic; it will turn out that once this is solved fixed-point arithmetic can be built on top of the integer arithmetic. We wish to process an arithmetic circuit *over the integers* where the input encrypted integers, and scalars, come from multiple ranges $[-L_i, \ldots, L_i]$ ($L_i \geq 0$). Allowing different ranges for different inputs and scalars will result in more accurate bounds when we come to consider the FFT algorithm later. Clearly as the circuit is computed the bound on the size of the integers increases, and it is this growth in size which we need to deal with if we are to be able to cope with integers encrypted via our SHE scheme.

As a warm up we consider the simpler case where we wish to compute a "regular" integer circuit which consists of at most $A \geq 0$ additions at each "level" in the circuit, and then, at each level, a layer of multiplications are performed.

The multiplicative depth of the circuit will be denoted by $M \geq 1$. In addition, to simplify this initial discussion, we assume all scalars and inputs are in the same range, i.e. we fix $L_i = L$ for all i. Clearly the output values from such a circuit will have absolute value bounded by

$$L_{max}^{A,M} := \left(2^{\sum_{i=1}^{M} 2^i \cdot A}\right) \cdot L^{2^M} = 2^{A\left(2^{M+1}-2\right)} \cdot L^{2^M}. \tag{1}$$

As explained in the introduction, natively the SHE scheme will encrypt polynomials modulo p, with degree bounded by d. The obvious natural encoding for integers is the *scalar encoding* method. In this encoding method an integer is encoded as the constant polynomial, then integer addition and multiplication become addition and multiplication modulo p. To ensure correctness we then require that $p > 2 \cdot L_{max}^{A,M}$, and hence p has to be very large indeed. This would make the SHE scheme highly inefficient, even for very low depth circuits.

2.1   Representing Integers as Polynomials

This led some authors, e.g. [11], to introduce the following method of encoding an integer, which we call the *non-balanced base-B encoding* method. We encode integers as an integer polynomial in base B, for some base value B to be determined. The polynomial will have negative coefficients for negative integers, and positive coefficients for positive integers. Thus we encode the integer as a polynomial with coefficients in the range $[-(B-1), \ldots, (B-1)]$. In particular this implies an integer in the range $[-L_i, \ldots, L_i]$ on input is encoded as a polynomial of degree at most

$$d_i^{\mathsf{non-Bal}} = \lfloor \log L_i / \log B \rfloor.$$

We are interested in how the infinity norm, and degree, of the polynomials increases as we pass through the circuit. Where for a polynomial $P(X) = p_0 + p_1 \cdot X + \cdots p_d \cdot X^d$ we have $\|P\|_\infty = \max_{i=0,\ldots,d} |p_i|$. Thus for this input/scalar integer at circuit level 0 the infinity norm of our polynomials is bounded by $B_{i,0}^{\mathsf{non-Bal}} = B - 1$.

Another method, considered in [6], is the *balanced base-B encoding*. The integer is now encoded as a polynomial with coefficients in the range $[-(B-1)/2, \ldots, (B-1)/2]$ for an odd integer $B \geq 3$. Any polynomial can now have both non-negative and negative coefficients. This method overcomes a limitation of the previous method that wasted part of the plaintext space by allowing only polynomials with coefficients of the same sign. At level 0, our integer is encoded as a polynomial of degree at most

$$d_i^{\mathsf{Bal}} = \lceil \log(2 \cdot L_i + 1) / \log B \rceil - 1. \tag{2}$$

The infinity norm of our input polynomials is bounded by $B_{i,0}^{\mathsf{Bal}} = (B-1)/2$.

In a later section we outline how to obtain bounds on the degree and infinity norm of the polynomials as we perform a calculation via an integer circuit. It will turn out that the optimal choice in the above two polynomial representations is to use the balanced base $B = 3$ representation, so in particular we select $B_{i,0}^{\mathsf{Bal}} = 1$ for the rest of this paper.

3   Fixed-Point Arithmetic

In this section we present two encoding methods for fixed-point arithmetic, introduced in [6], we then show that these two representations are isomorphic. To illustrate the techniques, we will use the two fixed-point numbers below throughout

$$y = 6.370370\ldots = \frac{172}{27} \text{ and } y' = 2.6666666\ldots = \frac{8}{3},$$

which in balanced base $B = 3$ representation are given by

$$y = 1\bar{1}0.101 \text{ and } y' = 10.\bar{1},$$

where $\bar{1} = -1$. The first method represents the fixed-point number as an integer, along with a "scaling parameter". Thus the fixed point number y is represented as the integer 172, along with a scaling factor of -3. The integer 172 being encoded as a polynomial via the balanced base-B encoding of the previous section.

The second encoding method takes the integer and fractional part of the fixed point number, seperately; it then encodes each part as polynomial (via the balance base-B representation of the associated integer) and then finally encoding the integer part in the lower plaintext coefficients, and the fractional part in the upper plaintext coefficients.

3.1   Balanced Base-B Encoding

The first method we use to represent a fixed-point number uses two integers, one representing the number and the one representing by which power of B one needs to decode. Thus this method requires a level of book keeping in order to keep track of the second integer. Let y be a real fixed-point number, and denote by $y = y^+.y^-$ its integer and fractional parts (upto desired precision) in balanced base-B representation. We then let I^+ be one less than the number of integer digits and I^- be equal to the number of fractional digits; thus we can write

$$y^+ = b_{I^+} \cdot B^{I^+} + b_{I^+-1} \cdot B^{I^+-1} + \cdots + b_1 \cdot B + b_0,$$
$$y^- = b_{-I^-} \cdot B^{-I^-} + b_{-I^-+1} \cdot B^{-I^-+1} + \cdots + b_{-2} \cdot B^{-2} + b_{-1} \cdot B^{-1}$$

where $b_i \in [-(B-1)/2, \ldots, (B-1)/2]$. Thus we can express y as

$$y = \sum_{i=-I^-}^{I^+} b_i \cdot B^i.$$

We then represent y as the pair of integers $(y \cdot B^{I^-}, I^-) = (\hat{y}, i)$. The integer $\hat{y}$ can then be represented by a polynomial $q(X)$, by replacing B in the above expression by X, to obtain the final representation (q, i). Thus we have

$$q_0(X) = b_{I^+} \cdot X^{I^+} + b_{I^+-1} \cdot X^{I^+-1} + \cdots + b_1 \cdot X + b_0,$$
$$q_1(X) = b_{-I^-} + b_{-I^-+1} \cdot X + \cdots + b_{-2} \cdot X^{I^--2} + b_{-1} \cdot X^{I^--1},$$
$$q(X) = q_0(X) \cdot X^i + q_1(X).$$

The degree of the polynomial $q(X)$ above is $\deg(q) = I^- + I^+$, and to recover the fixed-point number y from a pair (q, i) we compute $y = q(B) \cdot B^{-i}$. For our two example fixed-point numbers above we have $y \equiv (q, i)$ and $y' \equiv (q', i')$ where $i = 3$ and $i' = 1$ and

$$q(X) = (X^2 - X) \cdot X^3 + (X^2 + 1) = X^5 - X^4 + X^2 + 1,$$
$$q'(X) = X \cdot X - 1 = X^2 - 1.$$

Given this encoding we can now define how to perform basic arithmetic on the encoding.

Addition: Suppose we have two pairs (q, i) and (q', i') encoding the fixed-point numbers y and y', respectively. Write them as above, namely $q(X) = q_0(X) \cdot X^i + q_1(X)$ and similarly for $q'(X)$. Now if $i \neq i'$, this means that the encodings are not at the same "fixed-point level"[2] and thus the numbers they represent are expressed with a different number of significant digits. Thus, before adding two encodings we must ensure that they are at the same level, by multiplying one by a suitable power of X. Thus if we let $I = \max(i, i')$, we have that $(q, i) + (q', i') = (Q, I)$, where

$$(Q, I) = \begin{cases} (q + q' \cdot X^{I-i'}, i) & \text{if } i > i' \\ (q' + q \cdot X^{I-i}, i') & \text{if } i' \geq i. \end{cases}$$

To see that this indeed corresponds to fixed-point addition, notice that, assuming $i \geq i'$, that

$$Q(B) \cdot B^{-I} = (q + q' \cdot B^{I-i'}) \cdot B^{-I} = q \cdot B^{-I} + q' \cdot B^{I-i'} \cdot B^{-I}$$
$$= q \cdot B^{-i} + q' \cdot B^{-i'} = y + y'.$$

For our two example numbers we have, $i = 3 > i' = 1$, so that

$$Q = q + q' \cdot X^2 = (X^5 - X^4 + X^2 + 1) + (X^2 - 1) \cdot X^2 = X^5 + 1,$$

and $I = \max(3, 1) = 3$. To check correctness, notice that $Q(B) \cdot B^{-3} = B^2 + B^{-3} = 9 + 1/27 = 9.037037\ldots$ as required.

Multiplication: Multiplication is more straightforward, we simply perform

$$(q, i) \cdot (q', i') = (q \cdot q', i + i') = (Q, I),$$

with correctness being obvious. For our two example fixed-point numbers we have the product representation being given by

$$Q = (X^5 - X^4 + X^2 + 1) \cdot (X^2 - 1)$$
$$= X^7 - X^6 - X^5 + 2 \cdot X^4 - 1$$

and $I = i + i' = 3 + 1 = 4$. To check the correctness we note that $Q(B) \cdot B^{-4} = 1376/3^4 = 16.987654\ldots$ as required.

---

[2] Not to be confused with the associated level in the SHE scheme once we encrypt the polynomial.

The Ring $\mathfrak{R}_1$**:** We now define a ring $\mathfrak{R}_1$ out of the above operations. We define the underlying ring as pairs (q, i) where $q \in \mathbb{Z}[X]/\Phi_m(X)$ and $i \in \mathbb{Z}/\phi(m)\mathbb{Z}$, where $\phi(\cdot)$ denotes the Euler's totient function, where in practice we will take m to be a power of two. We define addition and multiplication as above, but now take the resulting pair modulo $\Phi_m(X)$ and $\phi(m)$.

Theorem 1. *With the above definitions* $\mathfrak{R}_1$ *is a ring.*

Proof. The additive identity in $\mathfrak{R}_1$ is the pair $(0, 0)$, which corresponds to the fixed-point number 0. The additive inverse of any element $(q, i) \in \mathfrak{R}_1$ is $(-q, i)$. It is clear that these two elements sum up to $(0, 0)$. Thus $\mathfrak{R}_1$ is an additive group; the fact that it is abelian is immediate.

The multiplicative identity is $(1, 0)$, corresponding to the fixed-point number 1. The associativity of the multiplication is trivially implied by associativity of (modular) polynomial multiplication and (modular) integer addition. We show that distributivity of multiplication over addition holds, thus completing the proof.

Let $(q_1, i_1), (q_2, i_2)$ and (q_3, i_3) be three elements of $\mathfrak{R}_1$. Without loss of generality, assume that $i_2 \geq i_3$, then

$$
\begin{aligned}
(q_1, i_1) \cdot \big((q_2, i_2) + (q_3, i_3)\big) &= (q_1, i_1) \cdot (q_2 + q_3 \cdot X^{i_2 - i_3}, i_2) \\
&= (q_1 \cdot q_2 + q_1 \cdot q_3 \cdot X^{i_2 - i_3}, i_1 + i_2) \\
&= (q_1 \cdot q_2 + q_1 \cdot q_3 \cdot X^{i_1 + i_2 - i_1 - i_3}, \\
&\qquad \max(i_1 + i_2, i_1 + i_3)) \\
&= (q_1 \cdot q_2, i_1 + i_2) + (q_1 \cdot q_3, i_1 + i_3) \\
&= (q_1, i_1) \cdot (q_2, i_2) + (q_1, i_1) \cdot (q_3, i_3). \qquad \square
\end{aligned}
$$

This representation of fixed-point numbers in the ring $\mathfrak{R}_1$ enables us to bound the degree of the polynomial and the coefficients, after a number of homomorphic operations, relatively easily, using the techniques in the next section. Of course it also implies that if we perform too many operations the degree of q will become too large and the polynomial will wrap around modulo $\Phi_m(X)$. Thus the complexity of the operations one performs not only provides a lower bound on p, i.e. an upper bound on the polynomial coefficients, but also a lower bound on the ring degree. These bounds enable us to set parameters for the SHE scheme. However, in performing homomorphic operations we not only need, for each pair (q, i), to keep the ciphertext corresponding to the plaintext q, but we also need to keep track (in the clear) of the value i.

3.2   Fractional Encoding

The second method we use to represent fixed-point numbers dispenses with the need to keep the second component i of our first representation. On the other hand it requires us to work in the cyclotomic ring $R = \mathbb{Z}[X]/(X^n + 1)$, where n is a power of two. Again we let $y = y^+ . y^-$ denote the fixed-point number as above, written in balanced base-B representation with $I^+ + 1$ digits in y^+ and I^- digits in y^-. We again write

$$y^+ = b_{I^+} \cdot B^{I^+} + b_{I^+-1} \cdot B^{I^+-1} + \cdots + b_1 \cdot B + b_0,$$
$$y^- = b_{-I^-} \cdot B^{-I^-} + b_{-I^-+1} \cdot B^{-I^-+1} + \cdots + b_{-2} \cdot B^{-2} + b_{-1} \cdot B^{-1},$$

where $b_i \in [-(B-1)/2, \ldots, (B-1)/2]$. We then encode the fixed-point number y in the ring R by the polynomial

$$p = \sum_{i \leq I^+} X^i b_i - \sum_{0 < i \leq I^-} X^{n-i} b_{-i}$$
$$= p_0(X) + p_1(X) \cdot X^{n-\partial_1}, \tag{3}$$

where $p_0(X) = \sum_{i \leq I^+} X^i b_i$ and $p_1(X) = -\sum_{0 < i \leq I^-} b_{-i} \cdot X^{I^--i}$, with ∂_0 and $\partial_1 - 1$ being the degrees of $p_0(X)$ and $p_1(X)$, respectively. Thus $\partial_0 = I^+$ is one less than the number of digits in the integer part y^+ and $\partial_1 = I^-$ is the number of digits in the fractional part y^-.

Given a polynomial $q(X)$ of this form we can recover the fixed-point number it represents. We will need to know an upper bound for our calculation on $p_0(X)$, which can be easily calculated from the formulae below. We then take $p(X)$ and split it into two polynomials p_0 and p_1 as in Eq. 3 (using the upper bound on the degree of $p_0(X)$ to resolve any ambiguity). We can then recover y by setting

$$y = p_0(B) - p_1(B) \cdot B^{-\partial_1},$$

where we utilize the ring equation $X^n + 1 = 0$.

For our two example numbers $y = 6.370370\ldots$ and $y' = 2.666666\ldots$ we have y represented by p, and y' represented by p', where

$$p = (X^2 - X) - (X^2 + 1) \cdot X^{n-3}$$
$$p' = X - (-1) \cdot X^{n-1}.$$

In both the cases above we have that, in terms of the representation $(q = q_0 \cdot X^i + q_1, i)$ of, say, y from Sect. 3.1, we have $p_0 = q_0$ and $p_1 = q_1$. We have $\partial_0 = 2$, $\partial_0' = 1$, $\partial_1 = 3$ and $\partial_1' = 1$.

Our second ring $\mathfrak{R}_2$ is the representation above, i.e. the set of polynomials modulo $X^n + 1$, which is trivially a ring. We now show that addition and multiplication in this ring corresponds to addition and multiplication of the encoded fixed point values.

Addition: Let $p(X) = p_0(X) + p_1(X) \cdot X^{n-\partial_1}$ and $p'(X) = p_0'(X) + p_1'(X) \cdot X^{n-\partial_1'}$ be two polynomials as described above, encoding y and y', respectively. To perform addition we simply add the associated polynomials as follows, without loss of generality, assume that $\partial_1 \geq \partial'_1$,

$$p + p' = (p_0 + p_1 \cdot X^{n-\partial_1}) + (p'_0 + p'_1 \cdot X^{n-\partial'_1})$$
$$= (p_0 + p'_0) + P_1 \cdot X^{n-\partial_1} = P_0 + P_1 \cdot X^{n-\partial_1},$$

where P_0 has degree $\max(\partial_0, \partial'_0)$ and P_1 has degree $\max(\partial_1, \partial'_1)$. The polynomial P_1 will in fact be $P_1 = p_1 + p'_1 \cdot X^{\partial_1 - \partial'_1}$.

For our two example numbers, their addition therefore has the encoding

$$p + p' = \left((X^2 - X) - (X^2 + 1) \cdot X^{n-3}\right) + \left(X - (-1) \cdot X^{n-1}\right)$$
$$= X^2 - X^{n-1} - X^{n-3} + X^{n-1} = X^2 - X^{n-3},$$

which agrees with the numerical value of their sum.

Multiplication: Let $p(X) = p_0(X) + p_1(X) \cdot X^{n-\partial_1}$ and $p'(X) = p'_0(X) + p'_1(X) \cdot X^{n-\partial'_1}$ be as above. We write $p_0 \cdot p'_1 = r_0 + r_1 \cdot X^{\partial'_1}$ and $p'_0 \cdot p_1 = r'_0 + r'_1 \cdot X^{\partial_1}$, where $\deg(r_0) \leq \partial'_1 - 1$, $\deg(r_1) \leq \partial_0 + \partial'_1 - \partial'_1 = \partial_0$, $\deg(r'_0) \leq \partial_1 - 1$, and $\deg(r'_1) \leq \partial'_0 + \partial_1 - \partial_1 = \partial'_0$, Then the product $y \cdot y'$ is encoded by the product of the two polynomials modulo $X^n + 1$,

$$p \cdot p' = (p_0 + p_1 \cdot X^{n-\partial_1}) \cdot \left(p'_0 + p'_1 \cdot X^{n-\partial'_1}\right)$$
$$= p_0 \cdot p'_0 + p_0 \cdot p'_1 \cdot X^{n-\partial'_1} + p'_0 \cdot p_1 \cdot X^{n-\partial_1} + p_1 \cdot p'_1 \cdot X^{2n-\partial_1-\partial'_1}$$
$$= p_0 \cdot p'_0 + p_1 \cdot p'_1 \cdot X^{2n-\partial_1-\partial'_1}$$
$$\quad + (r_0 + r_1 \cdot X^{\partial'_1}) \cdot X^{n-\partial'_1} + (r'_0 + r'_1 \cdot X^{\partial_1}) \cdot X^{n-\partial_1}$$
$$= p_0 \cdot p'_0 + p_1 \cdot p'_1 \cdot X^{n-\partial_1-\partial'_1} \cdot X^n$$
$$\quad + r_0 \cdot X^{n-\partial'_1} + r_1 \cdot X^n + r'_0 \cdot X^{n-\partial_1} + r'_1 \cdot X^n$$
$$= (p_0 \cdot p'_0 - r_1 - r'_1) + \left(-p_1 \cdot p'_1 + r_0 \cdot X^{\partial'_1} + r'_0 \cdot X^{\partial'_1}\right) \cdot X^{n-\partial_1-\partial'_1}$$
$$= P_0(X) + P_1(X) \cdot X^{n-\partial_2},$$

where $\deg(P_0) = \max(\deg(p_0 \cdot p'_0), \deg r_1, \deg r'_1) = \max(\partial_0 + \partial'_0, \partial_0, \partial'_0) = \partial_0 + \partial'_0$, and $\deg(P_1) \leq \partial_2 = \max(\deg(p_1 \cdot p'_1), \partial_1 + \deg r_0, \partial'_1 + \deg r'_0) = \max(\partial_1 + \partial'_1, \partial_1 + \partial'_1 - 1, \partial'_1 + \partial_1 - 1) = \partial_1 + \partial'_1$.

For our two example numbers, we have

$$p \cdot p' = \left((X^2 - X) - (X^2 + 1) \cdot X^{n-3}\right) \cdot \left(X - (-1) \cdot X^{n-1}\right)$$
$$= (X^3 - X^2) + (X^2 - X) \cdot X^{n-1} + (-X^3 - X) \cdot X^{n-3} + (-X^2 - 1) \cdot X^{2 \cdot n - 4}$$
$$= (X^3 - X^2) + (X^5 - X^4) \cdot X^{n-4} + (-X^4 - X^2) \cdot X^{n-4} + (X^2 + 1) \cdot X^{n-4}$$
$$= (X^3 - X^2) + (X - 1) \cdot X^n - X^n - X^2 \cdot X^{n-4} + (X^2 + 1) \cdot X^{n-4}$$
$$= (X^3 - X^2 - X + 2) + X^{n-4}$$
$$= P_0 + P_1 \cdot X^{n-\partial_2},$$

where $\partial_2 = \partial_1 + \partial_1' = 3 + 1 = 4$. To check this gives the correct value we note that

$$P_0(3) - P_1(3) \cdot 3^{-4} = \frac{1376}{81}.$$

3.3 Relating $\mathfrak{R}_1$ to $\mathfrak{R}_2$

The ring representation of fixed-point numbers in the ring $\mathfrak{R}_1$ allows us to bound the resulting degree and infinity norm of the associated polynomials encoding the fixed-point numbers (see the next section). In addition, it allows a wide choice of underlying rings, which could enable SIMD computation of specific fixed-point operations. However, it requires the "bookkeeping" of the base power that is needed to map the encoded integer into a fixed-point number.

The ring $\mathfrak{R}_2$ on the other hand requires no such bookkeeping, although limited book keeping is needed to ensure decoding after decryption works correctly. Additionally, it requires that we work in the ring defined by polynomial arithmetic modulo $X^n + 1$, where n is a power of two. A major drawback seems to be that one cannot derive obvious bounds on the degree and coefficients in the fractional representation, something which is crucial in order to set parameters of the SHE scheme. However, such bounds can be derived for the fractional representation, since this representation is isomorphic to the representation using the ring $\mathfrak{R}_1$, and the isomorphism presents a one-to-one direct relationship between the coefficients of the polynomials in each representation.

Let ϕ be as follows (from now on),

$$\phi : \begin{cases} \quad \mathfrak{R}_1 & \rightarrow \quad \mathfrak{R}_2 \\ (q = q_0 \cdot X^i + q_1, i) & \mapsto q_0 - q_1 \cdot X^{n-i} \end{cases}$$

Theorem 2. *If R is defined by $Z[X]/(X^n + 1)$, then ϕ is a ring isomorphism.*

Proof. First note that

1. $\phi(1_{\mathfrak{R}_1}) = \phi(1, 0) = \phi(1 \cdot X^0 + 0) = 1 - 0 \cdot X^0 = 1 = 1_{\mathfrak{R}_2}$.
2. Let (q, i) and (q', i') in $\mathfrak{R}_1$; without loss of generality assume $i \geq i'$. Then $(q, i) + (q', i') = q + q' \cdot X^{i-i'} =: (Q, i)$. Then

$$\begin{aligned} \phi(Q, i) &= \phi(q + q' \cdot X^{i-i'}, i) \\ &= \phi\big(q_0 \cdot X^i + q_1 + (q_0' \cdot X^{i'} + q_1') \cdot X^{i-i'}, i\big) \\ &= \phi\big((q_0 + q_0') \cdot X^i + (q_1 + q_1' \cdot X^{i-i'}), i\big) \\ &= (q_0 + q_0') + (q_1 + q_1' \cdot X^{i-i'}) \cdot X^{n-i} \\ &= q_0 + q_1 \cdot X^{n-i} + q_0' + q_1' \cdot X^{n-i'} \\ &= \phi(q, i) + \phi(q', i'). \end{aligned}$$

Notice that in the above, we have implicitly made use of additive property of $\mathfrak{R}_2$.

3. Let q, q' be as above.

$$\phi(q, i) \cdot \phi(q', i') = (q_0 - q_1 \cdot X^{n-i}) \cdot (q'_0 - q'_1 \cdot X^{n-i'})$$
$$= q_0 \cdot q'_0 - q_0 \cdot q'_1 \cdot X^{n-i'} - q'_0 \cdot q_1 \cdot X^{n-i} + q_1 \cdot q'_1 \cdot X^{n-I},$$

where $I = i + i'$. Now computing $(q, i) \cdot (q', i')$ first,

$$q \cdot q' = q_0 \cdot q'_0 \cdot X^I + q_0 \cdot q'_1 \cdot X^i + q'_0 \cdot q_1 \cdot X^{i'} + q_1 \cdot q'_1.$$

Now viewing this as the pair $(Q = q \cdot q' \mod X^n + 1, i + i' \mod n) = ((q_0 \cdot q'_0 + q_1 \cdot q'_1 \cdot X^{n-i-i'}) \cdot X^{i+i'}) + (q_0 \cdot q'_1 \cdot X^i + q'_0 \cdot q_1 \cdot X^{i'}), i + i')$, we obtain the following.

$$\phi(q \cdot q', I) = \phi((q_0 \cdot q'_0 + q_1 \cdot q'_1 \cdot X^{n-i-i'}) \cdot X^{i+i'}$$
$$+ (q_0 \cdot q'_1 \cdot X^i + q'_0 \cdot q_1 \cdot X^{i'}), I)$$
$$= q_0 \cdot q'_0 + q_1 \cdot q'_1 \cdot X^{n-i-i'} - (q_0 \cdot q'_1 \cdot X^i + q'_0 \cdot q_1 \cdot X^{i'}) \cdot X^{n-I}$$
$$= q_0 \cdot q'_0 - q_0 \cdot q'_1 \cdot X^{n-i'} - q'_0 \cdot q_1 \cdot X^{n-i} + q_1 \cdot q'_1 \cdot X^{n-I}$$
$$= \phi(q, i) \cdot \phi(q', i'),$$

so that ϕ is indeed a homomorphism between $\mathfrak{R}_1$ and $\mathfrak{R}_2$.

To finish the proof, we show that ϕ is bijective. For any $y = q_0 + q_1 \cdot X^{n-\eth_1}$ in $\mathfrak{R}_2$, we have that $(q, \eth_1) = (q_0 \cdot X^{\eth_1} + q_1, \eth_1)$ maps to y so that the mapping is surjective. To see that it is injective, suppose for $p, p' \in \mathfrak{R}_1$ we have that $\phi(p) = \phi(p') = z \in \mathfrak{R}_2$. Remember that both the rings contain encoding of fractional numbers written in balanced base B. Recall also that we recover the integers by simply evaluating (in our case) $z(B) = a \in \mathbb{Q}$, and since this is well-defined, a is unique. Now encode a in the ring $\mathfrak{R}_1$; the encoding operation (for both rings) is well-defined, therefore a will have an unique image in the ring $\mathfrak{R}_1$ and thus $p = p'$. It follows that ϕ is an isomorphism. $\qquad\square$

4   Bounds on Integer Arithmetic

Considering the balanced base B method for encoding integers as polynomials we need to estimate, for a given calculation, a lower bound on p and d. This is to determine parameters our SHE scheme needs to enable a given calculation to be performed correctly. In previous works this problem was not addressed. In this section we provide a methodology to produce tight bounds on the size of p, for any given computation.

To perform our analysis, we first note that as we pass through a general integer circuit each encrypted polynomial expression we are processing will be of the form

$$\sum_{d=0}^{M} \left(\sum_{d_1 < d_2 < \ldots < d_t} \left(\sum_{e_1 + e_2 + \cdots + e_k = d} \left(c_* \prod_{i=1}^{t} p_{d_i, *}^{e_i} \right) \right) \right).$$

where t is the number of distinct ranges $[-L_i, \ldots, L_i]$ for input/scalar values. $p_{d_i,*}$ is a polynomial of degree d_i with infinity norm $B_{i,0}^{\mathsf{Bal}} = 1$. The c_* are some constants and the value M is the maximal depth. Here we count scalar multiplication as consuming one level of depth. If we wish to determine the infinity norm of such a term we can simplify the discussion by just considering terms of the form

$$\prod_{i=1}^{t}(1 + x + x^2 + \ldots + x^{d_i})^{e_i}.$$

Indeed we define

$$c_{[(d_1,e_1),\ldots,(d_t,e_t)]} = \left\| \prod_{i=1}^{t}(1 + x + x^2 + \ldots + x^{d_i})^{e_i} \right\|_{\infty}.$$

In what follows, to ease discussion, the subscript indices are ordered such that

$$d_i \cdot e_i \leq (d_{i+1} \cdot e_{i+1}) \text{ and in the case of equality } d_i < d_{i+1}.$$

For two terms of the form $c_{[(d_1,e_1),\ldots,(d_t,e_t)]}$ and $c_{[(d_1,e_1'),\ldots,(d_t,e_t')]}$ we define

$$c_{[(d_1,e_1),\ldots,(d_t,e_t)]} \otimes c_{[(d_1,e_1'),\ldots,(d_t,e_t')]} = c_{[(d_1,e_1+e_1'),\ldots,(d_t,e_t+e_t')]}.$$

We can now bound the infinity norm of any polynomial P obtained in evaluating the integer circuit by an expression of the form

$$L_P = \sum_{e_1,\ldots,e_t} a_{[(d_1,e_1),\ldots,(d_t,e_t)]} \cdot c_{[(d_1,e_1),\ldots,(d_t,e_t)]},$$

where $a_{[(d_1,e_1),\ldots,(d_t,e_t)]}$ are constants depending on the precise polynomial P, and we think of this (for now) as a formal sum in the variables $c_{[(d_1,e_1),\ldots,(d_t,e_t)]}$. For an input or scalar value from the range $[-L_i, \ldots, L_i]$ the infinity norm of the polynomial P_0 is bounded by

$$L_{P_0} = c_{[(d_1,0),\ldots,(d_{i-1},0),\ (d_i,1),\ (d_{i+1},0),\ldots,(d_t,0)]}.$$

We can derive upper bounds on the infinity norm of the polynomials as we pass through the integer circuit using the following rules. Given upper bounds on the infinity norm of polynomials P and P' in this form given by

$$L_P = \sum_{e_1,\ldots,e_t} a_{[(d_1,e_1),\ldots,(d_t,e_t)]} \cdot c_{[(d_1,e_1),\ldots,(d_t,e_t)]},$$

$$L_{P'} = \sum_{e_1',\ldots,e_t'} a_{[(d_1,e_1'),\ldots,(d_t,e_t')]} \cdot c_{[(d_1,e_1'),\ldots,(d_t,e_t')]},$$

we can derive upper bounds on the infinity norm of the sum and the product of these polynomials terms via the equations

$$L_{P+P'} = L_P + L_{P'},$$

$$L_{P.P'} = \sum_{e_1,\ldots,e_t,e_1',\ldots,e_t'} \left(a_{[(d_1,e_1),\ldots,(d_t,e_t)]} \cdot a_{[(d_1,e_1'),\ldots,(d_t,e_t')]} \right)$$

$$\cdot \left(c_{[(d_1,e_1),\ldots,(d_t,e_t)]} \otimes c_{[(d_1,e_1'),\ldots,(d_t,e_t')]} \right).$$

Is it clear that the degree of the sum of two polynomials is the maximum of the degrees, and the degree of the product is the sum of the degrees.

4.1   Bounding $c_{[(d_1,e_1),\ldots,(d_t,e_t)]}$

To use these bounds we eventually obtain a formal expression for infinity norm of the output of the circuit consisting of a linear polynomial in the terms $c_{[(d_1,e_1),\ldots,(d_t,e_t)]}$. We thus are left with simply bounding $c_{[(d_1,e_1),\ldots,(d_t,e_t)]}$. We perform this bounding at the end, rather than as we go, as these leads to much tighter bounds on the infinity norm of the output polynomial.

We first present some basic facts on the case of a single pair of terms (d, e). Let $d, e \geq 0$ be integers, and define a_i for $0 \leq i \leq d \cdot e$ as

$$(1 + x + x^2 + \ldots + x^d)^e = \sum_{i=0}^{d \cdot e} a_i \cdot x^i. \tag{4}$$

We then define

$$c_{d,e} = \left\| (1 + x + x^2 + \ldots + x^d)^e \right\|_\infty = \max_{0 < i < d \cdot e} a_i.$$

Naively we can obtain upper and lower bounds on $c_{d,e}$ as follows:

$$\frac{(d+1)^e}{d \cdot e + 1} \leq c_{d,e} \leq (d+1)^e.$$

The upper bound is obtained by evaluating (4) at $x = 1$ and the lower bound is obtained from the upper bound by noting that there are only $d \cdot e + 1$ coefficients a_i in (4). We have the trivial bounds $c_{d,0} = c_{d,1} = 1$ and $c_{d,2} = (d+1)$.

The parameter $c_{d,e}$ is also of interest in probability theory and bounds on its value have been previously analysed [1,10]. The following upper bound follows from the main theorem in [10] (see also [1] for a relation between the parameter $c_{m,n}$ and the main parameter studied in [10]).

Theorem 3. *If $e \neq 2$ or $d \in \{1, 2, 3\}$, then*

$$c_{d,e} < \sqrt{\frac{6}{\pi \cdot d \cdot e \cdot (d+2)}} \cdot (d+1)^e. \tag{5}$$

The above upper bound is optimal in the following sense [10, Remark (a)].

Corollary 1. $\lim_{e \to \infty} \frac{\sqrt{e} \cdot c_{d,e}}{(d+1)^e} = \sqrt{\frac{6}{\pi \cdot d \cdot (d+2)}}.$

Although it is unknown whether the above convergence is uniform as d varies as well.

Given this bound on terms $c_{d,e}$ we can now derive bounds on our terms $c_{[(d_1,e_1),\ldots,(d_t,e_t)]}$ as follows. Recalling our ordering of the pairs in the subscript of $d_i \cdot e_i \leq (d_{i+1} \cdot e_{i+1})$ and in the case of equality, $d_i < d_{i+1}$. We (recursively) use the following bound, where d_k is the first value of d_i in the subscript for which the associated e_k value is non-zero,

$$c_{[(d_1,e_1),\ldots,(d_t,e_t)]} \leq (d_k \cdot e_k + 1) \cdot c_{d_k,e_k} \cdot c_{[(d_1,e_1'),\ldots,(d_t,e_t')]}, \tag{6}$$

where $e_i' = e_i$ except that $e_k' = 0$.

4.2 Applying the Bounds

We can now estimate the size of p and d needed to ensure correctness when evaluating our example balanced integer circuit that consists of M levels and A additions per level. The infinity norm bound on our polynomials becomes

$$B_M = c_{d,2^M} \cdot 2^{A(2^{M+1}-2)},$$

assuming the input values are in the range $[-L, \ldots, L]$ and using a balanced base-3 representation of the input values, so $d = d^{\mathsf{Bal}} = \lceil \log(2 \cdot L + 1)/\log 3 \rceil - 1$. The degree bound for our circuit output value is $d_{\mathsf{out}} = 2^M \cdot d$. From Theorem 3, a sharp upper bound on B_M (for $M > 1$, or $d > 3$ if $M = 1$) is

$$B_M < \sqrt{\frac{6}{\pi \cdot 2^M \cdot d(d+2)}} \cdot (d+1)^{2^M} \cdot 2^{A(2^{M+1}-2)}.$$

To ensure correctness, when we encrypt and manipulate these polynomials homomorphically, we need to ensure that our SHE scheme supports a plaintext with $p > 2 \cdot B_M$ and $\deg(R) > d_M$. The most stringent constraint is that on p, and we give examples in Subsect. 4.3 below.

Of course given a specific circuit we could derive other values of d_M and B_M, the above are just examples in the case of our regular circuit with multiplicative depth M and A additions per level. See the appendix for an application where our more general analysis becomes applicable.

4.3 Lower Bounds on p for Regular Circuits

Tables 2, 3 and 4 list the size in bits of the smallest prime satisfying the above b:unds and also the degree bound $d_M = 2^M \cdot d_0$ for small values of A and M for balanced base encoding with $B = 3$, 5 and 7 and $L = 2^{19}$. For the sake of comparison, we give also give Table 1 that suggests the size of the primes for the non-balanced base encoding for $B = 2$ and $L = 2^{19}$. It is evident that using balanced base encoding with $B = 3$ yields the smallest primes, although large multiplicative depth is hard to support in any method.

It should be noted that with current SHE schemes a ciphertext modulus over 256 bits in length seems currently infeasible for moderately sized circuits

Table 1. Size (in bits) of the smallest p and the degree bounds for *non-balanced* encoding with $B = 2$ and $L = 2^{19}$.

M	1	2	3	4	5	6	7	8	9	10
$A = 0$	6	14	31	65	133	271	547	1100	2206	4418
$A = 1$	8	20	45	95	195	397	801	1610	3228	6464
$A = 2$	10	26	59	125	257	523	1055	2120	4250	8510
$A = 3$	12	32	73	155	319	649	1309	2630	5272	10556
$A = 4$	14	38	87	185	381	775	1563	3140	6294	12602
$A = 5$	16	44	101	215	443	901	1817	3650	7316	14648
$A = 6$	18	50	115	245	505	1027	2071	4160	8338	16694
$A = 7$	20	56	129	275	567	1153	2325	4670	9360	18740
$A = 8$	22	62	143	305	629	1279	2579	5180	10382	20786
$A = 9$	24	68	157	335	691	1405	2833	5690	11404	22832
$A = 10$	26	74	171	365	753	1531	3087	6200	12426	24878
d_M	38	76	152	304	608	1216	2432	4864	9728	19456

Table 2. Size (in bits) of the smallest p and the degree bounds for *balanced* encoding with $B = 3$ and $L = 2^{19}$.

M	1	2	3	4	5	6	7	8	9	10
$A = 0$	5	12	26	55	114	232	468	942	1888	3783
$A = 1$	7	18	40	85	176	358	722	1452	2910	5829
$A = 2$	9	24	54	115	238	484	976	1962	3932	7875
$A = 3$	11	30	68	145	300	610	1230	2472	4954	9921
$A = 4$	13	36	82	175	362	736	1484	2982	5976	11967
$A = 5$	15	42	96	205	424	862	1738	3492	6998	14013
$A = 6$	17	48	110	235	486	988	1992	4002	8020	16059
$A = 7$	19	54	124	265	548	1114	2246	4512	9042	18105
$A = 8$	21	60	138	295	610	1240	2500	5022	10064	20151
$A = 9$	23	66	152	325	672	1366	2754	5532	11086	22197
$A = 10$	25	72	166	355	734	1492	3008	6042	12108	24243
d_M	24	48	96	192	384	768	1536	3072	6144	12288

to be evaluated. Thus it is clear that if anything but small values of M are to be considered one needs a different way of encoding fixed-point numbers. One such possibility is via multiple encryptions using different plaintext moduli, and then to use the Chinese Remainder Theorem to recover the final plaintext polynomial.

Table 3. Size (in bits) of the smallest p and the degree bounds for *balanced* encoding with $B = 5$ and $L = 2^{19}$.

M	1	2	3	4	5	6	7	8	9	10
$A = 0$	7	14	31	64	130	263	529	1062	2129	4264
$A = 1$	9	20	45	94	192	389	783	1572	3151	6310
$A = 2$	11	26	59	124	254	515	1037	2082	4173	8356
$A = 3$	13	32	73	154	316	641	1291	2592	5195	10402
$A = 4$	15	38	87	184	378	767	1545	3102	6217	12448
$A = 5$	17	44	101	214	440	893	1799	3612	7239	14494
$A = 6$	19	50	115	244	502	1019	2053	4122	8261	16540
$A = 7$	21	56	129	274	564	1145	2307	4632	9283	18586
$A = 8$	23	62	143	304	626	1271	2561	5142	10305	20632
$A = 9$	25	68	157	334	688	1397	2815	5652	11327	22678
$A = 10$	27	74	171	364	750	1523	3069	6162	12349	24724
d_M	16	32	64	128	256	512	1024	2048	4096	8192

Table 4. Size (in bits) of the smallest p and the degree bounds for *balanced* encoding with $B = 7$ and $L = 2^{19}$.

M	1	2	3	4	5	6	7	8	9	10
$A = 0$	8	16	34	70	143	289	582	1169	2342	4689
$A = 1$	10	22	48	100	205	415	836	1679	3364	6735
$A = 2$	12	28	62	130	267	541	1090	2189	4386	8781
$A = 3$	14	34	76	160	329	667	1344	2699	5408	10827
$A = 4$	16	40	90	190	391	793	1598	3209	6430	12873
$A = 5$	18	46	104	220	453	919	1852	3719	7452	14919
$A = 6$	20	52	118	250	515	1045	2106	4229	8474	16965
$A = 7$	22	58	132	280	577	1171	2360	4739	9496	19011
$A = 8$	24	64	146	310	639	1297	2614	5249	10518	21057
$A = 9$	26	70	160	340	701	1423	2868	5759	11540	23103
$A = 10$	28	76	174	370	763	1549	3122	6269	12562	25149
d_M	14	28	56	112	224	448	896	1792	3584	7168

Acknowledgements. This work has been supported in part by ERC Advanced Grant ERC-2010-AdG-267188-CRIPTO and by the European Union's H2020 Programme under grant agreement number ICT-644209 (HEAT). The authors would like to thank Carl Ek for input on image processing algorithms and Daniel P. Martin for valuable inputs throughout.

A Homomorphic Image Processing via the Fourier Transform

A standard image processing pipeline is to take an image (consisting of n pixels), pass it into the frequency domain by applying the Fourier transform, apply an operation in the Fourier domain, and then map back to the image space by applying the inverse Fourier transform. The operation in the Fourier domain in its simplest form could be the Hadamard component wise multiplication of the data by a fixed matrix. For example this is used when applying Gabor filters, which feature prominently in applications that are motivated by biological vision.

In this section we examine the application of our fixed-point analysis to the case of image processing in which both the initial image and the Hadamard transformation data are encrypted using a SHE scheme. It is well known that the Fourier transform is a linear operation, and hence only requires (in theory) an additively homomorphic encryption scheme to obtain an encrypted version. However, our requirement that the processing in the frequency domain is also unknown to the evaluator implies that our overall operation is non-linear.

Previous authors have examined homomorphic evaluation of the Fourier transform [2,3]. Indeed by exploiting the linear nature of the calculation they utilized an encoding of fixed-point numbers via scaled integers. Then they used the additively homomorphic Paillier encryption algorithm to perform the homomorphic evaluation of the Fourier transform. This has a number of disadvantages. Firstly by encoding in a purely integer manner the Paillier plaintext modulus space N increases dramatically if one is to perform an FFT, followed by a linear map, followed by an inverse FFT. In addition it requires all homomorphic operations in an application to be linear.

For means of comparison of parameters with prior work [2,3], which used Paillier encryption and only processed a single FFT operation, we also provide a comparison of parameters in that case.

A.1 The Mixed Fourier Transform Algorithm

The standard method to apply the (radix-2) Fourier transform[3] is to use the Fast Fourier Transform (FFT) which is a recursive algorithm requiring $O(\log n)$ depth of scalar multiplications and a total of $O(n \cdot \log n)$ scalar multiplications in total. As we have seen the need to perform a large depth of scalar multiplications will imply a large plaintext modulus for our SHE scheme. The naive method of performing the Fourier transform is to simply apply a matrix-vector product. This requires only depth one of scalar multiplications but on the other hand requires $O(n^2)$ scalar multiplications. We will refer to this method as the Naive Fourier Transform (NFT).

There is an obvious balance to be struck here, which we present in Fig. 1. This is an algorithm, which we dub the Mixed Fourier Transform (MFT) algorithm.

---

[3] Other FFT's, e.g. the radix-4 method, can be analysed using similar techniques to those in this paper.

A. Costache et al.

$$MFT(\mathbf{x}, n, \mathfrak{B})$$
$\quad$ **if** $n \le \mathfrak{B}$ **then**
$\qquad$ **for** $0 \le k \le n-1$ **do**
$\qquad\quad$ $\mathbf{y}_k \leftarrow \sum_{j=0}^{n-1} x_j \cdot \exp(-2 \cdot \pi \cdot \sqrt{-1} \cdot j \cdot k/n).$
$\qquad$ **end for**
$\quad$ **else**
$\qquad$ $m \leftarrow n/2.$
$\qquad$ $z_0, \cdots, z_{n/2-1} \leftarrow MFT((x_0, x_2, x_4, \ldots, x_{n-2}), m, \mathfrak{B}).$
$\qquad$ $z_{n/2}, \cdots, z_n \leftarrow MFT((x_1, x_3, x_5, \ldots, x_{n-1}), m, \mathfrak{B}).$
$\qquad$ **for** $0 \le k \le n/2 - 1$ **do**
$\qquad\quad$ $s \leftarrow \exp(-2 \cdot \pi \cdot \sqrt{-1} \cdot k/n) \cdot z_{k+n/2}.$
$\qquad\quad$ $t \leftarrow z_k.$
$\qquad\quad$ $\mathbf{y}_k \leftarrow t + s.$
$\qquad\quad$ $\mathbf{y}_{k+n/2} \leftarrow t - s.$
$\qquad$ **end for**
$\quad$ **end if**
$\quad$ **return y**

Fig. 1. The Mixed Fourier Transform algorithm

It executes standard recursive FFT algorithm down to a given depth $\lfloor \log_2(\mathfrak{B}) \rfloor$, and then at this lower level executes the naive Fourier transform method.

When we execute $MFT(\mathbf{x}, n, 1)$ we perform the full traditional Fast Fourier Transform method, while when we execute $MFT(\mathbf{x}, n, n)$ we perform the Naive Fourier Transform method. All values of $\mathfrak{B}$ in between execute a hybrid approach. By varying $\mathfrak{B}$ we can trade a reduced depth of scalar multiplications for an increased total number of multiplications. It is obvious that the depth of scalar multiplications required is given by

$$\mathsf{depth}(n, \mathfrak{B}) = \log_2(n) - \log_2(\mathfrak{B}) + 1.$$

Computing the total number of scalar multiplications requires a little more thought. For $n = 2^N$ and $\mathfrak{B} = 2^B$, the first level of the FFT operation has

$$\mathsf{mults}(n, \mathfrak{B}) = 2 \cdot \mathsf{mults}(n/2, \mathfrak{B}) + 2^{N-1}$$

multiplications. Doing FFT until we reach $\mathfrak{B}$ gives

$$\mathsf{mults}(n, \mathfrak{B}) = 2^{N-B} \cdot \mathsf{mults}(\mathfrak{B}, \mathfrak{B}) + (N - B) \cdot 2^{N-1}.$$

Solving this yields

$$\mathsf{mults}(n, \mathfrak{B}) = n \cdot B + (\log_2(n) - \log_2(\mathfrak{B})) \cdot \frac{n}{2}$$

as the number of multiplications performed in a MFT circuit.

A.2 Comparison with Prior Work

In [2,3] the authors present work on implementing a radix-2 FFT in the encrypted domain using the Paillier encryption algorithm. As a means of comparison of their work with ours we examine how their Paillier parameters would compare to our Ring-LWE parameters in their setting. The first key aspect is the precision of the input values, the roots of unity and the output precision. Both [2,3] and ourselves use a fixed-point encoding in which precision is never lost. But if one implemented FFT on a machine with b bits of floating point precision one would loose precision as the calculation proceeds. This means that to obtain the same output as running in the clear on a standard machine using floating point arithmetic, we can adapt the precision of the roots of unity.

In particular, we let b_1 denote the bits of precision in the input data (which is typically eight), b_2 denote the bits of precision in the roots of unity and b denote the bits of equivalent output bits of precision in an in-the-clear implementation. Then [2,3] show that for a single iteration of the FFT algorithm on data of size 2^v, one can take

$$b_2 = \left\lceil b - \frac{v}{2} + \frac{1}{2} \right\rceil.$$

Using this they are able to implement the FFT in the encrypted domain using a Paillier modulus of bit size

$$n_P \geq v + \alpha \cdot b_2 + b_1 + 4,$$

where $\alpha = 1$ for the Naive Fourier Transform, and $\alpha = v - 2$ for the full FFT; they do not consider a Mixed Fourier Transform.

As a means of comparison we look at the same situation using our polynomial encoding for use in the Ring-LWE system. The degrees of the associated polynomials to encode the input data and the roots of unity, in balanced base-3 encoding, are

$$d_i = \lceil \log(2 \cdot 2^{b_i} + 1)/\log 3 \rceil - 1.$$

Applying the analysis from Sect. 4 to a single Fourier Transform execution, we can obtain formulae for the infinity norm of the resulting polynomials via a computer algebra system in the form of a linear sum of terms the following form

$$c_{[(d_1,1),(d_2,e_2)]},$$

where $0 \leq e_2 \leq \mathsf{depth}(n, \mathfrak{B})$. Note that $e_1 = 1$ as we are only executing a single FFT operation.

Then using (5) and (6) and the fact that $c_{d,1} = 1$ we can give an upper bound on this quantity

$$c_{[(d_1,1),(d_2,e_2)]} \leq c \cdot (d_1 + 1) \cdot (d_2 + 1)^{e_2},$$

where

$$c = \sqrt{\frac{6}{\pi \cdot d_2 \cdot e_2 \cdot (d_2 + 2)}}.$$

Hence, we can upper bound the linear sum and so lower bound the plaintext modulus p needed for the SHE scheme to ensure correctness. A similar method allows us to upper bound the degree of the resulting polynomials. This itself leads to a lower bound on the ring dimension $\deg(R)$ needed for the SHE scheme. We summarize the results in Table 5 for emulating $b = 32$ bits of floating point precision and $b_1 = 8$ bit inputs.

Table 5. Comparing Paillier vs Ring-LWE encoding parameters for a single NFT/FFT execution for $b = 32$

n	b_2	d_1	d_2	FFT			NFT		
				$\log_2 p \geq$	$\deg(R) \geq$	$n_P \geq$	$\log_2 p \geq$	$\deg(R) \geq$	$n_P \geq$
64	30	5	19	35	138	138	11	24	48
256	29	5	18	45	167	194	13	23	49
1024	28	5	18	56	203	246	15	23	50

A.3   FFT-Hadamard-iFFT Pipeline

We now turn to investigating the FFT-Hadamard-iFFT standard image processing pipeline. Since we apply two Fourier transforms the precision of the roots of unity we take to be

$$b_2 = \left\lceil b - v + \frac{1}{2} \right\rceil,$$

in order to retain the same precision as b bits of floating point precision on a standard machine.

Applying the analysis from Sect. 4 again, we obtain formulae for the infinity norm of the resulting polynomials in the form of a linear sum of terms of the following form

$$c_{[(d_1,2),(d_2,e_2)]},$$

where $0 \leq e_2 \leq \mathsf{depth}(n, \mathfrak{B})$. Then using Eqs. 5 and 6, and the fact that $c_{d,2} = (d+1)$ we now upper bound this quantity via

$$c_{[(5,2),(10,e_2)]} \leq \begin{cases} 36 & \text{If } e_2 = 1, \\ c \cdot (2 \cdot 5 + 1) \cdot (5 + 1) \cdot (10 + 1)^{e_2} & \text{Otherwise,} \end{cases}$$

where

$$c = \begin{cases} \cdot \sqrt{\frac{6}{\pi \cdot 10 \cdot e_2 \cdot (10 + 2)}} & \text{If } e_2 > 2, \\ 1 & \text{Otherwise.} \end{cases}$$

Hence, we can upper bound the linear sum and so lower bound the plaintext modulus p needed for the SHE scheme to ensure correctness. This results in the parameters given in Table 6.

We then took these bounds and instantiated an SHE system to evaluate the pipeline using the HElib library [9]. The HElib library implements the BGV [4,8] Somewhat Homomorphic Encryption scheme, but restricts the plaintext modulus to be at most 64 bits in length. Hence, our experiments are limited to this reduced size of plaintext space.

In this scheme a plaintext $m \in R_p$ is encrypted as a pair of elements in $(c_0, c_1) \in R_q^2$, such that

$$c_0 - \mathfrak{st} \cdot c_1 = m + p \cdot \epsilon \pmod{q},$$

Table 6. Parameters for the FFT-Hadamard-iFFT pipeline

n	b_2	d_1	d_2	FFT $\mathfrak{B} = 1$		$\mathfrak{B} = \sqrt{n}$		NFT $\mathfrak{B} = n$	
				$\log_2 p \geq$	$\deg(R) \geq$	$\log_2 p \geq$	$\deg(R) \geq$	$\log_2 p \geq$	$\deg(R) \geq$
16	29	5	18	54	190	37	118	25	46
64	27	5	17	74	248	49	146	29	44
256	25	5	16	93	298	61	170	33	42
1024	23	5	15	112	340	72	190	37	40

where $\mathfrak{sk}$ is the secret key (a short element in R_q) and ϵ is a short "noise" element in R_q. As homomorphic operations progress the value q of the ciphertext is reduced, until it can be reduced no more. At this point, operations cease to be possible. The reduction in q enables the noise value to be controlled, and each reduction in q is said to consume a homomorphic "level". Note, that the HElib library due to its choice of moduli for each level actually consumes multiple "internal levels" for each of these "external levels".

In Table 7 we present our implementation results using the HElib. In each case we used the plaintext modulus size derived from the Table 6. We note that in all cases HElib selects a ring dimension for security reasons which is much larger than we need for our application. This last fact means that by careful choice of the plaintext modulus one can process many such operations in parallel using standard SIMD tricks; with the amortization constant being (roughly) the actual degree of R divided by the lower bound from Table 6. We note that we cannot obtain results for the larger plaintext spaces as HElib has a restriction of 60 bits on the plaintext modulus. In future work we aim to remove this restriction by utilizing a different SHE library. All run times measure the time in seconds to evaluate the FFT-Hadamard-iFFT pipeline in the homomorphic domain, and they are obtained on a machine with six Intel Xeon E5 2.7 GHz processors, and with 64 GB RAM.

Table 7. Results for homomorphically evaluating a full image processing pipeline

n	$\mathfrak{B}$	$\deg(R)$	$\log_2 q$	HElib levels	Amortization amount	CPU time	Amortized time
16	1	32768	710	33	172	188	1.09
16	4	32768	451	19	277	147	0.53
16	16	16384	192	9	356	106	0.3
64	8	32768	622	30	224	1500	6.69
64	64	16384	192	10	372	1582	4.25
256	256	16384	278	11	390	34876	89.4

References

1. Belbachir, H.: Determining the mode for convolution powers of discrete uniform distribution. Probab. Eng. Inf. Sci. **25**, 469–475 (2011)
2. Bianchi, T., Piva, A., Barni, M.: Comparison of different FFT implementations in the encrypted domain. In: 2008 16th European Signal Processing Conference, EUSIPCO 2008, Lausanne, Switzerland, 25–29 August 2008, pp. 1–5. IEEE (2008)
3. Bianchi, T., Piva, A., Barni, M.: On the implementation of the discrete fourier transform in the encrypted domain. IEEE Trans. Inf. Forensics Secur. **4**(1), 86–97 (2009)
4. Brakerski, Z., Gentry, C., Vaikuntanathan, V.: (Leveled) fully homomorphic encryption without bootstrapping. In: Goldwasser, S. (ed.) ITCS, pp. 309–325. ACM (2012)
5. Damgård, I., Pastro, V., Smart, N., Zakarias, S.: Multiparty computation from somewhat homomorphic encryption. In: Safavi-Naini, R., Canetti, R. (eds.) CRYPTO 2012. LNCS, vol. 7417, pp. 643–662. Springer, Heidelberg (2012). doi:10. 1007/978-3-642-32009-5_38
6. Dowlin, N., Gilad-Bachrach, R., Laine, K., Lauter, K., Naehrig, M., Wernsing, J.: Manual for using homomorphic encryption for bioinformatics (2015). http:// research.microsoft.com/pubs/258435/ManualHEv2.pdf. Accessed 05 May 2016 at 23:00
7. Gentry, C.: A fully homomorphic encryption scheme. Ph.D. thesis, Stanford University (2009). http://crypto.stanford.edu/craig. Accessed 05 May 2016 at 23:00
8. Gentry, C., Halevi, S., Smart, N.P.: Homomorphic evaluation of the AES circuit. In: Safavi-Naini, R., Canetti, R. (eds.) CRYPTO 2012. LNCS, vol. 7417, pp. 850–867. Springer, Heidelberg (2012). doi:10.1007/978-3-642-32009-5_49
9. Halevi, S., Shoup, V.: Algorithms in HElib. In: Garay, J.A., Gennaro, R. (eds.) CRYPTO 2014. LNCS, vol. 8616, pp. 554–571. Springer, Heidelberg (2014). doi:10. 1007/978-3-662-44371-2_31
10. Mattner, L., Roos, B.: Maximal probabilities of convolution powers of discrete uniform distributions. Stat. Prob. Lett. **78**(17), 2992–2996 (2008)
11. Naehrig, M., Lauter, K.E., Vaikuntanathan, V.: Can homomorphic encryption be practical? In: Cachin, C., Ristenpart, T. (eds.) Proceedings of the 3rd ACM Cloud Computing Security Workshop, CCSW 2011, Chicago, IL, USA, 21 October 2011, pp. 113–124. ACM (2011)
12. Smart, N.P., Vercauteren, F.: Fully homomorphic SIMD operations. Des. Codes Crypt. **71**(1), 57–81 (2014)
13. Torralba, A., Fergus, R., Freeman, W.T.: 80 million tiny images: a large data set for nonparametric object and scene recognition. IEEE Trans. Pattern Anal. Mach. Intell. **30**(11), 1958–1970 (2008)

A Full RNS Variant of FV Like Somewhat Homomorphic Encryption Schemes

Jean-Claude Bajard[1], Julien Eynard[2], M. Anwar Hasan[2(✉)],
and Vincent Zucca[1]

[1] Sorbonne Universités, UPMC, CNRS, LIP6, Paris, France
{jean-claude.bajard,vincent.zucca}@lip6.fr
[2] Department of Electrical and Computer Engineering,
University of Waterloo, Waterloo, Canada
{jeynard,ahasan}@uwaterloo.ca

Abstract. Since Gentry's breakthrough work in 2009, homomorphic cryptography has received a widespread attention. Implementation of a fully homomorphic cryptographic scheme is however still highly expensive. Somewhat Homomorphic Encryption (SHE) schemes, on the other hand, allow only a limited number of arithmetical operations in the encrypted domain, but are more practical. Many SHE schemes have been proposed, among which the most competitive ones rely on Ring Learning With Errors (RLWE) and operations occur on high-degree polynomials with large coefficients. This work focuses in particular on the Chinese Remainder Theorem representation (a.k.a. Residue Number Systems) applied to the large coefficients. In SHE schemes like that of Fan and Vercauteren (FV), such a representation remains hardly compatible with procedures involving coefficient-wise division and rounding required in decryption and homomorphic multiplication. This paper suggests a way to entirely eliminate the need for multi-precision arithmetic, and presents techniques to enable a full RNS implementation of FV-like schemes. For dimensions between 2^{11} and 2^{15}, we report speed-ups from $5\times$ to $20\times$ for decryption, and from $2\times$ to $4\times$ for multiplication.

Keywords: Lattice-based cryptography · Homomorphic encryption · FV · Residue Number Systems · Software implementation

1 Introduction

Cryptographers' deep interests in lattices are for multiple reasons. Besides possessing highly desirable post-quantum security features, lattice-based cryptography relies on simple structures, allowing efficient asymptotic complexities, and is quite flexible in practice. In addition to encryption/signature schemes [7,15,18,21,22,25], identity-based encryption [8], multilinear maps [10,16], lattices are also involved in homomorphic encryption (HE). The discovery of this property by Gentry in 2009 [12], through the use of ideal rings, is a major

© Springer International Publishing AG 2017
R. Avanzi and H. Heys (Eds.): SAC 2016, LNCS 10532, pp. 423–442, 2017.
https://doi.org/10.1007/978-3-319-69453-5_23

breakthrough which has opened the door to many opportunities in terms of applications, especially when coupled with cloud computing.

HE is generally composed of a basic layer, which is a Somewhat Homomorphic Encryption scheme (SHE). Such a scheme allows us to compute a limited number of additions and multiplications on ciphertexts. This can be explained by the fact that any ciphertext contains an inherent noise which increases after each homomorphic operation. Beyond a certain limit, the noise becomes too large to allow a correct decryption. This drawback may be tackled by using bootstrapping, which however constitutes a bottleneck in terms of efficiency. Further improvements of noise management [5,6] have been suggested so that, in practice, and given an applicative context, it may be wiser to select an efficient SHE with parameters enabling a sufficient number of operations. For instance, schemes like FV [9] and YASHE [4] have been implemented and tested for evaluating the SIMON Feistel Cipher [17]. Among the today's more practical SHE schemes, FV is arguably one of the most competitive. This scheme is being currently considered by major stakeholders, such as the European H2020 HEAT consortium [26], as a viable candidate for practical homomorphic encryption.

Our Contribution. This work focuses on practical improvement of SHE schemes, in particular FV. Despite the fact that the security of YASHE has been called into question recently [1], this scheme can also benefit from the present work. These schemes handle elements of a polynomial ring $\mathbb{Z}_q[X]/(X^n+1)$. The modulus q can be chosen as the product of some small moduli fitting with practical hardware requirements (machine word, etc.). This enables us to avoid the need of multi-precision arithmetic in almost the whole scheme. However, this CRT representation (a.k.a. Residue Number Systems, or RNS) is hardly compatible with a couple of core operations: coefficient-wise division and rounding, occurring in multiplication and decryption, and a noise management technique within homomorphic multiplication, relying on the access to a positional number system.

We show how to efficiently avoid any switch between RNS and the positional system for performing these operations. We present a full RNS variant of FV and analyse the new bounds on noise growth. A software implementation highlights the practical benefits of the new RNS variant.

It is important to note that this work is related to the arithmetic at the coefficient level. Thus, the security features of the original scheme are not modified.

Outline. Section 2 provides some preliminaries about FV and RNS. Section 3 provides a full RNS variant of decryption. Section 4 gives a full RNS variant of homomorphic multiplication. Results of a software implementation are presented in Sect. 5. Finally, some conclusions are drawn.

2   Preliminaries

Proofs of lemmas, propositions and theorems of this article can be found in its extended version [2].

Context. High-level operations occur in a polynomial ring $\mathcal{R} = \mathbb{Z}[X]/(X^n + 1)$ with n a power of 2. $\mathcal{R}$ is one-to-one mapped to integer polynomials of degree strictly smaller than n. Most of the time, elements of $\mathcal{R}$ are denoted by lower-case boldface letters and identified by their coefficients. Polynomial arithmetic is done modulo $(X^n + 1)$. The infinity norm of $\boldsymbol{a} = (a_0, \ldots, a_{n-1}) \in \mathcal{R}$ is defined by $\|\boldsymbol{a}\|_\infty = \max_{0 \leqslant i \leqslant n-1}(|a_i|)$. Ciphertexts will be managed as polynomials (of degree 1) in $\mathcal{R}[Y]$. For $\mathsf{ct} \in \mathcal{R}[Y]$, we note $\|\mathsf{ct}\|_\infty = \max_i \|\mathsf{ct}[i]\|_\infty$, $\mathsf{ct}[i]$ being the coefficient of degree i in Y. The multiplicative law of $\mathcal{R}[Y]$ is denoted by $\star$.

Behind lattice-based cryptosystems in general, and FV in particular, lies the principle of noisy encryption. Additionally to the plaintext, a ciphertext contains a noise (revealed by using the secret key) which grows after each homomorphic operation. Since the homomorphic multiplication involves multiplications in $\mathcal{R}$, it is crucial that the size of a product in $\mathcal{R}$ does not increase too much. This increase is related to the ring constant $\delta = \sup\{\|\boldsymbol{f} \cdot \boldsymbol{g}\|_\infty / \|\boldsymbol{f}\|_\infty \cdot \|\boldsymbol{g}\|_\infty : (\boldsymbol{f}, \boldsymbol{g}) \in (\mathcal{R} \setminus \{\boldsymbol{0}\})^2\}$. It means that $\|\boldsymbol{f} \cdot \boldsymbol{g}\|_\infty \leqslant \delta \|\boldsymbol{f}\|_\infty \cdot \|\boldsymbol{g}\|_\infty$. For the specific ring $\mathcal{R}$ used here, δ is equal to n.

For our subsequent discussions on decryption and homomorphic multiplication, we denote the "Division and Rounding" in $\mathcal{R}[Y]$ (depending on parameters t, q which are defined thereafter) by:

$$\mathsf{DR}_i : \mathsf{ct} = \sum_{j=0}^{i} \mathsf{ct}[j] Y^j \in \mathcal{R}[Y] \mapsto \sum_{j=0}^{i} \left\lfloor \frac{t}{q} \mathsf{ct}[j] \right\rceil Y^j \in \mathcal{R}[Y]. \qquad (1)$$

The notation $\lfloor \frac{t}{q} \boldsymbol{c} \rceil$, for any $\boldsymbol{c} \in \mathcal{R}$ (e.g. $\mathsf{ct}[j]$ in (1)), means a coefficient-wise division-and-rounding.

Plaintext and Ciphertext Spaces. The plaintext space is determined by an integer parameter t ($t \geqslant 2$). A message is an element of $\mathcal{R}_t = \mathcal{R}/(t\mathcal{R})$, i.e. a polynomial of degree at most $n - 1$ with coefficients in $\mathbb{Z}_t$. The notation $[\boldsymbol{m}]_t$ (resp. $|\boldsymbol{m}|_t$) means that coefficients lie in $[-t/2, t/2)$ (resp. $[0, t)$). Ciphertexts will lie in $\mathcal{R}_q[Y]$ with q a parameter of the scheme. On one side, some considerations about security imply a relationship between q and n which, for a given degree n, establish an upper bound for $\log_2(q)$ (cf. (6) in [9]). On the other side, the ratio $\Delta = \lfloor \frac{q}{t} \rfloor$ will basically determine the maximal number of homomorphic operations which can be done in a row to ensure a correct decryption.

RNS Representation. Beyond the upper bound on $\log_2(q)$ due to security requirements, the composition of q has no restriction. So, q can be chosen as a product of small pairwise coprime moduli $q_1 \ldots q_k$. The reason for such a choice is the Chinese Remainder Theorem (CRT) which offers a ring isomorphism $\mathbb{Z}_q \xrightarrow{\sim} \prod_{i=1}^{k} \mathbb{Z}_{q_i}$. Thus, the CRT implies the existence of a non-positional number system (RNS) in which large integers (mod q) are mapped to sets of small residues. Furthermore, the arithmetic modulo q over large integers can be substituted by k independent arithmetics in the small rings $\mathbb{Z}_{q_i}$. The isomorphism can be naturally extended to polynomials: $\mathcal{R}_q \simeq \mathcal{R}_{q_1} \times \ldots \times \mathcal{R}_{q_k}$. It means that RNS can be used at the coefficient level to accelerate the arithmetic in $\mathcal{R}_q$.

In the rest of the paper, the letter q may refer either to the product $q_1 \ldots q_k$ or to the "RNS base" $\{q_1, \ldots, q_k\}$. Symbol ν denotes the "width" of the moduli. In other words, from now on, any modulus m (should it belong to q or to any other RNS base) is assumed to verify $m < 2^\nu$.

Asymmetric Keys. The *secret key* s is picked up in $\mathcal{R}$ according to a discrete distribution χ_{key} on $\mathcal{R}$ (in practice, bounded by $B_{key} = 1$, i.e. $\|s\|_\infty \leqslant 1$).

For creating the public key, an "error" distribution χ_{err} over $\mathcal{R}$ is used. In practice, this is a discrete distribution statistically close to a Gaussian (with mean 0 and standard deviation σ_{err}) truncated at B_{err} (e.g. $B_{err} = 6\sigma_{err}$). χ_{err} is related to the hardness of the underlying (search version of) RLWE problem (for which the purpose is, given samples $([-(a_i s + e_i)]_q, a_i)$ with $e_i \leftarrow \chi_{err}$ and $a \leftarrow \mathcal{U}(\mathcal{R}_q)$, to find s; $\mathcal{U}(\mathcal{R}_q)$ is the uniform distribution on $\mathcal{R}_q$). The *public key* pk is created as follows: sample $a \leftarrow \mathcal{U}(\mathcal{R}_q)$ and $e \leftarrow \chi_{key}$, then output $\text{pk} = (p_0, p_1) = ([-(as + e)]_q, a)$.

Encryption, Addition, Inherent Noise of a Ciphertext. Encryption and homomorphic addition are already fully compliant with RNS arithmetic. They are recalled hereafter:

- $\text{Enc}_{FV}([m]_t)$: from samples $e_1, e_2 \leftarrow \chi_{err}$, $u \leftarrow \chi_{key}$, output
  $\text{ct} = (\text{ct}[0], \text{ct}[1]) = ([\Delta[m]_t + p_0 u + e_1]_q, [p_1 u + e_2]_q)$.
- $\text{Add}_{FV}(\text{ct}_1, \text{ct}_2)$: output $([\text{ct}_1[0] + \text{ct}_2[0]]_q, [\text{ct}_1[1] + \text{ct}_2[1]]_q)$.

By definition, the *inherent noise* of ct (encrypting $[m]_t$) is the polynomial v such that $[\text{ct}(s)]_q = [\text{ct}[0] + \text{ct}[1]s]_q = [\Delta[m]_t + v]_q$. Thus, it is revealed by evaluating $\text{ct} \in \mathcal{R}_q[Y]$ on the secret key s.

Elementary Operations. A basic word will fit in ν bits. In RNS, an "inner modular multiplication" (IMM) in a small ring like $\mathbb{Z}_m$ is a core operation. If EM stands for an elementary multiplication of two words, in practice an IMM is costlier than an EM. But it can be well controlled. For instance, the moduli provided in NFLlib library [19] (cf. Sect. 5) enable a modular reduction which reduces to one EM followed by a multiplication modulo 2^ν. Furthermore, the cost of an inner reduction can be limited by using lazy reduction, e.g. during RNS base conversions used throughout this paper. NTT and invNTT denote the Number Theoretic Transform and its inverse in a ring $\mathcal{R}_m$ for a modulus m. They enable an efficient polynomial multiplication (NTT, invNTT $\in \mathcal{O}(n \log_2(n))$).

3 Towards a Full RNS Decryption

This section deals with the creation of a variant of the original decryption function Dec_{FV}, which will only involve RNS representation. The definition of Dec_{FV} is recalled hereafter.

- $\text{Dec}_{FV}(\text{ct} = (c_0, c_1) \in \mathcal{R}_q[Y])$: compute $[\text{DR}_0([\text{ct}(s)]_q)]_t = \left[\left\lfloor \frac{t}{q}[c_0 + c_1 s]_q \right\rceil\right]_t$.

The idea is that computing $[c_0 + c_1 s]_q = [\Delta[m]_t + v]_q$ reveals the noise. If this noise is small enough, and given that $[m]_t$ has been scaled by Δ, the function DR_0 allows to cancel the noise while scaling down $\Delta[m]_t$ to recover $[m]_t$. Concretely, decryption is correct as long as $\|v\|_\infty < (\Delta - |q|_t)/2$, i.e. the size of the noise should not go further this bound after homomorphic operations.

The division-and-rounding operation makes Dec_{FV} hardly compatible with RNS at a first sight. Because RNS is of non positional nature, only exact integer division can be naturally performed (by multiplying by a modular inverse). But it is not the case here. And the rounding operation involves comparisons which require to switch from RNS to another positional system anyway, should it be a classical binary system or a mixed-radix one [11]. To get an efficient RNS variant of Dec_{FV}, we use an idea of [3]. To this end, we introduce relevant RNS tools.

3.1 Fast RNS Base Conversion

At some point, the decryption requires, among others, a polynomial to be converted from $\mathcal{R}_q$ to $\mathcal{R}_t$. To achieve such kind of operations as efficiently as possible, we suggest to use a "fast base conversion". In order to convert residues of $x \in [0, q)$ from base q to a base $\mathcal{B}$ (e.g. $\{t\}$) coprime to q, we compute:

$$\text{FastBconv}(x, q, \mathcal{B}) = \left(\sum_{i=1}^{k} \left| x_i \frac{q_i}{q} \right|_{q_i} \times \frac{q}{q_i} \bmod m \right)_{m \in \mathcal{B}}. \tag{2}$$

This conversion is relatively fast. This is because the sum should ideally be reduced modulo q in order to provide the exact value x; instead, (2) provides $x + \alpha_x q$ for some integer $\alpha_x \in [0, k-1]$. Computing α_x requires costly operations in RNS. So this step is by-passed, at the cost of an approximate result.

FastBconv naturally extends to polynomials of $\mathcal{R}$ by applying it coefficient-wise.

3.2 Approximate RNS Rounding

The above mentioned fast conversion allows us to efficiently compute an approximation of $\lfloor \frac{t}{q}[c_0 + c_1 s]_q \rceil$ modulo t. The next step consists in correcting this approximation.

First, we remark that $|\text{ct}(s)|_q$ can be used instead of $[\text{ct}(s)]_q$. Indeed, the difference between these two polynomials is a multiple of q. So, the division-and-rounding turns it into a polynomial multiple of t, which is cancelled by the last reduction modulo t. Second, a rounding would involve, at some point, a comparison. This is hardly compatible with RNS, so it is avoided. Therefore, we propose to simplify the computation, albeit at the price of possible errors, by replacing rounding by flooring. To this end, we use the following formula:

$$\left\lfloor \frac{t}{q}|ct(s)|_q \right\rfloor = \frac{t|ct(s)|_q - |t \cdot ct(s)|_q}{q}.$$

The division is now exact, so it can be done in RNS. Since this computation has to be done modulo t, the term $t|ct(s)|_q$ cancels. Furthermore, the term $(|t \cdot ct(s)|_q \bmod t)$ is obtained through a fast conversion.

Lemma 1 sums up the strategy by replacing $|ct(s)|_q$ by $\gamma|ct(s)|_q$, where γ is an integer which will help in correcting the approximation error.

Lemma 1. *Let ct be such that $[ct(s)]_q = \Delta[m]_t + v + qr$, and let $v_c := tv - [m]_t|q|_t$. Let γ be an integer coprime to q. Then, for $m \in \{t, \gamma\}$, the following equalities hold modulo m:*

$$FastBconv(|\gamma t \cdot ct(s)|_q, q, \{t, \gamma\}) \times |-q^{-1}|_m = \left\lfloor \gamma\frac{t}{q}[ct(s)]_q \right\rfloor - e$$

$$= \gamma([m]_t + tr) + \left\lfloor \gamma\frac{v_c}{q} \right\rfloor - e \tag{3}$$

where each integer coefficient of the error polynomial $e \in \mathcal{R}$ lies in $[0, k]$.

The error e is due to the fast conversion and the replacement of rounding by flooring. It is the same error for residues modulo t and γ. The residues modulo γ will enable a fast correction of it and of the term $\lfloor \gamma\frac{v_c}{q} \rceil$ at a same time. Also, note that r vanishes since it is multiplied by both t and γ.

3.3   Correcting the Approximate RNS Rounding

The next step is to show how γ in (3) can be used to correct the term $(\lfloor \gamma\frac{v_c}{q} \rceil - e)$. It can be done efficiently when the polynomial v_c is such that $\|v_c\|_\infty \leqslant q(\frac{1}{2} - \varepsilon)$, for some real number $\varepsilon \in (0, 1/2]$.

Lemma 2. *Let $\|v_c\|_\infty \leqslant q(\frac{1}{2} - \varepsilon)$, $e \in \mathcal{R}$ with coefficients in $[0, k]$, and γ an integer. Then,*

$$\gamma\varepsilon \geqslant k \Rightarrow \left[\left\lfloor \gamma\frac{v_c}{q} \right\rceil - e \right]_\gamma = \left\lfloor \gamma\frac{v_c}{q} \right\rceil - e. \tag{4}$$

Lemma 2 enables an efficient and correct RNS rounding as long as $k(\frac{1}{2} - \frac{\|v_c\|_\infty}{q})^{-1} \sim \gamma$ has the size of a modulus. Concretely, one computes (3) and uses the centered remainder modulo γ to obtain $\gamma([m]_t + tr)$ modulo t, which reduces to $\gamma[m]_t \bmod t$. And it remains to multiply by $|\gamma^{-1}|_t$ to recover $[m]_t$.

3.4 A Full RNS Variant of $\mathtt{Dec_{FV}}$

The new variant of the decryption is detailed in Algorithm 1. The main modification for the proposed RNS decryption is due to Lemma 2. As stated by Theorem 1, given a γ, the correctness of rounding requires a new bound on the noise to make the γ-correction technique successful.

Theorem 1. *Let* $ct(s) = \Delta[m]_t + v \pmod{q}$. *Let* γ *be a positive integer coprime to* t *and* q *such that* $\gamma > 2k/(1 - \frac{t|q|_t}{q})$. *For Algorithm 1 returning* $[m]_t$, *it suffices that* v *satisfies the following bound:*

$$\|v\|_\infty \leqslant \frac{q}{t}\left(\frac{1}{2} - \frac{k}{\gamma}\right) - \frac{|q|_t}{2}. \tag{5}$$

There is a trade-off between the size of γ and the bound in (5). Ideally, $\gamma \sim 2k$ at the price of a (*a priori*) quite small bound on the noise. But by choosing $\gamma \sim 2^{p+1}k$ for $p < \nu - 1 - \lceil \log_2(k) \rceil$ (i.e. $\gamma < 2^\nu$ is a standard modulus), the bound $(\Delta(1 - 2^{-p}) - |q|_t)/2$ for a correct decryption should be close to the original bound $(\Delta - |q|_t)/2$ for practical values of ν. A concrete estimation of γ in Sect. 5.1 will show that γ can be chosen very close to $2k$ in practice, and thus fitting on a basic word by far.

Algorithm 1. $\mathtt{Dec_{RNS}}(ct, s, \gamma)$

Require: ct an encryption of $[m]_t$, and s the secret key, both in base q; an integer γ
  coprime to t and q
Ensure: $[m]_t$
  1: **for** $m \in \{t, \gamma\}$ **do**
  2:     $s^{(m)} \leftarrow \mathtt{FastBconv}(|\gamma t \cdot ct(s)|_q, q, \{m\}) \times |-q^{-1}|_m \bmod m$
  3: **end for**
  4: $\tilde{s}^{(\gamma)} \leftarrow [s^{(\gamma)}]_\gamma$
  5: $m^{(t)} \leftarrow [(s^{(t)} - \tilde{s}^{(\gamma)}) \times |\gamma^{-1}|_t]_t$
  6: **return** $m^{(t)}$

3.5 Staying in RNS is Asymptotically Better

In any decryption technique, $(ct(s) \bmod q)$ has to be computed first. To optimize this polynomial product, one basically performs $k\mathtt{NTT} \to kn\mathtt{IMM} \to k\mathtt{invNTT}$. For next steps, a simple strategy is to compute $(\lfloor \frac{t}{q}[ct(s)]_q \rceil \bmod t)$ by doing an RNS-to-binary conversion in order to perform the division and rounding. By denoting $x_i = |ct(s)\frac{q_i}{q}|_{q_i}$, one computes $\sum_{i=1}^k x_i \frac{q}{q_i} \bmod q$, compares it to $q/2$ so as to center the result, and performs division and rounding. Hence, the division-and-rounding requires $\mathcal{O}(k^2n)\mathtt{EM}$. In practice, security analysis (cf. e.g. [4,9,17]) requires that $k\nu = \lceil \log_2(q) \rceil \in \mathcal{O}(n)$. So, the cost of leaving RNS to access a positional system is dominant in the asymptotic computational complexity.

Staying in RNS enables to get a better asymptotic complexity. Indeed, it is easy to see that Algorithm 1 requires $\mathcal{O}(kn)$ operations (excluding the polynomial product). Thus, the cost of $\mathtt{NTT}$ is dominant in this case. By considering

$k \in \mathcal{O}(n)$, we deduce $\mathcal{C}(\text{Dec}_{\text{FV}}) \in \mathcal{O}(n^3)$, while $\mathcal{C}(\text{Dec}_{\text{RNS}}) \in \mathcal{O}(n^2 \log_2(n))$. But the hidden constant in "$k \in \mathcal{O}(n)$" is small, and the NTT, common to both variants, should avoid any noticeable divergence (cf. Sect. 5.4) for practical ranges for parameters.

In order to provide optimized RNS variants of decryption, we make two remarks. First, the reduction modulo q is unnecessary. Indeed, any extra multiple of q in the sum $\sum_{i=1}^{k} x_i \frac{q}{q_i}$ is multiplied by $\frac{t}{q}$, making the resulting term a multiple of t, which is not affected by the rounding and is finally cancelled modulo t. Second, it is possible to precompute $\frac{t}{q}$ as a multiprecision floating point number in order to avoid a costly integer division. But given the first remark, it suffices to precompute the floating point numbers $\mathcal{Q}_i \sim \frac{t}{q_i}$ with $2\nu + \log_2(k) - \log_2(t)$ bits (~ 2 words) of precision. In this case, using standard double or quadruple (depending on ν) precision is sufficient. Finally, it is sufficient to compute $\lfloor \sum_{i=1}^{k} x_i \mathcal{Q}_i \rceil \bmod t$. This represents about $2kn$EM. Reducing modulo t is nearly free of cost when t is a power of 2.

A second optimized RNS variant, with only integer arithmetic, is based on Algorithm 1, in which γ is assumed to be coprime to t. It is possible to be slightly more efficient by noticing that the coprimality assumption can be avoided. This is because the division by γ is exact. To do it, the `for` loop can be done modulo $\gamma \times t$. For instance, even if t is a power of 2, one can choose γ as being a power of 2 too, and use the following lemma to finish the decryption efficiently.

Lemma 3. *Let γ be a power of 2. Let $z := |\gamma[m]_t + \lfloor \gamma \frac{v_c}{q} \rceil - e|_{\gamma t}$ coming from (3) when computed modulo γt. If γ satisfies (4), then ($\gg$ denotes the right bit-shifting, and v_1 is the polynomial with all its coefficients being equal to 1)*

$$[(z + \tfrac{\gamma}{2} v_1) \gg \log_2(\gamma)]_t = [m]_t. \tag{6}$$

Lemma 3 can be adapted to other values for γ. The important remark is that $[m]_t$ is contained in the $\lfloor \log_2(t) \rfloor + 1$ most significant bits of $(z + \frac{\gamma}{2} v_1) \bmod \gamma t$. So, by choosing γ as a power of 2, a simple bit shifting enables to recover these bits. Finally, as soon as γt fits in 1 word, the cost of such variant (besides the polynomial product) reduces to knIMM, or simply to knEM modulo $2^{\log_2(\gamma t)}$ whenever t is a power of 2.

Remark 1. In previous discussion, the product γt is assumed fitting in one machine word to simplify complexity analysis. However, for some applications, the plaintext modulus t can be bigger than a machine word (e.g. homomorphic neural networks [13], where $t > 2^{80}$). In such cases, either the plaintexts directly lie in $\mathcal{R}_t$, or t can be decomposed in a product of smaller moduli $t_1, \ldots, t_\ell$, enabling the use of RNS for encoding plaintexts (and then allowing better homomorphic multiplicative depth for a given dimension n). In the first case, the optimized RNS decryption (given by Lemma 3) remains available, but the residues modulo t should be handled with several words. In the second case, a plaintext is recovered by decrypting its residue modulo each of the t_i. These ℓ decryptions can be done as in Lemma 3, by using γ as a power of 2 (whatever the t_i's are).

Finally, the plaintext is reconstructed from residues modulo the t_i's by using a classical RNS to binary conversion. However, this conversion is only related to the way the plaintexts are encoded. This is not handled by RNS decryption described in this paper, which only deals with representation of ciphertexts (i.e. modulo q).

4   Towards a Full RNS Homomorphic Multiplication

4.1   Preliminaries About $\mathtt{Mult_{FV}}$

Below, we recall the main mechanisms of the homomorphic multiplication $\mathtt{Mult_{FV}}$ from [9]. More precisely, we focus on the variant with version 1 for relinearisation step. First, two functions, of which the purpose is to limit a too rapid noise growth during a multiplication, are recalled (these functions will be denoted as in [4]). They are applicable to any $a \in \mathcal{R}$, for any radix ω, and with the subsequent parameter $\ell_{\omega,q} = \lfloor \log_\omega(q) \rfloor + 1$. $\mathcal{D}_{\omega,q}$ is a decomposition in radix base ω, while $\mathcal{P}_{\omega,q}$ gets back powers of ω which are lost within the decomposition process.

$$\forall a \in \mathcal{R}, \begin{cases} \mathcal{D}_{\omega,q}(a) = ([a]_\omega, [\lfloor a\omega^{-1}\rfloor]_\omega, \ldots, [\lfloor a\omega^{-(\ell_{\omega,q}-1)}\rfloor]_\omega) \in \mathcal{R}_\omega^{\ell_{\omega,q}} \\ \mathcal{P}_{\omega,q}(a) = ([a]_q, [a\omega]_q, \ldots, [a\omega^{\ell_{\omega,q}-1}]_q) \in \mathcal{R}_q^{\ell_{\omega,q}} \end{cases} . \quad (7)$$

In particular, for any $(a, b) \in \mathcal{R}^2$, $\langle \mathcal{D}_{\omega,q}(a), \mathcal{P}_{\omega,q}(b)\rangle \equiv ab \bmod q$.

Next, $\mathtt{Mult_{FV}}$ is built as follows ($\mathtt{rlk_{FV}}$ is a public relinearisation key):

- $\mathtt{rlk_{FV}} = ([\mathcal{P}_{\omega,q}(s^2) - (\overrightarrow{e} + s\overrightarrow{a})]_q, \overrightarrow{a})$ where $\overrightarrow{e} \leftarrow \chi_{err}^{\ell_{\omega,q}}$, $\overrightarrow{a} \leftarrow \mathcal{U}(\mathcal{R}_q)^{\ell_{\omega,q}}$,
- $\mathtt{Relin_{FV}}(c_0, c_1, c_2, \mathtt{rlk_{FV}})$:
  compute $([c_0 + \langle \mathcal{D}_{\omega,q}(c_2), \mathtt{rlk_{FV}}[0]\rangle]_q, [c_1 + \langle \mathcal{D}_{\omega,q}(c_2), \mathtt{rlk_{FV}}[1]\rangle]_q)$,
- $\mathtt{Mult_{FV}}(\mathtt{ct}_1, \mathtt{ct}_2)$: denote $\mathtt{ct}_\star = \mathtt{ct}_1 \star \mathtt{ct}_2$ (degree-2 element of $\mathcal{R}[Y]$),
  - Step 1: $\widetilde{\mathtt{ct}}_{mult} = [\mathrm{DR}_2(\mathtt{ct}_\star)]_q = ([\mathrm{DR}_0(\mathtt{ct}_\star[i])]_q)_{i\in\{0,1,2\}}$,
  - Step 2: $\mathtt{ct}_{mult} = \mathtt{Relin_{FV}}(\widetilde{\mathtt{ct}}_{mult})$.

There are two main obstacles to a full RNS variant. First, the three calls to DR_0 in Step 1, for which the context is different than for the decryption. While in the decryption we are working with a noise whose size can be controlled, and while we are reducing a value from q to $\{t\}$, here the polynomial coefficients of the product $\mathtt{ct}_1 \star \mathtt{ct}_2$ have kind of random size modulo q (for each integer coefficient) and have to be reduced towards q. Second, the function $\mathcal{D}_{\omega,q}$ (in $\mathtt{Relin_{FV}}$) requires, by definition, an access to a positional system (in radix base ω), which is hardly compatible with RNS.

4.2   Auxiliary RNS Bases

Step 1 requires to use enough moduli to contain any product, in $\mathcal{R}[Y]$ (i.e. on $\mathbb{Z}$), of degree-1 elements from $\mathcal{R}_q[Y]$. So, we need an auxiliary base $\mathcal{B}$, additionally to the base q. We assume that $\mathcal{B}$ contains ℓ moduli (while q owns k elements). A sufficient size for ℓ will be given later. An extra modulus m_{sk} is added to $\mathcal{B}$

to create an extended base $\mathcal{B}_{sk}$. It will be used for a transition between the new steps 1 and 2. Computing the residues of ciphertexts in $\mathcal{B}_{sk}$ is done through a fast conversion from q. In order to reduce the extra multiples of q (called "q-overflows" in further discussions) this conversion can produce, a single-modulus base $\tilde{m}$ is introduced. All these bases are assumed to be pairwise coprime.

Reducing (mod q) a Ciphertext in $\mathcal{B}_{sk}$. A FastBconv from q can create q-overflows (i.e. unnecessary multiples of q) in the output. To limit the impact on noise growth (because of division by q in step 1), we give an efficient way to reduce a polynomial $c + qu$ in $\mathcal{B}_{sk}$. It should be done prior to each multiplication. For that purpose, we use the residues modulo $\tilde{m}$ as it is described in Algorithm 2.

Algorithm 2. $\mathsf{SmMRq}_{\tilde{m}}((c''_m)_{m \in \mathcal{B}_{sk} \cup \{\tilde{m}\}})$: Small Montgomery Reduction mod q

---

Require: c'' in $\mathcal{B}_{sk} \cup \{\tilde{m}\}$
Ensure: c' in $\mathcal{B}_{sk}$, with $c' \equiv c'' \tilde{m}^{-1} \bmod q$, $\|c'\|_\infty \leqslant \frac{\|c''\|_\infty}{\tilde{m}} + \frac{q}{2}$
1: $r_{\tilde{m}} \leftarrow [-c''_{\tilde{m}}/q]_{\tilde{m}}$
2: **for** $m \in \mathcal{B}_{sk}$ **do**
3:     $c'_m \leftarrow |(c''_m + qr_{\tilde{m}})\tilde{m}^{-1}|_m$
4: **end for**
5: **return** c' in $\mathcal{B}_{sk}$

Lemma 4. *On input $c''_m = |[\tilde{m}c]_q + qu|_m$ for all $m \in \mathcal{B}_{sk} \cup \{\tilde{m}\}$, with $\|u\|_\infty \leqslant \tau$, and given a parameter $\rho > 0$, then Algorithm 2 returns c' in $\mathcal{B}_{sk}$ with $c' \equiv c \bmod q$ and $\|c'\|_\infty \leqslant \frac{q}{2}(1 + \rho)$ if $\tilde{m}$ satisfies:*

$$\tilde{m}\rho \geqslant 2\tau + 1. \tag{8}$$

To use this fast reduction, the ciphertexts have to be handled in base q through the Montgomery [20] representation with respect to $\tilde{m}$ (i.e. $|\tilde{m}c|_q$ instead of $|c|_q$). This can be done for free of cost during the base conversions (in (2), multiply residues of c by precomputed $|\frac{\tilde{m}q_i}{q}|_{q_i}$ instead of $|\frac{q_i}{q}|_{q_i}$). Since $\{\tilde{m}\}$ is a single-modulus base, conversion of $r_{\tilde{m}}$ from $\{\tilde{m}\}$ to $\mathcal{B}_{sk}$ (l. 3 of Algorithm 2) is a simple copy-paste when $\tilde{m} < m_i$. Finally, if $\mathsf{SmMRq}_{\tilde{m}}$ is performed right after a FastBconv from q (for converting $|\tilde{m}c|_q$), τ is nothing but k.

4.3 Adapting the First Step

We recall that originally this step is the computation of $[\mathsf{DR}_2(\mathsf{ct}_\star)]_q$. Unlike the decryption, a γ-correction technique does not guarantee an exact rounding. Indeed, for the decryption we wanted to get $\mathsf{DR}_0([\mathsf{ct}(s)]_q)$, and through s we had access to the noise of ct, on which we have some control. In the present context, we cannot ensure a condition like $\|[t \cdot \mathsf{ct}_\star]_q\|_\infty \leqslant q(\frac{1}{2} - \varepsilon)$, for some $\varepsilon^{-1} \sim 2^\nu$, which would enable the use of an efficient γ-correction. Thus, we suggest to perform a simple uncorrected RNS flooring. For that purpose, we define:

$$\forall a \in \mathcal{R}, \mathsf{fastRNSFloor}_q(a, m) := (a - \mathsf{FastBconv}(|a|_q, q, m)) \times |q^{-1}|_m \bmod m.$$

Algorithm 2 should be executed first. Consequently, by Lemma 4, if $\tilde{m}$ satisfies the bound in (8) for a given parameter $\rho > 0$, we assume having, in $\mathcal{B}_{sk}$, the residues of ct_i' ($\equiv ct_i \bmod q$) such that:

$$\|ct_\star'\| := ct_1' \star ct_2'\|_\infty \leqslant \delta \frac{q^2}{2}(1+\rho)^2. \tag{9}$$

The parameter ρ is determined in practice. Notice that, in base q, ct_i' and ct_i are equal.

Lemma 5. *Let the residues of $ct_i' \equiv ct_i \bmod q$ be given in base $q \cup \mathcal{B}_{sk}$, with $\|ct_i'\|_\infty \leqslant \frac{q}{2}(1+\rho)$ for $i \in \{1,2\}$. Let $ct_\star' = ct_1' \star ct_2'$. Then, for $j \in \{0,1,2\}$,*

$$fastRNSFloor_q(t \cdot ct_\star'[j], \mathcal{B}_{sk}) = \left\lfloor \frac{t}{q} ct_\star'[j] \right\rceil + b_j \text{ in } \mathcal{B}_{sk}, \text{ with } \|b_j\|_\infty \leqslant k. \tag{10}$$

A first part of the noise growth is detailed in the following proposition.

Proposition 1. *Let $\widetilde{ct}_{mult} = DR_2(ct_\star')$ with (9) satisfied, and $r_\infty := \frac{1+\rho}{2}(1 + \delta B_{key}) + 1$. Let v_i be the inherent noise of ct_i'. Then $\widetilde{ct}_{mult}(s) = \Delta[m_1 m_2]_t + \tilde{v}_{mult} (\bmod q)$ with:*

$$\|\tilde{v}_{mult}\|_\infty < \delta t(r_\infty + \tfrac{1}{2})(\|v_1\|_\infty + \|v_2\|_\infty) + \tfrac{\delta}{2} \min \|v_i\|_\infty + \delta t|q|_t(r_\infty + 1)$$
$$+ \tfrac{1}{2}(3 + |q|_t + \delta B_{key}(1 + \delta B_{key})). \tag{11}$$

4.4 Transitional Step

Lemma 5 states that we have got back $DR_2(ct_\star') + b$ in $\mathcal{B}_{sk}$ so far, where we have denoted (b_0, b_1, b_2) by b. To perform the second step of multiplication, we need to convert it in base q. However, the conversion has to be exact because extra multiples of $M = m_1 \ldots m_\ell$ cannot be tolerated. m_{sk} allows us to perform a complete Shenoy and Kumaresan like conversion [24]. The next lemma describes such kind of conversion for a more general context where the input can be either positive or negative, and can be larger, in absolute value, than M.

Lemma 6. *Let $\mathcal{B}$ be an RNS base and m_{sk} be a modulus coprime to $M = \prod_{m \in \mathcal{B}} m$. Let x be an integer such that $|x| < \lambda M$ (for some real number $\lambda \geqslant 1$) and whose residues are given in $\mathcal{B}_{sk}$. Let's assume that m_{sk} satisfies $m_{sk} \geqslant 2(|\mathcal{B}| + \lceil \lambda \rceil)$. Let $\alpha_{sk,x}$ be the following integer:*

$$\alpha_{sk,x} := \left[(FastBconv(x, \mathcal{B}, \{m_{sk}\}) - x_{sk})M^{-1}\right]_{m_{sk}}. \tag{12}$$

Then, for x being either positive or negative, the following equality holds:

$$FastBconvSK(x, \mathcal{B}_{sk}, q) := (FastBconv(x, \mathcal{B}, q) - \alpha_{sk,x}M) \bmod q = x \bmod q. \tag{13}$$

Consequently, since $\|DR_2(ct_\star') + b\|_\infty \leqslant \delta t \frac{q}{2}(1+\rho)^2 + \frac{1}{2} + k$, we can establish the following proposition.

Proposition 2. *Given a positive real number* λ, *let* m_{sk} *and* $\mathcal{B}$ *be such that:*

$$\lambda M > \delta t \frac{q}{2}(1+\rho)^2 + \frac{1}{2} + k, \quad m_{sk} \geqslant 2(|\mathcal{B}| + \lceil \lambda \rceil). \tag{14}$$

Let's assume that $DR_2(\mathsf{ct}'_\star) + b$ *is given in* $\mathcal{B}_{sk}$, *with* $\|b\|_\infty \leqslant k$. *Then,*

$$FastBconvSK(DR_2(\mathsf{ct}'_\star) + b, \mathcal{B}_{sk}, q) = (DR_2(\mathsf{ct}'_\star) + b) \bmod q.$$

4.5   Adapting the Second Step

At this point, $\widetilde{\mathsf{ct}}_{mult} + b = (\overline{c}_0, \overline{c}_1, \overline{c}_2)$ is known in base q $(\widetilde{\mathsf{ct}}_{mult} := DR_2(\mathsf{ct}'_\star))$. We recall that the original second step of homomorphic multiplication would be done as follows:

$$\mathsf{ct}_{mult} = \left([\overline{c}_0 + \langle \mathcal{D}_{\omega,q}(\overline{c}_2), \mathcal{P}_{\omega,q}(s^2) - (\overrightarrow{e} + s\overrightarrow{a})\rangle]_q, [\overline{c}_1 + \langle \mathcal{D}_{\omega,q}(\overline{c}_2), \overrightarrow{a}\rangle]_q\right) \tag{15}$$

where $\overrightarrow{e} \leftarrow \chi_{err}^{\ell_{\omega,q}}$, $\overrightarrow{a} \leftarrow \mathcal{U}(\mathcal{R}_q)^{\ell_{\omega,q}}$. The decomposition of $\overline{c}_2$ in radix ω enables a crucial reduction of the noise growth due to the multiplications by the terms $e_i + sa_i$. It cannot be done directly in RNS as is. Indeed, it would require a costly switch between RNS and positional representation in radix ω. However, we can do something very similar. We recall that we can write $\overline{c}_2 = \sum_{i=1}^k |\overline{c}_2 \frac{q_i}{q}|_{q_i} \times \frac{q}{q_i} (\bmod q)$. If ω has the same order of magnitude than 2^ν (size of moduli in q), we obtain a similar limitation of the noise growth by using the vectors $\xi_q(\overline{c}_2) = (|\overline{c}_2 \frac{q_1}{q}|_{q_1}, \ldots, |\overline{c}_2 \frac{q_k}{q}|_{q_k})$ and $\mathcal{P}_{RNS,q}(s^2) = (|s^2 \frac{q}{q_1}|_q, \ldots, |s^2 \frac{q}{q_k}|_q)$, both in $\mathcal{R}^k$. This is justified by the following lemma.

Lemma 7. *For any* $c \in \mathcal{R}$, $\langle \xi_q(c), \mathcal{P}_{RNS,q}(s^2) \rangle \equiv cs^2 \bmod q$.

Thus, we replace the public relinearisation key rlk_{FV} by the following one: $\mathsf{rlk}_{RNS} = ([\mathcal{P}_{RNS,q}(s^2) - (\overrightarrow{e} + s\overrightarrow{a})]_q, \overrightarrow{a})$. The next lemma helps for providing a bound on the extra noise introduced by this step.

Lemma 8. *Let* $\overrightarrow{e} \leftarrow \chi_{err}^k$, $\overrightarrow{a} \leftarrow \mathcal{U}(\mathcal{R}_q)^k$, *and* $c \in \mathcal{R}$. *Then,*

$$\| (\langle \xi_q(c), -(\overrightarrow{e} + \overrightarrow{a}s) \rangle + \langle \xi_q(c), \overrightarrow{a} \rangle s) \bmod q \|_\infty < \delta B_{err} k 2^\nu. \tag{16}$$

Remark 2. It is still possible to add a second level of decomposition (like in original approach, but applied on the residues) to limit a bit more the noise growth. Furthermore, Sect. 4.6 details how the size of rlk_{RNS} can be reduced in a similar way that rlk_{FV} could be through the method described in ([4], Sect. 5.4).

Finally, the output of the new variant of multiplication is the following one:

$$\mathsf{ct}_{mult} = \left([\overline{c}_0 + \langle \xi_q(\overline{c}_2), \mathcal{P}_{RNS,q}(s^2) - (\overrightarrow{e} + \overrightarrow{a}s)\rangle]_q, [\overline{c}_1 + \langle \xi_q(\overline{c}_2), \overrightarrow{a}\rangle]_q\right). \tag{17}$$

Proposition 3. *Let* ct_{mult} *be as in (17), and* v_{mult} *(resp.* $\widetilde{v}_{mult}$*) the inherent noise of* ct_{mult} *(resp.* $\widetilde{\mathsf{ct}}_{mult}$*). Then* $\mathsf{ct}_{mult}(s) = \Delta [m_1 m_2]_t + v_{mult} (\bmod q)$ *with:*

$$\|v_{mult}\|_\infty < \|\widetilde{v}_{mult}\|_\infty + k(1 + \delta B_{key}(1 + \delta B_{key})) + \delta k B_{err} 2^{\nu+1}. \tag{18}$$

Algorithm 3 depicts the scheme of the new full RNS variant Mult_{RNS}.

Algorithm 3. RNS homomorphic multiplication $\texttt{Mult}_{\texttt{RNS}}$

Require: $\texttt{ct}_1, \texttt{ct}_2$ in q

Ensure: $\texttt{ct}_{mult}$ in q

  S0: Convert fast $\texttt{ct}_1$ and $\texttt{ct}_2$ from q to $\mathcal{B}_{\texttt{sk}} \cup \{\tilde{m}\}$: $\rightsquigarrow \texttt{ct}_i'' = \texttt{ct}_i + q$-overflows

  S1: Reduce q-overflows in $\mathcal{B}_{\texttt{sk}}$: $(\texttt{ct}_i'$ in $\mathcal{B}_{\texttt{sk}}) \leftarrow \texttt{SmMRq}_{\tilde{m}}(((\texttt{ct}_i'')_m)_{m \in \mathcal{B}_{\texttt{sk}} \cup \{\tilde{m}\}})$

  S2: Compute the product $\texttt{ct}_\star' = \texttt{ct}_1' \star \texttt{ct}_2'$ in $q \cup \mathcal{B}_{\texttt{sk}}$

  S3: Convert fast from q to $\mathcal{B}_{\texttt{sk}}$ to achieve first step (approximate rounding) in $\mathcal{B}_{\texttt{sk}}$:
  $$(\widetilde{\texttt{ct}}_{mult} + \mathbf{b} = \texttt{DR}_2(\texttt{ct}_\star') + \mathbf{b} \text{ in } \mathcal{B}_{\texttt{sk}}) \leftarrow \ldots \leftarrow \texttt{FastBconv}(t \cdot \texttt{ct}_\star', q, \mathcal{B}_{\texttt{sk}})$$

  S4: Convert exactly from $\mathcal{B}_{\texttt{sk}}$ to q to achieve transitional step:
  $$(\widetilde{\texttt{ct}}_{mult} + \mathbf{b} \text{ in } q) \leftarrow \texttt{FastBconvSK}(\widetilde{\texttt{ct}}_{mult} + \mathbf{b}, \mathcal{B}_{\texttt{sk}}, q)$$

  S5: Perform second step (relinearisation) in q:
  $$\texttt{ct}_{mult} \leftarrow \texttt{Relin}_{\texttt{RNS}}(\widetilde{\texttt{ct}}_{mult} + \mathbf{b}) \bmod (q_1, \ldots, q_k)$$

4.6   Reducing the Size of the Relinearization Key $\texttt{rlk}_{\texttt{RNS}}$

In [4], Sect. 5.4, a method to significantly reduce the size of the public evaluation key $\texttt{evk}$ is described (by truncating the ciphertext) and it is applicable to the original FV scheme. We provide an efficient adaptation of such kind of optimization to the RNS variant of the relinearisation step.

We recall that the relinearisation is applied to a degree-2 ciphertext denoted here by (c_0, c_1, c_2). The initial suggestion was to set to zero, say, the i lowest significant components of the vector $\mathcal{D}_{\omega,q}(c_2)$. Doing so is equivalent to replacing c_2 by $c_2' = \omega^i \lfloor c_2 \omega^{-i} \rfloor = c_2 - |c_2|_{\omega^i}$. Thus, only the $\ell_{\omega,q} - i$ most significant components of $\texttt{rlk}_{\texttt{FV}}[0]$ (and then of $\texttt{rlk}_{\texttt{FV}}[1]$) are required (in other words, when $\texttt{rlk}_{\texttt{FV}}[0]$ is viewed as an $(\ell_{q,\omega}, k)$ RNS matrix by decomposing each component in base q, ik entries are set to zero like this). This optimization causes a greater noise than the one in Lemma 4 of [4]. Given (c_0, c_1, c_2) decryptable under s, the relinearisation step provides the following ciphertext:

$$(\tilde{c}_0, \tilde{c}_1) := (c_0 + \langle \mathcal{D}_{\omega,q}(c_2'), \mathcal{P}_{\omega,q}(s^2) - (\overrightarrow{e} + \overrightarrow{a} s) \rangle, \; c_1 + \langle \mathcal{D}_{\omega,q}(c_2'), \overrightarrow{a} \rangle).$$

Thus, $(\tilde{c}_0, \tilde{c}_1)(s) = c_0 + c_1 s + c_2' s^2 - \langle \mathcal{D}_{\omega,q}(c_2'), \overrightarrow{e} \rangle \bmod q$. Consequently, the extra noise comes from the following term:

$$\| - |c_2|_{\omega^i} s^2 - \langle \mathcal{D}_{\omega,q}(c_2'), \overrightarrow{e} \rangle \|_\infty = \| - |c_2|_{\omega^i} s^2 - \sum_{j=i}^{\ell_{\omega,q}-1} \mathcal{D}_{\omega,q}(c_2)_j e_j \|_\infty \\ < \delta^2 \omega^i B_{key}^2 + (\ell_{\omega,q} - i) \delta \omega B_{err}. \tag{19}$$

In the present RNS variant, the computation of $\lfloor c_2 \omega^{-i} \rfloor$ is not straightforward. This could be replaced by $\lfloor c_2 (q_1 \ldots q_i)^{-1} \rfloor$ through a Newton like interpolation (also known as mixed-radix conversion [11]). Though the result would be quite similar to the original optimization in terms of noise growth, its efficiency is not satisfying. Indeed, despite ik entries of the RNS matrix $\texttt{rlk}_{\texttt{RNS}}[0]$ can be set to zero like this, such a Newton interpolation is intrinsically sequential, while the division by ω^i followed by a flooring is simply achieved by an immediate zeroing of the lowest significant coefficients in radix ω representation.

For our approach, we rely on the fact that $\texttt{rlk}_{\text{RNS}}$ contains the RLWE-encryptions of the polynomials $|s^2 \frac{q}{q_j}|_q$. Then, we notice that only the jth-residue of $|s^2 \frac{q}{q_j}|_q$ can be non zero. So, let's assume that we want to cancel ik entries in $\texttt{rlk}_{\text{RNS}}[0]$ (as it has been done in $\texttt{rlk}_{\text{FV}}$ with the previous optimization). Then we choose, for each index j, a subset of index-numbers $\mathcal{I}_j \subseteq [1, k] \setminus \{j\}$ with cardinality i (i.e. at line j of $\texttt{rlk}_{\text{RNS}}$, choose i columns, except the diagonal one; these terms will be set to zero). Next, for each j, we introduce an RLWE-encryption of $|s^2 \frac{q}{q_j q_{\mathcal{I}_j}}|_q$, where $q_{\mathcal{I}_j} = \prod_{s \in \mathcal{I}_j} q_s$, which is $(|s^2 \frac{q}{q_j q_{\mathcal{I}_j}} - (e_j + sa_j)|_q, a_j)$. So far, the underlying security features are still relevant. Now, it remains to multiply this encryption by $q_{\mathcal{I}_j}$, which gives in particular $|s^2 \frac{q}{q_j} - q_{\mathcal{I}_j}(e_j + sa_j)|_q$. This is the jth-line of the new matrix $\texttt{rlk}'_{\text{RNS}}[0]$. It is clear that this line contains zeros at columns index-numbered by $\mathcal{I}_j$. $\texttt{rlk}_{\text{RNS}}[1] = (a_1, \ldots, a_k)$ is replaced by $\texttt{rlk}'_{\text{RNS}}[1] = (|q_{\mathcal{I}_1} a_1|_q, \ldots, |q_{\mathcal{I}_k} a_k|_q)$.

Let's analyse the new noise growth. By evaluating in s the output of relinearisation with this new $\texttt{rlk}'_{\text{RNS}}$, we obtain:

$$c_0 + \langle \xi_q(c_2), \texttt{rlk}'_{\text{RNS}}[0] \rangle + (c_1 + \langle \xi_q(c_2), \texttt{rlk}'_{\text{RNS}}[1] \rangle) s$$

$$= c_0 + \sum_{j=1}^{k} |c_2 \tfrac{q_j}{q}|_{q_j} \left(s^2 \tfrac{q}{q_j} - q_{\mathcal{I}_j}(e_j + sa_j) \right) + \left(c_1 + \sum_{j=1}^{k} |c_2 \tfrac{q_j}{q}|_{q_j} q_{\mathcal{I}_j} a_j \right) s$$

$$= c_0 + c_1 s + c_2 s^2 - \sum_{j=1}^{k} |c_2 \tfrac{q_j}{q}|_{q_j} q_{\mathcal{I}_j} e_j.$$

Consequently, the cancellation of ik terms in the public matrix $\texttt{rlk}_{\text{RNS}}[0]$ by using this method causes an extra noise growth bounded by (this can be fairly compared to (19) in the case where $\omega = 2^\nu$, i.e. $k = \ell_{\omega, q}$):

$$\left\| \sum_{j=1}^{k} |c_2 \tfrac{q_j}{q}|_{q_j} q_{\mathcal{I}_j} e_j \right\|_\infty < \sum_{j=1}^{k} \delta q_j q_{\mathcal{I}_j} B_{err} < \delta k 2^{\nu(i+1)} B_{err}.$$

Therefore, the truncation of ciphertexts can be efficiently adapted to RNS representation without causing more significant noise growth.

4.7   About Computational Complexity

In a classical multi-precision (MP) variant, for the purpose of efficiency the multiplication can involve NTT-based polynomial multiplication to implement the ciphertext product (e.g. [23] for such kind of implementation on FPGA). This approach requires the use of a base $\mathcal{B}'$ (besides q) with $|\mathcal{B}'| = k + 1$ for storing the product. Notice that, in RNS variant, we also have $|\mathcal{B}_{sk}| = k + 1$ (the same amount of information is kept, but used differently). Thus, it is easy to show that RNS and MP variants involve the same number of NTT and invNTT operations in the case $\ell_{\omega, q} = k$ (e.g. the keys $\texttt{rlk}_{\text{FV}}$ and $\texttt{rlk}_{\text{RNS}}$ have the same size in this situation). In other words, the same number of polynomial products is performed in both cases when $\omega = 2^\nu$.

Above all, the RNS variant enables us to decrease the cost of other parts of the computation. Despite the fact that the asymptotic computational complexity of these parts remains identical for both variants, i.e. $\mathcal{O}(k^2 n)$ elementary multiplications, the RNS variant only involves single-precision integer arithmetic.

To sum up, because of a complexity of $\mathcal{O}(k^2 n \log_2(n))$ due to the NTT's, we keep the same asymptotic computational complexity $\mathcal{C}(\texttt{Mult}_{\text{FV}}) \underset{n \to +\infty}{\sim} \mathcal{C}(\texttt{Mult}_{\text{RNS}})$. However, the most important fact is that multi-precision multiplications within MP variant are replaced in RNS by fast base conversions, which are simple matrix-vector products. Thus, $\texttt{Mult}_{\text{RNS}}$ retains all the benefits of RNS properties and is highly parallelizable.

5   Software Implementation

The C++ NFLlib library [19] was used for efficiently implementing the arithmetic in $\mathcal{R}_q$. It provides an efficient NTT-based product in $\mathcal{R}_q$ for q a product of 30 or 62-bit prime integers, and with degree n a power of 2, up to 2^{15}.

5.1   Concrete Examples of Parameter Settings

In this part, we analyse which depth can be reached in a multiplicative tree, and for which parameters.

The initial noise is at most $V = B_{err}(1 + 2\delta B_{key})$ [17]. The output of a tree of depth L has a noise bounded by $C_{\text{RNS},1}^L V + L C_{\text{RNS},1}^{L-1} C_{\text{RNS},2}$ (cf. [4], Lemma 9) with, for the present RNS variant:

$$
\begin{cases}
C_{\text{RNS},1} = 2\delta^2 t \frac{(1+\rho)}{2} B_{key} + \delta t(4 + \rho) + \frac{\delta}{2}, \\
C_{\text{RNS},2} = (1 + \delta B_{key})(\delta t|q|_t \frac{1+\rho}{2} + \delta B_{key}(k + \frac{1}{2})) + 2\delta t|q|_t + k(\delta B_{err} 2^{\nu+1} + 1) \\
\qquad\quad + \frac{1}{2}(3 + |q|_t).
\end{cases}
\tag{20}
$$

We denote by $L_{\text{RNS}} = \max\{L \in \mathbb{N} \mid C_{\text{RNS},1}^L V + L C_{\text{RNS},1}^{L-1} C_{\text{RNS},2} \leqslant \frac{q}{t}(\frac{1}{2} - \frac{k}{\gamma}) - \frac{|q|_t}{2}\}$ the depth allowed by $\texttt{Mult}_{\text{RNS}}$, when $\texttt{Dec}_{\text{RNS}}$ is used for decryption.

Table 1. Examples of parameter settings, by using 30-bit moduli of NFLlib.

n	k	t	L_{RNS} (L_{std})	ρ	$\tilde{m}$	$\lceil \log_2(m_{sk}) \rceil$	γ
2^{11}	3	2	2 (2)	5	(no need)	18	7
		2^{10}	1 (1)	5	(no need)	27	7
2^{12}	6	2	5 (6)	11	(no need)	21	13
		2^{10}	4 (4)	10	2	29	54
2^{13}	13	2	13 (13)	$\frac{1}{3}$	81	15	36
		2^{10}	9 (9)	13	3	31	58
2^{14}	26	2	25 (25)	$\frac{1}{2}$	106	17	53
		2^{10}	19 (19)	1	53	27	53
2^{15}	53	2	50 (50)	$\frac{1}{16}$	1712	20	753
		2^{10}	38 (38)	$\frac{1}{2}$	214	30	107

For an 80-bit security level and parameters $B_{key} = 1$, $\sigma_{err} = 8$, $B_{err} = 6\sigma_{err}$, we consider the security analysis in [17], which provides ranges for $(\log_2(q), n)$ (cf. [17], Table 2). We analyse our parameters by using the moduli given in NFLlib, because those were used for concrete testing. For a 32-bit (resp. 64) implementation, a set of 291 30-bit (resp. 1000 62-bit) moduli is available. These moduli are chosen to enable an efficient modular reduction (cf. [19], Algorithm 2).

Table 1 lists parameters when q and $\mathcal{B}$ are built with the 30-bit moduli of NFLlib. These parameters were determined by choosing the largest ρ (up to $2k - 1$) allowing to reach the depth L_{RNS}. L_{std} corresponds to the bounds given in [17] for a classical approach. Sufficient sizes for γ and m_{sk} (allowing, for m_{sk}, to have $|\mathcal{B}| = k$ through (14) after having chosen for q the k greatest available moduli) are provided. For these specific parameters, the new bounds on the noise for RNS variant cause a smaller depth in only one case.

5.2   Influence of $\tilde{m}$ Over Noise Growth

After a fast conversion from q, ciphertexts in $\mathcal{B}_{\mathrm{sk}}$ can contain q-overflows and verify $\|\mathrm{ct}_i'\|_\infty < \frac{q}{2}(1 + \tau)$. In a multiplicative tree without any addition, $\tau \leqslant 2k - 1$. By applying Algorithm 2, this bound decreases to $\frac{q}{2}(1 + \rho)$, for some $0 < \rho \leqslant 2k - 1$. Having $\rho = 2k - 1$ in Table 1 means it is unnecessary to use $\mathrm{SmMRq}_{\tilde{m}}$ to reach the best depth. It happens only three times. Most of the time, doing such reduction is necessary before a multiplication so as to reach the highest depth. Moreover, choosing a lower ρ (i.e. higher $\tilde{m}$) than necessary can decrease the size of γ and m_{sk} (as shown in Table 1, one set of parameters leads to $\lceil \log_2(m_{\mathrm{sk}}) \rceil = 31$, avoiding the use of a 30-bit modulus; this can be solved by taking a larger $\tilde{m}$).

To illustrate the impact of $\mathrm{SmMRq}_{\tilde{m}}$, Fig. 1 depicts the noise growth for $\tilde{m} \in \{0, 2^8, 2^{16}\}$. Given Table 1, $\tilde{m} = 2^8$ is sufficient in such scenario to reach $L_{\mathrm{RNS}} = 13$. Against a computation with no reduction at all ($\tilde{m} = 0$, implying $L_{\mathrm{RNS}} = 11$ in this case), choosing $\tilde{m} = 2^8$ implies an average reduction of 25%. By using $\tilde{m} = 2^{16}$, we gain around 32%. Therefore, $\mathrm{SmMRq}_{\tilde{m}}$ has been systematically integrated within the implementation of $\mathrm{Mult}_{\mathrm{RNS}}$ for timing measurements in next part.

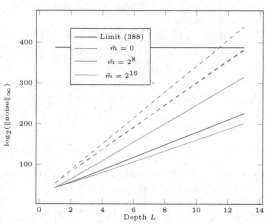

Fig. 1. Example of noise growth; $n = 2^{13}$, $\log_2(q) = 390$ ($\nu = 30, k = 13$), $t = 2$, $\sigma_{err} = 8$, $B_{key} = 1$ (dashed line: bound using (20); plain line: measurements).

5.3   Some Remarks

Convenient $\tilde{m}$ and γ. Given values of ρ in Table 1, $\tilde{m} = 2^8$ (resp. $\tilde{m} = 2^{16}$) satisfies, by far, any set of analysed parameters. This enables an efficient and straightforward modular arithmetic through standard types like `uint8_t` (resp. `uint16_t`) and a type conversion towards the signed `int8_t` (resp. `int16_t`) immediately gives the centered remainder. Parameter analysis with such $\tilde{m}$ shows that $\gamma = 2^8$ is sufficient to ensure a correct decryption for all configurations. A reduction modulo γ can then be achieved by a simple type conversion to `uint8_t`.

Tested Algorithms. The code[1] we compared with was implemented in the context of HEAT [26]. It is based on NFLlib too. Multi-precision arithmetic is handled with GMP 6.1.0 [14], and multiplications by $\frac{t}{q}$ are performed with integer divisions. Mult$_{\mathtt{MP}}$ and Dec$_{\mathtt{MP}}$ denote functions from this code.

Mult$_{\mathtt{RNS}}$ was implemented as described by Algorithm 3. Could the use of SmMRq$_{\tilde{m}}$ be avoided to reach the maximal theoretical depth, it was however systematically used. Its cost is negligible and it enables a noticeable decrease of noise growth.

Two variants of Dec$_{\mathtt{RNS}}$ (cf. Sect. 3.5) have been implemented. Depending on ν, the one with floating point arithmetic (named Dec$_{\mathtt{RNS-flp}}$ thereafter) uses double (resp. long double) for double (resp. quadruple) precision, and then does not rely on any other external library at all.

5.4 Results

The tests have been run on a laptop, running under Fedora 22 with Intel® Core™ i7-4810MQ CPU @ 2.80 GHz and using GNU compiler g++ version 5.3.1. Hyper-Threading and Turbo Boost were deactivated. The timings have been measured on 2^{12} decryptions/multiplications for each configuration.

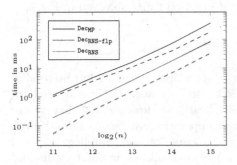
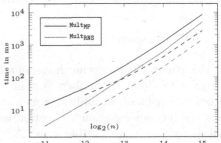

Fig. 2. Decryption time ($t = 2^{10}$), with $\nu = 30$ (plain lines) and $\nu = 62$ (dashed lines).

Fig. 3. Multiplication time ($t = 2^{10}$), with $\nu = 30$ (plain lines) and $\nu = 62$ (dashed lines).

Figure 2 presents timings for Dec$_{\mathtt{MP}}$, Dec$_{\mathtt{RNS}}$ and Dec$_{\mathtt{RNS-flp}}$, and Fig. 3 depicts timings for Mult$_{\mathtt{MP}}$ and Mult$_{\mathtt{RNS}}$. Both figures gather data for two modulus sizes: $\nu = 30$ and $\nu = 62$. Step 2 of Mult$_{\mathtt{MP}}$ uses a decomposition in radix-base $\omega = 2^{32}$ when $\nu = 30$, and $\omega = 2^{62}$ when $\nu = 62$. The auxiliary bases $\mathcal{B}_{sk}$ and $\mathcal{B}'$ involved in Mult$_{\mathtt{RNS}}$ and Mult$_{\mathtt{MP}}$ contain $k + 1$ moduli each. Table 2 shows which values of k have been tested (depending on n). Multiplication timing for $(n, \nu, k) = (2^{11}, 62, 1)$ is not given since $L = 1$ already causes decryption failures.

It has to be noticed that the performance are mainly due to NFLlib. The contributions appear in the speed-ups between RNS and MP variants.

---

[1] https://github.com/CryptoExperts/FV-NFLlib.

Table 2. Parameter k used in the tests (i.e. $\lceil \log_2(q) \rceil = k\nu$).

$\log_2(n)$	11	12	13	14	15
k ($\nu = 30$)	3	6	13	26	53
k ($\nu = 62$)	1	3	6	12	25

In Fig. 3, the convergence of complexities of Mult$_{\mathtt{RNS}}$ and Mult$_{\mathtt{MP}}$ (as explained in Sect. 4.7) is noticeable. The new algorithm presented in this paper allows speed-ups from 4.3 to 1.7 for degree n from 2^{11} to 2^{15} when $\nu = 30$, and from 3.6 to 1.9 for n from 2^{12} to 2^{15} when $\nu = 62$.

In Fig. 2, the two variants of decryption described in Sect. 3.5 are almost equally fast. Indeed, they perform the same number of elementary (floating point or integer) operations. Between degree 2^{11} and 2^{15}, the RNS variants allow speed-ups varying from 6.1 to 4.4 when $\nu = 30$, and from 20.4 to 5.6 when $\nu = 62$. All the implemented decryption functions take as input a ciphertext in NTT representation. Thus, only one invNTT is performed (after the product of residues) within each decryption. As explained (cf. Sect. 3.5), despite a better asymptotic computational complexity for RNS decryption, the efficiency remains in practice highly related to this invNTT procedure, even maybe justifying the slight convergence between MP and RNS decryption times observed in Fig. 2.

6   Conclusion

In this paper, the somewhat homomorphic encryption scheme FV has been fully adapted to Residue Number Systems. Prior to this work, RNS was used to accelerate polynomial additions and multiplications. However, the decryption and the homomorphic multiplication involve operations at the coefficient level which are hardly compatible with RNS, such as division and rounding.

Our proposed solutions overcome these incompatibilities, without modifying the security features of the original scheme. As a consequence, we have provided a SHE scheme which only involves RNS arithmetic. It means that only single-precision integer arithmetic is required, and the new variant fully benefits from the properties of RNS, such as parallelization.

The proposed scheme has been implemented in software using C++. Because arithmetic on polynomials (in particular polynomial product) is not concerned by the new optimizations provided here, the implementation has been based on the NFLlib library, which embeds a very efficient NTT-based polynomial product. Our implementation has been compared to a classical version of FV (based on NFLlib, and GMP). For degrees from 2^{11} to 2^{15}, the new decryption (resp. homomorphic multiplication) offers speed-ups from 20 to 5 (resp. 4 to 2) folds for cryptographic parameters.

Further work should demonstrate the high potential of the new variant by exploiting all the concurrency properties of RNS, in particular through dedicated hardware implementations.

Acknowledgement. We thank anonymous reviewers for their helpful comments and remarks.

This work was partially supported by the European Union's H2020 Programme under grant agreement # ICT-644209, and by the Natural Sciences and Engineering Research Council of Canada.

References

1. Albrecht, M., Bai, S., Ducas, L.: A Subfield Lattice Attack on Overstretched NTRU Assumptions: Cryptanalysis of some FHE and Graded Encoding Schemes. Cryptology ePrint Archive, Report 2016/127 (2016). http://eprint.iacr.org/2016/127

2. Bajard, J.-C., Eynard, J., Hasan, M.A., Zucca, V.: A Full RNS Variant of FV like Somewhat Homomorphic Encryption Schemes. Cryptology ePrint Archive, Report 2016/510 (2016). http://eprint.iacr.org/2016/510

3. Bajard, J.-C., Eynard, J., Merkiche, N., Plantard, T.: RNS arithmetic approach in lattice-based cryptography: accelerating the "rounding-off" core procedure. In: 22nd IEEE Symposium on Computer Arithmetic, ARITH 2015, Lyon, France, 22–24 June 2015, pp. 113–120. IEEE (2015)

4. Bos, J.W., Lauter, K., Loftus, J., Naehrig, M.: Improved security for a ring-based fully homomorphic encryption scheme. In: Stam, M. (ed.) IMACC 2013. LNCS, vol. 8308, pp. 45–64. Springer, Heidelberg (2013). doi:10.1007/978-3-642-45239-0_4

5. Brakerski, Z.: Fully Homomorphic Encryption without Modulus Switching from Classical GapSVP. Cryptology ePrint Archive, Report 2012/078 (2012). http://eprint.iacr.org/2012/078

6. Brakerski, Z., Gentry, C., Vaikuntanathan, V.: (Leveled) fully homomorphic encryption without bootstrapping. In Goldwasser, S. (ed.) Innovations in Theoretical Computer Science 2012, Cambridge, MA, USA, 8–10 January 2012, pp. 309–325. ACM (2012)

7. Ducas, L., Durmus, A., Lepoint, T., Lyubashevsky, V.: Lattice signatures and bimodal Gaussians. In: Canetti, R., Garay, J.A. (eds.) CRYPTO 2013. LNCS, vol. 8042, pp. 40–56. Springer, Heidelberg (2013). doi:10.1007/978-3-642-40041-4_3

8. Ducas, L., Lyubashevsky, V., Prest, T.: Efficient identity-based encryption over NTRU lattices. In: Sarkar, P., Iwata, T. (eds.) ASIACRYPT 2014. LNCS, vol. 8874, pp. 22–41. Springer, Heidelberg (2014). doi:10.1007/978-3-662-45608-8_2

9. Fan, J., Vercauteren, F.: Somewhat Practical Fully Homomorphic Encryption. Cryptology ePrint Archive, Report 2012/144 (2012). http://eprint.iacr.org/2012/144

10. Garg, S., Gentry, C., Halevi, S.: Candidate multilinear maps from ideal lattices. In: Johansson, T., Nguyen, P.Q. (eds.) EUROCRYPT 2013. LNCS, vol. 7881, pp. 1–17. Springer, Heidelberg (2013). doi:10.1007/978-3-642-38348-9_1

11. Garner, H.L.: The residue number system. Papers Presented at the the 3–5 March 1959, Western Joint Computer Conference, IRE-AIEE-ACM 1959 (Western), pp. 146–153. ACM, New York (1959)

12. Gentry, C.: Fully homomorphic encryption using ideal lattices. In: Proceedings of the Forty-First Annual ACM Symposium on Theory of Computing, STOC 2009, pp. 169–178. ACM, New York (2009)

13. Gilad-Bachrach, R., Dowlin, N., Laine, K., Lauter, K.E., Naehrig, M., Wernsing, J.: CryptoNets: applying neural networks to encrypted data with high throughput and accuracy. In: Balcan, M.-F., Weinberger, K.Q. (eds.) Proceedings of the 33nd International Conference on Machine Learning, ICML 2016, New York City, NY, USA, 19–24 June 2016, JMLR Workshop and Conference Proceedings, vol. 48, pp. 201–210. JMLR.org (2016)
14. Granlund, T., The GMP Development Team: GNU MP: The GNU Multiple Precision Arithmetic Library, 6.1.0 edn. (2015). http://gmplib.org/
15. Hoffstein, J., Pipher, J., Silverman, J.H.: NTRU: a ring-based public key cryptosystem. In: Buhler, J.P. (ed.) ANTS 1998. LNCS, vol. 1423, pp. 267–288. Springer, Heidelberg (1998). doi:10.1007/BFb0054868
16. Langlois, A., Stehlé, D., Steinfeld, R.: GGHLite: more efficient multilinear maps from ideal lattices. In: Nguyen, P.Q., Oswald, E. (eds.) EUROCRYPT 2014. LNCS, vol. 8441, pp. 239–256. Springer, Heidelberg (2014). doi:10.1007/978-3-642-55220-5_14
17. Lepoint, T., Naehrig, M.: A comparison of the homomorphic encryption schemes FV and YASHE. In: Pointcheval, D., Vergnaud, D. (eds.) AFRICACRYPT 2014. LNCS, vol. 8469, pp. 318–335. Springer, Cham (2014). doi:10.1007/978-3-319-06734-6_20
18. Lyubashevsky, V.: Lattice signatures without trapdoors. In: Pointcheval, D., Johansson, T. (eds.) EUROCRYPT 2012. LNCS, vol. 7237, pp. 738–755. Springer, Heidelberg (2012). doi:10.1007/978-3-642-29011-4_43
19. Aguilar-Melchor, C., Barrier, J., Guelton, S., Guinet, A., Killijian, M.-O., Lepoint, T.: NFLlib: NTT-based fast lattice library. In: Sako, K. (ed.) CT-RSA 2016. LNCS, vol. 9610, pp. 341–356. Springer, Cham (2016). doi:10.1007/978-3-319-29485-8_20
20. Montgomery, P.L.: Modular multiplication without trial division. Math. Comput. 44(170), 519–521 (1985)
21. Oder, T., Pöppelmann, T., Güneysu, T.: Beyond ECDSA and RSA: lattice-based digital signatures on constrained devices. In: DAC, pp. 110:1–110:6. ACM (2014)
22. Peikert, C.: Lattice cryptography for the internet. In: Mosca, M. (ed.) PQCrypto 2014. LNCS, vol. 8772, pp. 197–219. Springer, Cham (2014). doi:10.1007/978-3-319-11659-4_12
23. Roy, S.S., Järvinen, K., Vercauteren, F., Dimitrov, V., Verbauwhede, I.: Modular hardware architecture for somewhat homomorphic function evaluation. In: Güneysu, T., Handschuh, H. (eds.) CHES 2015. LNCS, vol. 9293, pp. 164–184. Springer, Heidelberg (2015). doi:10.1007/978-3-662-48324-4_9
24. Shenoy, A.P., Kumaresan, R.: Fast base extension using a redundant modulus in RNS. IEEE Trans. Comput. 38(2), 292–297 (1989)
25. Stehlé, D., Steinfeld, R., Tanaka, K., Xagawa, K.: Efficient public key encryption based on ideal lattices. In: Matsui, M. (ed.) ASIACRYPT 2009. LNCS, vol. 5912, pp. 617–635. Springer, Heidelberg (2009). doi:10.1007/978-3-642-10366-7_36
26. European Union: Homomorphic Encryption, Applications and Technology (HEAT). https://heat-project.eu. H2020-ICT-2014-1, Project reference: 644209

Security Considerations for Galois Non-dual RLWE Families

Hao Chen[1]([⊠]), Kristin Lauter[2], and Katherine E. Stange[3]

[1] University of Washington, Seattle, USA
chenh123@uw.edu
[2] Microsoft Research, Redmond, USA
klauter@microsoft.com
[3] University of Colorado, Boulder, USA
kstange@math.colorado.edu

Abstract. We explore further the hardness of the non-dual discrete variant of the Ring-LWE problem for various number rings, give improved attacks for certain rings satisfying some additional assumptions, construct a new family of vulnerable Galois number fields, and apply some number theoretic results on Gauss sums to deduce the likely failure of these attacks for 2-power cyclotomic rings and unramified moduli.

1 Introduction

Lattice-based cryptography was introduced in the mid 1990s in two different forms, independently by Ajtai-Dwork [1] and Hoffstein-Pipher-Silverman [12]. Thanks to the work of Stehlé-Steinfeld [19], we now understand the NTRU cryptosystem introduced by Hoffstein-Pipher-Silverman to be a variant of a cryptosystem which has security reductions to the Ring Learning With Errors (RLWE) problem. The RLWE problem was introduced in [14] as a version of the LWE problem [17]: both problems have reductions to hard lattice problems and thus are interesting for practical applications in cryptography. RLWE depends on a number ring R, a modulus q, and an error distribution. As such, it has added structure (the ring), which allows for greater efficiency, but also in some cases additional attacks.

The hardness of RLWE is crucial to cryptography, in particular as the basis of numerous homomorphic encryption schemes [2–6,13,19]. One main theoretical result in this direction is the security reduction theorem in [14], which reduces certain GapSVP problems in ideal lattices over R to RLWE, when the RLWE error distribution is sufficiently large and of a prescribed form. Although so far in practical cryptographic applications only cyclotomic rings are used, it is important to study the hardness of RLWE for general number rings, moduli and error distributions, so as to understand the boundaries of security in the parameter space. Recently, new attacks on the so-called *non-dual discrete* variant of the RLWE problem for certain number rings, error distributions, and special moduli were introduced [7–11]. The RLWE problem reduces to its discrete variant; and

© Springer International Publishing AG 2017
R. Avanzi and H. Heys (Eds.): SAC 2016, LNCS 10532, pp. 443–462, 2017.
https://doi.org/10.1007/978-3-319-69453-5_24

the non-dual RLWE problem is equivalent to the dual problem up to a change in the error distribution, so that non-dual RLWE may be viewed simply as a certain choice of error distribution in the parameter space of RLWE. The term *RLWE* is sometimes reserved for spherical Gaussian distributions.

This paper is an extension of [9], and here we explore further the hardness of the non-dual discrete variant of the RLWE problem for various number rings. We:

1. construct a new family of vulnerable Galois number fields,
2. improve the runtime of the attacks for certain rings satisfying some additional assumptions, and
3. apply some number theoretic results on Gauss sums to deduce the likely failure of these attacks for 2-power cyclotomic rings.

In cryptographic applications, it is most efficient to sample the error distribution coordinate-wise according to a polynomial basis for the ring. For 2-power cyclotomic rings, which are monogenic with a well-behaved power basis, it is justified to sample the RLWE error distribution directly in the polynomial basis for the ring, according to results in [5,10,14], where this error distribution choice is called Polynomial Learning With Errors (PLWE). Precisely, the PLWE (polynomial error), RLWE (meaning a spherical Gaussian), and non-dual RLWE problems are equivalent up to a scaling and rotation of the error distribution for 2-power cyclotomic fields. However, in general number rings the error distribution may be distorted by a general linear transformation when moving from one problem to another [11]. For certain choices of ring and modulus, efficient attacks on PLWE were presented in [10]. In [11], these attacks were extended to apply to the decision version of the non-dual RLWE problem in certain rings, and in [8,9], attacks on the search version of the RLWE problem for certain choices of ring and modulus were presented.

1.1 Summary of Contributions

- In Sect. 3, we present an improvement to the attack in [9, Sect. 4] and use it to dramatically cut down the runtime of the attacks on the weak instances found in [9, Sect. 5].
- In Sect. 4, we present a new infinite family of Galois number fields vulnerable to our attack in [9, Sect. 4], where the relative standard deviation parameter is allowed to grow to infinity, and we give a table of examples.
- In Sect. 5, we analyze the security of 2-power cyclotomic fields with unramified moduli under our attack. We prove Theorem 3, which gives an upper bound on the statistical distance between an approximated non-dual RLWE error distribution, reduced modulo a prime ideal $\mathfrak{q}$, and the uniform distribution on $R/\mathfrak{q}$. We conclude that the 2-power cyclotomic rings are safe against our attack when the modulus q is unramified with small residue degree (1 or 2), and is not too large ($q < m^2$).

2   Background

2.1   Discrete Gaussian on Lattices

Recall that a *lattice* in $\mathbb{R}^n$ is a discrete subgroup of $\mathbb{R}^n$ of rank n. For $r > 0$, let $\rho_r(x) = e^{-||x||^2/r^2}$.

Definition 1. *For a lattice $\Lambda \subset \mathbb{R}^n$ and $r > 0$, the discrete Gaussian distribution on Λ with width r is:*

$$D_{\Lambda,r}(x) = \frac{\rho_r(x)}{\sum_{y \in \Lambda} \rho_r(y)}, \ \forall x \in \Lambda.$$

2.2   Non-dual RLWE

A non-dual discrete RLWE instance is specified by a ring R, a positive integer q and an error distribution χ over R. Here R is normally taken to be the ring of integers of some number field K of degree n. The integer q, called the *modulus*, is often taken to be a prime number. We then fix an element $s \in R/qR$ called the *secret*.

Let $\iota : K \to \mathbb{R}^n$ be the *adjusted canonical embedding* defined as follows. Suppose $\sigma_1, \ldots, \sigma_{r_1}, \sigma_{r_1+1}, \ldots, \sigma_n$ are the distinct embeddings of K, such that $\sigma_1, \cdots, \sigma_{r_1}$ are the real embeddings and $\sigma_{r_1+r_2+j} = \overline{\sigma_{r_1+j}}$ for $1 \leq j \leq r_2$. We define $\iota : K \to \mathbb{R}^n$ by

$$x \mapsto (\sigma_1(x), \cdots, \sigma_{r_1}(x), \sqrt{2}\mathrm{Re}(\sigma_{r_1+1}(x)), \sqrt{2}\mathrm{Im}(\sigma_{r_1+1}(x)), \cdots,$$
$$\sqrt{2}\mathrm{Re}(\sigma_{r_1+r_2}(x)), \sqrt{2}\mathrm{Im}(\sigma_{r_1+r_2}(x))).$$

Then the *non-dual discrete RLWE error distribution* is the discrete Gaussian distribution $D_{\iota(R),r}$.

Definition 2. *Fix R, q, r as above. Let R_q denote the quotient ring R/qR. Then a non-dual RLWE sample is a pair*

$$(a, b = as + e) \in R_q \times R_q,$$

where the first coordinate a is chosen uniformly at random in R_q, and e is a sampled from the discrete Gaussian $D_{\iota(R),r}$, considered modulo q.

Definition 3 (Non-dual Search RLWE). *Given arbitrarily many non-dual RLWE samples, determine the secret s.*

Definition 4 (Non-dual Decision RLWE). *Given arbitrarily many samples in $R_q \times R_q$, which are either non-dual RLWE samples for a fixed secret s, or uniformly random samples, determine which.*

2.3 Comparing RLWE with Non-dual RLWE

In the original work [14], the RLWE problem is introduced using the dual ring $R^\vee$. Specifically, for the discrete variant, $s \in R_q^\vee := R^\vee/qR^\vee$, and an RLWE sample is taken to be of the form

$$(a, b = as + e) \in R_q \times R_q^\vee,$$

where e is sampled from $D_{\iota(R^\vee),r}$, then considered modulo q.

If the dual ring $R^\vee$ is principal as a fractional ideal, i.e., $R^\vee = tR$, then each non-dual instance is equivalent to a dual instance, by mapping a sample (a, b) to (a, tb), and vice versa. If $R^\vee$ is not principal, there are still inclusions $R^\vee \subset t_1 R$ and $R \subset t_2 R^\vee$, so that one can reduce dual and non-dual versions of the problem to one another. In either case, the reduction comes at the cost of distorting the error distribution.

For the infinite family constructed in Sect. 4, the dual ring $R^\vee$ is indeed principal (see Lemma 3 in Sect. 4). Note that multiplying by this field element t changes a spherical Gaussian to an elliptical Gaussian, so the two equivalent instances will have different error shapes.

Elliptical Gaussians are the most important class of error distributions for general rings, since in [14, Theorem 4.1], the reduction from hard lattice problems is to a class of RLWE problems where the distributions are elliptical Gaussians. Theorem 5.2 of [14] provides a further security reduction to decision RLWE with spherical Gaussian errors, but it is only stated for cyclotomic rings.

2.4 Comparing Discrete and Continuous Errors

Restricting now to the non-dual setup, there are still two variants of RLWE based on the form of the spherical errors: the *continuous* variant samples errors from spherical Gaussian on the space $K_\mathbb{R} = \iota(K \otimes_\mathbb{Q} \mathbb{R})$ (here we extend ι linearly), so that samples have the form

$$(a, b = as + e) \in R_q \times K_\mathbb{R}/qR,$$

whereas the *discrete* variant samples from a discrete Gaussian $D_{\iota R,r}$ on the lattice R, as defined above.

There is no known equivalence between the discrete problem and its continuous counterpart in general. However, the continuous problem reduces to the discrete one. Specifically, given a continuous sample $(a, b) \in R_q \times K_\mathbb{R}/qR$, one can perform a rounding on the second coordinate to get a discrete sample $(a, [b]) \in R_q \times R_q$. However, there is no obvious map in the reverse direction.

2.5 Search and Decision RLWE Problems

Let $\mathfrak{q}$ be a prime ideal of K lying above q; then the RLWE problem modulo $\mathfrak{q}$ means discovering $s \mod \mathfrak{q}$ from arbitrarily many RLWE samples. In [14] the authors gave a polynomial time reduction from search to decision for cyclotomic number fields and totally split primes, using the RLWE modulo $\mathfrak{q}$ as an intermediate problem. Their proof can be applied to prove a similar search-to-decision reduction for non-dual RLWE, when the underlying number field is Galois and the modulus q is unramified [9,10]. Moreover, the search-to-decision is most efficient when the residue degree of q is small. What is important in our paper is that for the instances in Sects. 3 and 4, our attacks on RLWE modulo $\mathfrak{q}$ could be efficiently transferred to attack the search problem.

2.6 Comparing Non-dual RLWE with PLWE for 2-Power Cyclotomic Fields

For cryptographic applications, it is perhaps natural to consider the PLWE error distribution on R: assuming the ring R is monogenic, i.e., $R = \mathbb{Z}[x]/(f(x))$, then a sample from the PLWE error distribution is $e = \sum_{i=0}^{n-1} e_i x^i$, where the e_i are "small errors", sampled independently from some error distribution over $\mathbb{Z}$ (e.g. a discrete Gaussian distribution).

In general number fields, a PLWE distribution differs greatly from the non-dual RLWE distribution (see [ELOS] for an effort to quantify the distance between the two distributions using spectral norms). However, for 2-power cyclotomic fields it turns out that the two error distributions are equivalent up to a factor of $\sqrt{n}$. Since this fact is used in Sect. 5, we give a proof below.

Lemma 1. *Let $m = 2^d$ be a power of 2 and let $R = \mathbb{Z}[\zeta_m]$. Consider the PLWE error distribution on R, i.e. samples $e = \sum_{i=0}^{n-1} e_i \zeta_m^i$, where $n = m/2$ and each e_i follows the discrete Gaussian $D_{\mathbb{Z},r}$. Then this PLWE distribution is equal to the non-dual RLWE distribution $D_{\iota(R), r\sqrt{n}}$.*

Proof. For an element $x = \sum_{i=0}^{n-1} x_i \zeta_m^i \in R$, the probability of x being sampled by the PLWE distribution is proportional to $\prod_{i=0}^{n-1} \rho_r(x_i) = \prod_{i=0}^{n-1} e^{-x_i^2/r^2} = e^{-||x||^2/r^2}$. On the other hand, one checks that $||\iota(x)|| = \sqrt{n}||x||$. So the above probability is proportional to $e^{-||\iota(x)||^2/nr^2}$, which is the exactly the same for the distribution $D_{\iota(R), r\sqrt{n}}$. This completes the proof.

2.7 Scaling Factors

As pointed out in [11], when analyzing the non-dual RLWE error distribution, one needs to take into account the sparsity of the lattice $\iota(R)$, measured by its covolume in $\mathbb{R}^n$. This covolume is equal to $|\operatorname{disc}(K)|^{1/2}$. In light of this, we define the scaled error width to be

$$r_0 = \frac{r}{|\operatorname{disc}(K)|^{\frac{1}{2n}}}.$$

2.8 Overview of Attack

We briefly review the method of attack in Sect. 4 of [9]. The basic principle of this family of attacks is to find a homomorphism

$$\rho : R_q \to F$$

to some small finite field F, such that the error distribution on R_q is transported by ρ to a non-uniform distribution on F. In this case, errors can be distinguished from elements uniformly drawn from R_q by a statistical test in F, for example, by a χ^2-test. The existence (or non-existence) of such a homomorphism depends on the parameters of the field, prime, and distribution in the setup of RLWE. In this section, we will describe parameters under which such a map exists.

Once such a map is known, the basic method of attack on Decision RLWE is as follows:

1. Apply ρ to samples (a, b) in $R_q \times R_q$, to obtain samples in $F \times F$.
2. Guess the image of the secret $\rho(s)$ in F, calling the guess g.
3. Compute the distribution of $\rho(b) - \rho(a)g$ for all the samples. If $g = \rho(s)$, this is the image of the distribution of the errors. Otherwise it is the image of a uniform distribution.
4. If the image looks uniform, try another guess g until all are exhausted. If any non-uniform distribution is found, the samples are RLWE samples. Otherwise they are not.

Whenever $\mathfrak{q}$ is a prime ideal lying above q, then reduction modulo $\mathfrak{q}$ is a valid map

$$\rho : R_q \to R_\mathfrak{q}$$

for the attack above. This attack targets the RLWE modulo $\mathfrak{q}$ problem for some prime $\mathfrak{q}$ lying above q, and as noted above, it can be turned into an attack on the search variant of the problem, whenever q is unramified and K is Galois.

2.9 Comparison to Related Works

In an independent preprint [7] which appeared on eprint around the same time as our preprint, Castryck et al. also constructed an infinite family of vulnerable Galois number fields, where the error width can be taken to be $O(|\operatorname{disc}(K)|^{\frac{1}{n} - \epsilon})$ for any $\epsilon > 0$. The asymptotic error width they obtained is wider than in our infinite family in Sect. 2. However, the method of attack is an errorless LWE linear algebra attack (based on short vectors), whereas our family is not susceptible to a linear algebra attack, and requires the novel techniques presented here and in [9].

3 An Improved Attack Using Cosets

In this section, we describe an improvement to our chi-square attack on RLWE mod $\mathfrak{q}$ outlined in Sect. 2.8 for a special case. As a result, we have an updated version of [9, Table 1], where we attacked each instance in the table in much shorter time. Note that the complexity of the previous attack in this special case is $O(nq^3)$. In contrast, our new attack has complexity $O(nq^2)$.

To clarify, the special case we consider in this section is characterised by the following assumptions (we need not be in the special family of the next section):

- The modulus q is a prime of residue degree 2 in the number field K.
- There exists a prime ideal $\mathfrak{q}$ above q such that the map $\rho : R_q \to R_{\mathfrak{q}}$ satisfies the following property: Let $e \in R_q$ be taken from the discrete RLWE error distribution. The probability that $\rho(e)$ lies in the prime subfield $\mathbb{F}_q$ of $\mathbb{F}_{q^2}$ is computationally distinguishable from $1/q$.

Granting these assumptions, we can distinguish the distribution of the "reduced error" $\rho(e)$ from the uniform distribution on $\mathbb{F}_{q^2}$. More precisely, the attack in [9] works exactly as we described in Sect. 4: with access to $\Omega(q)$ samples, one loops over all q^2 possible values of $\rho(s)$. It detects the correct guess $\rho(s)$ based on a chi-square test with two bins $\mathbb{F}_q$ and $\mathbb{F}_{q^2} \backslash \mathbb{F}_q$.

The distinguishing feature of the improved attack is to loop over the cosets of $\mathbb{F}_q$ of $\mathbb{F}_{q^2}$ instead of the whole space. Fix $t_1, \cdots t_q$ to be a set of coset representatives for the additive group $\mathbb{F}_{q^2}/\mathbb{F}_q$. Recall that s denotes the secret and $\rho : R_q \to R_{\mathfrak{q}} \cong \mathbb{F}_{q^2}$ is a reduction map modulo some fixed prime ideal $\mathfrak{q}$ lying above q. Then there exists a unique index i such that $\rho(s) = s_0 + t_i$ for some $s_0 \in \mathbb{F}_q$. Our improved attack will recover s_0 and t_i separately.

We start with an identity $b = as + e$, where $a, b, s, e \in \mathbb{F}_{q^2}$. We will regard s as fixed and a, b, e as random variables, such that a is uniformly distributed in $\mathbb{F}_{q^2} \backslash \mathbb{F}_q$ and b is uniformly distributed in $\mathbb{F}_{q^2}$. The reason why a is not taken to be uniform will become clear later in this section. We use a bar to denote the Frobenius automorphism, i.e.,

$$\bar{a} \overset{def}{=} a^q, \ \forall a \in \mathbb{F}_{q^2}.$$

Then $\bar{b} = \bar{a}\bar{s} + \bar{e}$. Using the identity $s = s_0 + t_i$ and subtracting, we obtain $\bar{b} - b - \overline{at_i} + at_i = s_0(\bar{a} - a) + \bar{e} - e$. Since $a \neq \bar{a}$, we can divide through by $\bar{a} - a$ and get

$$\frac{\bar{b} - b - \overline{at_i} + at_i}{\bar{a} - a} = s_0 + \frac{\bar{e} - e}{\bar{a} - a}. \tag{$**$}$$

Now for each $1 \leq j \leq q$, we can compute

$$m_j(a, b) := \frac{\bar{b} - b - \overline{at_j} + at_j}{\bar{a} - a}$$

with access to a and b, but without knowledge of s or s_0. Note that m_j is in the prime field $\mathbb{F}_q$ by construction.

Proposition 1. *For each* $1 \leq j \leq q$,

(1) If $j \neq i$, *then* $m_j(a, b)$ *is uniformly distributed in* $\mathbb{F}_q$, *for RLWE samples* (a, b).

(2) If $j = i$, then $m_j(a, b) = s_0 + \frac{\bar{e}-e}{\bar{a}-a}$.

We postpone the proof of Proposition 1 until the end of this section. Assuming the proposition, our improved attack works as follows: for $1 \leq j \leq q$, we compute a set of m_j from the samples. To avoid dividing by zero, we ignore the samples with $\rho(a) \in \mathbb{F}_q$ (which happens with probability $1/q$ since $\rho(a)$ is uniformly distributed). We then run a chi-square test on the m_j values. If $j \neq i$, then the distribution should be uniform; if $j = i$, then $P(m_i = s_0) = P(e \in \mathbb{F}_q)$, which by our assumption is larger than $1/q$. Hence if we plot the histogram of the m_i computed from the samples, we will see a spike at s_0. So we could recover s_0 as the element with the highest frequency, and output $\rho(s) = s_0 + t_i$. We give the pseudocode of the attack below.

Algorithm 1. Improved chi-square attack on RLWE modulo q)

Input: K – a number field; R – the ring of integers of K; $\mathfrak{q}$ – a prime ideal in K above q with residue degree 2; $\mathcal{S}$ – a collection of M RLWE samples; $\beta > 0$ – the parameter used for comparing χ^2 values.

Output: a guess of the value s (mod $\mathfrak{q}$), or **NOT-RLWE**, or **INSUFFICIENT-SAMPLES**

Let $\mathcal{G} \leftarrow \emptyset$.
for j in $1, \ldots, q$ do
    $\mathcal{E}_j \leftarrow \emptyset$.
    for a, b in $\mathcal{S}$ do
        $\bar{a}, \bar{b} \leftarrow a \pmod{\mathfrak{q}}, b \pmod{\mathfrak{q}}$.
        $m_j \leftarrow \frac{\bar{b}-b-\overline{at_j}+at_j}{\bar{a}-a}$.
        add m_j to $\mathcal{E}_j$.
    end for
    Run a chi-square test for uniform distribution on $\mathcal{E}_j$.
    if $\chi^2(\mathcal{E}_j) > \beta$ then
        $s_0 :=$ the element(s) in $\mathcal{E}_j$ with highest frequency.
        $s \leftarrow s_0 + t_j$, add s to $\mathcal{G}$.
    end if
end for
if $\mathcal{G} = \emptyset$ then
    return **NOT-RLWE**
else if $\mathcal{G} = \{s\}$ is a singleton then
    return s
else
    return **INSUFFICIENT-SAMPLES**
end if

We analyze the complexity of our improved attack. There are q iterations, each operating on $O(q)$ samples, and reduction of each sample is $O(n)$. So our new attack has complexity $O(nq^2)$.

3.1 Examples of Successful Attacks

To illustrate the idea, we apply our improved attack to the instances in Table 1 of [9]. Comparing the last column with the current Table 1, we see that the runtime has been improved significantly.

Table 1. RLWE instances under our improved attack

n	q	f	r_0	No. samples	Old runtime (in minutes)	New runtime (in minutes)
40	67	2	2.51	22445	209	3.5
60	197	2	2.76	3940	63	2.4
60	617	2	2.76	12340	8.2×10^5 (est.)	21.3
80	67	2	2.51	3350	288.6	0.5
90	2003	2	3.13	60090	6.6×10^4 (est.)	305
96	521	2	2.76	15630	4.5×10^3 (est.)	21.7
100	683	2	2.76	20490	1.6×10^4 (est.)	36.5
144	953	2	2.51	38120	342.6	114.5

3.2 Proof of Proposition 1

For notational convenience, we let A_q denote the set $\mathbb{F}_{q^2} \backslash \mathbb{F}_q$.

Lemma 2. *Let the random variable a be uniformly distributed in A_q. Suppose e is a random variable with value in $\mathbb{F}_{q^2}$ independent of a. Fix $\delta \in A_q$ and $s_0 \in \mathbb{F}_q$. Then*

$$m_\delta = g_\delta + s_0 + \frac{\bar{e} - e}{\bar{a} - a}$$

is uniformly distributed in $\mathbb{F}_q$. Here

$$g_\delta = \frac{\overline{a\delta} - a\delta}{\bar{a} - a}.$$

Proof. Since the uniform distribution is invariant under translation, we may assume $s_0 = 0$. We introduce a new set $V = \{x \in \mathbb{F}_{q^2} : \bar{x} = -x\}$. We claim that for any $c, d \in V$ with $c \neq 0$, we have $P(\bar{a} - a = c, \overline{a\delta} - a\delta = d) = \frac{1}{q(q-1)}$. To prove the claim, note that V is an $\mathbb{F}_q$-vector space of dimension one, and we have the following $\mathbb{F}_q$-linear map $f_\delta : \mathbb{F}_{q^2} \to V^2$.

$$f_\delta : a \mapsto (\bar{a} - a, \overline{a\delta} - a\delta).$$

First we show f_δ is injective: if $f_\delta(a) = 0$, then $a \in \mathbb{F}_q$ and thus $a(\bar{\delta} - \delta) = 0$, so $a = 0$. By dimension counting, f_δ is an isomorphism. Restricting to A_q, we see that $f_\delta|_{A_q}$ gives an isomorphism between A_q and $(V \backslash \{0\}) \times V$. This proves the claim.

Let $e' = \frac{\bar{e}-e}{\bar{a}-a}$. For any $z \in \mathbb{F}_q$, we have

$$P(g_\delta + e' = z)$$

$$= \sum_{x+y=z} P(g_\delta = x, e' = y)$$

$$= \sum_{x+y=z} \sum_{c \in V \setminus \{0\}} P(\bar{a}\delta - a\delta = xc, \bar{e} - e = yc, \bar{a} - a = c)$$

$$= \sum_{x+y=z, c \in V \setminus \{0\}} P(\bar{a}\delta - a\delta = xc, \bar{a} - a = c)P(\bar{e} - e = yc)$$

$$= \frac{1}{q(q-1)} \sum_{y \in \mathbb{F}_q, c \in V \setminus \{0\}} P(\bar{e} - e = yc)$$

$$= \frac{1}{q(q-1)} \cdot (q-1) \sum_{c' \in V} P(\bar{e} - e = c')$$

$$= \frac{1}{q}.$$

Proof (of Proposition 1). The second claim follows directly from (1). For the first claim, let $\delta = t_i - t_j$. Then $m_j \sim g_\delta + s_0 + \frac{\bar{e}-e}{\bar{a}-a}$, where $g_\delta = \frac{\bar{a}\delta - a\delta}{\bar{a}-a}$. Now the first claim is precisely Lemma 2.

4   Infinite Family of Vulnerable Galois RLWE Instances

Recall that a number field K of degree n is *Galois* if it has exactly n automorphisms. In this section, we describe Galois number fields which are vulnerable to the attack outlined in Sect. 2.8. In contrast to the vulnerable instances found by computer search in Sect. 5 of [9], in this section we explicitly construct infinite families of such fields with flexible parameters. Furthermore, the attacks of [9] were successful only on instances where the size of the distribution (in the form of the scaled standard deviation) is a small constant, where as in this paper the scaled standard deviation parameter can be taken to be $o(|d|^{1/4})$, where d is an integer parameter and can go to infinity.

To set up, let p be an odd prime and let $d > 1$ be a squarefree integer such that d is coprime to p and $d \equiv 2, 3 \mod 4$. We choose an odd prime q such that

(1) $q \equiv 1 \pmod{p}$.

(2) $\left(\frac{d}{q}\right) = -1$ (equivalently, the prime q is inert in $\mathbb{Q}(\sqrt{d})$).

Remark 1. Fix a pair (p, d) that satisfies the conditions described above. By quadratic reciprocity, condition (2) on q above is a congruence condition modulo $4d$. So by Dirichlet's theorem on primes in arithmetic progressions, there exists infinitely many primes q satisfying both (1) and (2).

Let $M = \mathbb{Q}(\zeta_p)$ be the p-th cyclotomic field and $L = \mathbb{Q}(\sqrt{d})$. Let $K = M \cdot L$ be the composite field and let $\mathcal{O}_K$ denote its ring of integers.

Theorem 1. *Let K and q be as above, and R_q defined as in the preliminaries in terms of K and q. Suppose $\mathfrak{q}$ is a prime ideal in K lying over q. We consider the reduction map $\rho : R/qR \to R/\mathfrak{q}R \cong \mathbb{F}_{q^f}$, where f is the residue degree. Suppose $\mathcal{D}$ is the RLWE error distribution with error width r such that $r < 2\sqrt{\pi d}$. Let*

$$\beta = \min\left\{ \left(\frac{\sqrt{4\pi ed}}{r} e^{-\frac{2\pi d}{r^2}} \right)^n, 1 \right\}.$$

Then, for $x \in R_q$ drawn according to $\mathcal{D}$, we have $\rho(x) \in \mathbb{F}_q$ with probability at least $1 - \beta$.

Example 1. As a sample application of the theorem, we take $d = 4871, r = 68.17$ and $p = 43$. Then we computed $\beta = 0.11\ldots$ So if $x \in R_q$ is drawn from the error distribution, then $\rho(x) \in \mathbb{F}_q$ with probability at least 0.88.

Lemma 3. *Under the notation above, we have*

(1) $K/\mathbb{Q}$ is a Galois extension.
(2) $[K : \mathbb{Q}] = [M : \mathbb{Q}][L : \mathbb{Q}] = 2(p - 1)$.
(3) The prime q has residue degree 2 in K.
(4) $\mathcal{O}_K = \mathcal{O}_M \cdot \mathcal{O}_L = \mathbb{Z}[\zeta_p, \sqrt{d}]$.
(5) $|\operatorname{disc}(\mathcal{O}_K)| = p^{2(p-2)}(4d)^{(p-1)}$.

Proof. (1) follows from the fact that K is a composition of Galois extensions M and L; (2) is equivalent to $M \cap L = \mathbb{Q}$, which holds because $L/\mathbb{Q}$ is unramified away from primes dividing $2d$ and $M/\mathbb{Q}$ is unramified away from p; for (3), note that our assumptions imply that q splits completely in M and is inert in L, hence the claim. The claims (4) and (5) follow directly from [15, II. Theorem 12], and the fact that $\operatorname{disc}(\mathcal{O}_M) = p^{p-2}$ and $\operatorname{disc}(\mathcal{O}_L) = 4d$ are coprime.

The following lemma is a standard upper bound on the Euclidean lengths of samples from discrete Gaussians. It can be deduced directly from [16, Lemma 2.10].

Lemma 4. *Suppose $\Lambda \subseteq \mathbb{R}^n$ is a lattice. Let $D_{\Lambda,r}$ denote the discrete Gaussian over Λ of width r. Suppose c is a positive constant such that $c > \frac{r}{\sqrt{2\pi}}$. Let v be a sample from $D_{\Lambda,r}$. Then*

$$\operatorname{Prob}(\|v\|_2 > c\sqrt{n}) \leq C^n_{c/r},$$

where $C_s = s\sqrt{2\pi e} \cdot e^{-\pi s^2}$.

Proof (of Theorem). Part (3) of Lemma 3 implies that

$$1, \zeta_p, \ldots, \zeta_p^{p-2}; \sqrt{d}, \ldots, \zeta_p^{p-2}\sqrt{d} \tag{$*$}$$

is an integral basis of $R = \mathcal{O}_K$. By our assumptions, we have $R/\mathfrak{q}R \cong \mathbb{F}_{q^2}$, the finite field of q^2 elements. Under the map ρ, the first $(p-1)$ elements of the basis reduce to $\mathbb{F}_q$, and the rest reduce to the complement $\mathbb{F}_{q^2}\backslash\mathbb{F}_q$, because d is not a square modulo q.

Let $n = p-1$ be the degree of M over $\mathbb{Q}$. Then the extension $K/\mathbb{Q}$ has degree $2n$. We denote the elements in (*) by $v_1, \ldots, v_n$ and $w_1, \ldots, w_n$. Then $||\iota(v_i)|| = \sqrt{2n}$, while $||\iota(w_i)|| = \sqrt{2nd}$. We compute the root volume $c := (vol(R))^{1/n}$. It is a general fact that $vol(R) = |\operatorname{disc}(R)|^{\frac{1}{2}}$, so we have

$$c = |\operatorname{disc}(R)|^{\frac{1}{2n}} = \sqrt{2}p^{\frac{p-2}{2(p-1)}}d^{\frac{1}{4}}.$$

So when $d \gg p$, we have $||v_i|| \ll c \ll ||w_i||$. We have a decomposition $R = V\oplus W$, where V and W are free abelian groups with bases $v_1, \ldots, v_n$ and $w_1, \ldots, w_n$, respectively. The embeddings of V and W are orthogonal subspaces, because $\operatorname{Tr}(v_i\bar{w}_j) = 0$ for all i, j. For any element $e \in R$, we can write $e = e_1 + e_2\sqrt{d}$ where e_1, e_2 are elements of $\mathbb{Z}[\zeta_p]$, and it follows that $||e||^2 = ||e_1||^2 + d||e_2||^2$. In particular, if $e_2 \neq 0$, then $||e|| \geq \sqrt{2nd}$.

By applying Lemma 4 with $c = \sqrt{2d}$, the assumptions in the statement of our theorem imply that the probability that the discrete Gaussian $D_{\iota(R),r}$ will output a sample with $e_2 \neq 0$ is less than β. So the statement of theorem follows, since $e_2 = 0$ implies $\rho(e) \in \mathbb{F}_q$, i.e., the image of e lies in the prime subfield.

Therefore, we can specialize the general attack in this situation as follows. Given a set S of samples $(a, b) \in (R/qR)^2$, we loop through all q^2 possible guesses g of the value $s \mod \mathfrak{q}$ and compute $e_g = \rho(b) - g\rho(a)$. We then perform a chi-square test on the set $\{e_g : (a, b) \in S\}$, using two bins $\mathbb{F}_q$ and $\mathbb{F}_{q^2}\backslash\mathbb{F}_q$. If the samples are not taken from the RLWE distribution, or if the guess is incorrect, we expect to obtain uniform distributions; for the correct guess, we have $e_g = \rho(e)$, and by the above analysis, if the error parameter r_0 is sufficiently small, then the chi-square test might detect non-uniformness, since the portion of elements that lie in $\mathbb{F}_q$ might be larger than $1/q$.

The theoretical time complexity of our attack is $O(nq^3)$: the loop runs through q^2 possible guesses. In each passing of the loop, the number of samples we need for the chi-square test is $O(q)$, and the complexity of computing the map ρ on one sample is $O(n)$. Note that using the techniques in Sect. 3 of this paper, we could reduce the complexity to $O(nq^2)$.

Remark 2. It is easy to verify that if a triple (p, q, d) satisfies our assumptions, then so does $(p, q, d + 4kq)$ for any integer k, as long as $d + 4kq$ is square free. This shows one infinite family of Galois fields vulnerable to our attack.

4.1 Examples

Table 2 records some of the successful attacks we performed on the instances described previously. In each row of Table 2, the degree of the number field is $2(p - 1)$. Note that the runtimes are computed based on the improved version of the attack described in Sect. 3 of this paper. Also, by varying the parameters p and d, we can find vulnerable instances with $r_0 \to \infty$. For example, any $r_0 = o(d^{1/4}/\sqrt{p})$ will suffice.

Remark 3. From Table 2, we see that the attack in practice seems to work better (i.e., we can attack larger width r) than what is predicted in Theorem 1. As a possible explanation, we remark that in proving the theorem we bounded the probability of $e_2 = 0$ from below. However, the condition $e_2 = 0$ is sufficient but not necessary for $\rho(e)$ to lie in $\mathbb{F}_q$, so our estimation may be a very loose one.

Table 2. New vulnerable Galois RLWE instances

p	d	q	r_0	r	No. samples	Runtime (in seconds)
31	4967	311	8.94	592.94	3110	144.92
43	4871	173	8.97	694.94	1730	6.44
61	4643	367	8.84	815.11	3670	205.28
83	4903	167	8.94	963.84	1670	5.74
103	4951	619	8.94	1076.32	6190	579.77
109	4919	1091	8.94	1105.44	10910	1818.82
151	100447	907	14.08	4356.02	9070	1394.18
181	100267	1087	14.11	4777.17	10870	1973.47

4.2   Remarks on Other Possible Attacks

First, we note that the instances we found in this section are not directly attackable using linear algebra, as in the recent paper [8]. The reason is that although the last $n/2$-coordinates of the error e under the basis (*) are small integers, they are nonzero most of the time, so it is not clear how one can extract exact linear equations from the samples. On the other hand, note that for linear equations with small errors, there is the attack on the search RLWE problem proposed by Arora and Ge. However, the attack requires $O(n^{d-1})$ samples and solving a linear system in $O(n^d)$ variables. Here d is the width of the discrete error: for example, if the error can take values $0, 1, 2, -1, -2$, then $d = 5$. Thus the attack of Arora and Ge becomes impractical when n is larger than 10^2 and $d \geq 5$, say. In contrast, the complexity of our attack depends linearly on n and quadratically on q. In particular, it does not depend on the error size (although the success rate does depend on the error size).

5   Security of 2-Power Cyclotomic Rings with Unramified Moduli

In this section we provide some numerical evidence that for 2-power cyclotomic rings, the image of a fairly narrow RLWE error distribution modulo an unramified prime ideal $\mathfrak{q}$ of residue degree one or two is practically indistinguishable from uniform, implying that the 2-power cyclotomic rings are protected against the family of attacks in this paper.

We restrict ourselves to 2-power cyclotomic rings because the geometry is simple, namely the discrete Gaussian distribution $D_{\iota(R),\sqrt{n}r}$ over the ring is equivalent to a PLWE distribution, where each coefficient of the error is sampled independently from a discrete Gaussian $D_{\mathbb{Z},r}$ over the integers.

To further aid the analysis, we make another simplifying assumption by replacing $D_{\mathbb{Z},r}$ in the PLWE distribution described above by a "shifted binomial distribution". This allows a closed form formula for a bound on the statistical distance, and hence eases the analysis.

Let $m = 2^d$ for some integer $d \geq 1$ and let $K = \mathbb{Q}(\zeta_m)$ be the m-th cyclotomic field, with degree $n = m/2$. Let q be a prime such that $q \equiv 1 \pmod{m}$. Finally, let $\mathfrak{q}$ be a prime ideal above q.

Now we introduce a class of "shifted binomial distributions".

Definition 5. *For an even integer $k \geq 2$, let $\mathcal{V}_k$ denote the distribution over $\mathbb{Z}$ such that for every $t \in \mathbb{Z}$,*

$$Prob(\mathcal{V}_k = t) = \begin{cases} \frac{1}{2^k}\binom{k}{t+\frac{k}{2}} & \text{if } |t| \leq \frac{k}{2} \\ 0 & \text{otherwise} \end{cases}$$

We will abuse notation and also use $\mathcal{V}_k$ to denote the reduced distribution $\mathcal{V}_k$ (mod q) over $\mathbb{F}_q$, and let ν_k denote its probability density function. Figure 1 shows a plot of ν_8.

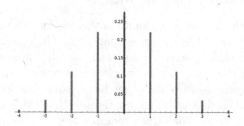

Fig. 1. Probability density function of $\mathcal{V}_8$

Definition 6. *Let $k \geq 2$ be an even integer. Then a sample from the distribution $P_{m,k}$ is*

$$e = \sum_{i=0}^{n-1} e_i \zeta_m^i,$$

where the coefficients e_i are sampled independently from $\mathcal{V}_k$.

5.1   Bounding the Distance from Uniform

We recall the definition and key properties of Fourier transform over finite fields. Suppose f is a real-valued function on $\mathbb{F}_q$. The *Fourier transform* of f is defined as

$$\widehat{f}(y) = \sum_{a \in \mathbb{F}_q} f(a)\overline{\chi_y(a)},$$

where $\chi_y(a) := e^{2\pi i a y/q}$.

Let u denote the probability density function of the uniform distribution over $\mathbb{F}_q$, that is $u(a) = \frac{1}{q}$ for all $a \in \mathbb{F}_q$. Let δ denote the characteristic function of the one-point set $\{0\} \subseteq \mathbb{F}_q$. Recall that the convolution of two functions $f, g : \mathbb{F}_q \to \mathbb{R}$ is defined as $(f * g)(a) = \sum_{b \in \mathbb{F}_q} f(a - b)g(b)$. We list without proof some basic properties of the Fourier transform.

1. $\widehat{\delta} = qu$; $\widehat{u} = \delta$.
2. $\widehat{f * g} = \widehat{f} \cdot \widehat{g}$.
3. $f(a) = \frac{1}{q}\sum_{y \in \mathbb{F}_q} \widehat{f}(y)\chi_y(a)$ (the Fourier inversion formula).

The following is a standard result.

Lemma 5. *Suppose F and G are independent random variables with values in $\mathbb{F}_q$, having probability density functions f and g. Then the density function of $F + G$ is equal to $f * g$. In general, suppose $F_1, \ldots, F_n$ are mutually independent random variables in $\mathbb{F}_q$, with probability density functions $f_1, \ldots, f_n$. Let f denote the density function of the sum $F = \sum F_i$, then $f = f_1 * \cdots * f_n$.*

The Fourier transform of ν_k has a nice closed-form formula, as below.

Lemma 6. *For all even integers $k \geq 2$, $\widehat{\nu_k}(y) = \cos\left(\frac{\pi y}{q}\right)^k$.*

Proof. We have

$$2^k \cdot \widehat{\nu_k}(y) = \sum_{m \in \mathbb{Z}/q\mathbb{Z}} \left(\sum_{a \in \mathbb{Z} : |aq+m| \leq k/2} \binom{k}{aq + m + \frac{k}{2}} \right) e^{-2\pi i y m/q}$$

$$= \sum_{m=-\frac{k}{2}}^{\frac{k}{2}} \binom{k}{m + \frac{k}{2}} e^{2\pi i y m/q}$$

$$= e^{-\pi i y k/q} \sum_{m'=0}^{k} \binom{k}{m'} e^{2\pi i y m'/q}$$

$$= e^{-\pi i y k/q}(1 + e^{2\pi i y/q})^k = (2\cos(\pi y/q))^k.$$

Dividing both sides by 2^k gives the result.

Next, we concentrate on the "reduced distribution" $P_{m,k} \pmod{\mathfrak{q}}$. Note that there is a one-to-one correspondence between primitive m-th roots of unity in $\mathbb{F}_q$ and the prime ideals above q in $\mathbb{Q}(\zeta_m)$. Let α be the root corresponding to our choice of $\mathfrak{q}$. Then a sample from $P_{m,k} \pmod{\mathfrak{q}}$ is of the form

$$e_\alpha = \sum_{i=0}^{n-1} \alpha^i e_i \pmod{q},$$

where the coordinates e_i are independently sampled from $\mathcal{V}_k$. We abuse notations and use e_α to denote its own probability density function.

Lemma 7

$$\widehat{e_\alpha}(y) = \prod_{i=0}^{n-1} \cos\left(\frac{\alpha^i \pi y}{q}\right)^k.$$

Proof. This follows directly from Lemma 6 and the independence of the coordinates e_i.

Lemma 8. *Let* $f : \mathbb{F}_q \to \mathbb{R}$ *be a function such that* $\sum_{a \in \mathbb{F}_q} f(a) = 1$. *Then for all* $a \in \mathbb{F}_q$, *the following holds.*

$$|f(a) - 1/q| \leq \frac{1}{q} \sum_{y \in \mathbb{F}_q, y \neq 0} |\hat{f}(y)|. \tag{1}$$

Proof. For all $a \in \mathbb{F}_q$,

$$\begin{aligned}
f(a) - 1/q &= f(a) - u(a) \\
&= \frac{1}{q} \sum_{y \in \mathbb{F}_q} (\hat{f}(y) - \hat{u}(y))\chi_y(a) \\
&= \frac{1}{q} \sum_{y \in \mathbb{F}_q} (\hat{f}(y) - \delta(y))\chi_y(a) \\
&= \frac{1}{q} \sum_{y \in \mathbb{F}_q, y \neq 0} \hat{f}(y)\chi_y(a). \quad \text{(since } \hat{f}(0) = 1)
\end{aligned}$$

Now the result follows from taking absolute values on both sides, and noting that $|\chi_y(a)| \leq 1$ for all a and all y.

Taking $f = e_\alpha$ in Lemma 8, we immediately obtain

Theorem 2. *The statistical distance between* e_α *and* u *satisfies*

$$\Delta(e_\alpha, u) \leq \frac{1}{2} \sum_{y \in \mathbb{F}_q, y \neq 0} |\widehat{e_\alpha}(y)|. \tag{2}$$

Now let $\epsilon(m, q, k, \alpha)$ denote the right hand side of (2), i.e.,

$$\epsilon(m, q, k, \alpha) = \frac{1}{2} \sum_{y \in \mathbb{F}_q, y \neq 0} \prod_{i=0}^{n-1} \cos\left(\frac{\alpha^i \pi y}{q}\right)^k.$$

To take into account all prime ideals above q, we let α run through all primitive m-th roots of unity in $\mathbb{F}_q$ and define

$$\epsilon(m, q, k) := \max\{\epsilon(m, q, k, \alpha) : \alpha \text{ has order } m \text{ in } (\mathbb{F}_q)^*\}.$$

If $\epsilon(m, q, k)$ is negligibly small, then the distribution $P_{m,k} \pmod{\mathfrak{q}}$ will be computationally indistinguishable from uniform. We will prove the following theorem.

Theorem 3. *Let q, m be positive integers such that q is a prime, m is a power of 2, $q \equiv 1 \mod m$ and $q < m^2$. Let $\beta = \frac{1+\frac{\sqrt{q}}{m}}{2}$; then $0 < \beta < 1$ and*

$$\epsilon(m, q, k) \leq \frac{q-1}{2} \beta^{\frac{km}{4}}.$$

In particular, if $\beta^{k/4} < \frac{1}{2}$, then the theorem says that $\epsilon(m, q, k) = O(q2^{-m})$ as $m \to \infty$.

Corollary 1. *The statistical distance between $P_{m,k}$ modulo $\mathfrak{q}$ and a uniform distribution is bounded above, independently of the choice of $\mathfrak{q}$ above q, by*

$$\frac{q-1}{2} \left(\frac{1 + \frac{\sqrt{q}}{m}}{2} \right)^{\frac{km}{4}}.$$

To prepare proving the theorem, we set up some notations of Shparlinski in [18]. Let $\Omega = (\omega_j)_{j=1}^{\infty}$ be a sequence of real numbers and let m be a positive integer. We define the following quantities:

- $L_\Omega(m) = \prod_{j=1}^{m}(1 - \exp(2\pi i \omega_j))$

- $S_\Omega(m) = \sum_{j=1}^{m} \exp(2\pi i \omega_j)$.

The following lemma is a special case of [18, Theorem 2.4].

Lemma 9
$$|L_\Omega(m)| \leq 2^{m/2}(1 + |S_\Omega(m)|/m)^{m/2}.$$

Proof (of Theorem 3). We specialize the above discussion to our situation, where m is a power of 2 and $n = m/2$. We fix $\omega_k = \frac{\alpha^{k-1}y}{q} + 1/2$, where we abuse notations and let α denote a lift of $\alpha \in \mathbb{F}_q$ to $\mathbb{Z}$.

Lemma 10. *We have*

$$|L_\Omega(n)| = 2^n \left| \prod_{j=0}^{n-1} \cos\left(\frac{\alpha^j \pi y}{q} \right) \right|$$

and $|L_\Omega(m)| = |L_\Omega(n)|^2$.

Proof. We have $L_\Omega(n) = \prod_{j=1}^{n}(1 - e^{2\pi i(\alpha^{j-1}y/q+1/2)}) = \prod_{j=0}^{n-1}(1 + e^{2\pi i\alpha^{j}y/q})$. So $|L_\Omega(n)| = \prod_{j=0}^{n-1}|e^{-\pi i\alpha^{j}y/q} + e^{\pi i\alpha^{j}y/q}| = \prod_{j=0}^{n-1} 2|\mathrm{Re}(e^{\pi i\alpha^{j}y/q})| = 2^m|\prod_{j=0}^{n-1} \cos(\alpha^{j}\pi y/q)|$. A similar argument with n replaced by m shows that $|L_\Omega(m)| = 2^m|\prod_{j=0}^{m-1} \cos(\alpha^{j}\pi y/q)|$. Since $\alpha^n \equiv -1 \mod q$ we have $\cos(\alpha^{j+n}\pi y/q) = \cos(\alpha^{j}\pi y/q)$ for $0 \le j \le n-1$. The claim now follows.

On the other hand, we have $S_\Omega(m) = -\sum_{j=0}^{m-1}\exp\left(\frac{2\pi i\alpha^{j}y}{q}\right)$, and standard bound on Gauss sums says that $|S_\Omega(m)| \le q^{1/2}$. Now combining Lemmas 9 and 10, we get

$$\left|\prod_{i=0}^{n-1}\cos\left(\frac{\alpha^{i}\pi y}{q}\right)\right| \le \beta^{n/2}$$

for β as defined in the statement of the theorem and for any nonzero $y \in \mathbb{F}_q$. Our result in the theorem now follows from taking both sides to k-th power and summing over y.

5.2   Numerical Distance from Uniform

We have computed $\epsilon(m, q, k)$ for various choices of parameters. Smaller values of ϵ imply that the error distribution looks more uniform when transferred to $R/\mathfrak{q}$, rendering the instance of RLWE invulnerable to the attacks in [9].

The data in Table 3 shows that when $n \ge 100$ and the size of the modulus q is polynomial in n, the statistical distances between $P_{m,k}$ (mod $\mathfrak{q}$) and the uniform distribution are negligibly small. Also, note that we fixed $k = 2$, and the epsilon values becomes even smaller when k increases.

For each instance in Table 3, we also generated the actual RLWE samples (where we fixed $r_0 = \sqrt{2\pi}$) and ran the chi-square attack of [9] using confidence level $\alpha = 0.99$. The column labeled "χ^2" contains the χ^2 values we obtained, and the column labeled "uniform?" indicates whether the reduced errors are uniform. We can see from the data how the practical situation agrees with our analysis on the approximated distributions.

Table 3. Values of $\epsilon(m, q, 2)$ and the χ^2 values

m ($n = m/2$)	q	$-[\log_2(\epsilon(m, q, 2))]$	χ^2	Uniform?
64	193	40	167.6	Yes
128	1153	97	1125.6	Yes
256	3329	194	3350.0	Yes
512	10753	431	10732.8	Yes

It is possible to generalize our discussion in this section to primes of arbitrary residue degree f, in which case the Fourier analysis will be performed over the field $\mathbb{F}_{q^f}$. The only change in the definitions would be $\chi_y(a) = e^{\frac{2\pi i \mathrm{Tr}(ay)}{q}}$. Here $\mathrm{Tr} : \mathbb{F}_{q^f} \to \mathbb{F}_q$ is the trace function. Similarly, we have

$$\widehat{e'_\alpha}(y) = \prod_{i=1}^{n} \cos\left(\frac{\pi \mathrm{Tr}(\alpha^i y)}{q}\right)^k.$$

Table 4 contains some data for primes of degree two.

Table 4. Values of $\epsilon(m, q, 2)$ for primes of degree two

m ($n = m/2$)	q	$-[\log_2(\epsilon(m, q, 2))]$
64	383	31
128	1151	54
256	1279	159
512	5583	341

Acknowledgements. We thank Chris Peikert, Igor Shparlinski, Léo Ducas and Ronald Cramer for helpful discussions.

References

1. Ajtai, M., Dwork, C.: A public-key cryptosystem with worst-case/average-case equivalence. In: Proceedings of the Twenty-Ninth Annual ACM Symposium on Theory of Computing, pp. 284–293. ACM (1997)
2. Bos, J.W., Lauter, K., Loftus, J., Naehrig, M.: Improved security for a ring-based fully homomorphic encryption scheme. In: Stam, M. (ed.) IMACC 2013. LNCS, vol. 8308, pp. 45–64. Springer, Heidelberg (2013). doi:10.1007/978-3-642-45239-0_4
3. Brakerski, Z.: Fully homomorphic encryption without modulus switching from classical GapSVP. In: Safavi-Naini, R., Canetti, R. (eds.) CRYPTO 2012. LNCS, vol. 7417, pp. 868–886. Springer, Heidelberg (2012). doi:10.1007/978-3-642-32009-5_50
4. Brakerski, Z., Gentry, C., Vaikuntanathan, V.: (Leveled) fully homomorphic encryption without bootstrapping. In: Proceedings of the 3rd Innovations in Theoretical Computer Science Conference, pp. 309–325. ACM (2012)
5. Brakerski, Z., Vaikuntanathan, V.: Fully homomorphic encryption from ring-lwe and security for key dependent messages. In: Rogaway, P. (ed.) CRYPTO 2011. LNCS, vol. 6841, pp. 505–524. Springer, Heidelberg (2011). doi:10.1007/978-3-642-22792-9_29
6. Brakerski, Z., Vaikuntanathan, V.: Efficient fully homomorphic encryption from (standard) LWE. SIAM J. Comput. **43**(2), 831–871 (2014)
7. Castryck, W., Iliashenko, I., Vercauteren, F.: On error distributions in Ring-based LWE. Cryptology ePrint Archive, Report 2016/240 (2016). http://eprint.iacr.org/2016/240

8. Castryck, W., Iliashenko, I., Vercauteren, F.: Provably weak instances of ring-LWE revisited. In: Fischlin, M., Coron, J.-S. (eds.) EUROCRYPT 2016. LNCS, vol. 9665, pp. 147–167. Springer, Heidelberg (2016). doi:10.1007/978-3-662-49890-3_6

9. Chen, H., Lauter, K., Stange, K.E.: Attacks on search-RLWE. Cryptology ePrint Archive, Report 2015/971 (2015). http://eprint.iacr.org/

10. Eisenträger, K., Hallgren, S., Lauter, K.: Weak instances of PLWE. In: Joux, A., Youssef, A. (eds.) SAC 2014. LNCS, vol. 8781, pp. 183–194. Springer, Cham (2014). doi:10.1007/978-3-319-13051-4_11

11. Elias, Y., Lauter, K.E., Ozman, E., Stange, K.E.: Provably weak instances of ring-LWE. In: Gennaro, R., Robshaw, M. (eds.) CRYPTO 2015. LNCS, vol. 9215, pp. 63–92. Springer, Heidelberg (2015). doi:10.1007/978-3-662-47989-6_4

12. Hoffstein, J., Pipher, J., Silverman, J.H.: An Introduction to Mathematical Cryptography, vol. 1. Springer, Heidelberg (2008). doi:10.1007/978-0-387-77993-5

13. López-Alt, A., Tromer, E., Vaikuntanathan, V.: On-the-fly multiparty computation on the cloud via multikey fully homomorphic encryption. In: Proceedings of the Forty-Fourth Annual ACM Symposium on Theory of Computing, pp. 1219–1234. ACM (2012)

14. Lyubashevsky, V., Peikert, C., Regev, O.: On ideal lattices and learning with errors over rings. J. ACM (JACM) **60**(6), 43 (2013)

15. Marcus, D.A.: Number Fields, vol. 18. Springer, Heidelberg (1977). doi:10.1007/978-1-4684-9356-6

16. Micciancio, D., Regev, O.: Worst-case to average-case reductions based on Gaussian measures. SIAM J. Comput. **37**(1), 267–302 (2007)

17. Regev, O.: On lattices, learning with errors, random linear codes, and cryptography. J. ACM (JACM) **56**(6), 34 (2009)

18. Shparlinski, I.E.: On some characteristics of uniformity of distribution and their applications. In: Bosma, W., van der Poorten, A. (eds.) Computational Algebra and Number Theory. Mathematics and Its Applications, pp. 227–241. Springer, Dordrecht (1995). doi:10.1007/978-94-017-1108-1_16

19. Stehlé, D., Steinfeld, R.: Making NTRU as secure as worst-case problems over ideal lattices. In: Paterson, K.G. (ed.) EUROCRYPT 2011. LNCS, vol. 6632, pp. 27–47. Springer, Heidelberg (2011). doi:10.1007/978-3-642-20465-4_4

Efficient Classical Public
Key Cryptography

Fast, Uniform Scalar Multiplication for Genus 2 Jacobians with Fast Kummers

Ping Ngai Chung[1], Craig Costello[2], and Benjamin Smith[3(✉)]

[1] University of Chicago, Chicago, USA
briancpn@math.uchicago.edu
[2] Microsoft Research, Redmond, USA
craigco@microsoft.com
[3] Inria and Laboratoire D'Informatique de l'École polytechnique (LIX),
Palaiseau, France
smith@lix.polytechnique.fr

Abstract. We give one- and two-dimensional scalar multiplication algorithms for Jacobians of genus 2 curves that operate by projecting to Kummer surfaces, where we can exploit faster and more uniform pseudo-multiplication, before recovering the proper "signed" output back on the Jacobian. This extends the work of López and Dahab, Okeya and Sakurai, and Brier and Joye to genus 2, and also to two-dimensional scalar multiplication. The technique is especially interesting in genus 2, because Kummer surfaces can outperform comparable elliptic curve systems.

Keywords: Kummer surface · Genus 2 · Scalar multiplication · Signatures · Pseudomultiplication · Uniform · Constant-time

1 Introduction

In this article we show how to exploit Gaudry's fast, uniform Kummer surface arithmetic [14] to carry out full scalar multiplications on genus 2 Jacobians. This brings the speed and side-channel security of Kummers, so far only used for Diffie–Hellman implementations, to implementations of other discrete-log-based cryptographic protocols including signature schemes.

To make things precise, let $\mathcal{J}_\mathcal{C}$ be the Jacobian of a genus 2 curve $\mathcal{C}$ over a finite field $\mathbb{F}_q$ of characteristic >3 (with $\oplus$ denoting the group law on $\mathcal{J}_\mathcal{C}$, and $\ominus$ the inverse). We want to compute scalar multiplications

$$(m, P) \longmapsto [m]P := \underbrace{P \oplus \cdots \oplus P}_{m \text{ times}} \quad \text{for } m \in \mathbb{Z}_{\geq 0} \text{ and } P \in \mathcal{J}_\mathcal{C}(\mathbb{F}_q)$$

which are at the heart of all discrete logarithm and Diffie–Hellman problem-based cryptosystems. If the scalar m is secret, then $[m]P$ must be computed in a *uniform* and *constant-time* way to protect against even the most elementary side-channel attacks. This means that the execution path of the algorithm must be independent of the scalar m (we may assume that the bitlength of m is fixed).

© Springer International Publishing AG 2017
R. Avanzi and H. Heys (Eds.): SAC 2016, LNCS 10532, pp. 465–481, 2017.
https://doi.org/10.1007/978-3-319-69453-5_25

The quotient *Kummer surface* $\mathcal{K}_C := \mathcal{J}_C / \langle \pm 1 \rangle$ identifies group elements with their inverses (this is the genus-2 analogue of projecting elliptic curve points onto the x-coordinate). If P is a point on $\mathcal{J}_C$, then $\pm P$ denotes its image in $\mathcal{K}_C$. Scalar multiplication on $\mathcal{J}_C$ induces a well-defined *pseudomultiplication*

$$(m, \pm P) \longmapsto \pm[m]P \quad \text{for } m \in \mathbb{Z}_{\geq 0} \text{ and } P \in \mathcal{J}_C(\mathbb{F}_q),$$

which can be computed using differential addition chains in the exact analogue of x-only arithmetic for elliptic curves. This suffices for implementing protocols like Diffie–Hellman key exchange which *only* involve scalar multiplication, as Bernstein's Curve25519 software did for elliptic curves [1]. But we emphasize that $\mathcal{K}_C$ is *not* a group, and its lack of a group operation prevents us instantiating many group-based protocols in $\mathcal{K}_C$ (see [25, Sect. 5]).

It has long been known that x-only pseudomultiplication can be used for full scalar multiplication on elliptic curves: López and Dahab [19] (followed by Okeya and Sakurai [22] and Brier and Joye [4]) showed that the auxiliary values computed by the x-only Montgomery ladder can be used to recover the missing y-coordinate, and hence to compute full scalar multiplications on elliptic curves. The main innovation of this paper is to extend this technique from elliptic curves to genus 2, and from one- to two-dimensional scalar multiplication. This allows cryptographic protocols instantiated in genus-2 Jacobians to delegate their scalar multiplications to faster, more uniform Kummer surfaces.

In the abstract, our algorithms follow the same common pattern:

1. **Project** the inputs from $\mathcal{J}_C$ to $\mathcal{K}_C$;
2. **Pseudomultiply** in $\mathcal{K}_C$ using a differential addition chain, such as the Montgomery ladder [21] or Bernstein's binary chain [2];
3. **Recover** the correct preimage for the full scalar multiplication in $\mathcal{J}_C$ from the outputs of the pseudomultiplication, using our new Algorithm 2.

More concretely, if $\mathcal{J}_C$ is a genus-2 Jacobian admitting a fast Kummer surface as in Sect. 2, and $\mathcal{B} \subset \mathcal{J}_C(\mathbb{F}_q)$ is the set of Definition 1, then our main results are

Theorem 1 (Project + Montgomery ladder + Recover): If P is a point in $\mathcal{J}_C(\mathbb{F}_q) \setminus \mathcal{B}$ then for any β-bit integer m, Algorithm 3 computes $[m]P$ in $(7\beta + 115)\mathbf{M} + (12\beta + 8)\mathbf{S} + (12\beta + 4)\mathbf{m_c} + (32\beta + 79)\mathbf{a} + 2\mathbf{I}$.

Theorem 2 (Project + Bernstein's binary chain + Recover): If P and Q are points in $\mathcal{J}_C(\mathbb{F}_q) \setminus \mathcal{B}$ with $P \oplus Q$ and $P \ominus Q$ not in $\mathcal{B}$ and m and n are positive β-bit integers, then Algorithm 4 computes $[m]P \oplus [n]Q$ in $(14\beta + 203)\mathbf{M} + (20\beta + 16)\mathbf{S} + (16\beta + 16)\mathbf{m_c} + (56\beta + 138)\mathbf{a} + 3\mathbf{I}$.

Both algorithms are uniform with respect to their scalars. The two-dimensional multiscalar multiplications of Theorem 2 appear explicitly in many cryptographic protocols (such as Schnorr signature verification), but they are also a key ingredient in endomorphism-accelerated one-dimensional scalar multiplication techniques like GLV [13] and its descendants.[1]

---

[1] Our techniques should readily extend to the higher-dimensional differential addition chains described by Brown [5]. We do not investigate this here.

There are two key benefits to this approach: speed and uniformity. For speed, we note that Gaudry's Kummer arithmetic is markedly faster than full Jacobian arithmetic, and competitive Diffie–Hellman implementations have shown that Kummer-based scalar multiplication software can outperform its elliptic equivalent [3]. Our results bring this speed to a wider range of protocols, such as ElGamal and signature schemes. Indeed, the methods described below (including Algorithms 2 and 3) have already been successfully put into practice in a fast and compact implementation of Schnorr signatures for microcontrollers [23], but without any proof of correctness or explanation of the algorithms[2]; this article provides that proof, and detailed algorithms to enable further implementations.

The second benefit is side-channel protection. Fast, uniform, constant-time algorithms for elliptic curve scalar multiplication are well-known and widely-used. In contrast, for genus 2 Jacobians, the uniform and constant-time requirements are problematic: conventional Cantor arithmetic [6] and its derivatives [16] are highly susceptible to simple side-channel attacks. The explicit formulæ derived for generic additions in Jacobians fail to compute correct results when one or both of the inputs are so-called "special" points (essentially, those corresponding to degree-one divisors on $\mathcal{C}$). While special points are rare enough that random scalar multiplications never encounter them, they are plentiful enough that attackers can easily mount exceptional procedure attacks [17], forcing software into special cases and using timing variations to recover secret data. It has appeared impossible to implement traditional genus 2 arithmetic in a uniform way without abandoning all hope of competitive efficiency [10]. The Jacobian point recovery method we present in Sect. 3 solves the problem of uniform genus 2 arithmetic (at least for scalar multiplication): rather than wrestling with the special cases of Cantor's algorithm on $\mathcal{J}_\mathcal{C}$, we can pseudomultiply on the Kummer and then recover the correct image on $\mathcal{J}_\mathcal{C}$.

Remark 1. Robert and Lubicz [20] use similar techniques to speed up their arithmetic for general abelian varieties based on theta functions, viewing the results of the Montgomery ladder on a g-dimensional Kummer variety K as a point on the corresponding abelian variety A embedded in K^2. In contrast to our method, Robert and Lubicz cannot treat A as a Jacobian (since general abelian varieties of dimension $g > 3$ are not Jacobians); so in the case of genus $g = 2$, there is no explicit connection with any curve $\mathcal{C}$, and the starting and finishing points do not involve the Mumford representation. Kohel [18] explores similar ideas for elliptic curves, leading to an interesting interpretation of Edwards curve arithmetic.

Remark 2. Since our focus here is on fast cryptographic implementations, for lack of space, in this article we restrict our attention to curves and Jacobians whose Kummer surfaces have so-called "fast" models (see Sect. 2). This implies that all of our Jacobians have full rational 2-torsion. Our techniques generalize without any difficulty to more general curves and Kummer surfaces, and then replacing the fast Kummer operations described in Appendix A with more

---

[2] The implementation in [23] was based on a much longer draft version of this paper.

general methods wherever they appear in Algorithms 3 and 4 yields efficient, uniform scalar multiplication algorithms for any genus 2 Jacobian.

Notation. As usual, M, S, I, and a denote the costs of one multiplication, squaring, inversion, and addition in $\mathbb{F}_q$, respectively; for simplicity, we assume subtraction and unary negation in $\mathbb{F}_q$ also cost a. We let m_c denote the cost of multiplication by the theta constants a, b, c, d, A, B, C, D of Sect. 2 and their inverses (we aim to make these as small as possible). We assume we have efficient constant-time conditional selection and swap routines: $\texttt{SELECT}(b, (X_0, X_1))$ returns X_b, and $\texttt{SWAP}(b, (X_0, X_1))$ returns (X_b, X_{1-b}) (see Appendix B for sample code).

2   Genus 2 Jacobians with fast Kummer Surfaces

Suppose we have a, b, c, and d in $\mathbb{F}_q \setminus \{0\}$ such that if we set

$$A := a + b + c + d \qquad\qquad B := a + b - c - d$$
$$C := a - b + c - d \qquad\qquad D := a - b - c + d$$

then $abcdABCD \neq 0$ and $CD/(AB) = \alpha^2$ for some α in $\mathbb{F}_q$. Setting

$$\lambda := a/b \cdot c/d \qquad \mu := c/d \cdot (1+\alpha)/(1-\alpha) \qquad \nu := a/b \cdot (1+\alpha)/(1-\alpha)$$

we define an associated genus 2 curve $\mathcal{C}$ in Rosenhain form:

$$\mathcal{C} : y^2 = f(x) = x(x-1)(x-\lambda)(x-\mu)(x-\nu)$$

so $f(x) = x^5 + f_4 x^4 + f_3 x^3 + f_2 x^2 + f_1 x$ with $f_4 = -(\lambda + \mu + \nu + 1)$, $f_3 = \lambda\mu + \lambda\nu + \lambda + \mu\nu + \mu + \nu$, $f_2 = -(\lambda\mu\nu + \lambda\mu + \lambda\nu + \mu\nu)$, $f_1 = \lambda\mu\nu$.

Elements of $\mathcal{J}_\mathcal{C}(\mathbb{F}_q)$ are presented in their standard Mumford representation:

$$P \in \mathcal{J}_\mathcal{C}(\mathbb{F}_q) \longleftrightarrow \langle a(x) = x^2 + a_1 x + a_0, b(x) = b_1 x + b_0 \rangle$$

where a_1, a_0, b_1, and b_0 are in $\mathbb{F}_q$ and $b(x)^2 \equiv f(x) \pmod{a(x)}$. The group law on $\mathcal{J}_\mathcal{C}$ is typically computed using Cantor's algorithm, specialized to genus 2. Here we suppose we have a function $\texttt{JacADD} : (P, Q) \mapsto P \oplus Q$ which computes the group law as in [16, Eq. (12)] at a cost of $22\mathsf{M} + 2\mathsf{S} + 1\mathsf{I} + 27\mathsf{a}$.

The *fast Kummer surface* for $\mathcal{C}$ is the quartic surface $\mathcal{K}_\mathcal{C}^{\text{fast}} \subset \mathbb{P}^3$ defined by

$$\mathcal{K}_\mathcal{C}^{\text{fast}} : \left(\begin{matrix} (X^2 + Y^2 + Z^2 + T^2) \\ -F(XT + YZ) - G(XZ + YT) - H(XY + ZT) \end{matrix} \right)^2 = EXYZT \quad (1)$$

where

$$F = \frac{a^2 - b^2 - c^2 + d^2}{ad - bc}, \quad G = \frac{a^2 - b^2 + c^2 - d^2}{ac - bd}, \quad H = \frac{a^2 + b^2 - c^2 - d^2}{ab - cd},$$

and $E = 4abcd \left(ABCD/((ad - bc)(ac - bd)(ab - cd))\right)^2$. These surfaces were algorithmically developed by the Chudnovskys [8], and introduced in cryptography by Gaudry [14]; here we use the "squared-theta" model of [9, Chap. 4].

Cryptographic parameters for genus-2 Jacobians equipped with fast Kummers can be (and have been) computed: the implementation of [23] uses the parameters from [15] in the algorithms presented below.

The map $\texttt{Project} : \mathcal{J}_\mathcal{C} \to \mathcal{K}_\mathcal{C}^{\text{fast}}$ mapping P to $\pm P$ is classical (cf. [9, Sect. 5.3]), and implemented by Algorithm 1. It is not uniform or constant-time, but it does not need to be: in most applications the input points are already public.

Algorithm 1. Project: $\mathcal{J}_\mathcal{C} \to \mathcal{K}_\mathcal{C}^{\text{fast}}$.

Input: $P \in \mathcal{J}_\mathcal{C}(\mathbb{F}_q)$
Output: $\pm P \in \mathcal{K}_\mathcal{C}^{\text{fast}}(\mathbb{F}_q)$
Cost: $8\text{M} + 1\text{S} + 4\text{m}_c + 14\text{a}$, assuming precomputed $\lambda\mu$, $\lambda\nu$.

1 **if** $P = 0$ **then return** $(a : b : c : d)$
2 **else if** $P = \langle x - u, v \rangle$ **then**
3 $\quad$ $(t_1, t_2, t_3, t_4) \leftarrow (u - 1, u - \lambda, u - \mu, u - \nu)$ $\qquad\qquad$ `// 4a`
4 $\quad$ **return** $(a \cdot t_1 \cdot t_3 : b \cdot t_2 \cdot t_4 : c \cdot t_1 \cdot t_4 : d \cdot t_2 \cdot t_3)$ $\qquad$ `// 4M+4m_c`
5 **else** (generic case $P = \langle x^2 + a_1 x + a_0, b_1 x + b_0 \rangle$)
6 $\quad$ $(t_1, t_2, t_3) \leftarrow (a_1 + \lambda, a_1 + 1, b_0^2)$ $\qquad\qquad\qquad$ `// 1S+2a`
7 $\quad$ $(t_4, t_5) \leftarrow (a_0 \cdot (a_0 - \mu) \cdot (t_1 + \nu), a_0 \cdot (a_0 - \lambda\nu) \cdot (t_2 + \mu))$ $\quad$ `// 4M+4a`
8 $\quad$ $(t_6, t_7) \leftarrow (a_0 \cdot (a_0 - \nu) \cdot (t_1 + \mu), a_0 \cdot (a_0 - \lambda\mu) \cdot (t_2 + \nu))$ $\quad$ `// 4M+4a`
9 $\quad$ **return** $(a \cdot t_4 + t_3, b \cdot t_5 + t_3, c \cdot t_6 + t_3, d \cdot t_7 + t_3)$ $\qquad$ `// 4m_c+4a`

Table 1 summarizes the key standard operations on $\mathcal{K}_\mathcal{C}^{\text{fast}}$ and their costs (for detailed pseudocode, see Appendix A). The pseudo-doubling $\texttt{xDBL}$ is correct on all inputs; the pseudo-additions $\texttt{xADD}^*$, $\texttt{xADD}$ and combined pseudo-double-and-add $\texttt{xDBLADD}$ are correct for all inputs provided the difference point has no coordinate equal to zero. Since almost all difference points are fixed in our algorithms, and these "bad" points are extremely rare (there are only $O(q)$ of them, versus $O(q^2)$ other points), we simply prohibit them as input: Definition 1 identifies their preimages in $\mathcal{J}_\mathcal{C}$ for easy identification and rejection.

Definition 1. *Let $\mathcal{B} \subset \mathcal{J}_\mathcal{C}(\mathbb{F}_q)$ be the set of elements P whose images $\pm P$ in $\mathcal{K}_\mathcal{C}^{\text{fast}}$ have a zero coordinate; or equivalently, $P = \langle x^2 + a_1 x + a_0, b_1 x + b_0 \rangle$ with*

1. $(\mu a_1 + a_0)(1(a_1 + \lambda + \nu) + a_0) + (\lambda\mu - \lambda\nu + \mu\nu - 1\mu)a_0 + \lambda\mu\nu = 0$, *or*
2. $(\nu a_1 + a_0)(\lambda(a_1 + 1 + \mu) + a_0) - (\lambda\nu - \mu\nu + 1\mu - 1\nu)a_0 + \lambda\mu\nu = 0$, *or*
3. $(\nu a_1 + a_0)(1(a_1 + \lambda + \mu) + a_0) - (\lambda\mu - \lambda\nu - \mu\nu + 1\nu)a_0 + \lambda\mu\nu = 0$, *or*
4. $(\mu a_1 + a_0)(\lambda(a_1 + 1 + \nu) + a_0) - (\lambda\mu - \mu\nu - 1\mu + 1\nu)a_0 + \lambda\mu\nu = 0$.

To optimize pseudo-additions, we define a mapping $\texttt{Wrap} : (x : y : z : t) \mapsto (x/y, x/z, x/t)$ (for $(x : y : z : y)$ not in $\mathcal{B}$). To $\texttt{Wrap}$ one Kummer point costs $7\text{M} + 1\text{I}$, but saves 7M in every subsequent pseudo-addition with that point as its difference. In Algorithm 4 we need to $\texttt{Wrap}$ four points; $\texttt{Wrap4}$ does this with a single shared inversion, for a total cost of $37\text{M} + 1\text{I}$.

Table 1. Operations on $\mathcal{K}_{\mathcal{C}}^{\text{fast}}$ and $\mathcal{J}_{\mathcal{C}}$. All but JacADD are uniform. The operations xADD*, xADD, and xDBLADD require $P \ominus Q \notin \mathcal{B}$.

Algorithm	Operation: Input $\mapsto$ Output	M	S	m_c	a	I
JacADD	$(P, Q) \mapsto P \oplus Q$	22	2	0	27	1
xDBL	$\pm P \mapsto \pm[2]P$	0	8	8	16	0
xADD*	$(\pm P, \pm Q, \pm(P \ominus Q)) \mapsto \pm(P \oplus Q)$	14	4	4	24	0
xADD	$(\pm P, \pm Q, \text{Wrap}(\pm(P \ominus Q))) \mapsto \pm(P \oplus Q)$	7	4	4	24	0
xDBLADD	$(\pm P, \pm Q, \text{Wrap}(\pm(P \ominus Q))) \mapsto (\pm[2]P, \pm(P \oplus Q))$	7	12	12	32	0
Wrap	$(x : y : z : t) \mapsto (x/y, x/z, x/t)$	7	0	0	0	1
Wrap4	$(\pm P_i)_{i=1}^4 \mapsto (\text{Wrap}(\pm P_i))_{i=1}^4$	37	0	0	0	1

3 Point Recovery in Genus 2

Our aim is to compute scalar multiplications $(m, P) \mapsto R = [m]P$ on $\mathcal{J}_{\mathcal{C}}$. Projecting to $\mathcal{K}_{\mathcal{C}}$ yields $\pm P$, and then pseudomultiplication (which we will describe below) gives $\pm R = \pm[m]P$; but it can also produce $\pm(R \oplus P)$ as an auxiliary output. We will reconstruct R from this data, by defining a map

$$\textbf{Recover} : (P, \pm P, \pm R, \pm(R \oplus P)) \longmapsto R \quad \text{for } P \text{ and } R \in \mathcal{J}_{\mathcal{C}}.$$

The map $\mathcal{J}_{\mathcal{C}} \to \mathcal{K}_{\mathcal{C}}^{\text{fast}}$ factors through the "general Kummer" $\mathcal{K}_{\mathcal{C}}^{\text{gen}}$, another quartic surface in $\mathbb{P}^3$ defined (as in [7, Chap. 3], taking $f_6 = f_0 = 0$ and $f_5 = 1$, and using coordinates ξ_i, to avoid confusion with $\mathcal{K}_{\mathcal{C}}^{\text{fast}}$) by

$$\mathcal{K}_{\mathcal{C}}^{\text{gen}} : K_2(\xi_1, \xi_2, \xi_3)\xi_4^2 + K_1(\xi_1, \xi_2, \xi_3)\xi_4 + K_0(\xi_1, \xi_2, \xi_3) = 0 \tag{2}$$

where $K_2 = \xi_2^2 - 4\xi_1\xi_3$, $K_1 = -2(f_1\xi_1^2 + f_3\xi_1\xi_3 + f_5\xi_3^2)\xi_2 - 4\xi_1\xi_3(f_2\xi_1 + f_4\xi_3)$, and $K_0 = (f_1\xi_1^2 - f_3\xi_1\xi_3 + f_5\xi_3^2)^2 - 4\xi_1\xi_3(f_1\xi_2 + f_2\xi_3)(f_4\xi_1 + f_5\xi_2)$. While fast Kummers offer significant gains in performance and uniformity, this comes at the price of full rational 2-torsion: hence, not every Kummer can be put in fast form. But the general Kummer exists for all genus 2 curves, not just those admitting a fast Kummer; roughly speaking, $\mathcal{K}_{\mathcal{C}}^{\text{gen}}$ is the analogue of the x-line of the Weierstrass model of an elliptic curve, while $\mathcal{K}_{\mathcal{C}}^{\text{fast}}$ corresponds to the x-line of a Montgomery model.[3] As such, $\mathcal{K}_{\mathcal{C}}^{\text{gen}}$ is much more naturally related to the Mumford model of $\mathcal{J}_{\mathcal{C}}$; so it makes sense to map our recovery problem from $\mathcal{K}_{\mathcal{C}}^{\text{fast}}$ into $\mathcal{K}_{\mathcal{C}}^{\text{gen}}$ and then recover from $\mathcal{K}_{\mathcal{C}}^{\text{gen}}$ to $\mathcal{J}_{\mathcal{C}}$.

---

[3] The use of $\mathcal{K}_{\mathcal{C}}^{\text{gen}}$ in cryptography was investigated by Smart and Siksek [25] and Duquesne [12]. The polynomials defining pseudo-operations on $\mathcal{K}_{\mathcal{C}}^{\text{gen}}$ (see [7, Sect. 3.4]) are hard to evaluate quickly, and do not offer competitive performance. However, they are completely compatible with our Project-pseudomultiply-Recover pattern, and we could use them to construct uniform and constant-time scalar multiplication algorithms for genus 2 Jacobians that do *not* admit fast Kummers.

The map $\pi : \mathcal{J}_\mathcal{C} \to \mathcal{K}_\mathcal{C}^{\mathrm{gen}}$ is described in [7, Eqs. (3.1.3–5)]; it maps generic points $\langle x^2 + a_1 x + a_0, b_1 x + b_0 \rangle$ in $\mathcal{J}_\mathcal{C}$ to $(\xi_1 : \xi_2 : \xi_3 : \xi_4)$ in $\mathcal{K}_\mathcal{C}^{\mathrm{gen}}$, where

$$(\xi_1 : \xi_2 : \xi_3 : \xi_4) = (1 : -a_1 : a_0 : b_1^2 + (a_1^2 - a_0)a_1 + a_1(f_3 - f_4 a_1) - f_2). \quad (3)$$

Projecting onto the $(\xi_1 : \xi_2 : \xi_3)$-plane yields a natural double cover $\rho : \mathcal{K}_\mathcal{C}^{\mathrm{gen}} \to \mathbb{P}^2$; comparing with (3), we see that $\rho \circ \pi$ corresponds to projecting onto the a-polynomial of the Mumford representation.

Proposition 1. *Suppose* $P = \langle x^2 + a_1^P x + a_0^P, b_1^P x + b_0^P \rangle$ *and* $R = \langle x^2 + a_1^R x + a_0^R, b_1^R x + b_0^R \rangle$ *are in* $\mathcal{J}_\mathcal{C}(\mathbb{F}_q)$. *Let* $(\xi_1^R : \xi_2^R : \xi_3^R : \xi_4^R) = \pi(R)$ *in* $\mathcal{K}_\mathcal{C}^{\mathrm{gen}}$, *and let* $(\xi_1^\oplus : \xi_2^\oplus : \xi_3^\oplus) = \rho(\pi(P \oplus R))$ *and* $(\xi_1^\ominus : \xi_2^\ominus : \xi_3^\ominus) = \rho(\pi(P \ominus R))$ *in* $\mathbb{P}^2$. *Let* $Z_1 = \xi_2^R + a_1^P \xi_1^R$, $Z_2 = \xi_3^R - a_0^P \xi_1^R$, *and* $Z_3 = -(a_1^P \xi_3^R + a_0^P \xi_2^R)$. *Then*

$$(\xi_1^R)^2 (b_1^R, b_0^R) = (G_3, G_4) \begin{pmatrix} \xi_2^R Z_1 - \xi_1^R Z_2 & -\xi_1^R Z_1 \\ -\xi_3^R Z_1 & \xi_1^R Z_2 \end{pmatrix} \quad (4)$$

where G_3 *and* G_4 *satisfy*

$$C(G_3, G_4) = D(G_1, G_2) \begin{pmatrix} \xi_1^\oplus \xi_3^\ominus - \xi_3^\oplus \xi_1^\ominus & \xi_2^\oplus \xi_3^\ominus - \xi_3^\oplus \xi_2^\ominus \\ \xi_2^\oplus \xi_1^\ominus - \xi_1^\oplus \xi_2^\ominus & \xi_3^\oplus \xi_1^\ominus - \xi_1^\oplus \xi_3^\ominus \end{pmatrix} \quad (5)$$

where $\xi_1^R D = Z_2^2 - Z_1 Z_3$ *and* G_1 *and* G_2 *satisfy*

$$D(G_1, G_2) = (b_1^P, b_0^P) \begin{pmatrix} Z_2 & a_0^P Z_1 \\ Z_1 & -a_1^P Z_1 - Z_2 \end{pmatrix} \quad (6)$$

and

$$C = \frac{-2D \left(2\xi_1^\oplus \xi_1^\ominus G_2^2 - (\xi_2^\oplus \xi_1^\ominus + \xi_1^\oplus \xi_2^\ominus) G_1 G_2 + (\xi_3^\oplus \xi_1^\ominus + \xi_1^\oplus \xi_3^\ominus) G_1^2 \right)}{G_1^2 + G_3^2}.$$

Proof. This is a disguised form of the geometric group law on $\mathcal{J}_\mathcal{C}$ (cf. [7, Sect. 1.2]). The points P and R correspond to unique degree-2 divisor classes on $\mathcal{C}$: say,

$$P \longleftrightarrow [(u_P, v_P) + (u_P', v_P')] \quad \text{and} \quad R \longleftrightarrow [(u_R, v_R) + (u_R', v_R')].$$

(We do not compute the values of $u_P, v_P, u_P', v_P', u_R, v_R, u_R'$, and v_R', which are generally in $\mathbb{F}_{q^2}$; they are purely formal devices here.) Let

$$E_1 = \frac{v_P}{(u_P - u_P')(u_P - u_R)(u_P - u_R')}, \quad E_2 = \frac{v_P'}{(u_P' - u_P)(u_P' - u_R)(u_P' - u_R')},$$

$$E_3 = \frac{v_R}{(u_R - u_P')(u_R - u_P)(u_R - u_R')}, \quad E_4 = \frac{v_R'}{(u_R' - u_P)(u_R' - u_P')(u_R' - u_R)}.$$

The functions $G_1 := E_1 + E_2$, $G_2 := u_P' E_1 + u_P E_2$, $G_3 := E_3 + E_4$, and $G_4 := u_R' E_3 + u_R E_4$ are functions of P and R, because they are symmetric with respect to $(u_P, v_P) \leftrightarrow (u_P', v_P')$ and $(u_R, v_R) \leftrightarrow (u_R', v_R')$. Now, the geometric expression of the group law on $\mathcal{J}_\mathcal{C}$ states that the cubic polynomial[4]

$$
\begin{aligned}
l(x) &= E_1(x - u_P')(x - u_R)(x - u_R') + E_2(x - u_P)(x - u_R)(x - u_R') \\
&\quad + E_3(x - u_P)(x - u_P')(x - u_R') + E_4(x - u_P)(x - u_P')(x - u_R) \\
&= (G_1 x - G_2)(x^2 + a_1^R x + a_0^R) + (G_3 x - G_4)(x^2 + a_1^P x + a_0^P)
\end{aligned}
$$

satisfies $\ell(x) \equiv b(x) \bmod a(x)$ when $\langle a(x), b(x) \rangle$ is any of P, R or $\ominus(R \oplus P)$. Together with $b(x)^2 \equiv f(x) \pmod{a(x)}$, which is satisfied by every $\langle a(x), b(x) \rangle$ in $\mathcal{J}_\mathcal{C}$, this gives (after some tedious symbolic manipulations, or, alternatively, by Littlewood's principle) the relations (4), (5), and (6). □

The two Kummers are related by a linear projective isomorphism $\tau : \mathcal{K}_\mathcal{C}^{\text{fast}} \xrightarrow{\sim} \mathcal{K}_\mathcal{C}^{\text{gen}}$, which maps $(X : Y : Z : T)$ to $(\xi_1 : \xi_2 : \xi_3 : \xi_4) = (X : Y : Z : T) M_\tau$ where

$$
M_\tau = \begin{pmatrix}
1 & \frac{\lambda - \mu\nu}{\lambda - \nu} & \frac{\lambda\nu(1-\mu)}{\lambda - \nu} & \frac{\lambda\nu(\lambda - \mu\nu)}{\lambda - \nu} \\
\frac{a(1-\mu)}{b(\lambda-\nu)} & \frac{a(\lambda - \mu\nu)}{b(\lambda - \nu)} & \frac{a}{b}\mu & \frac{a\mu(\lambda - \mu\nu)}{b(\lambda - \nu)} \\
\frac{a(\mu - \lambda)}{c(\lambda - \nu)} & \frac{a(\mu\nu - \lambda)}{c(\lambda - \nu)} & \frac{a\lambda\mu(\nu - 1)}{c(\lambda - \nu)} & \frac{a\lambda\mu(\mu\nu - \lambda)}{c(\lambda - \nu)} \\
\frac{a(\nu - 1)}{d(\lambda - \nu)} & \frac{a(\mu\nu - \lambda)}{d(\lambda - \nu)} & \frac{a\nu(\mu - \lambda)}{d(\lambda - \nu)} & \frac{a\nu(\mu\nu - \lambda)}{d(\lambda - \nu)}
\end{pmatrix}.
$$

The map $\rho \circ \tau : \mathcal{K}_\mathcal{C}^{\text{fast}} \to \mathbb{P}^2$ is defined by the matrix M_τ' formed by the first three columns of M_τ. The inverse isomorphism $\tau^{-1} : \mathcal{K}_\mathcal{C}^{\text{gen}} \to \mathcal{K}_\mathcal{C}^{\text{fast}}$ is defined by any scalar multiple of M_τ^{-1}, and then $\pm P = \tau^{-1}(\pi(P))$ for all P in $\mathcal{J}_\mathcal{C}$.

Proposition 2. *Let P and R be in $\mathcal{J}_\mathcal{C}(\mathbb{F}_q)$. Given $(P, \pm P, \pm R, \pm(R \oplus P))$, Algorithm 2 computes R in $107M + 11S + 4m_c + 81a + 1I$.*

Proof. We have $a_1^R = -\xi_2^R$ and $a_0^R = \xi_3^R$; it remains to compute b_1^R and b_0^R using Proposition 1, maintaining the notation of its proof. Let $E := \xi_1^R((DG_1)^2 + (DG_3)^2)$, $\Delta := D^2(2\xi_1^\oplus \xi_1^\ominus G_2^2 - (\xi_2^\oplus \xi_1^\ominus + \xi_1^\oplus \xi_2^\ominus)G_1 G_2 + (\xi_3^\oplus \xi_1^\ominus + \xi_1^\oplus \xi_3^\ominus)G_1^2)$, and $F := -2(\xi_1^R)^2 D\Delta$. Note that $C = F/(\xi_1^R E)$ and $\xi_1^R(DG_3)^2 = \xi_1^R(\xi_4^R D + f_1 Z_1 Z_2 + f_2 Z_2^2 + f_3 Z_2 Z_3 + f_4 Z_3^2) + (\xi_3^R Z_2 + \xi_2^R Z_3)Z_3$. Now, to Algorithm 2: Lines 1–4 compute $\pi(R)$, $\rho(\pi(P \oplus R))$, and $\rho(\pi(P \ominus R))$.[5] Then Lines 5–6 compute $D(G_1, G_2)$ from (b_1^P, b_0^P); Lines 7–8 compute $C(G_3, G_4)$ from $D(G_1, G_2)$; Lines 9–13 compute $F(b_1^P, b_0^P)$ from $EC(G_3, G_4)$. Finally, Lines 14–19 compute F and its inverse and renormalize, yielding R. □

---

[4] The cubic curve $y = \ell(x)$ is analogous to the line through P, R, and $\ominus(R \oplus P)$ in the classic elliptic curve group law.

[5] Okeya and Sakurai noticed that the formulæ for y-coordinate recovery on Montgomery curves are simpler if $\pm(R \ominus P)$ is also known [22, pp. 129–130]; here, we take advantage of an analogous simplification in genus 2.

Algorithm 2. Recover: Recovery from $\mathcal{K}_{\mathcal{C}}^{\text{fast}}$ to $\mathcal{J}_{\mathcal{C}}$.

Input: $(P, \pm P, \pm R, \pm(R \oplus P)) \in \mathcal{J}_{\mathcal{C}} \times (\mathcal{K}_{\mathcal{C}}^{\text{fast}})^3$ for P and (unknown) R in $\mathcal{J}_{\mathcal{C}}$.
Output: $R \in \mathcal{J}_{\mathcal{C}}$.
Cost: $107M + 11S + 4m_c + 81a + 1I$, assuming precomputed M_τ.

1 $\pm(R \ominus P) \leftarrow \texttt{xADD}^*(\pm R, \pm P, \pm(R \oplus P))$ // $14M + 8S + 4m_c + 24a$

2 $(\xi_1^R : \xi_2^R : \xi_3^R : \xi_4^R) \leftarrow \pm R \cdot M_\tau$ // $15M + 12a$

3 $(\xi_1^\oplus : \xi_2^\oplus : \xi_3^\oplus) \leftarrow \pm(R \oplus P) \cdot M_\tau'$ // $11M + 9a$

4 $(\xi_1^\ominus : \xi_2^\ominus : \xi_3^\ominus) \leftarrow \pm(R \ominus P) \cdot M_\tau'$ // $11M + 9a$

5 $(Z_1, Z_2, Z_3) \leftarrow (a_1^P \cdot \xi_1^R + \xi_2^R, \xi_3^R - a_0^P \cdot \xi_1^R, -(a_0^P \cdot \xi_2^R + a_1^P \cdot \xi_3^R))$ // $4M + 4a$

6 $(DG_1, DG_2) \leftarrow (Z_2 \cdot b_1^P + Z_1 \cdot b_0^P, (Z_1 \cdot a_0^P \cdot b_1^P - Z_1 \cdot a_1^P + Z_2) \cdot b_0^P)$ // $6M + 3a$

7 $(Y_{13}, Y_{21}, Y_{23}) \leftarrow (\xi_1^\oplus \cdot \xi_3^\ominus - \xi_3^\oplus \cdot \xi_1^\ominus, \xi_2^\oplus \cdot \xi_1^\ominus - \xi_1^\oplus \cdot \xi_2^\ominus, \xi_2^\oplus \cdot \xi_3^\ominus - \xi_3^\oplus \cdot \xi_2^\ominus)$ // $6M + 3a$

8 $(CG_3, CG_4) \leftarrow (DG_1 \cdot Y_{13} + DG_2 \cdot Y_{21}, DG_1 \cdot Y_{23} - DG_2 \cdot Y_{13})$ // $4M + 2a$

9 $\text{xiD} \leftarrow Z_2^2 - Z_1 \cdot Z_3$ // $1M + 1S + 1a$

10 $E \leftarrow \xi_1^R \cdot ((f_3 \cdot Z_3 + f_2 \cdot Z_2 + f_1 \cdot Z_1) \cdot Z_2 + DG_1^2) + \xi_4^R \cdot \text{xiD}$ // $6M + 1S + 4a$

11 $E \leftarrow E + Z_3 \cdot (Z_3 \cdot (f_4 \cdot \xi_1^R + \xi_2^R) + Z_2 \cdot \xi_3^R)$ // $4M + 3a$

12 $\text{xiFb}_1 \leftarrow E \cdot ((Z_1 \cdot \xi_2^R - Z_2 \cdot \xi_1^R) \cdot CG_3 - Z_1 \cdot \xi_3^R \cdot CG_4)$ // $6M + 2a$

13 $\text{xiFb}_0 \leftarrow E \cdot (Z_2 \cdot \xi_1^R \cdot CG_4 - Z_1 \cdot \xi_1^R \cdot CG_3)$ // $5M + 1a$

14 $\text{Delta} \leftarrow DG_1 \cdot (CG_3 + 2\xi_1^\oplus \cdot (DG_1 \cdot \xi_3^\ominus + DG_2 \cdot \xi_2^\ominus)) + 2DG_2^2 \cdot \xi_1^\oplus \cdot \xi_1^\ominus$ // $6M + 1S + 5a$

15 $F \leftarrow -2\text{xiD} \cdot \xi_1^R \cdot \text{Delta}$ // $2M + 2a$

16 $\text{invxiF} \leftarrow 1/(F \cdot \xi_1^R)$ // $1M + 1I$

17 $\text{invxi} \leftarrow F \cdot \text{invxiF}$ // $1M$

18 $(a_1^R, a_0^R, b_1^R, b_0^R) \leftarrow (-\text{invxi} \cdot \xi_2^R, \text{invxi} \cdot \xi_3^R, \text{invxiF} \cdot \text{xiFb}_1, \text{invxiF} \cdot \text{xiFb}_0)$ // $4M + 1a$

19 **return** $\langle x^2 + a_1^R x + a_0^R, b_1^R x + b_0^R \rangle$

Remark 3. Algorithm 2 assumes that P is not a special point in $\mathcal{J}_{\mathcal{C}}$, and that $\pm R$ is not the image of a special point $R \in \mathcal{J}_{\mathcal{C}}$. This assumption is reasonable for all cryptographic intents and purposes, since P is typically an input point to a scalar multiplication routine (that, if special, can be detected and rejected), and R is a secret multiple of P (that will be special with negligible probability). For completeness, we note that if either or both of P or R is special, then we can still use Algorithm 2 by translating the input points by a well-chosen 2-torsion point, and updating the output appropriately by the same translation (we recall that on the fast Kummer, all 16 of the two-torsion points are rational, which gives us plenty of choice here). A fully-fledged implementation could be made to run in constant-time (for *all* input and output points) by always performing these translations and choosing the correct inputs and outputs using bitmasks.

Remark 4. Gaudry computes the preimages in $\mathcal{J}_{\mathcal{C}}$ for points in $\mathcal{K}_{\mathcal{C}}^{\text{fast}}$ in [14, Sect. 4.3]; but this method (which is analogous to computing (x, y) and $(x, -y)$ on an elliptic curve given x and $y^2 = x^3 + ax + b$) cannot tell us which of the two preimages is the correct image for a given scalar multiplication on $\mathcal{J}_{\mathcal{C}}$.

4   Uniform One-Dimensional Scalar Multiplication

We are finally ready for scalar multiplication. Algorithm 3 lifts the Montgomery ladder [21] pseudomultiplication $(m, \pm P) \mapsto \pm[m]P$ on $\mathcal{K}_{\mathcal{C}}^{\text{fast}}$ to a full scalar multiplication $(m, P) \mapsto [m]P$ on $\mathcal{J}_{\mathcal{C}}$, generalizing the methods of [19,22], and [4]. It is visibly uniform with respect to (fixed-length) m.

Algorithm 3. One-dimensional uniform scalar multiplication on $\mathcal{J}_{\mathcal{C}}$ via `Project`, the Montgomery ladder, and `Recover`

Input: An integer $m = \sum_{i=0}^{\beta-1} m_i 2^i \geq 0$, with $m_{\beta-1} \neq 0$; a point $P \in \mathcal{J}_{\mathcal{C}}(\mathbb{F}_q) \setminus \mathcal{B}$
Output: $[m]P$
Cost: $(7\beta + 115)\mathrm{M} + (12\beta + 8)\mathrm{S} + (12\beta + 4)\mathrm{m_c} + (32\beta + 79)\mathrm{a} + 2\mathrm{I}$

1 $\pm P \leftarrow \mathtt{Project}(P)$ // 8M+1S+4m_c+14a
2 $\mathsf{x}P \leftarrow \mathtt{Wrap}(\pm P)$ // 7M+1I
3 $(t_1, t_2) \leftarrow (\pm P, \mathtt{xDBL}(\pm P))$ // 8S+8m_c+16a
4 **for** $i = \beta - 2$ *down to* 0 **do**
5      $(t_1, t_2) \leftarrow \mathtt{SWAP}(m_i, (t_1, t_2))$
6      $(t_1, t_2) \leftarrow \mathtt{xDBLADD}(t_1, t_2, \mathsf{x}P)$ // 7M+12S+12m_c+32a
7      $(t_1, t_2) \leftarrow \mathtt{SWAP}(m_i, (t_1, t_2))$
8 **end**
9 **return** $\mathtt{Recover}(P, \pm P, t_1, t_2)$ // 107M+11S+4m_c+81a+1I

Theorem 1 (`Project` + Montgomery ladder + `Recover`). *Let $m > 0$ be a β-bit integer, and P a point in $\mathcal{J}_{\mathcal{C}}(\mathbb{F}_q)$. Algorithm 3 computes $[m]P$ using one* `Project`, *one* `Wrap`, *one* `xDBL`, $\beta - 1$ `xDBLADD`s, *and one* `Recover`; *that is, in* $(7\beta + 115)\mathrm{M} + (12\beta + 8)\mathrm{S} + (12\beta + 4)\mathrm{m_c} + (32\beta + 79)\mathrm{a} + 2\mathrm{I}$.

Proof. Lines 3–7 are the Montgomery ladder; after each of the $\beta-1$ iterations we have $t_1 = \pm[\lfloor m/2^i \rfloor]P$ and $t_2 = \pm[\lfloor m/2^i \rfloor +1]P$, so $(t_1, t_2) = (\pm[m]P, \pm[m+1]P)$ at Line 8, and $\mathtt{Recover}(P, \pm P, t_1, t_2) = [m]P$. $\qquad\qquad\square$

If the base point P is fixed then we can precompute Lines 1–3 in Algorithm 3, thus saving $15\mathrm{M} + 9\mathrm{S} + 10\mathrm{m_c} + 30\mathrm{a} + 1\mathrm{I}$ in subsequent calls.

5   Uniform Two-Dimensional Scalar Multiplication

Algorithm 4 defines a uniform two-dimensional scalar multiplication for computing $[m]P \oplus [n]Q$, where P and Q (and $P \oplus Q$ and $P \ominus Q$) are in $\mathcal{J}_{\mathcal{C}} \setminus \mathcal{B}$ and $m = \sum_{i=0}^{\beta-1} m_i 2^i$ and $n = \sum_{i=0}^{\beta-1} n_i 2^i$ are β-bit scalars (with $m_{\beta-1}$ and/or $n_{\beta-1}$ not zero). The inner pseudomultiplication on $\mathcal{K}_{\mathcal{C}}^{\text{fast}}$ is based on Bernstein's binary differential addition chain [2, Sect. 4].[6] It is visibly uniform with respect

---

[6] The elliptic curve x-line version of this pseudomultiplication was used in [11].

to (fixed-length) multiscalars (m, n); while this is unnecessary for signature verification, where multiscalars are public, it is useful for GLV-style endomorphism-accelerated scalar multiplication with secret scalars.

Recall the definition of Bernstein's chain: for each pair of non-negative integers (A, B), we have two differential chains $C_0(A, B)$ and $C_1(A, B)$ with

$$C_0(0, 0) = C_1(0, 0) := ((0, 0), (1, 0), (0, 1), (1, -1)),$$

and then defined mutually recursively for $A \neq 0$ and/or $B \neq 0$ by

$$C_D(A, B) := C_d(\lfloor A/2 \rfloor, \lfloor B/2 \rfloor) \parallel (O, E, M)$$

where $\parallel$ is concatenation, $d = (D+1)(A - \lfloor A/2 \rfloor + 1) + D(B - \lfloor B/2 \rfloor) \pmod 2$, and O, E, and M (the "odd", "even", and "mixed" pairs) are

$$O := (A + (A + 1 \bmod 2), B + (B + 1 \bmod 2)), \tag{7}$$
$$E := (A + (A + 0 \bmod 2), B + (B + 0 \bmod 2)), \tag{8}$$
$$M := (A + (A + D \bmod 2), B + (B + D + 1 \bmod 2)). \tag{9}$$

By definition, (O, E, M) contains three of the four pairs (A, B), $(A + 1, B)$, $(A, B + 1)$, and $(A + 1, B + 1)$; the missing pair is $(A + (A + D + 1 \bmod 2), B + (B + D \bmod 2))$. The differences $M - O$, $M - E$, and $O - E$ depend only on D and the parities of A and B, as shown in Table 2.

Table 2. The differences between M, O, and E as functions of D and $A, B \pmod 2$.

$A \pmod 2$	$B \pmod 2$	$O - E$	$M - O$	$M - E$
0	0	$(1, 1)$	$(D - 1, -D)$	$(D, 1 - D)$
0	1	$(1, -1)$	$(D - 1, D)$	$(D, D - 1)$
1	0	$(-1, 1)$	$(1 - D, -D)$	$(-D, 1 - D)$
1	1	$(-1, -1)$	$(1 - D, D)$	$(-D, D - 1)$

Theorem 2 (Project + Bernstein's binary chain + Recover). *Let P and Q be in $\mathcal{J}_C(\mathbb{F}_q)$; let m and n be positive integers, with β the bitlength of $\max(m, n)$. Algorithm 4 computes $[m]P \oplus [n]Q$ using one JacADD, three Projects, one Wrap4, one xADD*, $\beta - 1$ xADDs, β xDBLADDs, and one Recover; that is,*
$$(14\beta + 203)\mathsf{M} + (20\beta + 16)\mathsf{S} + (16\beta + 16)\mathsf{m_c} + (56\beta + 138)\mathsf{a} + 3\mathsf{I}.$$

Proof. Consider $C_{m_0}(m, n) = C_0(0, 0) \parallel (O_{\beta-1}, E_{\beta-1}; M_{\beta-1}) \parallel \cdots \parallel (O_0, E_0, M_0)$. It follows from (7), (8), and (9) that (m, n) is one of O_0, E_0, or M_0 (and parity tells us which one). On the other hand, we have

$$C_{d_i}(\lfloor m/2^i \rfloor, \lfloor n/2^i \rfloor) = C_{d_{i+1}}(\lfloor m/2^{i+1} \rfloor, \lfloor n/2^{i+1} \rfloor) \parallel (O_i, E_i, M_i) \tag{10}$$

Algorithm 4. Two-dimensional uniform scalar multiplication on $\mathcal{J}_C$ via Project, Bernstein's two-dimensional "binary" differential addition chain, and Recover.

Input: $m = \sum_{i=0}^{\beta-1} m_i 2^i$ and $n = \sum_{i=0}^{\beta-1} n_i 2^i$ with $m_{\beta-1} n_{\beta-1} \neq 0$;
   $P, Q \in \mathcal{J}_C(\mathbb{F}_q) \setminus \mathcal{B}$ such that $P \oplus Q \notin \mathcal{B}$ and $P \ominus Q \notin \mathcal{B}$

Output: $[m]P \oplus [n]Q$

Cost: $(14\beta + 203)\mathrm{M} + (20\beta + 16)\mathrm{S} + (16\beta + 16)\mathrm{m_c} + (56\beta + 138)\mathrm{a} + 3\mathrm{I}$

```
1  S ← JacADD(P, Q)                                                    // 28M+2S+35a+1I
2  (±P, ±Q, ±S) ← (Project(P), Project(Q), Project(S))                 // 24M+3S+12m_c+42a
3  ±D ← xADD*(±P, ±Q, ±S)                                              // 14M+8S+4m_c+24a
4  (xP, xQ, xS, xD) ← Wrap4(±P, ±Q, ±S, ±D)                            // 37M+1I
5  d_0 ← m_0
6  for i ← 1 up to β − 1 do  d_i ← d_{i-1} + (d_{i-1} + 1)(m_{i-1} + m_i) + d_{i-1}(n_{i-1} + n_i)
7  U_0 ← SELECT(n_{β-1}, (xP, xQ))
8  U_1 ← SELECT(m_{β-1} n_{β-1}, (U_0, xS))
9  (U_2, U_3) ← SWAP(d_{β-1}, (xP, xQ))
10 (U_4, U_5) ← SELECT(d_{β-1}(m_{β-1} + n_{β-1}) + m_{β-1} + 1, ((xP, U_3), (xQ, xD)))
11 (U_6, U_7) ← SELECT(m_{β-1}(n_{β-1} + 1), ((xS, U_2), (U_4, xS)))
12 (E_{β-1}, U_8) ← xDBLADD(U_1, U_7, U_5)                             // 7M+12S+12m_c+32a
13 (O_{β-1}, M_{β-1}) ← SWAP(d_{β-1}(m_{β-1} + n_{β-1}) + m_{β-1} + 1, (U_6, U_8))
14 for i ← β − 2 down to 0 do
15 │  O_i ← xADD(O_{i+1}, E_{i+1}, SELECT(m_i + n_i, (xS, xD)))         // 7M+8S+4m_c+24a
16 │  V_0 ← SELECT((d_i + 1)(m_{i+1} + m_i) + d_i(n_{i+1} + n_i), (O_{i+1}, E_{i+1}))
17 │  (V_1, V_2) ← SWAP(m_i + m_{i+1} + n_i + n_{i+1}, (V_0, M_{i+1})))
18 │  (E_i, M_i) ← xDBLADD(V_1, V_2, SELECT(d_i, (xP, xQ)))            // 7M+12S+12m_c+32a
19 end
20 (W_0, W_1) ← SWAP(m_0, (O_0, E_0))
21 (W_2, W_3, W_4, W_5) ← SELECT(m_0 + n_0, ((S, xS, W_0, W_1), (P, xP, M_0, W_0)))
22 return Recover(W_2, W_3, W_4, W_5)                                   // 107M+11S+4m_c+81a+1I
```

for $0 \leq i \leq \beta - 2$, where the bits d_i are defined by $d_0 = m_0$ and $d_i := d_{i-1} + (d_{i-1} + 1)(m_{i-1} + m_i) + d_{i-1}(n_{i-1} + n_i)$ for $i > 0$. The definition of the chains, Table 2, and considerations of parity yield the following relations which allow us to construct each triple (O_i, E_i, M_i) from its antecedent $(O_{i+1}, E_{i+1}, M_{i+1})$:

1. $O_i = O_{i+1} + E_{i+1}$, with $O_{i+1} - E_{i+1} = \pm(1, 1)$ if $m_i = n_i$ and $\pm(1, -1)$ if $m_i \neq n_i$.
2. $E_i = 2E_{i+1}$ if $(m_i, n_i) = (m_{i+1}, n_{i+1})$; or $2O_{i+1}$ if $m_{i+1} \neq m_i$ and $n_{i+1} \neq n_i$; or $2M_{i+1}$ otherwise.
3. If $d_i = 0$ then $M_i = M_{i+1} + X$, where $X = E_{i+1}$ if $m_{i+1} = m_i$, or O_{i+1} if $m_{i+1} \neq m_i$; and $M_{i+1} - X = \pm(0, 1)$.
4. If $d_i = 1$ then $M_i = M_{i+1} + X$, where $X = E_{i+1}$ if $n_{i+1} \neq n_i$, or O_{i+1} if $n_{i+1} = n_i$; and $M_{i+1} - X = \pm(1, 0)$.

We can therefore compute $\pm R = \pm([m]P \oplus [n]Q)$ by mapping each pair (a, b) in $C_{m_0}(m, n)$ to $\pm([a]P \oplus [b]Q)$. Lines 1–4 (pre)compute the required difference points $\pm P$, $\pm Q$, $\pm S = \pm(P \oplus Q)$, and $\pm D = \pm(P \ominus Q)$. Lines 5–6 compute all of the d_i. After initializing the first nontrivial segment $(O_{\beta-1}, E_{\beta-1}, M_{\beta-1})$ in

Lines 7–13, the main loop (Lines 14–18) derives the following segments using the rules above. Table 3 gives the state of the final segment (O_0, E_0, M_0) immediately after the loop. In each case, we can recover $[m]P \oplus [n]Q$ using the call to Recover specified by the corresponding row, as is done in Lines 19–21. □

Table 3. The state of Algorithm 4 after the main loop.

(m_0, n_0)	O_0	E_0	M_0 if $d_0 = 0$	M_0 if $d_0 = 1$	$R = [m]P \oplus [n]Q$
$(0,0)$	$\pm(R \oplus S)$	$\pm R$	$\pm(R \oplus Q)$	$\pm(R \oplus P)$	Recover$(S, \pm S, E_0, O_0)$
$(0,1)$	$\pm(R \oplus P)$	$\pm(R \oplus Q)$	$\pm R$	$\pm(R \oplus S)$	Recover$(P, \pm P, M_0, O_0)$
$(1,0)$	$\pm(R \oplus Q)$	$\pm(R \oplus P)$	$\pm(R \oplus S)$	$\pm R$	Recover$(P, \pm P, M_0, E_0)$
$(1,1)$	$\pm R$	$\pm(R \oplus S)$	$\pm(R \oplus P)$	$\pm(R \oplus S)$	Recover$(S, \pm S, O_0, E_0)$

If the points P and Q are fixed then we can precompute Lines 1–4 in Algorithm 4, thus saving $103\text{M} + 13\text{S} + 16\text{m}_c + 101\text{a} + 2\text{I}$ in subsequent calls.

Remark 5. There are faster two-dimensional differential addition chains that are non-uniform, such as Montgomery's PRAC Algorithm [26, Chap. 3], which might be preferred in scenarios where the multiscalars are not secret (such as signature verification). However, PRAC is not well-suited to our recovery technique, because its outputs do not "differ" by an element with known preimage in $\mathcal{J}_C$.

A   Fast Kummer Arithmetic

We recall the formulæ for operations on fast Kummers from [14, Sect. 3.2]. To simplify the presentation of our algorithms, we define three operations on points in $\mathbb{P}^3$ (or more precisely, on 4-tuples of elements of $\mathbb{F}_q$). First, $\mathcal{M} : \mathbb{P}^3 \times \mathbb{P}^3 \to \mathbb{P}^3$ multiplies the corresponding coordinates of a pair of points:

$$\mathcal{M} : ((x_1 : y_1 : z_1 : t_1), (x_2 : y_2 : z_2 : t_2)) \longmapsto (x_1x_2 : y_1y_2 : z_1z_2 : t_1t_2),$$

costing 4M. The special case $(x_1 : y_1 : z_1 : t_1) = (x_2 : y_2 : z_2 : t_2)$ is denoted by

$$\mathcal{S} : (x : y : z : t) \longmapsto (x^2 : y^2 : z^2 : t^2),$$

costing 4S. Finally, the Hadamard transform[7] is defined by

$$\mathcal{H} : (x : y : z : t) \longmapsto (x' : y' : z' : t') \quad \text{where} \quad \begin{cases} x' = x + y + z + t, \\ y' = x + y - z - t, \\ z' = x - y + z - t, \\ t' = x - y - z + t. \end{cases}$$

The Hadamard transform can easily be implemented with 8a.

---

[7] Note $(A : B : C : D) = \mathcal{H}((a : b : c : d))$; dually, $(a : b : c : d) = \mathcal{H}((A : B : C : D))$.

Algorithm 5. xADD*: Differential addition on $\mathcal{K}_{\mathcal{C}}^{\text{fast}}$.

Input: $(\pm P, \pm Q, (x_\ominus : y_\ominus : z_\ominus : t_\ominus) = \pm(P \ominus Q))$ for some P, Q in $\mathcal{J}_\mathcal{C}(\mathbb{F}_q)$ with $P \ominus Q \notin \mathcal{B}$.
Output: $\pm(P \oplus Q) \in \mathcal{K}_{\mathcal{C}}^{\text{fast}}$.
Cost: $14M + 8S + 4m_c + 24a$

1 $(V_1, V_2) \leftarrow (\mathcal{H}(\mathcal{S}(\pm P)), \mathcal{H}(\mathcal{S}(\pm Q)))$	// 8S+16a
2 $V_3 \leftarrow \mathcal{M}(V_1, V_2)$	// 4M
3 $V_4 \leftarrow \mathcal{H}(\mathcal{M}(V_3, (1/A : 1/B, 1/C, 1/D))$	// 4m_c+8a
4 $(C_1, C_2) \leftarrow (x_\ominus \cdot y_\ominus, z_\ominus \cdot t_\ominus)$	// 2M
5 **return** $\mathcal{M}(V_4, (y_\ominus \cdot C_2, x_\ominus \cdot C_2, t_\ominus \cdot C_1, z_\ominus \cdot C_1))$	// 8M

Algorithm 6. xDBL: Pseudo-doubling on $\mathcal{K}_{\mathcal{C}}^{\text{fast}}$.

Input: $\pm P$ in $\mathcal{K}_{\mathcal{C}}^{\text{fast}}$ for P in $\mathcal{J}_\mathcal{C}(\mathbb{F}_q)$.
Output: $\pm[2]P$
Cost: $8S + 8m_c + 16a$

1 $V_1 \leftarrow \mathcal{H}(\mathcal{S}(\pm P))$	// 4S+8a
2 $V_2 \leftarrow \mathcal{S}(V_1)$	// 4S
3 $V_3 \leftarrow \mathcal{H}(\mathcal{M}(V_2, (1/A : 1/B : 1/C : 1/D)))$	// 4m_c+8a
4 **return** $\mathcal{M}(V_4, (1/a : 1/b : 1/c : 1/d))$	// 4m_c

---

Algorithm 7. Wrap: $(x : y : z : t) \mapsto (x/y, x/z, x/t)$.

Input: $(x_P : y_P : z_P : t_P) = \pm P$ for P in $\mathcal{J}_\mathcal{C}(\mathbb{F}_q) \setminus \mathcal{B}$
Output: $(x/y, x/z, x/t) \in \mathbb{F}_q^3$.
Cost: $7M + 1I$

1 $V_1 \leftarrow y \cdot z$	// 1M
2 $V_2 \leftarrow x/(V_1 \cdot t)$	// 2M+1I
3 $V_3 \leftarrow V_2 \cdot t$	// 1M
4 **return** $(V_3 \cdot z, V_3 \cdot y, V_1 \cdot V_2)$	// 3M

The basic (unoptimized) pseudo-addition operation is xADD* (Algorithm 5). The pseudo-doubling operation is xDBL (Algorithm 6).

Lines 4 and 5 of Algorithm 5 compute the point $(y_\ominus z_\ominus t_\ominus : x_\ominus z_\ominus t_\ominus : x_\ominus y_\ominus t_\ominus : x_\ominus y_\ominus z_\ominus)$, which is projectively equivalent to $(1/x_\ominus : 1/y_\ominus : 1/z_\ominus : 1/t_\ominus)$, but requires no inversions (note that this is generally *not* a point on $\mathcal{K}_\mathcal{C}$). This is the only point in our pseudoarithmetic where the third argument $(x_\ominus : y_\ominus : z_\ominus : t_\ominus)$ appears. In practice, the pseudoadditions used in our scalar multiplication all use a fixed third argument, so it makes sense to precompute this "inverted" point and to scale it by $x_\ominus$ so that the first coordinate is 1, thus saving 7M in each subsequent pseudo-addition for a one-off cost of 1I. The resulting data can be stored as the 3-tuple $(x_\ominus/y_\ominus, x_\ominus/z_\ominus, x_\ominus/t_\ominus)$, ignoring the trivial first coordinate: this is the *wrapped* form of $\pm(P \ominus Q)$. The function Wrap (Algorithm 7) applies this transformation; we also include Wrap4 (Algorithm 8), which simultaneously Wraps four points using a single shared inversion.

Algorithm 8. Wrap4: four simultaneous Kummer point wrappings

Input: $(\pm P, \pm Q, \pm S, \pm D)$ for P, Q, S, D in $\mathcal{J}_C(\mathbb{F}_q) \setminus \mathcal{B}$
Output: $\text{Wrap}(\pm P), \text{Wrap}(\pm Q), \text{Wrap}(\pm S), \text{Wrap}(\pm D)$
Cost: $37M + 1I$

1 $(c_1, c_2, c_3, c_4) \leftarrow (y^P \cdot z^P, y^Q \cdot z^Q, y^S \cdot z^S, y^D \cdot z^D)$	// 4M
2 $(f_1, f_2, f_3, f_4) \leftarrow (c_1 \cdot t^P, c_2 \cdot t^Q, c_3 \cdot t^S, c_4 \cdot t^D)$	// 4M
3 $(g_1, g_2) \leftarrow (f_1 \cdot f_2, f_3 \cdot f_4)$	// 2M
4 $I \leftarrow 1/(g_1 \cdot g_2)$	// 1M+1I
5 $(h_1, h_2) \leftarrow (g_1 \cdot I, g_2 \cdot I)$	// 2M
6 $(e_1, e_2, e_3, e_4) \leftarrow (x^P \cdot f_2 \cdot h_2, x^Q \cdot f_1 \cdot h_2, x^S \cdot f_4 \cdot h_1, x^D \cdot f_3 \cdot h_1)$	// 8M
7 $(r_1, r_2, r_3, r_4) \leftarrow (e_1 \cdot t^P, e_2 \cdot t^Q, e_3 \cdot t^S, e_4 \cdot t^D)$	// 4M
8 **return** $(r_1 \cdot z^P, r_1 \cdot y^P, c_1 \cdot e_1), (r_2 \cdot z^Q, r_2 \cdot y^Q, c_2 \cdot e_2), (r_3 \cdot z^S, r_3 \cdot y^S, c_3 \cdot e_3),$	
$\quad (r_4 \cdot z^D, r_4 \cdot y^D, c_4 \cdot e_4)$	// 12M

Algorithm 9. xADD: Differential addition on $\mathcal{K}_C^{\text{fast}}$ with wrapped difference.

Input: $(\pm P, \pm Q, (x_\ominus/y_\ominus, x_\ominus/z_\ominus, x_\ominus/t_\ominus) = \text{Wrap}(\pm(P \ominus Q)))$ for P, Q in
$\quad \mathcal{J}_C(\mathbb{F}_q)$ with $P \ominus Q \notin \mathcal{B}$
Output: $\pm(P \oplus Q) \in \mathcal{K}_C^{\text{fast}}$.
Cost: $7M + 8S + 4m_c + 24a$

1 $(V_1, V_2) \leftarrow (\mathcal{H}(\mathcal{S}(\pm P)), \mathcal{H}(\mathcal{S}(\pm Q)))$	// 8S+16a
2 $V_3 \leftarrow \mathcal{M}(V_1, V_2)$	// 4M
3 $V_4 \leftarrow \mathcal{H}(\mathcal{M}(V_3, (1/A : 1/B, 1/C, 1/D))$	// 4m_c+8a
4 **return** $\mathcal{M}(V_4, (1 : x_\ominus/y_\ominus, x_\ominus/z_\ominus, x_\ominus/t_\ominus))$	// 3M

Algorithm 10. xDBLADD: Combined differential double-and-add on $\mathcal{K}_C^{\text{fast}}$.

Input: $(\pm P, \pm Q, (x_\ominus/y_\ominus, x_\ominus/z_\ominus, x_\ominus/t_\ominus) = \text{Wrap}(\pm(P \ominus Q)))$ for P, Q in
$\quad \mathcal{J}_C(\mathbb{F}_q)$ with $P \ominus Q \notin \mathcal{B}$.
Output: $(\pm[2]P, \pm(P \oplus Q))$
Cost: $7M + 12S + 12m_c + 32a$

1 $(V_1, V_2) \leftarrow (\mathcal{H}(\mathcal{S}(\pm P)), \mathcal{H}(\mathcal{S}(\pm Q)))$	// 8S + 16a
2 $(V_1, V_2) \leftarrow (\mathcal{S}(V_1), \mathcal{M}(V_1, V_2))$	// 4M + 4S
3 $(V_1, V_2) \leftarrow (\mathcal{M}(V_1, (\frac{1}{A} : \frac{1}{B} : \frac{1}{C} : \frac{1}{D})), \mathcal{M}(V_2, (\frac{1}{A} : \frac{1}{B} : \frac{1}{C} : \frac{1}{D})))$	// 8m_c
4 $(V_1, V_2) \leftarrow (\mathcal{H}(V_1), \mathcal{H}(V_2))$	// 16a
5 **return** $(\mathcal{M}(V_1, (\frac{1}{a} : \frac{1}{b} : \frac{1}{c} : \frac{1}{d})), \mathcal{M}(V_2, (1 : \frac{x_\ominus}{y_\ominus} : \frac{x_\ominus}{y_\ominus} : \frac{x_\ominus}{t_\ominus})))$	// 3M + 4m_c

---

We can now define xADD (Algorithm 9), an optimized pseudo-addition using a Wrapped third argument, and xDBLADD (Algorithm 10), which is an optimized combined pseudo-doubling-and-addition.

B   Constant-Time Conditional Swaps and Selects

Our algorithms are designed to be a basis for uniform and constant-time implementations. As such, to avoid branching, we require constant-time conditional swap and selection routines. These are standard techniques, and can be implemented in many ways; Algorithms 11 and 12 give example pseudocode as an illustration of these techniques.

Algorithm 11. SWAP: Constant-time conditional swap.

Input: $b \in \{0,1\}$ and a pair (X_0, X_1) of objects encoded as n-bit strings
Output: (X_b, X_{1-b})
1 $b \leftarrow (b, \ldots, b)_n$
2 $V \leftarrow b$ and $(X_0 \text{ xor } X_1)$ // bitwise and, xor; do not short-circuit and
3 return $(X_0 \text{ xor } V, X_1 \text{ xor } V)$

Algorithm 12. SELECT: Constant-time conditional selection.

Input: $b \in \{0,1\}$ and a pair (X_0, X_1) of objects encoded as n-bit strings
Output: X_b
1 $b \leftarrow (b, \ldots, b)_n$
2 $V \leftarrow b$ and $(X_0 \text{ xor } X_1)$ // bitwise and, xor; do not short-circuit and
3 return $X_0 \text{ xor } V$

References

1. Bernstein, D.J.: Curve25519: new Diffie-Hellman speed records. In: Yung, M., Dodis, Y., Kiayias, A., Malkin, T. (eds.) PKC 2006. LNCS, vol. 3958, pp. 207–228. Springer, Heidelberg (2006). doi:10.1007/11745853_14
2. Bernstein, D.J.: Differential addition chains, preprint (2006)
3. Bernstein, D.J., Chuengsatiansup, C., Lange, T., Schwabe, P.: Kummer strikes back: new DH speed records. In: Sarkar and Iwata [24], pp. 317–337
4. Brier, É., Joye, M.: Weierstraß elliptic curves and side-channel attacks. In: Naccache, D., Paillier, P. (eds.) PKC 2002. LNCS, vol. 2274, pp. 335–345. Springer, Heidelberg (2002). doi:10.1007/3-540-45664-3_24
5. Brown, D.R.L.: Multi-dimensional montgomery ladders for elliptic curves (2006). http://eprint.iacr.org/2006/220
6. Cantor, D.G.: Computing in the Jacobian of a hyperelliptic curve. Math. Comput. **48**(177), 95–101 (1987)
7. Cassels, J.W.S., Flynn, E.V.: Prolegomena to a Middlebrow Arithmetic of Curves of Genus 2, vol. 230. Cambridge University Press, Cambridge (1996)
8. Chudnovsky, D.V., Chudnovsky, G.V.: Sequences of numbers generated by addition in formal groups and new primality and factorization tests. Adv. in Appl. Math. **7**, 385–434 (1986)
9. Cosset, R.: Applications of theta functions for hyperelliptic curve cryptography. Ph.D thesis. Université Henri Poincaré - Nancy I, November 2011
10. Cosset, R., Arene, C.: Construction of a k-complete addition law on Jacobians of hyperelliptic curves of genus two. Contemp. Math. **574**, 1–14 (2012)

11. Costello, C., Hisil, H., Smith, B.: Faster compact Diffie–Hellman: endomorphisms on the x-line. In: Nguyen, P.Q., Oswald, E. (eds.) EUROCRYPT 2014. LNCS, vol. 8441, pp. 183–200. Springer, Heidelberg (2014). doi:10.1007/978-3-642-55220-5_11

12. Duquesne, S.: Montgomery scalar multiplication for genus 2 curves. In: Buell, D. (ed.) ANTS 2004. LNCS, vol. 3076, pp. 153–168. Springer, Heidelberg (2004). doi:10.1007/978-3-540-24847-7_11

13. Gallant, R.P., Lambert, R.J., Vanstone, S.A.: Faster point multiplication on elliptic curves with efficient endomorphisms. In: Kilian, J. (ed.) CRYPTO 2001. LNCS, vol. 2139, pp. 190–200. Springer, Heidelberg (2001). doi:10.1007/3-540-44647-8_11

14. Gaudry, P.: Fast Genus 2 arithmetic based on theta functions. J. Math. Cryptol. **1**(3), 243–265 (2007)

15. Gaudry, P., Schost, E.: Genus 2 point counting over prime fields. J. Symb. Comput. **47**(4), 368–400 (2012)

16. Hisil, H., Costello, C.: Jacobian coordinates on genus 2 curves. In: Sarkar and Iwata [24], pp. 338–357

17. Izu, T., Takagi, T.: Exceptional procedure attack on elliptic curve cryptosystems. In: Desmedt, Y.G. (ed.) PKC 2003. LNCS, vol. 2567, pp. 224–239. Springer, Heidelberg (2003). doi:10.1007/3-540-36288-6_17

18. Kohel, D.: Arithmetic of split kummer surfaces: montgomery endomorphism of edwards products. In: Chee, Y.M., Guo, Z., Ling, S., Shao, F., Tang, Y., Wang, H., Xing, C. (eds.) IWCC 2011. LNCS, vol. 6639, pp. 238–245. Springer, Heidelberg (2011). doi:10.1007/978-3-642-20901-7_15

19. López, J., Dahab, R.: Fast multiplication on elliptic curves over $GF(2^m)$ without precomputation. In: Koç, Ç.K., Paar, C. (eds.) CHES 1999. LNCS, vol. 1717, pp. 316–327. Springer, Heidelberg (1999). doi:10.1007/3-540-48059-5_27

20. Lubicz, D., Robert, D.: Arithmetic on Abelian and Kummer varieties. Finite Fields Appl. **39**, 130–158 (2016)

21. Montgomery, P.L.: Speeding the pollard and elliptic curve methods of factorization. Math. Comput. **48**(177), 243–264 (1987)

22. Okeya, K., Sakurai, K.: Efficient elliptic curve cryptosystems from a scalar multiplication algorithm with recovery of the y-coordinate on a montgomery-form elliptic curve. In: Koç, Ç.K., Naccache, D., Paar, C. (eds.) CHES 2001. LNCS, vol. 2162, pp. 126–141. Springer, Heidelberg (2001). doi:10.1007/3-540-44709-1_12

23. Renes, J., Schwabe, P., Smith, B., Batina, L.: μkummer: efficient hyperelliptic signatures and key exchange on microcontrollers. IACR Cryptology ePrint Archive, Report 2016/366 (2016). http://eprint.iacr.org/2016/366

24. Sarkar, P., Iwata, T. (eds.): ASIACRYPT 2014. LNCS, vol. 8873. Springer, Heidelberg (2014). doi:10.1007/978-3-662-45611-8

25. Smart, N.P., Siksek, S.: A fast Diffie-Hellman protocol in Genus 2. J. Cryptol. **12**(1), 67–73 (1999)

26. Stam, M.: Speeding up subgroup cryptosystems. Technische Universiteit Eindhoven (2003)

PhiRSA: Exploiting the Computing Power of Vector Instructions on Intel Xeon Phi for RSA

Yuan Zhao[1,2,3], Wuqiong Pan[1,2(✉)], Jingqiang Lin[1,2], Peng Liu[4],
Cong Xue[1,2,3], and Fangyu Zheng[1,2]

[1] State Key Laboratory of Information Security,
Institute of Information Engineering, Chinese Academy of Sciences, Beijing, China
{yzhao,wqpan,linjq,cxue13,fyzheng}@is.ac.cn
[2] Data Assurance and Communication Security Research Center,
Chinese Academy of Sciences, Beijing, China
[3] University of Chinese Academy of Sciences, Beijing, China
[4] College of Information Sciences and Technology,
The Pennsylvania State University, State College, USA
pliu@ist.psu.edu

Abstract. Efficient implementations of public-key cryptographic algorithms on general-purpose computing devices, facilitate the applications of cryptography in communication security. Existing solutions work in two different directions: implementations on GPUs achieve high throughput but great latency, while those on CPUs are with low throughput and small latency. Intel Xeon Phi is the first highly parallel coprocessor of Many Integrated Core (MIC) architecture, with up to 61 cores and one 512-bit Vector Processing Unit (VPU) in each core, which offers the potential to achieve both high throughput and small latency. In this paper, we propose a vector-oriented Montgomery multiplication design based on *vector carry propagation chain* (VCPC) method to fully exploit the computing power of vector instructions on Intel Xeon Phi. Two key features of our design sharply reduce the number of instructions: (1) organizing the additions in Montgomery multiplication to be four VCPCs for saving the overhead of handling carry bits; (2) computing the intermediate scalar variable q in every round without breaking the flow of VCPCs. Furthermore, we offer the optimal Montgomery multiplication implementation of our design on Intel Xeon Phi, which make VPUs fully pipelined and maintain carry bits in vector mask registers. Based on the above, we implement RSA named *PhiRSA* and evaluate it on Intel Xeon Phi 7120P. For 1024, 2048 and 4096-bit RSA, PhiRSA performs 258,370, 41,803 and 5,358 decryptions per second, and the latencies are 0.94, 5.84 and 45.54 ms, respectively. These results achieve 4.1 to 8.5 times performance of the existing RSA implementations on Intel Xeon Phi, exhibit high throughput comparable to those on GPUs but with much less parallel tasks, and small latency comparable to those on CPUs.

Y. Zhao—This work was partially supported by National 973 Program under award No. 2014CB340603 and No. 2013CB338001, Strategy Pilot Project of Chinese Academy of Sciences under award No. XDA06010702.

R. Avanzi and H. Heys (Eds.): SAC 2016, LNCS 10532, pp. 482–500, 2017.
https://doi.org/10.1007/978-3-319-69453-5_26

Keywords: Intel Xeon Phi · Vectorization · Montgomery multiplication · RSA · Performance

1   Introduction

The computing power of general-purpose processors is enhanced by different parallelism designs. Firstly, single-instruction-multiple-data (SIMD) enables the elements of a vector to be processed in parallel. General-purpose CPUs are usually equipped with vector instruction extensions, such as Intel MMX/SSE/AVX, ARM NEON and AMD 3DNow. Graphics processing units (GPUs) follow a different parallelism structure, single-instruction-multiple-thread (SIMT), where thousands of independent threads execute the same instructions concurrently. Finally, simultaneous-multithreading (SMT) is adopted by both CPUs and GPUs, to enable instructions from multiple threads (in a GPU thread block or a CPU core) to be executed in any given pipeline stage at a time.

The GPUs' potential on public-key cryptographic computing has been investigated for several years. Thread-level parallelism and thousands of scalar stream processors in GPUs, produce very high throughput on a great number of simultaneous tasks, but greater latency than the scalar-instruction cryptographic implementations in CPUs [22]. Note that the frequency of GPUs is much lower than that of general-purpose CPUs, for example, Intel Core i7 CPU reaches up to 3.5 GHz while NVIDIA Tesla K20 is only 706 MHz [30]. The deficiency on latency limits the applications of GPUs as public-key cryptographic engines in many scenarios.

In 2012 November, Intel announced the first product family of Many Integrated Core (MIC) architectures, named Intel Xeon Phi. Xeon Phi provides an opportunity to implement public-key algorithms in a high-throughput and low-latency way. For example, Xeon Phi 7120P consists of 61 cores, and each core is shipped with (a) 512-bit SIMD unit, 16-way 32-bit vector instructions, and (b) 4-way SMT unit, 4 hyperthreads on one core for instruction pipelining. Intel Xeon Phi, with the computing power in tera floating-point operations per second (FLOPS), has been applied in the fields of supercomputing, such as molecular dynamics in [25], sparse matrix multiplication in [27] and large integer arithmetic in [8,16]. In fact, similar 512-bit SIMD units are supported in Intel Xeon Skylake and Skylake-E CPUs and will be in Intel Cannonlake CPUs.

This paper presents the first implementation of public-key cryptographic algorithms with 512-bit SIMD instructions on Xeon Phi, called *PhiRSA*. In particular, we evaluate 1024-bit, 2048-bit and 4096-bit RSA on vector instructions. PhiRSA fully exploits the computing power of Xeon Phi 7120P with the following designs. Firstly, to perform 512-bit Montgomery multiplication (see Algorithm 1 for details), the most expensive step of RSA, the intermediate products are organized into four 512-bit vectors; then, these vectors are added using the vector-add-with-carry instruction **vpadcd** in each round of the Montgomery multiplication's main loop. After n rounds, the corresponding 512-bit vector in each round composes a *vector carry propagation chain* (VCPC). This design exploits the vector mask registers and does not need to handle the carry bits after each addition in a round. Secondly, we exploit vector instructions to compute q (see Algorithm 3 for details), without breaking the flow of VCPCs. When a vector is used to compute q, the carry bit takes effect as the write-mask which is read-only in the operation; therefore, the correct q is obtained in the each round of VCPCs but does not break the chains.

The features of SIMD are fully exploited in PhiRSA, as our design magnifies the advantages of vector instruction extensions of Xeon Phi. Our method outperforms greatly the commonly-used redundant representation method in [3,5,10–12,21]. To avoid handling the carry bits after large-integer addition during Montgomery multiplication, redundant representation stores only 29-bit operands in each 64-bit element of vectors; then, every product of two elements multiplication is 58-bit and the additional 6 bits are used to hold addition carries without overflow. So, it requires extra instructions and vectors to finish the computations.

We implement 1024/2048-bit Montgomery multiplication (and then 2048/4096-bit RSA) based on 512-bit vectors. Two (or four) 512-bit vectors compose a 1024-bit (or 2048-bit) large integer, and the specific vector instruction *valignd* is used to right shift multiple 512-bit vectors of the large integer during the main loop of Montgomery multiplication. The operations of right shift and assignment are performed in only one vector instruction, for each 512-bit vector.

Meanwhile, the benefit of SMT is also kept in PhiRSA. The execution order of vector instructions is manually optimized to fully activate the pipeline of vector processing units (VPUs). When 4 threads are launched to perform RSA computations, the VPU utilization exceeds 90%, that is, almost one instruction is executed in each cycle.

Our contributions are as follows. Firstly, the vector-oriented designs are proposed to fully exploit the computing power of vector instructions for RSA. Secondly, we implement these designs on Intel Xeon Phi 7120P efficiently. To the best of our knowledge, this is the first implementation of public-key cryptography on Intel Xeon Phi. The experimental results exhibit both high throughput and low latency: for 1024-bit, 2048-bit and 4096-bit RSA, PhiRSA achieves the throughput of 258370, 41803 and 5358 decryptions per second with 244 parallel tasks, and the latency of 0.94 ms, 5.84 ms and 45.54 ms, respectively. This throughput is about 40 times of OpenSSL [23] on a single core of Intel Haswell i7-4770R, and the latency is about 5 times. Our throughput is higher than the best implementation [32] on GPUs [32], and the latency is reduced to about 25% only.

The rest of the paper is organized as follows. Section 2 is the related work. The preliminaries about Intel Xeon Phi and Montgomery multiplication are presented in Sect. 3. Section 4 describes the design of our Montgomery multiplication. In Sect. 5, we show how to implement Montgomery multiplication and RSA on Intel

Xeon Phi. In Sect. 6, performance results of our Montgomery multiplication and RSA implementations are given and compared with other works. We conclude in Sect. 7.

2   Related Work

There have been amount of studies using vector instructions to implement large integer multiplication, Montgomery multiplication and public-key cryptography. These works can be classified into three groups. The first group and also the main choice is storing the large integers in vectors horizontally for fine-grained parallel. Intel SSE2 instruction set has been exploited for large integer multiplication in [21] and cryptographic pairing computation in [11]. Redundant representation method proposed in [21] is widely used in vector implementations to help delay the carry propagation. Intel AVX2 instruction set is also applied to modular exponentiation in [12] and Curve25519 implementation in [10]. ARM NEON instruction set is explored to implement Montgomery multiplication in [28], Curve41417 in [3] and RSA in [29]. On Cell platform, an approach to implement Montgomery multiplication is described in [5]. The second group is splitting the Montgomery multiplication into two parts to compute in parallel. This approach is studied in [6] for 2-way vector instruction sets like Intel SSE2 and ARM NEON. The third group is using the vector instructions to carry out multiple tasks in parallel. Computing multiple Montgomery multiplications simultaneously in vector elements is investigated on Intel SSE2 instruction set in [24] and the Cell processor in [4].

Many previous studies have proved that GPUs are suitable for asymmetric cryptography. Most of them are based on the integer computing power of GPU, such as [1,31]. The floating-pointing power is also explored in [2,32]. For 2048-bit RSA GPU implementation, the highest throughput is reported in [32] and the lowest latency is obtained by [31].

Intel Xeon Phi is launched as a brand-new high performance computing platform, which performance has been evaluated in [9]. Large integer multiplication is firstly evaluated on Intel Xeon Phi in [16]. This work implements multiplication by using the usual redundant representation method described in [12]. While the study in [7] firstly implements multiplication and RSA based on the idea of carry propagation and endeavors to minimize memory footprints for reducing memory accesses. However, the results of these two studies are barely satisfactory and the computing power of Intel Xeon Phi has not been fully exploited.

3   Preliminaries

3.1   Overview of Intel Xeon Phi

Intel Xeon Phi comprises of up to 61 cores and every core possesses arithmetic logic units (ALUs) and one 512-bit VPU which provides the major computing power. The cores are in-order and pipelined. Each core supports four hyperthreads to keep the execution units busy and hide memory access latencies.

If the instructions are fully pipelined, the throughput of VPUs gets up to one vector instruction per cycle. There are L1 cache and L2 cache in each core and GDDR5 memory on board. The coprocessor communicates with the host through Peripheral Component Interconnect Express (PCIe) interface. The coprocessor OS based on an open-source Linux kernel runs on the coprocessor to manage resources and process applications. There are two predominant programming models for Intel Xeon Phi, offload execution mode and native execution mode. In native execution mode, the application is cross-compiled and runs directly on the coprocessor OS.

Intel Xeon Phi Instruction Set Architecture [13] introduces 512-bit vector instructions operating on thirty-two 512-bit vector registers (zmm0-zmm31), and offers eight 16-bit vector mask registers (k0-k7) for conditional operations on data elements within vector registers. One vector register consists on either sixteen 32-bit elements or eight 64-bit elements while the vector instructions executive operations on each element. The vector mask registers have many applications, the major is playing as write-mask to protect elements in the destination from updates during the execution of any operations. If a write mask bit is zero, the corresponding destination element is not modified. Vector mask registers can also be used for keeping carry bits, borrow bits and comparison results. Intel Xeon Phi does not support MMX, SSE and AVX instruction set, but introduces amount of novel vector instructions. For example, the vector-add-with-carry instruction *vpadcd* is extremely useful in large integer arithmetic, presented as follows.

$$vpadcd \quad (zmm2/memory), \quad k2, \quad zmm1\{k1\}$$

This instruction performs an element-by-element three-input addition between int32 vector zmm1, a int32 vector in memory or int32 vector zmm2, and the carry bits in k2. The result is written into zmm1, and the carry bits produced by the addition are written into k2. The instruction performing is controlled by the write-mask k1. Some other vector instructions are used in this paper. The instruction *vpmulhud* and *vpmulld* perform element-by-element multiplications between int32 vectors and store the high 32-bit result or the low 32-bit result respectively. The instruction *vpermd* performs an element permutation by using int32 vector elements as source indices. The instruction *valignd* concatenates and shifts right several 32-bit elements from two vectors.

3.2 Montgomery Multiplication

The major computations of RSA are modular multiplication. The modular reduction would be very costly if performing division operations. Montgomery multiplication [20] is proposed to replace division operations by cheaper multiplication and shifting operations. Let M be an odd modulus, $R = 2^n$ and $M < R$, Montgomery multiplication is defined as $MontMul(A, B) = A \cdot B \cdot R^{-1} \pmod{M}$. The process of calculating $A \cdot B \pmod{M}$ based on Montgomery multiplication can be computed as follows: $\widetilde{A} = MontMul(A, R^2)$, $\widetilde{B} = MontMul(B, R^2)$, then $\widetilde{C} = MontMul(\widetilde{A}, \widetilde{B})$, finally $C = MontMul(\widetilde{C}, 1)$, C is the result. If executing a sequence of modular multiplications, such as the modular exponentiation, one modular multiplication only needs to perform one Montgomery multiplication. Koç et al. proposed an interleaved Montgomery multiplication,

---

Algorithm 1. Montgomery multiplication CIOS method [19]

Input: Modulus M, $R = 2^{nw}$, $R > M$, $\gcd(M, R) = 1$, 2^w is radix, n is digits number
$\quad 0 \leqslant A, B < M$, $B = \sum_{i=0}^{n-1} b_i 2^{iw}$, $\mu = -M^{-1} \bmod 2^w$
Output: $S = A \cdot B \cdot R^{-1} \pmod{M}$, $0 \leqslant S < M$.
1: $S \leftarrow 0$
2: **for** i from 0 to $n-1$ **do**
3: $\quad S \leftarrow S + A \cdot b_i$
4: $\quad q \leftarrow S[0] \cdot \mu \bmod 2^w$
5: $\quad S \leftarrow S + M \cdot q$
6: $\quad S \leftarrow S/2^w$
7: **end for**
8: **if** $S \geqslant M$ **then**
9: $\quad S \leftarrow S - M$
10: **end if**
11: **return** S

---

named *Coarsely Integrated Operand Scanning* (CIOS) method [19] described in Algorithm 1. This method interleaves multiplication and Montgomery reduction, which is suitable to be implemented by vector instructions.

4   Montgomery Multiplication Design

In this section, we describes Montgomery multiplication design based on vector carry propagation chain (VCPC) method and the computation the intermediate scalar variable q. Then, we analyse the expected performance of our design and compare with the redundant representation method in [12]. Especially, we give out the vector length Montgomery multiplication (Algorithm 3) in Section Appendix as an example.

4.1   Vector Carry Propagation Chain Method

(1) Four VCPCs. As described in Algorithm 1, the main computations of Montgomery Multiplication CIOS Method are $S \leftarrow S + A \cdot b_i$ and $S \leftarrow (S + M \cdot q)$. Note that, this two formulas perform the same computation with different operands. The computing process is to multiply a vector with an element, then add the multiplication product to the sum vector S. We exploit Intel Xeon Phi vector instructions to carry out this computation. The logic instructions are used in this section for better clarification. We use *Mullow*, *Mulhigh* and *Vadc* standing for *vpmulhud*, *vpmulld* and *vpadcd*. We assume that the length of S is equal to one vector. The formula $S \leftarrow S + A \cdot b_i$ is computed in the following steps.

$$T \leftarrow Broadcast(b_i)$$
$$L \leftarrow Mullow(A, T)$$
$$H \leftarrow Mulhigh(A, T)$$
$$S \leftarrow S + L$$
$$S \leftarrow S + H$$

The step 1 and step 2 are easy to be carried out, but the products L and H are not the final multiplication product and the least significant element of H is aligned with the 2nd less significant element of L. The step 3 and step 4 must be computed by using *Vadc* instruction. We focus on L and S, extract the corresponding computations from Montgomery multiplication in Algorithm 1. As the integer L in i-th loop is higher one element than the integer L in (i-1)-th loop, and also S shift to the right for a element, so L is aligned to S at all times. We use *Vadc* instruction to overwrite the upper computing process 1 and will demonstrate computing process 2 completes the same calculation as computing process 1:

$$for \ \ i \ \ from \ \ 0 \ \ to \ \ n-1$$
$$T \leftarrow Broadcast(b_i)$$
$$L \leftarrow Mullow(A, T)$$
$$S \leftarrow Vadc(S, \ k1, \ L)$$
$$S \leftarrow Rshift(Zero, \ S, \ 1)$$

$k1$ is a vector mask register for storing carry. *Rshift* is used to stand for *valignd* which concatenates two vector and shifts the whole long vector to the right in 32-bit elements, stores the lower vector to the destination register. We observe that in each loop before performing *Vadc*, S, $k1$ and L are all aligned, so they can be add together by using *Vadc* directly. After performing *Vadc*, the carry bits in $k1$ are propagated forward in a element. While after performing *Rshift* in this loop and *Mullow* in the next loop, L and S are also higher a element than before. So when it is to perform *Vadc* in the next loop, L, S and $k1$ are all aligned, and can be add together by using *Vadc* directly. So computing process 2 has completed the calculation in computing process 1, except has not added the carry k1 back to S after the last round of the loop.

Based on all the above observation, we propose the notion *Vector Carry Propagation Chain* (VCPC), which describes a process: a group of vectors like S, L and a carry like $k1$ are added together in a chain to propagate $k1$ forward in a element after each round. Propagating carry $k1$ only need one *Vadc* instruction in a round of VCPC and adding $k1$ back to S is only performed at the end of VCPC (also known as handing carry). So VCPC very efficiently works out the carry propagation problem (the carry needs to propagate to the higher element). The strategy of VCPC is just propagating carry forward, delaying to handle it.

We propose a design as named VCPC method to overwrite major computations of Algorithm 1 by using VCPC.

The computing process 3 comprises four VCPCs. The VCPC 1 is $Vadc(S, k0, L)$, L is the product of $Mullow(A, b_i)$. The VCPC 2 is $Vadc(S, k1, L)$, L is the product of $Mullow(M, q)$. The VCPC 3 is $Vadc(S, k2, H)$, H is the product of $Mulhigh(A, b_i)$. The VCPC 4 is $Vadc(S, k3, H)$, H is the product of $Mulhigh(M, q)$. 4 VCPCs uses 4 vector mask registers k0, k1, k2 and k3 to propagating 4 carries respectively.

$$for\ i\ from\ 0\ to\ n-1$$
$$L \leftarrow Mullow(A,\ b_i)$$
$$S \leftarrow Vadc(S,\ k0,\ L)$$
$$L \leftarrow Mullow(M,\ q)$$
$$S \leftarrow Vadc(S,\ k1,\ L)$$
$$S \leftarrow Rshift(Zero,\ S,\ 1)$$
$$H \leftarrow Mulhigh(A,\ b_i)$$
$$S \leftarrow Vadc(S,\ k2,\ H)$$
$$H \leftarrow Mulhigh(M,\ q)$$
$$S \leftarrow Vadc(S,\ k3,\ H)$$

VCPC 1 and VCPC 2 are performed before S shifting to the right, because the L computed in two VCPCs are aligned to S before shift. While H computed in VCPC 3 and VCPC 4 are aligned to S after shift, so VCPC 3 and VCPC 4 are performed after S shifting to the right. At the start of each round, there are vector S and four carries $k0$, $k1$, $k2$ and $k3$, and $k0$ and $k1$ are aligned to S, the lowest bit of $k2$ and $k3$ are aligned to the second less significant element of S. In the end of each round, S and carries $k0$, $k1$, $k2$ and $k3$ are all move to more significant position in a element, and also maintain that $k0$ and $k1$ are aligned to S, the lowest bit of $k2$ and $k3$ are aligned to the 2nd less significant element of S. So the four VCPCs in computing process 3 can be maintained to end of the loop.

(2) Handling Tail. At the end of our Montgomery multiplication design, handling tail must be performed, which includes two steps: handling carry and reducing S. Handling carry is used to add all the carry vectors produced by VCPCs to the sum vectors S. First, we add carry vectors to a vector which initial value is zero. Then, we add this vector to S. As Intel Xeon Phi does not have the instruction to shift vector mask register, we use the $LMove$ to copy the carry to the general purpose register and perform left shift, then copy the carry back to the vector mask register. In the worst case, it will need to perform $s-1$ rounds move-shift-move operations, but usually only need one round. As presented in Algorithm 1, reducing S is used to ensure the output of Montgomery multiplication is smaller than modulus M. For the constant running time, reducing S is always performed. We use vector-sub-with-borrow instruction *vpsbbd* to perform subtraction just like addition. Handling borrow requires $s - 1$ rounds in the worst case, and performs two rounds in a greater chance. For performance reasons, the rounds of move-shift-move are not constant. While attackers can hardly get useful information from the running time of move-shift-move.

4.2   Computing q

The intermediate scalar variable q is produced and used for multiplications in Algorithm 1. q is computed as $q \leftarrow S[0] \cdot \mu \bmod 2^w$, which is very easy to be computed in other Montgomery multiplication design, such as redundant representation method. But in our Montgomery multiplication design, computing q is not easy as the carries are maintaining for propagation.

Note that, q computation is carried out by $Q \leftarrow Mullow(S, U)$, $U = Broadcast(\mu)$, $q = Q[0]$. But if we only perform this operation, q may be not correct. As we analyse the relations between q, four VCPCs and four carries, we can see that VCPC 2 and VCPC 4 need q to compute multiplications, and VCPC 2 propagates $k1$. If $k1$ has not be added to S ($k1$ propagates forward), the $S[0]$ may not be right since without being added carry bit $k1[0]$. So there is a contradiction that computing q requires $k1$ propagation forward, while propagating $k1$ (VCPC2) requires computing q first. If we add zero vector, $k1$ to S to obtain the right S, the $k1$ would propagate forward without adding with the product of $Mullow(M, q)$, which will break the VCPC 2, also destruct the VCPC method.

The obvious solution is trial addition which performs an addition to acquire the right $S[0]$ and does not modify $k1$. As depicted in Sect. 3, *vpadcd* is a three-operand instruction, carry k2 (see Sect. 3) is the source operand also the destination operand which means that the old value in $k2$ will be destructed by the new value. So k1 in VCPC2 must be copied for trial addition which need two operations copying of $k1$ and adding the new carry register to S. The drawbacks of trial addition are requiring an extra copy instruction, what's worse, an additional vector mask register (only 8 vector mask registers in a core [26]).

We propose an artistic method to compute q by using write-mask vector. The right counting process of q in VCPC method is $q \leftarrow (S[0] + k1[0]) \cdot \mu$, we rewrite it as $q \leftarrow S[0] \cdot \mu + k1[0] \cdot \mu$. We implements this formula by following two instructions:

$$Q \leftarrow Mullow(S, U)$$
$$Q \leftarrow Vadd(Q, U)\{k1\}$$

Vadd is a normal vector addition instruction without carry. We use k1 as write-mask for *Vadd*. if $k1[0]$ is 1, $Q[0]$ will add a μ; if $k1[0]$ is 0, $Q[0]$ will not be modified, so the value of q is correct. As the write mask is read-only, it would not be modified. Our method does not require an extra move instruction, also no need for an extra vector mask register. The most interesting idea of our method is using the carry vector as write-mask vector.

4.3   Performance Analysis

In this section, we analyse the performance of our Montgomery multiplication design (VCPC method) and compare with Redundant Representation (RR) method which is presented in [12].

We assume the length of element is w, the number of elements in a vector is n, the length of a vector is s ($s = w * n$) and the length of arguments is l. So the number of rounds in our design is $\lceil l/w \rceil$. And the number of vectors is $\lceil l/s \rceil$. So in every rounds, it needs to perform $2 * \lceil l/s \rceil$ *Mullow*, $2 * \lceil l/s \rceil$ *Mulhigh*, $4 * \lceil l/s \rceil$ *Vadc*, $\lceil l/s \rceil$ *RShift*, 2 operations to compute the intermediate scalar variable q and 2 *Broadcast* (b_i and q), which is equal to $9 * \lceil l/s \rceil + 4$. Therefore, for VCPC method, the total number of instructions is about $\lceil l/w \rceil * (9 * \lceil l/s \rceil + 4)$ (not including handling tail, which does not need many instructions).

RR method has two drawbacks: the first is need double spaces to store all the arguments and temporal variables, the second is the several high bits needed to be reserved for storing carries which can not involve in multiplications. For example, for 1024-bit Montgomery multiplication, RR method divides all the arguments and temporal variables into 29-bit parts for remaining high 6-bit (in 64-bit element) to maintain carries. So RR method need more than double vectors to store the arguments and variables than VCPC method. For RR method, we assume the number of reserve bits is t (in 32-bit element), which generally meets $2 * \lceil l/w \rceil \leqslant 2^t$, otherwise it needs to perform cleanup operation during Montgomery multiplication. The number of rounds in RR method is $\lceil l/(w-t) \rceil$. The number of vectors is $2 * \lceil l/((w-t) * n) \rceil = 2 * \lceil l/(s-t * n) \rceil$. So in every rounds, it need to carry out $4 * \lceil l/(s-t * n) \rceil$ multiplications, $4 * \lceil l/(s-t * n) \rceil$ additions, $2 * \lceil l/(s-t * n) \rceil$ *RShift*, 2 operations to compute q and 2 *Broadcast*, which is equal to $10 * \lceil l/(s-t*n) \rceil + 4$. Hence, for RR method, the total number of instructions is $\lceil l/(w-t) \rceil * (10 * \lceil l/(s-t * n) \rceil + 4)$.

Table 1. Comparison with redundant representation method

	RR method	VCPC method
Vector number	$2 * \lceil l/(s-t * n) \rceil$	$\lceil l/s \rceil$
Instructions/round	$10 * \lceil l/(s-t * n) \rceil + 4$	$9 * \lceil l/s \rceil + 4$
Round	$\lceil l/(w-t) \rceil$	$\lceil l/w \rceil$
Instructions	$\lceil l/(w-t) \rceil * (10 * \lceil l/(s-t * n) \rceil + 4)$	$\lceil l/w \rceil * (9 * \lceil l/s \rceil + 4)$

As presented in Table 1, compared with RR method, VCPC method needs less rounds and less instructions in each round. Consequently, VCPC method needs less instructions than RR method. For example, we want to compute 1024-bit Montgomery multiplication on Intel Xeon Phi. The element length w is 32, element number in a vector n is 16, the vector length s is 512, the length of arguments l is 1024, the number of reserve bit t is 3 (in 32-bit element). So VCPC method requires 32 rounds, 22 instructions in each round, so that requires about 704 instructions. RR method requires 36 rounds, 34 instructions in each round, so that requires about 1224 instructions. VCPC method only needs a factor of 0.58 instructions than RR method.

5   Implementation

In this section, we describe the implementations of Montgomery multiplication and RSA on Intel Xeon Phi. We choose assembly language instead of intrinsics in C language to implement Montgomery multiplication for fully controlling registers. Besides, we choose native execution mode [26] for our implementations as the ultimate performance can be evaluated in this mode.

5.1   Montgomery Multiplication Implementation

We implement 512-bit, 1024-bit and 2048-bit Montgomery multiplication on Intel Xeon Phi. Although our design provides the scheme with minimal instruction number, the implementation must be optimized to make the execution cycle approach to the instruction number. Two implementation issues are mainly concerned: making VPUs fully pipelined and maintaining carry bits in vector mask registers.

(1) Making VPUs Fully Pipelined. Data-dependencies in the instruction flow may cause pipeline stalls of VPUs. If an instruction about to be executed has to wait for the operands written by the previous instruction for several cycles, in the meantime no other instructions enter the pipeline, the cycles of VPUs will be wasted and performance will be compromised. First of all, we need to investigate the latency of instructions we used. As presented in [15], most vector instructions are four-cycle latency. We measure vector instruction latency by ourselves. The assessment results are presented in Table 2.

Table 2. The latencies of vector instructions on Intel Xeon Phi

Instruction	vpmulhud	vpmulld	vpadcd	vpermd	valignd
Cycles	4	4	4	6	7

As every core of Intel Xeon Phi has four hyperthreads, the four-cycle instructions (*vpmulhud*, *vpmulld* and *vpadcd*) can be fully pipelined, even though the instructions are data-dependent. But data-dependent *vpermd* and *valignd* are not easily pipelined. We observe that if *vpermd* and *valignd* do not use the data produced by the prior instruction, they can be fully pipelined. As the cores of Intel Xeon Phi are in-order, every vector instruction will be performed in terms of the sequence in the assembly code. So we manually adjust the sequence of instructions in our Montgomery multiplication implementation. The assembly code of one round in 512-bit Montgomery multiplication is presented in ASM Code 1. We insert the data-independent instructions (green ones) into the positions of pipeline stalls (red ones). Then we carry out four threads on one core as each thread executes ASM Code 1 repetitively, the result shows that the execution only requires 12.2 cycles per round which means the utilization of VPUs reaches 98%.

ASM Code 1	Adjusted
1: *vpmulld*	%zmm2{aaaa}, %zmm1, %zmm10
2: *vpadcd*	%zmm10, %k0, %zmm0
3: *vpmulld*	%zmm0, %zmm4, %zmm6
4: *vpaddd*	%zmm6, %zmm4, %zmm6{%k2}
5: *vpmulhud*	%zmm2{aaaa}, %zmm1, %zmm11
6: *vpermd*	%zmm6, %zmm5, %zmm6
7: *vpmulld*	%zmm6, %zmm3, %zmm12
8: *vpadcd*	%zmm12, %k2, %zmm0
9: *vpmulhud*	%zmm6, %zmm3, %zmm13
10: *valignd*	$1, %zmm0, %zmm5, %zmm0
11: *vpadcd*	%zmm11, %k1, %zmm0
12: *vpadcd*	%zmm13, %k3, %zmm0

(2) Maintaining Carry Bits in Vector Mask Registers. As 2048-bit Montgomery multiplication has sixteen VCPCs, it will produce sixteen carry vector every round. However, each core of Intel Xeon Phi has only eight vector mask registers. Although the instruction *kmov* can move data between vector mask registers and general purpose registers, frequent exchanging data will rouse gigantic performance loss. So maintaining carry bits in vector mask registers is essential. We split 2048-bit Montgomery multiplication implementation into four parts and every parts is similar to 1024-bit Montgomery multiplication implementation but without handling tail phase. Outside of these four parts, we need to handle carry bits two times. As every parts have eight VCPCs and the computation of scalar variable q will not break the flow of VCPCs, all carry bits in every round can be kept in vector mask registers.

5.2   RSA Implementation

We apply our Montgomery multiplication implementation to realize PhiRSA based on CRT method [18], which computes m-bit RSA by performing two $(m/2)$-bit Montgomery exponentiations. We also utilize m-ary method [17] to accelerate Montgomery exponentiations with the precomputed table. 2^5-ary method is applied for 1024-bit RSA and 2^6-ary method is applied for 2048-bit RSA and 4096-bit RSA. To complete CRT method, we implement a schoolbook multiplication, addition and subtraction on Intel Xeon Phi. The differences between implementations of multiplication and Montgomery multiplication are that the multiplication implementation has only two VCPCs, don't need to shift right every round and must save the double size product. Our multiplication implementation is very efficient as it also make VPUs fully pipelined.

6   Experimental Results

In this section, we conduct the experiments to evaluate our Montgomery multiplication implementations and RSA implementations on Intel Xeon Phi 7120P processor (1.33 GHz), and compare with the other studies on Xeon Phi, CPUs and GPUs. The configurations of our evaluation platform are described as follows: the coprocessor is Intel Xeon Phi 7120P, the host CPU is Intel Xeon E5 2697v2, the operating system is RedHat 6.4, and the compiler is Intel Composer XE 2013.

6.1   Implementation Result

We execute 244 threads running on 61 cores and bind four threads to each core for 4-way hyper-threading to avoid the performance loss of the thread migration. *Vector Instruction Number* in table indicates the number of assembly instructions and *Execution Cycles* denotes the real execution time. If VPUs reach the maximum performance that one instruction per cycle, *VPU Utilization* is 100%. We also evaluate the throughput and the latency. *Throughput/Thread* denotes the performance of one thread, which is equal to *Throughput*/244. Table 3 summarizes the performance of 512-bit, 1024-bit and 2048-bit Montgomery multiplication implementations. It shows that VPU Utilizations of all the implementations are above 92%. Note that, *Execution Cycles* are almost four times of *Vector Instruction Number* which dues to four threads performing on one core for pipelining. So *VPU Utilization* are equal to $4(VectorInstructionNumber)/(ExecutionCycles)$. Table 4 shows evaluation results of 1024-bit, 2048-bit and 4096-bit RSA. VPU Utilizations of RSA implementations are also above 90%.

6.2   Comparisons with the Previous Works on Intel Xeon Phi

(1) Comparison with the Implementation of Redundant Representation. The work in [16] applies redundant representation method to implement multiplication, which only provides the number of instructions. As the computation of the schoolbook multiplication is about one half of Montgomery multiplication, we double the instruction number in [16] for a rough comparison.

Table 5 shows that our implementation needs no more than one tenth of instructions compared with their implementation. The major reason is that our Montgomery multiplication design requires less instructions than redundant representation method inherently. Another reason is this generation Intel Xeon Phi (Knights Corner, KNC) does not support the multiplication instruction like *vpmuludq* in AVX2 [14] which needed in redundant representation method. So our design is not only better than redundant representation method but more suitable for Intel Xeon Phi (KNC).

Table 3. Performance of Montgomery multiplication on Intel Xeon Phi

	Montgomery multiplication		
	512-bit	1024-bit	2048-bit
Thread number	244	244	244
Core number	61	61	61
Vector instruction number	218	724	2797
Execution cycles	948	3076	12211
VPU utilization	92%	94%	92%
Throughput ($10^6/s$)	343.78	105.73	26.64
Throughput/thread ($10^6/s$)	1.41	0.43	0.11
Latency (μs)	0.71	2.31	9.16

Table 4. Performance of RSA decryption on Intel Xeon Phi

	RSA decryption					
	1024-bit		2048-bit		4096-bit	
	Window size: 5		Window size: 6		Window size: 6	
Thread number	1	244	1	244	1	244
Core number	1	61	1	61	1	61
Vector instruction number (10^6/op)	0.28	0.28	1.82	1.82	13.7	13.7
Execution cycles (10^6/op)	0.91	1.26	3.97	7.78	29.71	60.66
VPU utilization	31%	90%	46%	94%	46%	90%
Throughput (/s)	1466	258370	336	41803	45	5358
Throughput/thread (/s)	1466	1059	336	171	45	22
Latency (ms)	0.68	0.94	2.98	5.84	22.29	45.54

Table 5. Comparisons with the implementation of redundant representation on Intel Xeon Phi

	512-bit MontMul (instructions)	1024-bit MontMul (instructions)	2048-bit MontMul (instructions)
Keliris et al. [12] (Scaled)	3846	9498	28776
Our VCPC method	218	724	2797

(2) Comparison with the Implementation of Carry Propagation.
The work in [7] firstly uses carry propagation to implement multiplication and RSA on Intel Xeon Phi 5110P. As described in Table 6, the throughput in [7] is scaled to Intel Xeon Phi 7120P. Our implementations achieve 4.1 to 8.5 times performance of the scaled results. There are three possible reasons: (1) the *Extract* and *Store* operations are very cost; (2) using multiplication to compute Montgomery multiplication is not the best way; (3) intrinsics in C language can not fully control registers.

Table 6. Comparison with the implementation of carry propagation on Intel Xeon Phi

	512-bit RSA-1024 throughput (/s)	1024-bit RSA-2048 throughput (/s)	2048-bit RSA-4096 throughput (/s)
Chang et al. [7] (Scaled)	1310	7217	30282
Our VCPC method	5358	41803	258370

6.3 Comparisons with the Implementations on CPUs and GPUs

Table 7 shows the comparisons with the best implementations on CPUs and GPUs. Compared with CPU implementation in OpenSSL [23] which evaluated on Intel i7 4770R, the throughput of our implementation is about 40 times of a single CPU core, and the latency is about 5 times. The throughput of one core on Intel Xeon Phi is about a factor of 0.6 compared with one CPU core, which is due to the higher frequency of CPU core (3.2 GHz). On GPU platform, the integer implementation in [31] has the lowest latency so far which evaluated on NVIDIA GT 750m, and the floating-pointing implementation in [32] has the highest throughput until now which evaluated on NVIDIA GTX Titan. Compared with [31], the throughput of our implementation is about 7 times, and the latency is no more than 90%. And compared with [32], the throughput of our implementation is about 1.07 times, and the latency is only 26%. So PhiRSA has the advantage on achieving high throughput and small latency simultaneously.

Table 7. Comparisons with the implementations on CPUs and GPUs

RSA decryption	OpenSSL 1.0.1f [23]	Yang et al. [31]	Zheng et al. [32]	Our native implementations
Platform	Intel Haswell i7 4770R	NVIDIA GT 750m	NVIDIA GTX Titan	Intel Xeon Phi 7120P
Core number	4	384	2688	61
Frequency (GHz)	3.2	0.967	0.836	1.33
Computing power (SP GFLOPS)	410	743	4500	2600
RSA-1024 throughput (/s)	25850	34981	-	234981
RSA-2048 throughput (/s)	3427	5244	38975	41803
RSA-4096 throughput (/s)	485	-	-	5358
RSA-1024 latency (ms)	0.16	2.6	-	1.04
RSA-2048 latency (ms)	1.17	6.5	22.47	5.84
RSA-4096 latency (ms)	8.26	-	-	45.54

7 Conclusions

In this contribution, we propose a novel vector-oriented Montgomery multiplication design and implementation to fully exploit the computing power of vector instructions on Intel Xeon Phi. Based on the above, we implement RSA named PhiRSA. PhiRSA is much better than the existing RSA implementations on Intel Xeon Phi which attains 4.1 to 8.5 times performance. Our results also demonstrate that Intel Xeon Phi can be used to achieve both high throughput and small latency for RSA. On Intel Xeon Phi 7120P, PhiRSA achieves high throughput comparable to the implementations on GPUs but with much less parallel tasks, and small latency comparable to the implementations on CPU. PhiRSA and our Montgomery multiplication implementation can be applied to implement other cryptographic algorithms as primitives. We will also integrate PhiRSA into OpenSSL in the future.

A Appendix

Algorithm 3 is vector length Montgomery multiplication.

Algorithm 3. Vector length Montgomery multiplication

Input: 2^w is radix, n is element number, vector size is $s = n * w$
$R = 2^s$, Modulus M is s-bit number, $M < R$,
$\gcd(M, R) = 1$, $\mu = -M^{-1} \bmod 2^w$
A, B are s-bit number, $0 \leqslant A, B < M$, $B = \sum_{i=0}^{n-1} b_i 2^{iw}$
Output: $S = A \cdot B \cdot R^{-1} \pmod{M}$, $0 \leqslant S < M$
1: $k0 \leftarrow 0$, $k1 \leftarrow 0$, $k2 \leftarrow 0$, $k3 \leftarrow 0$
2: $S \leftarrow 0$, $Zero \leftarrow 0$,
3: $U \leftarrow Broadcast(\mu)$
   /* **VCPC   Phase** */
4: **for** i from 0 to $n - 1$ **do**
5:     $T \leftarrow Broadcast(b_i)$
       /* VCPC 1: $S + k0 = S + k0 + Low(A \cdot b_i)$ */
6:     $L \leftarrow Mullow(A, T)$
7:     $(S, k0) \leftarrow Vadc(S, k0, L)$
       /* $q \leftarrow (S[0] + k1[0]) \cdot \mu$ */
8:     $Q \leftarrow Mullow(S, U)$
9:     $Q \leftarrow Vadd(Q, U)\{k1\}$
10:    $Q \leftarrow Broadcast(Q[0])$
       /* VCPC 2: $S + k1 = S + k1 + Low(M \cdot q)$ */
11:    $L \leftarrow Mullow(M, Q)$
12:    $(S, k1) \leftarrow Vadc(S, k1, L)$
       /* Right Shift: $S = S \gg 1$ element */
13:    $S \leftarrow RShift(Zero, S, 1)$
       /* VCPC 3: $S + k2 = S + k2 + High(A \cdot b_i)$ */
14:    $H \leftarrow Mulhigh(A, T)$
15:    $(S, k2) \leftarrow Vadc(S, k2, H)$
       /* VCPC 4: $S + k3 = S + k3 + High(M \cdot q)$ */
16:    $H \leftarrow Mulhigh(M, Q)$
17:    $(S, k3) \leftarrow Vadc(S, k3, H)$
18: **end for**
    /* **Tail   Phase** */
    /* Handling carry */
19: $(T, k0) \leftarrow Vadc(Zero, k0, Zero)$
20: $(S, k1) \leftarrow Vadc(S, k1, T)$
21: $(T, k1) \leftarrow Vadc(Zero, k1, Zero)$
22: $(T, k2) \leftarrow Vadc(T, k2, Zero)$
23: $H \leftarrow Rshift(Zero, S, 1)$
24: $(H, k3) \leftarrow Vadc(H, k3, T)$
25: **for** i from 0 to $n - 2$ **do**
26:    **if** $k3 = 0$ **then**
27:        $BREAK$
28:    **end if**
29:    $(H, k3) \leftarrow Vadc(H, k3, Zero)$
30:    $k3 \leftarrow Lmove(k3)$
31: **end for**
32: $S \leftarrow Rshift(S, zero, 1)$
33: $S \leftarrow Rshift(H, S, n - 1)$
    /* Reducing */
34: **if** $H[n - 1] = 1$ **then**
35:    $(S, k0) \leftarrow Vsbb(S, k0, M)$
36:    $H \leftarrow Rshift(Zero, H, n - 1)$
37:    $H \leftarrow Rshift(H, S, 1)$
38:    **for** i from 0 to n-2 **do**
39:        **if** $k0 = 0$ **then**
40:            $BREAK$
41:        **end if**
42:        $(H, k0) \leftarrow Vsbb(H, k0, Zero)$
43:        $k0 \leftarrow Lmove(k0)$
44:    **end for**
45: **end if**
46: $S \leftarrow Rshift(S, zero, 1)$
47: $S \leftarrow Rshift(H, S, n - 1)$
48: **return** S.

References

1. Bernstein, D.J., Chen, H.-C., Chen, M.-S., Cheng, C.-M., Hsiao, C.-H., Lange, T., Lin, Z.-C., Yang, B.-Y.: The billion-mulmod-per-second PC. Workshop rec. SHARCS **9**, 131–144 (2009)
2. Bernstein, D.J., Chen, T.R., Cheng, C.M., Lange, T., Yang, B.-Y.: ECM on graphics cards. In: Joux, A. (ed.) EUROCRYPT 2009. LNCS, vol. 5479, pp. 483–501. Springer, Heidelberg (2009). doi:10.1007/978-3-642-01001-9_28
3. Bernstein, D.J., Chuengsatiansup, C., Lange, T.: Curve41417: Karatsuba revisited. In: Batina, L., Robshaw, M. (eds.) CHES 2014. LNCS, vol. 8731, pp. 316–334. Springer, Heidelberg (2014). doi:10.1007/978-3-662-44709-3_18
4. Bos, J.W.: High-performance modular multiplication on the cell processor. In: Hasan, M.A., Helleseth, T. (eds.) WAIFI 2010. LNCS, vol. 6087, pp. 7–24. Springer, Heidelberg (2010). doi:10.1007/978-3-642-13797-6_2
5. Bos, J.W., Kaihara, M.E.: Montgomery multiplication on the cell. In: Wyrzykowski, R., Dongarra, J., Karczewski, K., Wasniewski, J. (eds.) PPAM 2009. LNCS, vol. 6067, pp. 477–485. Springer, Heidelberg (2010). doi:10.1007/978-3-642-14390-8_50
6. Bos, J.W., Montgomery, P.L., Shumow, D., Zaverucha, G.M.: Montgomery multiplication using vector instructions. In: Lange, T., Lauter, K., Lisoněk, P. (eds.) SAC 2013. LNCS, vol. 8282, pp. 471–489. Springer, Heidelberg (2014). doi:10.1007/978-3-662-43414-7_24
7. Chang, C., Yao, S., Yu, D.: Vectorized big integer operations for cryptosystems on the Intel mic architecture. In: 2015 IEEE 22nd International Conference on High Performance Computing (HiPC), pp. 194–203. IEEE (2015)
8. Chen, J., Watson, W., Chen, M.F.: Efficient GCD computation for big integers on Xeon Phi coprocessor. In: 2014 9th IEEE International Conference on Networking, Architecture, and Storage (NAS), pp. 113–117. IEEE (2014)
9. Fang, J., Varbanescu, A.L., Sips, H., Zhang, L., Che, Y., Xu C.: An empirical study of Intel Xeon Phi. arXiv preprint arXiv:1310.5842 (2013)
10. Faz-Hernández, A., López, J.: Fast implementation of curve25519 using AVX2. In: Lauter, K., Rodríguez-Henríquez, F. (eds.) LATINCRYPT 2015. LNCS, vol. 9230, pp. 329–345. Springer, Cham (2015). doi:10.1007/978-3-319-22174-8_18
11. Grabher, P., Großschädl, J., Page, D.: On software parallel implementation of cryptographic pairings. In: Avanzi, R.M., Keliher, L., Sica, F. (eds.) SAC 2008. LNCS, vol. 5381, pp. 35–50. Springer, Heidelberg (2009). doi:10.1007/978-3-642-04159-4_3
12. Gueron, S., Krasnov, V.: Software implementation of modular exponentiation, using advanced vector instructions architectures. WAIFI **12**, 119–135 (2012)
13. Intel: Intel Xeon Phi coprocessor instruction set architecture reference manual (2012). https://software.intel.com/sites/default/files/forum/278102/327364001en.pdf
14. Intel: Intel 64 and IA-32 architectures software developer's manual, vol. 2 (2a, 2b and 2c): Instruction set reference, a-z (2015)
15. Jeffers, J., Reinders, J.: Intel Xeon Phi coprocessor high-performance programming. Newnes (2013)
16. Keliris, A., Maniatakos, M.: Investigating large integer arithmetic on Intel Xeon Phi SIMD extensions. In: 2014 9th IEEE International Conference on Design and Technology of Integrated Systems in Nanoscale Era (DTIS), pp. 1–6. IEEE (2014)
17. Knuth, D.E.: Seminumerical algorithms, the art of computer programming, vol. 2 (1981)

18. Koç, Ç.K.: High-speed RSA implementation. Technical report, RSA Laboratories (1994)
19. Koç, Ç.K., Acar, T., Kaliski Jr., B.S.: Analyzing and comparing montgomery multiplication algorithms. IEEE Micro **16**(3), 26–33 (1996)
20. Montgomery, P.L.: Modular multiplication without trial division. Math. comput. **44**(170), 519–521 (1985)
21. Moore, S.: Using streaming SIMD extensions (SSE2) to perform big multiplications. Application note AP-941, Intel Corporation 2000, version 2.0. Order (248606-001) (2000)
22. NVIDIA. Cuda C programming guide 7.5 (2015). http://docs.nvidia.com/cuda/cuda-c-programming-guide/
23. OpenSSL. The open source toolkit for SSL/TLS (2015)
24. Page, D., Smart, N.P.: Parallel cryptographic arithmetic using a redundant montgomery representation. IEEE Trans. Comput. **53**(11), 1474–1482 (2004)
25. Pennycook, S.J., Hughes, C.J., Smelyanskiy, M., Jarvis, S.A.: Exploring SIMD for molecular dynamics, using Intel Xeon Processors and Intel Xeon Phi coprocessors. In: 2013 IEEE 27th International Symposium on Parallel and Distributed Processing (IPDPS), pp. 1085–1097. IEEE (2013)
26. Rahman, R.: Intel Xeon Phi Coprocessor Architecture and Tools: The Guide for Application Developers. Apress, New York (2013)
27. Saule, E., Kaya, K., Çatalyürek, Ü.V.: Performance evaluation of sparse matrix multiplication kernels on Intel Xeon Phi. In: Wyrzykowski, R., Dongarra, J., Karczewski, K., Waśniewski, J. (eds.) PPAM 2013. LNCS, vol. 8384, pp. 559–570. Springer, Heidelberg (2014). doi:10.1007/978-3-642-55224-3_52
28. Seo, H., Liu, Z., Großschädl, J., Choi, J., Kim, H.: Montgomery modular multiplication on ARM-NEON revisited. In: Lee, J., Kim, J. (eds.) ICISC 2014. LNCS, vol. 8949, pp. 328–342. Springer, Cham (2015). doi:10.1007/978-3-319-15943-0_20
29. Seo, H., Liu, Z., Kim, H.: Efficient arithmetic on arm-neon and its application for high-speed RSA implementation
30. Wikipedia. List of NVIDIA graphics processing units (2015). https://en.wikipedia.org/wiki/List_of_Nvidia_graphics_processing_units
31. Yang, Y., Guan, Z., Sun, H., Chen, Z.: Accelerating RSA with fine-grained parallelism using GPU. In: Lopez, J., Wu, Y. (eds.) ISPEC 2015. LNCS, vol. 9065, pp. 454–468. Springer, Cham (2015). doi:10.1007/978-3-319-17533-1_31
32. Zheng, F., Pan, W., Lin, J., Jing, J., Zhao, Y.: Exploiting the floating-point computing power of GPUs for RSA. In: Chow, S.S.M., Camenisch, J., Hui, L.C.K., Yiu, S.M. (eds.) ISC 2014. LNCS, vol. 8783, pp. 198–215. Springer, Cham (2014). doi:10.1007/978-3-319-13257-0_12

FourQNEON: Faster Elliptic Curve Scalar Multiplications on ARM Processors

Patrick Longa[✉]

Microsoft Research, Redmond, USA
plonga@microsoft.com

Abstract. We present a high-speed, high-security implementation of the recently proposed elliptic curve FourQ (ASIACRYPT 2015) for 32-bit ARM processors with NEON support. Exploiting the versatile and compact arithmetic of this curve, we design a vectorized implementation that achieves high-performance across a large variety of ARM platforms. Our software is fully protected against timing and cache attacks, and showcases the impressive speed of FourQ when compared with other curve-based alternatives. For example, one single variable-base scalar multiplication is computed in about 235,000 Cortex-A8 cycles or 132,000 Cortex-A15 cycles which, compared to the results of the fastest genus 2 Kummer and Curve25519 implementations on the same platforms, offer speedups between 1.3x–1.7x and between 2.1x–2.4x, respectively. In comparison with the NIST standard curve K-283, we achieve speedups above 4x and 5.5x.

Keywords: Elliptic curves · FourQ · ARM · NEON · Vectorization · Efficient software implementation · Constant-time.

1 Introduction

In 2013, ARM surpassed the 50 billion mark of processors shipped worldwide, consolidating its hegemony as the most widely used architecture in terms of quantity [22]. One of the main drivers of this success has been the explosive growth of the mobile market, for which the Cortex-A and Cortex-M architectures based on the ARMv7 instruction set became key technologies. In particular, the Cortex-A series include powerful yet power-efficient processors that have successfully hit the smartphone/tablet/wearable mass market. For example, Cortex-A7 based SOCs power the Samsung Gear S2 (2015) smartwatch and the Microsoft Lumia 650 (2016) smartphone; Cortex-A8 and Cortex-A9 cores can be found in the Motorola Moto 360 (2014) smartwatch and the Samsung Galaxy Light (2013) smartphone, respectively; and Cortex-A15/Cortex-A7 (big.LITTLE) based SOCs power the Samsung Galaxy S5 (2014) and the Samsung Galaxy A8 (2015) smartphones. Many of these Cortex-A microarchitectures come equipped with a NEON engine, which provides advanced 128-bit Single Instruction Multiple Data (SIMD) vector instructions. Thus, these low-power

© Springer International Publishing AG 2017
R. Avanzi and H. Heys (Eds.): SAC 2016, LNCS 10532, pp. 501–519, 2017.
https://doi.org/10.1007/978-3-319-69453-5_27

RISC-based ARM platforms with NEON support have become an attractive platform for deploying and optimizing cryptographic computations.

Costello and Longa [12] recently proposed a highly efficient elliptic curve, dubbed FourQ, that provides around 128 bits of security and enables the fastest curve-based scalar multiplications on x64 software platforms by combining a four-dimensional decomposition based on endomorphisms [16], the fastest twisted Edwards formulas [18], and the efficient yet compact Mersenne prime $p = 2^{127} - 1$. In summary, the results from [12] show that, when computing a single variable-base scalar multiplication, FourQ is more than 5 times faster than the widely used NIST curve P-256 and more than 2 times faster than Curve25519 [4]. In comparison to other high-performance alternatives such as the genus 2 Kummer surface proposed by Gaudry and Schost [17], FourQ is, in most cases, more than 1.2x faster on x64 processors. For all of these comparisons, Costello and Longa's FourQ implementation (i) does not exploit vector instructions (in contrast to Curve25519 and Kummer implementations that do [7,11]), and (ii) is only optimized for x64 platforms. Therefore, the deployment and evaluation of FourQ on 32-bit ARM processors with NEON support, for which the use of vector instructions pose a different design paradigm, is still missing.

In this work, we engineer an efficient NEON-based implementation of FourQ targeting 32-bit ARM Cortex-A microarchitectures that are based on the widely used ARMv7 instruction set. Our design, although intended for high-performance applications, is not exclusive to only *one* microarchitecture; we analyze the different features from multiple Cortex-A microarchitectures and come up with an implementation that performs well across a wide range of ARM platforms. Specifically, our analysis includes *four* popular ARM processor cores: Cortex-A7, A8, A9 and A15. In addition, our implementation runs in constant-time, i.e., it is protected against timing and cache attacks [19], and supports the *three* core elliptic curve-based computations found in most cryptographic protocols (including Diffie-Hellman key exchange and digital signatures): variable-base, fixed-base and double-scalar multiplication. By considering these design decisions and functionality, we expect to ultimately produce practical software that can be used in real-world applications. Our code has been made publicly available as part of version 2.0 of FourQlib [13].

Our benchmark results extend FourQ's top performance to 32-bit ARM processors with NEON, and demonstrate for the first time FourQ's vectorization potential. For example, on a 2.0 GHz Odroid XU3 board powered by a Cortex-A15 CPU, our software computes a variable-base scalar multiplication in only 132,000 cycles (or 66 μs for a throughput above 15,150 operations/second). This result is about 1.7x faster than the Kummer implementation from [7], about 1.8x faster than the GLV+GLS based implementation from [15], about 2.4x faster than the Curve25519 implementation from [9], and about 5.6x faster than the implementation of the standardized NIST curve K-283 from [10]. As in our case, all of these implementations are state-of-the-art, exploit NEON instructions and are protected against timing and cache attacks. See Sect. 5 for complete benchmark results.

The paper is organized as follows. In Sect. 2, we provide relevant details about FourQ. In Sect. 3, we describe the 32-bit ARM architecture using NEON with focus on the targeted Cortex-A processors. We describe our vectorized NEON design and optimizations in Sect. 4 and, finally, Sect. 5 concludes the paper with the analysis and benchmark results.

2   The FourQ Curve

This section describes FourQ, where we adopt the notation from [12] for the most part. FourQ [12] is defined as the complete twisted Edwards [6] curve given by

$$\mathcal{E}/\mathbb{F}_{p^2} : \; - x^2 + y^2 = 1 + dx^2 y^2, \tag{1}$$

where the quadratic extension field $\mathbb{F}_{p^2} = \mathbb{F}_p(i)$ for $i^2 = -1$ and $p = 2^{127} - 1$, and $d = 125317048443780598345676279555970305165 \cdot i + 4205857648805777768770$.

The $\mathbb{F}_{p^2}$-rational points lying on the curve Eq. (1) form an abelian group for which the neutral element is $\mathcal{O}_{\mathcal{E}} = (0, 1)$ and the inverse of a point (x, y) is $(-x, y)$. The cardinality of this group is $\#\mathcal{E}(\mathbb{F}_{p^2}) = 392 \cdot N$, where N is a 246-bit prime; thus, the prime-order subgroup $\mathcal{E}(\mathbb{F}_{p^2})[N]$ can be used to build cryptographic systems.

FourQ is equipped with *two* efficiently computable endomorphisms, ψ and ϕ, which give rise to a four-dimensional decomposition $m \mapsto (a_1, a_2, a_3, a_4) \in \mathbb{Z}^4$ for any integer $m \in [1, 2^{256})$ such that $0 \le a_i < 2^{64}$ for $i = 1, \ldots, 4$ and such that a_1 is odd. This decomposition enables a four-dimensional variable-base scalar multiplication with the form

$$[m]P = [a_1]P + [a_2]\phi(P) + [a_3]\psi(P) + [a_4]\phi(\psi(P)),$$

for any point $P \in \mathcal{E}(\mathbb{F}_{p^2})[N]$.

The details of FourQ's variable-base scalar multiplication based on the four-dimensional decomposition are shown in Algorithm 1. The curve arithmetic is based on Hisil et al. explicit formulas that use *extended twisted Edwards coordinates* [18]: any projective tuple $(X : Y : Z : T)$ with $Z \ne 0$ and $T = XY/Z$ corresponds to an affine point $(x, y) = (X/Z, Y/Z)$. Note that these formulas are also *complete* on $\mathcal{E}$, which means that they work without exceptions for all points in $\mathcal{E}(\mathbb{F}_{p^2})$.

The execution of Algorithm 1 begins with the computation of the endomorphisms ψ and ϕ, and the computation of the 8-point precomputed table (Steps 1–2). These precomputed points are stored in coordinates $(X+Y, Y-X, 2Z, 2dT)$ for efficiency. Scalar decomposition and multiscalar recoding are then applied to the input scalar m at Steps 3 and 4 as described in [12, Proposition 5] and [12, Algorithm 1], respectively. Finally, the main loop consists of 64 iterations each computing a point doubling (Step 7) and a point addition with a point from the precomputed table (Step 8). Following [12], the next coordinate representations are used throughout the algorithm: $\mathbf{R_1} : (X, Y, Z, T_a, T_b)$, such that $T = T_a \cdot T_b$, $\mathbf{R_2} : (X+Y, Y-X, 2Z, 2dT)$, $\mathbf{R_3} : (X+Y, Y-X, Z, T)$ and

---

Algorithm 1. Fourℚ's scalar multiplication on $\mathcal{E}(\mathbb{F}_{p^2})[N]$ (from [12]).

Input: Point $P \in \mathcal{E}(\mathbb{F}_{p^2})[N]$ and integer scalar $m \in [0, 2^{256})$.
Output: $[m]P$.
   Compute endomorphisms:
  1: Compute $\phi(P)$, $\psi(P)$ and $\psi(\phi(P))$.
   Precompute lookup table:
  2: Compute $T[u] = P + [u_0]\phi(P) + [u_1]\psi(P) + [u_2]\psi(\phi(P))$ for $u = (u_2, u_1, u_0)_2$ in
     $0 \le u \le 7$.
     Write $T[u]$ in coordinates $(X+Y, Y-X, 2Z, 2dT)$.
   Scalar decomposition and recoding:
  3: Decompose m into the multiscalar (a_1, a_2, a_3, a_4) as in [12, Proposition 5].
  4: Recode (a_1, a_2, a_3, a_4) into $(d_{64}, \ldots, d_0)$ and $(m_{64}, \ldots, m_0)$ using [12, Algorithm 1].
     Write $s_i = 1$ if $m_i = -1$ and $s_i = -1$ if $m_i = 0$.
   Main loop:
  5: $Q = s_{64} \cdot T[d_{64}]$
  6: **for** $i = 63$ **to** 0 **do**
  7:    $Q = [2]Q$
  8:    $Q = Q + s_i \cdot T[d_i]$
  9: **return** Q

$\mathbf{R_4} : (X, Y, Z)$. In the main loop, point doublings are computed as $\mathbf{R_1} \leftarrow \mathbf{R_4}$ and point additions as $\mathbf{R_1} \leftarrow \mathbf{R_1} \times \mathbf{R_2}$, where the input using $\mathbf{R_1}$ comes from the output of a doubling (after ignoring coordinates T_a and T_b) and the input using $\mathbf{R_2}$ is a precomputed point from the table.

3 ARM NEON Architecture

The 32-bit RISC-based ARM architecture, which includes ARMv7, is the most popular architecture in mobile devices. It is equipped with 16 32-bit registers (r0–r15) and an instruction set supporting 32-bit operations or, in the case of Thumb and Thumb2, a mix of 16- and 32-bit operations. Many ARM cores include NEON, a powerful 128-bit SIMD engine that comes with 16 128-bit registers (q0–q15) which can also be viewed as 32 64-bit registers (d0–d31). NEON includes support for 8-, 16-, 32- and 64-bit signed and unsigned integer operations. For more information, refer to [2].

The following is a list of basic data processing instructions that are used in our NEON implementation. Since our design is based on a signed integer representation (see Sect. 4), most instructions below are specified with signed integer datatypes (denoted by .sXX in the instruction mnemonic). All of the timings provided correspond to Cortex-A8 and A9 (see [1,3]).

– vmull.s32 performs 2 signed 32×32 multiplications and produces 2 64-bit products. When there are no pipeline stalls, the instruction takes 2 cycles.

- `vmlal.s32` performs 2 signed 32×32 multiplications, produces 2 64-bit products and accumulates the results with 2 64-bit additions. It has a cost similar to `vmull.s32`, which means that additions for accumulation are for free.
- `vadd.s64` and `vsub.s64` perform 1 or 2 signed 64-bit additions (resp. subtractions). When there are no pipeline stalls, the instruction takes 1 cycle.
- `vshl.i32` performs 2 or 4 32-bit logical shifts to the left by a fixed value. When there are no pipeline stalls, the instruction takes 1 cycle.
- `vshr.s64` and `vshr.u64` perform 1 or 2 64-bit arithmetic and logical (resp.) shifts to the right by a fixed value. It has a cost similar to `vshl.i32`.
- `vand.u64` performs a bitwise logical **and** operation. It has a cost similar to `vadd.s64`.
- `vbit` inserts each bit from the first operand into the destination operand if the corresponding bit in the second operand is 1. Otherwise, the destination bit remains unchanged. When there are no pipeline stalls, the instruction takes 1 cycle if operands are 64-bit long and 2 cycles if operands are 128-bit long.

Multiply and multiply-and-add instructions (`vmull.s32` and `vmlal.s32`) have latencies of 6 cycles. However, if a multiply-and-add follows a multiply or another multiply-and-add that depends on the result of the first instruction then a special forwarding enables the issuing of these instructions back-to-back. In this case, a series of multiply and multiply-and-add instructions can achieve a throughput of *two* cycles per instruction.

3.1 Targeted Platforms

Our implementation is optimized for the 32-bit Cortex-A series with ARMv7 support, with a special focus on Cortex-A7, A8, A9 and A15 cores. Next, we describe the most relevant architectural features that are considered in the design of our NEON-based software to achieve a consistent performance across microarchitectures.

Cortex-A7. This microarchitecture has in-order execution, partial dual-issue and a NEON engine capable of executing (at most) one NEON arithmetic operation per cycle.

Cortex-A8. This microarchitecture has the NEON pipeline logically behind the integer pipeline. Once NEON instructions flow through the integer pipeline, they are stored in a queue getting ready for execution. This queue accumulates instructions faster than what it takes to execute them, which means that the integer pipeline can execute ARM instructions in the background while the NEON unit is busy. This is exploited in Sect. 4.2 to boost the performance of the $\mathbb{F}_{p^2}$ implementation by mixing ARM and NEON instructions. Cortex-A8 also has highly flexible dual-issue capabilities that support many combinations of ARM and NEON instructions; for example, Cortex-A8 can issue a NEON load/store instruction with a NEON arithmetic instruction in one cycle. The NEON engine has one execution port for arithmetic instructions and one execution port for load/store/permute instructions; this enables back-to-back execution of pairs of NEON {load/store, arithmetic} instructions (see Sect. 4.2).

Cortex-A9. This microarchitecture no longer has the NEON unit (with a detached NEON queue) behind the integer unit as in Cortex-A8. This significantly reduces the cost of NEON to ARM data transfers, but also reduces the efficiency gain obtained by mixing ARM and NEON instructions. In addition, NEON on Cortex-A9 has some limitations: load/store instructions have longer latency, and there is no dual-issue support. To minimize the inefficiency of the load/store port it is possible to interleave these instructions with other instructions, as detailed in Sect. 4.2.

Cortex-A15. This microarchitecture has out-of-order execution on both ARM and NEON units. The NEON engine, which is fully integrated with the ARM core, is capable of executing two operations per cycle. The ARM and NEON load/store ports are also integrated. In many platforms, Cortex-A15 cores are paired with Cortex-A7 cores to form powerful heterogeneous systems (a.k.a. big.LITTLE).

4   Vectorization Using NEON

The basic feature to consider in a NEON-based design is that vector multiplication is capable of working over *two pairs* of 32-bit values to produce *one pair* of 64-bit products.

In our preliminary analysis, we considered two approaches for vectorization:

– Vectorization across different $\mathbb{F}_{p^2}$ multiplications and squarings inside point formulas.
– Vectorization across different field multiplications inside $\mathbb{F}_{p^2}$ multiplications and squarings.

The first option has the disadvantage that pairing of $\mathbb{F}_{p^2}$ operations inside point addition and doubling formulas is not perfect and would lead to sub-optimal performance. E.g., the 3 squarings in the doubling formula would be computed either as 2 pairs of squarings (increasing the cost in 1 squaring) or as 1 pair of squarings and 1 pair of multiplications, using any available multiplication (degrading the speed of 1 squaring). This approach also puts extra pressure on register allocation, which can potentially lead to a high number of memory accesses. In contrast, the second approach can benefit from the availability of independent operations over $\mathbb{F}_p$ inside the $\mathbb{F}_{p^2}$ arithmetic. Both multiplications and squarings over $\mathbb{F}_{p^2}$ naturally contain *pairs* of field multiplications; all multiplications are independent from each other and, therefore, can be optimally paired for NEON vector multiplication.

We chose the second vectorization option for our implementation, which is described next.

4.1   Vectorization of $\mathbb{F}_{(2^{127}-1)^2}$ Arithmetic

For our design we use radix $t = 2^{26}$ and represent a quadratic extension field element $c = a + b \cdot i \in \mathbb{F}_{p^2}$ using $a = a_0 + a_1 t + a_2 t^2 + a_3 t^3 + a_4 t^4$ and $b = b_0 + b_1 t + b_2 t^2 + b_3 t^3 + b_4 t^4$. In a similar fashion to Naehrig et al.'s interleaving strategy [21], in our implementation the *ten-coefficient vector* representing element c is stored "interleaved" as $(b_4, a_4, b_3, a_3, b_2, a_2, b_1, a_1, b_0, a_0)$ in little endian format, i.e., a_0 and b_4 are stored in the lowest and highest memory addresses, respectively. Each coefficient is *signed* and occupies 32 bits in memory; however, when fully reduced, coefficients $a_0, b_0, \ldots, a_3, b_3$ have values in the range $[0, 2^{26})$ and coefficients a_4 and b_4 have values in the range $[0, 2^{23})$.

Using the representation above, addition and subtraction of two elements in $\mathbb{F}_{p^2}$ are simply done with 2 128-bit vector addition instructions (resp. subtractions) and 1 64-bit vector addition instruction (resp. subtraction) using the NEON instruction vadd.s32 (resp. vsub.s32). The corresponding results are immediately produced in the interleaved representation.

For the case of multiplication and squaring, we base the implementation on a schoolbook-like multiplication that includes the reduction modulo $p = 2^{127} - 1$. Given two field elements $a = a_0 + a_1 t + a_2 t^2 + a_3 t^3 + a_4 t^4$ and $b = b_0 + b_1 t + b_2 t^2 + b_3 t^3 + b_4 t^4$, multiplication modulo $(2^{127} - 1)$ can be computed by

$$\begin{aligned}
c_0 &= a_0 b_0 + 8(a_1 b_4 + a_4 b_1 + a_2 b_3 + a_3 b_2) \\
c_1 &= a_0 b_1 + a_1 b_0 + 8(a_2 b_4 + a_4 b_2 + a_3 b_3) \\
c_2 &= a_0 b_2 + a_2 b_0 + a_1 b_1 + 8(a_3 b_4 + a_4 b_3) \\
c_3 &= a_0 b_3 + a_3 b_0 + a_1 b_2 + a_2 b_1 + 8(a_4 b_4) \\
c_4 &= a_0 b_4 + a_4 b_0 + a_1 b_3 + a_3 b_1 + a_2 b_2.
\end{aligned} \quad (2)$$

Next, we show how to use (2) in the vectorized computation of multiplication and squaring over $\mathbb{F}_{p^2}$. Note that the operation sequences below are designed to maximize performance and to fit all intermediate computations in the 16 128-bit NEON registers at our disposal.

Multiplication in $\mathbb{F}_{p^2}$. Let $A = (b_4, a_4, b_3, a_3, b_2, a_2, b_1, a_1, b_0, a_0)$ and $B = (d_4, c_4, d_3, c_3, d_2, c_2, d_1, c_1, d_0, c_0)$ be coefficient vectors that represent elements $(a + b \cdot i) \in \mathbb{F}_{p^2}$ and $(c + d \cdot i) \in \mathbb{F}_{p^2}$, respectively. To multiply these two elements, we first shift A to the left by 3 bits to obtain

$$t_1 = (8b_4, 8a_4, \ldots, 8b_1, 8a_1, 8b_0, 8a_0),$$

which requires 1 64-bit and 2 128-bit vector shifts using vshl.i32.

We then compute the first *three* terms of the multiplications $bc = b \times c$ and $bd = b \times d$, by multiplying in pairs $(b_0 d_0, b_0 c_0)$, $(8b_1 d_4, 8b_1 c_4)$, $(8b_4 d_1, 8b_4 c_1)$ and so on, and accumulating the intermediate values to produce

$$\begin{aligned}
(bd_0, bc_0) &= (b_0 d_0 + 8b_1 d_4 + \ldots + 8b_3 d_2, \ b_0 c_0 + 8b_1 c_4 + \ldots + 8b_3 c_2) \\
(bd_1, bc_1) &= (b_0 d_1 + b_1 d_0 + \ldots + 8b_3 d_3, \ b_0 c_1 + b_1 c_0 + \ldots + 8b_3 c_3) \\
(bd_2, bc_2) &= (b_0 d_2 + b_2 d_0 + \ldots + 8b_4 d_3, \ b_0 c_2 + b_2 c_0 + \ldots + 8b_4 c_3).
\end{aligned}$$

The computation above is executed using (2). In total (including the missing two terms that are computed later on), it requires 25 vector multiplications: 5 are computed using `vmull.s32` and 20 are computed using `vmlal.s32`. Additions are not counted because they are virtually absorbed by the multiply-and-add instructions.

Then, we compute the *five* terms of the multiplications $ac = a \times c$ and $ad = a \times d$. Similarly to above, we compute pairwise multiplications (a_0d_0, a_0c_0), $(8a_1d_4, 8a_1c_4)$, $(8a_4d_1, 8a_4c_1)$ and so on, and accumulate the intermediate values to produce

$$(ad_0, ac_0) = (a_0d_0 + 8a_1d_4 + \ldots + 8a_3d_2, a_0c_0 + 8a_1c_4 + \ldots + 8a_3c_2)$$
$$(ad_1, ac_1) = (a_0d_1 + a_1d_0 + \ldots + 8a_3d_3, a_0c_1 + a_1c_0 + \ldots + 8a_3c_3)$$
$$(ad_2, ac_2) = (a_0d_2 + a_2d_0 + \ldots + 8a_4d_3, a_0c_2 + a_2c_0 + \ldots + 8a_4c_3)$$
$$(ad_3, ac_3) = (a_0d_3 + a_3d_0 + \ldots + 8a_4d_4, a_0c_3 + a_3c_0 + \ldots + 8a_4c_4)$$
$$(ad_4, ac_4) = (a_0d_4 + a_4d_0 + \ldots + a_2d_2, a_0c_4 + a_4c_0 + \ldots + a_2c_2).$$

As before, this vectorized schoolbook computation requires 25 multiplications: 5 computed using `vmull.s32` and 20 computed using `vmlal.s32`.

The intermediate values computed so far are subtracted and added to obtain the first *three* terms of the results $r = ac - bd$ and $s = ad + bc$. This requires 3 64-bit vector additions using `vadd.s64` and 3 64-bit vector subtractions using `vsub.s64`:

$$(s_0, r_0) = (ad_0 + bc_0, ac_0 - bd_0)$$
$$(s_1, r_1) = (ad_1 + bc_1, ac_1 - bd_1)$$
$$(s_2, r_2) = (ad_2 + bc_2, ac_2 - bd_2).$$

We then compute the remaining *two* terms in the computation of $bc = b \times c$ and $bd = b \times d$ (i.e., (bd_3, bc_3) and (bd_4, bc_4)) as follows

$$(bd_3, bc_3) = (b_0d_3 + b_3d_0 + \ldots + 8b_4d_4, b_0c_3 + b_3c_0 + \ldots + 8b_4c_4)$$
$$(bd_4, bc_4) = (b_0d_4 + b_4d_0 + \ldots + b_2d_2, b_0c_4 + b_4c_0 + \ldots + b_2c_2).$$

Finally, we complete the computation with the last *two* terms of the results $r = ac - bd$ and $s = ad + bc$. This involves 2 64-bit vector additions using `vadd.s64` and 2 64-bit vector subtractions using `vsub.s64`:

$$(s_3, r_3) = (ad_3 + bc_3, ac_3 - bd_3)$$
$$(s_4, r_4) = (ad_4 + bc_4, ac_4 - bd_4).$$

The coefficients in the resulting vector $(s_4, r_4, \ldots, s_0, r_0)$ need to be reduced before they are used by subsequent multiplications or squarings. We explain this process below, after discussing squarings in $\mathbb{F}_{p^2}$.

Squaring in $\mathbb{F}_{p^2}$. Let $A = (b_4, a_4, b_3, a_3, b_2, a_2, b_1, a_1, b_0, a_0)$ be a coefficient vector representing an element $(a + b \cdot i)$ in $\mathbb{F}_{p^2}$. To compute the squaring of this element we first shift its coefficients to the right to obtain

$$t_1 = (0, b_4, 0, b_3, 0, b_2, 0, b_1, 0, b_0),$$

which requires 1 64-bit and 2 128-bit vector shifts using `vshr.u64`.

Then, A is subtracted and added with t_1 to obtain

$$t_2 = (b_4, a_4 - b_4, b_3, a_3 - b_3, b_2, a_2 - b_2, b_1, a_1 - b_1, b_0, a_0 - b_0)$$
$$t_3 = (b_4, a_4 + b_4, b_3, a_3 + b_3, b_2, a_2 + b_2, b_1, a_1 + b_1, b_0, a_0 + b_0),$$

which requires 1 64-bit and 2 128-bit vector additions using `vadd.s32` and 1 64-bit and 2 128-bit vector subtractions using `vsub.s32`.

We then shift A to the left by one bit with 1 64-bit and 2 128-bit vector shifts using `vshl.i32`, as follows

$$t_4 = (2a_4, 0, 2a_3, 0, 2a_2, 0, 2a_1, 0, 2a_0, 0).$$

We perform a bitwise selection over t_2 and t_4 using 3 `vbit` instructions to obtain

$$t_5 = (2a_4, a_4 - b_4, 2a_3, a_3 - b_3, 2a_2, a_2 - b_2, 2a_1, a_1 - b_1, 2a_0, a_0 - b_0).$$

We then shift the result by 3 bits to the left using 1 64-bit and 2 128-bit vector shifts with `vshr.u64`, as follows

$$t_6 = (16a_4, 8(a_4 - b_4), 16a_3, 8(a_3 - b_3), 16a_2, 8(a_2 - b_2), 16a_1, 8(a_1 - b_1), 16a_0, 8(a_0 - b_0)).$$

We then compute the *five* terms of the multiplications $r = (a + b) \times (a - b)$ and $s = 2a \times b$. As before, we compute pairwise multiplications $(2a_0b_0, (a_0 - b_0)(a_0 + b_0))$, $(16a_1b_4, 8(a_1 - b_1)(a_4 + b_4))$, $(16a_4b_1, 8(a_4 - b_4)(a_1 + b_1))$ and so on, and accumulate the intermediate values to produce

$$(s_0, r_0) = (2a_0b_0 + \ldots + 16a_3b_2, (a_0 - b_0)(a_0 + b_0) + \ldots + 8(a_3 - b_3)(a_2 + b_2))$$
$$(s_1, r_1) = (2a_0b_1 + \ldots + 16a_3b_3, (a_0 - b_0)(a_1 + b_1) + \ldots + 8(a_3 - b_3)(a_3 + b_3))$$
$$(s_2, r_2) = (2a_0b_2 + \ldots + 16a_4b_3, (a_0 - b_0)(a_2 + b_2) + \ldots + 8(a_4 - b_4)(a_3 + b_3))$$
$$(s_3, r_3) = (2a_0b_3 + \ldots + 16a_4b_4, (a_0 - b_0)(a_3 + b_3) + \ldots + 8(a_4 - b_4)(a_4 + b_4))$$
$$(s_4, r_4) = (2a_0b_4 + \ldots + 2a_2b_2, (a_0 - b_0)(a_4 + b_4) + \ldots + (a_2 - b_2)(a_2 + b_2)).$$

As before, this computation follows (2) and involves 5 multiplications using `vmull.s32` and 20 multiplications using `vmlal.s32`. The reduction procedure that needs to be applied to the output vector $(s_4, r_4, \ldots, s_0, r_0)$ before subsequent multiplications or squarings is described next.

Coefficient Reduction. After computing a multiplication or squaring over $\mathbb{F}_{p^2}$, resulting coefficients must be reduced to avoid overflows in subsequent operations. Given a coefficient vector $(s_4, r_4, \ldots, s_0, r_0)$, coefficient reduction can be accomplished by applying a chain of shifts, additions and logical **and** instructions using the flow $(s_0, r_0) \rightarrow (s_1, r_1) \rightarrow (s_2, r_2) \rightarrow (s_3, r_3) \rightarrow (s_4, r_4) \rightarrow (s_0, r_0)$. In total, this requires 7 vector shifts using `vshr.s64`, 6 vector **and** operations using `vand.u64`, and 6 vector additions using `vadd.s64`. This chain of operations, however, introduces many data hazards that can stall the pipeline for several cycles. In our implementation, for computations in which instruction rescheduling is unable to eliminate most of these data hazards, we switch to a different alternative that consists of splitting the computation in the following *two* propagation chains: $(s_0, r_0) \rightarrow (s_1, r_1) \rightarrow (s_2, r_2) \rightarrow (s_3, r_3) \rightarrow (s_4, r_4)$, and $(s_3, r_3) \rightarrow (s_4, r_4) \rightarrow (s_0, r_0)$. Even though this approach increases the operation count in 1 vector shift, 1 vector addition and 1 vector **and**, it allows to speed up the overall computation because both chains can be interleaved, which eliminates all data hazards.

Vector-Instruction Count. Based on the operation description above, multiplication over $\mathbb{F}_{p^2}$ involves 11 shifts, 7 logical **and** instructions, 17 additions and 50 multiplications. Similarly, squaring over $\mathbb{F}_{p^2}$ involves 17 shifts, 7 logical **and** instructions, 3 bit-selection instructions, 13 additions and 25 multiplications. These counts include coefficient reduction.

4.2   Additional Optimizations to the $\mathbb{F}_{p^2}$ Implementation

As explained in Sect. 3.1, the ARM architecture with NEON support opens the possibility of optimizing software by exploiting the instruction-level parallelism between ARM and NEON instruction sets. We remark, however, that the capability of boosting performance by exploiting this feature strongly depends on the specifics of the targeted microarchitecture and application. For example, microarchitectures such as Cortex-A8 have a relatively large NEON instruction queue that keeps the NEON execution units busy once it is filled; when this happens the ARM core can execute ARM instructions virtually *in parallel*. In contrast, other microarchitectures such as Cortex-A7 and Cortex-A15 exhibit a more continuous flow of instructions to the NEON execution ports, which means that gaining efficiency from mixing ARM and NEON instructions gets significantly more challenging. This is especially true for implementations that rely on the full power of vector instructions. We note, however, that the technique could still be beneficial for implementations that generate enough pipeline stalls. In this case, NEON pipeline stalls could give enough room for ARM instructions to run while the NEON engine recovers (e.g., see [15]).

In the case of our NEON implementation, careful scheduling of instructions was effective in dealing with most latency problems inside the $\mathbb{F}_{p^2}$ functions and, thus, we were able to minimize the occurrence of pipeline stalls. We verified experimentally that this makes very difficult to obtain any additional speedup by mixing ARM and NEON instructions on microarchitectures such as Cortex-A7 and A15. In the case of microarchitectures that are more favorable to the instruction mixing technique (e.g., Cortex-A8 and A9), we applied the following approach. We use NEON to perform the relatively expensive multiplications and squarings over $\mathbb{F}_{p^2}$, and ARM to execute the simpler additions and subtractions (or any combination of these operations). To do this, we inserted add/sub ARM code into the larger NEON-based functions, carefully interleaving NEON and ARM instructions.

We verified that instantiating NEON-based multiplications and squarings that include ARM-based additions or subtractions effectively reduces the cost of these smaller operations. We do even better by suitably merging additions and subtractions inside NEON functions. Specifically, we have identified and implemented the following combinations of operations over $\mathbb{F}_{p^2}$ after analyzing twisted Edwards point doubling and addition formulas:

- `MulAdd`: multiplication $a \times b$ using NEON, addition $c + d$ using ARM.
- `MulSub`: multiplication $a \times b$ using NEON, subtraction $c - d$ using ARM.
- `MulDblSub`: multiplication $a \times b$ using NEON, doubling/subtraction $2 \times c - d$ using ARM.
- `MulAddSub`: multiplication $a \times b$ using NEON, addition $c + d$ and subtraction $c - d$ using ARM.
- `SqrAdd`: squaring a^2 using NEON, addition $c + d$ using ARM.
- `SqrAddSub`: squaring a^2 using NEON, addition $c + d$ and subtraction $c - d$ using ARM.

In our software, the use of these functions is optional. Users can enable this optimization by setting a command flag called "MIX_ARM_NEON". Following the details above, we suggest turning this flag on for Cortex-A8 and A9, and turning it off for Cortex-A7 and A15. See Appendix A for details about the use of these functions inside point doubling and addition.

Additionally, we improve the performance of multiplication and squaring over $\mathbb{F}_{p^2}$ even further by interleaving load/store operations with arithmetic operations. As explained in Sect. 3.1, microarchitectures such as Cortex-A8 are capable of executing one load or store instruction and one arithmetic instruction back-to-back. On the other hand, Cortex-A9 load/store instructions suffer from longer latencies. It is quite fortunate that, in both cases, suitable interleaving of load/store instructions with other non-memory instructions does benefit performance (albeit under different circumstances). We observed experimentally that some newer processors such as Cortex-A15 are negatively impacted by such interleaving. Since in our code input loading and output storing only occur at the very top and at the very bottom of $\mathbb{F}_{p^2}$ arithmetic functions, resp., it was straightforward to create two different execution paths with minimal impact to code size. The path selection is done at compile time: users can enable the optimization by setting a command flag called "INTERLEAVE". We suggest turning this flag on for Cortex-A7, A8 and A9, and turning it off for Cortex-A15.

4.3 Putting Pieces Together

We now describe the implementation details of other necessary operations, and explain how these and our vectorized functions are assembled together to compute Algorithm 1.

Since our vectorization approach is applied at the $\mathbb{F}_{p^2}$ level, most high-level functions in the scalar multiplication remain unchanged for the most part (relative to a non-vectorized implementation). Hence, in our software, endomorphism and point formulas, which are used for table precomputation and in the main loop of Algorithm 1 (Steps 1–2, 7–8), are implemented with only a few minor modifications in comparison with the original explicit formulas. Refer to Appendix A for the modified doubling and addition formulas used in our implementation.

The functions for scalar decomposition and recoding (Steps 3–4) are directly implemented as detailed in [12, Proposition 5] and [12, Algorithm 1], respectively. To extract points from the precomputed table, which is required at Step 8, we carry out a linear pass over the full content of the table performing bitwise selections with vbit instructions. At each step, a mask computed in constant-time determines if a given value is "taken" or not. Inversion over $\mathbb{F}_{p^2}$, which is required for final conversion to affine coordinates at the very end of Algorithm 1, involves the computation of field multiplications and squarings. For these operations, we represent a field element a as a coefficient vector $(a_4, a_3, a_2, a_1, a_0)$, and apply the schoolbook computation (2) (exploiting the typical savings for the case of squarings). In this case, vector multiplications are applied over pairs of internal integer multiplications. This pairing is not optimal, but the effect over the overall cost is relatively small.

Finally, we implemented straightforward functions to convert back and forth between our $\mathbb{F}_{p^2}$ vector representation and the canonical representation. These functions are required just once at the beginning of scalar multiplication to convert the input point to vector representation, and once at the very end to convert the output point to canonical representation. In addition, we need to perform one conversion from $\mathbb{F}_{p^2}$ to $\mathbb{F}_p$ vector representation (and one conversion back) when computing a modular inversion during the final conversion to affine coordinates.

In order to protect against timing and cache attacks, our implementation does not contain branches that depend on secret data and does not use secret memory addresses. For the most part, the elimination of secret branches is greatly facilitated by the regular structure of FourQ's algorithms [12], whereas the elimination of secret memory addresses is done by performing linear passes over the full content of tables in combination with some masking technique.

5   Implementation and Results

In this section, we carry out a theoretical analysis on the core scalar multiplication operations and then present benchmark results on a large variety of ARM Cortex-A based platforms: a 0.9 GHz Raspberry Pi 2 with a Cortex-A7 processor, a 1.0 GHz BeagleBone board with a Cortex-A8 processor, a 1.7 GHz Odroid X2 with a Cortex-A9 processor and a 2.0 GHz Odroid XU3 with a Cortex-A15 processor. All of these ARM-based devices come equipped with a NEON vector unit. The software was compiled with GNU GCC v4.7.2 for the case of

Raspberry Pi and BeagleBone, and with GNU GCC v4.8.2 for the case of the Odroid devices. We report the average of 10^4 operations which were measured with the clock_gettime() function and scaled to clock cycles using the processor frequency.

Next, we analyze the different scalar multiplications when using FourQ.

Variable-Base Scalar Multiplication. Following Algorithm 1, this operation involves the computation of 1 ϕ endomorphism, 2 ψ endomorphisms and 7 points additions in the precomputation stage; 64 doublings, 64 additions and 65 constant-time 8-point table lookups (denoted by **lut8**) in the evaluation stage; and, finally, 1 inversion and 2 multiplications over $\mathbb{F}_{p^2}$ for converting the final result to affine coordinates. This represents a cost of 1**i** + 842**m** + 283**s** + 950.5**a** + 65**lut8** or 3948**M** + 128**S** + 4436**A** + 65**lut8** (considering that 1**m** = 4**M** + 2**A** using schoolbook multiplication and that 1**s** = 2**M** + 3**A**[1]). This operation count does not include other relatively small computations, such as decomposition and recoding. We consider that field inversion of an element a is computed as $a^{2^{127}-3}$ mod $(2^{127} - 1)$ using a fixed chain consisting of 12 modular multiplications and 126 modular squarings.

Fixed-Base Scalar Multiplication. We implemented this operation using the mLSB-set comb method proposed by Faz-Hernández, Longa and Sánchez (see [14, Algorithm 5]). Recall that scalars are in the range $[0, 2^{256})$. By applying a relatively inexpensive Montgomery reduction, a given input scalar can be reduced to the range $[0, N)$ and, thus, fix the maximum scalar bitlength to $t = 249$. As an example, consider the table parameters $w = 5$ and $v = 5$. In this case, the mLSB-set comb method costs $\lceil \frac{249}{w \cdot v} \rceil - 1 = 9$ doublings and $v\lceil \frac{249}{w \cdot v} \rceil - 1 = 49$ mixed additions using $v \cdot 2^{w-1} = 80$ points computed *offline*. Since precomputed points are stored in coordinates $(x + y, y - x, 2t)$ the storage requirement is 7.5 KB and the operation cost is roughly given by 1**i** + 372**m** + 36**s** + 397**a** + 49**lut16** or 1574**M** + 128**S** + 1648**A** + 49**lut16**. This estimate does not include the cost of scalar recoding and conversion.

Double-Scalar Multiplication. We implemented this operation using width-w non-adjacent form (wNAF) with interleaving [16]. Given a computation with the form $[k]P + [l]Q$, scalars k and l can be split in *four* 64-bits sub-scalars each using FourQ's decomposition algorithm. After converting the eight sub-scalars to w-NAF, we compute an 8-way multiscalar multiplication as the main operation. As an example, consider window parameters $w_P = 8$ and $w_Q = 4$ (this assumes that the point P is known in advance, which typically happens in signature verification algorithms). In this case, the computation involves 4

---

[1] **I**, **M**, **S** and **A** represent the cost of modular inversion, multiplication, squaring and addition using $p = 2^{127} - 1$ (resp.); **i**, **m**, **s** and **a** represent the cost of inversion, multiplication, squaring and addition over $\mathbb{F}_{(2^{127}-1)^2}$ (resp.).

doublings and $4 \cdot (2^{w_Q-2} - 1) = 12$ additions (for the *online* precomputation), $4 \cdot (\frac{64}{w_P+1}) = 32$ mixed additions, $4 \cdot (\frac{64}{w_Q+1}) - 1 = 52$ additions and 63 doublings (for the evaluation stage) using $4 \cdot 2^{w_P-2} = 256$ points computed *offline*. Again, we store points in coordinates $(x+y, y-x, 2t)$. This fixes the storage requirement to 24 KB; the operation cost is roughly $1i + 951m + 268s + 1034a$ or $4354M + 128S + 4776A$. This estimate does not include the cost of 2 scalar decompositions and 8 recordings to wNAF. E.g., assuming that $1S = 0.8M$ and $1A = 0.1M$, double-scalar multiplication is expected to be roughly 10% more expensive than variable-base on FourQ.

5.1   Results

Table 1 includes benchmark results of our vectorized FourQ implementation for computing all of the core scalar multiplication operations. The results highlight the efficiency gain that can be obtained through the use of fixed-base scalar multiplications (e.g., during signature generation or ephemeral Diffie-Hellman key generation) using a relatively small amount of precomputation. Most notably, these results show for the first time the potential of using FourQ for signature verification: one double-scalar multiplication is, in most cases, less than 15% more expensive than single variable-base scalar multiplication.

Table 1. Performance results (in terms of thousands of cycles) of core scalar multiplication operations on FourQ with protection against timing and cache attacks on various ARM Cortex-A processors with NEON support. Results were rounded to the nearest 10^3 clock cycles. For this benchmark, fixed-base scalar multiplication used a precomputed table of 80 points (7.5 KB of storage) and double-scalar multiplication used a precomputed table of 256 points (24 KB of storage).

Scalar multiplication	Cortex-A7	Cortex-A8	Cortex-A9	Cortex-A15
$[k]P$, variable base	373	235	256	132
$[k]P$, fixed base	204	144	145	84
$[k]P + [l]Q$	431	269	290	155

In Table 2, we compare our results for variable-base scalar multiplication with other NEON-based implementations in the literature. We include results for the twisted Edwards GLV+GLS curve defined over $\mathbb{F}_{(2^{127}-5997)^2}$ that was proposed by Longa and Sica [20] and the genus 2 Kummer surface defined over $\mathbb{F}_{2^{127}-1}$ that was proposed by Gaudry and Schost [17]. These two curves, which we refer to as "GLV+GLS" and "Kummer", were the previous speed-record holders before the advent of FourQ. Our comparisons also include the popular Montgomery curve known as "Curve25519", which is defined over $\mathbb{F}_{2^{255}-19}$ [4], and two binary curve

Table 2. Performance results (expressed in terms of thousands of clock cycles) of state-of-the-art implementations of various curves targeting the 128-bit security level for computing variable-base scalar multiplication on various ARM Cortex-A processors with NEON support. Results were rounded to the nearest 10^3 clock cycles. The benchmarks for FourQ were done on a 0.9 GHz Raspberry Pi 2 (Cortex-A8), a 1.0 GHz BeagleBone (Cortex-A8), a 1.7 GHz Odroid X2 (Cortex-A9) and a 2.0 GHz Odroid XU3 (Cortex-A15). Cortex-A8 and A9 benchmarks for the Kummer implementation [7] and Cortex-A8, A9 and A15 benchmarks for the Curve25519 implementation [9] were taken from eBACS [8] (computers "h7beagle", "h7green" and "sachr"), while Cortex-A7 benchmarks for Kummer and Curve25519 and Cortex-A15 benchmarks for Kummer were obtained by running eBACS' SUPERCOP toolkit on the corresponding targeted platform. The benchmarks for the GLV-GLS curve were taken directly from [15], and the benchmarks for the binary Koblitz curve K-283 and the binary Edwards curve B-251 were taken directly from [10].

Work	Curve	Cortex A7	Cortex A8	Cortex A9	Cortex A15
This work	FourQ	**373**	**235**	**256**	**132**
Bernstein et al. [7]	Kummer	580	305	356	224
Faz-Hernández et al. [15]	GLV+GLS	-	-	417	244
Bernstein et al. [9]	Curve25519	926	497	568	315
Câmara et al. [10]	B-251	-	657	789	511
Câmara et al. [10]	K-283	-	934	1,148	736

alternatives: the binary Edwards curve defined over $\mathbb{F}_{2^{251}}$ [5], referred to as "B-251", and the NIST's standard Koblitz curve K-283 [23], which is defined over the binary field $\mathbb{F}_{2^{283}}$.

Using the operation counts above and those listed in [12, Table 2], one can determine that FourQ's variable-base scalar multiplication is expected to be roughly 1.28 times faster than Kummer's ladder computation (assuming that $1\mathbf{I} = 115\mathbf{M}$, $1\mathbf{S} = 0.8\mathbf{M}$, $1\mathbf{A} = 0.1\mathbf{M}$ and 1 word-mul $= 0.25\mathbf{M}$). Our actual results show that FourQ is between 1.3 and 1.7 times faster than Bernstein et al.'s Kummer implementation [7] on different ARM microarchitectures. Therefore, FourQ performs even better than expected, demonstrating that its efficient and compact arithmetic enable vector-friendly implementations. These results also highlight the effectiveness of the techniques described in Sect. 4.2.

In comparison to Curve25519, our NEON implementation is between 2.1 and 2.5 times faster when computing variable-base scalar multiplication. Our implementation is also significantly faster than state-of-the-art NEON implementations using binary curves; e.g., it is between 4 and 5.6 times faster than the implementation based on the NIST's standard K-283 curve.

In some cases, even larger speedups are observed for scenarios in which one can exploit precomputations. For example, for signature signing one can leverage the efficiency of fixed-base scalar multiplications to achieve between factor-2.1 and factor-2.8 speedups in comparison to the Kummer surface from [17], which does not support these efficient operations that exploit precomputations.

Acknowledgments. We thank Craig Costello for his valuable comments.

A   Algorithms for Point Operations

The basic point doubling and addition functions used in the NEON implementation are shown in Algorithms 2 and 3, respectively. When selector "MIX_ARM_NEON" is enabled, the algorithms use the functions with the labels on the right (`MulAddSub`, `SqrAdd`, etc.), which mix ARM and NEON instructions as described in Sect. 4.2.

Algorithm 2. Point doubling using homogeneous/extended homogeneous coordinates on $\mathcal{E}$.

Input: $P = (X_1, Y_1, Z_1) \in \mathcal{E}(\mathbb{F}_{p^2})$.
Output: $2P = (X_2, Y_2, Z_2, T_{2,a}, T_{2,b}) \in \mathcal{E}(\mathbb{F}_{p^2})$.

1: **if** MIX_ARM_NEON = *true* **then**
2:    $t_1 = X_1^2$, $X_2 = X_1 + Y_1$ {SqrAdd}
3:    $t_2 = Y_1^2$
4:    $Z_2 = Z_1^2$, $T_{2,b} = t_1 + t_2$, $t_1 = t_2 - t_1$ {SqrAddSub}
5:    $T_{2,a} = X_2^2$
6:    $Y_2 = t_1 \times T_{2,b}$, $t_2 = 2Z_2 - t_1$ {MulDblSub}
7:    $Z_2 = t_1 \times t_2$, $T_{2,a} = T_{2,a} - T_{2,b}$ {MulSub}
8:    $X_2 = t_2 \times T_{2,a}$
9: **else**
10:    $t_1 = X_1^2$
11:    $t_2 = Y_1^2$
12:    $X_2 = X_1 + Y_1$
13:    $T_{2,b} = t_1 + t_2$
14:    $t_1 = t_2 - t_1$
15:    $t_2 = Z_1^2$
16:    $T_{2,a} = X_2^2$
17:    $t_2 = t_2 + t_2$
18:    $t_2 = t_2 - t_1$
19:    $T_{2,a} = T_{2,a} - T_{2,b}$
20:    $Y_2 = t_1 \times T_{2,b}$
21:    $X_2 = t_2 \times T_{2,a}$
22:    $Z_2 = t_1 \times t_2$
23: **endif**
24: **return** $2P = (X_2, Y_2, Z_2, T_{2,a}, T_{2,b})$.

Algorithm 3. Point addition using extended homogeneous coordinates on $\mathcal{E}$.

Input: $P, Q \in \mathcal{E}(\mathbb{F}_{p^2})$ such that $P = (X_1, Y_1, Z_1, T_{1,a}, T_{1,b})$ and $Q = (X_2 + Y_2, Y_2 - X_2, 2Z_2, 2dT_2)$.
Output: $P + Q = (X_3, Y_3, Z_3, T_{3,a}, T_{3,b}) \in \mathcal{E}(\mathbb{F}_{p^2})$.

1: **if** MIX_ARM_NEON $= true$ **then**
2:    $T_{3,a} = T_{1,a} \times T_{1,b}$, $T_{3,b} = X_1 + Y_1$, $Y_3 = Y_1 - X_1$ {MulAddSub}
3:    $t_1 = 2Z_2 \times Z_1$
4:    $Z_3 = (2dT_2) \times T_{3,a}$
5:    $X_3 = (X_2 + Y_2) \times T_{3,b}$, $t_2 = t_1 - Z_3$, $t_1 = t_1 + Z_3$ {MulAddSub}
6:    $Y_3 = (Y_2 - X_2) \times Y_3$
7:    $Z_3 = t_1 \times t_2$, $T_{3,a} = X_3 + Y_3$, $T_{3,b} = X_3 - Y_3$
8:    $X_3 = T_{3,b} \times t_2$
9:    $Y_3 = T_{3,a} \times t_1$
10: **else**
11:    $t_1 = X_1 + Y_1$
12:    $t_2 = Y_1 - X_1$
13:    $t_3 = T_{1,a} \times T_{1,b}$
14:    $t_4 = 2Z_2 \times Z_1$
15:    $Z_3 = (2dT_2) \times t_3$
16:    $X_3 = (X_2 + Y_2) \times t_1$
17:    $Y_3 = (Y_2 - X_2) \times t_2$
18:    $T_{3,a} = X_3 + Y_3$
19:    $T_{3,b} = X_3 - Y_3$
20:    $t_3 = t_1 - Z_3$
21:    $t_1 = t_1 + Z_3$
22:    $X_3 = T_{3,b} \times t_3$
23:    $Z_3 = t_3 \times t_4$
24:    $Y_3 = T_{3,a} \times t_4$
25: **endif**
26: **return** $P + Q = (X_3, Y_3, Z_3, T_{3,a}, T_{3,b})$.

References

1. ARM Limited. Cortex-A8 technical reference manual, 2006–2010. http://infocent er.arm.com/help/topic/com.arm.doc.ddi0344k/DDI0344K_cortex_a8_r3p2_trm.pdf

2. ARM Limited. ARM architecture reference manual: ARMv7-A and ARMv7-R edition, 2007–2014. https://silver.arm.com/download/ARM_and_AMBA_Architect ure/AR570-DA-70000-r0p0-00rel2/DDI0406C_C_arm_architecture_reference_manu al.pdf

3. ARM Limited. ARM Cortex-A9 technical reference manual, 2008–2016. http://info center.arm.com/help/topic/com.arm.doc.100511_0401_10_en/arm_cortexa9_trm_10 0511_0401_10_en.pdf

4. Bernstein, D.J.: Curve25519: new Diffie-Hellman speed records. In: Yung, M., Dodis, Y., Kiayias, A., Malkin, T. (eds.) PKC 2006. LNCS, vol. 3958, pp. 207–228. Springer, Heidelberg (2006). doi:10.1007/11745853_14

5. Bernstein, D.J.: Batch binary Edwards. In: Halevi, S. (ed.) CRYPTO 2009. LNCS, vol. 5677, pp. 317–336. Springer, Heidelberg (2009). doi:10.1007/978-3-642-03356-8_19

6. Bernstein, D.J., Birkner, P., Joye, M., Lange, T., Peters, C.: Twisted Edwards curves. In: Vaudenay, S. (ed.) AFRICACRYPT 2008. LNCS, vol. 5023, pp. 389–405. Springer, Heidelberg (2008). doi:10.1007/978-3-540-68164-9_26

7. Bernstein, D.J., Chuengsatiansup, C., Lange, T., Schwabe, P.: Kummer strikes back: new DH speed records. In: Sarkar, P., Iwata, T. (eds.) ASIACRYPT 2014. LNCS, vol. 8873, pp. 317–337. Springer, Heidelberg (2014). doi:10.1007/978-3-662-45611-8_17

8. Bernstein, D.J., Lange, T.: eBACS: ECRYPT Benchmarking of Cryptographic Systems. http://bench.cr.yp.to/results-dh.html. Accessed 15 May 2016

9. Bernstein, D.J., Schwabe, P.: NEON crypto. In: Prouff, E., Schaumont, P. (eds.) CHES 2012. LNCS, vol. 7428, pp. 320–339. Springer, Heidelberg (2012). doi:10.1007/978-3-642-33027-8_19

10. Câmara, D., Gouvêa, C.P.L., López, J., Dahab, R.: Fast software polynomial multiplication on ARM processors using the NEON engine. In: Cuzzocrea, A., Kittl, C., Simos, D.E., Weippl, E., Xu, L. (eds.) CD-ARES 2013. LNCS, vol. 8128, pp. 137–154. Springer, Heidelberg (2013). doi:10.1007/978-3-642-40588-4_10

11. Chou, T.: Sandy2x: new Curve25519 speed records. In: Dunkelman, O., Keliher, L. (eds.) SAC 2015. LNCS, vol. 9566, pp. 145–160. Springer, Cham (2016). doi:10.1007/978-3-319-31301-6_8

12. Costello, C., Longa, P.: FourQ: four-dimensional decompositions on a Q-curve over the Mersenne prime. In: Advances in Cryptology – ASIACRYPT 2015. LNCS, vol. 9452, pp. 214–235. Springer, Heidelberg (2015). https://eprint.iacr.org/2015/565

13. Costello, C., Longa, P.: FourQlib (2015). http://research.microsoft.com/en-us/projects/fourqlib/

14. Faz-Hernández, A., Longa, P., Sánchez, A.H.: Efficient and secure algorithms for GLV-based scalar multiplication and their implementation on GLV-GLS curves (extended version). J. Cryptogr. Eng. 5(1), 31–52 (2015)

15. Faz-Hernández, A., Longa, P., Sánchez, A.H.: Efficient and secure algorithms for GLV-based scalar multiplication and their implementation on GLV-GLS curves. In: Benaloh, J. (ed.) CT-RSA 2014. LNCS, vol. 8366, pp. 1–27. Springer, Cham (2014). doi:10.1007/978-3-319-04852-9_1

16. Gallant, R.P., Lambert, R.J., Vanstone, S.A.: Faster point multiplication on elliptic curves with efficient endomorphisms. In: Kilian, J. (ed.) CRYPTO 2001. LNCS, vol. 2139, pp. 190–200. Springer, Heidelberg (2001). doi:10.1007/3-540-44647-8_11

17. Gaudry, P., Schost, E.: Genus 2 point counting over prime fields. J. Symb. Comput. 47(4), 368–400 (2012)

18. Hisil, H., Wong, K.K.-H., Carter, G., Dawson, E.: Twisted Edwards curves revisited. In: Pieprzyk, J. (ed.) ASIACRYPT 2008. LNCS, vol. 5350, pp. 326–343. Springer, Heidelberg (2008). doi:10.1007/978-3-540-89255-7_20

19. Kocher, P.C.: Timing attacks on implementations of Diffie-Hellman, RSA, DSS, and other systems. In: Koblitz, N. (ed.) CRYPTO 1996. LNCS, vol. 1109, pp. 104–113. Springer, Heidelberg (1996). doi:10.1007/3-540-68697-5_9

20. Longa, P., Sica, F.: Four-dimensional Gallant-Lambert-Vanstone scalar multiplication. In: Wang, X., Sako, K. (eds.) ASIACRYPT 2012. LNCS, vol. 7658, pp. 718–739. Springer, Heidelberg (2012). doi:10.1007/978-3-642-34961-4_43

21. Naehrig, M., Niederhagen, R., Schwabe, P.: New software speed records for cryptographic pairings. In: Abdalla, M., Barreto, P.S.L.M. (eds.) LATINCRYPT 2010. LNCS, vol. 6212, pp. 109–123. Springer, Heidelberg (2010). doi:10.1007/978-3-642-14712-8_7

22. Robinson, T.: 50 Billion ARM Processors shipped. ARM Connected Community, News Blog (2014). http://armdevices.net/2014/02/26/50-billion-arm-processors-shipped/
23. U.S. Department of Commerce/National Institute of Standards and Technology. Digital Signature Standard (DSS). FIPS-186-4 (2013). http://nvlpubs.nist.gov/nistpubs/FIPS/NIST.FIPS.186-4.pdf

Cryptanalysis of Asymmetric Primitives

Sieving for Closest Lattice Vectors (with Preprocessing)

Thijs Laarhoven[(✉)]

IBM Research, Rüschlikon, Switzerland
mail@thijs.com

Abstract. Lattice-based cryptography has recently emerged as a prime candidate for efficient and secure post-quantum cryptography. The two main hard problems underlying its security are the shortest vector problem (SVP) and the closest vector problem (CVP). Various algorithms have been studied for solving these problems, and for SVP, lattice sieving currently dominates in terms of the asymptotic time complexity: one can heuristically solve SVP in time $2^{0.292d+o(d)}$ in high dimensions d [Becker–Ducas–Gama–Laarhoven, SODA'16]. Although several SVP algorithms can also be used to solve CVP, it is not clear whether this also holds for heuristic lattice sieving methods. The best time complexity for CVP is currently $2^{0.377d+o(d)}$ [Becker–Gama–Joux, ANTS'14].

In this paper we revisit sieving algorithms for solving SVP, and study how these algorithms can be modified to solve CVP and its variants as well. Our first method is aimed at solving one problem instance and minimizes the overall time complexity for a single CVP instance with a time complexity of $2^{0.292d+o(d)}$. Our second method minimizes the amortized time complexity for several instances on the same lattice, at the cost of a larger preprocessing cost. Using nearest neighbor searching with a balanced space-time tradeoff, with this method we can solve the closest vector problem with preprocessing (CVPP) with $2^{0.636d+o(d)}$ space and preprocessing, in $2^{0.136d+o(d)}$ time, while the query complexity can be further reduced to $2^{0.059d+o(d)}$ at the cost of $2^{d+o(d)}$ space and preprocessing, or even to $2^{\varepsilon d+o(d)}$ for arbitrary $\varepsilon > 0$, at the cost of preprocessing time and memory complexities of $(1/\varepsilon)^{O(d)}$.

For easier variants of CVP, such as approximate CVP and bounded distance decoding (BDD), we further show how the preprocessing method achieves even better complexities. For instance, we can solve approximate CVPP with large approximation factors κ with polynomial-sized advice in polynomial time if $\kappa = \Omega(\sqrt{d/\log d})$. This heuristically closes the gap between the decision-CVPP result of [Aharonov–Regev, FOCS'04] (with equivalent κ) and the search-CVPP result of [Dadush–Regev–Stephens-Davidowitz, CCC'14] (which required larger κ).

Keywords: Lattices · Sieving algorithms · Approximate nearest neighbors · Shortest vector problem (SVP) · Closest vector problem (CVP) · Bounded distance decoding (BDD)

© Springer International Publishing AG 2017
R. Avanzi and H. Heys (Eds.): SAC 2016, LNCS 10532, pp. 523–542, 2017.
https://doi.org/10.1007/978-3-319-69453-5_28

1   Introduction

Hard lattice problems. Lattices are discrete subgroups of $\mathbb{R}^d$. More concretely, given a basis $B = \{b_1, \ldots, b_d\} \subset \mathbb{R}^d$, the lattice $\mathcal{L} = \mathcal{L}(B)$ generated by B is defined as $\mathcal{L}(B) = \left\{ \sum_{i=1}^{d} \lambda_i b_i : \lambda_i \in \mathbb{Z} \right\}$. Given a basis of a lattice $\mathcal{L}$, the Shortest Vector Problem (SVP) asks to find a shortest non-zero vector in $\mathcal{L}$ under the Euclidean norm, i.e., a non-zero lattice vector s of norm $\|s\| = \lambda_1(\mathcal{L}) := \min_{v \in \mathcal{L} \setminus \{0\}} \|v\|$. Given a basis of a lattice and a target vector $t \in \mathbb{R}^d$, the Closest Vector Problem (CVP) asks to find a vector $s \in \mathcal{L}$ closest to t under the Euclidean distance, i.e. such that $\|s - t\| = \min_{v \in \mathcal{L}} \|v - t\|$.

These two hard problems are fundamental in the study of lattice-based cryptography, as the security of these schemes is directly related to the hardness of SVP and CVP in high dimensions. Various other hard lattice problems, such as Learning With Errors (LWE) and the Shortest Integer Solution (SIS) problem are closely related to SVP and CVP, and many reductions between these and other hard lattice problems are known; see e.g. [LvdPdW12, Fig. 3.1] or [Ste16] for an overview. These reductions show that being able to solve CVP efficiently implies that almost all other lattice problems can also be solved efficiently in the same dimension, which makes the study of the hardness of CVP even more important for choosing parameters in lattice-based cryptography.

Algorithms for SVP and CVP. Although SVP and CVP are both central in the study of lattice-based cryptography, algorithms for SVP have received somewhat more attention, including a benchmarking website to compare different algorithms [SG15]. Various SVP methods have been studied which can solve CVP as well, such as enumeration (see e.g. [Kan83, FP85, GNR10, MW15]), discrete Gaussian sampling [ADRS15, ADS15], constructing the Voronoi cell of the lattice [AEVZ02, MV10a], and using a tower of sublattices [BGJ14]. On the other hand, for the asymptotically fastest method in high dimensions for SVP[1], lattice sieving, it is not known how to solve CVP with similar costs as SVP.

After a series of theoretical works on constructing efficient heuristic sieving algorithms [NV08, MV10b, WLTB11, ZPH13, Laa15a, LdW15, BGJ15, BL16, BDGL16] as well as practical papers studying how to speed up these algorithms even further [MS11, Sch11, Sch13, BNvdP14, FBB+14, IKMT14, MTB14, MODB14, MLB15, MB16, MLB16], the best time complexity for solving SVP currently stands at $2^{0.292d + o(d)}$ [BDGL16, MLB16]. Although for various other methods the complexities for solving SVP and CVP are similar [GNR10, MV10a, ADS15], one can only guess whether the same holds for lattice sieving methods.

---

[1] To obtain provable guarantees, sieving algorithms are commonly modified to facilitate a somewhat artificial proof technique, which drastically increases the time complexity beyond e.g. the discrete Gaussian sampler and the Voronoi cell algorithm [AKS01, NV08, PS09, MV10b]. On the other hand, if some natural heuristic assumptions are made to enable analyzing the algorithm's behavior, then sieving clearly outperforms these methods. We focus on heuristic sieving in this paper.

To date, the best heuristic time complexity for solving CVP in high dimensions stands at $2^{0.377d+o(d)}$, due to Becker–Gama–Joux [BGJ14].

1.1 Contributions

In this paper we revisit heuristic lattice sieving algorithms, as well as the recent trend to speed up these algorithms using nearest neighbor searching, and we investigate how these algorithms can be modified to solve CVP and its generalizations. We present two different approaches for solving CVP with sieving, each of which we argue has its own merits.

Adaptive sieving. In *adaptive sieving*, we adapt the entire sieving algorithm to the problem instance, including the target vector. As the resulting algorithm is tailored specifically to the given CVP instance, this leads to the best asymptotic complexity for solving a single CVP instance out of our two proposed methods: $2^{0.292d+o(d)}$ time and space. This method is very similar to solving SVP with lattice sieving, and leads to equivalent asymptotics on the space and time complexities as for SVP. The corresponding space-time tradeoff is illustrated in Fig. 1, and equals that of [BDGL16] for solving SVP.

Non-adaptive sieving. Our main contribution, *non-adaptive sieving*, takes a different approach, focusing on cases where several CVP instances are to be solved on the same lattice. The goal here is to minimize the costs of computations depending on the target vector, and spend more time on preprocessing the lattice, so that the amortized time complexity per instance is smaller when solving many CVP instances on the same lattice. This is very closely related to the Closest Vector Problem with Preprocessing (CVPP), where the difference is that we allow for exponential-size preprocessed space. Using nearest neighbor techniques with a balanced space-time tradeoff, we show how to solve CVPP with $2^{0.636d+o(d)}$ space and preprocessing, in $2^{0.136d+o(d)}$ time. A continuous tradeoff between the two complexities can be obtained, where in the limit we can solve CVPP with $(1/\varepsilon)^{O(d)}$ space and preprocessing, in $2^{\varepsilon d+o(d)}$ time. This tradeoff is depicted in Fig. 1.

A potential application of non-adaptive sieving is as a subroutine within enumeration methods. As described in e.g. [GNR10], at any given level in the enumeration tree, one is attempting to solve a CVP instance in a lower-dimensional sublattice of $\mathcal{L}$, where the target vector is determined by the path chosen from the root to the current node in the tree. That means that if we can preprocess this sublattice such that the amortized time complexity of solving CVPP is small, then this could speed up processing the bottom part of the enumeration tree. This in turn might help speed up the lattice basis reduction algorithm BKZ [Sch87,SE94,CN11], which commonly uses enumeration as its SVP subroutine, and is key in assessing the security of lattice-based schemes. As the preprocessing needs to be performed once, CVPP algorithms with impractically large preprocessing costs may not be useful, but we show that with sieving the preprocessing costs can be quite small.

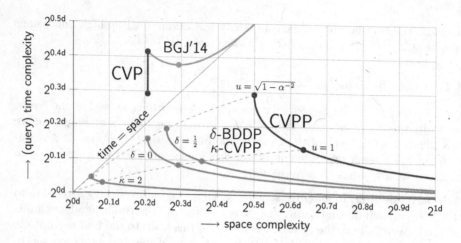

Fig. 1. Heuristic complexities for solving the Closest Vector Problem (CVP), the Closest Vector Problem with Preprocessing (CVPP), Bounded Distance Decoding with Preprocessing (δ-BDDP), and the Approximate Closest Vector Problem with Preprocessing (κ-CVPP). The red curve shows CVP complexities of Becker–Gama–Joux [BGJ14]. The left blue curve denotes CVP complexities of adaptive sieving. The right blue curve shows exact CVPP complexities using non-adaptive sieving. Purple curves denote relaxations of CVPP corresponding to different parameters δ (BDD radius) and κ (approximation factor). Note that exact CVPP corresponds to δ-BDDP with $\delta = 1$ and to κ-CVPP with $\kappa = 1$. (Color figure online)

Outline. The remainder of the paper is organized as follows. In Sect. 2 we describe some preliminaries, such as sieving algorithms and a useful result on nearest neighbor searching. Section 3 describes adaptive sieving and its analysis for solving CVP without preprocessing. Section 4 describes the preprocessing approach to solving CVP, with complexity analyses for exact CVP and some of its relaxations.

2   Preliminaries

2.1   Lattice Sieving for Solving SVP

Heuristic lattice sieving algorithms for solving the shortest vector problem all use the following basic property of lattices: if $v, w \in \mathcal{L}$, then their sum/difference $v \pm w \in \mathcal{L}$ is a lattice vector as well. Therefore, if we have a long list L of lattice vectors stored in memory, we can consider combinations of these vectors to obtain new, shorter lattice vectors. To make sure the algorithm makes progress in finding shorter lattice vectors, L needs to contain a lot of lattice vectors; for vectors $v, w \in \mathcal{L}$ of similar norm, the vector $v - w$ is shorter than v, w iff the angle between v, w is smaller than $\pi/3$, which for random vectors v, w occurs with probability $(3/4)^{d/2+o(d)}$. The expected space complexity of heuristic sieving algorithms follows directly from this observation: if we draw $(4/3)^{d/2+o(d)}$ random vectors from the unit sphere, we expect a large number of pairs of vectors to have angle less than $\pi/3$, leading to many short difference vectors. This is exactly the heuristic assumption used in analyzing these sieving algorithms: when normalized, vectors in L follow the same distribution as vectors sampled uniformly at random from the unit sphere.

Heuristic 1. *When normalized, the list vectors $w \in L$ behave as i.i.d. uniformly distributed random vectors from the unit sphere $\mathcal{S}^{d-1} := \{x \in \mathbb{R}^d : \|x\| = 1\}$.*

Therefore, if we start by sampling a list L of $(4/3)^{d/2+o(d)}$ long lattice vectors, and iteratively consider combinations of vectors in L to find shorter vectors, we expect to keep making progress. Note that naively, combining pairs of vectors in a list of size $(4/3)^{d/2+o(d)} \approx 2^{0.208d+o(d)}$ takes time $(4/3)^{d+o(d)} \approx 2^{0.415d+o(d)}$.

The Nguyen-Vidick sieve. The heuristic sieve algorithm of Nguyen and Vidick [NV08] starts by sampling a list L of $(4/3)^{d/2+o(d)}$ long lattice vectors, and uses a *sieve* to map L, with maximum norm $R := \max_{v \in L} \|v\|$, to a new list L', with maximum norm at most γR for $\gamma < 1$ close to 1. By repeatedly applying this sieve, after $\text{poly}(d)$ iterations we expect to find a long list of lattice vectors of norm at most $\gamma^{\text{poly}(d)} R = O(\lambda_1(\mathcal{L}))$. The final list is then expected to contain a shortest vector of the lattice. Algorithm 3 in Appendix A describes a sieve equivalent to Nguyen-Vidick's original sieve, to map L to L' in $|L|^2$ time.

Micciancio and Voulgaris' GaussSieve. Micciancio and Voulgaris used a slightly different approach in the GaussSieve [MV10b]. This algorithm reduces the memory usage by immediately *reducing* all pairs of lattice vectors that are sampled. The algorithm uses a single list L, which is always kept in a state where for all $w_1, w_2 \in L$, $\|w_1 \pm w_2\| \geq \|w_1\|, \|w_2\|$, and each time a new vector $v \in \mathcal{L}$ is sampled, its norm is reduced with vectors in L. After the norm can no longer be reduced, the vectors in L are reduced with v. Modified list vectors are added to a stack to be processed later (to maintain the pairwise reduction-property of L), and new vectors which are pairwise reduced with L are added to L. Immediately reducing all pairs of vectors means that the algorithm uses less time and memory in practice, but at the same time Nguyen and Vidick's heuristic proof technique does not apply here. However, it is commonly believed that the same bounds $(4/3)^{d/2+o(d)}$ and $(4/3)^{d+o(d)}$ on the space and time complexities hold for the GaussSieve. Pseudocode of the GaussSieve is given in Algorithm 4 in Appendix A.

2.2   Nearest Neighbor Searching

Given a data set $L \subset \mathbb{R}^d$, the nearest neighbor problem asks to preprocess L such that, when given a query $t \in \mathbb{R}^d$, one can quickly return a nearest neighbor $s \in L$ with distance $\|s - t\| = \min_{w \in L} \|w - t\|$. This problem is essentially identical to CVP, except that L is a finite set of unstructured points, rather than the infinite set of all points in a lattice $\mathcal{L}$.

Locality-Sensitive Hashing/Filtering (LSH/LSF). A celebrated technique for finding nearest neighbors in high dimensions is Locality-Sensitive Hashing (LSH) [IM98, WSSJ14], where the idea is to construct many random partitions of the space, and store the list L in hash tables with buckets corresponding to regions. Preprocessing then consists of constructing these hash tables, while a query t is answered by doing a lookup in each of the hash tables, and searching for a nearest neighbor in these buckets. More details on LSH in combination with sieving can be found in e.g. [Laa15a, LdW15, BGJ15, BL16].

Similar to LSH, Locality-Sensitive Filtering (LSF) [BDGL16, Laa15b] divides the space into regions, with the added relaxation that these regions do not have to form a partition; regions may overlap, and part of the space may not be covered by any region. This leads to improved results compared to LSH when L has size exponential in d [BDGL16, Laa15b]. Below we restate one of the main results of [Laa15b] for our applications. The specific problem considered here is: given a data set $L \subset \mathcal{S}^{d-1}$ sampled uniformly at random, and a random query $t \in \mathcal{S}^{d-1}$, return a vector $w \in L$ such that the angle between w and t is at most θ. The following result further assumes that the list L contains $n = (1/\sin\theta)^{d+o(d)}$ vectors.

Lemma 1. [Laa15b, Corollary 1] Let $\theta \in (0, \frac{1}{2}\pi)$, and let $u \in [\cos\theta, 1/\cos\theta]$. Let $L \subset \mathcal{S}^{d-1}$ be a list of $n = (1/\sin\theta)^{d+o(d)}$ vectors sampled uniformly at random from $\mathcal{S}^{d-1}$. Then, using spherical LSF with parameters $\alpha_q = u\cos\theta$ and $\alpha_u = \cos\theta$, one can preprocess L in time $n^{1+\rho_u+o(1)}$, using $n^{1+\rho_u+o(1)}$ space, and with high probability answer a random query $t \in \mathcal{S}^{d-1}$ correctly in time $n^{\rho_q+o(1)}$, where:

$$n^{\rho_q} = \left(\frac{\sin^2\theta\,(u\cos\theta + 1)}{u\cos\theta - \cos 2\theta}\right)^{d/2}, \quad n^{\rho_u} = \left(\frac{\sin^2\theta}{1 - \cot^2\theta\,(u^2 - 2u\cos\theta + 1)}\right)^{d/2}. \tag{1}$$

Applying this result to sieving for solving SVP, where $n = \sin(\frac{\pi}{3})^{-d+o(d)}$ and we are looking for pairs of vectors at angle at most $\frac{\pi}{3}$ to perform reductions, this leads to a space and preprocessing complexity of $n^{0.292d+o(d)}$, and a query complexity of $2^{0.084d+o(d)}$. As the preprocessing in sieving is only performed once, and queries are performed $n \approx 2^{0.208d+o(d)}$ times, this leads to a reduction of the complexities of sieving (for SVP) from $2^{0.208d+o(d)}$ space and $2^{0.415d+o(d)}$ time, to $2^{0.292d+o(d)}$ space and time [BDGL16].

3   Adaptive Sieving for CVP

We present two methods for solving CVP using sieving, the first of which we call *adaptive sieving* – we adapt the entire sieving algorithm to the particular CVP instance, to obtain the best overall time complexity for solving one instance. When solving several CVP instances, the costs roughly scale linearly with the number of instances.

Algorithm 1. The adaptive Nguyen-Vidick sieve for finding closest vectors

---

Require: Lists $L_0, L_t \subset \mathcal{L}$ containing $(4/3)^{d/2+o(d)}$ vectors at distance $\leq R$ from $\mathbf{0}, t$
Ensure: Lists $L_0', L_t' \subset \mathcal{L}$ contain $(4/3)^{d/2+o(d)}$ vectors at distance $\leq \gamma R$ from $\mathbf{0}, t$
1: Initialize empty lists L_0', L_t'
2: **for each** $(\boldsymbol{w}_1, \boldsymbol{w}_2) \in L_0 \times L_0$ **do**
3:     **if** $\|\boldsymbol{w}_1 - \boldsymbol{w}_2\| \leq \gamma R$ **then**
4:         Add $\boldsymbol{w}_1 - \boldsymbol{w}_2$ to the list L_0'
5: **for each** $(\boldsymbol{w}_1, \boldsymbol{w}_2) \in L_t \times L_0$ **do**
6:     **if** $\|(\boldsymbol{w}_1 - \boldsymbol{w}_2) - t\| \leq \gamma R$ **then**
7:         Add $\boldsymbol{w}_1 - \boldsymbol{w}_2$ to the list L_t'
8: **return** (L_0', L_t')

Using one list. The main idea behind this method is to translate the SVP algorithm by the target vector t; instead of generating a long list of lattice vectors reasonably close to $\mathbf{0}$, we generate a list of lattice vectors close to t, and combine lattice vectors to find lattice vectors even closer vectors to t. The final list then hopefully contains a closest vector to t.

One quickly sees that this does not work, as the fundamental property of lattices does not hold for the lattice coset $t + \mathcal{L}$: if $\boldsymbol{w}_1, \boldsymbol{w}_2 \in t + \mathcal{L}$, then $\boldsymbol{w}_1 \pm \boldsymbol{w}_2 \notin t + \mathcal{L}$. In other words, two lattice vectors close to t can only be combined to form lattice vectors close to $\mathbf{0}$ or $2t$. So if we start with a list of vectors close to t, and combine vectors in this list as in the Nguyen-Vidick sieve, then after one iteration we will end up with a list L' of lattice vectors close to $\mathbf{0}$.

Using two lists. To make the idea of translating the whole problem by t work for the Nguyen-Vidick sieve, we make the following modification: we keep track of two lists $L = L_0$ and L_t of lattice vectors close to $\mathbf{0}$ and t, and construct a sieve which maps two input lists L_0, L_t to two output lists L_0', L_t' of lattice vectors slightly closer to $\mathbf{0}$ and t. Similar to the original Nguyen-Vidick sieve, we then apply this sieve several times to two initial lists (L_0, L_t) with a large radius R, to end up with two lists L_0 and L_t of lattice vectors at distance at most approximately $\sqrt{4/3} \cdot \lambda_1(\mathcal{L})$ from $\mathbf{0}$ and t^2. The argumentation that this algorithm works is almost identical to that for solving SVP, where we now make the following slightly different heuristic assumption.

Heuristic 2. *When normalized, the list vectors L_0 and L_t in the modified Nguyen-Vidick sieve both behave as i.i.d. uniformly distributed random vectors from the unit sphere.*

The resulting algorithm, based on the Nguyen-Vidick sieve, is presented in Algorithm 1.

---

[2] Observe that by the Gaussian heuristic, there are $(4/3)^{d/2+o(d)}$ vectors in $\mathcal{L}$ within any ball of radius $\sqrt{4/3} \cdot \lambda_1(\mathcal{L})$. So the list size of the NV-sieve will surely decrease below $(4/3)^{d/2}$ when $R < \sqrt{4/3} \cdot \lambda_1(\mathcal{L})$.

Main result. As the (heuristic) correctness of this algorithm follows directly from the correctness of the original NV-sieve, and nearest neighbor techniques can be applied to this algorithm in similar fashion as well, we immediately obtain the following result. Note that space-time tradeoffs for SVP, such as the one illustrated in [BDGL16, Fig. 1], similarly carry over to solving CVP, and the best tradeoff for SVP (and therefore CVP) is depicted in Fig. 1.

Theorem 1. *Assuming Heuristic 2 holds, the adaptive Nguyen-Vidick sieve with spherical LSF solves CVP in time* T *and space* S, *with*

$$S = (4/3)^{d/2+o(d)} \approx 2^{0.208d+o(d)}, \quad T = (3/2)^{d/2+o(d)} \approx 2^{0.292d+o(d)}. \quad (2)$$

An important open question is whether these techniques can also be applied to the faster GaussSieve algorithm to solve CVP. The GaussSieve seems to make even more use of the property that the sum/difference of two lattice vectors is also in the lattice, and operations in the GaussSieve in $\mathcal{L}$ cannot as easily be *mimicked* for the coset $t + \mathcal{L}$. Solving CVP with the GaussSieve with similar complexities is left as an open problem.

4   Non-adaptive Sieving for CVPP

Our second method for finding closest vectors with heuristic lattice sieving follows a slightly different approach. Instead of focusing only on the total time complexity for one problem instance, we split the algorithm into two phases:

– Phase 1: Preprocess the lattice $\mathcal{L}$, without knowledge of the target t;
– Phase 2: Process the query t and output a closest lattice vector $s \in \mathcal{L}$ to t.

Intuitively it may be more important to keep the costs of Phase 2 small, as the preprocessed data can potentially be reused later for other instances on the same lattice. This approach is essentially equivalent to the Closest Vector Problem with Preprocessing (CVPP): preprocess $\mathcal{L}$ such that when given a target vector t later, one can quickly return a closest vector $s \in \mathcal{L}$ to t. For CVPP however the preprocessed space is usually restricted to be of polynomial size, and the time used for preprocessing the lattice is often not taken into account. Here we will keep track of the preprocessing costs as well, and we do not restrict the output from the preprocessing phase to be of size poly(d).

Algorithm description. To minimize the costs of answering a query, and to do the preprocessing independently of the target vector, we first run a standard SVP sieve, resulting in a large list L of almost all short lattice vectors. Then, after we are given the target vector t, we use L to reduce the target. Finally, once the resulting vector $t' \in t + \mathcal{L}$ can no longer be reduced with our list, we hope that this reduced vector t' is the shortest vector in the coset $t + \mathcal{L}$, so that 0 is the closest lattice vector to t' and $s = t - t'$ is the closest lattice vector to t.

Algorithm 2. Non-adaptive sieving (Phase 2) for finding closest vectors

---

Require: A list $L \subset \mathcal{L}$ of $\alpha^{d/2+o(d)}$ vectors of norm at most $\alpha \cdot \lambda_1(\mathcal{L})$, and $t \in \mathbb{R}^d$
Ensure: The output vector s is the closest lattice vector to t (w.h.p.)
 1: Initialize $t' \leftarrow t$
 2: **for each** $w \in L$ **do**
 3:    **if** $\|t' - w\| \leq \|t'\|$ **then**
 4:        Replace $t' \leftarrow t' - w$ and restart the **for**-loop
 5: **return** $s = t - t'$

The first phase of this algorithm consists in running a sieve and storing the resulting list in memory (potentially in a nearest neighbor data structure for faster lookups). For this phase either the Nguyen-Vidick sieve or the GaussSieve can be used. The second phase is the same for either method, and is described in Algorithm 2 for the general case of an input list essentially consisting of the $\alpha^{d+o(d)}$ shortest vectors in the lattice. Note that a standard SVP sieve would produce a list of size $(4/3)^{d/2+o(d)}$ corresponding to $\alpha = \sqrt{4/3}$.

List size. We first study how large L must be to guarantee that the algorithm succeeds. One might wonder why we do not fix $\alpha = \sqrt{4/3}$ immediately in Algorithm 2. To see why this choice of α does not suffice, suppose we have a vector $t' \in t + \mathcal{L}$ which is no longer reducible with L. This implies that t' has norm approximately $\sqrt{4/3} \cdot \lambda_1(\mathcal{L})$, similar to what happens in the GaussSieve. Now, unfortunately the fact that t' cannot be reduced with L anymore, does *not* imply that the closest lattice point to t' is 0. In fact, it is more likely that there exists an $s \in t + \mathcal{L}$ of norm slightly more than $\sqrt{4/3} \cdot \lambda_1(\mathcal{L})$ which is closer to t', but which is not used for reductions.

By the Gaussian heuristic, we expect the distance from t and t' to the lattice to be $\lambda_1(\mathcal{L})$. So to guarantee that 0 is the closest lattice vector to the reduced vector t', we need t' to have norm at most $\lambda_1(\mathcal{L})$. To analyze and prove correctness of this algorithm, we will therefore prove that, under the assumption that the input is a list of the $\alpha^{d+o(d)}$ shortest lattice vectors of norm at most $\alpha \cdot \lambda_1(\mathcal{L})$ for a particular choice of α, w.h.p. the algorithm reduces t to a vector $t' \in t + \mathcal{L}$ of norm at most $\lambda_1(\mathcal{L})$.

To study how to set α, we start with the following elementary lemma regarding the probability of reduction between two uniformly random vectors with given norms.

Lemma 2. *Let $v, w > 0$ and let $\boldsymbol{v} = v \cdot \boldsymbol{e}_v$ and $\boldsymbol{w} = w \cdot \boldsymbol{e}_w$. Then:*

$$\mathbb{P}_{\boldsymbol{e}_v, \boldsymbol{e}_w \sim \mathcal{S}^{d-1}} \left(\|\boldsymbol{v} - \boldsymbol{w}\|^2 < \|\boldsymbol{v}\|^2 \right) \sim \left[1 - \left(\tfrac{w}{2v} \right)^2 \right]^{d/2+o(d)}. \tag{3}$$

Proof. Expanding $\|\boldsymbol{v} - \boldsymbol{w}\|^2 = v^2 + w^2 - 2vw \langle \boldsymbol{e}_v, \boldsymbol{e}_w \rangle$ and $\|\boldsymbol{v}\|^2 = v^2$, the condition $\|\boldsymbol{v} - \boldsymbol{w}\|^2 < \|\boldsymbol{v}\|^2$ equals $\tfrac{w}{2v} < \langle \boldsymbol{e}_v, \boldsymbol{e}_w \rangle$. The result follows from [BDGL16, Lemma 2.1].

Under Heuristic 1, we then obtain a relation between the choice of α for the input list and the expected norm of the reduced vector t' as follows.

Lemma 3. *Let $L \subset \alpha \cdot \mathcal{S}^{d-1}$ be a list of $\alpha^{d+o(d)}$ uniformly random vectors of norm $\alpha > 1$, and let $v \in \beta \cdot \mathcal{S}^{d-1}$ be sampled uniformly at random. Then, for high dimensions d, there exists a $w \in L$ such that $\|v - w\| < \|v\|$ if and only if*

$$\alpha^4 - 4\beta^2 \alpha^2 + 4\beta^2 < 0. \tag{4}$$

Proof. By Lemma 2 we can reduce v with $w \in L$ with probability similar to $p = [1 - \frac{\alpha^2}{4\beta^2}]^{d/2 + o(d)}$. Since we have $n = \alpha^{d+o(d)}$ such vectors w, the probability that none of them can reduce v is $(1-p)^n$, which is $o(1)$ if $n \gg 1/p$ and $1 - o(1)$ if $n \ll 1/p$. Expanding $n \cdot p$, we obtain the given Eq. (4), where $\alpha^4 - 4\beta^2 \alpha^2 + 4\beta^2 < 0$ implies $n \gg 1/p$. □

Note that in our applications, we do not just have a list of $\alpha^{d+o(d)}$ lattice vectors of norm $\alpha \cdot \lambda_1(\mathcal{L})$; for any $\alpha_0 \in [1, \alpha]$ we expect L to contain $\alpha_0^{d+o(d)}$ lattice vectors of norm at most $\alpha_0 \cdot \lambda_1(\mathcal{L})$. To obtain a reduced vector t' of norm $\beta \cdot \lambda_1(\mathcal{L})$, we therefore obtain the condition that for *some* value $\alpha_0 \in [1, \alpha]$, it must hold that $\alpha_0^4 - 4\beta^2 \alpha_0^2 + 4\beta^2 < 0$.

From (4) it follows that $p(\alpha^2) = \alpha^4 - 4\beta^2 \alpha^2 + 4\beta^2$ has two roots $r_1 < 2 < r_2$ for α^2, which lie close to 2 for $\beta \approx 1$. The condition that $p(\alpha_0^2) < 0$ for some $\alpha_0 \leq \alpha$ is equivalent to $\alpha > r_1$, which for $\beta = 1 + o(1)$ implies that $\alpha^2 \geq 2 + o(1)$. This means that asymptotically we must set $\alpha = \sqrt{2}$, and use $n = 2^{d/2 + o(d)}$ input vectors, to guarantee that w.h.p. the algorithm succeeds. A sketch of the situation is also given in Fig. 2a.

Modifying the first phase. As we will need a larger list of size $2^{d/2 + o(d)}$ to make sure we can solve CVP exactly, we need to adjust Phase 1 of the algorithm as well. Recall that with standard sieving, we reduce vectors iff their angle is at most $\theta = \frac{\pi}{3}$, resulting in a list of size $(\sin \theta)^{-d+o(d)}$. As we now need the output list of the first phase to consist of $2^{d/2 + o(d)} = (\sin \theta')^{-d+o(d)}$ vectors for $\theta' = \frac{\pi}{4}$, we make the following adjustment: only reduce v and w if their common angle is less than $\frac{\pi}{4}$. For unit length vectors, this condition is equivalent to reducing v with w iff $\|v - w\|^2 \leq (2 - \sqrt{2}) \cdot \|v\|^2$. This further accelerates nearest neighbor techniques due to the smaller angle θ. Pseudocode for the modified first phase is given in Appendix B.

Main result. With the algorithm in place, let us now analyze its complexity for solving CVP. The first phase of the algorithm generates a list of size $2^{d/2 + o(d)}$ by combining pairs of vectors, and naively this can be done in time $T_1 = 2^{d+o(d)}$ and space $S = 2^{d/2 + o(d)}$, with query complexity $T_2 = 2^{d/2 + o(d)}$. Using nearest neighbor searching (Lemma 1), the query and preprocessing complexities can be further reduced, leading to the following result.

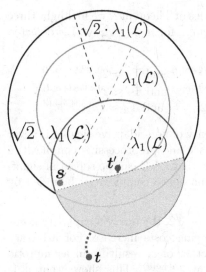 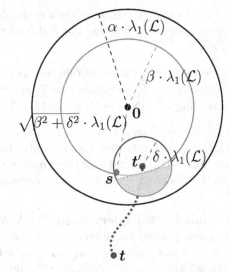

(a) For solving **exact CVP**, we must reduce the vector t to a vector $t' \in t + \mathcal{L}$ of norm at most $\lambda_1(\mathcal{L})$. The nearest lattice point to t' lies in a ball of radius approximately $\lambda_1(\mathcal{L})$ around t' (blue), and almost all the mass of this ball is contained in the (black) ball around 0 of radius $\sqrt{2} \cdot \lambda_1(\mathcal{L})$. So if $s \in \mathcal{L} \setminus \{0\}$ had lain closer to t' than 0, we would have reduced t' with s, since $s \in L$.

(b) For **variants of CVP**, a choice α for the list size implies a norm $\beta \cdot \lambda_1(\mathcal{L})$ of t'. The nearest lattice vector s to t' lies within $\delta \cdot \lambda_1(\mathcal{L})$ of t' ($\delta = 1$ for approx-CVP), so with high probability s has norm approximately $(\sqrt{\beta^2 + \delta^2}) \cdot \lambda_1(\mathcal{L})$. For δ-BDD, if $\sqrt{\beta^2 + \delta^2} \leq \alpha$ then we expect the nearest point s to be in the list L. For κ-CVP, if $\beta \leq \kappa$, then the lattice vector $t - t'$ has norm at most $\kappa \cdot \lambda_1(\mathcal{L})$.

Fig. 2. Comparison of the list size complexity analysis for CVP (left) and BDD/approximate CVP (right). The point t represents the target vector, and after a series of reductions with Algorithm 2, we obtain $t' \in t + \mathcal{L}$. Blue balls around t' depict regions in which we expect the closest lattice point to t' to lie, where the blue shaded area indicates a negligible fraction of this ball [BDGL16, Lemma 2]. (Color figure online)

Theorem 2. *Let* $u \in (\frac{1}{2}\sqrt{2}, \sqrt{2})$. *Using non-adaptive sieving, we can solve CVP with preprocessing time* T_1, *space complexity* S, *and query time complexity* T_2 *as follows:*

$$S = T_1 = \left(\frac{1}{u(\sqrt{2} - u)} \right)^{d/2 + o(d)} \quad , \qquad T_2 = \left(\frac{\sqrt{2} + u}{2u} \right)^{d/2 + o(d)} . \tag{5}$$

Proof. These complexities follow from Lemma 1 with $\theta = \frac{\pi}{4}$, noting that the first phase can be performed in time and space $T_1 = S = n^{1 + \rho_u}$, and the second phase in time $T_2 = n^{\rho_q}$.

To illustrate the time and space complexities of Theorem 2, we highlight three special cases u as follows. The full tradeoff curve for $u \in (\frac{1}{2}\sqrt{2}, \sqrt{2})$ is depicted in Fig. 1.

- Setting $u = \frac{1}{2}\sqrt{2}$, we obtain $S = T_1 = 2^{d/2+o(d)}$ and $T_2 \approx 2^{0.2925d+o(d)}$.
- Setting $u = 1$, we obtain $S = T_1 \approx 2^{0.6358d+o(d)}$ and $T_2 \approx 2^{0.1358d+o(d)}$.
- Setting $u = \frac{1}{2}(\sqrt{2}+1)$, we get $S = T_1 = 2^{d+o(d)}$ and $T_2 \approx 2^{0.0594d+o(d)}$.

The first result shows that the query complexity of non-adaptive sieving is never worse than for adaptive sieving; only the space and preprocessing complexities are worse. The second and third results show that CVP can be solved in significantly less time, even with preprocessing and space complexities bounded by $2^{d+o(d)}$.

Minimizing the query complexity. As $u \to \sqrt{2}$, the query complexity keeps decreasing while the memory and preprocessing costs increase. For arbitrary $\varepsilon > 0$, we can set $u = u_\varepsilon \approx \sqrt{2}$ as a function of ε, resulting in asymptotic complexities $S = T_1 = (1/\varepsilon)^{O(d)}$ and $T_2 = 2^{\varepsilon d+o(d)}$. This shows that it is possible to obtain a slightly subexponential query complexity, at the cost of superexponential space, by taking $\varepsilon = o(1)$ as a function of d.

Corollary 1. *For arbitrary $\varepsilon > 0$, using non-adaptive sieving we can solve CVPP with preprocessing time and space complexities $(1/\varepsilon)^{O(d)}$, in time $2^{\varepsilon d+o(d)}$. In particular, we can solve CVPP in $2^{o(d)}$ time, using $2^{\omega(d)}$ space and preprocessing.*

Being able to solve CVPP in subexponential time with superexponential preprocessing and memory is neither trivial nor quite surprising. A naive approach to the problem, with this much memory, could for instance be to index the entire fundamental domain of $\mathcal{L}$ in a hash table. One could partition this domain into small regions, solve CVP for the centers of each of these regions, and store all the solutions in memory. Then, given a query, one looks up which region t is in, and returns the answer corresponding to that vector. With a sufficiently fine-grained partitioning of the fundamental domain, the answers given by the look-ups are accurate, and this algorithm probably also runs in subexponential time.

Although it may not be surprising that it is possible to solve CVPP in subexponential time with (super)exponential space, it is not clear what the complexities of other methods would be. Our method presents a clear tradeoff between the complexities, where the constants in the preprocessing exponent are quite small; for instance, we can solve CVPP in time $2^{0.06d+o(d)}$ with less than $2^{d+o(d)}$ memory, which is the same amount of memory/preprocessing of the best provable SVP and CVP algorithms [ADRS15, ADS15]. Indexing the fundamental domain may well require much more memory than this.

4.1   Bounded Distance Decoding with Preprocessing

We finally take a look at specific instances of CVP which are easier than the general problem, such as when the target t lies unusually close to the lattice.

This problem naturally appears in practice, when a private key consists of a *good basis* of a lattice with short basis vectors, and the public key is a *bad basis* of the same lattice. An encryption of a message could then consist of the message being mapped to a lattice point $v \in \mathcal{L}$, and a small error vector e being added to v ($t = v + e$) to hide v. If the noise e is small enough, then with a good basis one can decode t to the closest lattice vector v, while someone with the bad basis cannot decode correctly. As decoding for arbitrary t (solving CVP) is known to be hard even with knowledge of a good basis [Mic01, FM02, Reg04, AKKV05], e needs to be very short, and t must lie unusually close to the lattice.

So instead of assuming target vectors $t \in \mathbb{R}^d$ are sampled at random, suppose that t lies at distance at most $\delta \cdot \lambda_1(\mathcal{L})$ from $\mathcal{L}$, for $\delta \in (0,1)$. For adaptive sieving, recall that the list size $(4/3)^{d/2+o(d)}$ is the minimum initial list size one can hope to use to obtain a list of short lattice vectors; with fewer vectors, one would not be able to solve SVP.[3] For non-adaptive sieving however, it may be possible to reduce the list size below $2^{d/2+o(d)}$.

List size. Let us again assume that the preprocessed list L contains almost all $\alpha^{d+o(d)}$ lattice vectors of norm at most $\alpha \cdot \lambda_1(\mathcal{L})$. The choice of α implies a maximum norm $\beta_\alpha \cdot \lambda_1(\mathcal{L})$ of the reduced vector t', as described in Lemma 3. The nearest lattice vector $s \in \mathcal{L}$ to t' lies within radius $\delta \cdot \lambda_1(\mathcal{L})$ of t', and w.h.p. $s - t'$ is approximately orthogonal to t'; see Fig. 2b, where the shaded area is asymptotically negligible. Therefore w.h.p. s has norm at most $(\sqrt{\beta_\alpha^2 + \delta^2}) \cdot \lambda_1(\mathcal{L})$. Now if $\sqrt{\beta_\alpha^2 + \delta^2} \le \alpha$, then we expect the nearest vector to be contained in L, so that ultimately 0 is nearest to t'. Substituting $\alpha^4 - 4\beta^2\alpha^2 + 4\beta^2 = 0$ and $\beta^2 + \delta^2 \le \alpha^2$, and solving for α, this leads to the following condition on α.

$$\alpha^2 \ge \tfrac{2}{3}(1 + \delta^2) + \tfrac{2}{3}\sqrt{(1 + \delta^2)^2 - 3\delta^2}. \tag{6}$$

Taking $\delta = 1$, corresponding to exact CVP, leads to the condition $\alpha \ge \sqrt{2}$ as expected, while in the limiting case of $\delta \to 0$ we obtain the condition $\alpha \ge \sqrt{4/3}$. This matches experimental observations using the GaussSieve, where after finding the shortest vector, newly sampled vectors often cause *collisions* (i.e. being reduced to the 0-vector). In other words, Algorithm 2 often reduces target vectors t which essentially lie *on* the lattice ($\delta \to 0$) to the 0-vector when the list has size $(4/3)^{d/2+o(d)}$. This explains why collisions in the GaussSieve are common when the list size grows to size $(4/3)^{d/2+o(d)}$.

Main result. To solve BDD with a target t at distance $\delta \cdot \lambda_1(\mathcal{L})$ from the lattice, we need the preprocessing to produce a list of almost all $\alpha^{d+o(d)}$ vectors of norm at most $\alpha \cdot \lambda_1(\mathcal{L})$, with α satisfying (6). Similar to the analysis for CVP, we can produce such a list by only doing reductions between two vectors if their

---

[3] The recent paper [BLS16] discusses how to use less memory in sieving, by using triple- or tuple-wise reductions, instead of the standard pairwise reductions. These techniques may also be applied to adaptive sieving to solve CVP with less memory, at the cost of an increase in the time complexity.

angle is less than θ, where now $\theta = \arcsin(1/\alpha)$. Combining this with Lemma 2, we obtain the following result.

Theorem 3. *Let α satisfy (6) and let $u \in (\sqrt{\frac{\alpha^2-1}{\alpha^2}}, \sqrt{\frac{\alpha^2}{\alpha^2-1}})$. Using non-adaptive sieving, we can heuristically solve BDD for targets t at distance $\delta \cdot \lambda_1(\mathcal{L})$ from the lattice, with preprocessing time T_1, space complexity S, and query time complexity T_2 as follows:*

$$ S = \left(\frac{1}{1 - (\alpha^2 - 1)(u^2 - \frac{2u}{\alpha}\sqrt{\alpha^2 - 1} + 1)} \right)^{d/2+o(d)}, \tag{7} $$

$$ T_1 = \max\left\{ S, \, (3/2)^{d/2+o(d)} \right\}, \qquad T_2 = \left(\frac{\alpha + u\sqrt{\alpha^2 - 1}}{2\alpha - \alpha^3 + \alpha^2 u\sqrt{\alpha^2 - 1}} \right)^{d/2+o(d)}. \tag{8} $$

Proof. These complexities directly follow from applying Lemma 1 with $\theta = \arcsin(1/\alpha)$, and again observing that Phase 1 can be performed in time $T_1 = n^{1+\rho_u}$ and space $S = n^{1+\rho_u}$, while Phase 2 takes time $T_2 = n^{\rho_q}$. Note that we cannot combine vectors whose angles are larger than $\frac{\pi}{3}$ in Phase 1, which leads to a lower bound on the preprocessing time complexity T_1 based on the costs of solving SVP.

Theorem 3 is a generalization of Theorem 2, as the latter can be derived from the former by substituting $\delta = 1$ above. To illustrate the results, Fig. 1 considers two special cases:

- For $\delta = \frac{1}{2}$, we find $\alpha \approx 1.1976$, leading to $S \approx 2^{0.2602d+o(d)}$ and $T_2 = 2^{0.1908d+o(d)}$ when minimizing the space complexity.
- For $\delta \to 0$, we have $\alpha \to \sqrt{4/3} \approx 1.1547$. The minimum space complexity is therefore $S = (4/3)^{d/2+o(d)}$, with query complexity $T_2 = 2^{0.1610d+o(d)}$.

In the limit of $u \to \sqrt{\frac{\alpha^2}{\alpha^2-1}}$ we need superexponential space/preprocessing $S, T_1 \to 2^{\omega(d)}$ and a subexponential query time $T_2 \to 2^{o(d)}$ for all $\delta > 0$.

4.2 Approximate Closest Vector Problem with Preprocessing

Given a lattice $\mathcal{L}$ and a target vector $t \in \mathbb{R}^d$, approximate CVP with approximation factor κ asks to find a vector $s \in \mathcal{L}$ such that $\|s - t\|$ is at most a factor κ larger than the real distance from t to $\mathcal{L}$. For random instances t, by the Gaussian heuristic this means that a lattice vector counts as a solution iff it lies at distance at most $\kappa \cdot \lambda_1(\mathcal{L})$ from t.

List size. Instead of reducing t to a vector t' of norm at most $\lambda_1(\mathcal{L})$ as is needed for solving exact CVP, we now want to make sure that the reduced vector t' has norm at most $\kappa \cdot \lambda_1(\mathcal{L})$. If this is the case, then the vector $t - t'$

is a lattice vector lying at distance at most $\kappa \cdot \lambda_1(\mathcal{L})$, which w.h.p. qualifies as a solution. This means that instead of substituting $\beta = 1$ in Lemma 3, we now substitute $\beta = \kappa$. This leads to the condition that $\alpha_0^4 - 4\kappa^2\alpha_0^2 + 4\beta^2 < 0$ for some $\alpha_0 \leq \alpha$. By a similar analysis α^2 must therefore be larger than the smallest root $r_1 = 2\kappa(\kappa - \sqrt{\kappa^2 - 1})$ of this quadratic polynomial in α^2. This immediately leads to the following condition on α:

$$\alpha^2 \geq 2\kappa\left(\kappa - \sqrt{\kappa^2 - 1}\right). \tag{9}$$

A sanity check shows that $\kappa = 1$, corresponding to exact CVP, indeed results in $\alpha \geq \sqrt{2}$, while in the limit of $\kappa \to \infty$ a value $\alpha \approx 1$ suffices to obtain a vector t' of norm at most $\kappa \cdot \lambda_1(\mathcal{L})$. In other words, to solve approximate CVP with very large (constant) approximation factors, a preprocessed list of size $(1 + \varepsilon)^{d+o(d)}$ suffices.

Main result. Similar to the analysis of CVPP, we now take $\theta = \arcsin(1/\alpha)$ as the angle with which to reduce vectors in Phase 1, so that the output of Phase 1 is a list of almost all $\alpha^{d+o(d)}$ shortest lattice vectors of norm at most $\alpha \cdot \lambda_1(\mathcal{L})$. Using a smaller angle θ for reductions again means that nearest neighbor searching can speed up the reductions in both Phase 1 and Phase 2 even further. The exact complexities follow from Lemma 1.

Theorem 4. *Using non-adaptive sieving with spherical LSF, we can heuristically solve κ-CVP with similar complexities as in Theorem 3, where now α must satisfy (9).*

Note that only the dependence of α on κ is different, compared to the dependence of α on δ for bounded distance decoding. The complexities for κ-CVP arguably decrease *faster* than for δ-BDD: for instance, for $\kappa \approx 1.0882$ we obtain the same complexities as for BDD with $\delta = \frac{1}{2}$, while $\kappa = \sqrt{4/3} \approx 1.1547$ leads to the same complexities as for BDD with $\delta \to 0$. Two further examples are illustrated in Fig. 1:

- For $\kappa = 2$, we have $\alpha \approx 1.1976$, which for $u \approx 0.5503$ leads to S = T_1 = $2^{0.2602d+o(d)}$ and $T_2 = 2^{0.1908d+o(d)}$, and for $u = 1$ leads to S = T_1 = $2^{0.3573d+o(d)}$ and $T_2 = 2^{0.0971d+o(d)}$.
- For $\kappa \to \infty$, we have $\alpha \to 1$, i.e. the required preprocessed list size approaches $2^{o(d)}$ as κ grows. For sufficiently large κ, we can solve κ-CVP with a preprocessed list of size $2^{\varepsilon d+o(d)}$ in at most $2^{\varepsilon d+o(d)}$ time. The preprocessing time is given by $2^{0.2925d+o(d)}$.

The latter result shows that for any superconstant approximation factor $\kappa = \omega(1)$, we can solve the corresponding approximate closest vector problem with preprocessing in subexponential time, with an exponential preprocessing time complexity $2^{0.292d+o(d)}$ for solving SVP and generating a list of short lattice vectors, and a subexponential space complexity required for Phase 2.

In other words, even without superexponential preprocessing/memory we can solve CVPP with large approximation factors in subexponential time.

To compare this result with previous work, note that the lower bound on α from (9) tends to $1 + 1/(8\kappa^2) + O(\kappa^{-4})$ as κ grows. The query space and time complexities are further both proportional to $\alpha^{\Theta(d)}$. To obtain a polynomial query complexity and polynomial storage after the preprocessing phase, we can solve for κ, leading to the following result.

Corollary 2. *With non-adaptive sieving we can heuristically solve approximate CVPP with approximation factor κ in polynomial time with polynomial-sized advice iff $\kappa = \Omega(\sqrt{d/\log d})$.*

Proof. The query time and space complexities are given by $\alpha^{\Theta(d)}$, where $\alpha = 1 + \Theta(1/\kappa^2)$. To obtain polynomial complexities in d, we must have $\alpha^{\Theta(d)} = d^{O(1)}$, or equivalently:

$$1 + \Theta\left(\frac{1}{\kappa^2}\right) = \alpha = d^{O(1/d)} = \exp O\left(\frac{\log d}{d}\right) = 1 + O\left(\frac{\log d}{d}\right). \tag{10}$$

Solving for κ leads to the given relation between κ and d.

Apart from the heuristic assumptions we made, this is equivalent to a result of Aharonov and Regev [AR04], who previously showed that the decision version of CVPP with approximation factor $\kappa = \Omega(\sqrt{d/\log d})$ can provably be solved in polynomial time. This further improves upon results of [LLS90, DRS14], who are able to solve search-CVPP with polynomial time and space complexities for $\kappa = O(d^{3/2})$ and $\kappa = \Omega(d/\sqrt{\log d})$ respectively. Assuming the heuristic assumptions are valid, Corollary 2 closes the gap between these previous results for decision-CVPP and search-CVPP with a rather simple algorithm: (1) preprocess the lattice by storing all $d^{O(1)}$ shortest vectors of the lattice in a list; and (2) apply Algorithm 2 to this list and the target vector to find an approximate closest vector. Note that nearest neighbor techniques only affect leading constants; even without nearest neighbor searching this would heuristically result in a polynomial time and space algorithm for κ-CVPP with $\kappa = \Omega(\sqrt{d/\log d})$.

Acknowledgments. The author is indebted to Léo Ducas, whose initial ideas and suggestions on this topic motivated work on this paper. The author further thanks Vadim Lyubashevsky and Oded Regev for their comments on the relevance of a subexponential time CVPP algorithm requiring (super)exponential space. The author is supported by the SNSF ERC Transfer Grant CRETP2-166734 FELICITY.

A   Pseudocode of SVP Algorithms

Algorithms 3 and 4 present pseudo-code for the (sieve part of the) original Nguyen-Vidick sieve and the GaussSieve, respectively, as described in Sect. 2. For the Nguyen-Vidick sieve, the presented algorithm is a more intuitive but equivalent version of the original sieve; see [Laa15a, Appendix B] for details on this equivalence.

Algorithm 3. The quadratic Nguyen-Vidick sieve for finding shortest vectors

Require: An input list $L \subset \mathcal{L}$ of $(4/3)^{d/2+o(d)}$ vectors of norm at most R
Ensure: The output list $L' \subset \mathcal{L}$ has $(4/3)^{d/2+o(d)}$ vectors of norm at most $\gamma \cdot R$
 1: Initialize an empty list L'
 2: **for each** $w_1, w_2 \in L$ **do**
 3:     **if** $\|w_1 - w_2\| \le \gamma R$ **then**
 4:         Add $w_1 - w_2$ to the list L'
 5: **return** L'

Algorithm 4. The GaussSieve algorithm for finding shortest vectors

Require: A basis B of a lattice $\mathcal{L}(B)$
Ensure: The algorithm returns a shortest lattice vector
 1: Initialize an empty list L and an empty stack S
 2: **repeat**
 3:     Get a vector v from the stack (or sample a new one if $S = \emptyset$)
 4:     **for each** $w \in L$ **do**
 5:         **if** $\|v - w\| \le \|v\|$ **then**
 6:             Replace $v \leftarrow v - w$
 7:         **if** $\|w - v\| \le \|w\|$ **then**
 8:             Replace $w \leftarrow w - v$
 9:             Move w from the list L to the stack S (unless $w = 0$)
10:     **if** v has changed **then**
11:         Add v to the stack S (unless $v = 0$)
12:     **else**
13:         Add v to the list L (unless $v = 0$)
14: **until** v is a shortest vector
15: **return** v

B Pseudocode of Phase 1 for Non-adaptive Sieving

To generate a list of the $\alpha^{d+o(d)}$ shortest lattice vectors with the GaussSieve, rather than the $(4/3)^{d/2+o(d)}$ lattice vectors one would get with standard sieving, we relax the reductions: reducing if $\|v - w\| < \|v\|$ corresponds to an angle $\pi/3$ between v and w, leading to a list size $(1/\sin(\frac{\pi}{3}))^{d+o(d)} = (4/3)^{d/2+o(d)}$. To obtain a list of size $\alpha^{d+o(d)}$, we reduce vectors if their angle is less than $\theta = \arcsin(1/\alpha)$, which for vectors v, w of similar norm corresponds to the following condition:

$$\|v - w\| < \sqrt{2(1 - \cos\theta)} \cdot \|v\| = \sqrt{2 - \frac{2}{\alpha}\sqrt{\alpha^2 - 1}} \cdot \|v\|. \tag{11}$$

This leads to the modified GaussSieve described in Algorithm 5.

Algorithm 5. The non-adaptive GaussSieve (Phase 1) for finding closest vectors

---

Require: A basis B of a lattice $\mathcal{L}(B)$, a parameter $\alpha > 1$
Ensure: The output list L contains $\alpha^{d+o(d)}$ vectors of norm at most $\alpha \cdot \lambda_1(\mathcal{L})$
 1: Initialize an empty list L and an empty stack S
 2: Let $\alpha_0 = \max\{\alpha, \sqrt{4/3}\}$
 3: **repeat**
 4:     Get a vector v from the stack (or sample a new one if $S = \emptyset$)
 5:     **for each** $w \in L$ **do**
 6:         **if** $\|v - w\|^2 \leq (2 - \frac{2}{\alpha_0}\sqrt{\alpha_0^2 - 1}) \cdot \|v\|^2$ **then**
 7:             Replace $v \leftarrow v - w$
 8:         **if** $\|w - v\|^2 \leq (2 - \frac{2}{\alpha_0}\sqrt{\alpha_0^2 - 1}) \cdot \|w\|^2$ **then**
 9:             Replace $w \leftarrow w - v$
10:             Move w from the list L to the stack S (unless $w = 0$)
11:     **if** v has changed **then**
12:         Add v to the stack S (unless $v = 0$)
13:     **else**
14:         Add v to the list L (unless $v = 0$)
15: **until** v is a shortest vector
16: **return** L

References

[ADRS15] Aggarwal, D., Dadush, D., Regev, O., Stephens-Davidowitz, N.: Solving the shortest vector problem in 2^n time via discrete Gaussian sampling. In: STOC, pp. 733–742 (2015)

[ADS15] Aggarwal, D., Dadush, D., Stephens-Davidowitz, N.: Solving the closest vector problem in 2^n time - the discrete Gaussian strikes again! In: FOCS (2015)

[AEVZ02] Agrell, E., Eriksson, T., Vardy, A., Zeger, K.: Closest point search in lattices. IEEE Trans. Inf. Theory **48**(8), 2201–2214 (2002)

[AKKV05] Alekhnovich, M., Khot, S., Kindler, G., Vishnoi, N.: Hardness of approximating the closest vector problem with pre-processing. In: FOCS, pp. 216–225 (2005)

[AKS01] Ajtai, M., Kumar, R., Sivakumar, D.: A sieve algorithm for the shortest lattice vector problem. In: STOC, pp. 601–610 (2001)

[AR04] Aharonov, D., Regev, O.: Lattice problems in NP∩coNP. In: FOCS, pp. 362–371 (2004)

[BDGL16] Becker, A., Ducas, L., Gama, N., Laarhoven, T.: New directions in nearest neighbor searching with applications to lattice sieving. In: SODA, pp. 10–24 (2016)

[BGJ14] Becker, A., Gama, N., Joux, A.: A Sieve algorithm based on overlattices. In: ANTS, pp. 49–70 (2014)

[BGJ15] Becker, A., Gama, N., Joux, A.: Speeding-up lattice sieving without increasing the memory, using sub-quadratic nearest neighbor search. Cryptology ePrint Archive, Report 2015/522, pp. 1–14 (2015)

[BL16] Becker, A., Laarhoven, T.: Efficient (ideal) lattice sieving using cross-polytope LSH. In: Pointcheval, D., Nitaj, A., Rachidi, T. (eds.) AFRICACRYPT 2016. LNCS, vol. 9646, pp. 3–23. Springer, Cham (2016). doi:10.1007/978-3-319-31517-1_1

[BLS16] Bai, S., Laarhoven, T., Stehlé, D.: Tuple lattice sieving. In: ANTS (2016)

[BNvdP14] Bos, J.W., Naehrig, M., van de Pol, J.: Sieving for shortest vectors in ideal lattices: a practical perspective. Cryptology ePrint Archive, Report 2014/880, pp. 1–23 (2014)

[CN11] Chen, Y., Nguyên, P.Q.: BKZ 2.0: better lattice security estimates. In: Lee, D.H., Wang, X. (eds.) ASIACRYPT 2011. LNCS, vol. 7073, pp. 1–20. Springer, Heidelberg (2011). doi:10.1007/978-3-642-25385-0_1

[DRS14] Dadush, D., Regev, O., Stephens-Davidowitz, N.: On the closest vector problem with a distance guarantee. In: CCC, pp. 98–109 (2014)

[FBB+14] Fitzpatrick, R., Bischof, C., Buchmann, J., Dagdelen, Ö., Göpfert, F., Mariano, A., Yang, B.-Y.: Tuning GaussSieve for speed. In: Aranha, D.F., Menezes, A. (eds.) LATINCRYPT 2014. LNCS, vol. 8895, pp. 288–305. Springer, Cham (2015). doi:10.1007/978-3-319-16295-9_16

[FM02] Feige, U., Micciancio, D.: The inapproximability of lattice and coding problems with preprocessing. In: CCC, pp. 32–40 (2002)

[FP85] Fincke, U., Pohst, M.: Improved methods for calculating vectors of short length in a lattice. Math. Comput. **44**(170), 463–471 (1985)

[GNR10] Gama, N., Nguyên, P.Q., Regev, O.: Lattice enumeration using extreme pruning. In: Gilbert, H. (ed.) EUROCRYPT 2010. LNCS, vol. 6110, pp. 257–278. Springer, Heidelberg (2010). doi:10.1007/978-3-642-13190-5_13

[IKMT14] Ishiguro, T., Kiyomoto, S., Miyake, Y., Takagi, T.: Parallel Gauss Sieve algorithm: solving the SVP challenge over a 128-dimensional ideal lattice. In: PKC, pp. 411–428 (2014)

[IM98] Indyk, P., Motwani, R.: Approximate nearest neighbors: towards removing the curse of dimensionality. In: STOC, pp. 604–613 (1998)

[Kan83] Kannan, R.: Improved algorithms for integer programming and related lattice problems. In: STOC, pp. 193–206 (1983)

[Laa15a] Laarhoven, T.: Sieving for shortest vectors in lattices using angular locality-sensitive hashing. In: Gennaro, R., Robshaw, M. (eds.) CRYPTO 2015. LNCS, vol. 9215, pp. 3–22. Springer, Heidelberg (2015). doi:10. 1007/978-3-662-47989-6_1

[Laa15b] Laarhoven, T.: Tradeoffs for nearest neighbors on the sphere (2015)

[LdW15] Laarhoven, T., Weger, B.: Faster sieving for shortest lattice vectors using spherical locality-sensitive hashing. In: Lauter, K., Rodríguez-Henríquez, F. (eds.) LATINCRYPT 2015. LNCS, vol. 9230, pp. 101–118. Springer, Cham (2015). doi:10.1007/978-3-319-22174-8_6

[LLS90] Lagarias, J.C., Lenstra, H.W., Schnorr, C.-P.: Korkin-Zolotarev bases and successive minima of a lattice and its reciprocal lattice. Combinatorica **10**(4), 333–348 (1990)

[LvdPdW12] Laarhoven, T., van de Pol, J., de Weger, B.: Solving hard lattice problems and the security of lattice-based cryptosystems. Cryptology ePrint Archive, Report 2012/533, pp. 1–43 (2012)

[MB16] Mariano, A., Bischof, C.: Enhancing the scalability and memory usage of HashSieve on multi-core CPUs. In: PDP (2016)

[Mic01] Micciancio, D.: The hardness of the closest vector problem with preprocessing. IEEE Trans. Inf. Theory **47**(3), 1212–1215 (2001)

[MLB15] Mariano, A., Laarhoven, T., Bischof, C.: Parallel (probable) lock-free HashSieve: a practical sieving algorithm for the SVP. In: ICPP, pp. 590–599 (2015)

[MLB16] Mariano, A., Laarhoven, T., Bischof, C.: A parallel variant of LDSieve for the SVP on lattices (2016)

[MODB14] Mariano, A., Dagdelen, Ö., Bischof, C.: A comprehensive empirical comparison of parallel ListSieve and GaussSieve. In: Lopes, L., Žilinskas, J., Costan, A., Cascella, R.G., Kecskemeti, G., Jeannot, E., Cannataro, M., Ricci, L., Benkner, S., Petit, S., Scarano, V., Gracia, J., Hunold, S., Scott, S.L., Lankes, S., Lengauer, C., Carretero, J., Breitbart, J., Alexander, M. (eds.) Euro-Par 2014. LNCS, vol. 8805, pp. 48–59. Springer, Cham (2014). doi:10.1007/978-3-319-14325-5_5

[MS11] Milde, B., Schneider, M.: A parallel implementation of GaussSieve for the shortest vector problem in lattices. In: PACT, pp. 452–458 (2011)

[MTB14] Mariano, A., Timnat, S., Bischof, C.: Lock-free GaussSieve for linear speedups in parallel high performance SVP calculation. In: SBAC-PAD, pp. 278–285 (2014)

[MV10a] Micciancio, D., Voulgaris, P.: A deterministic single exponential time algorithm for most lattice problems based on Voronoi cell computations. In: STOC, pp. 351–358 (2010)

[MV10b] Micciancio, D., Voulgaris, P.: Faster exponential time algorithms for the shortest vector problem. In: SODA, pp. 1468–1480 (2010)

[MW15] Micciancio, D., Walter, M.: Fast lattice point enumeration with minimal overhead. In: SODA, pp. 276–294 (2015)

[NV08] Nguyên, P.Q., Vidick, T.: Sieve algorithms for the shortest vector problem are practical. J. Math. Cryptology $2(2)$, 181–207 (2008)

[PS09] Pujol, X., Stehlé, D.: Solving the shortest lattice vector problem in time $2^{2.465n}$. Cryptology ePrint Archive, Report 2009/605, pp. 1–7 (2009)

[Reg04] Regev, O.: Improved inapproximability of lattice and coding problems with preprocessing. IEEE Trans. Inf. Theory $50(9)$, 2031–2037 (2004)

[Sch87] Schnorr, C.-P.: A hierarchy of polynomial time lattice basis reduction algorithms. Theoret. Comput. Sci. $53(2-3)$, 201–224 (1987)

[Sch11] Schneider, M.: Analysis of Gauss-Sieve for solving the shortest vector problem in lattices. In: Katoh, N., Kumar, A. (eds.) WALCOM 2011. LNCS, vol. 6552, pp. 89–97. Springer, Heidelberg (2011). doi:10.1007/978-3-642-19094-0_11

[Sch13] Schneider, M.: Sieving for short vectors in ideal lattices. In: AFRICACRYPT, pp. 375–391 (2013)

[SE94] Schnorr, C.-P., Euchner, M.: Lattice basis reduction: improved practical algorithms and solving subset sum problems. Math. Program. $66(2-3)$, 181–199 (1994)

[SG15] Schneider, M., Gama, N.: SVP challenge (2015)

[Ste16] Stephens-Davidowitz, N.: Dimension-preserving reductions between lattice problems (2016). http://noahsd.com/latticeproblems.pdf

[WLTB11] Wang, X., Liu, M., Tian, C., Bi, J.: Improved Nguyen-Vidick heuristic sieve algorithm for shortest vector problem. In: ASIACCS, pp. 1–9 (2011)

[WSSJ14] Wang, J., Shen, H.T., Song, J., Ji, J.: Hashing for similarity search: a survey. arXiv:1408.2927 [cs.DS], pp. 1–29 (2014)

[ZPH13] Zhang, F., Pan, Y., Hu, G.: A three-level sieve algorithm for the shortest vector problem. In: Lange, T., Lauter, K., Lisoněk, P. (eds.) SAC 2013. LNCS, vol. 8282, pp. 29–47. Springer, Heidelberg (2014). doi:10.1007/978-3-662-43414-7_2

Key Recovery Attack on the Cubic ABC Simple Matrix Multivariate Encryption Scheme

Dustin Moody[1], Ray Perlner[1], and Daniel Smith-Tone[1,2(✉)]

[1] National Institute of Standards and Technology, Gaithersburg, MD, USA
{dustin.moody,ray.perlner,daniel.smith}@nist.gov
[2] Department of Mathematics, University of Louisville, Louisville, KY, USA

Abstract. In the last few years multivariate public key cryptography has experienced an infusion of new ideas for encryption. Among these new strategies is the ABC Simple Matrix family of encryption schemes which utilize the structure of a large matrix algebra to construct effectively invertible systems of nonlinear equations hidden by an isomorphism of polynomials. The cubic version of the ABC Simple Matrix Encryption was developed with provable security in mind and was published including a heuristic security argument claiming that an attack on the scheme should be at least as difficult as solving a random system of quadratic equations over a finite field.

In this work, we prove that these claims are erroneous. We present a complete key recovery attack breaking full sized instances of the scheme. Interestingly, the same attack applies to the quadratic version of ABC, but is far less efficient; thus, the enhanced security scheme is less secure than the original.

Keywords: Multivariate public key cryptography · Differential invariant · MinRank · Encryption

1 Introduction

Classical public key cryptography is mainly based on arithmetic constructions on Abelson groups. Since the discovery by Shor in the 1990s of efficient algorithms for factoring and computing discrete logarithms with quantum computers, see [1], there has been a growing interest in the international community in the task of constructing algorithms resistant to cryptanalysis with quantum computers. Indeed, in light of the announcement [?] by the National Institute of Standards and Technology (NIST) of an imminent call for proposals for post-quantum standards, the challenge of migrating from the homogeneous heritage of public key cryptography to a more diverse collection of tools has become mainstream.

One possible candidate for practical, efficient, and nonconforming solutions to some of the most consequential public key applications is Multivariate Public Key Cryptography (MPKC). Multivariate schemes are attractive in certain applications because of the maleability of the schemes. Different modifications

© Springer International Publishing AG 2017
R. Avanzi and H. Heys (Eds.): SAC 2016, LNCS 10532, pp. 543–558, 2017.
https://doi.org/10.1007/978-3-319-69453-5_29

of similar ideas can make a scheme more suited to lightweight architectures, enhance security, or parametrize various aspects of performance.

In addition, MPKC is one among a few serious candidates to have risen to prominence as post-quantum options. The fundamental problem of solving a system of quadratic equations is known to be NP-hard, and so in the worst case, solving a system of generic quadratic equations is unfeasible for a classical computer; neither is there any indication that the task is easier in the quantum computing paradigm.

MPKC has experienced a fair amount of success in the realm of digital signatures. Some trustworthy schemes that have survived for almost two decades include UOV [2], HFE- [3], and HFEv- [4]. Moreover, some of these schemes have optimizations which have strong theoretical support or have stood unbroken in the literature for some time. Specifically, UOV has a cyclic variant [5] which reduces the key size dramatically, and Gui, a new HFEv- scheme, see [6], has parameters far more appealing than QUARTZ due to greater confidence in the complexity of algebraically solving the underlying system of equations [7].

The situation with multivariate public key encryption is quite different, however. Many attempts at multivariate encryption, see [8,9] for example, have been shown to be weak based on rank or differential weaknesses. Recently, a few interesting attempts to achieve multivariate encryption have surfaced. ZHFE, see [10], and the ABC Simple Matrix Scheme, see [11], both use fundamentally new structures for the derivation of an encryption system. While it was shown that the best attack known on the Simple Matrix structure, see [12]—which relies on the differential invariant structure of the central map—supports the claimed security level of the scheme, a subset of the original authors proposed a cubic version of the scheme, [13], as a step towards provable security.

In this article, we present a key recovery attack on a full scale version of the Cubic Simple Matrix encryption scheme, having a complexity on the order of q^{s+2} for characteristic $p > 3$, q^{s+3} for characteristic 3 and q^{2s+6} for characteristic 2. Here s is the dimension of the matrices in the scheme, and q is the cardinality of the finite field used. This technique is an extension and augmentation of the technique of [12], and, similarly, exploits a differential invariant property of the core map to perform a key recovery attack. We can show that the attack uses a property which uniquely distinguishes the isomorphism class of the central map from that of a random collection of formulae.

Specifically, our attack breaks CubicABC ($q = 127$, $s = 7$), designed for 80-bit security, in approximately 2^{76} operations (or around 2^{80} if one pessimistically uses $\omega = 3$ as the linear algebra constant). More convincingly, our attack completely breaks CubicABC ($q = 127$, $s = 8$), designed for 100-bit security, in approximately 2^{84} operations (or 2^{88} for $\omega = 3$). Furthermore, the attack is fully parallelizable and requires very little memory; hence, the differential invariant attack is far more efficient than algebraic attacks, the basis for the original security estimation. Thus, the security claims in [13] are clearly unfounded; in fact, the cubic version of the scheme, whose security was claimed to be closely related to an NP-complete problem, is actually less secure than the quadratic case.

The paper is organized as follows. In the next section, we present the structure of the Cubic ABC Simple Matrix encryption scheme. In the following section, we recall differential invariants and present a natural extension of this notion to the case of cubic polynomials. The differential invariant structure of the ABC scheme is derived in the subsequent section and the effect of this structure on minrank calculations is determined. We next calculate the complexity of the full attack including the linear algebra steps required to extend the distinguisher into a key recovery mechanism. Finally, we review these results and discuss the surprising relationship between the practical security of the Cubic ABC scheme and its quadratic counterpart.

2 The Cubic ABC Matrix Encryption Scheme

In [13], the Cubic ABC Matrix encryption scheme is proposed. The motivation behind the scheme is to use a large matrix algebra over a finite field to construct an easily invertible cubic map. The construction uses matrix multiplication to combine random quadratic formulae and random linear formula into cubic formulae in a way that allows a user with knowledge of the structure of the matrix algebra and polynomial isomorphism used to compose the scheme to invert the map.

Let $k = \mathbb{F}_q$ be a finite field. Linear forms and variables over k will be denoted with lower case letters. Vectors of any dimension over k will be denoted with bold font, $\mathbf{v}$. Fix $s \in \mathbb{N}$ and set $n = s^2$ and $m = 2s^2$. An element of $M_s(k)$, $M_n(k)$ or $M_m(k)$, or the linear transformations they represent, will be denoted by upper case letters, such as M. When the entries of the matrix are being considered functions of a variable, the matrix will be denoted $M(\mathbf{x})$. Let ϕ represent the vector space isomorphism from $M_{s \times 2s}(k)$ to k^{2s^2} sending a matrix to the column vector consisting of the concatenation of its rows. The output of this map, being a vector, will be written with bold font; however, to indicate the relationship to its matrix preimage, it will be denoted with an upper case letter, such as $\mathbf{M}$.

The scheme utilizes an isomorphism of polynomials to hide the internal structure. Let $\mathbf{x} = [x_1, x_2, \ldots, x_n]^\top \in k^n$ denote plaintext while $\mathbf{y} = [y_1, \ldots, y_m] \in k^m$ denotes ciphertext. Fix two invertible linear transformations $T \in M_m(k)$ and $U \in M_n(k)$ (One may use affine transformations, but there is no security or performance benefit in doing so.) Denote the input and output of the central map by $\mathbf{u} = U\mathbf{x}$ and $\mathbf{v} = T^{-1}(\mathbf{y})$.

The construction of the central map is as follows. Define three $s \times s$ matrices A, B, and C in the following way:

$$A = \begin{bmatrix} p_1 & p_2 & \cdots & p_s \\ p_{s+1} & p_{s+2} & \cdots & p_{2s} \\ \vdots & \vdots & \ddots & \vdots \\ p_{s^2-s+1} & p_{s^2-s+2} & \cdots & p_{s^2} \end{bmatrix}, B = \begin{bmatrix} b_1 & b_2 & \cdots & b_s \\ b_{s+1} & b_{s+2} & \cdots & b_{2s} \\ \vdots & \vdots & \ddots & \vdots \\ b_{s^2-s+1} & b_{s^2-s+2} & \cdots & b_{s^2} \end{bmatrix},$$

and

$$C = \begin{bmatrix} c_1 & c_2 & \cdots & c_s \\ c_{s+1} & c_{s+2} & \cdots & c_{2s} \\ \vdots & \vdots & \ddots & \vdots \\ c_{s^2-s+1} & c_{s^2-s+2} & \cdots & c_{s^2} \end{bmatrix}.$$

Here the p_i are quadratic forms on $\mathbf{u}$ chosen independently and uniformly at random from among all quadratic forms and the b_i and c_i are linear forms on $\mathbf{u}$ chosen independently and uniformly at random from among all linear forms.

We define two $s \times s$ matrices $E_1 = AB$ and $E_2 = AC$. Since A is quadratic and B and C are linear in u_i, E_1 and E_2 are cubic in the u_i. The central map $\mathcal{E}$ is defined by

$$\mathcal{E} = \phi \circ (E_1 \| E_2).$$

Thus $\mathcal{E}$ is an m dimensional vector of cubic forms in $\mathbf{u}$. Finally, the public key is given by $\mathcal{F} = T \circ \mathcal{E} \circ U$.

Encryption with this system is standard: given a plaintext $(x_1, \ldots, x_n)$, compute $(y_1, \ldots, y_m) = \mathcal{F}(x_1, \ldots, x_n)$. Decryption is somewhat more complicated.

To decrypt, one inverts each of the private maps in turn: apply T^{-1}, invert $\mathcal{E}$, and apply U^{-1}. To "invert" $\mathcal{E}$, one assumes that $A(\mathbf{u})$ is invertible, and forms a matrix

$$A^{-1}(\mathbf{u}) = \begin{bmatrix} w_1 & w_2 & \cdots & w_s \\ w_{s+1} & w_{s+2} & \cdots & w_{2s} \\ \vdots & \vdots & \ddots & \vdots \\ w_{s^2-s+1} & w_{s^2-s+2} & \cdots & w_{s^2} \end{bmatrix},$$

where the w_i are indeterminants. Then using the relations $A^{-1}(\mathbf{u})E_1(\mathbf{u}) = B(\mathbf{u})$ and $A^{-1}(\mathbf{u})E_2(\mathbf{u}) = C(\mathbf{u})$, we have $m = 2s^2$ linear equations in $2n = 2s^2$ unknowns w_i and u_i. Using, for example, Gaussian elimination one can eliminate all of the variables w_i and most of the u_i. The resulting relations can be substituted back into $E_1(\mathbf{u})$ and $E_2(\mathbf{u})$ to obtain a large system of equations in very few variables which can be solved efficiently in a variety of ways.

3   Subspace Differential Invariants for Cubic Maps

Let $f : k^n \to k^m$ be an arbitrary fixed function on k^n. Consider the discrete differential $Df(\mathbf{a}, \mathbf{x}) = f(\mathbf{a} + \mathbf{x}) - f(\mathbf{a}) - f(\mathbf{x}) + f(\mathbf{0})$.

If f is quadratic, we can express the differential as an n-tuple of bilinear differential coordinate forms in the following way: $[Df(\mathbf{a}, \mathbf{x})]_i = \mathbf{a}^\top Df_i \mathbf{x}$, where Df_i is a symmetric matrix representation of the action on the ith coordinate of the bilinear differential. If the function f is cubic $Df(\mathbf{a}, \mathbf{x})$ is a symmetric bi-quadratic function. By the symmetry, it is well defined to compute a second differential $D^2 f(\mathbf{a}, \mathbf{b}, \mathbf{x})$ by computing the discrete differential of Df with respect to either $\mathbf{a}$ or $\mathbf{x}$. In this case, we may consider the second differential as an n-tuple of trilinear differential coordinate forms by letting $D^2 f_i$ be the symmetric 3-tensor representing the action on the ith coordinate of the trilinear differential.

In [12], the following definition of a subspace differential invariant was provided:

Definition 1. *A subspace differential invariant of a quadratic map* $f : k^n \to k^m$ *with respect to a subspace* $X \subseteq k^m$ *is a subspace* $V \subseteq k^n$ *with the property that there exists a* $W \subseteq k^n$ *of dimension at most* $dim(V)$ *such that simultaneously* $AV \subseteq W$ *for all* $A = \sum_{i=1}^m x_i Df_i$ *where* $(x_1, \ldots, x_m) \in X$, *i.e.* $A \in Span_X(Df_i)$.

This definition captures the idea of a subspace of the span of the public polynomials acting linearly on a subspace of the plaintext space in the same way. Such behavior is strange for quadratic maps in general. Furthermore, as shown in [12], this behavior is computable regardless of the rank of the maps involved.

A natural generalization of this definition is the following:

Definition 2. *A subspace differential invariant of a cubic map* $f : k^n \to k^m$ *with respect to a subspace* $X \subseteq k^m$ *is a pair of subspaces* $(V_1, V_2) \subseteq (k^n)^2$ *for which there exists a subspace* $W \subseteq k^n$ *with* $dim(W) \leq mindim(V_i)$ *such that for all* $A = \sum_{i=1}^m x_i D^2 f_i$ *where* $(x_1, \ldots, x_m) \in X$, *for all* $\mathbf{a} \in V_2$, *for all* $\mathbf{b} \in V_2$ *and for all* $\mathbf{x} \in W^\perp$ *we have that* $A(\mathbf{a}, \mathbf{b}, \mathbf{x}) = 0$.

This definition captures the notion of a subspace of the span of the public cubic polynomials acting quadratically on a subspace of the plaintext space in the same way. Such behavior is strange for cubic maps in general.

4   The Differential Invariant Structure of the Cubic ABC Scheme

4.1   Column Band Spaces

Each component of the central $\mathcal{E}(\mathbf{u}) = E_1(\mathbf{u}) \| E_2(\mathbf{u})$ map may be written as:

$$\mathcal{E}_{(i-1)s+j} = \sum_{l=1}^{s} p_{(i-1)s+l} b_{(l-1)s+j}, \tag{1}$$

for the E_1 equations, and likewise, for the E_2 equations:

$$\mathcal{E}_{s^2+(i-1)s+j} = \sum_{l=1}^{s} p_{(i-1)s+l} c_{(l-1)s+j} \tag{2}$$

where i and j run from 1 to s.

Consider the s sets of s polynomials that form the columns of E_1, i.e. for each $j \in \{1, \ldots, s\}$ consider $(\mathcal{E}_j, \mathcal{E}_{s+j}, \ldots, \mathcal{E}_{s^2-s+j})$. With high probability, the linear forms $b_j, b_{s+j}, \ldots, b_{s^2-s+j}$ are linearly independent, and if so the polynomials may be re-expressed, using a linear change of variables to $(u'_1, \ldots u'_{s^2})$ where $u'_i = b_{(i-1)s+j}$ for $i = 1, \ldots, s$. After the change of variables, the only cubic monomials contained in $(\mathcal{E}_j, \mathcal{E}_{s+j}, \ldots, \mathcal{E}_{s^2-s+j})$ will be those containing at least one factor of $u'_1, \ldots, u'_s$. We can make a similar change of variables to reveal structure in the s sets of s polynomials that form the columns of E_2: Setting $u'_i = c_{(i-1)s+j}$ for $i = 1, \ldots, s$ and a fixed j, the only cubic monomials contained in $(\mathcal{E}_{s^2+j}, \mathcal{E}_{s^2+s+j}, \ldots, \mathcal{E}_{2s^2-s+j})$ will be those containing at least one factor of $u'_1, \ldots, u'_s$.

More generally, we can make a similar change of variables to reveal structure in any of a large family of s dimensional subspaces of the span of the component polynomials of E_1 and E_2, which we will call column band spaces in analogy to the band spaces used to analyze the quadratic ABC cryptosystem in [12]. Each family is defined by a fixed linear combination, (β, γ), of the columns of E_1 and E_2:

Definition 3. *The column band space defined by the $2s$-dimensional linear form (β, γ) is the space of cubic maps, $\mathcal{B}_{\beta,\gamma}$, given by:*

$$\mathcal{B}_{\beta,\gamma} = Span(\mathcal{E}_{\beta,\gamma,1}, \ldots, \mathcal{E}_{\beta,\gamma,s})$$

where

$$\mathcal{E}_{\beta,\gamma,i} = \sum_{j=1}^{s} (\beta_j \mathcal{E}_{(i-1)s+j} + \gamma_j \mathcal{E}_{s^2+(i-1)s+j})$$

$$= \sum_{l=1}^{s} \left(p_{(i-1)s+l} \sum_{j=1}^{s} (\beta_j b_{(l-1)s+j} + \gamma_j c_{(l-1)s+j}) \right)$$

Theorem 1. *There is a pair of subspaces $(V_1, V_2) \in (k^n)^2$ which is a subspace differential invariant with respect to $\mathcal{B}_{\beta,\gamma}$ for all (β, γ). Moreover, there exists an $\mathbf{x}_1 \in k^n$ such that $rank(D^2\mathcal{E}(\mathbf{x}_1)) \leq 2s$ for all $\mathcal{E} \in \mathcal{B}_{\beta,\gamma}$.*

Proof. Note that under a change of variables $(x_1, \ldots, x_{s^2}) \xrightarrow{M} (u'_1, \ldots u'_{s^2})$, where $u'_i = \sum_{j=1}^{s} (\beta_j b_{(i-1)s+j} + \gamma_j c_{(i-1)s+j})$ for $i = 1, \ldots, s$, the only cubic monomials contained in the elements of $\mathcal{B}_{\beta,\gamma}$ will be those containing at least one factor of $u'_1, \ldots, u'_s$. In such a basis, the second differential of any map in $\mathcal{B}_{\beta,\gamma}$, and thus the second differential of $\mathcal{E}$ can be visualized as a 3-tensor with a special block form, see Fig. 1.

Let V be the $(s^2 - s)$-dimensional preimage $M^{-1}(Span(u'_1, \ldots, u'_s)^{\perp})$. This 3-tensor $D^2\mathcal{E}$ may be thought of as a bilnear map which takes two vectors $\mathbf{x}_1, \mathbf{x}_2 \in V$, i.e. of the form:

$$(0, \ldots, 0, u'_{s+1}(\mathbf{x}_k), \ldots, u'_{s^2}(\mathbf{x}_k))^{\top}$$

to a covector of the form:

$$(y(u'_1), \ldots, y(u'_s), 0, \ldots, 0).$$

Thus, in this basis $D^2\mathcal{E}(\mathbf{x}_1)$ is a symmetric matrix which is zero on $V \times V$. Therefore, $rank(D^2\mathcal{E}(\mathbf{x})) \leq 2s$. One checks that (V, V) is a subspace differential with respect to $\mathcal{B}_{\beta,\gamma}$ with $W := V^{\perp}$, since $dim(W) = s < s^2 - s = dim(V)$.

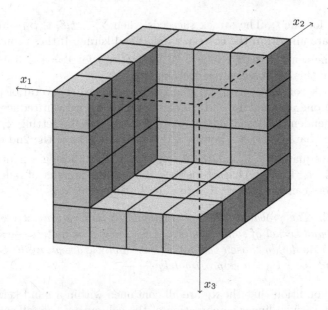

Fig. 1. 3-tensor structure of the second differential of a band space map. Solid regions correspond to nonzero coefficients. Transparent regions correspond to zero coefficients.

We will define the term "band-kernel" to describe the space of vectors of the same form as x_1 and x_2 in the proof above, i.e.:

Definition 4. *The band kernel of* $\mathcal{B}_{\beta,\gamma}$*, denoted* $\mathcal{BK}_{\beta,\gamma}$*, is the space of vectors,* x*, such that*

$$u_i' = \sum_{j=1}^{s} \left(\beta_j b_{(i-1)s+j}(x) + \gamma_j c_{(i-1)s+j}(x) \right) = 0$$

for $i = 1, \ldots, s$*.*

5 A Variant of MinRank Exploiting the Column Band Space Structure

A minrank-like attack may be used to locate the column band-space maps defined in the previous section. In this case, the attack proceeds by selecting s^2-dimensional vectors $\mathbf{x}_1$, $\mathbf{x}_2$, $\mathbf{x}_3$, and $\mathbf{x}_4$, setting

$$\sum_{i=1}^{2s^2} t_i D^2 \mathcal{E}_i(\mathbf{x}_1, \mathbf{x}_2) = 0$$

$$\sum_{i=1}^{2s^2} t_i D^2 \mathcal{E}_i(\mathbf{x}_3, \mathbf{x}_4) = 0,$$

(3)

and solving for the t_i. The attack succeeds when $\sum_{i=1}^{2s^2} t_i \mathcal{E}_i \in \mathcal{B}_{\beta,\gamma}$ and $\mathbf{x}_1$, $\mathbf{x}_2$, $\mathbf{x}_3$, and $\mathbf{x}_4$ are all within the corresponding band kernel. If these conditions are met, then the rank of the 2-tensor $\sum_{i=1}^{2s^2} t_i D^2\mathcal{E}_i(\mathbf{x}_k)$ for $k = 1, 2, 3, 4$ will be at most $2s$, and this will be easily detectable.

The attack complexity will be significantly reduced if several of the $\mathbf{x}_k$ are set equal to one another. In odd characteristic fields, we can reduce the number of independently chosen vectors to 2, (for example, by setting $\mathbf{x}_1 = \mathbf{x}_2$ and $\mathbf{x}_3 = \mathbf{x}_4$.) In characteristic 2, however, the antisymmetry of the 2nd differential prevents the equation $\sum_{i=1}^{2s^2} t_i D^2\mathcal{E}_i(\mathbf{x}_1, \mathbf{x}_1) = 0$ from imposing a nontrivial constraint on the t_i. Even in characteristic 2, though, the number of independently chosen vectors can be reduced to 3 (e.g. by setting $\mathbf{x}_1 = \mathbf{x}_4$).

Theorem 2. *The probability that 2 randomly chosen vectors, $\mathbf{x}_1$ and $\mathbf{x}_2$, are both in the band kernel of some band-space $\mathcal{B}_{\beta,\gamma}$ is approximately $\frac{1}{q-1}$; The probability that 3 randomly chosen vectors, $\mathbf{x}_1$, $\mathbf{x}_2$, and $\mathbf{x}_3$, are all in the band kernel of some band-space $\mathcal{B}_{\beta,\gamma}$ is approximately $\frac{1}{(q-1)q^s}$.*

Proof. The condition that the $\mathbf{x}_k$ are all contained within a band kernel is that there be a nontrivial linear combination of the columns of the following matrix equal to zero (i.e. that the matrix has nonzero column corank):

$$
\left[
\begin{array}{cccc|cccc}
b_1(\mathbf{x}_1) & b_2(\mathbf{x}_1) & \dots & b_s(\mathbf{x}_1) & c_1(\mathbf{x}_1) & c_2(\mathbf{x}_1) & \dots & c_s(\mathbf{x}_1) \\
b_{s+1}(\mathbf{x}_1) & b_{s+2}(\mathbf{x}_1) & \dots & b_{2s}(\mathbf{x}_1) & c_{s+1}(\mathbf{x}_1) & c_{s+2}(\mathbf{x}_1) & \dots & c_{2s}(\mathbf{x}_1) \\
\vdots & \vdots & \ddots & \vdots & \vdots & \vdots & \ddots & \vdots \\
b_{s^2-s+1}(\mathbf{x}_1) & b_{s^2-s+2}(\mathbf{x}_1) & \dots & b_{s^2}(\mathbf{x}_1) & c_{s^2-s+1}(\mathbf{x}_1) & c_{s^2-s+2}(\mathbf{x}_1) & \dots & c_{s^2}(\mathbf{x}_1) \\
b_1(\mathbf{x}_2) & b_2(\mathbf{x}_2) & \dots & b_s(\mathbf{x}_2) & c_1(\mathbf{x}_2) & c_2(\mathbf{x}_2) & \dots & c_s(\mathbf{x}_2) \\
b_{s+1}(\mathbf{x}_2) & b_{s+2}(\mathbf{x}_2) & \dots & b_{2s}(\mathbf{x}_2) & c_{s+1}(\mathbf{x}_2) & c_{s+2}(\mathbf{x}_2) & \dots & c_{2s}(\mathbf{x}_2) \\
\vdots & \vdots & \ddots & \vdots & \vdots & \vdots & \ddots & \vdots \\
b_{s^2-s+1}(\mathbf{x}_2) & b_{s^2-s+2}(\mathbf{x}_2) & \dots & b_{s^2}(\mathbf{x}_2) & c_{s^2-s+1}(\mathbf{x}_2) & c_{s^2-s+2}(\mathbf{x}_2) & \dots & c_{s^2}(\mathbf{x}_2) \\
\vdots & \vdots & & \vdots & \vdots & \vdots & & \vdots
\end{array}
\right]
$$

$$(4)$$

In the case with 2 randomly chosen vectors, the matrix is a uniformly random $2s \times 2s$ matrix, which has nonzero column corank with probability approximately $\frac{1}{q-1}$. In the case with 3 randomly chosen vectors, the matrix is a uniformly random $3s \times 2s$ matrix, which has nonzero column corank with probability approximately $\frac{1}{(q-1)q^s}$.

Theorem 3. *If $\mathbf{x}_1$, $\mathbf{x}_2$, $\mathbf{x}_3$, and $\mathbf{x}_4$ are chosen in such a way that all four vectors are in the band kernel of a column band space $\mathcal{B}_{\beta,\gamma}$ and also that the symmetric tensor products $\mathbf{x}_1 \odot \mathbf{x}_2$ and $\mathbf{x}_3 \odot \mathbf{x}_4$ are linearly independent from one another and statistically independent from the private quadratic forms, $p_{(i-1)s+j}$ in the matrix A, then the tensor products $\mathbf{x}_1 \otimes \mathbf{x}_2$ and $\mathbf{x}_3 \otimes \mathbf{x}_4$ are both in the kernel of some column band-space differential $D^2\mathcal{E} = \sum_{\mathcal{E}_{\beta,\gamma,i} \in \mathcal{B}_{\beta,\gamma}} \tau_i D^2\mathcal{E}_{\beta,\gamma,i}$ with probability approximately $\frac{1}{(q-1)q^s}$.*

Proof. A $D\mathcal{E}$ meeting the above condition exists iff there is a nontrivial solution to the following system of equations

$$\sum_{\mathcal{E}_{\beta,\gamma,i} \in \mathcal{B}_{\beta,\gamma}} \tau_i D^2 \mathcal{E}_{\beta,\gamma,i}(\mathbf{x}_1, \mathbf{x}_2) = 0,$$

$$\sum_{\mathcal{E}_{\beta,\gamma,i} \in \mathcal{B}_{\beta,\gamma}} \tau_i D^2 \mathcal{E}_{\beta,\gamma,i}(\mathbf{x}_3, \mathbf{x}_4) = 0. \tag{5}$$

Expressed in a basis (e.g. the u_i' basis used in Definition 4) where the first s basis vectors are chosen to be outside the band kernel, and the remaining $s^2 - s$ basis vectors are chosen from within the band kernel, the column band-space differentials, $D^2 \mathcal{E}_{\beta,\gamma,i}$ are 3-tensors of the form shown in Fig. 1.

Likewise $\mathbf{x}_1$, $\mathbf{x}_2$, $\mathbf{x}_3$, and $\mathbf{x}_4$ take the form $(0| \mathbf{x}_k)$. The 2-tensors $D^2 \mathcal{E}_{\beta,\gamma,i}(\mathbf{x}_k)$ can then be represented by matrices of the form:

$$D^2 \mathcal{E}_{\beta,\gamma,i}(\mathbf{x}_k) = \left[\begin{array}{c|c} S_{k,i} & R_{k,i} \\ \hline R_{k,i}^\top & 0 \end{array} \right] \tag{6}$$

where $R_{k,i}$ is a random $s \times s^2 - s$ matrix and $S_{k,i}$ is a random symmetric $s \times s$ matrix. Removing the redundant degrees of freedom we have the system of $2s$ equations in s variables:

$$\sum_{i=1}^{s} \tau_i R_{1,i} \mathbf{x}_2^\top = 0,$$

$$\sum_{i=1}^{s} \tau_i R_{3,i} \mathbf{x}_4^\top = 0. \tag{7}$$

This has a nontrivial solution precisely when the following $2s \times s$ matrix has nonzero column corank:

$$M = \left[\begin{array}{c|c} \begin{array}{cccc} | & | & & | \\ R_{1,1}\mathbf{x}_2^\top & R_{1,2}\mathbf{x}_2^\top & \dots & R_{1,s}\mathbf{x}_2^\top \\ | & | & & | \end{array} \\ \hline \begin{array}{cccc} | & | & & | \\ R_{3,1}\mathbf{x}_4^\top & R_{3,2}\mathbf{x}_4^\top & \dots & R_{3,s}\mathbf{x}_4^\top \\ | & | & & | \end{array} \end{array} \right] \tag{8}$$

This is a random matrix over $k = \mathbb{F}_q$, which has nonzero column corank with probability approximately $\frac{1}{(q-1)q^s}$, for practical parameters.

To verify that the conditions given in the theorem are sufficient to establish the randomness of the matrix M, we can give the following explicit expression for the matrix M, which is most easily derived by applying the product rule for the discrete differential to Definition 3:

$$M = \begin{bmatrix} Dp_1(\mathbf{x}_1, \mathbf{x}_2) & Dp_{s+1}(\mathbf{x}_1, \mathbf{x}_2) & \cdots & Dp_{s^2-s+1}(\mathbf{x}_1, \mathbf{x}_2) \\ Dp_2(\mathbf{x}_1, \mathbf{x}_2) & Dp_{s+2}(\mathbf{x}_1, \mathbf{x}_2) & \cdots & Dp_{s^2-s+2}(\mathbf{x}_1, \mathbf{x}_2) \\ \vdots & \vdots & \ddots & \vdots \\ Dp_s(\mathbf{x}_1, \mathbf{x}_2) & Dp_{2s}(\mathbf{x}_1, \mathbf{x}_2) & \cdots & Dp_{s^2}(\mathbf{x}_1, \mathbf{x}_2). \\ Dp_1(\mathbf{x}_3, \mathbf{x}_4) & Dp_{s+1}(\mathbf{x}_3, \mathbf{x}_4) & \cdots & Dp_{s^2-s+1}(\mathbf{x}_3, \mathbf{x}_4) \\ Dp_2(\mathbf{x}_3, \mathbf{x}_4) & Dp_{s+2}(\mathbf{x}_3, \mathbf{x}_4) & \cdots & Dp_{s^2-s+1}(\mathbf{x}_3, \mathbf{x}_4) \\ \vdots & \vdots & \ddots & \vdots \\ Dp_s(\mathbf{x}_3, \mathbf{x}_4) & Dp_{2s}(\mathbf{x}_3, \mathbf{x}_4) & \cdots & Dp_{s^2}(\mathbf{x}_3, \mathbf{x}_4) \end{bmatrix} \tag{9}$$

Combining the results of Theorems 2 and 3, we find that for each choice of the vectors $\mathbf{x}_k$, there is a column band-space map among the solutions of Eq. (3) with probability approximately $\frac{1}{(q-1)^2 q^{2s}}$ for even characteristic and $\frac{1}{(q-1)^2 q^s}$ for odd characteristic. Equation (3) is a system of $2s^2$ equations in $2s^2$ variables; one might expect it to generally have a 0-dimensional space of solutions. In some cases, however, there are linear dependencies among the equations, due to the fact that the $D^2\mathcal{E}_i$ are symmetric tensors. In even characteristic, we get 4 linear dependencies: $D^2\mathcal{E}_i(\mathbf{x}_1, \mathbf{x}_2)(\mathbf{x}_1) = 0$, $D^2\mathcal{E}_i(\mathbf{x}_1, \mathbf{x}_2)(\mathbf{x}_2) = 0$, $D^2\mathcal{E}_i(\mathbf{x}_3, \mathbf{x}_4)(\mathbf{x}_3) = 0$, and $D^2\mathcal{E}_i(\mathbf{x}_3, \mathbf{x}_4)(\mathbf{x}_4) = 0$, and an additional linear dependency when we reduce the number of independent vectors to 3 by setting $\mathbf{x}_1 = \mathbf{x}_4$: $D^2\mathcal{E}_i(\mathbf{x}_1, \mathbf{x}_2)(\mathbf{x}_3) + D^2\mathcal{E}_i(\mathbf{x}_3, \mathbf{x}_4)(\mathbf{x}_2) = 0$, resulting in a 5-dimensional space of solutions. In characteristic 3, reducing the number of independent vectors to 2 results in 2 linear dependencies among the equations: e.g. setting $\mathbf{x}_1 = \mathbf{x}_2$ and $\mathbf{x}_3 = \mathbf{x}_4$, we have $D^2\mathcal{E}_i(\mathbf{x}_1, \mathbf{x}_2)(\mathbf{x}_1) = 0$ and $D^2\mathcal{E}_i(\mathbf{x}_3, \mathbf{x}_4)(\mathbf{x}_3) = 0$. In higher characteristic, there are no linear dependencies imposed on the equations by setting $\mathbf{x}_1 = \mathbf{x}_2$ and $\mathbf{x}_3 = \mathbf{x}_4$.

For characteristic 2, finding the expected 1-dimensional space of band-space solutions in a 5-dimensional space costs $q^4 + q^3 + q^2 + q + 1$ rank operations, which in turn cost $(s^2)^\omega$ field operations, where $\omega \approx 2.373$ is the linear algebra constant. Likewise, for characteristic 3, finding the expected 1-dimensional space of band-space solutions in a 2-dimensional space costs $q+1$ rank operations. Thus the total cost of finding a column band-space map using our variant of MinRank is approximately $q^{2s+6}s^{2\omega}$ for characterisitc 2, $q^{s+3}s^{2\omega}$ for characteristic 3, and $q^{s+2}s^{2\omega}$ for higher characteristic.

6   Complexity of Invariant Attack

The detection of a low rank induced bilinear form $D^2\mathcal{E}(x)$ already constitutes a distinguisher from a random system of equations. Extending this calculation to a full key recovery requires further use of the differential invariant structure of the public key.

First, note that U is not a critical element of the scheme. If A is a random matrix of quadratic forms and B and C are random matrices of linear forms, so are $A \circ U$, $B \circ U$ and $C \circ U$ for any full rank map U. Thus, since clearly

$T \circ \phi(AB\|AC) \circ U = T \circ \phi((A \circ U)(B \circ U)\|(A \circ U)(C \circ U))$, we may absorb the action of U into A, B, and C, and consider the public key to be of the form:

$$P(\mathbf{x}) = T \circ \phi(AB\|AC)(\mathbf{x}).$$

Next, consider a trilinear form $D^2\mathcal{E}$ in the band space generated by $\mathcal{B}_{\beta,\gamma}$. Since the coefficients of $D^2\mathcal{E}$ are products of coefficients of A and coefficients of an element of $Im(B\|C)$, both of which are uniform i.i.d., there is a change of basis M in which $D^2\mathcal{E}$ has the form in Fig. 1 and the nonzero coefficients are uniform i.i.d.

Consider $D^2\mathcal{E}(\mathbf{x}_1)$ and $D^2\mathcal{E}(\mathbf{x}_2)$ for $\mathbf{x}_1, \mathbf{x}_2$ in the band kernel corresponding to $\mathcal{B}_{\beta,\gamma}$. Being maps from the same band space, there is a basis in which both $D^2\mathcal{E}(\mathbf{x}_1)$ and $D^2\mathcal{E}(\mathbf{x}_2)$ have the form in Fig. 2. Thus, with high probability for $s \geq 2$, the kernels of both maps are contained in the corresponding band kernel, $\mathcal{B}_{\beta,\gamma}$, and $\mathrm{span}(\ker(D^2\mathcal{E}(\mathbf{x}_1) \cup \ker(D^2\mathcal{E}(\mathbf{x}_2)) = \mathcal{B}_{\beta,\gamma}$.

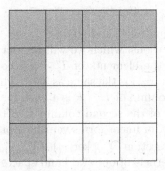

Fig. 2. Structure of the bilinear forms induced by cubic maps in the same band space.

Remark 1. *Here we have utilized a property which explicitly distinguishes differential invariant structure from rank structure.*

Given the basis for an $s^2 - s$ dimensional band kernel $\mathcal{BK}$, we may choose a basis $\{v_1, \ldots, v_s\}$ for the subspace of the dual space vanishing on $\mathcal{BK}$. We can also find a basis $\mathcal{E}_{v_1}, \ldots, \mathcal{E}_{v_s}$ for the band space itself by solving the linear system

$$\sum_{\mathcal{E}_i} \tau_i D^2\mathcal{E}_i(\mathbf{x}_{11}, \mathbf{x}_{12}, \mathbf{x}_{13}) = 0,$$

$$\sum_{\mathcal{E}_i} \tau_i D^2\mathcal{E}_i(\mathbf{x}_{21}, \mathbf{x}_{22}, \mathbf{x}_{23}) = 0,$$

$$\vdots = \vdots$$

$$\sum_{\mathcal{E}_i} \tau_i D^2\mathcal{E}_i(\mathbf{x}_{t1}, \mathbf{x}_{t2}, \mathbf{x}_{t3}) = 0,$$

where $t \approx 2s^2$ and $\mathbf{x}_{ij}$ is in the band kernel.

Since the basis $\mathcal{E}_{v_1}, \dots, \mathcal{E}_{v_s}$ is in a single band space, there exists an element $\begin{bmatrix} b'_1 & \cdots & b'_s \end{bmatrix}^\top \in ColSpace(B||C)$ and two matrices Ω_1 and Ω_2 such that

$$\Omega_1 A \left(\Omega_2 \begin{bmatrix} b'_1 \\ \vdots \\ b'_s \end{bmatrix} \right) =: A' \left(\begin{bmatrix} v_1 \\ \vdots \\ v_s \end{bmatrix} \right) = \begin{bmatrix} \mathcal{E}_{v_1} \\ \vdots \\ \mathcal{E}_{v_s} \end{bmatrix}.$$

Solving the above system of equations over $\mathbb{F}_q[x_1, \dots, x_{s^2}]$ uniquely determines A' in $\mathbb{F}_q[x_1, \dots, x_{s^2}]/\langle v_1, \dots, v_s \rangle$. To recover all of A', note that the above system is part of an equivalent key

$$\mathcal{F} = T' \circ A'(B'||C')$$

where $\begin{bmatrix} v_1 & \cdots & v_s \end{bmatrix}^\top$ is the first column of B'.

Applying T'^{-1} to both sides and inserting the information we know we may construct the system

$$A'(B'||C') = T'^{-1}\mathcal{F} \tag{10}$$

Solving this system of equations modulo $\langle v_1, \dots, v_s \rangle$ for B', C' and T'^{-1} we can recover a space of solutions, which we will restrict by arbitrarily fixing the value of T'^{-1}. Note that the elements of T'^{-1} are constant polynomials, and therefore $T'^{-1} (\mathrm{mod} \langle v_1, \dots, v_s \rangle)$ is the same as T'^{-1}. Thus, for any choice of T'^{-1} in this space, the second column of $T'^{-1}\mathcal{F}$ is a basis for a band space. Moreover, the elements $v'_{s+1}, \dots, v'_{2s}$ of the second column of $B' (\mathrm{mod} \langle v_1, \dots, v_s \rangle)$ are the image, modulo $\langle v_1, \dots, v_s \rangle$, of linear forms vanishing on the corresponding band kernel. Therefore, the intersection $\bigcap_{i=1}^{s} \ker(v_i) \cap \bigcap_{i=s+1}^{2s} \ker(v'_i)$ is the intersection $\mathcal{BK}_2 \cap \mathcal{BK}_1$ of the band kernels of our two band spaces.

We can reconstruct the full band kernel of this second band space using the same method we used to obtain our first band kernel: We take a map $\mathcal{E}_2$ from the second column of $T'^{-1}\mathcal{F}$, and two vectors x_a and x_b from $\mathcal{BK}_2 \cap \mathcal{BK}_1$, and we compute $\mathcal{BK}_2 = \mathrm{span}(\ker(D^2\mathcal{E}_2(\mathbf{x}_a) \cup \ker(D^2\mathcal{E}_2(\mathbf{x}_b))$. We can now solve for the second column of B', $\begin{bmatrix} v_{s+1} & \cdots & v_{2s} \end{bmatrix}^\top$, uniquely over $\mathbb{F}_q[x_1, \dots, x_{s^2}]$ (NOT modulo $\langle v_1, \dots, v_s \rangle$) by solving the following system of linear equations:

$$v_i \equiv v'_i (\mathrm{mod} \langle v_1, \dots, v_s \rangle)$$
$$v_i(\mathbf{x}_1) = 0$$
$$v_i(\mathbf{x}_2) = 0$$
$$\vdots = \vdots$$
$$v_i(\mathbf{x}_{s^2-s}) = 0$$

where $i = s+1, \dots, 2s$, and $(\mathbf{x}_1, \dots, \mathbf{x}_{s^2-s})$ is a basis for $\mathcal{BK}_2$. We can now solve for A' (again, uniquely over $\mathbb{F}_q[x_1, \dots, x_{s^2}]$) by solving:

$$A' \left(\begin{bmatrix} v_1 \\ \vdots \\ v_s \end{bmatrix} \right) \equiv \begin{bmatrix} \mathcal{E}_{v_1} \\ \vdots \\ \mathcal{E}_{v_s} \end{bmatrix} (\mathrm{mod} \langle v_1, \dots, v_s \rangle)$$

$$A'\left(\begin{bmatrix} v_{s+1} \\ \vdots \\ v_{2s} \end{bmatrix}\right) \equiv \begin{bmatrix} \mathcal{E}_{v_{s+1}} \\ \vdots \\ \mathcal{E}_{v_{2s}} \end{bmatrix} (\mathrm{mod}\, \langle v_{s+1}, \ldots, v_{2s} \rangle)$$

where $\begin{bmatrix} \mathcal{E}_{v_{s+1}} \cdots \mathcal{E}_{v_{2s}} \end{bmatrix}^{\top}$ is the second column of $T'^{-1}\mathcal{F}$. This allows us to solve Eq. 10 for the rest of B' and C', completing the attack.

The primary cost of the attack involves finding the band space map. The rest of the key recovery is additive in complexity and dominated by the band space map recovery; thus, the total complexity of the attack is of the same order as band space map recovery. Hence, the cost of private key extraction is approximately $q^{2s+6}s^{2\omega}$ for characteristic 2, $q^{s+3}s^{2\omega}$ for characteristic 3, and $q^{s+2}s^{2\omega}$ for higher characteristic. We note that with these parameters we can break full sized instances of this scheme with parameters chosen for the 80-bit and 100-bit security levels via the criteria presented in [13].

Specifically, our attack breaks CubicABC($q = 127$, $s = 7$), designed for 80-bit security, in approximately 2^{76} operations (or around 2^{80} if one pessimistically uses $\omega = 3$ as the linear algebra constant). More convincingly, our attack completely breaks CubicABC($q = 127$, $s = 8$), designed for 100-bit security, in approximately 2^{84} operations (or 2^{88} for $\omega = 3$). Furthermore, the attack is fully parallelizable and requires very little memory; hence, the differential invariant attack is far more efficient than algebraic attacks, the basis for the original security estimation. Thus, the security claims in [13] are clearly unfounded; in fact, the cubic version of the scheme, whose security was claimed to be closely related to an NP-complete problem, is actually less secure than the quadratic case.

We can explain this dramatic discrepancy on the fact that the parameters in [13] are derived by assuming that the algebraic attack is the most effective. In the case of the quadratic ABC scheme, for the proposed parameters, the attack of [12] was slower than the algebraic attack, though asymptotically faster. In the case of the Cubic scheme, the attack is actually more efficient, in asymptotics as well as for practical parameters.

7   Experiments

Using SAGE [14], we performed some minrank computations on small scale variants of the Cubic ABC scheme. The computations were done on a computer with a 64 bit quad-core Intel i7 processor, with clock cycle 2.8 GHz. We were interested in verifying our complexity estimates on the most costly step in the attack, the MinRank instance, rather than the full attack on the ABC scheme. Given as input the finite field size q, and the scheme parameter s, we computed the average number of vectors v required to be sampled in order for the rank of the 2-tensor $D^2\mathcal{E}(v)$ to fall to $2s$. As explained in Sect. 5, when the rank falls to this level, we have identified the subspace differential invariant structure of the scheme and can exploit this structure to attack the scheme. Our results for odd q are given in Table 1.

Table 1. Average number of vectors needed for the rank to fall to $2s$ (for odd q)

	$s = 3$	$(q-1)^2 q^s$	$s = 4$	$(q-1)^2 q^s$	$s = 5$	$(q-1)^2 q^s$
$q = 3$	14.75	108	333	324	952	972
$q = 5$	378	2000	9986	10000		
$q = 7$	1688	12348	72951	86436		
$q = 9$	606	46656				
$q = 11$	13574	133100				

For higher values of q and s the computations took too long to produce sufficiently many data points and obtain meaningful results with SAGE. When q is odd, our analysis predicted the number of vectors needed would be on the order of $(q-1)^2 q^s$. Table 1 shows the comparison between our experiments and the expected value. We see that for $s = 3$, the rank fell quicker than expected, while for $s > 3$ the results are quite near the predicted value. This is because when $s = 3$ our complexity estimates given in Sect. 5 are simply not accurate enough, which happens for small values of q and/or s.

For even q, we also ran some experiments. We found that for $s = 3$ and $q = 2, 4$, or 8, with high probability only a single vector was needed before the rank fell to $2s$. For $s = 4$ and $s = 5$, the computations were only feasible in SAGE for $q = 2$. The average number of vectors needed in the $s = 4$ case was 244, with the expected value being $(q-1)^2 q^{2s} = 256$. With $s = 5$, the average number in our experiments was 994 (although the number of trials was small), with the expected value 1024. For higher values of q and s the computations took too long to obtain meaningful results.

8   Conclusion

The ABC schemes are very interesting new ideas for multivariate public key schemes. Essentially all of MPKC can be bisected into big field schemes, utilizing the structure of an extension of the field used for public calculations, and small field schemes which require no such extension. (For the purpose of this comment we consider "medium" field schemes to be big field schemes.)

The ABC cryptosystems present a fundamentally new structure for the development of schemes. In fact, if we consider the structure of simple algebras over the public field (which are surely the only such structures we should consider for secure constructions) then "big field" and "big matrix algebra" complete the picture of possible large structure schemes.

It is interesting to note that the authors provide in [13] a heuristic security argument for the scheme and, as reinforced in the first presentation of the scheme at [15], suggest that with some work the scheme may be able to be shown provably secure. The idea behind their argument is at least somewhat reasonable, if not precise. Their argument essentially amounts to the following: every cubic polynomial in the public key is in the ideal generated by the quadratic forms in A under a certain basis; thus, one might expect the public key to contain a subset of the information one would obtain by applying one step of a Gröbner basis algorithm such as F4, see [16].

Unfortunately, this analysis is not very tight. In fact, we exploit the subspace differential invariant structure inherent to the ABC methodology to show that for odd characteristic the cubic scheme is less secure than its quadratic counterpart. We may therefore conclude that any attempt at a secure cubic "big matrix algebra" scheme must rely on the application of modifiers. The challenge, then, is to construct such a scheme which is still essentially injective for the purpose of encryption. Schemes such as this one can never compete with the secure multivariate options for digital signatures we already know.

We are thus left with the same lingering question that has been asked for the last two decades: Is secure multivariate encryption possible? Currently there is a small list of candidates none of which has both been extensively reviewed and has existed for longer than a few years. If we are to discover a secure multivariate encryption scheme with a convincing security proof or some other security metric, it will require some new techniques and new science. Only time will tell.

References

1. Shor, P.W.: Polynomial-time algorithms for prime factorization and discrete logarithms on a quantum computer. SIAM J. Sci. Stat. Comp. **26**, 1484 (1997)
2. Kipnis, A., Patarin, J., Goubin, L.: Unbalanced oil and vinegar signature schemes. In: Stern, J. (ed.) EUROCRYPT 1999. LNCS, vol. 1592, pp. 206–222. Springer, Heidelberg (1999). doi:10.1007/3-540-48910-X_15
3. Patarin, J., Goubin, L., Courtois, N.: C^*_{-+}, and HM: variations around two schemes of T. Matsumoto and H. Imai. In: Ohta, K., Pei, D. (eds.) ASIACRYPT 1998. LNCS, vol. 1514, pp. 35–50. Springer, Heidelberg (1998). doi:10.1007/3-540-49649-1_4
4. Patarin, J., Courtois, N., Goubin, L.: QUARTZ, 128-bit long digital signatures. In: Naccache, D. (ed.) CT-RSA 2001. LNCS, vol. 2020, pp. 282–297. Springer, Heidelberg (2001). doi:10.1007/3-540-45353-9_21
5. Petzoldt, A., Bulygin, S., Buchmann, J.: CyclicRainbow – a multivariate signature scheme with a partially cyclic public key. In: Gong, G., Gupta, K.C. (eds.) INDOCRYPT 2010. LNCS, vol. 6498, pp. 33–48. Springer, Heidelberg (2010). doi:10.1007/978-3-642-17401-8_4
6. Petzoldt, A., Chen, M., Yang, B., Tao, C., Ding, J.: Design principles for HFEv-based multivariate signature schemes. In: Iwata, T., Cheon, J.H. (eds.) ASIACRYPT 2015. LNCS, vol. 9452, pp. 311–334. Springer, Heidelberg (2015). doi:10.1007/978-3-662-48797-6_14
7. Ding, J., Yang, B.: Degree of regularity for HFEv and HFEv-. In: Gaborit, P. (ed.) PQCrypto 2013. LNCS, vol. 7932, pp. 52–66. Springer, Heidelberg (2013). doi:10.1007/978-3-642-38616-9_4
8. Goubin, L., Courtois, N.T.: Cryptanalysis of the TTM cryptosystem. In: Okamoto, T. (ed.) ASIACRYPT 2000. LNCS, vol. 1976, pp. 44–57. Springer, Heidelberg (2000). doi:10.1007/3-540-44448-3_4

9. Tsujii, S., Gotaishi, M., Tadaki, K., Fujita, R.: Proposal of a signature scheme based on STS trapdoor. In: Sendrier, N. (ed.) PQCrypto 2010. LNCS, vol. 6061, pp. 201–217. Springer, Heidelberg (2010). doi:10.1007/978-3-642-12929-2_15

10. Porras, J., Baena, J., Ding, J.: ZHFE, a new multivariate public key encryption scheme. In: Mosca, M. (ed.) PQCrypto 2014. LNCS, vol. 8772, pp. 229–245. Springer, Cham (2014). doi:10.1007/978-3-319-11659-4_14

11. Tao, C., Diene, A., Tang, S., Ding, J.: Simple matrix scheme for encryption. In: Gaborit, P. (ed.) PQCrypto 2013. LNCS, vol. 7932, pp. 231–242. Springer, Heidelberg (2013). doi:10.1007/978-3-642-38616-9_16

12. Moody, D., Perlner, R.A., Smith-Tone, D.: An asymptotically optimal structural attack on the ABC multivariate encryption scheme. In: Mosca, M. (ed.) PQCrypto 2014. LNCS, vol. 8772, pp. 180–196. Springer, Cham (2014). doi:10.1007/978-3-319-11659-4_11

13. Ding, J., Petzoldt, A., Wang, L.: The cubic simple matrix encryption scheme. In: Mosca, M. (ed.) PQCrypto 2014. LNCS, vol. 8772, pp. 76–87. Springer, Cham (2014). doi:10.1007/978-3-319-11659-4_5

14. Sage, S.: The Sage Mathematics Software System (Version x.y.z). (YYYY) http://www.sagemath.org

15. Mosca, M. (ed.): PQCrypto 2014. LNCS, vol. 8772. Springer, Cham (2014). doi:10.1007/978-3-319-11659-4

16. Faugere, J.C.: A new efficient algorithm for computing grobner bases (f4). J. Pure Appl. Algebra **139**, 61–88 (1999)

17. Gaborit, P. (ed.): PQCrypto 2013. LNCS, vol. 7932. Springer, Heidelberg (2013). doi:10.1007/978-3-642-38616-9

Solving Discrete Logarithms on a 170-Bit MNT Curve by Pairing Reduction

Aurore Guillevic[5,6]($\boxtimes$), François Morain[1,4], and Emmanuel Thomé[1,2,3]

1 Institut National de Recherche en Informatique et en Automatique (INRIA),
Villers-lès-Nancy and Saclay, France
emmanuel.thome@inria.fr
2 Université de Lorraine, Loria, UMR 7503, Vandoeuvre-lès-Nancy, France
3 CNRS, Loria, UMR 7503, Vandoeuvre-lès-Nancy, France
4 École Polytechnique/LIX, CNRS UMR 7161, Palaiseau, France
morain@lix.polytechnique.fr
5 University of Calgary, Alberta, Canada
aurore.guillevic@inria.fr
6 Pacific Institute for the Mathematical Sciences, CNRS UMI 3069,
Vancouver, Canada

Abstract. Pairing based cryptography is in a dangerous position following the breakthroughs on discrete logarithms computations in finite fields of small characteristic. Remaining instances are built over finite fields of large characteristic and their security relies on the fact the embedding field of the underlying curve is relatively large. How large is debatable. The aim of our work is to sustain the claim that the combination of degree 3 embedding and too small finite fields obviously does not provide enough security. As a computational example, we solve the DLP on a 170-bit MNT curve, by exploiting the pairing embedding to a 508-bit, degree-3 extension of the base field.

Keywords: Discrete logarithm · Finite field · Number Field Sieve · MNT elliptic curve

1 Introduction

Pairings were introduced as a constructive cryptographic tool in 2000 by Joux [31], who proposed a one-round three participants key-exchange. Numerous protocols also based on pairings have been developed since. Beyond efficient broadcast protocols, prominent applications include Identity-Based Encryption [13,35,36], or short signatures [14].

The choice of appropriate curves and pairing definitions in the context of pairing-based cryptography has been the topic of many research articles. An important invariant is the degree of the embedding field, which measures the complexity of evaluating pairings, but is also related to the security of systems (see Sect. 2 for more precisions). The first cryptographic setups proposed used pairings on supersingular curves of embedding degree 2 defined over a

© Springer International Publishing AG 2017
R. Avanzi and H. Heys (Eds.): SAC 2016, LNCS 10532, pp. 559–578, 2017.
https://doi.org/10.1007/978-3-319-69453-5_30

prime field $\mathbb{F}_p$, where p is 512-bit long, so that the pairing embeds into a 1024-bit finite field $\mathbb{F}_{p^2}$. Another early curve choice is a supersingular elliptic curve in characteristic three, defined over $\mathbb{F}_{3^{97}}$, of embedding degree 6 (used e.g. in [14], as well as various implementation proposals, e.g. [10]). More recent proposals define *pairing-friendly* ordinary curves over large characteristic fields, where constraining the embedding degree to selected values is a desired property [9, 15, 16, 19, 22, 23, 25, 42].

Cryptanalysis of pairings can be attempted via two distinct routes. Either attack the discrete logarithm problem *on the curve*, or in the embedding field of the pairing considered. The former approach is rarely successful, given that it is usually easy to choose curves which are large enough to thwart $O(\sqrt{N})$ attacks such as parallel collision search or Pollard rho. Note however that derived problems such as the discrete logarithm *with auxiliary inputs* are much easier to handle, as shown by [46].

Attacking pairings via the embedding field is a strategy known as the Menezes–Okamoto–Vanstone [41] or Frey–Rück [24] attack, depending on which pairing is considered. Successful cryptanalyses that follow this strategy have been described in small characteristic. In [29], for a supersingular curve over $\mathbb{F}_{3^{97}}$, the small characteristic allowed the use of the Function Field Sieve algorithm [1], and the composite extension degree was also a very useful property. More recently, following recent breakthroughs for discrete logarithm computation in small characteristic finite fields [7, 27], a successful attack has been reported on a supersingular curve over $\mathbb{F}_{2^{1223}}$, with degree-4 embedding [27]. The outcome of these more recent works is that curves in small characteristic are now definitively avoided for pairing-based cryptography.

As far as we know, there is no major record computation of discrete logarithms over pairing-friendly curves in large characteristic using a pairing reduction in the finite field. The pairing-friendly curves used in practice have a large embedding field of more than 1024 bits, where computing a discrete logarithm is still very challenging. A few curves in large characteristic have comparatively small embedding fields, and were identified as weak to this regard, although no practical computation to date demonstrated the criticality of this weakness. This includes the so-called MNT curves defined by Miyaji–Nakabayashi–Takano, e.g. [42, Example 1], an elliptic curve defined over a 170-bit prime p, and of 508-bit embedding field $\mathbb{F}_{p^3}$.

Despite the academic agreement on the fact that the pairing embedding fields for 170-bit MNT curves in general, and the one just mentioned in particular, are too small for cryptographic use, recent work like [2] has shown how cryptography relying on overly optimistic hardness assumptions can linger almost indefinitely in the wild. Demonstrating a practical break is key to really phasing out such outdated cryptographic choices. As far as we know, an MNT curve of low embedding degree 3 was never used in pairing-based cryptography, but was never attacked by a pairing reduction either. In this article, we present our attack over the weak[1] MNT curve [42, Example 1], with p of 170 bits and $n = 3$.

---

[1] Already described as weak in the paper by the authors.

We report a discrete logarithm computation in the group of points of this curve by a pairing reduction, using only a moderate amount of computing power.

In order to attack the discrete logarithm problem in the embedding field, appropriate variants of the Number Field Sieve must be used. The crucial point is the adequate choice of a polynomial pair defining the Number Field Sieve setup, among the various choices proposed in the literature [6,8,32,33,40]. It is also important to arrange for the computation to take advantage of Galois automorphisms when available, both within sieving and linear algebra. Last, some care is needed in order to efficiently compute individual logarithms of arbitrary field elements.

This article is organized as follows. Section 2 reviews some background and notations for MNT curves on the one hand, and the Number Field Sieve (NFS) as a general framework on the other hand. Section 3 discusses in more detail the various possible choices of polynomial selection techniques for NFS. Section 4 discusses the details of the discrete logarithm computation with NFS, while Sect. 4.3 defines and solves an arbitrary challenge on the MNT curve.

2   Background and Notations

2.1   Using Pairing Embedding to Break DLP

We follow [12, Chap. IX]. To fix notations, pairings are defined as follows, the map being bilinear, non-degenerate and computable in polynomial time in the size of the inputs.

$$e : \begin{cases} E(\mathbb{F}_p)[\ell] \times E(\mathbb{F}_{p^n})[\ell] \to \mu_\ell \subset \mathbb{F}_{p^n}^* \\ (P, Q) \qquad \mapsto e(P, Q). \end{cases} \tag{1}$$

Here, μ_ℓ is the subgroup of ℓ-th roots of unity, i.e. an element $u \in \mu_\ell$ satisfies $u^\ell = 1 \in \mathbb{F}_{p^n}^*$. The integer n is the so-called *embedding degree*, that is the smallest integer i for which the ℓ-torsion is contained in $\mathbb{F}_{p^i}$. It has a major impact on evaluating the difficulty of solving the DLP on the curve.

Let G_1 be a generator of $E(\mathbb{F}_p)[\ell]$ and P in the same group, whose discrete logarithm u is sought (so that $P = [u]G_1$). We choose a generator G_2 for $E(\mathbb{F}_{p^n})[\ell]$. We observe that

$$e(P, G_2) = e(G_1, G_2)^u$$

so that u can be recovered as the logarithm of $U = e(P, G_2)$ in base $T = e(G_1, G_2)$, where both elements belong to the subgroup of order ℓ of $\mathbb{F}_{p^n}^*$. Note that by construction, $\ell = O(p)$, so that the Number Field Sieve linear algebra phase has to be considered modulo ℓ, which is a priori much smaller than the largest prime order subgroup of $\mathbb{F}_{p^n}^*$, which has size $O(p^{\phi(n)})$.

2.2   MNT Curves

The Miyaji–Nakabayashi–Takano curves were designed in 2000 in [42] as the first example of *ordinary* curves with low embedding degree $n = 3$, 4, or 6. The curves were presented as a weak instance of ordinary elliptic curves that should be avoided in elliptic-curve cryptography because of the Menezes–Okamoto–Vanstone and Frey–Rück attacks [24,41] that embed the computation of a discrete logarithm from the group of points of the curve to the embedding field $\mathbb{F}_{p^n}$. At the 80-bit security level which was used in the 2000's, an elliptic curve of 160-bit prime order was considered safe, and of at least the same security as an 1024-bit RSA modulus. However for MNT curves over prime fields of 160 bits, the MOV and FR reduction attacks embed to finite fields of size 480, 640, or 960 bits, none of which should be considered as having a hard enough DLP. For these three cases and most of all for $n = 3$, computing a discrete logarithm in the embedding field is considerably easier than over the elliptic curve. The conclusion of the MNT paper was to advise developers to systematically check that the embedding degree of an elliptic curve is large enough, to avoid pairing reduction attacks. The authors also mentioned as a constructive use of their curves the prequel work of Kasahara, Ohgishi, and Sakai on identity-based encryption using pairings [35,36]. Some implementations using MNT curves exist, for example the Miracl Library proposes software on an MNT curve over a 170-bit prime, with embedding degree $n = 6$, providing a 80-bit security level (Table 1).

Table 1. MNT curves as pairing-friendly curves in the 2000's

Embedding degree n	$\log_2 p$ ($\#E(\mathbb{F}_p)$)	$n \log_2 p$ ($\#\mathbb{F}_{p^n}$)	80-bit security
3	170	510	No
4	170	680	No
6	170	1020	Yes

Construction of MNT curves. The parameters p, τ, ℓ (base field, trace, and number of points) of the curve are given by polynomials of degree at most two. For $n = 3$, 4, or 6, these are

Embedding degree n	$p = P(x)$	$\tau = \mathrm{Tr}(x)$	$\#E(\mathbb{F}_p) = p + 1 - \tau$
3	$12x^2 - 1$	$\pm 6x - 1$	$12x^2 \mp 6x + 1$
4	$x^2 + x + 1$	$-x$, or $x + 1$	$x^2 + 2x + 2$ or $x^2 + 1$
6	$4x^2 + 1$	$1 \pm 2x$	$4x^2 \mp 2x + 1$

To generate a curve, one needs to find an integer y of the appropriate size, such that $p = P(y)$ is prime and $\#E(\mathbb{F}_p)$ is also prime, or equal to a small cofactor times a large prime. To compute the coefficients of the curve equation, a Pell equation needs to be solved.

The Target curve. Our target will be the MNT curve given in [42, Example 1]. We recall that the curve parameters satisfy

$$y = -\text{0x732c8cf5f983038060466}$$
$$p = 12y^2 - 1 = \text{0x26dccacc5041939206cf2b7dec50950e3c9fa4827af} \text{ of } 170 \text{ bits}$$
$$\tau = 6y - 1 \text{ where } \tau \text{ is the trace of the curve}$$
$$\#E(\mathbb{F}_p) = p + 1 - \tau = 7^2 \cdot 313 \cdot \ell \text{ where } \ell \text{ is a 156-bit prime}$$
$$\ell = \text{0xa60fd646ad409b3312c3b23ba64e082ad7b354d}$$

The pairing embeds into the prime order ℓ subgroup of the cyclotomic subgroup of $\mathbb{F}_{p^3}$, where ℓ divides $p^2 + p + 1$.

2.3   A Brief Overview of NFS-DL

Our target field is $\mathbb{F}_{p^n}$. NFS-DL starts by selecting two irreducible integer polynomials f and g such that $\varphi = \gcd(f \bmod p, g \bmod p)$ is irreducible of degree n (construction of f and g is discussed in Sect. 3). We use the representation $\mathbb{F}_{p^n} = \mathbb{F}_p[x]/(\varphi(x))$. Let $K_f = \mathbb{Q}[x]/(f(x)) = \mathbb{Q}(\alpha)$, and $\mathcal{O}_f$ be its ring of integers. Note that because f is not necessarily monic, α might not be an algebraic integer. Let ρ_f be the map from K_f to $\mathbb{F}_{p^n}$, sending α to $T \bmod (p, \varphi(T))$. We define likewise $K_g = \mathbb{Q}(\beta)$, together with $\mathcal{O}_g$ and ρ_g. This installs the (typical) commutative diagram in Fig. 1.

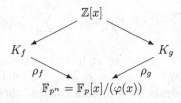

Fig. 1. NFS-DL diagram for $\mathbb{F}_{p^n}$

Given f and g, we choose a smoothness bound B and build factor bases $\mathcal{F}_f$ (resp. $\mathcal{F}_g$) consisting of prime ideals in $\mathcal{O}_f$ (resp. $\mathcal{O}_g$) of norm less than B, to which we add prime ideals dividing $\text{lc}(f)$ (resp. $\text{lc}(g)$) to take into account the fact that α and β are not algebraic integers. Then, we collect relations, that is polynomials $\phi(x) \in \mathbb{Z}[x]$ such that both ideals $\langle \phi(\alpha) \rangle$ and $\langle \phi(\beta) \rangle$ are *smooth*, namely factor completely over $\mathcal{F}_f$ (resp. $\mathcal{F}_g$). Smoothness is related to $\text{Norm}(\phi(\alpha))$, and in turn to $\text{Res}(f, \phi)$ since we have

$$\pm \text{lc}(f)^{\deg(\phi)} \text{Norm}(\phi(\alpha)) = \text{Res}(f, \phi).$$

When ϕ is such that the integers $\mathrm{Res}(f, \phi)$ and $\mathrm{Res}(g, \phi)$ are B-smooth (only prime factors below B), we have a relation:

$$\begin{cases} \phi(\alpha)\mathcal{O}_f = \prod_{\mathfrak{q} \in \mathcal{F}_f} \mathfrak{q}^{\mathrm{val}_{\mathfrak{q}}(\phi(\alpha))}, \\ \phi(\beta)\mathcal{O}_g = \prod_{\mathfrak{r} \in \mathcal{F}_g} \mathfrak{r}^{\mathrm{val}_{\mathfrak{r}}(\phi(\beta))} \end{cases}$$

that are transformed as linear relation between virtual logarithms of ideals [52], to which are added the so-called Schirokauer maps [51], labelled $\lambda_{f,i}$ for $1 \leq i \leq r_f$ where r_f is the unit rank of K_f (and the same for g).

To overcome the problem of dealing with fractional ideals instead of integral ideals, we use the following result from [43] (see also [20]).

Proposition 1. *Let $f(X) = \sum_{i=0}^{d} c_i X^i$ with coprime integer coefficients and α a root of f. Let*

$$J_f = \langle c_d, c_d\alpha + c_{d-1}, c_d\alpha^2 + c_{d-1}\alpha + c_{d-2}, \ldots, c_d\alpha^{d-1} + c_{d-1}\alpha^{d-2} + \cdots + c_1 \rangle.$$

Then $\langle 1, \alpha \rangle J_f = (1)$, J_f has norm $|c_d|$, and $J_f\langle a - b\alpha \rangle$ is an integral ideal for integers a and b.

If $\phi(X)$ has degree $k - 1$, we have $\mathrm{Norm}(J_f^{k-1}\langle\phi(\alpha)\rangle) = \pm\mathrm{Res}(f, \phi)$, so that we can read off the factorization of the integral $J_f^{k-1}\langle\phi(\alpha)\rangle$ directly from the factorization of its norm. A relation can now be written as:

$$(k-1)\,\mathrm{vlog}(J_f) + \sum_{\mathfrak{q} \in \mathcal{F}_f} \mathrm{val}_{\mathfrak{q}}(\phi(\alpha))\,\mathrm{vlog}(\mathfrak{q}) + \sum_{i=1}^{r_f} \lambda_{f,i}(\phi(\alpha))\,\mathrm{vlog}(\lambda_{f,i})$$

$$\equiv (k-1)\,\mathrm{vlog}(J_g) + \sum_{\mathfrak{r} \in \mathcal{F}_g} \mathrm{val}_{\mathfrak{r}}(\phi(\beta))\,\mathrm{vlog}(\mathfrak{r}) + \sum_{i=1}^{r_g} \lambda_{g,i}(\phi(\beta))\,\mathrm{vlog}(\lambda_{g,i}) \bmod \ell.$$

We select as many $\phi(x)$ of degree at most $k - 1$ (for $k \geq 2$ and very often $k = 2$) as needed to find $\#\mathcal{F}_f + \#\mathcal{F}_g + r_f + r_g + 2$ relations. Note that J_f and J_g are not always prime ideals. Nevertheless since all their prime divisors have a grouped contribution for each relation, we may count them as single columns. We may even replace the two columns by one, corresponding to $\mathrm{vlog}(J_f) - \mathrm{vlog}(J_g)$ (e.g. this is done in `cado-nfs`).

Given sufficiently many equations, the linear system in the virtual logarithms can be solved using sparse linear algebra techniques such as the Block Wiedemann algorithm [18]. When we want to compute the logarithm of a given target, we need to rewrite some power (or some multiple) of the target as a multiplicative combination of the images in $\mathbb{F}_{p^n}$ of the factor base ideals, and use the precomputed data base of computed logarithms. Section 4 will briefly discusses algebraic factorization in practice.

3   Polynomial Selection

The polynomial selection is the first step of the NFS algorithm. Polynomial selection is rather cheap, but care is needed since the quality of the polynomial pair it outputs conditions the running time of the three next steps. Section 3.1 below explains the two phases of polynomial selection. In a nutshell, we first decide from which family the polynomials are chosen, and then we search among possible solutions for "exceptionally good" polynomials. Note that because all degree n irreducible polynomials correspond to isomorphic finite fields $\mathbb{F}_{p^n}$, we are not constrained in the choice of $\mathrm{Res}(f, g)$. This degree of freedom allows to select good polynomials.

As of 2016, the available polynomial selection algorithms are:

- the Conjugation method (Conj) [6, Sect. 3.3], explained in Algorithm 1;
- the Generalized Joux–Lercier method (GJL) [6, Sect. 3.2] and [40] that produces polynomials of unbalanced coefficient sizes;
- the Joux–Lercier–Smart–Vercauteren method (JLSV$_1$) [32, Sect. 2.3], that produces two polynomials of degree n and coefficient size in $O(\sqrt{p})$ for both polynomials;
- the second proposition (JLSV$_2$) of the same paper [32, Sect. 3.2];
- the Joux–Pierrot (JP) method for pairing-friendly curves [33] which produces polynomials equivalent to the Conjugation method for MNT curves;
- the Tower-NFS method (TNFS) of Barbulescu, Gaudry and Kleinjung [8];
- the Sarkar–Singh method that combines and generalizes the GJL and Conjugation methods [49].

Remark 1 (Non-applicable methods.). The Extended-TNFS method of Kim and Kim-Barbulescu and its numerous variants [30,37,38,47,48,50] do not apply to finite fields of prime extension degree n such as $\mathbb{F}_{p^3}$. The TNFS method is not better than the best above methods for our practical case study, as shown in the paper [8, Sect. 5]. The Sarkar–Singh method [49] has two parameters (d, r): d is a divisor of n and $r \geq n/d$. Since n is prime, the pair $(d = 1, r \geq n)$ corresponds to the GJL method and the pair $(d = n, r = 1)$ to the Conjugation method. The pair $(d = n, r = 2)$ produces a polynomial f of degree 9 and small coefficients, and a polynomial g of degree 6 and coefficients in $O(p^{1/3})$. This is not competitive for our size of parameters $n = 3$ and p of 170 bits: the cross-over point between the Conjugation ($r = 1$) and their method ($r = 2$) is at $\log_2 p^3 = 9592$ bits.

Algorithm 1 presents the Conjugation method, which eventually provided the best yield. Pseudo-code describing the other methods can be found in Appendix A.

3.1   A First Comparison

The various methods above yield polynomial pairs whose characteristics differ significantly. Table 2 gives the expected degrees and coefficient sizes. From this

---

Algorithm 1. Polynomial selection with the Conjugation method [6, Sect. 3.3]

Input: p prime and n integer
Output: f, g, ψ with $f, g \in \mathbb{Z}[x]$ irreducible and $\psi = \gcd(f \bmod p, g \bmod p)$ in
$\mathbb{F}_p[x]$ irreducible of degree n

1 **repeat**
2 $\quad$ Select $g_1(x), g_0(x)$, two polynomials with small integer coefficients,
$\quad\quad \deg g_1 < \deg g_0 = n$
3 $\quad$ Select $a(y)$ a quadratic, monic, irreducible polynomial over $\mathbb{Z}$ with small
$\quad\quad$ coefficients
4 **until** $a(y)$ *has a root* y *in* $\mathbb{F}_p$ *and* $\psi(x) = g_0(x) + yg_1(x)$ *is irreducible in* $\mathbb{F}_p[x]$
5 $f \leftarrow \mathrm{Res}_y(a(y), g_0(x) + yg_1(x))$
6 $(u, v) \leftarrow$ a rational reconstruction of y
7 $g \leftarrow vg_0 + ug_1$
8 **return** (f, g, ψ)

---

data, we can derive bounds on the resultants on both sides of a relation (either using the coarse bound $(\deg f + \deg \phi)! \|f\|_\infty^{\deg \phi} \|\phi\|_\infty^{\deg f}$, or finer bounds such as [11, Theorem 7], as used in [8, Sect. 3.2]). These norms should be minimized in order to obtain the best running-time for the NFS algorithm. We obtain the plot of Fig. 2 for the bit-size of the product of norms, similar to [6, Fig. 3].

Table 2. Norm bound w.r.t. Q

Method	$\deg f$	$\|f\|_\infty$	$\deg g$	$\|g\|_\infty$
GJL	$D + 1 \geq n + 1$	$O(\log p)$	$D \geq n$	$O(Q^{1/(D+1)})$
JP or Conj	$2n$	$O(\log p)$	n	$O(Q^{1/(2n)})$
JLSV$_1$	n	$O(Q^{1/(2n)})$	n	$O(Q^{1/(2n)})$
JLSV$_2$	$D \geq n + 1$	$O(Q^{1/(D+1)})$	n	$O(Q^{1/(D+1)})$

Figure 2 suggests that the GJL method yields the smallest norms for $\log_2 Q = 508$. The norms produced with the Conjugation and JLSV$_1$ methods are not very far however so we compared more precisely these three methods for our 170-bit parameters. This entails finding competitive polynomial pairs for each method, and comparing their merits. Estimated bounds as well as experimental values for the products of norms for $\log_2 Q = 508$ are reported in Table 3. Results of sieving on one slide of special-q is reported in Table 4. The algorithms and computed polynomials are given in Appendix A. The theoretical bound $\|f\|_\infty$ equals one bit in the Conjugation and GJL methods whereas in practice to improve the smoothness properties of f, we have chosen a polynomial with moderately larger coefficients, and with better α and Murphy's E values (see [44, Sect. 5.2 Eq. (5.7)] on Murphy's E value). The coefficient size of g selected with the GJL, Conj and JLSV$_1$methods is a few bits larger than the theoretical bound because we

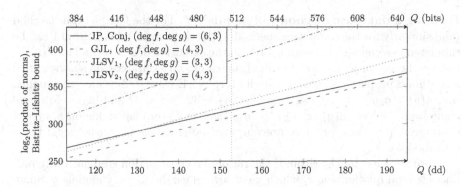

Fig. 2. Norm bound for four polynomial selection methods for $\mathbb{F}_{p^3}$

computed linear combinations of two distinct g, and of f and the initial g in the JLSV$_1$ case (since they are of same degree). The advantage of the hybrid Joux–Pierrot method (Algorithm 2) in the MNT case is that g can be monic, which does not allow for linear combinations.

Table 3. Norm bounds in bits for $\log Q = 508$ and $\log E = 25.25$: estimates based on Table 2, compared to experimental values with our selected polynomials.

Method	$\|f\|_\infty$		$\|g\|_\infty$		Norm bound f		Norm bound g		Product	
	Bound	Exp.	Bound	Exp.	Bound	Exp.	Bound	Exp.	Bound	Exp.
GJL	1	2	127	130	106	107	206	208	311	314
Conj	1	9	85	86	157	165	163	164	320	328
Hybrid JP	1	12	85	85	157	168	163	164	320	331
JLSV$_1$	85	85	85	86	163	163	163	164	326	327
JLSV$_2$	102	–	102	–	206	–	180	–	386	–

Galois actions. For small extension degrees $n \in \{3, 4, 6\}$ there exist families of polynomials producing number fields with cyclic Galois groups, and an easy-to-compute automorphism [21, Prop. 1.2]. Taking polynomials from these families yields a speed-up in the sieving part as well as in the linear algebra part for the JLSV$_1$ and Conjugation methods. We take $g = x^3 - y_0 x^2 - (y_0 + 3)x - 1$ for the Conjugation method, i.e. $g_0 = x^3 - 3x - 1$ and $g_1 = -x^2 - x$ in Algorithm 1. The Galois action is $\sigma(x) = (-x - 1)/x$ which is independent of the parameter y_0. In that case, given the factorization for $\langle a - b\alpha \rangle$, we can deduce that of

$$\sigma(\langle a - b\alpha \rangle) = \langle a - b\sigma(\alpha) \rangle = -\frac{1}{\alpha}(b - (-a - b)\alpha).$$

The same holds on the f side.

Forming a database of good polynomials f. For the Conjugation method (and similarly for the competing methods), the early steps in Algorithm 1 can be tabulated in some way, depending only on the extension degree n (and for $JLSV_1$, also on the size of p, but not its value): we can store a database of f's with good smoothness properties (low α and high Murphy's E values). Actually we searched over $a(y) = a_2 y^2 + a_1 y + a_0$, where $0 < a_2 < 32$, $|a_1| < 32$ and $|a_0| < 512$, and computed $f = \mathrm{Res}_y(a(y), x^3 - yx^2 - (y+3)x - 1)$. Later, depending on p, we can continue Algorithm 1 for these precomputed polynomials (test whether a has a root modulo p).

Note also that in Algorithm 1, the rational reconstruction step naturally produces several quotients u/v, which yield several candidate polynomials g. Small linear combinations of these polynomials can be tried, in order to improve on the Murphy's E value.

3.2   Probing the Sieving Yield

To finalize the comparison between the polynomials, we compared the relation yield for small special-q ranges sampled over the complete special-q space. Because the $JLSV_1$ and Conjugation methods feature balanced norms (see Table 3), we used similar large prime bounds (27 bits) on both sides in both cases, and allowed two large prime on each side. In contrast, for the GJL method, we allowed 28-bit large primes on the g side, and chose q to be only on that side. The Conjugation method (polynomial below) appeared as the best option based on the seconds/relation measure, given that the overall yield was sufficient. Results of this test are reported on Table 4.

$$
\begin{aligned}
f &= 28x^6 + 16x^5 - 261x^4 - 322x^3 + 79x^2 + 152x + 28 \\
\alpha(f) &= -2.94 \\
\log_2 \|f\|_\infty &= 8.33 \\
g &= {\scriptstyle 2475781518663919737044122}\, x^3 + {\scriptstyle 4080689704025368047177775183}\, x^2 \\
&\quad -{\scriptstyle 33466548519663911639551183}x - {\scriptstyle 2475781518663919737044122} \\
\alpha(g) &= -4.16 \\
\log_2 \|g\|_\infty &= 85.08, \text{ the optimal being } \tfrac{1}{2}\log_2 p = 85 \\
E(f, g) &= 1.31 \cdot 10^{-12}
\end{aligned}
\tag{2}
$$

4   Solving DLP over $\mathbb{F}_{p^3}$

4.1   Sieving and Linear Algebra

We took a smoothness bound of 50×10^6 on both sides; and all special-q in $[50 \times 10^6, 2^{27}]$, on both sides. This took roughly 660 core-days, normalized on the most common hardware used, namely 4-core Intel Xeon E5520 CPUs (2.27GHz). We collected 57070251 relations, out of which 34740801 were non duplicate. Filtering produced a 1982791×1982784 matrix M with weight 396558692. Taking into account the block of 7 Schirokauer maps S, the matrix $M\|S$ is square.

We computed 8 sequences in the Block Wiedemann algorithm, using the trick mentioned in [18, Sect. 8], as programmed in cado-nfs (rediscovered and

Table 4. Probed yield for special-q ranges. Cpu time on Intel Xeon E5520 (2.27GHz).

Method	Seconds/ relation	Relations/ special-q	Remarks
Generalized Joux–Lercier	3.48	4.96	0+3 large primes below 2^{28}
JLSV$_1$	1.31	4.24	2+2 large primes below 2^{27}, orbits of three special-q batched together
Conjugation	**0.91**	5.93	

further analyzed in [34]). All these sequences can be computed independently. Computation time for the 8 Krylov sequence was about 250 core-days (Xeon E5-2650, 2.4GHz, using four 16-core nodes per sequence). Finding the linear (matrix) generator for the matrices took 75 core-hours, parallelized over 64 cores. Building the solution cost some more 170 core-days. We reconstructed virtual logarithms for 15196345 out of the 15206761 factor base elements (99.9%). This was good enough to start looking for individual logarithms.

4.2 Computing Individual Discrete Logarithms in $\mathbb{F}_{p^3}$

From the linear algebra step, we know how to compute the logarithm modulo ℓ of any element of $\mathbb{F}_{p^3}$ whose lift in either K_f or K_g factors completely over the factor base. Lifting in K_f is often convenient because norms are smaller.

The tiny case. A particular element which lifts conveniently in K_f is the common root t of both polynomials. By construction, its lift $\alpha \in K_f$ generates a principal (fractional) ideal that factors as J_f^{-1} (see Proposition 1) times prime ideals of norm dividing 28, namely: $(\alpha) = I_{2,0}^2 I_{2,\infty}^{-2} I_{7,0} I_{7,\infty}^{-1}$, where $I_{2,\infty}^2 I_{7,\infty}$ corresponds to J_f and the prime ideals in the right-hand side can be made explicit. Its logarithm therefore writes as[2]

$$\log(t) = 2 \operatorname{vlog} I_{2,0} - 2 \operatorname{vlog} I_{2,\infty} + \operatorname{vlog} I_{7,0} - \operatorname{vlog} I_{7,\infty} + \sum_{i=1}^{5} \lambda_{f,i}(\alpha) \operatorname{vlog}(\lambda_{f,i}).$$

$\lambda_{f,1}(\alpha) = \text{0x3720106a3d368d7f731a0757b905778050ae327},$

$\lambda_{f,2}(\alpha) = \text{0x1dbeace7d0ec187712ae8afcd6ccdc4db06f781},$

$\lambda_{f,3}(\alpha) = \text{0x9c3109f7741d625869f135706be03fc09375450},$

$\lambda_{f,4}(\alpha) = \text{0x1e46653b287d99c502a5c6e12ab17a3dd10988c},$

$\lambda_{f,5}(\alpha) = \text{0x31628f3e0b491e622946b32f66292c1389a7427}.$

By construction the value $\log(t)$ above is invertible modulo ℓ, and we can freely normalize our virtual logarithm values so that it is equal to one.

---

[2] The convention in `cado-nfs` is to take coefficients of largest degree first in the Schirokauer maps computation $z \mapsto \frac{1}{\ell}(z^{\ell^m - 1} - 1)$ where $m = \operatorname{lcm}_{\mathfrak{l} \text{ prime, } \mathfrak{l} \mid \ell}[\mathfrak{l} : \ell]$. Here we have $m = 1$.

The tame case. Elements whose lifts do not factor completely over any of the factor base but have only moderate-size outstanding factors can be dealt with using a classical *descent* procedure. This finds recursively new relations involving smaller and smaller primes, until all primes involved belong to the factor base. Software achieving this exists, such as the `las_descent` program in `cado-nfs`.

The general case. For computing individual logarithms of arbitrary elements, we used the boot technique described in [28]. For each target, we compute a preimage in $\mathbb{Z}[x]$ represented by a polynomial of degree at most 5 and coefficients bounded by $p^{1/3}$. The norm in K_f of the preimage is $O(p^2) = O(Q^{2/3})$, of approximately 340 bits. The asymptotic complexity of this step is $L_Q[1/3, 1.26]$, and would be $L_Q[1/3, 1.132]$ with one early-abort test (see e.g. [45, Sect. 4.3] or [3, Chap. 4]). The optimal size of largest prime factors in the decomposition is given by the formula $L_Q[2/3, (e^2/3)^{1/3} \approx 0.529]$, where $e = 2/3$ (see [17, Sect. 4]). Applying it for $\log_2 Q = 508$ gives a bound of 68 bits and a running-time of approximately 2^{42} tests. In practice we found very easily initial splittings where B_1 is less than 64 bits, which eased the descent.

4.3   Solving the Challenge

Our main use case for individual logarithm computation in $\mathbb{F}_{p^3}$ is to solve a DLP challenge on the curve. The challenge definition procedure (described in the appendix[3], the Magma code is also available[4]) gives:

$$G_1 = \big(\text{0x106b415d7b4a2d71659ae97440cbb20a6de42d76d69}, \text{0x16d74a2a88e817f1821a1c40e220d34eec93e33cb83}\big),$$
$$P = \big(\text{0x15052ba45717710e6b0cbf8ed89c5c1a0a279480e26}, \text{0x8050f05a231ae1f13e56de1171c108294656052339}\big)$$

From Sect. 2.1, we need to compute $\log(G_T)$ and $\log(S)$, where $G_T = e(G_1, G_2)$ and $S = e(P, G_2)$ are given in the Magma verification script(see footnote 4) . We searched for randomized values G_T^r and $G_T^{r'}S$ which were amenable to the descent procedure. After 32 core-hours looking in the range $r \in [1, 64000]$, we selected the following element

$$G_T^{52154} = -\text{0x21d517d51512e9} - \text{0x95233b3af1b3c7}\,x + \text{0x8d324ebc7849bb}\,x^2$$
$$+ \text{0x18ff0d5ae0b52b}\,x^3 + \text{0x13f711fe92d63cd}\,x^4 - \text{0x15c778630d36920}\,x^5$$

whose straightforward lift in K_f has 59-bit smooth norm (resultant with f, more precisely):

0x87ac1a057df9772d1e08d4de56b3e6b5f208710437b5f92ac4a494c318c9781107e00364934e34efa87b26597771c

$= 2^2 \cdot 5 \cdot 7^2 \cdot 31 \cdot 193 \cdot 277 \cdot 1787 \cdot 12917 \cdot 125789 \cdot 142301513 \cdot 380646221 \cdot 2256567883$

$\cdot 132643203397 \cdot 1380194432565816569 \cdot 603094914193031251 \cdot 8010607393005538627$

---

[3] Sect. B.1 and Sect. B.2 of the pre-proceedings version available at https://hal.inria. fr/hal-01320496.

[4] http://www.lix.polytechnique.fr/~guillevic/discrete-log/SAC2016-mnt170-verificati on-script.mag.

Virtual logarithms for primes below $50 \cdot 10^6$ (25.57 bits) were known. The descent procedure took 13.4 h. Once all logarithms were computed, the value of $\log(G_T)$ could be deduced:

$$\log(G_T) = \texttt{0x8c58b66f0d8b2e99a1c0530b2649ec0c76501c3} \text{ (normalized to } \log t = 1).$$

Similarly, we selected

$$G_T^{35313} S \mapsto \quad \texttt{0x457449569db669} + \texttt{0x88c32ec54242fd}\, x - \texttt{0x2370c0f5914ba9}\, x^2$$
$$+ \texttt{0x14c7ccbafc20e2}\, x^3 + \texttt{0xde2e21c5f1a4c4}\, x^4 - \texttt{0x10b6bfd826db49c}\, x^5$$

whose lift in K_f has norm

$$-\texttt{0x44dafd6ec57c91e64567fa045187100da9a98c5c509b388cb61759f345b3ce27226a5e8520be0bd4559acbd538b90}$$

$$= -2^4 \cdot 5^2 \cdot 7 \cdot 643 \cdot 1483 \cdot 2693 \cdot 95617 \cdot 9573331 \cdot 33281579 \cdot 1608560119 \cdot 48867401441$$
$$\cdot 516931716361 \cdot 896237937459937 \cdot 16606283628226811 \cdot 19530910835315983$$

the largest factor having 54 bits, a very small size indeed (compared to the 68 bits predicted by theory). The descent procedure for other primes took 10.7 hours. We found that

$$\log(S) = \texttt{0x48a6bcf57cacca997658c98a0c196c25116a0aa} \text{ (normalized to } \log t = 1).$$

We eventually found that

$$\log_{G_1}(P) = \texttt{0x711d13ed75e05cc2ab2c9ec2c910a98288ec038} \bmod \ell.$$

5   Conclusion and Future Work

5.1   Consequences for Pairing-Based Cryptography

Our work showed that the choice of embedding degree n and finite field size $\log p^n$ should be done carefully. The size of $\mathbb{F}_{p^n}$ should be large enough to provide the desired level of security. We recall these sizes for $\mathbb{F}_{p^3}$. The recent improvements of Kim and Kim–Barbulescu [37,38] do not apply to $\mathbb{F}_{p^n}$ where n is prime, so $\mathbb{F}_{p^3}$ is not affected. The asymptotic complexity of the NFS algorithm for $\mathbb{F}_{p^3}$ is $\exp\left((c + o(1))(\log p^n)^{1/3}(\log\log p^n)^{2/3}\right) = L_{p^3}[1/3, (64/9)^{1/3}]$. Since there is a polynomial factor hidden in the notation $c + o(1)$, taking $\log_2 L_{p^3}[1/3, (64/9)^{1/3}]$ does not give the exact security level but only an approximation. We may compare our present record with previous records of same size for prime fields $\mathbb{F}_p$ and quadratic fields $\mathbb{F}_{p^2}$. Kleinjung in 2007 announced a record computation in a prime field $\mathbb{F}_p$ of 530 bits (160 decimal digits) [39]. Barbulescu, Gaudry, Guillevic and Morain in 2014 announced a record computation in $\mathbb{F}_{p^2}$ of 529 bits (160 decimal digits) [4]. We compare the timings in Table 5. The timings of relation collection and linear algebra were not balanced in Kleinjung record: 3.3 years compared to 14 years and moreover, this is a quite old record so it is not really possible to compare our record with this one directly. We can compare our record with the 529-bit $\mathbb{F}_{p^2}$ record computation of 2014 [4]. Our total running-time is 15.5 times longer whereas the finite field is 21 bit smaller.

Table 5. Comparison of running-time for discrete logarithm records in $\mathbb{F}_p$, $\mathbb{F}_{p^2}$ and $\mathbb{F}_{p^3}$ of 530, 529, 512 and 508 bits.

record	relation collection	linear algebra	individual log	total
Kleinjung [39] 530-bit field $\mathbb{F}_p$, 2007	3.3 CPU-years 3.2 GHz Xeon64	14 years 3.2 GHz Xeon64	few hours 3.2 GHz Xeon64	17.3 years
BGGM [4] 529-bit field $\mathbb{F}_{p^2}$, 2014	68 core-days=0.19y 2.0 GHz E5-2650	30.3 hours NVidia GTX 680 graphic card	few hours 2.0 GHz E5-2650	70 days = 0.2 year
BGGM [5] 512-bit field $\mathbb{F}_{p^3}$, 2015	850 core-days = 2.33 years	5500 core-days = 15 years	few days	17.3 years
	2.4 GHz Xeon E5-2650			
this work 508-bit field $\mathbb{F}_{p^3}$, 2016	660 core-days =1.81 years 2.27GHz 4-core Xeon E5520	423 days = 1.16 years 2.4 GHz Xeon E5-2650	2 days 2.27GHz 4-core Xeon E5520	1085 days = 2.97 years

5.2 Future Work

We have computed a DLP on an MNT curve with embedding degree 3. What are the next candidates? We could continue the series in two directions: increasing the size of p^n to 600 bits, in order to compare this new record to the previous records of the same size, in particular the $\mathbb{F}_{p^2}$ record of 600 bits [6]. We could conjecture, according to the present record and the size of the norms, that a DLP record in $\mathbb{F}_{p^3}$ of 600 bits will be more than 15 times harder than in a 600-bit field $\mathbb{F}_{p^2}$.

The second direction would be to continue the series of MNT curves, with $n = 4$. We found an MNT curve of embedding degree 4 in Miracl (file k4mnt.ecs). The curve was generated by Drew Sutherland for Mike Scott a long time ago.

$$y = \text{0xf19192168b16c1315d33}$$
$$p = y^2 + y + 1 = \text{0xe3f367d542c82027f33dc5f3245769e676a5755d}$$
$$\ell = \text{0x6b455e0a014f1e30eaef7300bd4bb4258290fc5}$$
$$\tau = y + 1 = \text{0xf19192168b16c1315d34}$$
$$\#E(\mathbb{F}_p) = y^2 + 1 = p + 1 - \tau = 2 \cdot 17 \cdot \ell$$

Since n is a prime power, we have to adapt the Kim–Barbulescu technique (dedicated to non-prime power n) to prime-power extension degrees[5]. We construct $\mathbb{F}_{p^4}$ as $\mathbb{F}_{p^2}[x]/(\varphi(x))$, where $\mathbb{F}_{p^2} = \mathbb{F}_p[s]/(h_1(s))$ and both h_1 and φ are of degree 2, and φ has coefficients in $\mathbb{F}_{p^2}$. As a consequence, the polynomials f and g will have coefficients in $\mathbb{Z}[s]/(h_1(s))$ instead of $\mathbb{Z}$. For example, one could take

---

[5] Right after the submission, several variants of Kim's Extended TNFS where proposed, that deal with any composite n, in particular prime power n, and generalize the Sarkar–Singh method [30,47,48,50].

$$h_1(s) = s^2 + 2,$$
$$h_2(x, t_0, s) = x^2 + s + t_0,$$
$$P(t_0) = t_0^2 + t_0 + 1,$$
$$f = \text{Res}_{t_0}(P(t_0), h_2(x, t_0, s)) = x^4 + (2s - 1)x^2 - s - 1,$$
$$g = h_2(x, y, s) = x^2 + s + \text{0xf19192168b16c1315d33}.$$

The major difference is that to be efficient, we have to sieve polynomials of degree 1 with coefficients in $\mathbb{Z}[s]/(h_1(s))$, that is elements of the form $(a_0 + a_1 s) + (b_0 + b_1 s)x$ where the a_i's and b_i's are small rational integers, say $|a_i|, |b_i| \leq A$. For instance, taking $\log_2(E) = 1.1(\log Q)^{1/3}(\log \log Q)^{2/3} \approx 28$, we obtain $A = E^{2/(2 \deg h)}$ of 14 bits. The upper bound on the norm would be of 120 bits on f-side and 219 bits on g-side, the total being roughly of 339 bits. This is 11 bits more than our present record for the 508-bit $n = 3$ MNT curve (328 bits, Table 3), but by far much less than with any previous technique applied to that $\mathbb{F}_{p^4}$. Norm estimates are provided in Table 6. From a practical point of view, we would need extensions of the work [26].

Table 6. Norm bound estimates for $\mathbb{F}_{p^4}$ of 640 bits.

Method	$\|f\|_\infty$	$\|g\|_\infty$	NB_f	NB_g	$NB_f + NB_g$
Extended TNFS+hybrid JP	1	80	120	219	339
GJL	1	128	144	243	387
JLSV$_1$	80	80	195	195	390
Sarkar-Singh, $d = 2, r = 2$	1	107	172	222	394
Hybrid JP–Conj	1	80	159	240	399
JLSV$_2$, $D = 6$ (D best choice)	91	91	264	206	470

Acknowledgements. The authors are grateful to Pierrick Gaudry for his help in running the computations.

A   Polynomial Selection Methods

We provide in this section the polynomials computed for our $\mathbb{F}_{p^3}$ record with the other competitive polynomial selection methods that we compared in Sect. 3.

Generalized Joux–Lercier method. The first step of the GJL polynomial selection algorithm is to choose a polynomial f of degree 4 in our context. We need f to factor as a linear polynomial times a degree 3 polynomial modulo p, hence we cannot allow for a degree two subfield, or any of the Galois groups C4, V4 or D4. We extracted from the Magma number field database the list of irreducible polynomials of degree 4 and Galois group A4 (of order 12), class number one and signature $(0, 2)$ (592 polynomials) and $(4, 0)$ (3101 polynomials).

In the GJL method, the LLL algorithm outputs four polynomials g_1, g_2, g_3 and g_4 with small coefficients. To obtain the smallest possible coefficients, we set the LLL parameters to $\delta = 0.99999$ and $\eta = 0.50001$. We compute linear combinations $g = \sum_{i=1}^{4} \lambda_i g_i$ with $|\lambda_i| \|g_i\|_\infty \leq 2^5 \cdot \min_{1 \leq i \leq 4} \|g_i\|_\infty$ (roughly speaking, $|\lambda_i| \leq 32$) so that the size of the coefficients of g do not increase too much, while we can obtain a polynomial g with a better Murphy's $\mathbb{E}$ value.

Then we run the GJL method with our modified post-LLL step for each polynomial f in our database and we selected the pair with the highest Murphy's $\mathbb{E}$ value. We obtained

$$
\begin{aligned}
f &= x^4 - 2x^3 + 2x^2 + 4x + 2 \\
\alpha(f) &= 1.2 \\
\log_2 \|f\|_\infty &= 2 \\
g &= 1337141023326143365636811811937049609555\, x^3 + 1738187069076994966689945593428022999969\, x^2 \\
&\quad + 878019651910536420352249995702628405053\, x - 1854039481155034984713783237852106058885 \\
\alpha(g) &= -2.1 \\
\log_2 \|g\|_\infty &= 129.37, \text{ the optimal being } \tfrac{3}{4} \log_2 p = 127.5 \\
\mathbb{E}(f, g) &= 5.08 \cdot 10^{-13}
\end{aligned}
$$

Joux-Lercier-Smart-Vercauteren method. The Joux-Lercier-Smart-Vercauteren method (JLSV1) is possibly the most straighforward polynomial selection method adapted to non-prime finite fields. It is possible to force this method to pick polynomials f within a specific family, in order to force nice Galois properties. For example, we may use the form $\psi = x^3 - tx^2 - (t + 3)x - 1$.

The enumeration was the largest for the JLSV$_1$ method: we searched over 2^{25} polynomials f in the cyclic family $x^3 - t_0 x^2 - (t_0 + 3)x - 1$, with a parameter t_0 of 84 up to 85 bits. We kept the polynomials whose α value was less than -3.0. We continued the JLSV$_1$ polynomial selection algorithm selectively for these good precomputed polynomials. The "initial" g (say g_0) produced by the method can be improved by using instead any linear combination $g = \lambda f + \mu g_0$ for small λ and μ, thereby improving the Murphy's $\mathbb{E}$ value. We set $|\lambda|, |\mu| \leq 2^5$.

$$
\begin{aligned}
f &= x^3 - 3014566310085793929634346\, x^2 - 3014566310085793929634349\, x - 1 \\
\alpha(f) &= -3.0 \\
\log_2 \|f\|_\infty &= 84.64 \\
g &= 3014566310085793929969540\, x^3 + 4684527414449598057831640 7\, x^2 \\
&\quad - 4359171515807783732078221 3\, x - 3014566310085793929969540 \\
\alpha(g) &= -2.8 \\
\log_2 \|g\|_\infty &= 85.28, \text{ the optimal being } \tfrac{1}{2} \log_2 p = 85 \\
\mathbb{E}(f, g) &= 1.02 \cdot 10^{-12}
\end{aligned}
\tag{3}
$$

Conjugation and Joux–Pierrot methods. The Joux-Pierrot method produces polynomials with the same degree and coefficient properties as the Conjugation method for MNT curves and that are moreover monic. The polynomials constructed with the Conjugation method allow a factor two speed-up thanks to a Galois automorphism. We propose here a hybrid variant in Algorithm 2 for pairing-friendly curves. The conjugation method, in Algorithm 1, is the one which eventually produced the best polynomial pair.

For the Conjugation method as well as the hybrid method of Algorithm 2, and similarly to the JLSV_1 method, it is possible to choose polynomials g of the form $\psi = x^3 - tx^2 - (t+3)x - 1$ to allow a Galois automorphism of degree 3.

Algorithm 2. Variant of Joux–Pierrot and Conjugation methods

Input: p prime, $p = P(y)$ where $\deg P \geq 2$ and P of tiny coefficients, n integer
Output: $f, g \in \mathbb{Z}[x]$ irreducible and $\psi = \gcd(f \bmod p, g \bmod p)$ in $\mathbb{F}_p[x]$
       irreducible of degree n

1 **repeat**
2   Select $g_1(x), g_0(x)$, two polynomials with small integer coefficients,
       $\deg g_1 < \deg g_0 = n$
3   Select small integers a, b, c, d
4     $\psi(x) = g_0(x) + \left(\dfrac{a + by}{c + dy} \bmod p \right) g_1(x)$
5     $f \leftarrow \text{Res}_Y\left(P(Y), (c + dY)g_0(x) + (a + bY)g_1(x) \right)$
6     $g \leftarrow (c + dy)g_0(x) + (a + by)g_1(x)$ // $g \equiv (c + dy)\psi(x) \bmod p$
7 **until** $\psi(x)$ *is irreducible in* $\mathbb{F}_p[x]$ *and* f, g *are irreducible in* $\mathbb{Z}[x]$
8 **return** (f, g, ψ)

In practice, in Algorithm 2 one might prefer to constrain $d = 0$, so that g has small leading coefficient c. Going further and requiring $c = 1$ so that g is monic reduces however too much the possibilities to find a good pair of polynomials.

The following example has been obtained with Algorithm 2, searching over all $(a + by)/c$ with $|a|, |b|, |c| \leq 256$.

$$y = -8702303353090049898316902 \text{ is the targeted MNT curve parameter}$$
$$f = 108x^6 + 1116x^5 + 3347x^4 + 2194x^3 - 613x^2 - 468x + 108$$
$$g = 6x^3 + 3480921341236019959326763 9\ x^2 + 348092134123601995932676 21\ x - 6$$
$$= 6x^3 - (4y - 31)x^2 - (4y - 13)x - 6$$
$$\varphi = \tfrac{1}{6}g \bmod p = x^3 + 15146016729840465134625816509459896150600476996648 1\ x^2$$
$$+ 1514601672984046513462581650945989615060047699664 78\ x - 1$$

References

1. Adleman, L.M., Huang, M.-D.: Function field sieve methods for discrete logarithms over finite fields. Inf. Comput. **151**(1), 5–16 (1999)
2. Adrian, D., Bhargavan, K., Durumeric, Z., Gaudry, P., Green, M., Halderman, J.A., Heninger, N., Springall, D., Thomé, E., Valenta, L., VanderSloot, B., Wustrow, E., Béguelin, S.Z., Zimmermann, P.: Imperfect forward secrecy: how Diffie-Hellman fails in practice. In: Ray, I., Li, N., Kruegel, C. (eds.) ACM CCS 15, pp. 5–17. ACM Press, October 2015
3. Barbulescu, R.: Algorithmes de logarithmes discrets dans les corps finis. Ph.D. thesis, Université de Lorraine (2013)

4. Barbulescu, R., Gaudry, P., Guillevic, A., Morain, F.: Discrete logarithms in GF(p^2) – 160 digits. Announcement on the Number Theory List, June 2014. https://listserv.nodak.edu/cgi-bin/wa.exe?A2=NMBRTHRY;2ddabd4c.1406

5. Barbulescu, R.,. Gaudry, P., Guillevic, A., Morain, F.: Discrete logarithms in GF(p^3) – 512 bits. Announcement at the CATREL workshop, October 2015. https://webusers.imj-prg.fr/razvan.barbaud/p3dd52.pdf

6. Barbulescu, R., Gaudry, P., Guillevic, A., Morain, F.: Improving NFS for the discrete logarithm problem in non-prime finite fields. In: Oswald, E., Fischlin, M. (eds.) EUROCRYPT 2015. LNCS, vol. 9056, pp. 129–155. Springer, Heidelberg (2015). doi:10.1007/978-3-662-46800-5_6

7. Barbulescu, R., Gaudry, P., Joux, A., Thomé, E.: A heuristic Quasi-polynomial algorithm for discrete logarithm in finite fields of small characteristic. In: Nguyen, P.Q., Oswald, E. (eds.) EUROCRYPT 2014. LNCS, vol. 8441, pp. 1–16. Springer, Heidelberg (2014). doi:10.1007/978-3-642-55220-5_1

8. Barbulescu, R., Gaudry, P., Kleinjung, T.: The tower number field sieve. In: Iwata, T., Cheon, J.H. (eds.) ASIACRYPT 2015. LNCS, vol. 9453, pp. 31–55. Springer, Heidelberg (2015). doi:10.1007/978-3-662-48800-3_2

9. Barreto, P.S.L.M., Naehrig, M.: Pairing-friendly elliptic curves of prime order. In: Preneel, B., Tavares, S. (eds.) SAC 2005. LNCS, vol. 3897, pp. 319–331. Springer, Heidelberg (2006). doi:10.1007/11693383_22

10. Beuchat, J., Brisebarre, N., Detrey, J., Okamoto, E., Shirase, M., Takagi, T.: Algorithms and arithmetic operators for computing the η_T pairing in characteristic three. IEEE Trans. Comput. **57**(11), 1454–1468 (2008)

11. Bistritz, Y., Lifshitz, A.: Bounds for resultants of univariate and bivariate polynomials. Linear Algebra Appl. **432**(8), 1995–2005 (2010). Special issue devoted to the 15th ILAS Conference at Cancun, Mexico, June 16–20, 2008

12. Blake, I.F., Seroussi, G., Smart, N.: Advances in Elliptic Curve Cryptography. London Mathematical Society Lecture Note Series, vol. 317. Cambridge University Press, Cambridge (2005)

13. Boneh, D., Franklin, M.K.: Identity-based encryption from the weil pairing. In: Kilian, J. (ed.) CRYPTO 2001. LNCS, vol. 2139, pp. 213–229. Springer, Heidelberg (2001). doi:10.1007/3-540-44647-8_13

14. Boneh, D., Lynn, B., Shacham, H.: Short signatures from the weil pairing. In: Boyd, C. (ed.) ASIACRYPT 2001. LNCS, vol. 2248, pp. 514–532. Springer, Heidelberg (2001). doi:10.1007/3-540-45682-1_30

15. Brezing, F., Weng, A.: Elliptic curves suitable for pairing based cryptography. Des. Codes Cryptogr. **37**(1), 133–141 (2005)

16. Cocks, C., Pinch, R.G.: ID-based cryptosystems based on the Weil pairing (2001, Unpublished manuscript)

17. Commeine, A., Semaev, I.: An algorithm to solve the discrete logarithm problem with the number field sieve. In: Yung, M., Dodis, Y., Kiayias, A., Malkin, T. (eds.) PKC 2006. LNCS, vol. 3958, pp. 174–190. Springer, Heidelberg (2006). doi:10.1007/11745853_12

18. Coppersmith, D.: Solving homogeneous linear equations over GF(2) via block Wiedemann algorithm. Math. Comput. **62**(205), 333–350 (1994)

19. Dupont, R., Enge, A., Morain, F.: Building curves with arbitrary small MOV degree over finite prime fields. J. Cryptol. **18**(2), 79–89 (2005)

20. Elkenbracht-Huizing, R.M.: An implementation of the number field sieve. Exp. Math. **5**(3), 231–253 (1996)

21. Foster, K.: HT90 and "simplest" number fields. Illinois J. Math. **55**(4), 1621–1655 (2011)

22. Freeman, D.: Constructing pairing-friendly elliptic curves with embedding degree 10. In: Hess, F., Pauli, S., Pohst, M. (eds.) ANTS 2006. LNCS, vol. 4076, pp. 452–465. Springer, Heidelberg (2006). doi:10.1007/11792086_32

23. Freeman, D., Scott, M., Teske, E.: A taxonomy of pairing-friendly elliptic curves. J. Cryptology **23**(2), 224–280 (2010)

24. Frey, G., Rück, H.G.: A remark concerning m-divisibility and the discrete logarithm in the divisor class group of curves. Math. Comput. **62**(206), 865–874 (1994)

25. Galbraith, S.D., McKee, J.F., Valença, P.C.: Ordinary abelian varieties having small embedding degree. Finite Fields Appl. **13**(4), 800–814 (2007)

26. Gaudry, P., Grémy, L., Videau, M.: Collecting relations for the number field sieve in $GF(p^6)$. LMS Journal of Computation and Mathematics, Special issue ANTS-XII, August 2016, to appear. https://eprint.iacr.org/2016/124

27. Granger, R., Kleinjung, T., Zumbrägel, J.: Breaking '128-bit secure' supersingular binary curves. In: Garay, J.A., Gennaro, R. (eds.) CRYPTO 2014. LNCS, vol. 8617, pp. 126–145. Springer, Heidelberg (2014). doi:10.1007/978-3-662-44381-1_8

28. Guillevic, A.: Computing individual discrete logarithms faster in $GF(p^n)$ with the NFS-DL algorithm. In: Iwata, T., Cheon, J.H. (eds.) ASIACRYPT 2015. LNCS, vol. 9452, pp. 149–173. Springer, Heidelberg (2015). doi:10.1007/978-3-662-48797-6_7

29. Hayashi, T., Shimoyama, T., Shinohara, N., Takagi, T.: Breaking pairing-based cryptosystems using η^T pairing over $GF(3^{97})$. In: Wang, X., Sako, K. (eds.) ASIACRYPT 2012. LNCS, vol. 7658, pp. 43–60. Springer, Heidelberg (2012). doi:10.1007/978-3-642-34961-4_5

30. Jeong, J., Kim, T.: Extended tower number field sieve with application to finite fields of arbitrary composite extension degree. Cryptology ePrint Archive, Report 2016/526 (2016). http://eprint.iacr.org/2016/526

31. Joux, A.: A one round protocol for tripartite Diffie–Hellman. In: Bosma, W. (ed.) ANTS 2000. LNCS, vol. 1838, pp. 385–393. Springer, Heidelberg (2000). doi:10.1007/10722028_23

32. Joux, A., Lercier, R., Smart, N., Vercauteren, F.: The number field sieve in the medium prime case. In: Dwork, C. (ed.) CRYPTO 2006. LNCS, vol. 4117, pp. 326–344. Springer, Heidelberg (2006). doi:10.1007/11818175_19

33. Joux, A., Pierrot, C.: The special number field sieve in $\mathbb{F}_{p^n}$. In: Cao, Z., Zhang, F. (eds.) Pairing 2013. LNCS, vol. 8365, pp. 45–61. Springer, Cham (2014). doi:10.1007/978-3-319-04873-4_3

34. Joux, A., Pierrot, C.: Nearly sparse linear algebra. Cryptology ePrint Archive, Report 2015/930 (2015). http://eprint.iacr.org/2015/930

35. Kasahara, M., Ohgishi, K., Sakai, R.: Notes on ID-based key sharing systems on elliptic curve. Technical report, IEICE (1999)

36. Kasahara, M., Ohgishi, K., Sakai, R.: Cryptosystems based on pairing. In: The 2000 Symposium on Cryptography and Information Security, volume SCIS2000-C20, January 2000

37. Kim, T.: Extended tower number field sieve: a new complexity for medium prime case. Cryptology ePrint Archive, Report 2015/1027, version 1, October 2015. http://eprint.iacr.org/eprint-bin/getfile.pl?entry=2015/1027&version=20151026:154834&file=1027.pdf

38. Kim, T., Barbulescu, R.: Extended tower number field sieve: a new complexity for the medium prime case. In: Robshaw, M., Katz, J. (eds.) CRYPTO 2016. LNCS, vol. 9814, pp. 543–571. Springer, Heidelberg (2016). doi:10.1007/978-3-662-53018-4_20

39. Kleinjung, T.: Discrete logarithms in GF(p) – 160 digits. Announcement on the Number Theory List, February 2007. https://listserv.nodak.edu/cgi-bin/wa.exe? A2=NMBRTHRY;1c737cf8.0702

40. Matyukhin, D.: Effective version of the number field sieve for discrete logarithms in the field GF(p^k) (in Russian). Tr. Diskr. Mat. **9**, 121–151 (2006). http://m.mathnet.ru/php/archive.phtml?wshow=paper&jrnid=tdm&paperid=144&option_lang=eng

41. Menezes, A., Okamoto, T., Vanstone, S.A.: Reducing elliptic curve logarithms to logarithms in a finite field. IEEE Trans. Inf. Theory **39**(5), 1639–1646 (1993)

42. Miyaji, A., Nakabayashi, M., Takano, S.: Characterization of elliptic curve traces under FR-reduction. In: Won, D. (ed.) ICISC 2000. LNCS, vol. 2015, pp. 90–108. Springer, Heidelberg (2001). doi:10.1007/3-540-45247-8_8

43. Montgomery, P.L.: Square roots of products of algebraic numbers (1997). Unpublished draft, dated 16 May 1997

44. Murphy, B.A.: Polynomial selection for the number field sieve integer factorisation algorithm . Ph.D. thesis, Australian National University (1999). http://maths-people.anu.edu.au/~brent/pd/Murphy-thesis.pdf

45. Pomerance, C.: Analysis and comparison of some integer factoring algorithms. In: Lenstra, H.W.J., Tijdeman, R. (eds.) Computational methods in number theory, part I. Mathematical Centre Tracts, vol. 154, pp. 89–139. Mathematisch Centrum, Amsterdam (1982). http://oai.cwi.nl/oai/asset/19571/19571A.pdf

46. Sakemi, Y., Hanaoka, G., Izu, T., Takenaka, M., Yasuda, M.: Solving a discrete logarithm problem with auxiliary input on a 160-bit elliptic curve. In: Fischlin, M., Buchmann, J., Manulis, M. (eds.) PKC 2012. LNCS, vol. 7293, pp. 595–608. Springer, Heidelberg (2012). doi:10.1007/978-3-642-30057-8_35

47. Sarkar, P., Singh, S.: A general polynomial selection method and new asymptotic complexities for the tower number field sieve algorithm. In: Cheon, J.H., Takagi, T. (eds.) ASIACRYPT 2016. LNCS, vol. 10031, pp. 37–62. Springer, Heidelberg (2016). doi:10.1007/978-3-662-53887-6_2

48. Sarkar, P., Singh, S.: A generalisation of the conjugation method for polynomial selection for the extended tower number field sieve algorithm. Cryptology ePrint Archive, Report 2016/537 (2016). http://eprint.iacr.org/2016/537

49. Sarkar, P., Singh, S.: New complexity trade-offs for the (multiple) number field sieve algorithm in non-prime fields. In: Fischlin, M., Coron, J.-S. (eds.) EUROCRYPT 2016. LNCS, vol. 9665, pp. 429–458. Springer, Heidelberg (2016). doi:10.1007/978-3-662-49890-3_17

50. Sarkar, P., Singh, S.: Tower number field sieve variant of a recent polynomial selection method. Cryptology ePrint Archive, Report 2016/401 (2016). http://eprint.iacr.org/2016/401

51. Schirokauer, O.: Discrete logarithms and local units. Philos. Trans. Roy. Soc. London Ser. A **345**(1676), 409–423 (1993)

52. Schirokauer, O.: Virtual logarithms. J. Algorithms **57**(2), 140–147 (2005)

Author Index

Printed in the United States
By Bookmasters